# The
# Reoviridae

# THE VIRUSES

Series Editors
HEINZ FRAENKEL-CONRAT, *University of California*
*Berkeley, California*
ROBERT R. WAGNER, *University of Virginia School of Medicine*
*Charlottesville, Virginia*

---

THE HERPESVIRUSES, Volumes 1, 2, 3, and 4
Edited by Bernard Roizman

THE REOVIRIDAE
Edited by Wolfgang K. Joklik

THE PARVOVIRUSES
Edited by Kenneth I. Berns

# The
# Reoviridae

Edited by
## WOLFGANG K. JOKLIK
*Duke University Medical Center*
*Durham, North Carolina*

PLENUM PRESS • NEW YORK AND LONDON

Library of Congress Cataloging in Publication Data

Main entry under title:

The Reoviridae.

  (Viruses)
  Includes bibliographical references and index.
  1. Reoviruses. I. Joklik, Wolfgang K. II. Series.
QR414.R46  1983                      576′.64                      83-6276
ISBN 0-306-42344-0

©1983 Plenum Press, New York
A Division of Plenum Publishing Corporation
233 Spring Street, New York, N.Y. 10013

Printed in the United States of America

# Contributors

**Guido Boccardo,** Istituto di Fitovirologia Applicata del C.N.R., 10135 Torino, Italy

**Bernard N. Fields,** Department of Microbiology and Molecular Genetics, Harvard Medical School, Boston, Massachusetts 02115; and Department of Medicine, Brigham and Women's Hospital, Boston, Massachusetts 02115

**R. I. B. Francki,** Department of Plant Pathology, Waite Agricultural Research Institute, The University of Adelaide, Adelaide 5064, South Australia

**Barry M. Gorman,** Queensland Institute of Medical Research, Bramston Terrace, Brisbane 4006, Australia

**Ian H. Holmes,** Department of Microbiology, University of Melbourne, Parkville, Victoria 3052, Australia

**Wolfgang K. Joklik,** Department of Microbiology and Immunology, Duke University Medical Center, Durham, North Carolina 27710

**Marilyn Kozak,** Department of Biological Sciences, University of Pittsburgh, Pittsburgh, Pennsylvania 15260

**Peter P. C. Mertens,** Animal Virus Research Institute, Pirbright, Woking, Surrey GU24 0NF, United Kingdom

**Stewart Millward,** Department of Biochemistry, McGill University, Montreal, Quebec, Canada H3G 1Y6

**Christopher C. Payne,** Glasshouse Crops Research Institute, Littlehampton, West Sussex BN16 3PU, United Kingdom

**Robert F. Ramig,** Department of Virology and Epidemiology, Baylor College of Medicine, Houston, Texas 77030

**Arlene H. Sharpe,** Department of Microbiology and Molecular Genetics, Harvard Medical School, Boston, Massachusetts 02115

**Aaron J. Shatkin,** Department of Cell Biology, Roche Institute of Mole- Biology, Nutley, New Jersey 07110

**Jill Taylor,** Queensland Institute of Medical Research, Bramston Terrace, Brisbane 4006, Australia

**Peter J. Walker,** Queensland Institute of Medical Research, Bramston Terrace, Brisbane 4006, Australia

**Helmut Zarbl,** Department of Biochemistry, McGill University, Montreal, Quebec, Canada  H3G 1Y6

# Preface

It is now just 20 years since Gomatos and his co-workers at the Rockefeller University showed that the nucleic acid in reovirus particles is double-stranded RNA (dsRNA). This discovery created great excitement, for dsRNA was at that time under intense investigation as the replicative form of viral genomes consisting of single-stranded RNA. An equally interesting and important finding followed soon after: it was found that the reovirus genome consists, not of a single nucleic acid molecule, but of 10 discrete "segments," each with its specific sequence content and each transcribed into its own messenger RNA. It is clear now that these segments are genes. Not surprisingly, the availability of a viral genome consisting of 10 unlinked genes has permitted some unique lines of investigation in molecular biology.

Mammalian and avian reoviruses proved to be but the first of several viruses recognized as sharing similarity in size and morphology and genomes consisting of 10, 11, or 12 separate genes. These viruses are distributed throughout living organisms; among the natural hosts of members of this virus family are vertebrates, insects, and plants.

Members of the Reoviridae family differ widely in the virulence that they exhibit toward their hosts. For example, the first discovered mammalian reovirus literally is, as the name signifies, a "respiratory enteric orphan" virus, that is, a virus unassociated with disease. In contrast, the most recently discovered members of this family, the rotaviruses, which were recognized as comprising a separate genus not much more than a decade ago, cause a disease, infantile gastroenteritis, that, while not severe in communities where health care is good, is one of the most important, if not the most important, cause of death in infants under the age of 2 years in less developed countries. Studies of this family of viruses from molecular, biological, and medical perspectives complement each other to an unusual degree because of the unique nature of the genomes of these viruses.

The ten chapters in this book describe what is currently known concerning the molecular biology of members of this virus family on the one hand and the pathogenesis that they exhibit toward their hosts on the other. About half the chapters are devoted to the mammalian reoviruses, which are the Reoviridae that have been studied most intensively in molecular and genetic terms; there are chapters that deal with the nature of reovirus particles and of the reovirus genome, its transcription and replication, the reovirus multiplication cycle, and reovirus genetics. There is also a chapter that describes reovirus pathogenesis in terms of the effects that various individual reovirus gene products exert on host cells. Other chapters deal with the orbiviruses, which include some important animal pathogens such as bluetongue virus; the rotaviruses, important human pathogens; the cytoplasmic polyhedrosis viruses of insects; and the plant reoviruses, which include some important plant pathogens.

Wolfgang K. Joklik

# Contents

Chapter 3

**Biochemical Aspects of Reovirus Transcription and Translation**

*Aaron J. Shatkin and Marilyn Kozak*

## Chapter 4

### The Reovirus Multiplication Cycle

*Helmut Zarbl and Stewart Millward*

## Chapter 5

### Genetics of Reoviruses

*Robert F. Ramig and Bernard N. Fields*

*Chapter 6*

**Pathogenesis of Reovirus Infection**

*Arlene H. Sharpe and Bernard N. Fields*

*Chapter 7*

**Orbiviruses**

*Barry M. Gorman, Jill Taylor, and Peter J. Walker*

*Chapter 8*

**Rotaviruses**

*Ian H. Holmes*

Chapter 9

## Cytoplasmic Polyhedrosis Viruses

*Christopher C. Payne and Peter P. C. Mertens*

*Chapter 10*

**The Plant Reoviridae**

*R. I. B. Francki and Guido Boccardo*

## CHAPTER 1

# The Members of the Family Reoviridae

WOLFGANG K. JOKLIK

## I. INTRODUCTION

The studies of Gomatos and collaborators in the early 1960s that indicated that the RNA in reovirus particles was double-stranded created a great deal of interest because although dsRNA was at that time under active investigation as the replicative form of viral genomes consisting of single-stranded RNA, they pinpointed for the first time a source of stable dsRNA (Gomatos *et al.*, 1962).

Reovirus was but the first of several viruses that proved to contain dsRNA. Gomatos and Tamm (1963) themselves very soon showed that the RNA of wound tumor virus was also double-stranded, as did Black and Markham (1963). Cytoplasmic polyhedrosis virus came next (Miura *et al.*, 1968), followed by bluetongue virus, which was purified in 1969 by Verwoerd (1969), who showed that its RNA was also double-stranded. Finally, in 1971, Welch (1971) showed that neonatal calf diarrhea virus, the first rotavirus to be studied, also contained dsRNA.

While possession of dsRNA was the primary reason for grouping these viruses into one family, it was by no means the only one. In fact, there are other viruses that contain dsRNA (see Section IV); but an initial proposal to group all dsRNA-containing viruses into one family, the Diplornaviridae, proved untenable, just as it would not be feasible to place all dsDNA-containing viruses into one family. The manner in which they manage their replication cycles and express the information encoded in their genomes is just too different.

---

WOLFGANG K. JOKLIK • Department of Microbiology and Immunology, Duke University Medical Center, Durham, North Carolina 27710.

TABLE I. Host Ranges of Members of the Family
Reoviridae

| Genus | Host range |
|---|---|
| Orthoreovirus | Vertebrates |
| Orbivirus | Vertebrates, insects |
| Rotavirus | Mammals |
| Cypovirus | Insects |
| Phytoreovirus | Plants, insects |
| Fijivirus | Plants, insects |
| Presently unassigned viruses | Insects, fish, oysters |

The viruses that are now classified together as members of the Reoviridae family are grouped into six genera, the prototypes of which are mammalian reovirus, bluetongue virus, human rotavirus, cytoplasmic polyhedrosis virus of the silkworm, wound tumor virus, and Fijivirus. These genera are the orthoreoviruses, orbiviruses, rotaviruses, cypoviruses, phytoreoviruses, and fijiviruses. The extremely wide collective host ranges of these viruses are listed in Table I. The members of these genera share certain common properties, but also possess distinctive features of their own. The properties that they share, and that therefore form the distinctive features of members of the Reoviridae family, are described in the following section.

## II. PROPERTIES OF MEMBERS OF THE FAMILY REOVIRIDAE

### A. Possession of Double-Stranded RNA

This has already been discussed. There are numerous criteria for double-strandedness, including possession of a very sharp melting profile, resistance to ribonuclease, susceptibility to ribonuclease III, density in $Cs_2SO_4$, base composition exhibiting equality of $A + U$ as well of $G + C$, and X-ray diffraction pattern. Demonstration that cytoplasmic inclusions in infected cells fluoresce pale green with acridine orange is also a criterion and was in fact the feature that first alerted Gomatos et al. (1962) to the fact that there was something unusual about reovirus. These criteria are discussed in greater detail in Chapter 2.

### B. Structure

All members of this family have a diameter of about 70 nm. In fact, it was primarily on the basis of particle size that the virus originally designated ECHO 10 was taken out of the picornavirus family and designated the prototype of a new group of viruses, the reoviruses; it was the original reovirus serotype 1.

There are three major structural patterns of Reoviridae particles. The first is exemplified by that of the orthoreoviruses and phytoreoviruses. This pattern is characterized by two distinct capsid shells, the capsomer arrangement of which has been difficult to identify, since there appears to be very extensive sharing of protein subunits between capsomers. Particles viewed along five-fold axes of symmetry exhibit 20 peripheral capsomers, while particles viewed along two- or three-fold axes of symmetry exhibit 36 peripheral capsomers (see Chapter 2). The inner capsid shell, or core, possesses 12 prominent projections or spikes situated as though on the vertices of a regular icosahedron.

The second pattern is that of the orbiviruses and rotaviruses, the structure of which is not identical, but similar. Orbiviruses display 32 large ring-shaped capsomers, hence the name (from Latin "orbis," ring). Rotaviruses exhibit a similar arrangement of capsomers surrounding holes that has also been interpreted as comprising 32 large capsomers, though other structural patterns, such as one based on 162 holes surrounded by extensively shared structural units, have also been suggested. In both, the characteristic capsomer arrangement is primarily that of the inner capsid shell or core; the outer shell, though undoubtedly present, does not have a capsomer structure as pronounced as that of orthoreoviruses and phytoreoviruses. This outer shell is more pronounced in rotaviruses than in orbiviruses and confers on the particles the appearance of a wheel, hence the name (from Latin "rota," wheel).

The third pattern is that of the cypoviruses and fijiviruses. These particles resemble reovirus cores; that is, they possess 12 large projections or spikes distributed icosahedrally on the particle surface. The question of the presence of an outer capsid shell in these particles has not been satisfactorily resolved; if there is one, it is morphologically even more featureless than that of orbiviruses and rotaviruses.

One of the reasons for the uncertainty concerning the outer capsid shells of some members of this family is that the shells tend to be unstable to a variety of physical and chemical agents. Thus, the reovirus and rotavirus outer capsid shells are sensitive to proteolytic enzymes, those of orbiviruses and phytoreoviruses to high salt, and those of phytoreoviruses and cypoviruses to fluorocarbons. Care must therefore be taken not to expose virus particles to these reagents during purification and isolation.

Whereas investigators have reported a wide spread in the diameters of these virus particles (60–80 nm), most workers agree that the diameter of the central cavity of members of the Reoviridae family is about 40 nm. This agrees with the fact that the genomes of all these viruses are about the same size (see Section II.C).

## C. Nature of the Genome

The fact that the genomes of Reoviridae are composed of dsRNA is not their only unusual feature; they are also segmented, each segment

TABLE II. Number of Genes in Members of the
Family Reoviridae

| Genus | Number of genes |
|-------|-----------------|
| *Orthoreovirus* | 10 |
| *Orbivirus* | 10 |
| *Rotavirus* | 11 |
| *Cypovirus* | 10 |
| *Phytoreovirus* | 12 |
| *Fijivirus* | 10 |

representing a single gene. In fact, it was with reovirus that it was first shown that the individual genome segments present in virus particles are unique RNA molecules: they do not hybridize with each other, and each is transcribed into specific and unique messenger RNA (mRNA) molecules (see Chapter 2).

The aggregate size of the genomes of all Reoviridae is about the same, namely, a molecular weight of about 15 million. In all cases, the largest gene has a molecular weight of 2.5–3 million [corresponding to about 4000 base pairs (bp)]. The size spread of the smallest genes is wider: the smallest reovirus gene has a molecular weight of about 600,000 (corresponding to about 1000 bp), but some others are up to 5-fold smaller (such as the smallest orbivirus gene).

An important feature of the segmented Reoviridae genome is the number of component genes (Table II). An anomalous situation is presented by Colorado tick fever virus, which is generally regarded as an orbivirus, but has recently been found to possess 11 genes.

Possession of 10–12 genes of dsRNA is the primary prerequisite for inclusion in the Reoviridae family.

## D. Lack of Complete Uncoating

Viruses that possess icosahedral symmetry are uncoated following infection; there is physical separation of protein coat and genome. The dsRNA of Reoviridae, however, is not uncoated following infection; only the outer shell (if there is one) of the virus particles is removed, and the viral genomes remain associated with so-called subviral particles that persist as such throughout the infection cycle (see Chapter 4).

## E. Possession of Enzymes That Transcribe the Double-Stranded RNA into Messenger RNA

The reason that the dsRNA genomes of members of this family are not completely uncoated is that they must be transcribed into mRNA, and since cells do not possess enzymes capable of transcribing dsRNA, such enzymes must be present in the virus particles themselves. Reoviridae particles therefore contain all five enzymes that are required to transcribe their genes into mRNA molecules capped at their 5′ termini. It is interesting to recall that the 5′-methylated mRNA cap was first demonstrated with cytoplasmic polyhedrosis virus and reovirus (see Chapters 2, 3, and 4). A very unusual feature of mRNA molecules of this virus family is that they are not polyadenylated at their 3′ termini.

## III. UNCLASSIFIED MEMBERS OF THE FAMILY REOVIRIDAE

Several viruses have been isolated and partially characterized that appear to be Reoviridae but do not fit into any of the six genera of this family. One of these has already been referred to: it is Colorado tick fever virus, which is regarded as an orbivirus, but which has recently been shown to possess 11 genes, rather than the 10 that are characteric of this genus.    Other potential members of the Reoviridae family are described below.

### A. *Drosophila* Virus F and *Ceratitis* Virus I

Reoviruslike viruses have been found repeatedly in both laboratory and wild populations of *Drosophila* (Brun and Plus, 1980) and have also been isolated from malignant *Drosophila* blood cell lines (Gateff *et al.*, 1980; Haars *et al.*, 1980). The prototype of these viruses, *Drosophila* F virus, possesses a morphology similar to that of reoviruses and a genome composed of ten dsRNA genes. In distinction to orthoreoviruses, but like phytoreoviruses, its outer capsid shell is resistant to proteolytic enzymes. The Ceratitis I virus possesses a similar morphology and also has ten genes. It is serologically unrelated to *Drosophila* F virus and does not multiply in *Drosophila* cells (Plus *et al.*, 1981). Both contain an RNA-dependent RNA polymerase (Plus, personal communication).

### B. Housefly Virus

Moussa (1978) isolated a virus (HFV) from the housefly (*Musca domestica*) that multiplied in the flies' hemocytes. The structure of HFV particles exhibits some similarities to that of mammalian reovirus, such as possession of outer and inner capsid shells; however, HFV particles also appear to contain a middle layer and a nucleoprotein core (Moussa, 1981). The virus possesses ten genes of dsRNA, the aggregate size of which is about 15% larger than that of the mammalian reovirus genome (Moussa, 1980). HFV appears to be unrelated serologically to other members of the family Reoviridae.

## C. Chum Salmon Virus

Recently a virus was isolated from adult chum salmon that possesses typical reovirus morphology (Winton *et al.*, 1981). Its outer capsid shell is susceptible to digestion by chymotrypsin to yield subviral particles with enhanced infectivity. The genome of this virus consists of eleven dsRNA segments (genes) that range in size from 2.5 to $0.37 \times 10^6$ (total genome size, $16 \times 10^6$ daltons). The virus, tentatively identified as the chum salmon virus (CSV), grows best in cells derived from salmonid fish.

A virus isolated from oysters (Myers, 1979) also contains eleven genes that fall into similar size classes (Winton, personal communication). It appears to be another member of the same group of viruses.

## IV. OTHER VIRUSES THAT CONTAIN DOUBLE-STRANDED RNA

Several other viruses also possess genomes consisting of dsRNA. Some, such as infectious bursal disease virus of chickens, infectious pancreatic necrosis virus of fish (IPNV), *Tellina* virus and oyster virus of bivalve molluscs, and *Drosophila* virus X, appear to possess two RNA segments (Dobos *et al.*, 1979) [or possibly only one (Revet and Delain, 1982)]; others, such as the bacteriophage $\phi6$, possess three genes; others still, like the mycoviruses and the viruslike yeast killer factor particles, probably contain only one RNA segment. However, even though at least some of these viruses also possess an RNA-dependent RNA polymerase and some resemble reoviruses morphologically (particularly those the hosts of which are animals, which are also naked icosahedral viruses with diameters of about 60 nm), there appear to be too many differences between them and bona fide members of the Reoviridae for them to be included in this virus family. For example, the RNA segment(s) of *Drosophila* virus X appear to be circularized by a 67-kilodalton terminal protein, and the two RNA segments of IPNV appear to be polycistronic (Mertens and Dobos, 1982), while the RNA segments of members of the Reoviridae family are, as far as is known, monocistronic.

## REFERENCES

Black, L.M., and Markham, R., 1963, Base pairing in the ribonucleic acid of wound tumor virus, *Neth. J. Plant Pathol.* **69**:215.
Brun, G., and Plus, N., 1980, The viruses of Drosophila, in: *The Genetics and Biology of Drosophila*, 2nd ed. (M. Ashburner and T.R.F. Wright, eds.), p. 625, Academic Press, New York.
Dobos, P., Hill, B.J., Hallett, R., Kells, D.T.C., Becht, H., and Teninges, D., 1979, Biophysical and biochemical characterization of five animal viruses with bisegmented double-stranded RNA genomes, *J. Virol.* **32**:593.

Gateff, E., Gissmann, L., Shrestha, R., Plus, N., Pfister, H., Schroder, J., and zur Hausen, H., 1980, Characterization of two tumorous blood cell lines of *Drosophila melanogaster* and the viruses they contain, in: *Invertebrate Systems in Vitro* (E. Kurstak, M. Maramorosch, and A. Dubendorfer, eds.), p. 517, Elsevier/North-Holland, Amsterdam.

Gomatos, P.J., and Tamm, I., 1963, Animal and plant viruses with double-helical RNA, *Proc. Natl. Acad. Sci. U.S.A.* **50**:878.

Gomatos, P.J., Tamm, I., Dales, S., and Franklin, R.M., 1962, Reovirus type 3: Physical characteristics and interactions with L cells, *Virology* **17**:441.

Haars, R., Zentgraf, H., Gateff, E., and Bautz, F.A., 1980, Evidence for endogenous reovirus-like particles in a tissue culture cell line from *Drosophila melanogaster, Virology* **101**:124.

Mertens, P.C., and Dobos, P., 1982, Messenger RNA of infectious pancreatic necrosis virus is polycistronic, *Nature (London)* **247**:243.

Meyers, T.R., 1979, A reo-like virus isolated from juvenile American oysters (*Crassostrea virginica*), *J. Gen. Virol.* **43**:203.

Miura, K.-I., Fujii, I., Sakaki, T., Fuke, M., and Kawase, S., 1968, Double-stranded ribonucleic acid from cytoplasmic polyhedrosis virus of the silkworm, *J. Virol.* **2**:1211.

Moussa, A.Y., 1978, A new virus disease in the housefly, *Musca domestica* (Diptera), *J. Invertebr. Pathol.* **31**:204.

Moussa, A.Y., 1980, The housefly virus contains double-stranded RNA, *Virology* **106**:173.

Moussa, A.Y., 1981, Studies of the housefly virus structure and disruption during purification procedures, *Micron* **12**:131.

Plus, N., Gissmann, L., Veyrunes, J.C., Pfister, H., and Gateff, E., 1981, Reoviruses of *Drosophila* and *Ceratitis* populations and of *Drosophila* cell lines: A possible new genus of the Reoviridae family, *Ann. Virol. (Inst. Pasteur)* **132E**:261.

Revet, B., and Delain, E., 1982, The *Drosophila* X virus contains a 1 μm double-stranded RNA circularized by a 67kd terminal protein: High resolution denaturation mapping of its genome, *Virology* **123**:29.

Verwoerd, D.W., 1969, Purification and characterization of bluetongue virus, *Virology* **38**:203.

Welch, A.B., 1971, Purification, morphology, and partial characterization of a reovirus-like agent associated with neonatal calf diarrhoea, *Can. J. Comp. Med.* **35**:195.

Winton, J.R., Lannan, C.N., Fryer, J.L. and Kimura, T., 1981, Isolation of a new reovirus from chum salmon in Japan, *Fish Pathology,* **15**:155.

CHAPTER 2

# The Reovirus Particle

WOLFGANG K. JOKLIK

## I. INTRODUCTION

The term reovirus (the prefix reo- being an acronym of respiratory enteric orphan, that is, not associated with any human disease) was proposed by Sabin (1959) as the group name for a number of related viruses that had been isolated in the 1950s. This was an extremely active period for virus isolation: tissue-culture methods permissive for virus growth *in vitro* were being developed rapidly, and under the influence of efforts to develop mass vaccination against poliomyelitis, interest in enteric viruses was intense. Many cytopathogenic agents that were neither poliovirus nor coxsackie virus were isolated from the alimentary tracts, not only of patients with mild febrile illness and diarrhea but also of healthy individuals (Robbins *et al.*, 1951; Ramos-Alvarez and Sabin, 1954). Most of these agents possessed rather similar properties and were therefore lumped together under the acronym echovirus (enteric cytopathogenic human orphan); these viruses are now classified as members of the *Enterovirus* genus of the Picornaviridae family. Some isolates, however, while meeting the original and appropriately loose definition of echoviruses, were different: they were much larger and produced different cytopathic effects. These were the viruses for which Sabin proposed to term reovirus. While morphologically identical, they could be subdivided into three serological subgroups or serotypes: the original echovirus 10 became the Lang strain of reovirus serotype 1, the D5 Jones strain became the prototype of reovirus serotype 2, and the Dearing strain became the prototype of reovirus serotype 3.

WOLFGANG K. JOKLIK • Department of Microbiology and Immunology, Duke University Medical Center, Durham, North Carolina 27710.

The mammalian reoviruses are ubiquitous in nature and have been isolated from a wide variety of species including primates, monkeys, cattle, mice (Rosen, 1962), dogs (Lou and Wenner, 1963), cats (Scott *et al.*, 1970), and even an isolated population of quokkas (Stanley and Leak, 1963). All these viruses belong to the three serotypes described above. For example, a virus originally isolated from man in suckling mice (Stanley *et al.*, 1953) and found to produce hepatoencephalomyelitis in mice turned out to be reovirus serotype 3 (Stanley, 1961), and two viruses previously isolated from *Macaca* monkeys and designated simian virus 12 (SV12) and SV59 were found to be reovirus serotype 1 and 2 isolates, respectively (Rosen, 1960). The wide range of reoviruses in nature has also been demonstrated by the presence of antibodies to them in camels, cats, cattle, chickens, dogs, guinea pigs, hares, horses, marsupials, mice, monkeys, pigs, quokkas, rabbits, rats, sheep, and others (Stanley, 1967). Collectively, all these viruses form the genus *Orthoreovirus*.

In addition to these mammalian hosts, orthoreoviruses (or reoviruses for short) have also been isolated from birds, and antibodies to reoviruses have been detected in birds, bats, and reptiles. While it seems clear, however, that the reoviruses that circulate in mammals all belong to serotype 1, 2, or 3, the situation is less clear for the reoviruses that can be isolated from other vertebrates. Those best characterized are the reoviruses isolated from birds. Such viruses are readily isolated from a wide variety of species, particularly chickens, turkeys, and ducks (Wooley *et al.*, 1972; Simmons *et al.*, 1972); see also Olson, 1968). Some are associated with respiratory or enteric illness (Kawamura *et al.*, 1965; Deshmukh *et al.*, 1968; Fahey and Crawley, 1954; Petek *et al.*, 1967); they can also be isolated from birds expressing clinical tenosynovitis/arthritis (van der Heide *et al.*, 1974; see also van der Heide, 1977). Like mammalian reoviruses, they can also be isolated readily from asymptomatic animals (Kawamura *et al.*, 1965). It has been suggested that they can be grouped into five serotypes by means of the neutralization test (Kawamura *et al.*, 1965), but a great deal of work remains to be done to place their serological relationship on a firm and systematic footing; certainly they share common antigens when examined by complement-fixation and agar gel precipitation tests (Kawamura and Tsubahara, 1966). Avian reoviruses are related serologically to mammalian reoviruses as judged by a variety of tests, some being more closely related to one particular mammalian reovirus serotype than to others (Deshmukh *et al.*, 1968).

An interesting virus is the Nelson Bay virus, a reovirus isolated from the flying fox Petropus poliocephalus (Gard and Compans, 1970; Gard and Marshall, 1973), that has characteristics that appear to be intermediate between reoviruses of mammalian and avian origin: it shares a common complement-fixing and immunofluorescent antigen with mammalian reoviruses, but possesses the cell fusion activity characteristic of avian reoviruses (Wilcox and Compans, 1982).

Avian reoviruses differ from mammalian ones by their lack of hemagglutinating ability, their ability to cause cell fusion (Kawamura *et al.*, 1965), and by their host range. Many avian reoviruses only grow in avian cells (Glass *et al.*, 1973), although the initial steps of infection do proceed in L cells and four genes are transcribed (Spandidos and Graham, 1976a). Interestingly, *all* genes of the S1133 strain of avian reovirus are transcribed in L cells that are simultaneously infected with mammalian reovirus serotype 3, which Spandidos and Graham attributed to the possible presence of a cellular repressor that is removed by infection with the mammalian virus. However, no progeny avian reovirus genes are formed even in doubly infected cells (Spandidos and Graham, 1976a). It is not known how universal this inability of avian reoviruses to grow in mammalian cells is, since it has been observed that some avian reoviruses are indeed able to multiply in L cells (Gaillard, personal communication).

Very little work of a molecular nature has been carried out on avian reoviruses apart from the work of Spandidos and Graham discussed above, which also included a physical and chemical characterization of the S1133 agent (Spandidos and Graham, 1976b), and a study of the relative migration rates in polyacrylamide gels of the genes of several avian reovirus strains by Gouvea and Schnitzer (1982) that demonstrated the same type of variability (polymorphism) as had also been observed by Hrdy *et al.* (1979) for mammalian reoviruses. Most of the work on avian reoviruses has been concerned with the nature of the pathogenesis of many different isolates and with some of their biological properties [for a review see van der Heide (1977)]. Therefore, this chapter deals almost exclusively with the mammalian reoviruses which are by far the most thoroughly studied members of the genus orthoreovirus.

There are three serotypes of mammalian reoviruses. As judged by measurement of the fraction of single-stranded regions in hybrid genomes the component plus and minus strands of which are those of different serotypes, naturally occurring strains of serotypes 1 and 3 are related to each other to the extent of about 70%, while strains of serotype 2 are related to those of the other two serotypes to the extent of only about 10% (Martinson and Lewandowski, 1975; Gaillard and Joklik, 1982). Strains of these three serotypes are indistinguishable morphologically. Their cognate genes and the proteins that they encode are similar in size but not identical; this permits gene and protein sizes to be used as genetic markers (Ramig *et al.*, 1977a; Sharpe *et al.*, 1978). Most studies of the biochemistry and molecular biology of reovirus have been carried out on one virus strain, the Dearing strain of serotype 3, and most of the discussion that follows will apply to it. Recently, the usefulness of recombinants for elucidating the functions of reovirus proteins has become more appreciated, and this has increasingly focused attention on the other two reovirus serotypes. The most commonly used strains of serotypes 1 and 2 are the Lang strain and the D5 Jones strain, respectively.

## II. MORPHOLOGY

### A. Reovirus Particles

Reovirus particles possess not one but two capsid shells, each composed exclusively of protein arranged in discrete morphological subunits or capsomers (Gomatos *et al.*, 1962; Jordan and Mayor, 1962; Loh *et al.*, 1965) (Fig. 1). The outside diameters of the outer and inner shells are about 76 and 52 nm, respectively; the diameter of the central cavity is about 38 nm (Luftig *et al.*, 1972). Thus, the outer and inner shells are about 24 and 14 nm thick, respectively, and account for 68 and 19,5% of the particle volume, respectively; the central cavity accounts for only 12.5%. The diameters given above are the "electron-microscopic" diameters, which are considerably smaller than the "hydrodynamic" diameters derived from measuring the diffusion coefficients of intact virus particles and particles from which the outer shell has been removed, which are 96 and 70 nm, respectively (Harvey *et al.*, 1974). The fact that the hydrodynamic diameters exceed the electron-microscopic diameters indicates that extensive shrinkage occurs during the fixation and dehydration procedures involved in preparing virus particles for electron-microscopic examination.

The particle that lacks the outer shell is known as the core. The outer shell is stable at high and physiological salt concentrations, but it

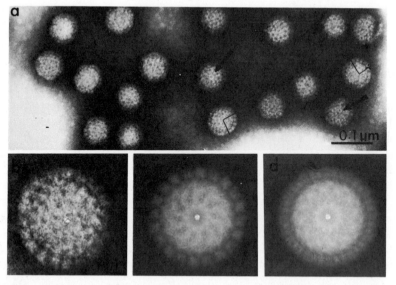

FIGURE 1. (a) Reovirus particles stained with 4% phosphotungstic acid. Note that two particles are situated such that they reveal 20 peripheral capsomers. (←) Structures that appear to be core projections or spikes that penetrate through the outer capsid shell to the surface of the particle. × 135,000. (b, c, d) These represent $n$ = 1-, 10-, and 12-fold rotations, respectively, for a selected particle. × 450,000. From Luftig *et al.* (1972).

loses capsomers on storage at low ionic strength (Amano *et al.*, 1971).
Under certain conditions, it is disrupted by heat (Mayor and Jordan, 1968)
and by NaCNS (Almedia *et al.*, 1979), and it is susceptible to digestion
by chymotrypsin, to which cores are completely resistant (Shatkin and
Sipe, 1968a; Smith *et al.*, 1969). Virus particles and cores are readily separated by density-gradient centrifugation: their sedimentation coefficients are about 630 and 470 S, and their buoyant densities in CsCl are
1.36 and 1.43 g/ml, respectively (Smith *et al.*, 1969). The average particle
densities (anhydrous molecular weight plus bound water/vol) of virus
particles and cores are 1.14 and 1.42 g/ml, and their water contents are
1.5 and 0.3 ml/g particle, respectively (Harvey *et al.*, 1974).

Although it is easy to see that both the outer and inner capsid shells
are composed of capsomers, either their arrangement or their nature is
such that it has so far been impossible to determine their spatial interrelationships with certainty. The original interpretation was that the
outer shell consists of 80 hexagonal and 12 pentagonal hollow columnar
or prismatic capsomers arranged according to $5:3:2$ icosahedral symmetry ($T = 9$, 4 capsomers along each edge) (Vasquez and Tournier, 1962;
Mayor *et al.*, 1965). Another interpretation favored the view that there
are 92 holes on the particle surface, 12 of them surrounded by five structural subunits and the rest by six, which are shared in such a manner
that the total is 180 (Vasquez and Tournier, 1964). Studies on virus particles stored in water provided support for the latter view, since this treatment causes loosening and disruption of the outer capsid shell with the
generation of structures that on six-fold rotation clearly revealed six subunits arranged hexagonally in the form of rosettes (Amano *et al.*, 1971).
Later, evidence was obtained that was difficult to reconcile with this
model (Luftig *et al.*, 1972), this evidence being that the outer capsid shell
appears to possess 20 peripheral capsomers, and not 18, as demanded by
either of the two interpretations described above. If there are 20 peripheral
capsomers, then there must be a total of $400/\pi$, or 127, capsomers if they
are all aligned exactly on a meridian, or slightly fewer if, as is likely, this
is not the case. On the basis of this model, the capsomer arrangement
was postulated to conform to the class of icosahedral deltahedra for which
$P = 3$ and $T = 12$ and the number of morphological units is 122 (Caspar
and Klug, 1962). Luftig *et al.* (1972) found that the capsomers appeared
to be either cylinders or truncated pyramids, 9 nm in length, and separated
by about 1 nm from their neighbors. This is quite different from the
hexagonal rosettes observed by Amano *et al.* (1971), which were 18–20
nm in diameter and had a hole 4–6 nm in diameter.

Palmer and Martin (1977) reexamined the fine structure of the reovirus outer capsid shell by negative-contrast electron microscopy and
enhancement of image detail by rotational analysis. They found the capsid to be composed of large capsomers measuring 18 nm in diameter.
Most of these capsomers were hexagonal units composed of six separate
wedge-shaped subunits that are in turn made up of three smaller subunits

and have the architecture of a $T = 9$ icosadeltahedron. The capsid itself seems to be composed of 32 morphological units according to a $T = 3$ morphology. A prominent feature of the architecture of the reovirus capsid is clearly a sharing of subunits, apparently a unique feature of all members of the Reoviridae family. It is this subunit sharing that makes precise assignment of the symmetry system so difficult.

Palmer and Martin (1977) also pointed out two additional novel features: First, they showed that the number of peripheral capsomers depended on the orientation of the virus particle, that is, on whether it is viewed along five-fold or along two- or three-fold axes of symmetry. In the former case, there are 20 large peripheral capsomers (enhanced by $n = 10$ rotation), whereas in the latter, there are 36 smaller peripheral subunits (enhanced by $n = 9$ rotation). They also found either five-fold or six-fold clustering of morphological units, depending on how virus particles were situated. Second, they found some virus particles that were positioned in such a way that peripheral indentations were evident. Five such indentations were visible when favorably oriented particles were viewed on an axis of five-fold symmetry, which suggests that there are 12 such indentations on the virus particle surface, distributed icosahedrally. The relevance of this finding will become apparent in Section II.B, where the structure of cores is discussed.

Very recently, Metcalf (1982) found whole $T = 13$ triangles among the capsid fragments that are often found as minor contaminants in negatively stained reovirus-particle preparations. Using a computer program to match spherical triangulated icosahedra with triangulation numbers between 9 and 16 with optically low-pass filtered images of reovirus particles prepared by the mica-sandwich freeze–etching technique, he showed that the $T = 13$, $z$ structure provides a better fit than any other. $T = 13$, $z$ triangulation has also been established for the bacteriophage T2 head (Branton and Klug, 1975) and for rotavirus (Roseto $et$ $al.$, 1979).

## B. Cores

In contrast to the rather labile outer shell, the core shell is an extremely stable structure that cannot be disrupted except by high concentrations of urea, guanidine, dimethylformamide, dimethylsulfoxide, sodium dodecyl sulfate, and the like. Its most obvious morphological feature is 12 projections or spikes that are located as though they were on the 12 five-fold vertices of an icosahedron (Luftig $et$ $al.$, 1972) (Fig. 2). They are about 10 nm in diameter, possess central channels about 5 nm wide, and project about 5.5 nm beyond the core surface, that is, about half the distance through the outer capsid shell. It is very likely that the location of these projections or spikes on the cores coincides with the indentations noted by Palmer and Martin (1977) in the outer capsid shell. In fact, there is evidence that the projections penetrate through the outer

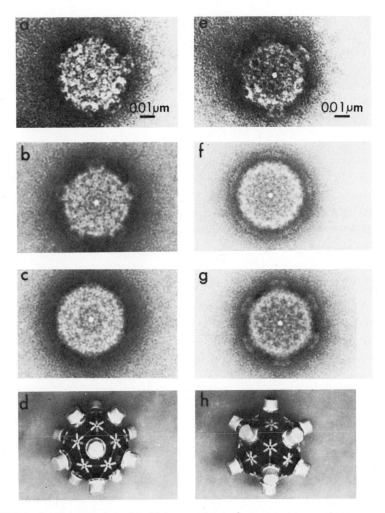

FIGURE 2. Application of the Markham rotation technique to two reovirus core particles the central axis of which was on either a presumptive five-fold (a) or three-fold vertex (e). Enhancement of the five peripheral spikes of (a) was achieved by an $n = 5$ rotation (b), but not by an $n = 6$ rotation (c). (d) Model that depicts the spike orientations when the central axis is through a five-fold vertex. Enhancement of the six peripheral spikes (e) was exhibited with an $n = 6$ rotation (g), but not an $n = 5$ rotation (f). (h) Model with the central axis through a three-fold vertex. The particles were stained with 2% uranyl acetate. ×450,000. From Luftig *et al.* (1972).

capsid shell to the surface of the reovirus particle; as discussed in Section IX, antibodies to the 140-kilodalton protein λ2, five molecules of which comprise a projection spike (Ralph *et al.*, 1980), are capable of reacting with intact reovirus particles, neutralizing their infectivity, preventing them from hemagglutinating red cells, and aggregating them (Hayes *et al.*, 1981). This suggests that there is no outer shell above the projections/spikes. It is known, however, that a minor outer-capsid shell component, protein σ1, is closely associated with protein λ2 on the virus particle surface and may partially cover it (Lee *et al.*, 1981b). Thus, it seems that in 12 icosahedrally distributed locations on the virus particle surface, through which pass the five-fold axes of symmetry, the structure of the particle surface is quite different from that of the remainder of the outer capsid shell. It is conceivable that in two virus particles shown in Fig. 1, the view is directly down into a projection/spike.

The remainder of the inner capsid shell is made up of morphological subunits that are smaller than those of the outer shell (4 nm in diameter) (Luftig *et al.*, 1972). Their overall arrangement appears to be similar to that of the capsomers of the outer shell, since here also suitably oriented particles clearly reveal 20 peripheral subunits. However, the precise arrangement of capsomers of the inner capsid shell has not been studied in detail.

## C. Top-Component Particles

The yield of icosahedral viruses often includes a small proportion of empty virus particles, that is, virus particles that contain no nucleic acid. These empty virus particles are known as top component, since their buoyant density is lower than that of virus particles. The relative amounts of empty and normal virus particles present in virus yields from infected cells can very widely; when virus yields are high, the number of empty particles is generally no more than 10% of that of normal virus particles, but when virus yields are low, the number of empty particles can exceed the number of normal virus particles that are formed. Omitting lysine from the medium (Loh and Oie, 1969) or adding ethidium bromide to it (M.-H.T. Lai *et al.*, 1973) increases the relative amount of empty virus particles. Empty and normal virus particles are formed concomitantly, as can be shown by harvesting cells at various stages of the multiplication cycle and purifying the yield. Empty particles are therefore not precursors of "full" particles, nor do they accumulate to any extent after most normal virus particles have formed. The morphology of empty virus particles is identical to that of normal virus particles except that they are penetrated by phosphotungstic acid stain (Smith *et al.*, 1969). Examination of empty particles shows very clearly that the outer capsid shell consists of relatively few large capsomers, whereas the inner shell consists of much smaller subunits and resembles a thin skin, rather like the heads of the T-even bacteriophages (Smith *et al.*, 1969). Interestingly, a temperature-sensitive (*ts*) mutant of reovirus, the group C mutant *ts* 447, forms, even at permissive temperatures, two kinds of top-component particles, one of which has the normal morphology, while the other lacks the inner capsid shell (Matsuhisa and Joklik, 1974). The reason for this appears to be a lesion in protein σ2, one of the major components of cores.

The average particle density of top-component particles is 1.11 g/ml, while their water content significantly exceeds that of virus particles (1.9 ml/g compared with 1.5 ml/g). As would be expected, the water content of top-component particles and cores is strikingly different, that of the former being more than 6 times greater (Harvey *et al.*, 1974).

## III. PURIFICATION

Reoviruses grow well in many cultured cell types; those most commonly used for the preparation of large amounts of virus are mouse L fibroblasts growing in suspension culture. Since infection at 37°C results in extensive cytopathic damage, infected cells are usually incubated at 33°C, a temperature at which cytopathic effects are much less marked and virus yields are therefore 3- to 5-fold larger (Smith *et al.*, 1969). Reovirus is not liberated readily from cells, but remains cell-associated until well after the virus has ceased to multiply; early purification procedures therefore employed chymotryptic digestion to dissociate viral from cellular components (Gomatos and Tamm, 1963). However, since it was subsequently found that the outer shell of reovirus particles is sensitive to chymotrypsin (Shatkin and Sipe, 1968a) (see Section IV.A), this practice has been replaced by homogenization with Freon 113 (Shatkin, 1965; Bellamy *et al.*, 1967; Smith *et al.*, 1969). Final purification is achieved by zone sedimentation in 20–40% sucrose density gradients, followed by equilibrium centrifugation in CsCl density gradients (Smith *et al.*, 1969). As a rule, two bands are obtained: a major band with a buoyant density of 1.36 g/ml that contains intact virus particles and a minor band with a buoyant density of 1.30 g/ml that contains empty virus particles (top component). Yields of purified virus as high as 30 mg/$10^9$ cells (300,000 virus particles/cell) can be obtained.

## IV. MEASUREMENT

### A. Plaque Formation

Reovirus replicate and produce cytopathic effects in human, monkey, hamster, and mouse cells, among others. For plaquing, monkey kidney cells (Rhim and Malnick, 1961) were used at first, but nowadays mouse L fibroblasts are preferred. The plaque size is characteristic of both the virus and the cell strain (McClain *et al.*, 1967). Among the commonly used strains of reovirus, strains Dearing and Abney (both type 3) produce plaques that are about 3 mm across after 4–5 days of incubation at 37°C, while those produced by strain Carter (type 3) , D5 Jones (type 2), and Lang (type 1) are about 1–2 mm in diameter after 5–10 days. Heating and cooling infectious culture fluids in 2 M $Mg^{2+}$ enhances the titer of reovirus (Wallis *et al.*, 1964); the mechanism involved is not known. The addition of proteolytic enzymes such as pancreatin or chymotrypsin to cell monolayers sometimes increases the size of plaques and therefore decreases the time before they can be counted (Wallis *et al.*, 1966). Another expression of this effect is probably the fact that the titers of unpurified preparations of some strains of reovirus, for example, strain Lang (type 1), are sometimes increased 10- to 50-fold by treatment with proteolytic enzymes such as chymotrypsin; the particles that become activated in this manner have been called "potentially infectious virus" (Spendlove *et al.*, 1966). The diameter of the virus particles appears to decrease at the same time, but they retain their hemagglutinating activity (Spendlove *et al.*, 1970). It is likely, therefore, that under certain conditions chymotryptic digestion of some strains of reovirus removes part of the outer capsid shell, thereby giving rise to particles with a specific infectivity greater than that of intact virus particles (see Section XII for a discussion of the mode of action of chymotrypsin on reovirus). This enhancement of infectivity may be mistaken for virus multiplication. An interesting example of this type was described by Whitcomb and Jensen (1968), who, in attempting to determine whether reovirus could multiply in leafhoppers, found that its titer increased following injection, but traced this effect to enhancement of infectivity by proteolytic enzymes present in insects, rather than to multiplication.

## B. Hemagglutination

Reovirus particles of all three serotypes agglutinate human erythrocytes of all four blood groups, the order of reactivity being A > AB > O > B (Brubaker *et al.*, 1964). Agglutination proceeds at both low (4°C) and elevated (37°C) temperatures. Serotype 3 strains also agglutinate bovine erythrocytes, but only at low temperatures (Eggers *et al.*, 1962).

## C. Enumeration of Virus Particles by Measurement of Optical Density

The total number of virus particles in purified reovirus preparations is best determined by measurement of optical density at 260 nm. The relationship between the optical density of suspensions of reovirus particles at 260 nm, the number of particles, and viral protein mass is 1 $ODU_{260nm} = 2.1 \times 10^{12}$ particles = 185 µg protein (Smith *et al.*, 1969).

Under optimal conditions of plaquing, the ratio of reovirus type 1 particles to plaque-forming units has been reported to be as low as 2:1 (Wallis *et al.*, 1964) and the ratio of chymotrypsin-activated reovirus particles to infectious units as measured by fluorescent cell count as low as 1:1 (Spendlove *et al.*, 1970). However, these ratios are 25–100 times lower than those commonly observed in routine assays of purified virus stocks.

## V. REOVIRUS GENOME

Studies of Gomatos and collaborators in the early 1960s that indicated that the RNA of reovirus particles was double-stranded (ds) created a great deal of interest, since they provided the first demonstration of the occurrence of dsRNA in nature. The original observation was that cytoplasmic inclusions in L cells infected with reovirus fluoresced pale green when treated with acridine orange (Gomatos et al., 1962). This is indicative of double-stranded nucleic acids, which bind only a small amount of the dye and therefore stain orthochromatically; single-stranded (ss) nucleic acids, which bind more dye, stain metachromatically, and fluoresce bright red. The inclusions proved to be resistant to deoxyribonuclease but susceptible to ribonuclease at low salt concentrations, and it was soon shown that the nucleic acid isolated from purified virus particles was indeed dsRNA (Gomatos and Tamm, 1963). The evidence rests on the following facts, among others: (1) The RNA exhibits very sharp melting profiles, with the $T_m$ depending on the ionic strength (Shatkin, 1965; Bellamy et al., 1967). (2) It is resistant to ribonuclease, the resistance depending on the concentration of both monovalent and divalent cations, as well as on the concentration of ribonuclease (Shatkin, 1965; Bellamy et al., 1967). (3) It is susceptible to ribonuclease III, which is specific for dsRNA. (4) Formaldehyde fails to induce hyperchromicity (Gomatos and Tamm, 1963). (5) Its density in $Cs_2SO_4$ is 1.61 g/ml, rather than 1.65 g/ml, which is characteristic of ssRNA (Shatkin, 1965; Iglewski and Franklin, 1967). (6) Its base composition indicates equality of A + U as well as of G + C (Gomatos and Tamm, 1963; Bellamy et al., 1967). (7) X-ray diffraction patterns are consistent with double-strandedness (Langridge and Gomatos, 1963; Arnott et al., 1966).

Attempts to extract this RNA from reovirus particles in the form of one long molecule failed, and evidence soon accumulated that the reovirus genome exists in the form of a collection of discrete and unique segments that proved to be genes. The first indication of this was the electron-microscopic demonstration that reovirus RNA consists of a population of molecules that exhibits a trimodal length distribution with maxima at 1.1, 0.6, and 0.35 μg, corresponding to molecular weights of about 2.5, 1.4, and 0.8 million (Gomatos and Stoeckenius, 1964; Dunnebacke and Kleinschmidt, 1967; Vasquez and Kleinschmidt, 1968). Clearly, even the largest of these molecules corresponded to only a portion of the genome, since reovirus particles had already been shown to contain about 14.6% RNA, corresponding to an aggregate molecular weight of at least $10 \times 10^6$ (Gomatos and Tamm, 1963). It was then shown that regardless of the means used to liberate it, reovirus RNA comprises three size classes of molecules when analyzed on sucrose-density gradients; these are the large (L), M (medium), and small (S) species of molecules, which sediment with 14, 12, and 10.5 S, corresponding to molecular weights of about 2.7, 1.4, and 0.7 million, respectively, or about 4500, 2300, and 1200 nucleotide base pairs (Shatkin, 1965; Bellamy et al., 1967); Watanabe and Graham, 1967). The molecules in these three size classes were shown to be discrete segments rather than random fragments of larger molecules because (1) they did not hybridize with each other and (2) they were transcribed into specific species of messenger RNA (mRNA) molecules within infected cells (Bellamy et al., 1967; Bellamy and Joklik, 1967a; Watanabe and Graham, 1967; Watanabe et al., 1967). These three size classes were then further separated by polyacrylamide gel electrophoresis into ten discrete and unique molecular species that are present in equimolar amounts and possess an aggregate molecular weight of about $15 \times 10^6$ (Shatkin et al., 1968). Figure 3 shows autoradiograms of polyacrylamide gels in which the genome segments (genes) of the Lang (serotype 1), D5 Jones (serotype 2), and Dearing (serotype 3) strains of reovirus had been electrophoresed. As pointed out in Section I, the three patterns are similar, but readily differentiable (Sharpe et al., 1978). Furthermore, none of the three patterns is invariate. Hrdy et al. (1979) examined the genome-segment patterns of 94 strains of reovirus isolates from human, cattle, and mice and found extensive variability. This variation was found in all three serotypes and involved all ten genome segments. Although a single pattern was present among several samples isolated from individuals and collected at a single time and place, multiple genetic variants of a single serotype were often present in a population. Samples isolated from widely different geographic origins or different mammalian hosts showed different patterns; samples from a single species from the same area over a period of time showed more limited variations. Electrophoretic migration rate can thus be used as a genetic marker.

Final proof of the segmented nature of the reovirus genome came with the demonstration that even prior to extraction, there are 20 3' termini in the RNA within each reovirus particle (Millward and Graham, 1970), and with the finding of Banerjee and Shatkin (1971) that all 10 dsRNA segments in reovirus particles possess ppGp at one of their 5' termini.

While there is no doubt that the ten reovirus genes are discrete molecules, the nature of their arrangement within virus particles is not clear. The two major questions are (1) how is the RNA packed within the central cavity and (2) are the genes linked in some regular linear arrangement? These two questions are important because the double-stranded reovirus genome must be transcribed into ssRNA (mRNA) to initiate infection, and it is transcribed without being uncoated, that is, while it is present within the core. Thus, movement of dsRNA sequences relative to the catalytic site of RNA polymerase must be possible within cores, and it is of interest to know how closely the dsRNA is packaged within the central cavity. Further, it is known that the ten genes are not transcribed with equal frequency but with a frequency that is inversely proportional to their size. Thus, it is of interest to learn the nature of the mechanism that is responsible for regulating the relative transcription frequencies.

With respect to the first of these questions, small-angle X-ray dif-

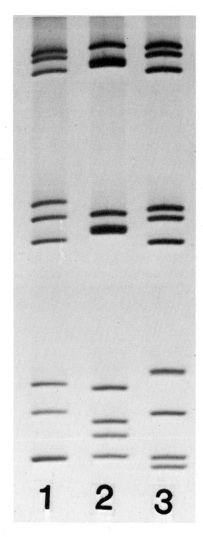

FIGURE 3. Genes of the Lang (serotype 1), D5 Jones (serotype 2), and Dearing (serotype 3) strains of reovirus. The genes were stained with ethidium bromide and the gels (7.5% polyacrylamide) photographed with UV light. The genes are denoted (starting from the top) *L1*, *L2*, and *L3*; *M1*, *M2*, and *M3*; and *S1*, *S2*, *S3*, and *S4*. Genes *S3* and *S4* of serotype 1 and *L2* and *L3* and *M2* and *M3* of serotype 2 are not resolved in this gel system, which is designed to resolve the genes of serotype 3. Kindly supplied by Dr. R.K. Gaillard.

fraction analysis has shown that the RNA exists within cores in a well-ordered packaging arrangement, with adjacent helices aligned locally parallel to each other and hexagonally in cross section, rather like the DNA of the T-even bacteriophages (Harvey *et al.*, 1981). If the entire reovirus genome were packed in this manner, it would pack into a sphere with a diameter of 47 nm, which is almost exactly the same as the observed value of 49 nm for the outer small-angle X-ray diffraction diameter of the RNA sphere within reovirus particles. Thus, the reovirus genome exists within virus particles in a closely packed condensed form, yet is transcribed within them. Evidently, the constraints of close packing prevent neither local strand separation nor movement either of genes past fixed enzyme sites or of enzyme molecules along immobile genes.

As for the second question, Granboulan and Niveleau (1967) were

able to liberate, with very low frequency and demonstrable only by electron-microscopic analysis, very long linear arrays of RNA from reovirus
particles; more recently, Kavenoff et al. (1975) were able to release the
RNA in the form of spiderlike arrangements with all genes connected
at a common focus. Such evidence for possible linkage of the ten genes,
which is very weak since the genes are released from cores far more often
in unlinked than in linked form, must be reconciled with the gene transcription-frequency pattern. Namely, individual genes are transcribed
with frequencies that are inversely proportional to their size; that is, all
genes are transcribed at the same rate, and the time interval between
successive rounds of transcription is either very short or proportionately
longer for $L$ than for $M$ and $S$ genes. This transcription-frequency pattern
is most readily accounted for by postulating that each gene is transcribed
independently of all others; that is, there is no linkage at all between the
genes. The only scheme that would account for the observed transcription-frequency pattern if there *is* linkage between the genes is for transcription to start at one end of a linear array in which the most and least
frequently transcribed genes are located proximally and distally respectively to the transcription origin and the likelihood of premature termination increases with increasing length of the transcribed sequence.
This is how relative frequency of transcription of rhabdovirus and par-
amyxovirus genes appears to be controlled (Abraham and Banerjee, 1976).
On balance, it seems that the ten reovirus genes are *not* linked in the
virus particle, but the evidence is of a negative rather than a positive
nature. As discussed in Section VI, the function of several species of proteins that are present in reovirus cores in the form of very few molecules
only is not known, and they could serve as linkers. Certainly, a highly
specific mechanism, most probably involving interaction with protein
molecules, must be responsible for assembling the ten discrete species
of RNA into each virus particle (Section X).

## VI. PROTEINS OF REOVIRUS

Protein composition of reovirus particles was first examined by Loh
and Shatkin (1968) and Smith et al. (1969). The nomenclature of reovirus
proteins [the $\lambda$, $\mu$, and $\sigma$ protein species being encoded by the $L$ (large),
$M$ (medium), and $S$ (small) size classes of genes] is due to the latter group
of workers.

Reovirus particles are composed of nine species of proteins (Fig. 4
and Table I). The outer capsid shell is composed of three species: the
major species $\mu 1C$ [a cleavage produce of protein $\mu 1$ (Zweerink and Joklik,
1970)] and $\sigma 3$, which together make up of over 60% of the mass of the
reovirus particle, and the minor protein species $\sigma 1$, which is present in
reovirus particles to the extent of only about 24 molecules. The core,
which comprises about one third of the reovirus particle protein mass,

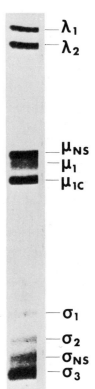

FIGURE 4. Proteins of the Dearing strain of reovirus serotype 3. The gel is of the cytoplasmic fraction of infected cells labeled with [$^{35}$S]methionine for 90 min at 10 hr after infection, when only reovirus-coded proteins are synthesized. Proteins λ3 and μ2 comigrate with proteins λ2 and μ1C, respectively.

TABLE I. Reovirus Proteins

| Species | Molecular weight | Percentage in virion | Approximate number of molecules per virus particle | Location |
|---|---|---|---|---|
| Structural proteins | | | | |
| λ1 | 155,000 | 15 | 105 | Core |
| λ2 | 140,000 | 11 | 90 | Core |
| λ3 | 135,000 | ≲2 | ≤12 | Core |
| μ1 | 80,000 | 2 | 20 | Core |
| μ1C | 72,000 | 35 | 550 | Outer shell |
| μ2 | 70,000 | ≤2 | ≤12 | Core |
| σ1 | 42,000 | 1 | 24 | Outer shell |
| σ2 | 38,000 | 7 | 200 | Core |
| σ3 | 34,000 | 28 | 900 | Outer shell |
| Nonstructural proteins | | | | |
| μNS | 75,000 | | | |
| σNS | 36,000 | | | |

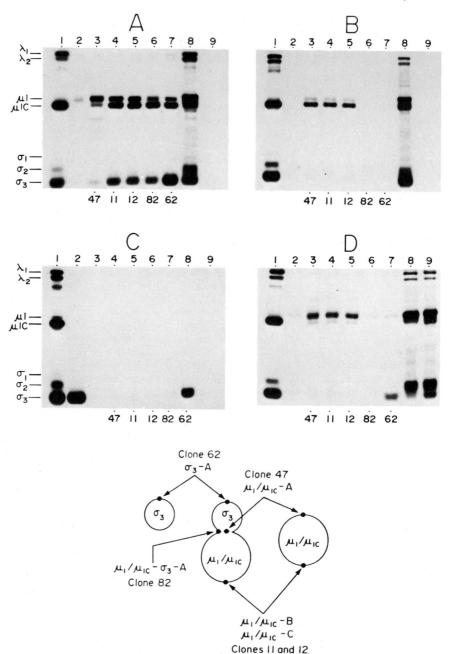

FIGURE 5. Demonstration that most of proteins μ1/μ1C and σ3 exist in infected cells complexed with each other. (A) The cytoplasmic fraction of L cells infected with reovirus serotype 3 and labeled with [$^{35}$S]methionine from 10.5 to 12 hr after infection and then harvested, centrifuged at 45,000 rpm for 1 hr, and preabsorbed with myeloma cell supernatant was mixed with five monoclonal antibodies (identified by the numbers beneath the gel). All five antibodies precipitated proteins μ1/μ1C (although with varying efficiencies),

contains six protein species: the major components λ1, λ1, and σ2 and the minor components λ3, μ1, and μ2.

With one exception, all reovirus proteins that have been examined possess blocked amino-terminal groups (Pett *et al.*, 1973). The exception is protein μ1C, the N-terminal amino acid sequence of which is $NH_2$-Pro-Gly-Gly-Val-Pro-. Thus, the peptide of molecular weight 8000 that is removed during the conversion of μ1 to μ1C is probably derived from the former's amino-terminal end (Pett *et al.*, 1973).

The structural, antigenic, and enzymatic functions of the various reovirus proteins, as presently known, are discussed in the following sections.

## A. Outer Capsid Shell

### 1. Proteins μ1C and σ3

Protein μ1C is a cleavage product of protein μ1, as was shown by Zweerink and Joklik (1970) by kinetic analysis of their relative rates of formation and by McCrae and Joklik (1978) by peptide mapping analysis. It is the principal component of the reovirus outer capsid shell, as well as of reovirus particles. Hayes *et al.* (1981) isolated four monoclonal antibodies against it; none neutralized infectivity or prevented hemagglutination by reovirus particles, but two were capable of precipitating/aggregating reovirus particles.

Protein σ3 is the other principal constituent of the outer capsid shell. Proteins σ3 and μ1C possess strong affinity for each other, as is shown by the fact that about 80% of the unassembled form of each (that is, not in the form of immature or mature virus particles) exists in the cytoplasm of infected cells complexed with each other. This has been shown in two ways: First, when S100 extracts of L cells infected with reovirus are centrifuged into sucrose density gradients, the unassembled reovirus-specified proteins become distributed according to their size; but the most rapidly sedimenting proteins are not, as expected, the λ species proteins, but a complex of μ1C and σ3 (Huismans and Joklik, 1976). Second, Lee *et al.*(1981a) isolated four monoclonal antibodies that precipitated both

---

and four also precipitated protein σ3. (B) The cell extract had been treated with 1% sodium dodecyl sulfate to dissociate the complex between proteins μ1/μ1C and σ3. Now, antibodies 47, 11, and 12 still precipitated proteins μ1/μ1C, but antibodies 82 and 62 no longer precipitated either protein. (C) The ability of the five antibodies to precipitate free σ3 was tested; only antibody 62 could do so. (D) The cell extract was preabsorbed with antibody from clone 82. Now, antibodies 47, 11, and 12 reacted only with uncomplexed μ1/μ1C (note that much more protein μ1 is now in complex form than μ1C, which suggests that μ1 is cleaved to μ1C in the process of complex formation), and antibody 62 now precipitated only free protein σ3. *Bottom:* The chart is of a model depicting the putative locations of the antigenic sites on the complex between proteins μ1/μ1C and σ3, and on free proteins μ1/μ1C and σ3. From Lee *et al.* (1981a).

μ1C and σ3. Two of these antibodies were also capable of precipitating free μ1C and one was also capable of precipitating free σ3, while one was incapable of precipitating either free μ1C or free σ3. These specificities were explained as illustrated in Fig. 5: the binding site for the first two antibodies is on μ1C and that for the third on σ3, while that for the fourth includes sequences of both μ1C and σ3. In fact, a fifth antibody was also isolated that precipitated only free μ1C, but not the complex; the antigenic binding site for that antibody was evidently on that portion of μ1C that was either at or close to the μ1C–σ3 binding site.

Since μ1C and σ3 forms a complex in their free form, they are presumably also intimately associated with each other in virus particles, in which they may exist in the form of capsomers with the constitution of $1n$ μ1C:$2n$ σ3 (with the value of $n$ perhaps being 2, since μ1C exists in virus particles as a disulfide-bonded dimer (Smith *et al.*, 1969). It should be noted, however, that (1) when chymotrypsin digests the outer capsid shell, σ3 is removed from virus particles before μ1C is degraded (Joklik, 1972); (2) following infection, reovirus particles are converted to subviral particles from which σ3 is removed completely, whereas μ1C loses only a polypeptide with a molecular weight of about 12,000, being converted to protein δ (Silverstein *et al.*, 1970, 1972; Chang and Zweerink, 1971); and (3) antibody to σ3 neutralizes infectivity and possesses hemagglutination-inhibition activity, while antibodies to μ1C do not (Hayes *et al.*, 1981; see also above). Thus, although closely associated, proteins σ3 and μ1C can also react independently.

Protein μ1C controls the susceptibility of the reovirus outer capsid shell to proteolytic digestion (Rubin and Fields, 1980). When chymotrypsin is added to reovirus, proteins σ3 and σ1 are removed, and protein μ1C is degraded via a series of intermediates that remain more or less transiently associated with the virus particle (see Section XII). Conditions are readily arranged such that the outer shell of virus particles of serotype 3 is degraded completely, that of particles of serotype 2 much less so, and that of particles of serotype 1 not at all. By examining the susceptibility of the outer capsid shells of recombinants of the three serotypes, Drayna and Fields (1982) were able to show that such susceptibility is controlled by the *M2* gene, which encodes protein μ1C (see Section VII).

Protein σ3 possesses the remarkable property of having strong affinity for double-stranded RNA (dsRNA), and that portion of it that occurs in free form (that is, not complexed with μ1C) in the cytoplasm of infected cells can be isolated by adsorption to and elution from polyriboinosinict:polycytidylic acid (Huismans and Joklik, 1976). The significance of this property, a remarkable one for a protein that is a component of the outer capsid shell, is not known; perhaps σ3 has some function during reovirus morphogenesis. Protein σ3 is also the protein that is responsible for inhibiting cellular RNA and protein synthesis following infection (Sharpe and Fields, 1982).

## 2. Protein σ1

The third component of the reovirus outer capsid shell is a protein with a molecular weight of about 42,000 that is present in each virus particle to the extent of only a few molecules (Smith *et al.* (1969). Its location in the outer capsid shell has recently been pinpointed by the use of monoclonal antibodies. Lee *et al.* (1981b) found that antibody to protein λ2, but not antibody to either μ1C or σ3, can prevent antibody against σ1 from binding to it, which indicates that σ1 is located close to where the projections/spikes penetrate through the outer capsid shell to the outer particle surface. Since there are 12 of these projections/spikes and since the total number of σ1 molecules per virus particle is about 20 (Smith *et al.*, 1969), two σ1 molecules are presumably associated with each of these structures, and the total number of σ1 molecules in the intact reovirus outer capsid shell is probably 24.

Protein σ1 is the reovirus cell-attachment protein. It can adsorb to cells by itself; when an extract of infected radioactively labeled cells is added to uninfected cells, free protein σ1 attaches to cells via the same receptors as are used by intact virus particles (Lee *et al.*, 1981b).

Protein σ1 is the most type-specific of all reovirus proteins; the ability of antibodies against σ1 to precipitate homologous and heterologous σ1 molecules (that is, those of the other two serotypes), to neutralize virus, and to inhibit hemagglutination is almost completely type-specific (Lee *et al.*, 1981a; Gaillard and Joklik, 1980; Hayes *et al.*, 1981). Not surprisingly, the serotype 1, 2, and 3 *S1* genes, which encode protein σ1 (see Section VII), differ greatly in sequence (Li *et al.*, 1980a) and exhibit negligible ability to hybridize with each other [judging by the single-stranded (ss) region content of intergenic heteroduplexes (Gaillard and Joklik, 1982)]. However, the type specificity is not absolute. For example, antiserum to reovirus serotype 3 precipitates the σ1 proteins of serotypes 1 and 2 more than 100 times less efficiently than the homologous σ1, but antiserum to reovirus serotype 2 behaves differently: it precipitates the σ1 of serotypes 1 and 3 about one third and one tenth as efficiently as the homologous σ1. Thus, the specificity is neither absolute nor symmetrical (Gaillard and Joklik, 1980).

Since protein σ1 is the reovirus cell-attachment protein, it has the very important biological role of specifying how reovirus particles interact with host cells and with the host. Among the interactions for which it is responsible are the following: it is the reovirus hemagglutinin (Brubaker *et al.*, 1964; Eggers *et al.*, 1962; Weiner *et al.*, 1978), it elicits the formation of neutralizing antibody (Weiner and Fields, 1977) (but it is not the *only* reovirus protein that does so), it is responsible for the development of delayed-type hypersensitivity (Weiner *et al.*, 1980b) and for the generation of suppressor T cells (Fontana and Weiner, 1980; Greene and Weiner, 1980) and cytolytic T lymphocytes (Finberg *et al.*, 1979,

1981), and it specifies reovirus tissue tropism and virulence,* particularly neurovirulence (Weiner et al., 1977, 1980c) (see also Chapter 6).

Interestingly enough, protein σ1 not only profoundly influences how reovirus interacts with cells, but also has very important effects within the cell; for example, it determines the extent to which reovirus particles interact with microtubules (Babiss et al., 1979), and it appears to be the protein that inhibits cellular DNA replication (Sharpe and Fields, 1981). The fact that it is protein σ1 that is responsible for this effect was found by combining comparative virology with genetics. It is found that strains of reovirus serotype 3 inhibit DNA synthesis in mouse cells, whereas strains of serotype 1 do not. By using recombinant viruses that contain various assortments of genes of these two serotypes, it was shown that the inhibition of cellular DNA synthesis is a property of the S1 gene (which encodes protein σ1).

Evidence has recently been obtained for the presence of at least four distinct antigenic regions (epitopes) on protein σ1 of reovirus serotype 3. This was shown by analyzing the efficiency with which several monoclonal antibodies combine with σ1, neutralize infectivity, and inhibit neutralization. The ratios of these three activities differ widely among 15 monoclonal antibodies studied by Lee et al. (1981a,b) and Burstin et al. (1982), which suggests the existence of functional domains in protein σ1.

## B. Inner Capsid Shell: The Core

### 1. Proteins λ1, λ2, and σ2

The major components of reovirus cores are proteins λ1, λ2, and σ2. Protein λ2 is the major, and most likely the only, component of the core projections/spikes. White and Zweerink (1976) showed that when cores are incubated for 15 min at pH 11.8 at 4°C, the projections/spikes are removed, and so is protein λ2. Subsequently, Ralph et al. (1980) showed that each projection/spike is a pentamer of λ2. Thus, there should be 60 molecules of λ2 per core. This new estimate of the number of λ2 molecules, which seems quite firm and reasonable, conflicts with an earlier estimate that also appeared to be solidly based (Smith et al., 1969). The total molecular weights of the protein components of reovirus particles and cores are 71 and 39.5 million, respectively, as calculated using the

---

* For example, murine and human lymphocytes have a receptor for the σ1 of reovirus serotype 3 but not for that of serotype 1 (Weiner et al., 1980a), even though both use the same receptor on mouse L fibroblasts (Lee et al., 1981b). It should be noted, however, that tissue tropism is not controlled solely by ability to adsorb. Ability to initiate the subsequent stages of the infection cycle is also essential, and certain aspects of these reactions are controlled by the nature of protein μ1C (Rubin and Fields, 1980). The relative roles of proteins σ1 and μ1C in tissue tropism and virulence are discussed in Chapter 6.

molecular-weight and number-per-virus-particle estimates for the individual proteins reported by Smith *et al.* (1969) (and assuming the number of λ3 and μ2 molecules per virus particle to be 6 each). These figures are in excellent agreement with the values of 72 and 37 million, respectively, obtained by Farrell *et al.* (1974) on the basis of ultracentrifugation and dynamic light-scattering analyses. The problem is that Smith *et al.* (1969) calculated the number of λ2 molecules per virus particle to be 80, rather than 60. Clearly, further work will be required to resolve this discrepancy.

Since the structures of the outer and inner capsid shells of reovirus appear to be similar, both displaying 20 peripheral morphological subunits if viewed along appropriate axes of symmetry, the arrangement of protein subunits in the outer and inner capsid shells may also well be similar, if not identical. Thus, if outer capsid shell capsomers are made up of 1*n* molecules of μ1C and 2*n* molecules of σ3, the inner capsid shell capsomers may be made up of 1*n* molecules of λ1 and 2*n* molecules of σ2, with λ2 and σ1 being components of the projections/spikes. White and Zweerink (1976) showed that protein λ1 of cores is more readily iodinated than protein σ2, whereas both are equally efficiently iodinated in free form. Protein σ2 may therefore be located predominantly on the inner core shell surface, while protein λ1 may be located predominantly on the outer core shell surface.

## 2. Minor Components λ3, μ1, and μ2

Reovirus cores also contain small amounts (12 molecules or less) of three other protein species, namely, λ3, μ1, and μ2. The fact that protein μ1 is present in cores is easy to demonstrate (Smith *et al.*, 1969); the presence of proteins λ3 and μ2 is more difficult to document, since they migrate very close to proteins λ2 and μ1C, respectively, in the commonly used electrophoretic systems. In fact, both were first discovered as *in vitro* translation products of reovirus messenger RNA (mRNA) species (Both *et al.*, 1975; McCrae and Joklik, 1978). Their precise location within cores [whether components of the core shell (outer or inner surface) or free or associated with RNA in the central cavity] is not known. Conceivably, they may be responsible for some of the enzymatic activities exhibited by reovirus particles (see Section XIII), or they may serve to link the genes (see above).

A recent report by Drayna and Fields (1982) is very interesting in this regard. They found that the transcriptases of the three serotypes of reovirus have characteristic pH optima. Using recombinant viruses that possess some genes of one serotype and some of another, they found that the transcriptase pH optimum is specified by the *L1* gene, which codes for protein λ3. It will be very interesting to determine whether this minor core component is really part of the transcriptase molecule.

Reovirus particles also contain about 500 molecules of "*component viii*," a very small protein with a molecular weight of 5000–10,000 (Smith *et al.*, 1969). Since the number of component viii molecules per virus particle is about the same as that of protein μ1C molecules, and since the size of component viii is similar to the difference in the molecular weights of proteins μ1 and μ1C, it has been suggested that component viii may be the other cleavage product in the conversion of μ1 and μ1C. However, no direct evidence to this effect has yet been obtained.

It is clear that although the outline, in molecular terms, of reovirus structure is known, many important and essential pieces of information are not yet available. Some of these may derive from a recently developed approach. Ramig *et al.* (1977b) and Ramig and Fields (1979) have found that many spontaneous revertants of temperature-sensitive mutants are not true revertants, but are able to grow at temperatures nonpermissive for the mutant by virtue of possessing extragenic suppressor mutations (see also Chapter 5). In other words, proteins defective at elevated temperatures are enabled to resume normal function by virtue of an alteration in some other protein. Since such pairs of proteins would presumably have to interact in the virus particle, one would expect very specific "reactivation" patterns, which may provide important clues concerning reovirus morphogenesis and structure.

## C. Nonstructural Reovirus Proteins

The reovirus genome also codes for two proteins, μNS and σNS, that are not virus-particle components; they are nonstructural proteins (Zweerick *et al.*, 1971). Protein μNS exists in infected cells in two forms: in the form of the entire protein and in the form of a cleavage product, μNSC, that is about 5000 daltons smaller. Infected cells contain roughly equal amounts of μNS and μNSC (Lee *et al.*, 1981a).

The other nonstructural reovirus protein, σNS, possesses strong affinity for ssRNA and exists in infected cells as a complex with cellular and viral mRNA. It can be readily purified by passing extracts of infected cells through columns of poly(A) or poly(U) linked to cellulose and eluting with NaCl (Huismans and Joklik, 1976). Both reovirus nonstructural proteins are remarkable for being synthesized in infected cells in relatively large amounts; unlike other nonstructural virus-coded proteins that are often produced in small amounts only, proteins μNS and σNS are among the most abundantly produced reovirus-coded proteins in infected cells.

The functions of proteins μNS and σNS are not known. Among possible functions (for σNS) are functions during morphogenesis (that is, assembling the unique sets of ten plus strands for encapsidation) and control of the translation of reovirus mRNA species. Furthermore, Shelton *et al.* (1981) recently reported that σNS also binds to DNA *in vitro*, and Gomatos *et al.* (1980) have reported that particles with poly(C)-dependent RNA polymerase activity accumulate in reovirus-infected cells, and that the only protein that they contain is σNS. It will be interesting to determine the significance of these observations.

## D. Modification of Reovirus Proteins

### 1. Cleavage

Three reovirus proteins, namely, proteins λ2, μ1, and μNS, exist not only in the form in which they are translated from mRNAs, but also as cleavage products of such proteins. None is converted completely to its cleavage products; the ratio of cleavage product to uncleaved proteins that remains varies widely among them.

Protein μ1 is cleaved to μ1C to the extent of about 95%. About 90% of the form of the protein that occurs in free form in the cytoplasm is the uncleaved protein, but about 95% of the form that occurs in the cytoplasm complexed with σ3 (see above) is μ1C, and in virus particles also about 95% of the protein is in the form of μ1C (Lee et al., 1981a; Smith et al., 1969). Thus, protein μ1 appears to be cleaved to μ1C at the time when it forms a complex with σ3.

In the case of protein μNS, about 50% is converted to μNSC, which is about 5000 daltons smaller; in the case of protein λ2, less than 10% is converted to the cleavage product λ2C, which is about 15,000 daltons smaller (Lee et al. 1981a). Protein λ2C was first detected in extracts of cells infected with reovirus (Lee et al., 1981a). It is not known whether it is a component of reovirus particles, but autoradiograms of electropherograms of reovirus structural proteins do indeed often reveal a protein in the position occupied by λ2C.

### 2. Chemical Modification

The question of whether any of the reovirus proteins are modified chemically is unclear and unsettled. First, Krystal et al. (1976) reported that 10–20 of the 550 molecules of protein μ1C in virus particles appear to contain a tri- or tetrasaccharide linked O-glycosydically to a serine or threonine residue. However, not only is the significance of such a low degree of glycosylation of so few μ1C molecules not apparent, but also the fact that protein μ1C may be both polyadenylylated and ADP-ribosylated (see below) has thrown some doubt on the original preliminary and tentative identification of N-acetylneuraminic acid, N-acetylgalactosamine, and galactose in protein μ1C.

Second, Krystal et al. (1975) reported that when the proteins of reovirus grown in the presence of [$^{32}$P]orthophosphate are electrophoresed in sodium dodecyl sulfate (SDS)–polyacrylamide gels, label is associated with the protein μ1C band, and obtained evidence consistent with the presence of 1 (or more) phosphoserine residue(s) per protein molecule.

However, no studies to correlate the degree or extent of phosphorylation with infectivity were carried out. Further, Carter has reported that P may be associated with reovirus proteins in forms other than simple ortho-phosphate. She has found that when the proteins of reovirus grown in the presence of [³H]adenosine are subjected to electrophoresis in SDS–polyacrylamide gels, labeled material migrates with both the μ1C and the component viii bands, and she has suggested that these two proteins contain covalently bound oligoadenylate, that is, that they are polyaden-ylylated [through the 3' termini of oligo(A)] (Carter, 1979a; Carter et al., 1980). The number of AMP residues per poly(A) chain was estimated to range from 7 to 11, and their source was suggested to be the free oli-goadenylates that exist within reovirus particles (see Section XI).

Finally, Carter has also reported that proteins μ1C and viii are ADP-ribosylated, that is, that oligoadenylate is linked covalently to them through their ribose moiety (Carter et al., 1980). The average chain length of the oligo(ADP-ribose) is apparently about 1.3. It is clear that both oli-goadenylylated and oligo(ADP-ribosylated) μ1C would be labeled in reo-virus grown in the presence of [³²P]orthophosphate.

Neither the significance nor the function of these protein modifi-cations has been studied to date. All three, that is, phosphorylation, oli-goadenylylation, and oligo(ADP-ribosylation), could conceivably serve to regulate the transcriptase activity of cores, or the transcription of plus strands into minus strands when ds genes are formed during the initial stages of morphogenesis (see Chapter 4). It is conceivable in this con-nection that the significant modifications may be on protein μ1, a con-stituent of cores, rather than on μ1C, a constituent of the outer capsid shell. However that may be, it is clearly essential first to confirm and rigorously prove that reovirus proteins are in fact phosphorylated, poly-adenylylated and ADP-ribosylated, and to characterize the amino acid residues that are so modified.

## E. Mechanisms of Chemical and Physical Inactivation of Reovirus Particles

The three serotypes of reovirus differ remarkably in their response to a variety of chemical and physical inactivating agents. Drayna and Fields (1982b) used intertypic recombinants of serotypes 1 and 3, 2 and 3, and 1 and 2, that contained various combinations of genes derived from the parental serotypes, to study the basis of these differences; for the intertypic recombinants behaved like one parent or the other in the pres-ence of the inactivating agents, and therefore permitted determination of the genes responsible for each difference. They found that sensitivity to 2.5 M guanidine·HCl and pH 11 was determined by the S1 gene (that is, proteins σ1, which is removed from the outer capsid of reovirus ser-otype 2, but not from that of serotype 1, by treatment with pH 11, Drayna and Fields, 1982c); that sensitivity to 55° and 1 percent SDS is determined by the S4 gene (that is, protein σ3, which is removed from both serotype 1 and 2 particles by treatment with SDS (Drayna and Fields, 1982c)); and that sensitivity to 33 percent ethanol and 1 percent phenol is determined by the M2 gene (which encodes protein μ1/μ1C). Thus, as expected, the ultimate structural basis for the stability of reovirus toward harsh chemical and physical treatments lies in the nature of and interactions between its structural proteins.

## VII. INFORMATION CONTENT OF THE TEN REOVIRUS GENES

By the late 1960s and early 1970s, it had been established that the three size classes of reovirus genes, namely, the L, M, and S genes (Bellamy et al., 1967), code for the three size classes of reovirus proteins, namely the λ, μ, and σ size class proteins (Smith et al., 1969). It was realized, however, that the largest L gene did not necessarily code for the largest λ protein, and so on. Techniques for identifying the information content of the individual genes have recently become available. They came from two quite different directions, but yielded identical answers. The first involved translation in an in vitro protein synthesizing system. Attempts to translate reovirus messenger RNAs (mRNAs) in vitro had been made since the early 1970s, when McDowell and Joklik (1971) demonstrated that polyribosomes isolated from infected cells were capable of incorporating labeled amino acids into proteins with the electrophoretic mobility of authentic reovirus proteins. Shortly thereafter, McDowell et al. (1972) devised cell-free systems from several mammalian cells including HeLa cells, L cells, and Ehrlich ascites tumor cells that were capable of translating reovirus mRNA species transcribed in vitro by reovirus cores into σ, μ, and even λ size class proteins, and Both et al. (1975) devised a similar system from wheat germ that was capable of translating all ten species of reovirus mRNAs. In fact, it was in this system that the existence of the two minor protein species λ3 and μ2 was first demonstrated (not only are these two proteins present in virus particles and in extracts of infected cells in very small amounts only, but their electrophoretic migration rates are also very close to those of the major capsid protein species λ2 and μ1C, so that they are usually obscured by them). The straightforward way of determining the information content of reovirus genes by this approach would be to isolate the ten species of mRNA transcribed by reovirus cores in vitro, to hybridize them to the various genes to determine from which each was transcribed, and to translate them individually in an in vitro protein-synthesizing system. It is difficult technically, however, to isolate in pure form sufficient quantities of the larger mRNA species. The approach therefore adopted by McCrae and Joklik (1978) was to isolate the ten ds genes, denature them at 50° in 90%

DMSO, and dilute them into a protein-synthesizing system prepared from wheat germ. At the RNA concentration and ionic strength employed in these incubation mixtures reannealing of the plus and minus strands was sufficiently slow to permit the translation of even $l$ size class mRNA molecules into complete protein molecules. Identification of proteins was achieved not only by comparison of the electrophoretic mobilities of the translation products with those of authentic proteins, but also by Staphylococcus V8 protease peptide mapping. It was found that there was no absolute correspondence between relative gene and protein size as judged by electrophoretic migration rates. The gene-protein assignments that were found are summarized in Fig. 6. Recently Levin and Samuel (1980) confirmed these assignments for the S class genes by isolating their mRNAs and translating them in a wheat germ *in vitro* protein synthesizing system.

Mustoe *et al.* (1978) studied the same problem by a completely different method. As described in Section VIII, there are three serotypes of mammalian reovirus, namely, serotypes 1, 2, and 3. The sizes of the genes of virus strains belonging to these three serotypes, and of the proteins encoded by them, differ slightly but detectably. As would be expected, the genes of these three serotypes undergo extensive reassortment in mixedly infected cells. Thus, by examining polyacrylamide gel electrophoresis profiles of the RNA and protein species of cloned recombinants after pairwise mixed infection, it is possible to determine which protein

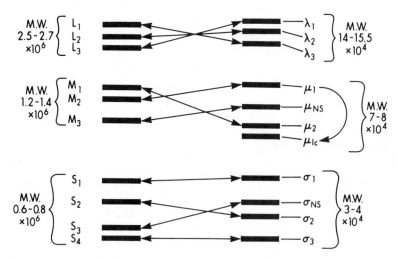

FIGURE 6. Summary of RNA coding assignments for reovirus serotype 3, strain Dearing. The positions of the RNA and protein species are those assumed in the Loening's buffer E (0.036 M Tris, 0.03 M phosphate, pH 7.8) and Tris–glycine systems, respectively. From McCrae and Joklik (1978).

changes accompany specific gene changes. The gene-protein assignments made by this method are exactly the same as those yielded by the direct *in vitro* translation method described above.

## VIII. GENETIC RELATEDNESS AMONG THE THREE SEROTYPES OF REOVIRUS

The numerous strains of reovirus that circulate in nature among mammalian hosts can be conveniently divided into three groups on the basis of the immunological properties of their σ1 proteins: there are three types of σ1 proteins among all these strains that cross-react immunologically only a little if at all, and they therefore define the three serotypes. Sequence analysis of the termini of the three *S1* genes that encode the σ1 proteins has confirmed the expectation that their sequences differ extensively (Li *et al.*, 1980a) (see Section X). It should be noted, however, that although the sequences of the three *S1* genes differ greatly, and the amino acid sequences of the three σ1 proteins are also quite different, such differences are nevertheless subject to restraints, since the three σ1 proteins have retained identical functions during evolution: all σ1 proteins fit into an outer capsid shell that has diverged much less than they have themselves; all are responsible for hemagglutinating red blood cells; all serve as the reovirus cell-attachment protein; and all control the interaction between reovirus particles and host cells in organisms, as described in Section VI.

The question then arises as to how closely related are the other genes that are associated in nature with the three forms of the *S1* gene. There are many ways of examining this problem. One is to compare the sequences of the various cognate gene species associated with the three types of S1 genes; this approach is discussed in Section X. Another is to inquire how many types, in functional terms, there are of each protein. For example, it is known that strains of serotype 2 are more temperature-sensitive than those of serotypes 1 and 3, that this is caused by a difference in gene *S1* (Ramig *et al.*, 1978), and that proteins μ1C of serotypes 3, 2, and 1 are digested very readily, moderately well, and not all by chymotrypsin under certain standard conditions (Drayna and Fields, 1982a). Finally, one may examine how closely related immunologically are the other nine reovirus proteins.

The problem has been approached from three angles.

1. Studies have been carried out to determine how closely related immunologically are the cognate proteins encoded by the three reovirus serotypes. This was studied by Leers *et al.* (1968), who used the technique of immunodiffusion on cellulose acetate, reacting extracts of cells infected with strains of the three serotypes with antisera against the three serotypes. Although they could not identify the individual proteins that

TABLE II. Quantitation of the Serologic Relatedness of Reovirus-Coded Proteins[a]

| Protein | Molecular weight | Approximate number of molecules per virus particle | Location | Coded by serotype | Antiserum against: | | |
|---|---|---|---|---|---|---|---|
| | | | | | Strain Abney serotype 1 | Strain D/5 Jones serotype 2 | Strain Dearing serotype 3 |
| λ1 | 155,000 | 100 | Core | 1 | 250 | 180 | 160 |
| | | | | 2 | 190 | 200 | 240 |
| | | | | 3 | 110 | 140 | 120 |
| λ2 | 140,000 | 100 | Core spike/ virion surface | 1 | 230 | 60 | 120 |
| | | | | 2 | 260 | 160 | 200 |
| | | | | 3 | 80 | 240 | 1000 |
| μ1 | 80,000 | 24 | Core? | 1 | 660 | 140 | 520 |
| | | | | 2 | 400 | 200 | 660 |
| | | | | 3 | 200 | 410 | 700 |
| μ1C | 72,000 | 550 | Virion surface | 1 | 1420 | 80 | 700 |
| | | | | 2 | 1000 | 1100 | 1230 |
| | | | | 3 | 700 | 1780 | 1100 |

| Protein | Mol. wt. | | Particle location | Type | 1 | 2 | 3 |
|---|---|---|---|---|---|---|---|
| μNS | 85,000 | — | Nonstructural | 1 | 40 | 30 | 30 |
| | | | | 2 | 40 | 40 | 30 |
| | | | | 3 | 60 | 30 | 40 |
| σ1 | 42,000 | 24 | Virion surface | 1 | 1240 | 60 | <10 |
| | | | | 2 | 100 | 200 | <10 |
| | | | | 3 | <10 | <10 | 2000 |
| σ2 | 38,000 | 200 | Core | 1 | 280 | 100 | 1100 |
| | | | | 2 | 280 | 1250 | 2300 |
| | | | | 3 | 320 | 400 | 820 |
| σ3 | 34,000 | 900 | Virion surface | 1 | 1000 | 260 | 1060 |
| | | | | 2 | 1060 | 2400 | 1700 |
| | | | | 3 | 860 | 1660 | 680 |
| σNS | 36,000 | — | Nonstructural | 1 | 40 | 40 | 45 |
| | | | | 2 | 30 | 45 | 30 |
| | | | | 3 | 50 | 40 | 50 |

[a] From Gaillard and Joklik (1980). The numbers are the reciprocals of the antiserum dilutions at which 50% of the maximum amount of each protein reacts. Boxed numbers indicate significant type specificity.

formed precipitation lines, they did show that reoviruses possess both group- and type-specific antigenic determinants distributed among several proteins. More recently, Gaillard and Joklik (1980) studied this problem by adding antisera to each of the three reovirus serotypes raised in pathogen-free rabbits (in which reovirus multiplies readily, so that antibodies to all reovirus-coded proteins are formed) to extracts of mouse L fibroblasts infected with a reovirus strain of serotype 1, 2, or 3. In each case, the proteins capable of reacting with antibody were identified and quantitated by a radioimmunoprecipitation procedure using *Staphylococcus* A protein. By using stepwise dilutions of antiserum, it was possible to determine the antiserum dilutions at which 50% of the maximum amount of each individual protein reacted. The reciprocals of these dilutions, which provide a quantitative measure of the avidity of each protein for homologous as well as heterologous antibodies, are listed in Table II.

The most significant results of this analysis were as follows:

a. The most unique, that is, type-specific, of all reovirus proteins is σ1. However, the type specificity is not absolute; for example, although antiserum against serotype 3 does not react with the σ1 proteins of serotypes 1 and 2 at all, antiserum to serotype 2 reacts with the serotype 1 σ1 almost one third as efficiently as the homologous σ1, and antiserum to serotype 1 reacts with serotype 2 σ1, although less than 10% as efficiently as the homologous σ1.

b. Four other proteins, namely λ2, μ1C, σ2, and σ3, also exhibit some degree of type specificity. Antiserum to serotype 2, not unexpectedly, displays a greater ability to discriminate among the antigenic determinants of the three serotypes than the other two antisera, and usually it discriminates against serotype 1 more than against serotype 3 (see polypeptides μ1C, σ2, and σ3). The only exception is λ2, where it is antiserum to serotype 3 that possesses significant type specificity.

c. Not infrequently, the affinity of a given antiserum is greater for a heterologous than for the homologous antigen. For example, antiserum against serotype 3 reacts with 50% of homologous σ3 at a dilution of 1:680, but reacts with 50% of serotype 2 σ3 at a dilution of 1:1700. Several other examples of this type can be found in Table II. Similar results were obtained when the specificities of individual monoclonal antibodies were examined, up to 4-fold greater affinities for heterologous as compared with homologous antigens being noted (Lee *et al.*, 1981a).

d. The most remarkable result of this analysis was that the antigenic determinants on proteins coded by serotype 2 are not significantly more different from those on serotype 1 and 3 proteins than those on serotype 1 and 3 proteins are among themselves. In other words, in most cases the antigenic determinants are highly conserved, even on proteins specified by serotype 2, which must differ markedly in amino acid sequence from those specified by serotypes 1 and 3 (since there is no more than 10% homology between their genes). Thus, the antigenic determinants on most of the proteins of the three reovirus serotypes have been highly conserved during evolution.

2. Gentsch and Fields (1981) compared the tryptic peptides of the three proteins of the outer capsid shell of serotypes 1, 2, and 3. They found that the tryptic peptides of the three μ1C proteins showed a very high degree of conservation: of 10 tryptic peptides that could be readily recognized, 8 appeared to be shared. The three M2 genes, which encode protein μ1C, thus appear to have diverged minimally during evolution. In the case of proteins σ3 and σ1, 15–20 tryptic peptides of each could be examined. In both cases, 4 or 5 tryptic peptides appeared to be common to the proteins of all three serotypes, and 5 or 6 were unique to the protein of one serotype or another, the remainder being shared by two. It is surprising, on the one hand, that, using this criterion, the degree of relatedness among the σ3 proteins is more similar to that among the (type-specific) σ1 proteins than among the μ1C proteins and, on the other, that the three σ1 proteins are so closely related. Two observations are relevant in this regard: (a) It is not certain that two tryptic peptides with similar chromatographic–electrophoretic properties are identical; for example, replacement of amino acids by similar amino acids may not be detectable, and this may exaggerate the apparent relatedness of proteins. (b) The fact that the three σ1 proteins are related although they are quite distinct immunologically and although the sequences at the termini of the three S1 genes are quite different (see Section X) can be explained by the necessity to preserve ability to carry out common functions, as argued above.

3. Gaillard and Joklik (1982) have compared the relatedness of the ten genes of the three serotypes by a hybridization analysis. Hybrid genes containing the plus strand of one serotype and the minus strand of another were examined for the extent of non-base-paired regions by two criteria: sensitivity to ribonuclease and reduction in electrophoretic migration rate [ss regions slow the migration rate of dsRNA molecules (Schuerch and Joklik, 1973)]. A summary of the results obtained using the ribonuclease sensitivity technique is presented in Table III (Gaillard and Joklik, 1982). It can be seen that the degree of relatedness among the genes of serotypes 1 and 3 ranges from 31 to 94% (with the sole exception of gene S1). Among the genes of serotype 2, on the one hand, and those of serotypes 1 and 3, on the other, the values are all between 3 and 23%. These estimates are compatible with estimates of the overall relatedness of these genomes, that is, about 70 and 10%, respectively (Martinson and Lewandowski, 1975; Gaillard and Joklik, 1982, see previous page). Interestingly, the most closely related groups of genes are the L genes and the two genes that code for nonstructural proteins; and the genes that code for core components (genes L1, L2, L3, M1, and S2) are more closely related than the genes that code for the proteins of the outer capsid shell (M2, S1, and S3). It is particularly interesting that the M2 genes are less closely related than the S4 genes, even though the μ1C proteins are more

TABLE III. Percentage Homology among the Genes of Reovirus Serotypes 1, 2, and 3[a]

| Gene | Hybrid genes[b] | | | | | |
|------|----------------|------|------|------|------|------|
|      | +(1)<br>−(2) | +(1)<br>−(3) | +(3)<br>−(1) | +(3)<br>−(2) | +(2)<br>−(1) | +(2)<br>−(3) |
| L1   | 15 | 86 | 84 | 18 | 21 | 11 |
| L2   | 23 | 85 | 94 | 13 | 12 | 7  |
| L3   | 12 | 88 | 79 | 13 | 15 | 14 |
| M1   | 9  | 66 | 69 | 5  | 8  | 3  |
| M2   | 5  | 31 | 42 | 9  | 11 | 8  |
| M3   | 10 | 63 | 67 | 5  | 6  | 8  |
| S1   | 4  | 12 | 6  | 1  | 1  | 4  |
| S2   | 14 | 56 | 44 | 9  | 6  | 13 |
| S3   | 12 | 69 | 81 | 14 | 8  | 12 |
| S4   | 11 | 57 | 51 | 11 | 11 | 15 |

[a] Percentage homology is the percentage of sequences resistant to digestion by ribonuclease under standard conditions relative to the ribonuclease sensitivity of homologous genes.
[b] Hybrid genes possess a + and a − strand of genes of serotypes 1, 2, or 3 as denoted by the numbers in parentheses.
(From Gaillard and Joklik, 1982.)

similar among themselves than the σ3 proteins (Gentsch and Fields, 1981) (however, see Gaillard and Joklik, 1980). Further implications of these results are discussed in Section X.

In summary, examination of three isolates of reovirus that possess the three alleles of the σ1 gene (that is, the Lang, D5 Jones and Dearing strains of serotypes 1, 2 and 3, respectively) has revealed that the *S1* gene is indeed the gene that has diverged to the greatest extent during evolution; all other genes have diverged far less. Further, in all cases the serotype 1 and 3 genes are related far more closely to each other than to the cognate serotype 2 gene. This suggests that the gene sets of reovirus serotype 1, 2 and 3 have evolved independently of each other. Apparently double infection of hosts with strains of two reovirus serotypes, which would very likely yield recombinants, occurs infrequently and/or recombinants have a lower survival advantage than strains containing "pure" gene sets (Gaillard and Joklik, 1982).

## IX. NATURE OF THE ANTIBODIES ELICITED BY REOVIRUS-SPECIFIED PROTEINS

The nature of the antibodies elicited by the four proteins on the outer surface of reovirus particles has been studied by the use of a series of 19 monoclonal antibodies isolated and characterized by Lee *et al.* (1981a). Their properties, as well as the properties of antibodies against the core component σ2 and the two nonstructural proteins μNS and σNS, are summarized in Tables IV and V. The following points are of interest:

1. Protein σ1 gives rise to antibodies with strong, completely type-specific hemagglutination-inhibition (HI) activity; strong, almost completely type-specific neutralizing activity; and ability to precipitate reovirus particles. The ratios of the first two of these activities differ widely among the various monoclonal antibodies, which suggests that they are directed against different parts of the σ1 molecule, which in turn implies the existence of functional domains in σ1, as suggested by Burstin et al. (1982).

2. Protein λ2 gives rise to antibodies with moderate, almost completely type-specific HI activity; strong, group-specific neutralizing activity; and ability to precipitate reovirus particles very efficiently. The finding that antibodies to protein λ2 are capable of interacting with reovirus particles was quite unexpected, since this protein was thought to be a core component. It seems that the 12 projections/spikes of reovirus cores, which are composed of protein λ2 (see Section II.B.), project through the outer capsid shell to the surface of the virus particle. The indentations

TABLE IV. Measurement of the Ability of 19 Hybridoma-Secreted IgGs to React with Proteins Specified by Reovirus Serotypes 1, 2, and 3

| IgG secreted by hybridoma cell line | Reovirus serotype[a] | | | Reovirus serotype[b] | | |
|---|---|---|---|---|---|---|
| | 1 | 2 | 3 | 1 | 2 | 3 |
| λ2-A | 42 | 29 | 75 | 150 | 39 | 100 |
| λ2-B | 42 | 26 | 64 | 66 | 41 | 100 |
| μl/μlC-A | 6060 | 1310 | 1600 | 380 | 82 | 100 |
| μl/μlC-B | 1960 | 225 | 555 | 350 | 41 | 100 |
| μl/μlC-C | 605 | 445 | 835 | 73 | 53 | 100 |
| μl/μlC-σ3-A | 2700 | 2060 | 910 | 265 | 225 | 100 |
| μNS-A | 2410 | 185 | 540 | 445 | 34 | 100 |
| μNS-B | 630 | 670 | 450 | 140 | 150 | 100 |
| μNS-C | 1300 | 1000 | 315 | 415 | 320 | 100 |
| μNS-D | 13 | 5 | 12 | 110 | 39 | 100 |
| σ1-A | 0 | 0 | 70 | 0 | 0 | 100 |
| σ1-B | 0 | 0 | 220 | 0 | 0 | 100 |
| σ1-C | 7 | 0 | 950 | 0 | 0 | 100 |
| σ1-D | 0 | 0 | 290 | 0 | 0 | 100 |
| σ2-A | 210 | 565 | 600 | 35 | 94 | 100 |
| σ2-B | 175 | 140 | 245 | 73 | 57 | 100 |
| σ3-A | 13 | 0 | 12 | 110 | 0 | 100 |
| σNS-A | 1820 | 870 | 3510 | 52 | 25 | 100 |
| σNS-B | 2245 | 2740 | 4345 | 52 | 63 | 100 |

[a] The figures in these three columns are the reciprocals of the dilutions of 1 mg/ml IgG solutions that reacted with 20% of the maximum amount of antigen. Zero signifies that no detectable amount of protein reacted even at an IgG concentration of 1 mg/ml.
[b] The figures in these three columns are those in the first three columns normalized with respect to the serotype 3 (the homologous) value.
(From Lee et al., 1981a)

TABLE V. Interactions of Hybridoma-Secreted IgGs (Monoclonal Antibodies) Directed against Reovirus Proteins with Reovirus Particles[a]

| Neutralization of homologous and heterologous reovirus serotypes by hybridoma IgGs | | | | HI of homologous and heterologous reovirus serotypes by hybridoma IgGs | | | | Radioimmunoprecipitation of reovirus serotype 3 by hybridoma IgGs | |
|---|---|---|---|---|---|---|---|---|---|
| IgG or antiserum | Sero-type 1 | Sero-type 2 | Sero-type 3 | IgG or antiserum | Sero-type 1 | Sero-type 2 | Sero-type 3 | IgG or antiserum | Precipitation (%) |
| λ2-A | 2,100 | 1,140 | 1,820 | λ2-A | 120 | 960 | 15,360 | λ2-A | 100 |
| λ2-B | 2,220 | 1,960 | 2,520 | λ2-B | — | 240 | 1,920 | λ2-B | 100 |
| μ1/μ1C-A | — | — | — | μ1/μ1C-A | — | — | — | μ1/μ1C-A | 29 |
| μ1/μ1C-B | — | — | — | μ1/μ1C-B | — | — | — | μ1/μ1C-B | 7 |
| μ1/μ1C-C | — | — | — | μ1/μ1C-C | — | — | — | μ1/μ1C-C | 5 |
| μ1/μ1C-σ3-A | — | — | — | μ1/μ1C-σ3-A | — | — | — | μ1/μ1C-σ3-A | 65 |
| μNS-A | — | — | — | μNS-A | — | — | — | μNS-A | 4 |
| μNS-B | — | — | — | μNS-B | — | — | — | μNS-B | 4 |
| μNS-C | — | — | — | μNS-C | — | — | — | μNS-C | 4 |
| μNS-D | — | — | — | μNS-D | — | — | — | μNS-D | 28 |
| σ1-A | — | — | 720 | σ1-A | — | — | 1,920 | σ1-A | 67 |
| σ1-B | — | — | 580 | σ1-B | — | — | 122,880 | σ1-B | 52 |
| σ1-C | 500 | — | 660 | σ1-C | 480 | — | 245,760 | σ1-C | 64 |
| σ1-D | — | — | 1,220 | σ1-D | — | — | 122,880 | σ1-D | 59 |

| Reagent | | | | |
|---|---|---|---|---|
| σ2-A | — | — | — | 105 |
| σ2-B | — | — | — | — |
| σ3-A | 40 | — | 60 | — |
| σNS-A | — | — | — | — |
| σNS-B | — | — | — | — |
| 45.6TG 1.7 | — | — | — | 105 |
| Rabbit antiserum against reovirus serotype 3 σ3 | 1,275 | 1,395 | 1,530 | — |
| Rabbit antiserum against reovirus serotype 3 λ1 | — | — | — | — |
| Rabbit antiserum against reovirus serotype 1 | 5,415 | 735 | 1,665 | 294,910 |
| Rabbit antiserum against reovirus serotype 2 | 990 | 15,310 | 2,190 | 1,200 |
| Rabbit antiserum against reovirus serotype 3 | 1,050 | 320 | 25,560 | 2,690 |

| Reagent | | | |
|---|---|---|---|
| σ2-A | — | — | 2 |
| σ2-B | — | — | 2 |
| σ3-A | 480 | — | 93 |
| σNS-A | — | — | 1 |
| σNS-B | — | — | 1 |
| 45.6TG 1.7 | — | 105 | 1 |
| Phosphate-buffered saline | 13,450 | 105 | 1 |
| Rabbit antiserum against reovirus serotype 3 (1:10) | — | — | 100 |
| Rabbit antiserum against reovirus serotype 3 λ1 (1:10) | 385 | 1,150 | 2 |
| Rabbit antiserum against reovirus serotype 3 σ3 (1:10) | 800 | 153,600 | 100 |
| Nonimmune rabbit serum (1:10) | 21,505 | 1,345 | 2 |

[a] From Hayes et al. (1981). The values are the reciprocals of the dilutions of 10 mg/ml IgG solutions or antisera that reduced the number of plaques by 50% or completely inhibited hemagglutination. In the rightmost section, the IgG concentration was 1 mg/ml.

in the outer capsid shell noted by Palmer and Martin (1977) may correspond to the locations of these projections (see Section II.A ) Although part of reovirus cores, protein λ2 is thus also an outer virus particle surface component.

3. Protein σ3 gives rise to antibodies with type-specific HI activity, partially type-specific neutralizing activity, and ability to precipitate reovirus particles very efficiently.

4. Antibodies to μ1C possess no HI activity and do not neutralize infectivity, but may or may not precipitate virus particles (two of four monoclonal antibodies were capable of precipitating reovirus particles, two were not).

These conclusions are derived from studies of the properties of a limited number of monoclonal antibodies (four against protein σ1, two against λ2, one against σ3, and four against μ1C). What one would really like to know is the nature of the antibody–antigen interactions that are responsible for HI, neutralization, and precipitation when a particular antiserum is added to a particular antigen or mixture of antigens, including reovirus particles. Clearly, this would be a function of the nature of the individual antibodies present in it (as exemplified by the sample listed in Table IV) and of the actual concentration of each of these antibodies [which may be gauged from the data presented in Table II) (Section VIII)]. For example, the antiserum against serotype 3 in Table II had a high titer against σ1, about one half this titer against proteins μ1C and λ2, moderately high activity against proteins σ2 and σ3, and weak activity against proteins λ1, μNS, and σNS. It is likely that the HI type specificity of this antiserum was due primarily to antibodies against protein σ1 and that most of its group-specific neutralizing activity was due to antibodies against λ2. The situation is likely to be different for the antiserum against serotype 2, which had a lower activity against σ1 than against σ3, σ2, and μ1C.

## X. SEQUENCES OF REOVIRUS GENES AND MESSENGER RNAs

Attempts to sequence reovirus genes started soon after it was realized that they were individual RNA molecules, the purpose being to test the possibility that the association of sets of genes during morphogenesis involves recognition of complementary sequences at their ends. Although at that time, in the early 1970s, methods were available for identifying only the terminal two or three residues, it was soon shown that all reovirus genes possessed identical terminal dinucleotide pairs, which ruled out the possibility of terminal complementarity (Miura et al., 1974b). It was then shown that both ends of reovirus genes are completely resistant to the single-stranded (ss) specific nuclease S1 (Muthukrishnan and Shatkin, 1975), which indicated that the plus and minus strands are exactly collinear. Such collinearity is a prerequisite for the model of reovirus gene replication proposed by Schonberg et al. (1971), who found that when reovirus particles infect cells, the double-stranded RNA (dsRNA)-containing genes are transcribed into single-stranded plus-stranded transcripts that function first as messenger RNA (mRNA) for several hours, are then assembled into sets of ten, and are then transcribed into minus strands with which they remain associated, thereby forming the progeny genes. Clearly, this mode of replication demands that the plus and minus strands of all genes be exact complements of each other.

In 1974, Miura et al. (1974a) discovered that the 5' ends of the plus strands of cytoplasmic polyhedrosis virus RNA are blocked. Shortly thereafter, it was found that the plus strands of all reovirus genes, as well as all reovirus mRNA molecules, are also blocked at their 5' termini (Miura et al., 1974b). The blocking groups were found to be what are now known as caps, that is, $^{m7}G(5')ppp(5')G^mpC$ . . . (Furuichi et al., 1975a,b; Chow and Shatkin, 1975). The four enzymes that are involved in capping reovirus gene transcripts, namely, the nucleoside triphosphate phosphohydrolase, the guanylyltransferase, and the two RNA methyltransferases, are all components of reovirus cores [since cores synthesize capped mRNAs (see Section XIII)] and are most probably virus-coded.

The cap described above is known as cap 1. During the second half of the virus multiplication cycle, the cap 1 on about one half of reovirus mRNA molecules in infected cells is modified by a host-cell methyltransferase to become cap 2, namely, $^{m7}G(5')ppp(5')G^mpC^mpU$ . . ., but the caps on the 5' termini of plus strands of dsRNA are almost exclusively cap 1 (Desrosiers et al., 1976). Antibody specific for cap 1 has been elicited in rabbits and used to enrich for capped mRNAs (Nakazato, 1980). The significance of capping for the translation of reovirus mRNA is discussed in Chapter 3.

The first serious attempt to sequence a reovirus RNA molecule was when Nichols et al. (1972a) found that under certain limiting conditions, reovirus cores synthesize only a single species of mRNA in vitro, namely, the s4 species (Darzynkiewicz and Shatkin, 1980), and sequenced its first 25 5'-terminal residues. This work was followed by the studies of Kozak and Shatkin, who in a series of elegant studies sequenced the regions in six of the seven s and m species of mRNA of the Dearing strain (serotype 3) that bound to and were protected by wheat germ 40 S ribosomal subunits and 80 S ribosomes (Kozak and Shatkin, 1976, 1977a,b; Kozak, 1977; see also Darzynkiewicz and Shatkin, 1980; Lazarowitz and Robertson, 1977). Recently, Kozak (1982) has extended this work to two of the three l mRNA species. In all cases, the 40 S subunits protected sequences 50–60 residues long that included the 5' terminus as well as the initiation codon. The sequences protected by intact 80 S ribosomes were subsets of these sequences that were only about one half as long and were centered around the initiation codons (Kozak, 1977). The eight sequences are shown in Fig. 7. Interesting features are that: (1) all 5' termini start with GCUA; (2) the distance between the 5' terminus and the first initiation

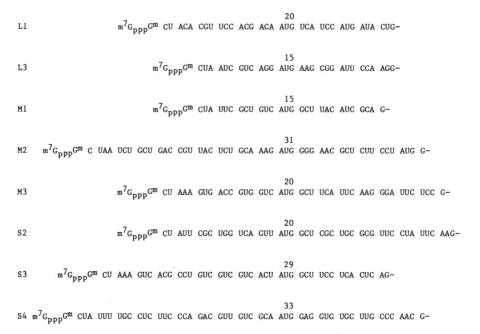

FIGURE 7. Sequences of the ribosome-binding sites of eight species of reovirus serotype 3 messenger RNA (Kozak, 1977, 1982). The gene assignments are from Darzynkiewicz and Shatkin (1980) and Antczak et al. (1982).

codon is short, only 15–33 residues; (3) the sequences between the 5' termini and the initiation codons are completely different; and (4) the sequences lack secondary structure features such as hairpins.

Recently, 50–90 residues at the 3' ends of both plus and minus strands of all ten genes of the serotype 3 strain Dearing and of the *L3*, *M3*, *S1*, and *S2* genes of strains of serotypes 1 and 2 have been sequenced (Li *et al.*, 1980a,b; Darzynkiewicz and Shatkin, 1980; McCrae, 1981; Antczak *et al.*, 1982; Gaillard *et al.*, 1982.) The results permit the conclusions discussed below.

## A. Sequence Identity of the Plus Strands of Reovirus Genes and of Their Transcripts

This identity was shown for the specific case of the *S2* gene of serotype 3, but is undoubtedly true for all genes. The background to this demonstration is as follows:

It is generally assumed that all forms of plus-stranded reovirus RNA are identical. There are three forms of such RNA: (1) the plus strands of viral genes (which consist of dsRNA), (2) the plus strands transcribed from these genes *in vitro* by transcriptase present in reovirus cores (see below), and (3) the single-stranded RNAs (ssRNAs) that are translated by the polyribosomes of infected cells. The evidence that these three forms of plus-stranded RNA are identical is good, but circumstantial: (1) following infection, parental dsRNA is transcribed into plus-stranded RNA, which is transcribed in turn into minus strands with which it remains associated, thereby generating progeny dsRNA molecules (Schonberg *et al.*, 1971); (2) plus-stranded RNA transcribed from virion dsRNA by cores *in vitro* is translated in *in vitro* protein-synthesizing systems into proteins that are indistinguishable from authentic reovirus-coded proteins synthesized in infected cells (Both *et al.*, 1975); and (3) the plus strands of reovirus genes can similarly be translated into proteins that can be identified as reovirus-coded (McCrae and Joklik, 1978).

Conclusive evidence concerning the identity of the various forms of reovirus plus-stranded RNA can come only from sequencing data. Li *et al.* (1980b) showed that the 79 3'-terminal residues of the ssRNA transcribed from the *S2* gene of reovirus serotype 3 (that is, its mRNA) and of the plus strand of this gene are identical and that the 3'-terminal sequence of the minus strand of this gene is exactly complementary to the ribosome-binding region of *s2* mRNA, which is known to encompass its 5' terminus (Li *et al.*, 1980b; Kozak, 1977). Taken together, these results prove that the sequences at both ends of plus-stranded transcripts of the *S2* gene of reovirus serotype 3 are *exactly* collinear with the plus strands of the gene itself, and there is no evidence to suggest that this situation does not hold for all reovirus genes.

## B. Sequences at the Termini of the Ten Genes of Reovirus Serotype 3

### 1. 5' Termini

All ten genes share a common tetranucleotide GCUA at the 5' terminus of their plus strands (Fig. 8). All genes contain an initiation codon within 13–33 residues of the 5' terminus; an initiation codon starting at position 6 in gene *L2* is inoperative, since its reading frame leads quickly to a termination codon. Several genes possess additional initiation codons, either in phase or out of phase, further downstream. It is impossible to know which of these initiation codons are used in the absence of further sequencing information, but since the recognition signal for ribosome attachment appears to be the 5'-terminal cap, it is very likely that it is indeed the first initiation codon that is used (Kozak and Shatkin, 1978).

The amino acid sequences of proteins starting at the first initiation codon (or the second, in gene *L2*) are shown beneath each sequence (Antczak *et al.*, 1982). For most genes, the number of hydrophobic and charged amino acid residues is about equal, and in most cases the net charge of the first 15–20 amino acid residues is either 0 or 1. The only exceptions

Gene

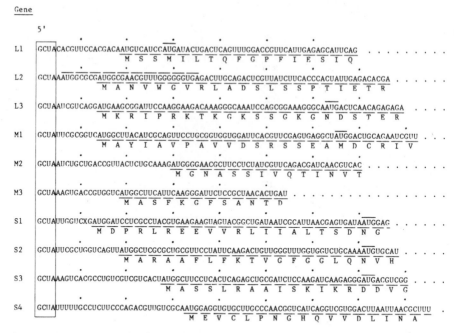

FIGURE 8. Sequences of the 5'-terminal regions of the ten genes of the Dearing strain of reovirus serotype 3. From Antczak *et al.* (1982).

are the proteins encoded by genes *M2* and *S2*, which encode proteins μ1/ μ1C and σ2, respectively, in which the net charge is +3 and +4, respectively, and the protein encoded by gene *L3*, protein λ1, a major core constituent. The nature of the first 22 amino acids of this protein is striking: there is only one hydrophobic amino acid residue and no fewer than nine basic amino acid residues and two acidic ones, which gives the N-terminal 22 amino acid residues a net charge of +7. The amino terminus of protein λ1 is thus very hydrophilic and basic.

It should be pointed out that the sequences presented in Fig. 8 differ markedly from those presented by McCrae (1981) for the same strain of reovirus. Only McCrae's sequence of gene *S2* agrees with the sequence presented in Fig. 8; in most other sequences, there are four or more differences, and in gene *L2*, there are nine differences in the first 22 residues. There are also numerous differences between the sequences in Fig. 8 and those found by Darzynkiewicz and Shatkin (1980) [which, incidentally, do not agree with those of McCrae (1981)]. These differences are especially marked in genes *S1*, *S4*, *S3*, and *M3*. By contrast, the sequences of Antczak *et al.* (1982) are very similar to the sequences of the ribosome-binding sites of eight species of reovirus mRNA (all of which include their 5' terminus) reported by Kozak and Shatkin (1977a,b) and Kozak (1977, 1982); the only differences are two brief regions of acknowledged ambiguity in the ribosome-binding site sequences of *m2* and *s2* mRNA and an extra U that Antczak *et al.* (1982) find in position 9 of *s*4 RNA. It is interesting to note that McCrae (1981) and Darzynkiewicz and Shatkin (1980) sequenced RNA by the use of enzymes, whereas Kozak and Shatkin (1977a,b) and Kozak (1977, 1982) used classic sequencing techniques and Antczak *et al.* (1982) used the chemical rapid RNA sequencing technique. It would appear that the latter is considerably more accurate than the technique using enzymes.

## 2. 3′ Termini

The 3′-terminal sequences of the plus strands of all ten genes are presented in Fig. 9. The only sequence shared by all ten genes is the 3′-terminal pentanucleotide UCAUC-3′. Multiple termination codons are present in all ten sequences; in fact, some genes, like *M1* and *M3*, possess as many as seven termination codons in the 70 3′-terminal residues, in all reading frames, and some adjacent to each other. Thus, most reovirus genes may contain sizeable stretches of untranslated sequence at their 3′ ends. It is of course not known which of the many termination codons are actually used. In gene *S2*, it may well be the codon finishing at position 55 from the 3′ terminus, since there are also termination codons terminating in positions 53 and 54 in the *S2* genes of serotypes 1 and 2, respectively, and the amino acid sequences upstream from it in the genes of serotypes 3 and 2 are IRALI and IAALI, with very different codon usage.

| Gene | |
|---|---|
| L1 | (CGUGUC)GUCACUUGGACAGAGAUUCAUGACUUGGAUGCGACAGGAAGGAGGUCAGCGUGAGUCAUCAUGGGUCGUGGUGCGUCAACUCAUC-OH 3′ |
| L2 | . . . . . UGAAU(C)AAUAAUUGUGCACCGGUGAUUAUACCUUGACUCCUCUGUAACUCACUGUCCCUCGUGAGCGCGCCUAAUUCAUC-OH 3′ |
| L3 | . . . . . . . . . . . . . . . . . . . . . . . . CAGGUGCUGGCAUUGG(C)CUCAAUGCUAGUUGGGCCGAUUCAUC-OH 3′ |
| M1 | . . . . . . GCGUGAUCGUGACAUGCGUAGUGUGAG(UAC)UAGGUCAAUGGGGGUAGGGGGCGGGCUAAGACUACGUACGCGCUUCAUC-OH 3′ |
| M2 | . . . . . . . . GGUGGUCAACGUGGGUACACACGUUAGGCCGCUC(CGC)CUGGUGACGCGGGGUUAAGGGAUGCAGGCAAAUCAUC-OH 3′ |
| M3 | . . . . . . . . . . CUGAUGAGUUGUAAAUAAUCCGUGUAGCAGUGUUGCCCUAAUCCCUUUAAGCCUUCCCGGGCCACAUUCAUC-OH 3′ |
| S1 | . . . . . . . . . GGUGCUGCUAGACCGUUUCGUACCCGCGUAGUUUCACGUGAGGAUXAGACCCACGCGGCACUGGGGCAUUUCAUC-OH 3′ |
| S2 | . . . . . . . . . CGAUUCGAGCCCUUAUCUGAUCAGUCUUGGAGUAGGGUCCCCCACACCCCUCACGACUGACCACACAUUCAUC-OH 3′ |
| S3 | . . . . . . . . . GUAAUGGGUGAGUGAGCUGAUGUGGUCGCCAAGACAUGUGCCGGUGUCUUGGUGGUGGGUGACGCCUAAUCAUC-OH 3′ |
| S4 | . . . . . . . . . . UAACUAAACACCCCAUCUUCACAGCGCCGGGCUUGACCAACCUGGUGUGACGUGGACAGGCUUCAUUCAUC-OH 3′ |

FIGURE 9. Sequences of the 3′-terminal regions of the ten genes of the Dearing strain of reovirus serotype 3. From Antczak *et al.* (1982).

## C. Examination of the 5′-Terminal and 3′-Terminal Sequences for Potentially Functional Features

The availability of sequence data concerning all ten reovirus genes presents the opportunity for asking several questions of fundamental importance. Among them are:

1. Can any evolutionary relationships be discerned among the ten genes (that is, do some genes appear to be derived from others)?
2. Do the genes possess features that may function in gene assortment during morphogenesis?
3. Do they possess features that may regulate translation efficiency?
4. Can binding recognition sites for RNA polymerases and capsid proteins be discerned?

The sequences were therefore examined for the following by means of computer programs: (1) homologies (i.e., the presence of *identical* or *very similar same-sense* sequences); (2) symmetries (i.e., the presence of *identical* or *very similar opposite-sense* sequences); (3) dyad symmetries (i.e., the presence of *complementary* sequences); (4) dinucleotide frequencies; and (5) *complementary* sequences between the 5′-terminal sequences of the mRNAs and the 3′-terminal consensus sequence of eukaryotic 18 S ribosomal RNA.

To assess whether any features found were significant, identical searches were carried out on 20 random sequences constructed using a random number table. The results discussed below were obtained (Antczak *et al.*, 1982).

### 1. Possible Evolutionary Relationships

Ten highly statistically significant *homologies* are present in the 5′- and 3′-terminal regions of the ten reovirus genes; none is present in the random sequences at this level of significance. Four of these homologies are among 5′ termini, one among 3′ termini, and five among 5′ termini and 3′ termini. The homologous regions are indicated in Fig. 10. The reovirus genes also contain eight highly statistically significant *symmetries*, that is, sequences identical or very similar to sequences in the opposite sense (inverted repeats if on the same strand). These sequences are also indicated in Fig. 10. One of these symmetries, in the 3′-terminal region of gene *L1*, is a 19-residue-long nearly perfect palindrome.

Antczak *et al.* (1982) concluded that the 5′- and 3′-terminal regions of reovirus genes possess 22 sequences 20–30 residues long that are very similar to other sequences, either in the same sense or in the opposite sense. These sequences are widely distributed among reovirus genes. Only gene *S1* does not possess at least 1; gene *L1* possesses no fewer than 6, and 11, 6, and 5 of these sequences are in *L*, *M*, and *S* genes, respectively. The most remarkable feature of these sequences is that no fewer than 7

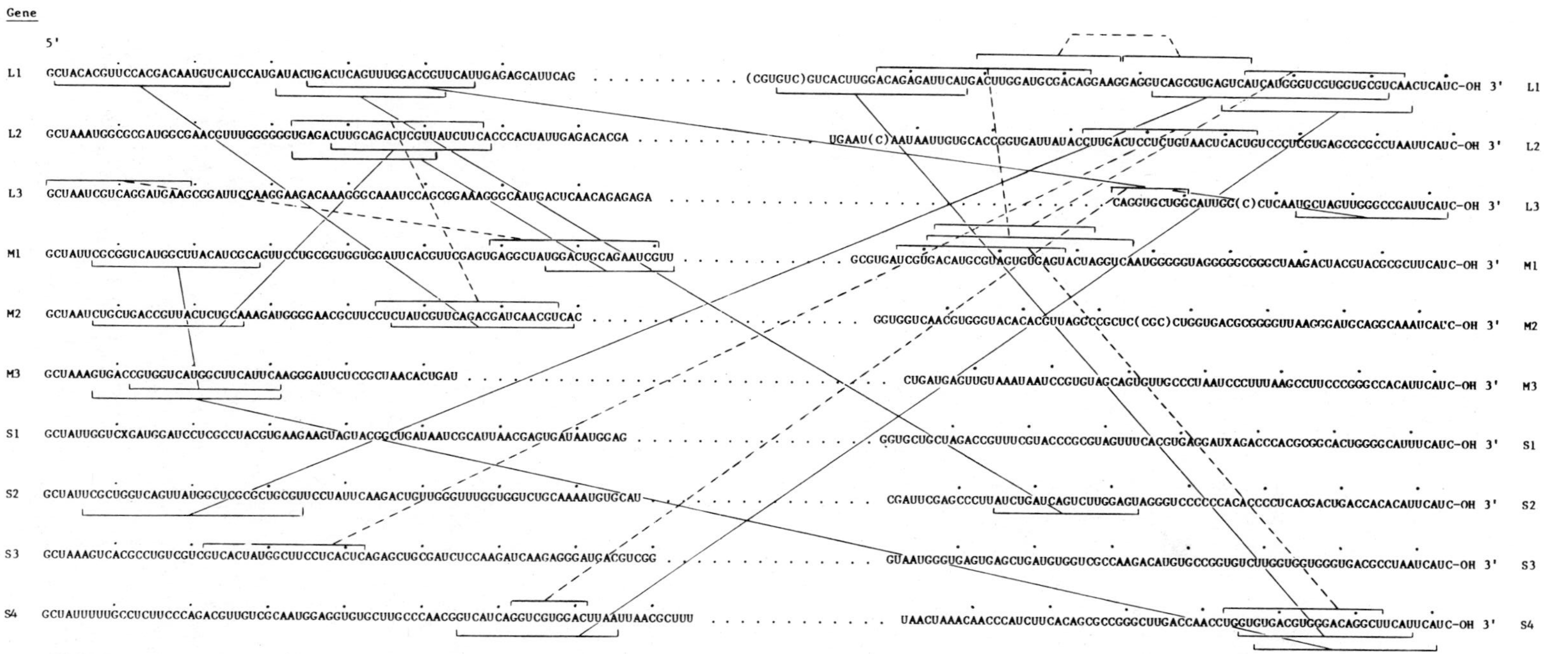

FIGURE 10. Location of homologies and symmetries in the genes of reovirus serotype 3. The limits of sequences exhibiting homology are indicated beneath each sequence, and are joined by solid lines; the limits of sequences exhibiting symmetry are indicated above each sequence, and they are joined by dashed lines. From Antczak *et al.* (1982).

exhibit either homology or symmetry not only to one but also to two and even three other sequences, that 4 sequences exhibit homology to one sequence and symmetry to another, and that 2 sequences exhibit homology to two sequences and symmetry to a third. This results in an extraordinarily complex network of closely related sequences the origin of which is not readily apparent but which probably represents vestiges of gene-duplication events and aberrant transcription of both plus and minus strands. It should be pointed out that the sequences involved in homologies and symmetries contain no distinctive base compositional features, nor are they surrounded by sequences that possess common distinctive features.

A possibly analogous situation has recently been described for influenza virus by Fields and Winter (1982), who found a small defective interfering RNA that is derived from five separate regions in one virion RNA segment and one in another. These regions, which are normally near the termini of the RNA segments, are arranged in the small RNA in an alternating fashion (that is, each region from near a 5′ terminus is adjacent to a region from near a 3′ terminus). Fields and Winter (1982) proposed that the small RNA is generated during plus strand synthesis as a result of viral polymerase pausing at U-rich sequences in the template and reinitiating synthesis at another site. A fascinating similarity between these regions and the reovirus regions that exhibit homology or symmetry or both is that they are not identical, but only *nearly* identical (homology upward of 80%), to the regions from which they appear to be derived (in the case of influenza virus) or to their partners (in the case of reovirus). This strengthens the hypothesis that they originate during transcription, rather than as a result of recombination.

## 2. Features That May Function in Gene Assortment

The 20 gene sequences have also been examined for signals that might explain how sets of 10 reovirus plus strands are assembled during the initial stage of reovirus morphogenesis (Antczak *et al.*, 1982). Numerous regions exhibiting complementarity and potentially capable of forming reasonably strong associations by means of hydrogen-bonding can be found; in fact, some interactions have free energies as low as $-25$ kcal and even lower that are highly significant energetically and thermodynamically very stable. The problem is that random sequences exhibit an equal frequency of such interactions. Thus, although the termini of reovirus genes are no doubt capable of interacting by base-pairing, such interactions are unlikely to possess the specificity necessary for such interactions to form the basis of a highly selective association process. In fact, no mechanism could be detected capable of directing the unique incorporation of one copy each of the 10 plus strands into an immature virus complex. In particular, the sequences were examined for ability to interact in unique arrangements such as 1 gene acting as a "collector" gene for the other 9, or for head-to-head, tail-to-tail, or head-to-tail and tail-to-head associations that would result in the formation of circles on the insertion of the 10th element. No evidence supporting such models could be detected (Antczak *et al.*, 1982). One must conclude either that recognition signals of this nature are not present in the terminal regions, or that the associations occur via mechanisms that do not depend on base-pairing, or that the assembly mechanism involves not only nucleic acid–nucleic acid but also nucleic acid–protein interactions.

### 3. Features That May Control Frequency of Translation

Attempts were also made to discern signals that might function to control the frequency of translation of the ten reovirus mRNAs. It is known that some mRNAs, particularly species like *m*1 and *s*1 mRNA, and probably also *l*1 mRNA, are translated about 25 times less frequently than others, such as species *l*2, *l*3, *m*2, and *m*3 mRNA (Joklik, 1973, 1981). Several factors could be eliminated readily. For example, almost none of the reovirus plus strands exhibit significant ability to form hairpin loops in the 5'-terminal regions; and although most of them can form rather stable, and even very stable, interactions between sequences in their 5'-terminal and 3'-terminal regions, thereby forming rings, they do so neither more nor significantly less frequently than random sequences, and the relative stability of such interactions does not correlate, positively or negatively, with relative translation frequency. Secondary structure considerations based on hairpin loop and ring formation therefore do not seem to enter into controlling frequency of translation. Nor does ability to base-pair with the consensus 3'-terminal region of mammalian ribosomal 18 S RNA appear to be a factor; such interactions are rather weak, do not occur more frequently with the 5'-terminal regions of reovirus messenger RNAs than with their 3'-terminal regions, and do not occur more or less frequently than with random sequences. Nor could frequency of translation be correlated with distance of the first initiation codon from the 5' terminus or with the presence of additional initiation codons. Two factors, however, do correlate with relative frequency of translation. First, Kozak (1981) has found that $^{A}_{G}$XXAUGG is a favored sequence for eukaryotic initiation sites; whereas most functional initiation codons are preceded by a purine, usually A, in position $-3$, and possess a G in position $+4$, most nonfunctional or little-used AUGs have pyrimidines in these positions. All reovirus mRNAs that are translated with high or medium efficiency, as well as one of the mRNAs (*m*1) translated with low efficiency, possess this consensus sequence, while two of the three mRNAs translated with low efficiency do not; one has a pyrimidine in position $-3$, the other a pyrimidine in position $+4$. It is of interest that of the six initiation codons downstream from the first initiation codon on genes *L1*, *L3*, *M1*, *S1*, *S2*, and *S3*, three have pyrimidines in either position $-3$ or $+4$. Second, reovirus mRNA species that are translated frequently have AG-rich regions upstream and surrounding the first initiation codon; mRNA species that are translated infrequently do not possess such regions. Neither the presence nor the absence of sequences rich in any other nucleotide base pair correlates with relative frequency of translation (Antczak et al., 1982).

### 4. Protein-Binding Recognition Sites

Apart from the terminal tetra- and pentanucleotide, the ends of the ten genes possess no *common* sequences sufficiently similar for them to be interpreted as representing common protein-binding sites. If such sites exist, they must share features other than similarity of base sequence.

## D. Sequences at the Termini of the L3, M3, S1, and S2 Genes of Reovirus Serotypes 1, 2, and 3

Li et al. (1980a) and Gaillard et al. (1982) sequenced the 3'-terminal regions of the plus and minus strands of the L3, M3, S1, and S2 genes of reovirus serotypes 1, 2, and 3, which provided the sequences of the regions at both ends of their plus strands. These sequences are presented in Figs. 11 and 12. This selection of genes provides information concerning genes of all three size classes, comparison between two genes of the S size class, and information concerning genes that encode either structural or nonstructural proteins. As for the antigenic determinants on these proteins, those on proteins λ1 and μNS appear to be group-specific, those on protein σ2 display some type specificity, and those on protein 1 are type-specific (Gaillard and Joklik, 1980; Lee et al., 1981b; Hayes et al., 1981).

At the 5' termini, the three L1 genes and the three M3 genes are extraordinarily similar. Where there are substitutions, they are mostly in third codon positions, so that the amino acid sequence remains the same. There is no sequence divergence in the first 18 codons of the three L3 genes, but the sequences of the serotype 1 and 2 M3 genes do diverge completely after the 15th codon. In the case of the S2 genes, the serotype 1 and 3 genes are almost identical, but in the serotype 2 gene, there are 7 changes in the first 59 residues, including deletions that change the reading frame, so that the amino acid sequence of the serotype 2 σ2 protein differs from that of the serotype 1 and 3 σ2 proteins in 5 of the first 12 amino acids. Finally, the three S1 genes are almost totally dissimilar, both in the regions upstream from the initiation codon and in their coding regions, so that the amino acid sequences of the three σ1 proteins are quite different.

At their 3' ends, the four sets of cognoate genes display similar patterns: the sets of L1, M3, and S2 genes are very similar, the three S1 genes much less so. Most genes possess termination codons at least 60 residues

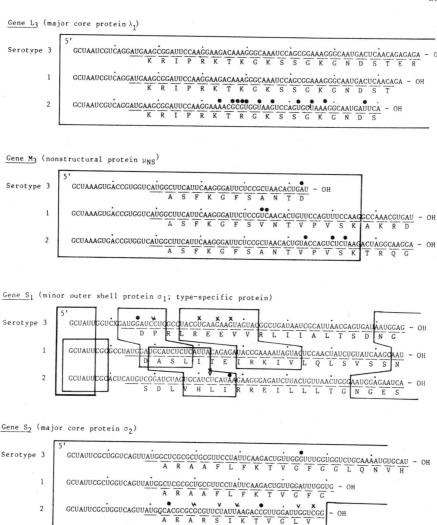

FIGURE 11. Sequences of the 5′-terminal regions of the plus strands of genes *L3*, *M3*, *S1*, and *S2* of reovirus serotypes 1, 2, and 3. The sequences shown here are the complements of the actually sequenced 3′-terminal regions of the corresponding minus strands. (●) Base changes (the odd base out of three being indicated); (**V**) deletions (that is, bases missing); (**×**) insertions (that is, bases inserted into the sequence of one gene, but not into that of the other two). From Gaillard *et al.* (1982).

from the 3′ terminus; many possess multiple termination codons in all reading frames. Thus, most reovirus genes appear to possess sizeable untranslated 3′-terminal regions.

Thus, examination of the sequences of four genes of all three reovirus serotypes reveal some features that were expected, others that were unexpected. Among the expected features are the fact that there is far less similarity among the three *S1* genes, which code for the type-specific σ1

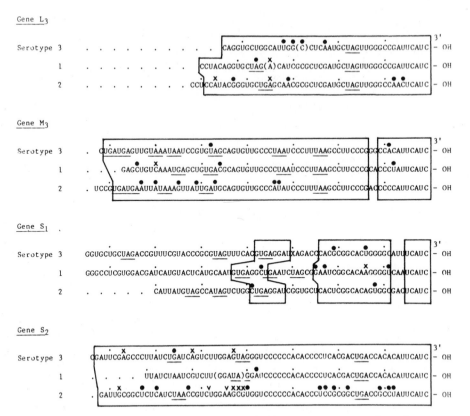

FIGURE 12. 3'-Terminal regions of the plus strands of genes *L3*, *M3*, *S1*, and *S2* of reovirus serotypes 1, 2, and 3. Symbols are defined in the Fig. 11 caption. From Gaillard *et al.* (1982).

protein, than among the sets of *L3*, *M3*, and *S2* genes. Among the unexpected features is the extent of similarity among the serotype 2 genes on the one hand and the serotype 1 and 3 genes on the other. Since the relatedness of serotype 2 genes to serotype 1 and 3 genes is no more than about 15% (Gaillard and Joklik, 1982), sequence divergence among these genes must be concentrated in their internal regions. This is curious, since the reason for terminal homology is clearly not the need for recognition signals for polymerases and encapsidation: the three *S1* genes show only very limited homology at their termini.

These results are also of interest from another point of view, namely, that related to the known antigenic properties of the reovirus proteins and the degree of genetic relatedness of the three reovirus serotypes. The background is as follows: First, mammalian reoviruses possess three types of *S1* genes that encode σ1 proteins that can be readily differentiated by immunological means; they are the type-specific antigens (Weiner and Fields, 1977; Weiner *et al.*, 1978; Gaillard and Joklik, 1980; Lee *et al.*, 1981b; Hayes *et al.*, 1981). This raises the question of how many types of other reovirus genes there are. There are clearly several possibilities. For example, *S1* could be the only gene that has diverged during evolution; all other genes could be identical. Indeed, this seems not unreasonable, because the antigenic determinants on most of the other nine reovirus proteins are predominantly group-specific; that is, antisera cannot readily differentiate among them. Alternatively, there could be several different types of each of the other genes, but because reovirus genes reassociate efficiently in multiply infected cells (Fields and Joklik, 1968), each of the three *S1* genes would be associated with a randomly distributed set of other genes. In either case, and even if the sets of cognate genes are quite dissimilar, the genomes of serotype 1, 2, and 3 strains would be related to each other to the *same* extent: closely in the first case considered above, and more distantly in the second. However, this is not what is found: as judged by the ability of the genomes of strains of the three serotypes to hybridize with each other, strains of serotypes 1 and 3 are related to the extent of about 70%, while strains of serotype 2 are related to strains of serotypes 1 and 3 only to the extent of about 10% (Martinson and Lewandowski, 1975; Gaillard and Joklik, 1980). This implies not only that the genes other than gene *S1* have also diverged widely during evolution, but also that reassociation among them in nature has not occurred; thus, the type 1 *S1* gene is still associated with type 1 *L1*, *L2*, etc., genes, rather than with a random set of genes.*

As already mentioned, the most striking feature of the sequence data of Gaillard *et al.* (1982) is not the difference between serotype 2 genes on the one hand and serotype 1 and 3 genes on the other, but rather the extraordinary similarity between them. Since individual serotype 1 and 3 genes (except gene *S1*) are related to the extent of about 50–90%, whereas serotype 2 genes (except gene *S1*) are related to serotype 1 and 3 genes to the extent of only about 5–15% (Gaillard and Joklik, 1982), it would seem that the sequence differences among the genes are not distributed uniformly along their double strands, but that they are localized, or rather than there are local regions of profound similarity that have presumably been conserved during evolution. Clearly, such regions of similarity exist at the two ends. They apparently also exist at those regions that encode the antigenic determinants of their proteins, for, as pointed out in Section VIII, these antigenic determinants are group-spe-

---

* Recently, Hrdy *et al.* (1979) examined a series of 94 reovirus isolates from humans, cattle, and mice, and found extensive variability in the patterns of the electrophoretic migration rates of their genes; that is, the migration rates of their *L1* genes were not the same, but varied detectably and significantly, and so on. The significance of this polymorphism vis-à-vis the postulate that the three sets of reovirus genes have evolved independently will become apparent only when the genetic relatedness of *individual* genes of several of these isolates is determined. So far, this has been done with only three reovirus isolates, namely, the Lang strain of serotype 1, the D5 Jones strain of serotype 2, and the Dearing strain of serotype 3 (Gaillard and Joklik, 1982). It was found that *all* ten genes of serotypes 1 and 3 are far more closely related to each other than to those of serotype 2. This supports the postulate of independent gene-set evolution.

```
              15            ATG    30                  45               60
GCT ATT CGC TGG TCA GTT [ATG] GCT CGC GCT GCG TTC CTA TTC AAG ACT GTT GGG TTT GGT
                         M    A   R   A   A   F   L   F   K   T   V   G   F   G

                     75              90                105               120
GGT CTG CAA AAT GTG CCA ATT AAC GAC GAA CTA TCT TCA CAT CTA CTC CGA GCT GGT AAT
 G   L   Q   N   V   P   I   N   D   E   L   S   S   H   L   L   R   A   G   N

                    135             150                165               180
TCA CCA TGG CAG TTA ACA CAG TTT TTA GAC TGG ATA AGC CTT GGG AGG GGT TTA GCT ACA
 S   P   W   Q   L   T   Q   F   L   D   W   I   S   L   G   R   G   L   A   T

                    195             210                225               240
TCG GCT CTC GTT CCG ACG GCT GGG TCA AGA TAC TAT CAA ATG AGT TGC CTT CTA AGT GGC
 S   A   L   V   P   T   A   G   S   R   Y   Y   Q   M   S   C   L   L   S   G

                    255             270                285               300
ACT CTC CAG ATT CCG TTC CGT CCT AAC CAC CGA TGG GGA GAC ATT AGG TTC TTA CGC TTA
 T   L   Q   I   P   F   R   P   N   H   R   W   G   D   I   R   F   L   R   L

                    315             330                345               360
GTG TGG TCA GCT CCT ACT CTC GAT GGA TTA GTC GTA GCT CCA CCA CAA GTT TTG GCT CAG
 V   W   S   A   P   T   L   D   G   L   V   V   A   P   P   Q   V   L   A   Q

                    375             390                405               420
CCC GCT TTG CAA GCA CAG GCA GAT CGA GTG TAC GAC TGC GAT GAT TAT CCA TTT CTA GCG
 P   A   L   Q   A   Q   A   D   R   V   Y   D   C   D   D   Y   P   F   L   A

                    435             450                465               480
CGT GAT CCA AGA TTC AAA CAT CGG GTG TAT CAG CAA TTG AGT GCT GTA ACT CTA CTT AAC
 R   D   P   R   F   K   H   R   V   Y   Q   Q   L   S   A   V   T   L   L   N

                    495             510                525               540
TTG ACA GGT TTT GGC CCG ATT TCC TAC GTT CGA GTG GAT GAA GAT ATG TGG AGT GGA GAT
 L   T   G   F   G   P   I   S   Y   V   R   V   D   E   D   M   W   S   G   D

                    555             570                585               600
GTG AAC CAG CTT CTC ATG AAC TAT TTC GGG CAC ACG TTT GCA GAG ATT GCA TAC ACA TTG
 V   N   Q   L   L   M   N   Y   F   G   H   T   F   A   E   I   A   Y   T   L

                    615             630                645               660
TGT CAA GCC TCG GCT AAT AGG CCT TGG GAA TAT GAC GGT ACA TAT GCT AGG ATG ACT CAG
 C   Q   A   S   A   N   R   P   W   E   Y   D   G   T   Y   A   R   M   T   Q

                    675             690                705               720
ATT GTG TTA TCC TTG TTC TGG CTA TCG TAT GTC GGT GTA ATT CAT CAG CAG AAT ACG TAT
 I   V   L   S   L   F   W   L   S   Y   V   G   V   I   H   Q   Q   N   T   Y

                    735             750                765               780
CGG ACA TTC TAT TTT CAG TGT AAT CGG CGA GGT GAC GCC GCT GAG GTG TGG ATT CTT TCT
 R   T   F   Y   F   Q   C   N   R   R   G   D   A   A   E   V   W   I   L   S

                    795             810                825               840
TGT TCG TTG AAC TAT TCC GCA CAA ATT AGA CCG GGT AAT CGT AGC TTA TTC GTT ATG CCA
 C   S   L   N   Y   S   A   Q   I   R   P   G   N   R   S   L   F   V   M   P

                    855             870                885               900
ACT AGC CCA GAT TGG AAC ATG GAC GTC AAT TTG ATC CTG AGT TCA ACG TTG ACG GGG TGT
 T   S   P   D   W   N   M   D   V   N   L   I   L   S   S   T   L   T   G   C

                    915             930                945               960
TTG TGT TCG GGT TCA CAG CTG CCA CTG ATT GAC AAT AAT TCA GTA CTG CAG TGT CGC GTA
 L   C   S   G   S   Q   L   P   L   I   D   N   N   S   V   L   Q   C   R   V

                    975             990               1005              1020
ACA TCC ATG GCT GGA CTG GTA GAG CTG GTA ACC AAT TGC ATG GGT TCC AGG TGA GAC GAA
 T   S   M   A   G   L   V   E   L   V   T   N   C   M   G   S   R    D   E

                   1035            1050               1065              1080
TGG TGA CTG AAT TTT GTG ACA GGT TGA GAC GCG ATG GTG TCA TGA CCC AAG CTC AGC AGA
 W    L   N   F   V   T   G    D   A   M   V   S     P   K   L   S   R

                   1095            1110               1125              1140
ATC AAG TTG AAG CGT TGG CAG ATC AGA CTC AAC AGT TTA AGA GGG ACA AGC TCG AAA CGT

                   1155            1170               1185              1200
GGG CGA GAG AAG ACG ATC AAT ATA ATC AGG CTC ATC CCA ACT CCA CAA TGT TCC GTA CGA

                   1215            1230               1245              1260
AGC CAT TTA CGA ATG CGC AAT GGG GAC GAG GTA ATA CGG GGG CGA CTA GTG CCG CGA TTG

                   1275            1290               1305              1320
CAG CCC TTA TCT GAT CGT CGT TGG AGT GAG GGT CCC CCC ACA CCC CTG ACG ACT GAC CAC

ACA TTC ATC
```

cific, that is, very similar, which is certainly very unexpected among proteins the genes of which are related only to the extent of 5–15% and the amino acid sequences of which must therefore be quite different.

Thus, the pattern that emerges is that of the existence of three sets of ten reovirus genes that have evolved independently, and that in nine of these genes regions exist that have been conserved during evolution, among them being the terminal regions and the region coding for the antigenic sites of the proteins that they encode. Why these regions have been conserved, and why they have not been conserved in the *S1* genes, remains to be determined.

There is no doubt that it will be possible to ask both better and more informed questions concerning these problems when the complete sequences of all these ten genes are known. Several laboratories are currently attempting to clone the genes of reovirus (as well as of other members of the Reoviridae family), and no doubt the cloned genes will be quickly sequenced. Examination of the sequences of entire reovirus genes along the lines described above should provide insight into the questions that have been discussed here.

## E. Cloning the Reovirus Genes: The Sequence of the Cloned *S2* Gene of Serotype 3

Cashdollar *et al.* (1982) have cloned the genes of the Dearing strain of reovirus serotype 3 in pBR322 by tailing both strands of each gene with poly(A), transcribing them with reverse transcriptase, annealing the cognate plus and minus cDNA strands, incubating them with *Escherichia coli* DNA polymerase I to ensure that they are complete, and cloning the double-stranded cDNA molecules by standard procedures. The DNA of insert-containing plasmids was then hybridized to reovirus genes immobilized on diazotized aminophenylthioether-paper in order to determine the identity of each. All ten genes were cloned, and all cloned genes were shown to be complete gene copies because the sequences at their termini were identical to those of the reovirus genes themselves as determined by Antczak *et al.* (1982).

One of the cloned genes, the *S2* gene, has been sequenced (Fig. 13). The gene is 1329 nucleotides long and possesses a single long open reading frame that starts at the first initiation codon (residue 19) and extends for 331 codons to the first of four in-phase TGA termination codons. This reading frame can code for a protein the same size as the known *S2* gene

←_____

FIGURE 13. Nucleotide sequence of the cloned *S2* gene of the Dearing strain of serotype 3. ■, Pst 1 restriction site; ▼, Alu I restriction sites; ▲, Taq 1 restriction sites. The long open reading frame starts at residue 19; the second (smaller) open reading frame starts at residue 1020. The former is terminated by four TGA codons in phase (●), and the latter is terminated by three (at positions 1272, 1287 and 1314). (From Cashdollar *et al.*, 1982).

product, the major core component σ2 ($M_r$, 38,000). Protein σ2 was found to possess 151 nonpolar hydrophobic, 128 uncharged polar, 23 acidic, and 29 basic amono acids. In agreement with the fact that reovirus mRNAs are not polyadenylylated, there is no polyadenylylation signal near the 3'-end of the S2 mRNA.

Interestingly, the S2 gene possesses another open reading frame in a different reading frame that starts at residue 1020, shortly after the first TGA of the long open reading frame, and terminates with 3 in-phrase TGA codons, the first of which is at residue 1272. This second open reading frame extends for 85 codons and could encode a protein with a molecular weight of about 10,000. However, there is no evidence that this reading frame is operative, as no such 10,000-dalton protein has yet been detected in cells infected with reovirus.

## XI. REOVIRUS OLIGONUCLEOTIDES

In addition to the ten genes, reovirus particles contain numerous small oligonucleotides; in fact, about 25% of the total RNA in reovirus particles is single-stranded RNA (Bellamy and Joklik, 1967b; Shatkin and Sipe, 1968b). This RNA is in the form of short molecules that are not breakdown products, since they possess ppp (as well as smaller amounts of pp and p) at their 5' ends. These molecules fall into two classes (Bellamy and Hole, 1970; Bellamy et al., 1972; Nichols et al., 1972b; Stoltzfus and Banerjee, 1972; Carter et al., 1974). Their composition is summarized in Table VI. About one third contain only A; these are the oligoadenylates. They range in length from 2 to 20 residues, and there are about 900 of them in each reovirus particle. The remainder fall into the following series: (p)ppGpC, (p)ppGpCpU, (p)ppGpCpUpA, (p)ppGpCpUpA(pU)$_{1-4}$, and (p)ppGpCpUpA(pA)$_{1-3}$ (Bellamy et al., 1970, 1972; Nichols et al., 1972b). About 10% are guanylylated (i.e., terminate in GpppGp), but none is methylated (Carter, 1977). These are the 5'-G-terminated oligonucleotides. These sequences are the same as those that are present at the 5' termini of all reovirus gene transcripts (see Section X.A.1.) (Kozak, 1977; Li et al., 1980a; Gaillard et al., 1982; Antczak et al., 1982). They are clearly products of abortive reovirus gene transcription. Interestingly, the relative amounts of double-stranded RNA (dsRNA), oligoadenylates, and 5'-G-terminated oligonucleotides in virus yields vary markedly depending on the temperature at which the virus was grown; at 31 and 40°C, the ratios of the amounts of oligoadenylates to the amounts of 5'-G-terminated oligonucleotides are 0.4 and 2.0, respectively, and the ratios of the amounts of dsRNA to the amounts of 5'-G-terminated oligonucleotides at 31 and 40°C, are 0.05 and 0.33, respectively (K.C. Lai and Bellamy, 1971). The origin of the reovirus oligonucleotides, and how they come to be sealed into reovirus particles, is discussed in Section XIII.

The oligonucleotides are released from infecting virus particles once

TABLE VI. Oligonucleotides Present in Reovirus Particles[a]

| Oligonucleotide(s) | Sequence | Chain length | Approximate number of molecules per virus particle |
|---|---|---|---|
| Oligoadenylates | $(p)(p)p(A)_{1-19}A_{OH}$ | 2–20 | 850 |
| 5'-G-terminated oligonucleotides | $(p)ppGC_{OH}$ | 2 | 50 |
| | $(p)ppGCU_{OH}$ | 3 | 900 |
| | $(p)(p)pGCUA_{OH}$ | 4 | 775 |
| | $(p)ppGCU(A)_{1-3}A_{OH}$ | 5–7 | 130 |
| | $(p)ppGCUA(U)_{1-4}U_{OH}$ | 6–9 | 130 |
| | Unknown | 2–8 | 350 |

[a] From Nichols et al. (1972b).

they have been converted to subviral particles. They are then capped and methylated and persist for at least 5 hr (Carter and Lin, 1979). Several workers have hypothesized that they may have some function during the infection cycle, but Carter et al. (1974) have shown that under special conditions, cores, which lack them, are infectious.

## XII. CONVERSION OF REOVIRUS PARTICLES TO CORES AND ACTIVATION* OF THE TRANSCRIPTASE

The conversion of reovirus particles to cores has been studied in some detail. It is effected by several proteolytic enzymes, that most commonly used being chymotrypsin. The conversion proceeds via several distinct steps that can be at least partially dissociated and therefore studied by manipulating the temperature of incubation, the concentration of reactants, and the composition of the suspending solutions. If the concentration of chymotrypsin is low (less than 100 μg/ml), $K^+$ or $Cs^+$ is present (at about 0.1 mM) and $Mg^{2+}$ is absent, and the Dearing strain of reovirus serotype 3 is used, the sequence of events is as follows (Joklik, 1972): The first stage consists of the removal of protein σ3 and yields particles that retain full infectivity, lose their oligonucleotide complement (see below) on prolonged incubation at 37°C, and possess no transcriptase activity. The second stage involves the removal of protein μ1C via a series of sequentially arising cleavage products that remain transiently associated with the particles. During this stage, the virus flocculates and increases in buoyant density, specific infectivity diminishes by about 5 logs, and the virus loses its oligonucleotide complement and develops transcriptase activity. During the final stage, the particles lose protein σ1. At high enzyme concentrations (more than 1000 μg/ml), a different end product is formed. This consists of particles that lack protein σ3 and a 12,000-

* The term "activation" is taken to signify development of the ability to transcribe complete transcripts (see below).

dalton fragment of $\mu 1C$, possess a buoyant density about 0.01 g/ml higher than that of virus particles, and possess full infectivity, the entire oligonucleotide complement, and no transcriptase activity.

Interestingly, the conversion of reovirus particles to cores is strongly influenced by virus concentration. For example, under the conditions outlined above, 10 $\mu$g/ml chymotrypsin causes a greater than 6 log decrease in the infectivity of the Dearing strain reovirus when the virus is at a concentration of 1000 $\mu$g/ml, but even double this enzyme concentration scarcely reduces infectivity when the virus concentration is 10 $\mu$g/ml (Joklik, 1972).

Borsa and his associates have confirmed and extended these results. They identified two sets of conditions for the chymotrypsin-catalyzed digestion of the outer shell of reovirus particles that results in the formation of cores and activation of the transcriptase.

1. If the reaction is carried out in the presence of $NH_4^+$, $Li^+$, $Na^+$, or $Mg^{2+}$, or at high concentrations of chymotrypsin, intermediate subviral particles (ISVPs) are formed that correspond to the second type of particle identified by Joklik (1972) (Borsa et al., 1973a–c). These particles are converted to cores in the absence of further proteolytic digestion provided that conditions are "facilitating" (the activation of transcriptase): such conditions include the presence of the "facilitating" ions $K^+$, $Rb^+$, or $Cs^+$, a chymotrypsin concentration that is not too high (Sargent et al., 1977) (since chymotrypsin and other proteases except trypsin reversibly block the second step of the conversion), and a virus concentration that is not too low [so as to effect "virus-concentration-mediated facilitation" (Borsa et al., 1974)]. It should be noted, as pointed out above, that ISVPs may be either infectious or noninfectious; both are formed in the absence of facilitating ions, the former at low, the latter at high, virus concentrations. In fact, in the presence of $Na^+$ or $Li^+$, chymotrypsin may even enhance the infectivity of dilute suspensions of reovirus (Borsa et al., 1973c), a phenomenon that had also been noted by Spendlove et al. (1970), and both Cox and Clinkscales (1976) and Borsa et al. (1981) noted that infection with infectious ISVPs proceeds with a significantly shortened eclipse phase and results in up to 3 times higher yields than infection with untreated reovirus particles. It has also been suggested that reovirus particles and ISVPs may be taken up into cells by different pathways, viropexis for virus particles and more rapid, direct penetration for ISVPs (Borsa et al., 1979). In either case, ISVPs with inactive transcriptase are converted to cores with transcriptase in the activated state when they are allowed to adsorb to and infect cells (Borsa et al., 1981).

2. If the reaction between chymotrypsin and reovirus particles is carried out in the presence of facilitating ions, the reaction proceeds directly to cores with transcriptase in the activated state, as described by Joklik (1972) (see above).

Another type of ISVP was described by Shatkin and LaFiandra (1972), who digested reovirus strain Dearing with chymotrypsin in the presence of nonfacilitating ions and phosphate-buffered saline. Like the ISVPs of Joklik (1972), these ISVPs retained the μ1C cleavage product protein δ and were infectious, but unlike them they possessed transcriptase activity. These ISVPs appeared to be very similar to the products of uncoating in infected cells (D.H. Levin et al., 1970; Chang and Zweerink, 1971) (see Chapter 4 for a more detailed treatment). When reovirus particles infect cells, they are uncoated to subviral particles (SVPs) that, like the ISVPs generated in vitro in the presence of chymotrypsin, possess polypeptide δ, are infectious [specific infectivity about one fifth that of virus particles (Astell et al., 1972)], and possess transcriptase activity.

Another approach to the conversion of reovirus particles to ISVPs and cores was taken by Carter (1979b), who correlated the protein composition of ISVPs, in particular with respect to the nature of protein μ1C cleavage products, with activation of several enzyme activities, such as transcriptase capable of synthesizing full-length transcripts, poly(A) polymerase, and oligonucleotide methyltransferase. Evidence was obtained that the nature of the μ1C cleavage products and of the enzymes that were activated were influenced by the presence of an ATP-generating system during digestion with chymotrypsin, which raises the possibility that protein modifications such as polyadenylylation and poly(ADP) ribosylation (see Section VI) may play a role in specifying the pathway of the reovirus particle to core conversion.

These results raise the question of why some ISVPs and the SVPs generated in infected cells are infectious, while others are not. The specific infectivity of "infectious" ISVPs and of SVPs is within an order of magnitude of that of virus particles (Joklik, 1972; Borsa et al., 1973c; Astell et al., 1972), while that of "noninfectious" ISVPs is 4–6 logs lower (Joklik, 1972; Borsa et al., 1973b). Furthermore, cores are "noninfectious," despite being able to adsorb to cells almost as efficiently as virus particles (M.-H.T. Lai and Joklik, 1973). The most likely explanation of these findings is that virus particles and infectious ISVPs are taken up into cells via different pathways from noninfectious ISVPs and cores, and that their fates within cells differ. It is conceivable that the crucial difference is the presence of protein σ1, which reovirus particles and infectious ISVPs possess (Joklik, 1972), but cores, and perhaps also noninfectious ISVPs, do not possess.

## XIII. ENZYMES IN REOVIRUS

In 1967, Kates and McAuslan (1967) discovered the first virus-coded virus-associated enzyme capable of catalyzing a synthetic reaction, namely, the DNA-dependent RNA polymerase of vaccinia virus. A year later, Borsa and Graham (1968) and Shatkin and Sipe (1968a) independently discovered an enzyme with a similar function, namely, transcribing the viral genome into messenger RNA (mRNA), in reovirus particles, or, more accurately, in reovirus cores: this is the double-stranded → single-stranded (ds→ss) RNA polymerase, or "transcriptase." Shortly thereafter, Kapuler et al. (1970) and Borsa et al. (1970) demonstrated the presence of nucleoside triphosphate phosphohydrolase activity in reovirus, and in 1975 Furuichi and Shatkin and their collaborators discovered that terminal guanylyltransferase and the enzymes that methylate the cap G and the ribose of the original 5'-terminal nucleotide are also present (Shatkin, 1974; Furuichi et al., 1975a; see also Wachsman et al., 1970). Together, these enzymes constitute a mechanism that enables reovirus to transcribe its ten genes into ten capped species of mRNA that are extruded through the 12 projections/spikes located on cores (Gillies et al., 1971).

Until recently, it was thought that the reovirus transcriptase and the four enzymes involved in capping are active only in cores, in subviral particles (SVPs) formed in cells following infection and in the intermediate SVPs (ISVPs) of Shatkin and LaFiandra (1972) (see Section XII), but not in virus particles. Yamakawa et al. (1982) have shown recently that this is not so. When intact reovirus particles are incubated with $Mn^{2+}$ (or $Mg^{2+}$), the four nucleoside triphosphates, $S$-adenosyl-L-methionine (AdoMet), and an ATP-generating system, they synthesize G-terminated oligonucleotides. The rate and extent of oligonucleotide synthesis by reovirus particles is about one quarter that exhibited by cores, but reovirus particles, in contrast to cores, do not synthesize complete mRNA molecules. Of these oligonucleotides, 98% are uncapped: about two thirds are ppGpC, one fifth are ppGpCpU, and most of the remainder are ppGpCpUpA. Although most of these oligonucleotides are uncapped, the four enzymes involved in capping are active in reovirus particles, albeit at lower levels than in cores (varying from 10 to 100%). The fact that nucleoside triphosphate phosphohydrolase is active in intact reovirus particles had already been noted by Borsa et al. (1970) and Kapuler et al. (1970), and Carter (1977, 1978) had shown that when reovirus particles are incubated with AdoMet, the endogenous oligonucleotides are methylated and acquire both $m^7GpppG$ and $m^7GpppG^m$ structures.

These five enzymes are also active in cores as well as in the SVPs to which reovirus particles are converted in infected cells, and in the in vitro–generated ISVPs described by Shatkin and LaFiandra (1972), which still retain the 60,000 μlC cleavage product δ. Not only is their activity up to 3-fold higher in such particles than in virus particles, but their transcriptase also generates complete transcripts (that is, mRNA molecules). However, oligonucleotides are also still formed by cores; in fact, both Zarbl et al. (1980) and Yamakawa et al. (1981) found that even in cores, transcription aborts in the great majority of initiation events before the transcripts are more than five residues long, so that even in cores a vast molar excess of oligonucleotides as compared with mRNA molecules is synthesized. Detailed mechanisms of transcription are discussed in Chapter 3.

Nothing is known concerning the nature of the transcriptase or of the four capping-function enzymes, since all enzyme activities are lost when cores or reovirus particles are disrupted. It is therefore thought that the enzymes may be components of the core shell and that the genes move past transcriptase catalytic sites fixed on the interior surface of the core shell, perhaps at the base of the spikes, so that the double-stranded templates remain within the cores, while transcripts are fed into and through the spikes. It this is correct, the transcriptase catalytic site may comprise sequences on both $\lambda 2$ and neighboring protein molecules ($\lambda 1$ or $\sigma 2$), while the capping functions would be on $\lambda 2$ in the projection/ spike channels. Indeed, when cores transcribe dsRNA in the presence of labeled pyridoxal phosphate, which is hypothesized to react with the reactive site of the enzyme, both $\lambda 1$ and $\lambda 2$ become labeled (Morgan and Kingsbury, 1980) and all four capping enzymes are inhibited, although pyridoxal phosphate does not bind to any of their active sites (except the AdoMet-binding sites of the methyltransferases) (Morgan and Kingsbury, 1981). This suggests that reovirus mRNA transcription and modification are accomplished by a topographically related group of enzyme molecules. It should be pointed out, however, that this is the only evidence that the transcriptase is located on the inner capsid shell. As discussed, no functions have been assigned to the three minor proteins $\lambda 3$, $\mu 1$, and $\mu 2$, which are also associated with cores, and these proteins, either free in the interior of the virus particle or attached to the inner core shell, may themselves possess transcriptase or capping activity or both. In fact, Drayna and Fields (1982) have shown that the pH optimum of the transcriptase is specified by gene $L1$, the gene that encodes protein $\lambda 3$ (see Section VII).

However this may be, the question remains as to why cores synthesize complete transcripts, while virus particles synthesize only abortive transcripts [although most of the transcription initiation events in cores are also abortive, so that chain *elongation* is clearly the rate-limiting step in reovirus mRNA synthesis, at least *in vitro* (Yamakawa *et al.*, 1981)]. The transcriptase thus has two distinct and separable phases: repetitive initiation, which is very active in both virus particles and cores, and elongation, which occurs only in structurally altered particles like cores (and SVPs). Presumably, loss of integrity of the outer capsid shell causes structural changes in the inner capsid shell as a result of which elongation is released from contraints that prevent it in virus particles. Interestingly, this process is reversible, for addition of $\sigma 3$ to SVPs causes inhibition of elongation, that is, loss of ability to synthesize mRNA (Astell *et al.*, 1972).

Nothing is known concerning the nature of the changes that are induced in cores as a result of which they become capable of transcribing complete transcripts. Presumably, movement of template relative to the catalytic site (or vice versa) is impossible if the outer capsid shell is intact, and this constraint is released when it is partially disrupted or completely removed. A remarkable conclusion that follows from these facts is that the appropriate ends of at least some of the genes are always in close juxtaposition to the catalytic sites of the transcriptase. Further insight into these mechanisms will have to await identification of the nature of the protein chains on which the catalytic site of the transcriptase is located. In particular, it will be interesting to determine whether it is really located on both proteins $\lambda 1$ and $\lambda 2$, as suggested by the experiments of Morgan and Kingsbury (1980).

The transcriptase activity of reovirus cores is very stable. Its optimum temperature is between 47 and 52°C, a range at which it is active for over 48 hr. It has an extraordinary temperature dependence, being 15 times more active at 44 than at 34°C (Kapuler, 1970). The transcription that it catalyzes has been studied extensively (Skehel and Joklik, 1969; D.H. Levin et al., 1970; Banerjee and Shatkin, 1970; Shatkin and Banerjee, 1970) (see also Chapter 3). It is totally asymmetric—strands with only one polarity (plus, that is, translatable) being synthesized—and conservative (that is, the first plus strand to be formed is already a transcript, not the plus strand of the double-stranded molecule being transcribed) (Skehel and Joklik, 1969). In the presence of optimum $Mg^{2+}$ concentrations, all genes are transcribed at the same rate; i.e., 4 and 2 times as many $s$ and $m$ class transcripts, respectively, are synthesized as $l$ class transcripts (Skehel and Joklik, 1969). In other words, the frequency of transcription initiation is not the same for each gene, but is inversely proportional to gene size, which argues strongly against the ten genes being transcribed as a linked complex (see Section V).

The transcripts that are formed during the early part of the reovirus multiplication cycle or in vitro in the presence of AdoMet are capped and methylated; that is, the structure of their 5' termini is $m^7G(5')ppp(5')$-$G^mpCp$ . . . (Shatkin, 1974; Furuichi et al., 1975a,b). However, neither capping nor methylation is tightly coupled to transcription; uncapped and unmethylated transcripts are both formed readily (Furuichi and Shatkin, 1976, 1977; Yamakawa et al., 1982), and AdoMet does not stimulate transcription. Interestingly, AdoMet stimulates the transcriptase of orbiviruses 2-fold (Van Dijk and Huismans, 1980) and that of cytoplasmic polyhedrosis viruses by up to 70-fold (Furuichi, 1981), the mechanism of stimulation being thought to be lowering of the $K_m$ of the initiating nucleotide at the promoter site. Another finding that points in this direction was reported by Nakashima et al. (1979), who showed that while irradiation with long-wave length UV light in the presence of 4'-substituted psoralens abolishes transcriptase activity—by damaging reovirus genes as templates through cross-linking their strands and, incidentally, also abolishing reovirus infectivity—guanylyltransferase and methyltransferase activities are not affected.

The reason oligonucleotides are present in reovirus particles, are released slowly from ISVPs and are not present in cores is not known with certainty. They are firmly associated with virus particles; they are, in essence, sealed in. Before it was known that the transcriptase is active in virus particles, it was thought that the $\lambda 2$ clusters that make up the projections/spikes may be closed in virus particles and open in cores. However, this is clearly not the case, because (1) oligonucleotides in virus particles can be methylated; i.e., AdoMet can enter (Carter, 1978); (2) reovirus particles synthesize oligonucleotides; i.e., nucleoside triphosphates can enter; and (3) the oligonucleotides synthesized by reovirus particles are mostly liberated into the medium. Why, then, are the endogenous oligonucleotides not released? Presumably, endogenous oligonucleotides *within* virus particles are associated with them in a way that differs from the way in which oligonucleotides *synthesized by* reovirus particles are associated with them. If the endogenous oligonucleotides are "sealed" in, this event probably happens during the final stages of morphogenesis, when transcription elongation is inhibited through association of σ3 with the penultimate form of immature virus particles (Astell *et al.*, 1972) (see above); that is, they are oligonucleotides generated within incomplete, rather than within complete, virus particles.

As for the oligoadenylates, their synthesis is catalyzed by intact virus particles in the presence of $Mn^{2+}$ but not of $Mg^{2+}$ (Stoltzfus *et al.*, 1974). Infectious SVPs also synthesize them, but cores do not. Within cells, their synthesis may be an expression of a partial activity of the transcriptase in the process of being immobilized, when its $K_m$ for ATP may greatly exceed that for other nucleoside triphosphates, which would result in its catalyzing the template-independent polymerization of ATP to short oligoadenylates (Bellamy *et al.*, 1972; Nichols *et al.*, 1972b). Indeed, Silverstein *et al.* (1974) found that oligoadenylates are formed within nascent virus particles during the final stages of morphogenesis and suggested that oligo(A) polymerase activity may be an alternative activity of the viral transcriptase that is regulated by outer capsid proteins. Further, Johnson *et al.* (1976) found that late temperature-sensitive mutants of reovirus do not synthesize oligoadenylates at nonpermissive temperatures, which again supports the conclusion that they are synthesized during the final stages of virus maturation.

Finally, the fact that the activities of the transcriptase and of the capping-function enzymes can be assayed with intact virus particles raises very interesting questions concerning their locations; even more surprising, cores (but not virus particles) can guanylylate and methylate exogenously added mRNA molecules (Furuichi and Shatkin, 1977; Yamakawa *et al.*, 1982). It seems clear that the catalytic sites of the transcriptase are located within the core [since transcripts can be seen to be extruded from cores (Gillies *et al.*, 1971)], but it is conceivable that the capping-function enzymes are located on portions of the core (either on the core "body" or on the projections/spikes) that are freely accessible to exogenously added substrates, which could thus react without having to pass through the projections/spikes into the interior of the core. Clearly, a great deal more is to be learned concerning the structure of reovirus particles and reovirus cores.

## XIV. REOVIRUS MUTANTS THAT CONTAIN FEWER THAN TEN GENES

Reovirus particles contain ten genes. The mechanism that ensures that each virus particle contains a complete set of all ten genes is not known; available evidence indicates that assortment of RNA species into sets proceeds at the level of single-stranded RNA which associates with several species of virus-specified proteins to form "ds RNA synthesizing structures" (Matsuhisa and Joklik, 1974). Within these structures, the (plus-stranded) ssRNA is transcribed once and once only into minus strands that remain associated with them, thereby generating the progeny dsRNA genes.

This mechanism obviously functions very efficiently, since most reovirus particles appear to contain all ten genes, as is evidenced by the fact that infectious/total particle ratios close to 1 have been reported (Spendlove et al., 1970). It is not infallible, however, for Nonoyama et al. (1970) and Nonoyama and Graham (1970) described reovirus particles that contained only nine segments, the largest, L1, being missing. Such defective particles are produced if the virus is passaged repeatedly at high multiplicity. They are readily removed from virus stocks by plaquing or passaging at low multiplicity.

These studies were extended by Schuerch et al. (1974), who found that yields of wt virus begin to decline and contain defective particles after about 8 passages at high multiplicity; the progeny then consists of about one quarter of nondefective particles and three quarters of particles that lack segment L1. On further passaging, virus particles that lack both L1 and L3 genes also begin to accumulate (Ahmed and Fields, 1981). Furthermore, when the group F mutant ts556 is passaged no more than 3 or 4 times at high multiplicity, virus yields decrease precipitously and consist of roughly equal numbers of particles that lack RNA segment L1 and particles that lack both segments L1 and L3; only a small proportion of the yield then consists of nondefective particles. Also, yields of the group C mutant ts447, even under conditions of low-multiplicity passage at permissive temperatures, consist of equal numbers of particles that contain all ten genes and particles that lack gene L3, and if this mutant is passaged under conditions of high multiplicity, particles that lack both genes L1 and L3 begin to be formed by the 8th passage, and particles that lack gene M1 are formed somewhat later (Ahmed and Graham, 1977). Ahmed and Graham (1977) also noted that by the 15th passage at high multiplicity, 90% of the yields of the group E mutant ts320 lack genes L1 and L3.

All defective reovirus particles contain the normal complement of capsid proteins and therefore adsorb normally and possess full transcriptase activity on activation (Schuerch et al., 1974). They are noninfectious and can replicate only in the presence of helper virus to provide the deleted gene function; either wt virus or temperature-sensitive (ts) mutants that do not carry lesions in the genes missing from the defective particles (Spandidos and Graham, 1975a,b) can complement. Recombination (by gene reassortment) has also been shown to occur between deletion mutants and *ts* mutants (Ahmed and Fields, 1981) (see also Chapter 5).

Defective reovirus particles are not easy to isolate, since the only difference between them and normal virus particles is lack of one or two genes; the resulting density differences are small (less than 0.01 g/ml) and sufficient for enrichment, but not for isolation (Ahmed and Fields, 1981). It has been shown, however, that practically pure preparations of defective virus particles can result when mixtures of defective virus particle preparations and certain *ts* mutants are passaged several times at a nonpermissive temperature (Spandidos and Graham, 1975a). It is easier to separate cores that contain complete and incomplete sets of genes (Nonoyama *et al.*, 1970; Schuerch *et al.*, 1974; Spandidos and Graham, 1975a), since their nucleic acid content is much higher than that of virus particles.

Finally, the question arises as to why reovirus deletion mutants that lack one or more genes interfere with the multiplication of complete virus. The fact that they interfere is undoubted, since on continued high-multiplicity passage the yield of infectious *wt* virus decreases as the proportion of defective virus particles among the yield increases. However, is it the *absence* of a gene that causes the virus particles to interfere [that is, to be defective interfering (DI) particles]? The most likely reason DI particles of viruses with nonsegmented genomes interfere with the multiplication of *wt* virus may be that their smaller genomes compete successfully for virus-coded replicases (Huang and Baltimore, 1977). The mechanism of interference by DI particles of viruses with segmented genomes is probably quite different. Ahmed and Fields (1981) found that the genomes of defective reovirus particles generally contain numerous mutations, such as *ts* mutations, plaque-size mutations, and so on. In fact, populations of defective reovirus particles provide a source of mutant genes and contribute to the emergence of genetic variants. These workers propose that the biological properties of defective reovirus particles are determined not by the genes that are deleted, but by mutations in the genes that are still present. Interference by DI particles of reovirus would then be due to the expression of mutant genes and the incorporation of defective proteins into progeny virus particles. Ahmed and Fields (1981) did, in fact, show that recombinant viruses containing mutant genes derived from DI particles are capable of interfering with the growth of *wt* reovirus. The genetic aspects of DI reovirus particles are discussed further in Chapter 5.

ACKNOWLEDGMENTS. I would like to thank numerous colleagues for their collaboration, advice, and illuminating discussions. Among them are Patrick Lee, Ed Hayes, Joe Li, Jack Keene, Jim Antczak, David Pickup, and

Dick Gaillard. Much of the work from my laboratory discussed in this chapter was supported by Grant R01 AI08909 from the U.S. Public Health Service.

## REFERENCES

Abraham, G., and Banerjee, A.K., 1976, Sequential transcription of the genes of vesicular stomatitis virus, *Proc. Natl. Acad. Sci. U.S.A.* **73**:1504.

Ahmed, R., and Fields, B.N., 1981, Reassortment of genome segments between reovirus defective interfering particles and infectious virus: Construction of temperature-sensitive and attenuated viruses by rescue of mutations from DI particles, *Virology* **111**:351.

Ahmed, R., and Graham, A.F., 1977, Persistent infections in L cells with temperature-sensitive mutants of reovirus, *J. Virol.* **23**:250.

Almeida, J.D., Bradburne, A.F., and Wreghitt, T.G., 1979, The effect of sodium thiocyanate on virus structure, *J. Med. Virol.* **4**:269.

Amano, Y., Katagiri, S., Ishida, N., and Watanabe, Y., 1971, Spontaneous degradation of reovirus capsid into subunits, *J. Virol.* **8**:805.

Antczak, J.B., Chmelo, R., Pickup, D.J., and Joklik, W.K., 1982, Sequences at both termini of the ten genes of reovirus serotype 3 (strain Dearing), *Virology* **121**:307.

Arnott, S., Hutchinson, F., Spencer, M., Wilkins, M.H.F., Fuller, W., and Langridge, R., 1966, X-ray diffraction studies of double helical ribonucleic acid, *Nature (London)* **211**:227.

Astell, C., Silverstein, S.C., Levin, D.H., and Acs, G., 1972, Regulation of the reovirus transcriptase by a viral capsomer protein, *Virology* **48**:648.

Babiss, L.E., Luftig, R.B., Weatherbee, J.A., Weihing, R.R., Ray, U.R., and Fields, B.N., 1979, Reovirus serotypes 1 and 3 differ in their *in vitro* association with microtubules, *J. Virol.* **30**:863.

Banerjee, A.K., and Shatkin, A.J., 1970, Transcription *in vitro* by reovirus-associated ribonucleic acid-dependent polymerase, *J. Virol.* **6**:1.

Banerjee, A.K., and Shatkin, A.J., 1971, Guanosine-5'-diphosphate at the 5' termini of reovirus RNA: Evidence for a segmented genome within the virion, *J. Mol. Biol.* **61**:643.

Bellamy, A.R., and Hole, L.V., 1970, Single-stranded oligonucleotides from reovirus type 3, *Virology* **40**:808.

Bellamy, A.R., and Joklik, W.K., 1967a, Studies of the reovirus RNA. II. Characterization of reovirus messenger RNA and of the genome RNA segments from which it is transcribed, *J. Mol. Biol.* **29**:19.

Bellamy, A.R., and Joklik, W.K., 1967b, Studies on the A-rich RNA of reovirus, *Proc. Natl. Acad. Sci. U.S.A.* **58**:1389.

Bellamy, A.R., Shapiro, L., August, J.T., and Joklik, W.K., 1967, Studies on reovirus RNA. I. Characterization of reovirus genome RNA, *J. Mol. Biol.* **29**:1.

Bellamy, A.R., Hole, L.V., and Baguley, B.C., 1970, Isolation of the trinucleotide pppGpCpU from reovirus, *Virology* **42**:415.

Bellamy, A.R., Nichols, J.L., and Joklik, W.K., 1972, Nucleotide sequences of reovirus oligonucleotides: Evidence for abortive RNA synthesis during virus maturation, *Nature (London) New Biol.* **238**:49.

Borsa, J., and Graham, A.F., 1968, Reovirus: RNA polymerase activity in purified virions, *Biochem. Biophys. Res. Commun.* **33**:896.

Borsa, J., Grover, J., and Chapman, J.D., 1970, Presence of nucleoside triphosphate phosphohydrolase activity in purified virions of reovirus, *J. Virol.* **6**:295.

Borsa, J., Sargent, M.D., Long, D.G., and Chapman, J.D., 1973a, Extraordinary effects of specific monovalent cations on activation of reovirus transcriptase by chymotrypsin *in vitro*, *J. Virol.* **11**:207.

Borsa, J., Copps, T.P., Sargent, M.D., Long, D.G., and Chapman, J.D., 1973b, New intermediate subviral particles in the *in vitro* uncoating of reovirus virions by chymotrypsin, *J. Virol.* **11**:552.

Borsa, J., Sargent, M.D., Copps, T.P., Long, D.G., and Chapman, J.D., 1973c, Specific monovalent cation effects on modification of reovirus infectivity by chymotrypsin digestion *in vitro*, *J. Virol.* **11**:1017.

Borsa, J., Long, D.G., Sargent, M.D., Copps, T.P., and Chapman, J.D., 1974, Reovirus transcriptase activation *in vitro*: Involvement of an endogenous uncoating activity in the second stage of the process, *Intervirology* **4**:171.

Borsa, J., Monash, B.D., Sargent, M.D., Copps, T.P., Lievaart, P.A., and Szekely, J.G., 1979, Two modes of entry of reovirus particles into cells, *J. Gen. Virol.* **45**:161.

Borsa, J., Sargent, M.D., Lievaart, P.A., and Copps, T.P., 1981, Reovirus: Evidence for a second step in the intracellular uncoating and transcriptase activation process, *Virology* **111**:191.

Both, G.W., Lavi, S., and Shatkin, A.J., 1975, Synthesis of all ten gene products of the reovirus genome *in vivo* and *in vitro*, *Cell* **4**:173.

Branton, D., and Klug, A., 1975, Capsid geometry of bacteriophage T2: A freeze–etching study, *J. Mol. Biol.* **92**:559; **98**:445.

Brubaker, M.M., West, B., and Ellis, R.J., 1964, Human blood group influence on reovirus hemagglutination titers, *Proc. Soc. Exp. Biol. Med.* **115**:1118.

Burstin, S.J., Spriggs, D.R., and Fields, B.N., 1982, Evidence for functional domains on the reovirus type 3 hemagglutinin, *Virology* **117**:146.

Carter, C.A., 1977, Methylation of reovirus oligonucleotides *in vivo* and *in vitro*, *Virology* **80**:249.

Carter, C.A., 1978, *In vitro* methylation of adenosine residues in reovirus RNA, *Virology* **88**:222.

Carter, C.A., 1979a, Polyadenylylation of protein in reovirions, *Proc. Natl. Acad. Sci. U.S.A.* **76**:3087.

Carter, C.A., 1979b, Activation of reovirion-associated poly(A) polymerase and oligomer methylase by cofactor-dependent cleavage of μ polypeptides, *Virology* **94**:417.

Carter, C.A., and Lin, B.Y., 1979, Conservation and modification of the pyrimidine-rich reovirus oligonucleotides after infection, *Virology* **93**:329.

Carter, C.A., Stoltzfus, C.M., Banerjee, A.K., and Shatkin, A.J., 1974, Origin of reovirus oligo(A), *J. Virol.* **13**:1331.

Carter, C.A., Lin, B.Y., and Metley, M., 1980, Polyadenylylation of reovirus proteins: Analysis of the RNA bound to structural proteins, *J. Biol. Chem.* **255**:6479.

Cashdollar, L.W., Esparza, J., Hudson, G.R., Chmelo, R., Lee, P.W.K., and Joklik, W.K., 1982, Cloning the double-stranded RNA genes of reovirus: Sequence of the cloned S2 gene, *Proc. Natl. Acad. Sci. U.S.A.* **79**:7644.

Caspar, D.L.D., and Klug, A., 1962, Physical properties in the construction of regular viruses, *Cold Spring Harbor Symp. Quant. Biol.* **27**:1.

Chang, C.-T., and Zweerink, H.J., 1971, Fate of parental reovirus in infected cells, *Virology* **46**:544.

Chow, N.-L., and Shatkin, A.J., 1975, Blocked and unblocked 5′ termini in reovirus genome RNA, *J. Virol.* **15**:1057.

Cox, D.C., and Clinkscales, C.W., 1976, Infectious reovirus subviral particles: Virus replication, cellular cytopathology, and DNA synthesis, *Virology* **74**:259.

Darzynkiewicz, E., and Shatkin, A.J., 1980, Assignment of reovirus mRNA ribosome binding sites to virion genome segments by nucleotide sequence analysis, *Nucleic Acids Res.* **8**:337.

Deshmukh, D.R., Sayed, H.I., and Pomeroy, D.S., 1968, Avian reoviruses. IV. Relationship to human reoviruses, *Avian Dis.* **13**:16.

Desrosiers, R.C., Sen, G.C., and Lengyel, P., 1976, Difference in 5′ terminal structure between the mRNA and the double-stranded virion RNA of reovirus, *Biochem. Biophys. Res. Commun.* **73**:32.

Drayna, D., and Fields, B.N., 1982a, Activation and characterization of the reovirus transcriptase: genetic analysis, *J. Virol.* **41**:110.

Drayna, D., and Fields, B.N., 1982b, Genetic studies on the mechanisms of chemical and physical interactions of reovirus, *J. Gen. Virol.* **63**:149.

Drayna, D., and Fields, B.N., 1982c, Biochemical studies on the mechanism of chemical and physical interaction of reovirus, *J. Gen. Virol.* **63**:161.

Dunnebacke, T.H., and Kleinschmidt, A.K., 1967, Ribonucleic acid from reovirus as seen in protein monolayers by electron microscopy, *Z. Naturforsch.* **22b**:159.

Eggers, H.J., Gomatos, P.J., and Tamm, I., 1962, Agglutination of bovine erythrocytes: A general characteristic of reovirus type 3, *Proc. Soc. Exp. Biol. Med.* **110**:879.

Fahey, J.E., and Crawley, J.F., 1954, Studies on chronic respiratory disease of chickens. II. Isolation of a virus, *Canad. J. Comp. Med.* **18**:13.

Farrell, J.A., Harvey, J.D., and Bellamy, A.R., 1974, Biophysical studies of reovirus type 3. 1. The molecular weights of reovirus and reovirus cores, *Virology* **62**:145.

Fields, B.N., and Joklik, W.K., 1969, Isolation and preliminary genetic and biochemical characterization of temperature-sensitive mutants of reovirus, *Virology* **37**:335.

Fields, S., and Winter, G., 1982, Nucleotide sequences of influenza virus segments 1 and 3 reveal mosaic structure of a small viral RNA segment, *Cell* **28**:303.

Finberg, R., Weiner, H.L., Fields, B.N., Benacerraf, B., and Burakoff, S.J., 1979, Generation of cytolytic T lymphocytes after reovirus infection: Role of S1 gene, *Proc. Natl. Acad. Sci. U.S.A.* **76**:442.

Finberg, R., Weiner, H.L., Burakoff, S.J., and Fields, B.N., 1981, Type-specific reovirus antiserum blocks the cytotoxic T-cell–target cell interaction: Evidence for the association of the viral hemagglutinin of a nonenveloped virus with the cell surface, *Infect. Immun.* **31**:646.

Fontana, A., and Weiner, H.L., 1980, Interaction of reovirus with cell surface receptors. II. Generation of suppressor T cells by the hemagglutinin of reovirus type 3, *J. Immunol.* **125**:2660.

Furuichi, Y., 1981, Allosteric stimulatory effect of *S*-adenosylmethionine on the RNA polymerase in cytoplasmic polyhedrosis virus: A model for the positive control of eukaryotic transcription, *J. Biol. Chem.* **256**:483.

Furuichi, Y., and Shatkin, A.J., 1976, Differential synthesis of blocked and unblocked 5′-termini in reovirus mRNA: Effect of pyrophosphate and pyrophosphatase, *Proc. Natl. Acad. Sci. U.S.A.* **73**:3448.

Furuichi, Y., and Shatkin, A.J., 1977, 5′-Termini of reovirus mRNA: Ability of viral cores to form caps post-transcriptionally, *Virology* **77**:566.

Furuichi, Y., Morgan, M., Muthukrishnan, S., and Shatkin, A.J., 1975a, Reovirus messenger RNA contains a methylated, blocked 5′-terminal structure: $m^7G(5')ppp(5')G^mpCp$-, *Proc. Natl. Acad. Sci. U.S.A.* **72**:362.

Furuichi, Y., Muthukrishnan, S., and Shatkin, A.J., 1975b, 5′-Terminal $m^7G(5')ppp(5')G^mp$ *in vivo*: Identification in reovirus genome RNA, *Proc. Natl. Acad. Sci. U.S.A.* **72**:742.

Gaillard, R.K., and Joklik, W.K., 1980, The antigenic determinants of most of the proteins coded by the three serotypes of reovirus are highly conserved during evolution, *Virology* **107**:533.

Gaillard, R.K., and Joklik, W.K., 1982, Quantitation of the relatedness of the genomes of reovirus serotypes 1, 2 and 3 at the gene level, *Virology* **123**:152.

Gaillard, R.K., Li, J.K.-K., Keene, J.D., and Joklik, W.K., 1982, The sequences at the termini of four genes of the three reovirus serotypes, *Virology* **121**:320.

Gard, G., and Compans, R.W., 1970, Structure and cytopathic effects of Nelson Bay Virus, *J. Virol.* **6**:100.

Gard, G.P., and Marshall, I.D., 1973, Nelson Bay Virus. A novel reovirus, *Arch. Virol.* **43**:34.

Gentsch, J.R., and Fields, B.N., 1981, Tryptic peptide analysis of outer capsid polypeptides of mammalian reovirus serotypes 1, 2, and 3, *J. Virol.* **38**:208.

Gillies, S., Bullivant, S., and Bellamy, A.R., 1971, Viral RNA polymerases: Electron microscopy of reovirus reaction cores, *Science* **174**:694.

Glass, S.E., Naqui, S.A., Hall, C.F., and Kerr, K.M., 1973, Isolation and characterization of virus associated with arthritis of chickens, *Avian Dis.* **17:**415.

Gomatos, P.J., and Stoeckenius, W., 1964, Electron microscope studies on reovirus, *Proc. Natl. Acad. Sci. U.S.A.* **52:**1449.

Gomatos, P.J., and Tamm, I., 1963, The secondary structure of reovirus RNA, *Proc. Natl. Acad. Sci. U.S.A.* **49:**707.

Gomatos, P.J., Tamm, I., Dales, S., and Franklin, R.M., 1962, Reovirus type 3: Physical characteristics and interactions with L cells, *Virology* **17:**441.

Gomatos, P.J., Stamatos, N.M., and Sarkar, N.H., 1980, Small reovirus-specific particle with polycytidylate-dependent RNA polymerase activity, *J. Virol.* **36:**556.

Gouvea, V., and Schnitzer, T.J., 1982, Polymorphism of the genomic RNAs among the avian reoviruses, *J. Gen. Virol.* **61:**87.

Granboulan, N., and Niveleau, A., 1967, Etude au microscope électronique du RNA de réovirus, *J. Microsc.* **6:**23.

Greene, M.I., and Weiner, H.L., 1980, Delayed hypersensitivity in mice infected with reovirus. II. Induction of tolerance and suppressor T cells to viral specific gene products, *J. Immunol.* **125:**283.

Harvey, J.D., Farrell, J.A., and Bellamy, A.R., 1974, Biophysical studies of reovirus type 3. II. Properties of the hydrated particle, *Virology* **62:**154.

Harvey, J.D., Bellamy, A.R., Earnshaw, W.C., and Schutt, C., 1981, Biophysical studies of reovirus type 3. IV. Low-angle X-ray diffraction studies, *Virology* **112:**240.

Hayes, E.C., Lee, P.W.K., Miller, S.E., and Joklik, W.K., 1981, The interaction of a series of hybridoma IgGs with reovirus particles: Demonstration that the core protein λ2 is exposed on the particle surface, *Virology* **108:**147.

Hrdy, D.B., Rosen, L., and Fields, B.N., 1979, Polymorphism of the migration of double-stranded RNA genome segments of reovirus isolates from humans, cattle, and mice, *J. Virol.* **31:**104.

Huang, A.S., and Baltimore, D., 1977, Defective interfering animal viruses, in: *Comprehensive Virology*, Vol. 10 (H. Fraenkel-Conrat and R.R. Wagner, eds.), pp. 73–116, Plenum Press, New York.

Huismans, H., and Joklik, W.K., 1976, Reovirus-coded polypeptides in infected cells: Isolation of two native monomeric polypeptides with high affinity for single-stranded and double-stranded RNA, respectively, *Virology* **70:**411.

Iglewski, W.J., and Franklin, R.M., 1967, Purification and properties of reovirus ribonucleic acid, *J. Virol.* **1:**302.

Johnson, R.B., Soeiro, R., and Fields, B.N., 1976, The synthesis of A-rich RNA by temperature-sensitive mutants of reovirus, *Virology* **73:**173.

Joklik, W.K., 1972, Studies on the effect of chymotrypsin on reovirions, *Virology* **49:**700.

Joklik, W.K., 1973, The transcription and translation of reovirus RNA, in: *Virus Research* (C.F. Fox and W.S. Robinson, eds.), pp. 105–126, Academic Press, New York.

Joklik, W.K., 1981, Structure and function of the reovirus genome, *Microbiol. Rev.* **45:**483.

Jordon, L.E., and Mayor, H.D., 1962, The fine structure of reovirus, a new member of the icosahedral series, *Virology* **17:**597.

Kapuler, A.M., 1970, An extraordinary temperature dependence of the reovirus transcriptase, *Biochemistry* **9:**4453.

Kapuler, A.M., Mendelsohn, N., Klett, H., and Acs, G., 1970, Four base-specific 5′-triphosphatases in the subviral core of reovirus, *Nature (London)* **225:**1209.

Kates, J.R., and McAuslan, B.R., 1967, Messenger RNA synthesis by a "coated" viral genome, *Proc. Natl. Acad. Sci. U.S.A.* **57:**314.

Kavenoff, R., Talcove, D., and Mudd, J.A., 1975, Genome-sized RNA from reovirus particles, *Proc. Natl. Acad. Sci. U.S.A.* **72:**4317.

Kawamura, H., and Tsubahara, H., 1966, Common antigenicity of avian reoviruses, *Natl. Inst. Anim. Health Q.* (Tokyo) **6:**187.

Kawamura, H., Shimizu, F., Maeda, M., and Tsubahara, H., 1965, Avian reovirus: Its properties and serological classification, *Natl. Inst. Anim. Health Q.* **5:**115.

Kozak, M., 1977, Nucleotide sequences of 5'-terminal ribosome-protected initiation regions from two reovirus messages, *Nature (London)* **269:**390.

Kozak, M., 1981, Possible role of flanking nucleotides in recognition of AUG initiator codon by eukaryotic ribosomes, *Nucleic Acids Res.* **9:**5233.

Kozak, M., 1982, Sequences of ribosome binding sites from the large size class of reovirus mRNA, *J. Virol.* **42:**467.

Kozak, M., and Shatkin, A.J., 1976, Characterization of ribosome-protected fragments from reovirus messenger mRNA, *J. Biol. Chem.* **251:**4259.

Kozak, M., and Shatkin, A.J., 1977a, Sequences of two 5'-terminal ribosome-protected fragments from reovirus messenger RNAs, *J. Mol. Biol.* **112:**75.

Kozak, M., and Shatkin, A.J., 1977b, Sequences and properties of two ribosome binding sites from the small size class of reovirus messenger RNA, *J. Biol. Chem.* **252:**6895.

Kozak, M., and Shatkin, A.J., 1978, Migration of 40S ribosomal subunits on messenger RNA in the presence of Edeine, *J. Biol. Chem.* **253:**6568.

Krystal, G., Wiin, P., Millward, S., and Sakuma, S., 1975, Evidence for phosphoproteins in reovirus, *Virology* **64:**505.

Krystal, G., Perrault, J., and Graham, A.F., 1976, Evidence for a glycoprotein in reovirus, *Virology* **72:**308.

Lai, K.C., and Bellamy, A.R., 1971, Factors affecting the amount of oligonucleotides in reovirus particles, *Virology* **45:**821.

Lai, H.-H.T., and Joklik, W.K., 1973, The induction of interferon by temperature-sensitive mutants of reovirus, UV-irradiated reovirus, and subviral reovirus particles, *Virology* **51:**191.

Lai, M.-H.T., Werenne, J., and Joklik, W.K., 1973, The preparation of reovirus top component and its effect on host DNA and protein synthesis, *Virology* **54:**237.

Langridge, R., and Gomatos, P.J., 1963, The structure of RNA: Reovirus RNA and transfer RNA have similar 3-dimensional structures, which differ from DNA, *Science* **141:**694.

Lazarowitz, S.G., and Robertson, H.D., 1977, Initiator regions from the small size class of reovirus messenger RNA protected by rabbit reticulocyte ribosomes, *J. Biol. Chem.* **252:**7842.

Lee, P.W.K., Hayes, E.C., and Joklik, W.K., 1981a, Characterization of antireovirus immunoglobulins secreted by cloned hybridoma cell lines, *Virology* **108:**134.

Lee, P.W.K., Hayes, E.C., and Joklik, W.K., 1981b, Protein σ1 is the reovirus cell attachment protein, *Virology* **108:**156.

Leers, W.-P., Rozee, K.R., and Wardlaw, A.C., 1968, Immunodiffusion and immunoelectrophoretic studies of reovirus antigens, *Can. J. Microbiol.* **14:**161.

Levin, D.H., Mendelsohn, N., Schonberg, M., Klett, H., Silverstein, S.C., Kapuler, A.M., and Acs, G., 1970, Properties of RNA transcriptase in reovirus subviral particles, *Proc. Natl. Acad. Sci. U.S.A.* **66:**890.

Levin, K.H., and Samuel, C.E., 1980, Biosynthesis of reovirus-specified polypeptides: Purification and characterization of the small-sized class mRNAs of reovirus type 3; coding assignments and translational efficiencies, *Virology* **106:**1.

Li, J.K.-K., Keene, J.D., Scheible, P.P., and Joklik, W.K., 1980a, Nature of the 3'-terminal sequence of the plus and minus strands of the S1 gene of reovirus serotypes 1, 2 and 3, *Virology* **105:**41.

Li, J.K.-K., Scheible, P.P., Keene, J.D., and Joklik, W.K., 1980b, The plus strands of reovirus genes S2 are identical with its *in vitro* transcripts, *Virology* **105:**282.

Loh, P.C., and Oie, H.K., 1969, Role of lysine in the replication of reovirus. I. Synthesis of complete and empty virions, *J. Virol.* **4:**890.

Loh, P.C., and Shatkin, A.J., 1968, Structural proteins of reovirus, *J. Virol.* **2:**1353.

Loh, P.C., Hohl, H.R., and Soergel, M., 1965, Fine structure of reovirus type 2, *J. Bacteriol.* **89:**1140.

Lou, T.Y., and Wenner, H.A., 1963, Natural and experimental infection of dogs with reovirus type 1: Pathogenicity of the strain for other animals, *Am. J. Hyg.* **77:**293.

Luftig, R.B., Kilham, S., Hay, A.J., Zweerink, H.J., and Joklik, W.K., 1972, An ultrastructure study of virions and cores of reovirus type 3, *Virology* **48**:170.

Martinson, H.G., and Lewandowski, L.J., 1975, Sequence homology studies between the double-stranded RNA genomes of cytoplasmic polyhedrosis virus, wound tumor virus and reovirus strains, 1, 2 and 3, *Intervirology* **4**:91.

Matsuhisa, T., and Joklik, W.K., 1974, Temperature-sensitive mutants of reovirus. V. Studies on the nature of the temperature-sensitive lesion of the group C mutant ts447, *Virology* **60**:380.

Mayor, H.D., and Jordon, L.E., 1968, Preparation and properties of the internal capsid components of reovirus, *J. Gen. Virol.* **3**:233.

Mayor, H.D., Jamison, R.M., Jordon, L.E., and Mitchell, M.V., 1965, Reoviruses. II. Structure and composition of the virion, *J. Bacteriol.* **89**:1548.

McClain, M.E., Spendlove, R.S., and Lennette, E.H., 1967, Infectivity assay of reoviruses: Comparison of immunofluorescent cell output and plaque methods, *J. Immunol.* **98**:1301.

McCrae, M.A., 1981, Terminal structure of reovirus RNAs, *J. Gen. Virol.* **55**:393.

McCrae, M.A., and Joklik, W.K., 1978, The nature of the polypeptide encoded by each of the ten double-stranded RNA segments of reovirus type 3, *Virology* **89**:578.

McDowell, M.J., and Joklik, W.K., 1971, An *in vitro* protein synthesizing system from mouse L fibroblasts infected with reovirus, *Virology* **45**:724.

McDowell, M.J., Joklik, W.K., Villa-Komaroff, L., and Lodish, H.F., 1972, Translation of reovirus messenger RNAs synthesized *in vitro* into reovirus polypeptides by several mammalian cell-free extracts, *Proc. Natl. Acad. Sci. U.S.A.* **69**:2649.

Metcalf, P., 1982, The symmetry of the reovirus outer shell, *J. Ultrastruct. Res.* **78**:292.

Millward, S., and Graham, A.F., 1970, Structural studies on reovirus: Discontinuities in the genome, *Proc. Natl. Acad. Sci. U.S.A.* **65**:422.

Miura, K.-I., Watanabe, K., and Sugiura, M., 1974a, 5′-Terminal nucleotide sequences of the double-stranded RNA of silkworm cytoplasmic polyhedrosis virus, *J. Mol. Biol.* **86**:31.

Miura, K.-I., Watanabe, K., Sugiura, M., and Shatkin, A.J., 1974b, The 5′-terminal nucleotide sequences of the double-stranded RNA of human reovirus, *Proc. Natl. Acad. Sci. U.S.A.* **71**:3979.

Morgan, E.M., and Kingsbury, D.W., 1980, Pyridoxal phosphate as a probe of reovirus transcriptase, *Biochemistry* **19**:484.

Morgan, E.M., and Kingsbury, D.W., 1981, Reovirus enzymes that modify messenger RNA are inhibited by perturbation of the lambda proteins, *Virology* **113**:565.

Mustoe, T.A., Ramig, R.F., Sharpe, A.H., and Fields, B.N., 1978, Genetics of reovirus: Identification of the ds RNA segments encoding the polypeptides of the μ and σ size classes, *Virology* **89**:594.

Muthukrishnan, S., and Shatkin, A.J., 1975, Reovirus genome RNA segments: Resistance to S1 nuclease, *Virology* **64**:96.

Nakashima, K., LaFiandra, A.J., and Shatkin, A.J., 1979, Differential dependence of reovirus-associated enzyme activities on genome RNA as determined by psoralen photosensitivity, *J. Biol. Chem.* **254**:8007.

Nakashima, K., Darzynkiewicz, E., and Shatkin, A.J., 1980, Proximity of mRNA 5′-regions and 18S rRNA in eukaryotic initiation complexes, *Nature* **286**:226.

Nakazato, H., 1980, Immunospecific binding of capped mRNA of reovirus by an antibody to a cap structure: m$^7$G(5′)pppGm, *Biochem. Biophys. Res. Commun.* **96**:400.

Nichols, J.L., Hay, A.J., and Joklik, W.K., 1972a, 5′-Terminal nucleotide sequence in reovirus mRNA synthesized *in vitro*, *Nature (London) New Biol.* **235**:105.

Nichols, J.L., Bellamy, A.R., and Joklik, W.K., 1972b, Identification of the nucleotide sequences of the oligonucleotides present in reovirions, *Virology* **49**:562.

Nonoyama, M., and Graham, A.F., 1970, Appearance of defective virions in clones of reovirus, *J. Virol.* **6**:693.

Nonoyama, M., Watanabe, Y., and Graham, A.F., 1970, Defective virions of reovirus, *J. Virol.* **6**:226.

Olson, N.O., 1978, Reovirus infections, in: *Diseases of Poultry* (M.S. Hofstad, ed), p. 641, Iowa State University Press, Ames.

Palmer, E.L., and Martin, M.L., 1977, The fine structure of the capsid of reovirus type 3, *Virology* **76**:109.

Petek, M., Fulluga, B., Borghi, G., and Baroni, A., 1967, The Crawley Agent: An avian reovirus, *Arch ges. Virusforschung* **21**:413.

Pett, D.M., Vanaman, T.C., and Joklik, W.K., 1973, Studies on the amino and carboxyl-terminal amino acid sequences of reovirus capsid polypeptides, *Virology* **52**:174.

Ralph, S.J., Harvey, J.D., and Bellamy, A.R., 1980, Subunit structure of the reovirus spike, *J. Virol.* **36**:894.

Ramig, R.F., and Fields, B.N., 1979, Revertants of temperature-sensitive mutants of reovirus: Evidence for frequent extragenic suppression, *Virology* **92**:155.

Ramig, R.F., Cross, R.K., and Fields, B.N., 1977a, Genome RNAs and polypeptides of reovirus serotypes 1, 2, and 3, *J. Virol.* **22**:726.

Ramig, R.F., White, R.M., and Fields, B.N., 1977b, Suppression of the temperature-sensitive phenotype of a mutant of reovirus type 3, *Science* **195**:406.

Ramig, R.F., Mustoe, T.A., Sharpe, A.H., and Fields, B.N., 1978, A genetic map of reovirus. II. Assignment of the double-stranded RNA-negative mutant groups C, D, and E to genome segments, *Virology* **85**:531.

Ramos-Alvarez, M., and Sabin, A.B., 1954, Characterization of poliomyelitis and other enteric viruses recovered in tissue culture from healthy American children, *Proc. Soc. Exp. Biol. Med.* **87**:655.

Rhim, J.S., and Melnick, J.L., 1961, Plaque formation by reoviruses, *Virology* **15**:80.

Robbins, F.C., Enders, J.F., Weller, T.H., and Florentino, G.L., 1951, Studies on the cultivation of poliomyelitis virus in tissue culture. V. The direct isolation and serologic identification of virus strains in tissue culture from patients with paralytic and nonparalytic poliomyelitis, *Am. J. Hyg.* **54**:56.

Rosen, L., 1960, Serologic groupings of reoviruses by hemagglutination-inhibition, *Am. J. Hyg.* **71**:242.

Rosen, L., 1962, Reoviruses in animals other than man, *Ann. N. Y. Acad. Sci.* **101**:461.

Roseto, A., Escaig, J., Delain, E., Cohen, J., and Scherrer, R., 1979, Structure of rotaviruses as studied by the freeze–drying technique, *Virology* **98**:471.

Rubin, D.H., and Fields, B.N., 1980, Molecular basis of reovirus virulence: Role of the M2 gene, *J. Exp. Med.* **152**:853.

Sabin, A.B., 1959, Reoviruses: A new group of respiratory and enteric viruses formerly classified as ECHO 10 is described, *Science* **130**:1387.

Sargent, M.D., Long, D.G., and Borsa, J., 1977, Functional analysis of the interactions between reovirus particles and various proteases *in vitro*, *Virology* **78**:354.

Schonberg, M., Silverstein, S.C., Levin, D.H., and Acs, G., 1971, Asynchronous synthesis of the complementary strands of the reovirus genome, *Proc. Natl. Acad. Sci. U.S.A.* **68**:505.

Schuerch, A.R., and Joklik, W.K., 1973, Temperature-sensitive mutants of reovirus. IV. Evidence that anomalous electrophoretic migration behavior of certain double-stranded RNA hybrid species is mutant group-specific, *Virology* **56**:218.

Schuerch, A.R., Matsuhisa, T., and Joklik, W.K., 1974, Temperature-sensitive mutants of reovirus. VI. Mutant ts447 and ts556 particles that lack either one or two genome RNA segments, *Intervirology* **3**:36.

Scott, F.W., Kahn, D.E., and Gillespie, J.H., 1970, Feline reovirus: Isolation, characterization and pathogenicity of a feline reovirus, *Am. J. Vet. Res.* **71**:11.

Sharpe, A.H., and Fields, B.N., 1981, Reovirus inhibition of cellular DNA synthesis: Role of the S1 gene, *J. Virol.* **38**:389.

Sharpe, A.H., and Fields, B.N., 1982, Reovirus inhibition of cellular RNA and protein synthesis: Role of the S4 gene, *Virology* **122**:381.

Sharpe, A.H., Ramig, R.F., Mustoe, T.A., and Fields, B.N., 1978, A genetic map of reovirus. I. Correlation of genome RNAs between serotypes 1, 2, and 3, *Virology* **84**:63.

Shatkin, A.J., 1965, Inactivity of purified reovirus RNA as a template for *E. coli* polymerases *in vitro*, *Proc. Natl. Acad. Sci. U.S.A.* **54**:1721.

Shatkin, A.J., 1974, Methylated messenger RNA synthesis *in vitro* by purified reovirus, *Proc. Natl. Acad. Sci. U.S.A.* **71**:3204.

Shatkin, A.J., and Banerjee, A.K., 1970, *In vitro* transcription of double-stranded RNA by reovirus-associated RNA polymerase, *Cold Spring Harbor Symp. Quant. Biol.* **35**:781.

Shatkin, A.J., and LaFiandra, A.J., 1972, Transcription by infectious subviral particles of reovirus, *J. Virol.* **10**:698.21.

Shatkin, A.J., and Sipe, J.D., 1968a, RNA polymerase activity in purified reoviruses, *Proc. Natl. Acad. Sci. U.S.A.* **61**:1462.

Shatkin, A.J., and Sipe, J.D., 1968b, Single-stranded adenine-rich RNA from purified reoviruses, *Proc. Natl. Acad. Sci. U.S.A.* **59**:246.

Shatkin, A.J., Sipe, J.D., and Loh, P.C., 1968, Separation of ten reovirus genome segments by polyacrylamide gel electrophoresis, *J. Virol.* **2**:986.

Shelton, I.H., Kasupski, G.J., Jr., Oblin, C., and Hand, R., 1981, DNA-binding of a nonstructural reovirus protein, *Can. J. Biochem.* **59**:122.

Silverstein, S.C., Levin, D.H., Schonberg, M., and Acs, G., 1970, The reovirus replicative cycle: Conservation of parental RNA and protein, *Proc. Natl. Acad. Sci. U.S.A.* **67**:275.

Silverstein, S.C., Astell, C., Levin, D.H., Schonberg, M., and Acs, G., 1972, The mechanisms of reovirus uncoating and gene activation *in vivo*, *Virology* **47**:797.

Silverstein, S.C., Astell, C., Christman, J., Klett, H., and Acs, G., 1974, Synthesis of reovirus oligoadenylic acid *in vivo* and *in vitro*, *J. Virol.* **13**:740.

Simmons, D.G., Colwell, W.M., Muse, K.E., and Brewer, C.E., 1972, Isolation and characterization of an enteric reovirus causing high mortality in turkey poults, *Avian Dis.* **16**:1094.

Skehel, J.J., and Joklik, W.K., 1969, Studies on the *in vitro* transcription of reovirus RNA catalyzed by reovirus cores, *Virology* **39**:822.

Smith, R.E., Zweerink, H.J., and Joklik, W.K., 1969, Polypeptide components of virions, top component and cores of reovirus type 3, *Virology* **39**:791.

Spandidos, D.A., and Graham, A.F., 1975a, Complementation of defective reovirus by ts mutants, *J. Virol.* **15**:954.

Spandidos, D.A., and Graham, A.F., 1975b, Complementation between temperature-sensitive and deletion mutants of reovirus, *J. Virol.* **16**:1444.

Spandidos, D.A., and Graham, A.F., 1976a, Nonpermissive infection of L cells by an avian reovirus: Restricted transcription of the viral genome, *J. Virol.* **19**:977.

Spandidos, D.A., and Graham, A.F., 1976b, Physical and chemical characterization of an avian reovirus, *J. Virol.* **19**:968.

Spendlove, R.S., Lennette, E.H., Knight, C.O., and Chin, J.N., 1966, Production in FL cells of infectious and potentially infectious reovirus, *J. Bacteriol.* **92**:1036.

Spendlove, R.S., McClain, M.E., and Lennette, E. H., 1970, Enhancement of reovirus infectivity by extracellular removal or alteration of the virus capsid by proteolytic enzymes, *J. Gen. Virol.* **8**:83.

Stanley, N.F., 1961, Relationship of hepatoencephalomyelitis virus and reoviruses, *Nature (London)* **189**:687.

Stanley, N.F., 1967, Reoviruses, *Br. Med. Bull.* **23**:150.

Stanley, N.F., and Leak, P.J., 1963, The serologic epidemiology of reovirus infection with special reference to the Rottnest Island quokka (*Setenix brachyurus*), *Am. J. Hyg.* **78**:82.

Stanley, N.F., Dorman, D.C., and Ponsford, J., 1953, Studies on the pathogenesis of a hitherto undescribed virus (hepato-encephalomyelitis) producing unusual symptoms in suckling mice, *Aust. J. Exp. Biol. Med. Sci.* **31**:147.

Stoltzfus, C.M., and Banerjee, A.K., 1972, Two oligonucleotide classes of single-stranded ribopolymers in reovirus A-rich RNA, *Arch. Biochem. Biophys.* **152**:733.

Stoltzfus, C.M., Morgan, M., Banerjee, A.K., and Shatkin, A.J., 1974, Poly(A) polymerase activity in reovirus, *J. Virol.* **13**:1338.

Van der Heide, L., Geissler, J., and Bryant, E.S., 1974, Infectious tenosynovitis: Serologic and histopathologic response after experimental infection with a Connecticut isolate, *Avian Dis.* **18:**289.

Van der Heide, L., 1977, Viral arthritis/tenosynovitis: A review, *Avian Pathol.* **6:**271.

Van Dijk, A.A., and Huismans, H., 1980, The *in vitro* activation and further characterization of the blue-tongue virus-associated transcriptase, *Virology* **104:**347.

Vasquez, C., and Kleinschmidt, A.K., 1968, Electron microscopy of RNA strands released from individual reovirus particle, *J. Mol. Biol.* **34:**137.

Vasquez, C., and Tournier, P., 1962, The morphology of reovirus, *Virology* **17:**503.

Vasquez, C., and Tournier, P., 1964, New interpretation of the reovirus structure, *Virology* **24:**128.

Wachsman, J.T., Levin, D.H., and Acs, G., 1970, Ribonucleoside triphosphate-dependent pyrophosphate exchange of reovirus cores, *J. Virol.* **6:**563.

Wallis, C., Smith, K.O., and Melnick, J.L., 1964, Reovirus activation by heating and inactivation by cooling in $MgCl_2$ solution, *Virology* **22:**608.

Wallis, C., Melnick, J.L., and Rapp, F., 1966, Effect of pancreatin on the growth of reovirus, *J. Bacteriol.* **92:**155.

Watanabe, Y., and Graham, A.F., 1967, Structural units of reovirus RNA and their possible functional significance, *J. Virol.* **1:**665.

Watanabe, Y., Prevec, L., and Graham, A.F., 1967, Specificity in transcription of the reovirus genome, *Proc. Natl. Acad. Sci. U.S.A.* **58:**1040.

Weiner, H.L., and Fields, B.N., 1977, Neutralization of reovirus: The gene responsible for the neutralization antigen, *J. Exp. Med.* **146:**1305.

Weiner, H.L., Drayna, D., Averill, D.R., Jr., and Fields, B.N., 1977, Molecular basis of reovirus virulence: Role of the S1 gene, *Proc. Natl. Acad. Sci. U.S.A.* **74:**5744.

Weiner, H.L., Ramig, R.F., Mustoe, T.A., and Fields, B.N., 1978, Identification of the gene coding for the hemagglutinin of reovirus, *Virology* **86:**581.

Weiner, H.L, Ault, K.A., and Fields, B.N., 1980a, Interaction of reovirus with cell surface receptors. I. Murine and human lymphocytes have a receptor for the hemagglutinin of reovirus type 3, *J. Immunol.* **124:**2143.

Weiner, H.L., Greene, M.I., and Fields, B.N., 1980b, Delayed hypersensitivity in mice infected with reovirus. I. Identification of host and viral gene products responsible for the immune reponse, *J. Immunol.* **125:**278.

Weiner, H.L., Powers, M.L., and Fields, B.N., 1980c, Absolute linkage of virulence and central nervous system cell tropism of reoviruses to viral hemagglutinin, *J. Infect. Dis.* **141:**609.

Whitcomb, R.F., and Jensen, D.D., 1968, Enhancement of infectivity of a mammalian virus preparation after injection into insects, *Virology* **34:**182.

White, C.K., and Zweerink, H.J., 1976, Studies on the structure of reovirus cores: Selective removal of polypeptide $\lambda 2$, *Virology* **70:**171.

Wilcox, G.E., and Compans, R.W., 1982, Cell fusion induced by Nelson Bay Virus, *Virology* **123:**312.

Wooley, R.E., Dees, T.A., Cromack, A.S., and Gratzek, J.B., 1972, Characterization of two reoviruses isolated by sucrose density gradient centrifugation from turkeys with infectious enteritis, *Am. J. Vet. Res.* **33:**165.

Yamakawa, M., Furuichi, Y., Nakashima, K., LaFiandra, A.J., and Shatkin, A.J., 1981, Excess synthesis of viral mRNA 5′-terminal oligonucleotides by reovirus transcriptase, *J. Biol. Chem.* **256:**6507.

Yamakawa, M., Furuichi, Y., and Shatkin, A.J., 1982, Reovirus transcriptase and capping enzymes are active in intact virions, *Virology* **118:**157.

Zarbl, H., Hastings, K.E.M., and Millward, S., 1980, Reovirus core particles synthesize capped oligonucleotides as a result of abortive transcription, *Arch. Biochem. Biophys.* **202:**348.

Zweerink, H.J., and Joklik, W.K., 1970, Studies on the intracellular synthesis of reovirus-specified proteins, *Virology* **41:**501.

Zweerink, H.J., McDowell, M.J., and Joklik, W.K., 1971, Essential and nonessential non-capsid reovirus proteins, *Virology* **45:**716.

CHAPTER 3

# Biochemical Aspects of Reovirus Transcription and Translation

Aaron J. Shatkin and Marilyn Kozak

## I. INTRODUCTION

It should be evident from Chapter 2 that reoviruses provide an excellent system for studying transcription and translation at the molecular level. Reoviruses replicate rapidly to high titers in a variety of tissue-culture cells. Progeny virions remain cell-associated and are readily isolated in yields of 10 mg/$10^9$ cells or higher. Purified reoviruses contain several stable enzymes including RNA polymerase and messenger (mRNA) capping activities. Consequently, under appropriate incubation conditions, they produce full-length transcripts of one strand of each of the ten segments of genome duplex RNA. Transcription *in vitro* continues for long periods, resulting in the accumulation of milligram amounts of ten different viral mRNA species. Reovirus mRNAs synthesized in the presence of the methyl donor *S*-adenosyl-L-methionine contain 5'-terminal m$^7$G(5')ppp(5')N, a structure identical to the "cap" found in most eukaryotic cellular and viral mRNAs. Viral mRNAs made *in vitro* and denatured genome RNAs can be translated into authentic viral polypeptides in cell-free protein-synthesizing systems, providing coding assignments for the genome segments. Reovirus genetics is also especially powerful because three distinct virus serotypes are available and the frequency of

AARON J. SHATKIN • Department of Cell Biology, Roche Institute of Molecular Biology, Nutley, New Jersey 07110.    MARILYN KOZAK • Department of Biological Sciences, University of Pittsburgh, Pittsburgh, Pennsylvania 15260.

genome-segment reassortment is high. These and other properties of reoviruses have facilitated studies of viral mRNA formation and function that have provided new insights into some general features of eukaryotic genetic expression.

## II. TRANSCRIPTION

### A. Discovery of Reovirus-Associated RNA Polymerase

Many new issues were raised by the finding that reoviruses contain double-stranded RNA (dsRNA) genomes (Gomatos and Tamm, 1963). RNA–RNA duplexes are highly stable, and protein synthesis is necessarily directed by single-stranded RNA (ssRNA) templates (Miura and Muto, 1965). Consequently, it was particularly intriguing to understand how the genetic information in reoviruses becomes expressed in order to initiate virus replication. DNA viruses that replicate in the nucleus can direct viral messenger (mRNA) synthesis by reprogramming the host-cell DNA-dependent RNA polymerase. First reports suggested that the same mechanism is used by reoviruses (Gomatos et al., 1964), although their replication is cytoplasmic (Rhim et al., 1962). Subsequent studies showed that the ssRNAs produced in vitro by incubating reovirus genome RNA with purified Escherichia coli DNA-dependent RNA polymerase were in fact templated by mouse-cell DNA that had persisted during virus purification (Shatkin, 1965b).

Precedent for a different mechanism of reovirus mRNA synthesis came from studies of vaccinia virus, which contains dsDNA but also replicates in the cytoplasm (for a review, see Moss, 1974). The dilemma about the origin of virus-specified early mRNAs from a cytoplasmic DNA virus led to the suggestion (Salzman et al., 1964) and subsequent demonstration (Kates and McAuslan, 1967; Munyon et al., 1967) that purified vaccinia virions contain RNA polymerase. Specific transcription of the vaccinia genome was mediated by the particle-associated RNA polymerase both in vitro and in infected cells. The presence of a similar kind of transcriptase activity in purified reoviruses was reported in 1968 by two groups. In one series of experiments, involvement of a pre-existing polymerase was predicted because a low level of reovirus early mRNA was produced in infected cells even when protein synthesis was blocked from the start of infection (Watanabe et al., 1968). Later, the preformed polymerase was found to be a virion-associated transcriptase that could be "activated" by heat shock treatment of purified particles (Borsa and Graham, 1968). A second, independent approach that led to demonstration of the reovirus transcriptase made use of reports that parental virions are partially degraded at the beginning of the replicative cycle by lysosomal proteases in infected mouse L cells (Silverstein and Dales, 1966) and that reovirus infectivity can be enhanced by mild protease treatment

(Spendlove and Schaffer, 1965). Digestion of purified virions with chymotrypsin removed the outer protein shell. The resulting viral cores were shown to contain a unique dsRNA-dependent RNA polymerase that synthesizes virus-specific transcripts (Shatkin and Sipe, 1968). Subsequent work verified that the virion-associated transcriptase synthesizes full-length copies of one strand of each of the ten genome RNA duplex segments by a fully conservative mechanism (Skehel and Joklik, 1969; Banerjee and Shatkin, 1970; D. H. Levin et al., 1970b; for a review, see Joklik, 1974).

## B. Reovirus Messenger RNA Synthesis in Infected Cells

Because reoviruses contain RNA-dependent RNA polymerase, parental virions are capable of initiating replication in infected cells directly, i.e., without inducing a new transcriptase or modifying a host-cell enzyme. Infecting reovirions attach to cell surface receptors via the outer shell protein σ1 (Weiner et al., 1980; Lee et al., 1981b). The particles enter the cytoplasm within phagocytic vacuoles that fuse with lysosomes (Silverstein and Dales, 1968). Proteolytic digestion follows, resulting in removal of outer-capsid proteins σ1 and σ3 and cleavage of μ1C (Chang and Zweerink, 1971; Silverstein et al., 1972). The RNA polymerase in the partially disrupted particles is highly active, and mRNA synthesis proceeds after their release into the cytoplasm of the infected cell.

Viral mRNA synthesis is most readily studied in cells that are treated with low levels of actinomycin to inhibit host-cell DNA-dependent transcription differentially and almost completely (Shatkin, 1965a; Kudo and Graham, 1965). Under these conditions, reovirus genome RNA-templated transcripts constitute the bulk of the RNA synthesized in infected mouse L cells. During the early stage of reovirus infection, virus-specific transcription is mainly from genome segments L1, M3, S3, and S4 (Nonoyama et al., 1974), but transcription of all segments is apparent within 6–8 hr in permissive cells (Zweerink and Joklik, 1970). Expansion of transcription to ten viral mRNA species is prevented in permissive cells by blocking protein synthesis with cycloheximide (Watanabe et al., 1968). Viral gene expression is similarly restricted in untreated virus–cell combinations that are nonpermissive for replication (Spandidos and Graham, 1976). Furthermore, subviral particles isolated from cycloheximide-treated, permissive cells synthesized ten viral mRNAs in vitro, but became restricted and transcribed segments L1, M3, S3, and S4 on reinfection of cycloheximide-treated cells (Shatkin and LaFiandra, 1972). These results suggest that one or more host proteins are involved in regulating viral transcription, but exactly how is not clear.

After the transcription of all ten genome segments has commenced in reovirus-infected mouse cells, the number of copies of each mRNA produced is regulated. The number varies considerably (Table I). To some

TABLE I. Approximate Relative Frequencies of Transcription and Translation of the Ten Reovirus Genes[a]

| Gene | Protein | Transcription frequency | Translation frequency | Frequency of translation/transcription |
|------|---------|------------------------|----------------------|----------------------------------------|
| L1 | λ3 | 0.05 | 0.03 | 0.6 |
| L2 | λ2 | 0.05 | 0.15 | 3 |
| L3 | λ1 | 0.05 | 0.1 | 2 |
| M1 | μ2 | 0.15 | 0.03 | 0.2 |
| M2 | μ1 | 0.3 | 1.0 | 3.3 |
| M3 | μNS | 0.5 | 0.5 | 1 |
| S1 | σ1 | 0.5 | 0.05 | 0.1 |
| S2 | σ2 | 0.5 | 0.2 | 0.4 |
| S3 | σNS | 1.0 | 0.3 | 0.3 |
| S4 | σ3 | 1.0 | 0.7 | 0.7 |

[a] From Joklik (1981).

extent, the size of the template RNA determines the relative frequency with which it is transcribed. Thus, in mouse L cells examined between 2 and 8 hr after infection, there were 20 times more s3 and s4 species of mRNA synthesized as compared to large-class transcripts (Zweerink and Joklik, 1970). Since the s mRNAs are approximately 25–35% the length of the large mRNA species, they must be initiated considerably more often and/or synthesized at a faster rate than the large transcripts. Synthesis of similarly disparate relative amounts of mRNAs can be demonstrated under some conditions with purified cores (Banerjee and Shatkin, 1970; Skehel and Joklik, 1969). Results of in vitro pulse-labeling experiments (Banerjee and Shatkin, 1970) indicate that the individual genome segments are simultaneously transcribed at synthetic rates in the order $s > m > l$ in agreement with the relative molar yields of the different mRNAs synthesized in vivo. These yields probably reflect the synthesis rates of the individual viral mRNAs as well as the size of the corresponding template RNAs, because there was no differential turnover of mRNA in infected cells (Watanabe et al., 1968). The net result is that many more reovirus small transcripts than large mRNAs are produced both in vivo and in vitro. This pattern of synthesis is consistent with transcription of individual genome segments rather than coordinate copying of a linked array of templates. The basis for selective transcription of a few segments early in infection and of preferential utilization of small-class genome RNAs during transcription of all ten templates at later times remains to be determined. Also unanswered are many other basic questions about reovirus transcription, for example, the number of RNA polymerase molecules per virion. Some particles in a population of active cores clearly transcribe all segments simultaneously, indicating a minimum of one polymerase molecule per genome segment (Bartlett et al., 1974).

## C. Properties of the Virion Polymerase

Much of the published work on cell-free reovirus mRNA synthesis has been carried out with chymotrypsin-digested purified virions. Several parameters, including the ratio of enzyme to virus, the concentration of virions during proteolysis, and the presence of monovalent cations, influence the extent of digestion of the outer-shell proteins (Shatkin and LaFiandra, 1972; Joklik, 1972; Borsa *et al.*, 1973). Complete removal of the outer capsid converts virions to viral cores. They have a markedly decreased specific infectivity but an active RNA polymerase capable of producing large amounts of ten full-length viral mRNAs. Under some conditions of chymotrypsin digestion, the outer coat of virions is only partially removed. The resulting subviral particles are highly infectious. They are also morphologically very similar to, and retain the same polypeptides as, particles formed in L cells from parental virions at early stages of infection (Silverstein *et al.*, 1972). Reports from different laboratories have disagreed about whether the transcriptase is active in subviral particles or whether enzyme activation requires conversion to cores by further modification or removal of the outer-capsid polypeptides (Borsa *et al.*, 1981). This may now be a moot point because recent results indicate that the polymerase and other mRNA-modifying enzymes are present in intact virions in a functional state (Yamakawa *et al.*, 1982a). In agreement with many previous reports, whole reovirions do not catalyze the incorporation of radiolabeled precursors into acid-precipitable products. However, virions incubated in a standard transcription reaction mixture (without protease) produce short oligonucleotides corresponding to the 5′-terminal sequence common to the ten viral mRNAs. Thus, virion "activation" by heat shock or proteolytic treatment is not required to engage the transcriptase activity. Instead, it appears to be necessary to loosen the enzyme–template complex to allow elongation of nascent, initiated short chains. It should be noted that cores also produce "initiator" oligonucleotides (Yamakawa *et al.*, 1981). They are made in high molar excess relative to complete mRNAs and presumably account for the single-stranded oligonucleotides in purified virions. Their formation *in vitro* in excess suggests that the transcriptase produces viral mRNA by a two-step process: reiterative initiation followed by elongation of a small fraction of the resulting oligonucleotides.

Consistent with a tight association between the RNA polymerase and duplex RNA templates, the synthesis of mRNA by viral cores continues for long periods if precursor ribonucleoside triphosphates are continually replenished (D. H. Levin *et al.*, 1970b). Another measure of the stability of the polymerase–template complex is its elevated temperature optimum for mRNA synthesis, approximately 45–50°C (Kapuler, 1970). Furthermore, brief exposure to approximately 60°C was one of the procedures originally used to induce mRNA synthesis by purified reovirus. While the high stability of viral cores is convenient for accumulating large amounts of authentic mRNAs, it also makes it virtually impossible to solubilize the RNA polymerase in an active form for detailed molecular studies. Of particular interest is the mechanism by which the reovirus polymerase repeatedly and correctly initiates, elongates, and terminates ten full-length mRNAs.

Some insights into the overall mechanism of reovirus transcription have been obtained by studying cores treated with chemical agents that alkylate the genome RNAs. Psoralens are photoreactive compounds that form covalent adducts mainly with pyrimidines in RNA and DNA. 4'-Substituted psoralen derivatives react particularly effectively with RNA and form interstrand crosslinks as well as monoadducts with the genome RNA in reovirus cores (Nakashima et al., 1979). Photoreaction of cores with 4'-aminomethyl-4,5',8-trimethylpsoralen (AMT) caused a progressive, time- and concentration-dependent inactivation of mRNA synthesis (Nakashima and Shatkin, 1978). The mechanism of psoralen inactivation of transcription as reflected by the nature of the RNA products was "all or none"; i.e., reduced amounts of full-length transcripts were produced by cores that had been 80% inactivated by psoralen photoreaction. It was predicted that if psoralen acts like short-wavelength UV irradiation and modifies polynucleotides at multiple and variable sites, the RNA products made by partially inactivated cores would consist of a random assortment of prematurely terminated transcripts. The contrary finding that decreased yields of full-length mRNAs were formed suggests that psoralen attachment to reovirus RNA may not be random. Photoreaction studies with $^3$H-radiolabeled AMT indicate that both strands of all genome segments of reovirus RNA are modified, predominantly at C residues. The levels of [$^3$H]adducts formed are approximately but not exactly proportional to RNA-segment molecular weight (Y. Furuichi, A. J. Shatkin, and R. E. Smith, unpublished results). Thus, it remains possible that the all-or-none effect on transcription is due to the presence of promoter-proximal psoralen "hot spots" in the reovirus genome RNAs.

An alternative explanation of the psoralen findings is that during transcription, the template RNA molecules move past enzymatic sites that are fixed in the viral capsid. Damage to a template RNA by psoralen-adduct formation at any site along its length would prevent transit through these fixed protein structures that include the RNA polymerase. The overall effect would be to limit transcription to initiation, a step that should require little or no linear movement of the template relative to the enzyme (or vice versa). In agreement with a moving template model, initiation measured as 5'-terminal oligonucleotide synthesis was essentially unaffected in viral cores after photoreaction with psoralen under conditions that extensively inactivated mRNA formation. Chain elongation as compared to initiation was also differentially inactivated by dimethylsulfate alkylation of the genome RNAs in viral cores (Yamakawa et al., 1981).

## D. Polypeptides in Transcribing Cores

Many investigators have tried unsuccessfully to purify the reovirus RNA-dependent RNA polymerase from virus-infected cell cultures or from isolated virions. Others attempted to prepare template-dependent enzyme by nuclease digestion of partially disrupted viral cores (Silverstein et al., 1976). In each case, loss of enzyme activity accompanied disruption of either the nucleic acid or the protein components of the nucleoprotein complex. Maintenance of an active enzyme conformation apparently requires tight association of the transcriptase with the dsRNA template.

Reovirus cores, the minimum structure capable of synthesizing viral mRNA in vitro, consist of the viral genome RNAs and five or six polypeptides. In the recent nomenclature, they are λ1, λ2, and λ3; μ2 and possibly μ1; and σ2 (Joklik, 1981). As discussed in Section II.E, cores contain in addition to RNA polymerase at least four other enzyme activities that are involved in the formation of mature viral mRNA (Furuichi et al., 1976). Assignment of the RNA polymerase or any of the other enzyme activities to one or more of the core polypeptides has not been achieved, but several experimental approaches to the problem have provided much new information about reovirus structure.

One potentially powerful tool for relating the reovirus transcriptase activity to the responsible core polypeptide(s) is specific antibody. However, rabbit antisera to purified reovirus type 3 (prepared by B. Fields) effectively neutralized virus infectivity and immunoprecipitated radiolabeled λ, μ, and σ polypeptides from extracts of [$^{35}$S]methionine-labeled infected L cells, but did not block mRNA synthesis by purified viral cores (A. LaFiandra and A. J. Shatkin, unpublished results). Furthermore, mouse antisera to reovirus cores failed to inhibit the core transcriptase as measured by mRNA synthesis or formation of short 5'-terminal oligonucleotides (A. LaFiandra, A. J. Shatkin, and S. M. Tahara, unpublished results). Presumably, the tight association between the enzyme and template RNA protects the polymerase catalytic site from inactivation by antibody binding. Monoclonal antibodies with specificity for several individual reovirus polypeptides including the major core constituents λ2 and σ2 have been reported recently (Lee et al., 1981a); they may prove useful for studying the core-associated enzyme activities.

Another standard procedure for identifying enzymes in protein mixtures is the use of compounds that resemble normal substrates but are sufficiently different to preclude enzymatic attack. Such compounds bind specifically but unproductively to the enzyme in question. Pyridoxal 5'-phosphate resembles nucleotide 5'-phosphates, and many polymerases including the reovirus transcriptase are inactivated by binding of this analogue in place of authentic RNA precursors (E. M. Morgan and Kingsbury, 1980). The aldehyde and phosphate groups in pyridoxal 5'-phosphate are both required for maximal inhibition of the reovirus core transcriptase—80–90% decrease in activity at 1 mM (E. M. Morgan and Kingsbury, 1980). However, kinetic analyses showed that the binding of the analogue and ribonucleotide precursors was noncompetitive, consistent with an indirect effect by analogue attachment to one or more core proteins that are required for mRNA synthesis but are not necessarily components of the transcriptase. It is of some interest that pyridoxal-5′-phosphate-mediated reductive labeling of cores yielded λ1 and λ2 as the only radioactive polypeptides (E. M. Morgan and Kingsbury, 1980). Furthermore, it was estimated that binding of the analogue to only ten molecules each of λ1 and λ2, i.e., approximately 10% of the potential λ sites for pyridoxal 5′-phosphate attachment, was sufficient to inactivate mRNA synthesis by cores. The agreement between the number of analogue-binding sites and ten genome template RNAs is striking. However, the implication that each reovirus genome segment is transcribed by a dimeric polymerase consisting of λ1 and λ2 subunits was made less attractive by more recent studies (E. M. Morgan and Kingsbury, 1981). Transcriptase inactivation by pyridoxal 5′-phosphate was accompanied by inhibition of all the mRNA-modifying enzymes in viral cores, and again the analogue acted noncompetitively with the substrates of most of the other core enzymes. This suggests that the analogue binds to one or more sites required in common by all the core activities. The postulated shared element may be the core spikes or "chimneys" that consist of polypeptide λ2 (White and Zweerink, 1976) and serve as channels for mRNA extrusion (Barlett et al., 1974). Thus, alteration of λ2 by pyridoxal 5′-phosphate linkage may block the RNA channels, resulting in feedback inhibition of all mRNA-forming activities in the core. Alternatively, two polypeptides, λ1 and λ2, may comprise a multifunctional complex that expresses the several activities necessary for synthesis of 5′-terminally blocked and methylated reovirus mRNAs.

The general notion that the transcriptase and other enzymes in reovirus cores constitute a complex of interrelated polypeptides is consistent with genetic studies. The reovirus genome segments have been assigned to ten reassortment groups (see Chapter 6) and mapped with respect to the polypeptides they encode (McCrae and Joklik, 1978; Mustoe et al., 1978). With this wealth of background information at hand, it was anticipated that analyses of cells infected with various temperature-sensitive mutants would allow assignment of each enzymatic activity to its respective core polypeptide. However, infection with mutants in either polypeptide σ2 or λ3 resulted in reduced amounts of viral RNAs at nonpermissive temperatures (Cross and Fields, 1972; Ito and Joklik, 1972) (see also Chapter 6). In these cases, lack of mRNA synthesis was probably due to failure to produce progeny particles containing active polymerases responsible for amplification of transcription. Thus, as noted for inactivation of in vitro transcription by pyridoxal 5′-phosphate, in vivo effects on viral mRNA synthesis in mutant-infected cells can also be indirect. In support of this point, it was observed recently that serotype-specific differences in chymotrypsin activation of reovirus mRNA synthesis are determined by genomic segment *M1* corresponding to polypeptide μ2 (Drayna and Fields, 1982). Another property of the transcriptase, a lower pH optimum in serotype 2 as compared to type 3 particles, is specified by the *L1* genome segment, which codes for polypeptide λ3. From the information currently available, no firm conclusion can be made about which polypeptide(s) catalyzes reovirus RNA synthesis. Perhaps it will be possible to identify the polymerase by *in vitro* reconstitution, for example, by using dsRNA and extracts of cells transfected with cloned complementary DNAs of the genes for the individual core polypeptides.

## E. Core Enzymes That Synthesize Messenger RNA 5′-Terminal Caps

A second reovirus-associated enzyme activity was detected shortly after the transcriptase was described. It converts nucleoside triphosphates to diphosphates (Borsa *et al.*, 1970; Kapuler *et al.*, 1970). This finding provided an explanation of the partial dependence of *in vitro* transcription on addition of a triphosphate-regenerating system (Shatkin and Sipe, 1968). The nucleotide phosphohydrolase also cleaves triphosphate ends on polynucleotides, accounting for failure to find radioactivity in GTP-initiated reovirus transcripts made in the presence of $[\gamma\text{-}^{32}P]$-GTP (Banerjee *et al.*, 1971). By contrast, 5′-end-labeled products were obtained with $[\beta,\gamma\text{-}^{32}P]$-GTP as precursor (D. H. Levin *et al.*, 1970a; Banerjee *et al.*, 1971). Furthermore, when the RNA was synthesized in the presence of *S*-adenosyl-L-methionine (AdoMet) under conditions of methylation, a large fraction of the radioactivity incorporated from $[\beta,\gamma\text{-}^{32}P]$-GTP was resistant to hydrolysis by alkaline phosphatase (Shatkin, 1974). This result constituted one of the first indications that mRNA made *in vitro*, like the plus strands in virion genome duplex RNAs (Miura *et al.*, 1974a; Chow and Shatkin, 1975), contains blocked 5′ termini.

Involvement of other enzyme activities in the formation of viral transcripts was suggested by the presence of 2′-*O*-methylated residues in the 5′-terminal sequences of genome RNA plus strands (Miura *et al.*, 1974a,b). Messenger RNA methylating activity was found associated with the transcriptase in insect cytoplasmic polyhedrosis virus (CPV) (Furuichi, 1974) and later in reovirus cores (Shatkin, 1974) and several other viruses (Wei and Moss, 1974; Banerjee, 1980). Reovirus-associated transferases sequentially methylate two different sites on nucleotidyl acceptors, a purine ring position followed by a ribose 2′-hydroxyl group, consistent with the presence of two distinct methyltransferase activities. As in eukaryotic cellular methylations, the viral enzymes utilize AdoMet as methyl donor. The transcriptase in CPV was activated by AdoMet and some analogues that prevent methylation (Wertheimer *et al.*, 1980). The situation differed for reovirus; AdoMet and *S*-adenosyl-L-homocysteine (AdoHcy) were without significant effect on overall RNA synthesis by reovirus cores.

During the course of studies on the synthesis and structure of reovirus and CPV RNAs, it was reported that eukaryotic cellular mRNAs are methylated (Perry and Kelley, 1974). Comparison of the modified residues in viral transcripts and cellular RNAs made it quickly apparent (Rottman *et al.*, 1974) that they contain similar methylated structures that resemble the blocked, methylated 5' ends of low-molecular-weight nuclear RNAs (Ro-Choi *et al.*, 1974). The general structure of the 5' ends of reovirus mRNA and most other eukaryotic viral and cellular messages is shown in Fig. 1. "Caps" consist of 7-methylguanosine in 5'–5' linkage through three phosphate groups to the initial nucleotide incorporated into the RNA by template-dependent transcription. In all ten species of reovirus mRNA, base 1 is guanine (Furuichi *et al.*, 1975a), and usually but not invariably the 5'-penultimate residue of other mRNAs is a purine. The guanine in position 1 but not the adjacent cytosine is 2'-*O*-methylated in the plus strands of reovirus genome RNA (Furuichi *et al.*, 1975b) and in mRNA made *in vitro*. Termini of this type have been defined previously as cap 1 structures (Shatkin, 1976). In infected mouse cells, some reovirus mRNAs contain cap 2 structures, i.e., $m^7GpppG^mpC^mp$ . . . (Desrosiers *et al.*, 1976). The second 2'-*O*-methylation is probably catalyzed by a cytoplasmic, cellular methyltransferase (Langberg and Moss, 1981).

The mechanism of capped mRNA synthesis by reovirus transcriptase and other core-associated enzymes has been worked out in detail. By analyzing the products of partial incubation mixtures containing GTP and CTP as the only ribonucleoside triphosphates, it was shown that capping can (and probably does) occur at the dinucleotide level (Furuichi *et al.*, 1976). Viral cores can also form caps on preformed reovirus mRNAs by the same series of reactions (Furuichi and Shatkin, 1977). They include the following:

$$\overset{*}{p}ppG + pppC \xrightarrow{\text{RNA polymerase}} \overset{*}{p}ppGpC + PP_i$$

$$\overset{*}{p}ppGpC \xrightarrow{\text{nucleotide phosphohydrolase}} \overset{*}{p}pGpC + P_i$$

$$pppG + \overset{*}{p}pGpC \xleftrightarrow{\text{guanylyltransferase}} Gp\overset{*}{p}pGpC + PP_i$$

$$Gp\overset{*}{p}pGpC + AdoMet \xrightarrow{\text{methyltransferase 1}} m^7Gp\overset{*}{p}pGpC + AdoHcy$$

$$m^7Gp\overset{*}{p}pGpC + AdoMet \xrightarrow{\text{methyltransferase 2}} m^7Gp\overset{*}{p}pG^mpC + AdoHcy$$

Although cellular mRNA capping probably occurs at an early stage of transcription by modification of nascent nuclear transcripts (Salditt-Georgieff *et al.*, 1980) by the same general mechanism (Venkatesan and Moss, 1982), there is one distinct difference as compared to reovirus mRNA capping—the ready reversibility of the viral guanylyltransferase

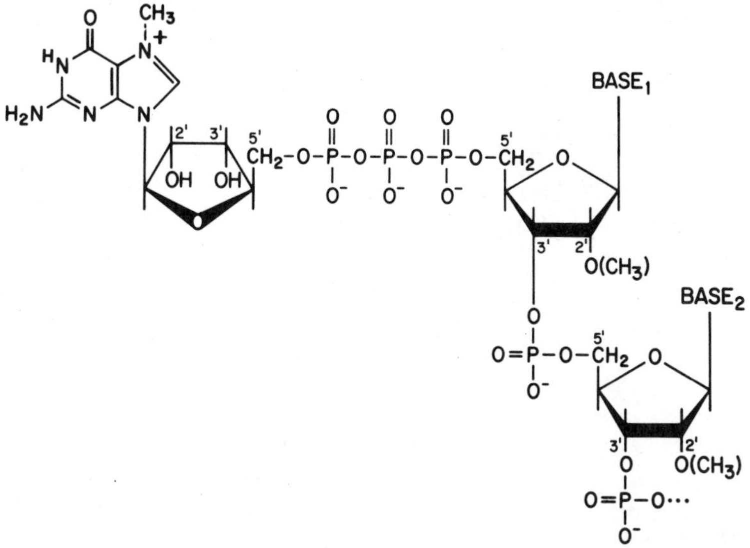

FIGURE 1. Structure of 5' cap.

reaction. In the presence of inorganic pyrophosphate, reovirus cores synthesize mRNA with predominantly 5'-terminal ppG (Furuichi and Shatkin, 1976). By including inorganic pyrophosphatase in transcriptase incubation mixtures (to cleave $PP_i$ generated during mRNA synthesis), it was possible to obtain mRNA containing mainly GpppG termini. In the presence of AdoMet, the mRNAs were fully modified and contained 5'-terminal $m^7GpppG^m$. Once methylated, the termini were resistant to pyrophosphorolysis by the viral guanylyltransferase (Furuichi et al., 1976). The corresponding pyrophosphate-promoted back-reaction catalyzed by the cellular guanylyltransferase was difficult to detect (Venkatesan and Moss, 1982).

Reovirus mRNAs containing other types of 5' termini have been obtained in vitro by using inhibitors of selected steps of cap synthesis. For example, RNA made by cores in the presence of GMP-P(NH)P (the nonhydrolyzable analogue of GTP) contained triphosphorylated 5' ends because the β,γ-imido bond was resistant to cleavage by the viral nucleotide phosphohydrolase (Furuichi and Shatkin, 1977). Similarly, incorporation of γ-S-GTP yielded transcripts with 5'-terminal S-pppGpC (Reeve et al., 1982). These results demonstrate that capping and transcription can be uncoupled in reovirus cores.

Other enzymes involved in capping also have demonstrable substrate and donor specificities. For example, the core guanylyltransferase preferred ppGpC over ppApG as a GMP acceptor, consistent with recognition of an oligonucleotide corresponding to the common 5'-terminal sequence of reovirus mRNA (Furuichi et al., 1976). GTP was effective as a nucleotide donor, but $m^7GTP$ was not used by the guanylyltransferase. This is the result predicted from the aforedescribed scheme for sequential modification of transcripts. Recently, it was found (Shatkin et al., 1983) that a reovirus λ polypeptide that comigrates with λ2 by gel electrophoresis is labeled by [α-$^{32}$P]-GTP, suggesting that the guanylyltransferase forms an enzyme–GMP stable intermediate like the corresponding enzymes from vaccinia virus (Shuman and Hurwitz, 1981) and HeLa cells (Venkatesan and Moss, 1982; Wang et al., 1982).

The reovirus core methyltransferases used GpppGpC in preference to GpppG to form the corresponding m$^7$G-containing oligonucleotides. Structures of the type GppG, GppppG, and GpppA were not active as methyl acceptors, and recent studies demonstrate that GppppG (but not GppG) blocks methylation but not transcription by reovirus cores (Yamakawa et al., 1982b). Methyltransferase donor specificity was somewhat less stringent, and AdoMet was utilized for both methyltransfer reactions (Furuichi et al., 1979). The resulting mRNAs contained 5'-terminal 7-ethylguanosine linked to 2'-O-ethylguanosine. These inhibitor studies provided the means to accumulate mRNAs with different types of 5' termini for functional tests. For example, transcripts with ethylated caps were actively translated in cell-free protein-synthesizing systems. The general role of the cap (and cap derivatives) in promoting cell-free translation of reovirus mRNAs is described in Section III.B.3.

In addition to its effect on translation, the presence of a blocked 5' end also enhances mRNA stability (Furuichi et al., 1977). Reovirus mRNAs containing 5'-terminal m$^7$GpppG$^m$ or GpppG were more stable than the corresponding transcripts with unblocked ppG ends when microinjected into Xenopus oocytes or incubated in cell-free translating extracts derived from wheat germ or mouse L cells. Since GpppG-terminated transcripts did not bind to ribosomes in wheat germ extract, the stabilizing effect was not due to protection by formation of translation initiation complex. Apparently, the presence of the 5'–5' linked terminal G was sufficient to protect the transcripts against exonucleolytic digestion by one or more nucleases in the protein-synthesizing systems. Comparisons of the structure, stability, and function of individual reovirus mRNAs may help to elucidate how efficiencies of eukaryotic transcription and translation are determined.

## III. TRANSLATION OF REOVIRUS MESSENGER RNAs

### A. Protein Synthesis in Virus-Infected Cells

Newly synthesized reovirus proteins can usually be detected by 2 hr post-infection. The precise kinetics of viral protein synthesis vary, of course, depending on the multiplicity of infection and the incubation temperature (Zweerink and Joklik, 1970). There appears to be no early-to-late switch operating at the level of translation, although transcriptional regulation (see above) causes restricted expression of the reovirus genome at early times. As explained below, translation of various reovirus messages is modulated in that some messenger RNA species are utilized far more efficiently than others, but ability to translate a given message does not change as the infection progresses. The quantity of viral proteins made *in vivo* appears to reflect the size of the RNA pool. Thus, late in infection when prodigious amounts of mRNA have accumulated (Walden *et al.*, 1981), viral proteins comprise 10–30% of the total labeled cytoplasmic proteins (Fields *et al.*, 1972; Walden *et al.*, 1981).

Reovirus-specified proteins have been studied by polyacrylamide gel electrophoresis following exposure of infected mouse L cells to radioactive amino acids. Although viral proteins are readily detected by this technique late in the infection, during the early hours of infection various manipulations are required to increase the sensitivity of the assay. One approach used to demonstrate viral protein synthesis at early times involved subtracting the polyacrylamide gel profile of uninfected-cell proteins from the corresponding gel pattern obtained with virus-infected cells (Zweerink and Joklik, 1970). Clearer evidence for viral protein synthesis was obtained by using actinomycin D to suppress host protein synthesis (Zweerink and Joklik, 1970; Gupta *et al.*, 1974) or by immunoprecipitating reovirus proteins with virus-specific antiserum (Cross and Fields, 1976; Lau *et al.*, 1975). These techniques permit at least four of the reovirus proteins to be detected during the early hours of infection. However, accurate determination of the number and relative amounts of reovirus-encoded proteins depends on analyses carried out at later stages of infection, when viral proteins are more abundant. The analytical techniques that were available several years ago permitted detection of eight reovirus-encoded proteins: four small ($\sigma$), two medium ($\mu$), and two large ($\lambda$) species (Zweerink *et al.*, 1971). Recent improvements in gel electrophoresis technology and use of high-specific-activity radioactive amino acids revealed two additional proteins, designated $\lambda 3$ and $\mu 2$ (Both *et al.*, 1975c; Cross and Fields, 1976). Thus, ten primary translation products have been identified at late times in virus-infected cells. The four proteins corresponding in size to the coding capacity of the small genomic RNA species are designated $\sigma 1$, $\sigma NS$, $\sigma 2$, and $\sigma 3$, in the order of their migration in a Laemmli Tris–glycine polyacrylamide gel system. In the medium-size range, three primary translation products have been identified and designated $\mu 1$, $\mu NS$, and $\mu 2$. Pulse-chase experiments revealed that a fourth medium-sized polypeptide ($\mu 1C$) derives from proteolytic cleavage of the primary translation product $\mu 1$ (Zweerink *et al.*, 1971), although not all the radioactivity that disappears from the $\mu 1$ band reappears in the $\mu 1C$ band (Cross and Fields, 1976). Resolution of the three large polypeptide species ($\lambda 1$, $\lambda 2$, and $\lambda 3$) requires electrophoresis conditions different from those used for the $\sigma$ and $\mu$ proteins (Both *et al.*, 1975c).

Because this set of ten viral proteins accounts for the expected coding capacity of the ten genome segments (see Chapter 2), there is a tendency to disregard the additional polypeptide bands that are frequently seen in variable amounts in gels of reovirus proteins (Both *et al.*, 1975c; Zweerink *et al.*, 1971; Lee *et al.*, 1981a; Cross and Fields, 1976). Although the "extra" polypeptide species are usually attributed to cleavage of one of the primary translation products, evidence supporting this interpretation is lacking except for the well-characterized $\mu 1-\mu 1C$ pair. Alternative explanations include the possibility that "extra" polypeptides result from utilizing two initiation sites within a given message (Section III.B.2) or from translation of a processed (i.e., spliced) message. The latter phenomenon occurs with at least one other RNA virus, influenza (Lamb and Lai, 1980). Further heterogeneity in reovirus-encoded polypeptides is introduced by posttranslational modifications such as glycosylation (Krystal *et al.*, 1976), phosphorylation (Krystal *et al.*, 1975; Carter and Shatkin, unpublished data), ADP-ribosylation, and polyadenylylation (Carter *et al.*, 1980). Although the shortened and chemically modified forms of reovirus proteins have not yet been shown to function differently from the corresponding primary translation products, it seems unlikely that such modifications are gratuitous. Thus, the number of functionally distinct polypeptide species in reovirus-infected cells might well exceed ten.

Accurate quantitation of reovirus polypeptides is complicated somewhat by the aforementioned modifications. Furthermore, not all viral proteins are immunoprecipitated with equal efficiency (Cross and Fields, 1976). The insolubility of reovirus structural proteins might also cause differential losses during extraction. Despite these potential problems in quantitating viral proteins, it is clear from the measurements that have been made that some reovirus mRNA species are translated far more efficiently than others. As can be seen in Table I (Section II.B), the relative amounts of reovirus proteins made *in vivo* do not simply reflect the proportion of each template in the mRNA pool. The mechanism by which some viral mRNA species (notably those that encode proteins σ3, σNS, $\mu 1$, $\mu$NS, λ1, and λ2) are translated more efficiently than others is not yet understood. The same phenomenon is observed when reovirus mRNAs are translated *in vitro*, however (Section III.B.4).

Attempts to study the polysomal distribution of reovirus mRNAs in infected cells have yielded curious results. The rough correlation between mRNA size and polyribosome size that one often detects in other systems (Mager and Planta, 1976) was not seen in early investigations of reovirus-infected cells. In other words, rather than finding small-sized reovirus mRNAs associated with relatively small polysomes, Ward *et al.*, (1972) found that polysomes of any given size contained similar amounts of large, medium, and small reovirus mRNAs. One possible interpretation is that reovirus mRNAs are linked to one another at the time they are being translated. Although this interesting possibility merits consideration, it is conceivable that the polysome fraction was aggregated, or contaminated with assembly intermediates. Indeed, more recent studies indicate that one of the reovirus σ proteins is translated on polysomes that are smaller than those that translate μ proteins (Walden *et al.*, 1981). Because it is difficult to resolve and identify the myriad of viral RNA-containing structures in reovirus-infected cells, the question of whether reovirus mRNAs associate with cellular proteins to form ribonucleoprotein (RNP) particles has not been studied extensively. Christman *et al.* (1973) have shown that reovirus mRNAs that accumulate in RNP structures when protein synthesis is inhibited can subsequently be recruited into polysomes.

There are hints that reovirus mRNAs somehow differ from the majority of cellular mRNAs. For example, the histidine analogue L-histidinol preferentially inhibits host protein synthesis (Warrington and Wratten, 1977), whereas interferon discriminates in the opposite direction (Gupta *et al.*, 1974). Both results imply that the translational machinery can distinguish between cellular and viral mRNAs, and consequently translation of host mRNAs *could* be preferentially shut off during virus infection. But it seems that in many cases host protein synthesis either is not shut off (Ensminger and Tamm, 1969; Walden *et al.*, 1981) or is shut off only very late in reovirus-infected cells (Zweerink and Joklik, 1970). In contrast with other investigators, Skup *et al.* (1981) have reported a drastic switch from host to viral protein synthesis during the course of reovirus infection. This phenomenon is further described in Chapter 4.

## B. Translation of Reovirus Messenger RNAs *in Vitro*

Because reovirions contain all the enzymes required for synthesis and modification of viral mRNA, highly purified mRNAs suitable for translation studies can be readily obtained from *in vitro* transcription reactions (Section II.E). An alternative source of single-stranded reovirus RNAs that are competent for translation is to utilize denatured genomic RNA segments (McCrae and Joklik, 1978). Reovirus mRNAs have been translated in cell-free extracts from various sources, including wheat germ (Both *et al.*, 1975c; McCrae and Joklik, 1978), rabbit reticulocytes (McDowell *et al.*, 1972), and mouse ascites cells (Brendler *et al.*, 1981a), as well as by microinjection into frog oocytes (McCrae and Woodland, 1981). Although synthesis of some reovirus proteins is more readily detected than that of others, success depends not so much on the type of cell-free extract as on certain other parameters. These include avoiding nuclease (which preferentially reduces the yield of large proteins), raising the salt concentration to facilitate completion of elongation [although initiation of translation proceeds better at low salt concentrations (see McCrae and Joklik, 1978)], and selection of electrophoresis conditions that are adequate to resolve all the products (Both *et al.*, 1975c). The set of ten reovirus proteins that are made *in vivo* (Section III.A) are also synthesized *in vitro*, under optimal conditions. The correspondence be-

tween *in vivo*– and *in vitro*–synthesized polypeptides was established by
coelectrophoresis (McDowell *et al.*, 1972; Both *et al.*, 1975c), immunoprecipitation (Graziadei *et al.*, 1973), and characterization of partial proteolytic digestion products (McCrae and Joklik, 1978). In addition to the
ten apparently authentic reovirus proteins, the polyacrylamide gel profiles usually contain additional bands that probably represent prematurely terminated chains. The possibility that some of the shorter polypeptides result from spurious initiation events within the interior of
intact mRNA chains seems unlikely, since analysis of the ribosome-protected sequences generally reveals only one initiation site per message
(Section III.B.2). But generation of mRNA fragments, by nucleolytic
cleavage during incubation in the lysate, might allow ribosomes to initiate at spurious internal sites (Kozak, 1980a).

The ability to translate reovirus mRNAs *in vitro* with high efficiency
has been exploited to answer a number of questions about the biology of
reoviruses, as well as more general questions about the mechanism of
protein synthesis in eukaryotes.

### 1. Protein Coding Assignments of Reovirus RNA Segments

The question of which mRNA species encodes which reovirus protein was difficult to approach because the single-stranded mRNAs within
a given size class are poorly resolved under the usual electrophoresis
conditions. McCrae and Joklik (1978) circumvented this problem by frac-
tionating the double-stranded genomic RNAs. Individual genomic RNA
species were then denatured and translated in a wheat germ lysate, permitting clear assignment of one reovirus polypeptide to each of the ten
genome segments (see Chapter 2). The identity of each polypeptide band
was carefully checked by comparing partial proteolytic digestion profiles
of *in vitro* translation products with those of authentic viral proteins.
Gene–protein assignments for the medium and small size classes have
been confirmed by an elegant analysis of recombinants between different
strains of reovirus (Mustoe *et al.*, 1978).

More recently, K. H. Levin and Samuel (1980) succeeded in fraction-
ating the four small-sized mRNAs, so that each could be translated individually. The resulting coding assignments (s1–σ1, s2–σ2, s3–σNS, s4–
σ3) are consistent with those just described.

### 2. Identification of Ribosome-Binding Sites in Nine Reovirus
### Messenger RNAs

The hypothesis that each reovirus mRNA encodes one protein gained
substantial support from finding ten viral-encoded polypeptides, the sizes
of which more or less corresponded to those of the mRNAs. The mon-

ocistronic character of reovirus mRNAs was confirmed (with one interesting exception) by demonstrating that each message yielded a single ribosome-protected initiation site. The ribosome protection assay devised many years ago to study coliphage MS2 translation (Steitz, 1969) was readily adapted to the reovirus system. High-specific-activity $^{32}$P-labeled reovirus mRNAs were incubated with wheat germ ribosomes in the presence of sparsomycin, which inhibits elongation, and the resulting 80 S initiation complexes were trimmed with ribonuclease. The nuclease-resistant mRNA fragments that cosedimented with 80 S ribosomes (i.e., the presumptive initiation sites for translation) were then purified and sequenced. In addition to the initiation sites protected by 80 S ribosomes, which typically encompass about 30 nucleotides, the somewhat larger mRNA fragments protected by 40 S ribosomal subunits were also analyzed whenever possible. Figure 2 shows the nucleotide sequences of ri-

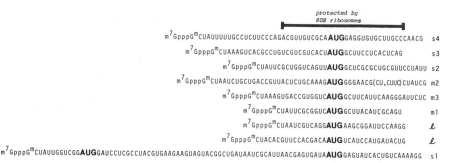

FIGURE 2. Ribosome-binding sites in nine mRNAs of reovirus serotype 3. The presumptive AUG initiator codon in each message is shown in boldface. The sequences shown for the first five mRNAs (s4, s3, s2, m2, and m3) are the regions protected by wheat germ 40 S ribosomal subunits (Kozak and Shatkin, 1977a,b; Kozak, 1977). The smaller region of each message that is protected by 80 S ribosomes is indicated across the top of the figure. [Protection by 80 S ribosomes usually extends about 12 nucleotides to the left (5′ side) of the AUG triplet and 12–14 nucleotides to the right. The exact site of cleavage depends, of course, on the specificity of the ribonuclease.] With m1 mRNA (Kozak, 1977) and the two l messages (Kozak, 1982a), only the 80 S ribosome-protected region was characterized and is shown. In m2 mRNA, the order of the two pyrimidine blocks enclosed in parentheses is not known. Near the 5′ terminus of s4 mRNA, there appears to be a run of five U residues (Hastings and Millward, 1978; Darzynkiewicz and Shatkin, 1980), rather than the four U residues that we initially reported (Kozak and Shatkin, 1977b). McCrae (1981) has confirmed the sequence CCUCUUCCC in position 11–19 of s4 mRNA. In the sequence originally reported for s2 mRNA (Kozak, 1977), there was an uncertainty in the order of nucleotides 25–30; this sequence was subsequently clarified by Li et al. (1980b) and by Darzynkiewicz and Shatkin (1980). The first 76 nucleotides shown for s1 mRNA are from Li et al. (1980a). The remainder of the s1 sequence is from Kozak (1982b). In the latter study, analysis of the 80 S ribosome-protected mRNA fragments revealed two initiation sites in s1 mRNA, centered respectively at the first and the second AUG triplets. All the data in this figure were obtained using wheat germ ribosomes. Initiation sites selected by reticulocyte (Lazarowitz and Robertson, 1977) and by mouse ascites cell ribosomes (Kozak and Shatkin, 1977a) have also been analyzed with some of the reovirus mRNAs, confirming the results obtained with wheat germ lysates.

bosome-binding sites from nine reovirus mRNAs that were characterized in this way. The relative yield of each ribosome-binding site corresponds roughly to the translational efficiency of reovirus mRNAs. Thus, initiation sites from $m2$, $m3$, $s3$, and $s4$ mRNAs were preferentially recovered (Kozak and Shatkin, 1977a,b).

As shown in Fig. 2, an AUG triplet is found approximately in the center of the 80 S ribosome-protected sequence from each reovirus mRNA. It seems likely that this AUG is the initiator codon, although that assumption must be confirmed by analyzing the N-terminal amino acid sequences of nascent reovirus proteins. The analyses cannot be carried out with proteins extracted from virions because most mature reovirus proteins have a blocked N-terminus (Pett et al., 1973). Samuel and Joklik (1976) reported preliminary characterization of $N$-formylmethionyl-containing peptides derived from reovirus proteins synthesized in vitro, but their data were not sufficient to deduce the N-terminal amino acid sequences. The ribosome-protected initiation sites in Fig. 2 have little in common apart from the 5'-terminal sequence $m^7GpppG^mC$-U-A (conservation of which is probably related to transcriptional requirements) and the sequence $^G_ANNAUGG$, which characterizes seven of the reovirus initiation sites and more than half the initiation sites that have been sequenced from other eukaryotic organisms. The reovirus ribosome-binding sites reveal no consistent pattern of complementarity with 3' end of 18 S ribosomal RNA. The distance of the $m^7G$ cap from the AUG codon varies from 13 to 33 nucleotides. In this respect, reovirus mRNAs fall at the short end of the spectrum when compared with other viral or cellular messages (Kozak, 1981a). Eight of the reovirus mRNAs represented in Fig. 2 follow the pattern observed with nearly all other eukaryotic mRNAs, in that ribosomes initiate exclusively at the AUG triplet that lies closest to the 5' end of the message. The interesting exception is $s1$ mRNA, which unexpectedly yielded two ribosome-protected initiation sites (Kozak, 1982b). In that message, ribosomes appear to initiate at the first and the second AUG codons. Inspection of the sequence flanking the 5'-proximal AUG triplet in $s1$ mRNA suggests a possible explanation. Although most functional initiator codons in eukaryotic mRNAs conform to the sequence $^{-3}_A$GNNAUGG (Kozak, 1981b), the first AUG triplet in $s1$ mRNA deviates from that pattern; it has a C in position $-3$. The unfavorable context surrounding the first AUG triplet might cause some 40 S ribosomal subunits to bypass that site and select instead the next AUG codon, which conforms to the consensus sequence for eukaryotic initiation sites. The significance of initiation at the first and the second AUG triplets (which lie in different reading frames) in $s1$ mRNA depends on what lies downstream. If both AUG triplets are followed by long open reading frames, it would seem that reovirus has found a way to produce two proteins from a single mRNA species. (Although this scenario predicts that one should detect five σ proteins instead of the observed set of four, it should be borne in mind that polyacrylamide gels of reovirus proteins do reveal "extra" protein bands, which are usually dismissed as artifacts. Moreover, polypeptides that are marginally smaller than σ3 would probably have been missed, since they would have migrated off the lower edge of the gel.) Reovirus s1 mRNA is unusual but not unique in permitting initiation at a second AUG codon in a second reading frame. A similar phenomenon seems to occur with the small mRNA of bunyavirus (Clerx-van Haaster et al., 1982).

3. Function of the m⁷G Cap in Translation of Reovirus Messenger RNAs

Shortly after Furuichi et al. (1975a) discovered that the 5′ terminus of reovirus mRNAs is blocked and methylated, Both et al. (1975a) showed that in vitro translation was highly dependent on the m⁷G cap. Glycerol gradient analyses of initiation complexes formed with ³²P-labeled reovirus mRNA revealed that wheat germ 40 S ribosomal subunits were unable to form stable initiation complexes with uncapped reovirus mRNAs (Both et al., 1975b). The facilitating effect of the m⁷G cap on translation, which was first seen in these studies in vitro with reovirus mRNA, has been observed subsequently with many other eukaryotic mRNAs (Shatkin, 1976). Although there is as yet no direct evidence that the m⁷G cap functions during translation of reovirus mRNAs in vivo, data obtained with globin and vesicular stomatitis virus mRNAs suggests that the methylated cap enhances translation in vivo (Lockard and Lane, 1978; de Ferra and Baglioni, 1981).

In view of the strong dependence of reovirus translation on the m⁷G cap, one might be surprised at the early success of several investigators in translating reovirus mRNAs that were synthesized in vitro without inclusion of S-adenosyl-L-methionine. Several factors might account for the successful in vitro translation experiments that were carried out prior to discovery of the cap. In some cases, the translatability of reovirus mRNA was enhanced by pretreating with formaldehyde (D.H. Levin et al., 1971)—a procedure that, we now know, obviates the m⁷G requirement, to some extent (Kozak and Shatkin, 1978a). The stringency of the m⁷G requirement depends on the source of the cell-free extract, with reticulocyte (Muthukrishnan et al., 1976; Lodish and Rose, 1977; Bergmann and Lodish, 1979) and ascites cell lysates (Samuel et al., 1977) showing less of a dependence on the m⁷G cap than wheat germ extracts. Moreover, extracts from wheat germ (Muthukrishnan et al., 1975; Both et al., 1975a; Keith et al., 1982) and ascites cells (K.H. Levin and Samuel, 1977) contain capping and methylating activities that, unless specifically inhibited by S-adenosyl-L-homocysteine, will add the required m⁷G moiety to the 5′ end of uncapped mRNA.

As noted in the preceding section, the 5′-terminal cap was included within the region of each reovirus mRNA protected by wheat germ 40 S ribosomal subunits (although 80 S ribosomes protected a small region of each message that often did not include the cap). It seems likely that protection by 40 S ribosomes of the $m^7G$ cap on reovirus mRNAs is merely a consequence of the AUG initiator codon lying quite close to the 5′ terminus. Thus, with other messages in which more than 60 nucleotides separate the cap from the AUG initiator codon, the cap was not protected by 40 S ribosomal subunits (Kozak and Shatkin, 1978b). The fact that the $m^7G$ cap facilitates ribosome-binding but is not necessarily part of the ribosome-protected region of the mRNA hints of a two-site initiation mechanism. This idea has been elaborated elsewhere (Kozak, 1981a).

The mechanism by which the cap facilitates initiation remains obscure. Enhanced translation of $m^7$GpppN-terminated mRNA cannot be attributed merely to protection from 5′-exonuclease activity, since unmethylated GpppG-terminated mRNA, which is resistant to 5′-exonuclease (Furuichi et al., 1977), is nevertheless translated poorly. K.H. Levin and Samuel (1977) have shown that unmethylated reovirus mRNA directs the same spectrum of polypeptides as methylated mRNA, albeit with lower efficiency. The region of each reovirus message protected by 40 S ribosomes is also the same with both forms of mRNA (Kozak and Shatkin, 1978a). Thus, a methylated cap is not required to identify the authentic AUG initiator codon. Although the methylated cap does not contribute to the fidelity of initiation-site selection, the $m^7G$ moiety dramatically enhances both the rate and the extent of binding of reovirus mRNAs (Both et al., 1975b) or fragments thereof (Kozak and Shatkin, 1978a). Situations in which the cap seems not to be needed might provide a clue to its mode of action. Denaturation of reovirus mRNAs by treatment with formaldehyde (Kozak and Shatkin, 1978a) or incorporation of inosine (Kozak, 1980b; M.A. Morgan and Shatkin, 1980) lessens the requirement for both the $m^7G$ cap and cap-binding protein (Sonenberg et al., 1981). One possible interpretation is that a cap-recognizing protein (Sonenberg et al., 1980) might bind to the $m^7G$ terminus of native mRNA and subsequently unwind a short stretch of the mRNA chain, thereby inviting a ribosome to bind. Unfolded mRNA would no longer require the facilitating (denaturing) effect of the cap and its associated protein.

Experiments carried out with modified cap structures have provided considerable information about what features are necessary for $m^7G$ to exert its facilitating effect on reovirus translation. The assay makes use of the observation that cap analogues such as 7-methylguanosine-5′-monophosphate ($m^7G^{5'}p$), $m^7G^{5'}$ pp, and $m^7G^{5'}$ ppp inhibit translation of capped mRNAs (Hickey et al., 1976). Adams et al. (1978) found that 7-ethylguanosine-5′-diphosphate and 7-benzylguanosine-5′-diphosphate inhibited translation of reovirus mRNAs as effectively as $m^7G^{5'}$pp, indicating that the nature of the 7-substituent is relatively unimportant.

On the other hand, ring-opened forms of the three aforementioned derivates were inactive, as was 8-hydro-$m^7G^{5'}$ pp; these compounds lack a plus charge on the imidazole moiety, which seems to be crucial for cap function. Electrostatic interaction between the positively charged imidazole moiety of the $N^7$-alkylated G residue and the negatively charged phosphate groups causes the cap to have a rigid structure. Although this may be necessary for cap function, it is not sufficient because the methyl ester of $m^7G^{5'}$ p, which retains the characteristic rigid conformation, is far less active than $m^7G^{5'}$ p as an inhibitor of translation (Darzynkiewicz et al., 1981).

## 4. Variation in Translational Efficiency among Reovirus Messenger RNAs

In Section III.A., we pointed out that some reovirus mRNA species are translated far more efficiently than others in infected cells. A similar phenomenon occurs in vitro. The most efficiently translated reovirus messages in the small-size class are s3 and s4 (McDowell et al., 1972), while m2 and m3 predominate in the medium-size class (Brendler et al., 1981b). Interestingly, the poor translatability of s1 and s2 mRNAs persists even in the absence of other competing mRNAs, as K.H. Levin and Samuel (1980) showed by translating each message individually. The s1 message has the peculiar property of allowing initiation at two AUG codons (Kozak, 1982b), but it remains to be seen whether this contributes to the low translational efficiency of that message.

In a recent series of experiments, Brendler et al. (1981a) studied reovirus translation under conditions of competition by an unrelated mRNA, namely, that of rabbit globin. Their competition studies confirm the high efficiency of mRNAs m3 and m2 (encoding polypeptides μNS and μ1, respectively), but, surprisingly, they did not detect marked differences in efficiency among the four small-sized reovirus mRNAs.

## IV. CLOSING THOUGHTS

We have avoided calling reovirus a "model system"—in part because that overworked term lacks clear meaning and in part because reovirus (with its double-stranded, segmented RNA genome) is hardly typical of any other viral or cellular system. But it is amusing to note that a system as atypical as reovirus has been at the center of some very fundamental discoveries concerning the structure and function of eukaryotic messenger RNAs. It goes without saying that major questions about the biochemistry of reovirus transcription and translation remain to be answered. One may hope that some of those answers will provide clues to the workings of other biological systems.

ACKNOWLEDGMENTS. We are grateful to Janet Hansen for her capable assistance in preparing the manuscript. Research carried out by M.K. was supported by U.S. Public Health Service Grants AI 16634 and AI 00380.

## REFERENCES

Adams, B.L., Morgan, M., Muthukrishnan, S., Hecht, S.M., and Shatkin, A.J., 1978, The effect of "cap" analogs on reovirus mRNA binding to wheat germ ribosomes, *J. Biol. Chem.* **253**:2589.

Banerjee, A.K., 1980, 5'-Terminal cap structure in eucaryotic messenger ribonucleic acids, *Microbiol. Rev.* **44**:175.

Banerjee, A.K., and Shatkin, A.J., 1970, Transcription *in vitro* by reovirus-associated ribonucleic acid-dependent polymerase, *J. Virol.* **6**:1.

Banerjee, A.K., Ward, R., and Shatkin, A.J., 1971, Initiation of reovirus mRNA synthesis *in vitro*, *Nature (London) New Biol.* **230**:169.

Bartlett, N.M., Gillies, S.C., Bullivant, S., and Bellamy, A.R., 1974, Electron microscopy study of reovirus reaction cores, *J. Virol.* **14**:315.

Bergmann, J.E., and Lodish, H.F., 1979, Translation of capped and uncapped vesicular stomatitis virus and reovirus mRNAs, *J. Biol. Chem.* **254**:459.

Borsa, J., and Graham, A.F., 1968, Reovirus: RNA polymerase activity in purified virions, *Biochem. Biophys. Res. Commun.* **33**:895.

Borsa, J., Grover, J., and Chapman, J.D., 1970, Presence of nucleoside triphosphate phosphohydrolase activity in purified virions of reovirus, *J. Virol.* **6**:295.

Borsa, J., Copps, T.P., Sargent, M.D., Long, D.G., and Chapman, J.D., 1973, New intermediate subviral particles in the *in vitro* uncoating of reovirus virions by chymotrypsin, *J. Virol.* **11**:552.

Borsa, J., Sargent, M.D., Lievaart, P.A., and Copps, T.P., 1981, Reovirus: Evidence for a second step in the intracellular uncoating and transcriptase activation process, *Virology* **111**:191.

Both, G.W., Banerjee, A.K., and Shatkin, A.J., 1975a, Methylation-dependent translation of viral messenger RNAs *in vitro*, *Proc. Natl. Acad. Sci. U.S.A.* **72**:1189.

Both, G.W., Furuichi, Y., Muthukrishnan, S., and Shatkin, A.J., 1975b, Ribosome binding to reovirus mRNA in protein synthesis requires 5' terminal 7-methylguanosine, *Cell* **6**:185.

Both, G.W., Lavi, S., and Shatkin, A.J., 1975c, Synthesis of all the gene products of the reovirus genome *in vivo* and *in vitro*, *Cell* **4**:173.

Brendler, T., Godefroy-Colburn, T., Carlill, R.D., and Thach, R.E., 1981a, The role of mRNA competition in regulating translation. II. Development of a quantitative *in vitro* assay, *J. Biol. Chem.* **256**:11747.

Brendler, T., Godefroy-Colburn, T., Yu, S., and Thach, R.E., 1981b, The role of mRNA competition in regulating translation. III. Comparison of *in vitro* and *in vivo* results, *J. Biol. Chem.* **256**:11755.

Carter, C.A., Lin, B.Y., and Metlay, M., 1980, Polyadenylylation of reovirus proteins: Analysis of the RNA bound to structural proteins, *J. Biol. Chem.* **255**:6479.

Chang, C.-T., and Zweerink, H.J., 1971, Fate of parental reovirus in infected cell, *Virology* **46**:544.

Chow, N.-L., and Shatkin, A.J., 1975, Blocked and unblocked 5' termini in reovirus genome RNA, *J. Virol.* **15**:1057.

Christman, J.K., Reiss, B., Kyner, D., Levin, D.H., Klett, H., and Acs, G., 1973, Characterization of a viral messenger ribonucleoprotein particle accumulated during inhibition of polypeptide chain initiation in reovirus-infected L cells, *Biochim. Biophys. Acta* **294**:153.

Clerx-van Haaster, C.M., Akashi, H., Auperin, D.D., and Bishop, D.H.L., 1982, Nucleotide sequence analyses and predicted coding of bunyavirus genome RNA species, *J. Virol.* **41**:119.

Cross, R.K., and Fields, B.N., 1972, Temperature-sensitive mutants of reovirus type 3: Studies on the synthesis of viral RNA, *Virology* **50**:799.

Cross, R.K., and Fields, B.N., 1976, Reovirus-specific polypeptides: Analysis using discontinuous gel electrophoresis, *J. Virol.* **19**:162.

Darzynkiewicz, E., and Shatkin, A.J., 1980, Assignment of reovirus mRNA ribosome binding sites to virion genome segments by nucleotide sequence analyses, *Nucleic Acids Res.* **8**:337.

Darzynkiewicz, E., Antosiewicz, J., Ekiel, I., Morgan, M.A., Tahara, S.M., and Shatkin, A.J., 1981, Methyl esterification of $m^7G^{5'}$ p reversibly blocks its activity as an analog of eukaryotic mRNA 5'-caps, *J. Mol. Biol.* **153**:451.

De Ferra, F., and Baglioni, C., 1981, Viral messenger RNA unmethylated in the 5'-terminal guanosine in interferon-treated HeLa cells infected with vesicular stomatitis virus, *Virology* **112**:426.

Desrosiers, R.C., Sen, G.C., and Lengyel, P., 1976, Difference in 5' terminal structure between the mRNA and the double-stranded virion RNA of reovirus, *Biochem. Biophys. Res. Commun.* **73**:32.

Drayna, D., and Fields, B.N., 1982, Activation and characterization of the reovirus transcriptase: Genetic analysis, *J. Virol.* **41**:110.

Ensminger, W.D., and Tamm, I., 1969, Cellular DNA and protein synthesis in reovirus-infected L cells, *Virology* **39**:357.

Fields, B.N., Laskov, R., and Scharff, M.D., 1972, Temperature-sensitive mutants of reovirus type 3: Studies on the synthesis of viral peptides, *Virology* **50**:209.

Furuichi, Y., 1974, "Methylation-coupled" transcription by virus-associated transcriptase of cytoplasmic polyhedrosis virus containing double-stranded RNA, *Nucleic Acids Res.* **1**:809.

Furuichi, Y., and Shatkin, A.J., 1976, Differential synthesis of blocked and unblocked 5'-termini in reovirus mRNA: Effect of pyrophosphate and pyrophosphatase, *Proc. Natl. Acad. Sci. U.S.A.* **73**:3448.

Furuichi, Y., and Shatkin, A.J., 1977, 5'-Termini of reovirus mRNA: Ability of viral cores to form caps post-transcriptionally, *Virology* **77**:566.

Furuichi, Y., Morgan, M., Muthukrishnan, S., and Shatkin, A.J., 1975a, Reovirus messenger RNA contains a methylated, blocked 5'-terminal structure: $m^7G(5')ppp(5')G^mpCp$-, *Proc. Natl. Acad. Sci. U.S.A.* **72**:362.

Furuichi, Y., Muthukrishnan, S., and Shatkin, A.J., 1975b, 5'-Terminal $m^7G(5')ppp(5')G^mp$ *in vivo*: Identification in reovirus genome RNA, *Proc. Natl. Acad. Sci. U.S.A.* **72**:742.

Furuichi, Y., Muthukrishnan, S., Tomasz, J., and Shatkin, A.J., 1976, Mechanism of formation of reovirus mRNA 5'-terminal blocked and methylated sequence, $m^7GpppG^mpC$, *J. Biol. Chem.* **251**:5043.

Furuichi, Y., LaFiandra, A., and Shatkin, A.J., 1977, 5'-Terminal structure and mRNA stability, *Nature (London)* **266**:235.

Furuichi, Y., Morgan, M.A., and Shatkin, A.J., 1979, Synthesis and translation of mRNA containing 5'-terminal 7-ethylguanosine cap, *J. Biol. Chem.* **254**:6732.

Gomatos, P.J., and Tamm, I., 1963, The secondary structure of reovirus RNA, *Proc. Natl. Acad. Sci. U.S.A.* **49**:707.

Gomatos, P.J., Krug, R.M., and Tamm, I., 1964, Enzymic synthesis of RNA with reovirus RNA as template. I. Characteristics of the reaction catalyzed by the RNA polymerase from *Escherichia coli*, *J. Mol. Biol.* **9**:193.

Graziadei, W.D., III, Roy, D., Konigsberg, W., and Lengyel, P., 1973, Translation of reovirus messenger ribonucleic acids synthesized *in vitro* into reovirus proteins in a mouse L cell extract, *Arch. Biochem. Biophys.* **158**:266.

Gupta, S.L., Graziadei, W.D., III, Weideli, H., Sopori, M.L., and Lengyel, P., 1974, Selective inhibition of viral protein accumulation in interferon-treated cells: Nondiscriminate inhibition of the translation of added viral and cellular messenger RNAs in their extracts, *Virology* **57**:49.

Hastings, K.E.M., and Millward, S., 1978, Nucleotide sequences at the 5′ termini of reovirus mRNA's, *J. Virol.* **28**:490.

Hickey, E.D., Weber, L.A., and Baglioni, C., 1976, Inhibition of initiation of protein synthesis by 7-methylguanosine-5′-monophosphate, *Proc. Natl. Acad. Sci. U.S.A.* **73**:19.

Ito, Y., and Joklik, W.K., 1972, Temperature-sensitive mutants of reovirus. I. Patterns of gene expression by mutants of groups C, D, and E, *Virology* **50**:189.

Joklik, W.K., 1972, Studies on the effect of chymotrypsin on reovirions, *Virology* **49**:700.

Joklik, W.K., 1974, Reproduction of reoviridae, in: *Comprehensive Virology* (H. Fraenkel-Conrat and R.R. Wagner, eds.), pp. 231–334, Plenum Press, New York.

Joklik, W.K., 1981, Structure and function of the reovirus genome, *Microbiol. Rev.* **45**:483.

Kapuler, A.M., 1970, An extraordinary temperature dependence of the reovirus transcriptase, *Biochemistry* **9**:4453.

Kapuler, A.M., Mendelsohn, N., Klett, H., and Acs, G., 1970, Four base-specific nucleoside 5′-triphosphatases in the subviral core of reovirus, *Nature (London)* **225**:1209.

Kates, J.A., and McAuslan, B.R., 1967, Poxvirus DNA-dependent RNA polymerase, *Proc. Natl. Acad. Sci. U.S.A.* **58**:134.

Keith, J.M., Venkatesan, S., Gershowitz, A., and Moss, B., 1982, Purification and characterization of the messenger ribonucleic acid capping enzyme GTP:RNA guanylyltransferase from wheat germ, *Biochemistry* **21**:327.

Kozak, M., 1977, Nucleotide sequences of 5′-terminal ribosome-protected initiation regions from two reovirus messages, *Nature (London)* **269**:390.

Kozak, M., 1980a, Binding of wheat germ ribosomes to fragmented viral mRNA, *J. Virol.* **35**:748.

Kozak, M., 1980b, Influence of mRNA secondary structure on binding and migration of 40S ribosomal subunits, *Cell* **19**:79.

Kozak, M., 1981a, Mechanism of mRNA recognition by eukaryotic ribosomes during initiation of protein synthesis, in: *Current Topics in Microbiology and Immunology*, Vol. 93 (A.J. Shatkin, ed.), pp. 81–123, Springer-Verlag, Berlin.

Kozak, M., 1981b, Possible role of flanking nucleotides in recognition of the AUG initiator codon by eukaryotic ribosomes, *Nucleic Acids Res.* **9**:5233.

Kozak, M., 1982a, Sequences of ribosome binding sites from the large size class of reovirus messenger RNA, *J. Virol.* **42**:467.

Kozak, M., 1982b, Analysis of ribosome binding sites from the s1 message of reovirus: Initiation at the first and second AUG codons, *J. Mol. Biol.* **156**:807.

Kozak, M., and Shatkin, A.J., 1977a, Sequences of two 5′-terminal ribosome-protected fragments from reovirus messenger RNAs, *J. Mol. Biol.* **112**:75.

Kozak, M., and Shatkin, A.J., 1977b, Sequences and properties of two ribosome binding sites from the small size class of reovirus messenger RNA, *J. Biol. Chem.* **252**:6895.

Kozak, M., and Shatkin, A.J., 1978a, Identification of features in 5′ terminal fragments from reovirus mRNA which are important for ribosome binding, *Cell* **13**:201.

Kozak, M., and Shatkin, A.J., 1978b, Migration of 40 S ribosomal subunits on messenger RNA in the presence of edeine, *J. Biol. Chem.* **253**:6568.

Krystal, G., Winn, P., Millward, S., and Sakuma, S., 1975, Evidence for phosphoproteins in reovirus, *Virology* **64**:505.

Krystal, G., Perrault, J., and Graham, A.F., 1976, Evidence for a glycoprotein in reovirus, *Virology* **72**:308.

Kudo, H., and Graham, A.F., 1965, Synthesis of reovirus ribonucleic acid in L cells, *J. Bacteriol.* **90**:936.

Lamb, R.A., and Lai, C.-J., 1980, Sequence of interrupted and uninterrupted mRNAs and cloned DNA coding for the two overlapping nonstructural proteins of influenza virus, *Cell* **21**:475.

Langberg, S.R., and Moss, B., 1981, Post-transcriptional modifications of mRNA: Purification and characterization of cap I and cap II RNA (nucleoside-2'-)-methyltransferases from HeLa cells, *J. Biol. Chem.* **256**:10054.

Lau, R.Y., Van Alstyne, D., Berckmans, R., and Graham, A.F., 1975, Synthesis of reovirus-specific polypeptides in cells pretreated with cycloheximide, *J. Virol.* **16**:470.

Lazarowitz, S.G., and Robertson, H.D., 1977, Initiator regions from the small size class of reovirus messenger RNA protected by rabbit reticulocyte ribosomes, *J. Biol. Chem.* **252**:7842.

Lee, P.W.K., Hayes, E.C., and Joklik, W.F., 1981a, Characterization of anti-reovirus immunoglobulins secreted by cloned hybridoma cell lines, *Virology* **108**:134.

Lee, P.W.K., Hayes, E.C., and Joklik, W.K., 1981b, Protein σ1 is the reovirus cell attachment protein, *Virology* **108**:156.

Levin, D.H., Acs, G., and Silverstein, S.C., 1970a, Chain initiation by reovirus RNA transcriptase *in vitro, Nature (London)* **227**:603.

Levin, D.H., Mendelsohn, N., Schonberg, M., Klett, H., Silverstein, S., Kapuler, A.M., and Acs, G., 1970b, Properties of RNA transcriptase in reovirus subviral particles, *Proc. Natl. Acad. Sci. U.S.A.* **66**:890.

Levin, D.H., Kyner, D., Acs, G., and Silverstein, S., 1971, Messenger activity in mammalian cell-free extracts of reovirus single-stranded RNA prepared *in vitro, Biochem. Biophys. Res. Commun.* **42**:454.

Levin, K.H., and Samuel, C.E., 1977, Biosynthesis of reovirus-specified polypeptides: Effect of methylation on the efficiency of reovirus genome expression *in vitro, Virology* **77**:245.

Levin, K.H., and Samuel, C.E., 1980, Biosynthesis of reovirus-specified polypeptides: Purification and characterization of the small-sized class mRNAs of reovirus type 3: Coding assignments and translational efficiencies, *Virology* **106**:1.

Li, J.K.-K., Keene, J.D., Scheible, P.P., and Joklik, W.K., 1980a, Nature of the 3'-terminal sequences of the plus and minus strands of the S1 gene of reovirus serotypes 1, 2 and 3, *Virology* **105**:41.

Li, J.K.-K., Scheible, P.P., Keene, J.D., and Joklik, W.K., 1980b, The plus strand of reovirus gene S2 is identical with its *in vitro* transcript, *Virology* **105**:282.

Lockard, R.E., and Lane, C., 1978, Requirement for 7-methylguanosine in translation of globin mRNA *in vivo, Nucleic Acids Res.* **5**:3237.

Lodish, H.F., and Rose, J.K., 1977, Relative importance of 7-methylguanosine in ribosome binding and translation of vesicular stomatitis virus mRNA in wheat germ and reticulocyte cell-free systems, *J. Biol. Chem.* **252**:1181.

Mager, W.H., and Planta, R.J., 1976, Yeast ribosomal proteins are synthesized on small polysomes, *Eur. J. Biochem.* **62**:193.

McCrae, M.A., 1981, Terminal structure of reovirus RNAs, *J. Gen. Virol.* **55**:393.

McCrae, M.A., and Joklik, W.K., 1978, The nature of the polypeptide encoded by each of the 10 double-stranded RNA segments of reovirus type 3, *Virology* **89**:578.

McCrae, M.A., and Woodland, H.R., 1981, Stability of non-polyadenylated viral mRNAs injected into frog oocytes, *Eur. J. Biochem.* **116**:467.

McDowell, M.J., Joklik, W.K., Villa-Komaroff, L., and Lodish, H.F., 1972, Translation of reovirus messenger RNAs synthesized *in vitro* into reovirus polypeptides by several mammalian cell-free extracts, *Proc. Natl. Acad. Sci. U.S.A.* **69**:2649.

Miura, K.-I., and Muto, A., 1965, Lack of messenger RNA activity of a double-stranded RNA, *Biochim. Biophys. Acta* **108**:707.

Miura, K.-I., Watanabe, K., Sugiura, M., and Shatkin, A.J., 1974a, The 5'-terminal nucleotide sequences of the double-stranded RNA of human reovirus, *Proc. Natl. Acad. Sci. U.S.A.* **71**:3979.

Miura, K.-I., Watanabe, K., and Sugiura, M., 1974b, 5'-Terminal nucleotide sequences of the double-stranded RNA of silkworm cytoplasmic polyhedrosis virus, *J. Mol. Biol.* **86**:31.

Morgan, E.M., and Kingsbury, D.W., 1980, Pyridoxal phosphate as a probe of reovirus transcriptase, *Biochemistry* **19**:484.

Morgan, E.M., and Kingsbury, D.W., 1981, Reovirus enzymes that modify messenger RNA are inhibited by perturbation of the lambda proteins, *Virology* **113**:565.

Morgan, M.A., and Shatkin, A.J., 1980, Initiation of reovirus transcription of inosine 5'-triphosphate and properties of 7-methylinosine-capped, inosine-substituted messenger ribonucleic acids, *Biochemistry* **19**:5960.

Moss, B., 1974, Reproduction of poxviruses, in: *Comprehensive Virology*, Vol. 3 (H. Fraenkel-Conrat and R.R. Wagner, eds.), pp. 405–474, Plenum Press, New York.

Munyon, W.E., Paoletti, E., and Grace, J.T., Jr., 1967, RNA polymerase activity in purified infectious vaccinia virus, *Proc. Natl. Acad. Sci. U.S.A.* **58**:2280.

Mustoe, T.A., Ramig, R.F., Sharpe, A.H., and Fields, B.N., 1978, Genetics of reovirus: Identification of the ds RNA segments encoding the polypeptides of the μ and σ size classes, *Virology* **89**:594.

Muthukrishnan, S., Both, G.W., Furuichi, Y., and Shatkin, A.J., 1975, 5'-Terminal 7-methylguanosine in eukaryotic mRNA is required for translation, *Nature (London)* **255**:33.

Muthukrishnan, S., Morgan, M., Banerjee, A.K., and Shatkin, A.J., 1976, Influence of 5'-terminal $m^7G$ and 2'-O-methylated residues on messenger ribonucleic acid binding to ribosomes, *Biochemistry* **15**:5761.

Nakashima, K., and Shatkin, A.J., 1978, Photochemical cross-linking of reovirus genome RNA *in situ* and inactivation of viral transcriptase, *J. Biol. Chem.* **253**:8680.

Nakashima, K., LaFiandra, A.J., and Shatkin, A.J., 1979, Differential dependence of reovirus-associated enzyme activities on genome RNA as determined by psoralen photosensitivity, *J. Biol. Chem.* **254**:8007.

Nonoyama, M., Millward, S., and Graham, A.F., 1974, Control of transcription of the reovirus genome, *Nucleic Acids Res.* **1**:373.

Perry, R.P., and Kelley, D.E., 1974, Existence of methylated messenger RNA in mouse L cells, *Cell* **1**:37.

Pett, D.M., Vanaman, T.C., and Joklik, W.K., 1973, Studies on the amino and carboxyl terminal amino acid sequences of reovirus capsid polypeptides, *Virology* **52**:174.

Reeve, A.E., Shatkin, A.J., and Huang, R.C.C., 1982, Guanosine-5'-O-(3-thiotriphosphate) inhibits capping of reovirus mRNA, *J. Biol. Chem.* **257**:7018.

Rhim, J.S., Jordan, L.E., and Mayor, H.D., 1962, Cytochemical, fluorescent-antibody and electron microscopic studies on the growth of reovirus (ECHO 10) in tissue culture, *Virology* **17**:342.

Ro-Choi, T.S., Reddy, R., Choi, Y.C., Raj, N.B., and Henning, D., 1974, Primary sequence of U1 nuclear RNA and unusual feature of 5'-end structure of LMWN RNAs, *Fed. Proc. Fed. Am. Soc. Exp. Biol.* **33**:1548 (Abstract 1832).

Rottman, F., Shatkin, A.J., and Perry, R.P., 1974, Sequences containing methylated nucleotides at the 5' termini of messenger RNAs: Possible implications for processing, *Cell* **3**:197.

Salditt-Georgieff, M., Harpold, M., Chen-Kiang, S., and Darnell, J.E., Jr., 1980, The addition of 5' cap structures occurs early in hnRNA synthesis and prematurely terminated molecules are capped, *Cell* **19**:69.

Salzman, N.P., Shatkin, A.J., and Sebring, E.D., 1964, The synthesis of a DNA-like RNA in the cytoplasm of HeLa cells infected with vaccinia virus, *J. Mol. Biol.* **8**:405.

Samuel, C.E., and Joklik, W.K., 1976, Biosynthesis of reovirus-specified polypeptides: Initiation of reovirus messenger RNA translation *in vitro* and identification of methionyl-X initiation peptides, *Virology* **74**:403.

Samuel, C.E., Farris, D.A., and Levin, K.H., 1977, Biosynthesis of reovirus-specified polypeptides: System-dependent effect of reovirus mRNA methylation on translation *in vitro* catalyzed by ascites tumor and wheat embryo cell-free extracts, *Virology* **81**:476.

Shatkin, A.J., 1965a, Actinomycin and the differential synthesis of reovirus and L cell RNA, *Biochem. Biophys. Res. Commun.* **19**:506.

Shatkin, A.J., 1965b, Inactivity of purified reovirus RNA as a template for *E. coli* polymerases *in vitro*, *Proc. Natl. Acad. Sci. U.S.A.* **54**:1721.

Shatkin, A.J., 1974, Methylated messenger RNA synthesis *in vitro* by purified reovirus, *Proc. Natl. Acad. Sci. U.S.A.* **71**:3204.

Shatkin, A.J., 1976, Capping of eucaryotic mRNAs, *Cell* **9**:645.

Shatkin, A.J., and LaFiandra, A.J., 1972, Transcription by infectious subviral particles of reovirus, *J. Virol.* **10**:698.

Shatkin, A.J., and Sipe, J.D., 1968, RNA polymerase activity in purified reoviruses, *Proc. Natl. Acad. Sci. U.S.A.* **61**:1462.

Shatkin, A.J., Furuichi, Y., LaFiandra, A.J., and Yamakawa, M., 1983, Initiation of mRNA synthesis and 5'-terminal modification of reovirus transcripts, in: *Proc. Symp. Double-stranded RNA viruses* (D.H.L. Bishop and R.W. Compans, eds.), Elsevier (in press).

Shuman, S., and Hurwitz, J., 1981, Mechanism of mRNA capping by vaccinia virus guanylyltransferase: Characterization of an enzyme-guanylate intermediate, *Proc. Natl. Acad. Sci. U.S.A.* **78**:187.

Silverstein, S., and Dales, S., 1966, Role of lysosomes in the penetration of reovirus RNA into strain L cells, in: Abstracts of the Sixth Annual Meeting of the American Society of Cell Biology, Abstract 217, p. 107A.

Silverstein, S.C., and Dales, S., 1968, The penetration of reovirus RNA and initiation of its genetic function in L-strain fibroblasts, *J. Cell Biol.* **36**:197.

Silverstein, S.C., Astell, C., Levin, D.H., Schonberg, M., and Acs, G., 1972, The mechanism of reovirus uncoating and gene activation *in vivo*, *Virology* **47**:797.

Silverstein, S.C., Christman, J.K., and Acs, G., 1976, The reovirus replicative cycle, *Annu. Rev. Biochem.* **45**:375.

Skehel, J.J., and Joklik, W.K., 1969, Studies on the *in vitro* transcription of reovirus RNA catalyzed by reovirus cores, *Virology* **39**:822.

Skup, D., Zarbl, H., and Millward, S., 1981, Regulation of translation in L-cells infected with reovirus, *J. Mol. Biol.* **151**:35.

Sonenberg, N., Trachsel, H., Hecht, S., and Shatkin, A.J., 1980, Differential stimulation of capped mRNA translation *in vitro* by cap binding protein, *Nature (London)* **285**:331.

Sonenberg, N., Guertin, D., Cleveland, D., and Trachsel, H., 1981, Probing the function of the eucaryotic 5' cap structure by using monoclonal antibody directed against cap-binding proteins, *Cell* **27**:563.

Spandidos, D.A., and Graham, A.F., 1976, Non-permissive infection of L cells by an avian reovirus: Restricted transcription of the viral genome, *J. Virol.* **19**:977.

Spendlove, R.S., and Schaffer, F.L., 1965, Enzymatic enhancement of infectivity of reovirus, *J. Bacteriol.* **89**:597.

Steitz, J.A., 1969, Polypeptide chain initiation: Nucleotide sequences of the three ribosomal binding sites in bacteriophage R17 RNA *Nature (London)* **224**:957.

Venkatesan, S., and Moss, B., 1982, Eukaryotic mRNA capping enzyme-guanylate covalent intermediate, *Proc. Natl. Acad. Sci. U.S.A.* **79**:340.

Walden, W.E., Godefroy-Colburn, T., and Thach, R.E., 1981, The role of mRNA competition in regulating translation. I. Demonstration of competition *in vivo*, *J. Biol. Chem.* **256**:11739.

Ward, R., Banerjee, A.K., LaFiandra, A., and Shatkin, A.J., 1972, Reovirus-specific ribonucleic acid from polysomes of infected L cells, *J. Virol.* **9**:61.

Warrington, R.C., and Wratten, N., 1977, Differential action of L-histidinol in reovirus-infected and uninfected L-929 cells, *Virology* **81**:408.

Watanabe, Y., Millward, S., and Graham, A.F., 1968, Regulation of transcription of the reovirus genome, *J. Mol. Biol.* **36**:107.

Wang, D., Furuichi, Y., and Shatkin, A.J., 1982, Covalent guanylyl intermediate formed by Hela cell mRNA capping enzyme, *Mol. Cell. Biol.* **2**:993.

Wei, C.M., and Moss, B., 1974, Methylation of newly synthesized viral messenger RNA by an enzyme in vaccinia virus, *Proc. Natl. Acad. Sci. U.S.A.* **71**:3014.

Weiner, H.L., Ault, K.A., and Fields, B.N., 1980, Interaction of reovirus with cell surface receptors. I. Murine and human lymphocytes have a receptor for the hemagglutinin of reovirus type 3, *J. Immunol.* **124**:2143.

Wertheimer, A.M., Chen, S.-Y., Borchardt, R.T., and Furuichi, Y., 1980, S-Adenosylme-
    thionine and its analogs: Structural features correlated with synthesis and methylation
    of mRNAs of cytoplasmic polyhedrosis virus, J. Biol. Chem. 255:5924.
White, C.K., and Zweerink, H.J., 1976, Studies on the structure of reovirus cores: Selective
    removal of polypeptide λ2, Virology 70:171.
Yamakawa, M., Furuichi, Y., Nakashima, K., LaFiandra, A.J., and Shatkin, A.J., 1981, Excess
    synthesis of viral mRNA 5'-terminal oligonucleotides by reovirus transcriptase, J. Biol.
    Chem. 256:6507.
Yamakawa, M., Furuichi, Y., and Shatkin, A.J., 1982a, Reovirus transcriptase and capping
    enzymes are active in intact virions, Virology 118:157.
Yamakawa, M., Furuichi, Y., and Shatkin, A.J., 1982b, Priming of reovirus transcription by
    GppppG and formation of CpG(5')pppp(5')GpC, Proc. Natl. Acad. Sci. U.S.A. 79:6142.
Zweerink, H.J., and Joklik, W.K., 1970, Studies on the intracellular synthesis of reovirus-
    specified proteins, Virology 41:501.
Zweerink, H.J., McDowell, M.J., and Joklik, W.K., 1971, Essential and nonessential non-
    capsid reovirus proteins, Virology 45:716.

CHAPTER 4

# The Reovirus Multiplication Cycle

Helmut Zarbl and Stewart Millward

## I. INTRODUCTION

Human reoviruses have been recovered from a variety of mammals (Stanley, 1961a; Rosen, 1968). Serological studies, which reveal the presence of reovirus specific antibodies, have indicated a ubiquitous distribution of reovirus among mammals (Stanley, 1961a,b; Rosen, 1962). The American Type Culture Collection has suggested at least 14 mammalian cell lines as suitable for the growth of reoviruses, including cell lines derived from 12 different species and 7 different tissue types (see *Catalogue of Strains II*, 3rd ed., 1981). As a result of this extensive host range, several aspects of the reovirus multiplication cycle have been studied at one time or another in a variety of mammalian cells. However, most of the studies concerning the molecular biology of the reovirus multiplication cycle have been carried out on mouse L fibroblasts infected with the Dearing strain of reovirus type 3. The reason for this is that mouse L cells grow readily in suspension culture and that the Dearing strain multiplies particularly well in these cells. Therefore, the discussion of the events in the reovirus multiplication cycle will of necessity be based primarily on studies of reovirus multiplication in mouse L cells. Wherever appropriate, the discussion will include the multiplication of all three serotypes of human reovirus in a variety of cell lines.

HELMUT ZARBL and STEWART MILLWARD • Department of Biochemistry, McGill University, Montreal, Quebec, Canada H3G 1Y6.

## II. PRODUCTION OF INFECTIOUS VIRIONS

Reovirus multiplies to very high titers in a large number of primary and continuous mammalian cell lines, in both monolayer and suspension cultures. The number of infectious virions produced by each infected cell depends on a number of factors including the pH of the culture medium (Fields and Eagle, 1973), the type of viral particle or subviral particle used for infection (Borsa et al., 1981), and the number of serial passages of the virus at high multiplicity of infection, which results in the generation of defective virus particles (Nonoyama and Graham, 1970; Nonoyama et al., 1970; Schuerch et al., 1974) and temperature-sensitive mutants (Ahmed et al., 1980). The temperature at which the infection is carried out also affects the yield of infectious virus (Smith et al., 1969). Infection at 38°C results in extensive cytopathic effects, whereas cytopathic effects are much reduced at lower temperatures. As a result, infections are usually done at 31 or 34°C, resulting in yields severalfold higher than at 37°C. Under optimal conditions for infection, yields of purified virus may be as high as 30 mg/$10^9$ cells (Joklik, 1974). This represents a burst size of 300,000 particles per cell, or about 3000 plaque-forming units (PFU) per cell.

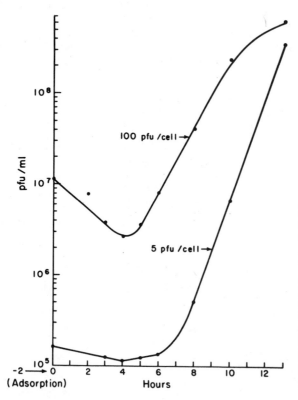

FIGURE 1. One-step growth curves of reovirus. Petri dishes, each containing 3 × $10^6$ L cells, were inoculated with either 5 or 100 PFU/cell. At the end of the adsorption period, the cells were overlaid with nutrient medium and incubated at 37°C. Duplicate plates were sampled at regular intervals and assayed for infectious progeny. Reprinted from Silverstein and Dales (1968), with permission.

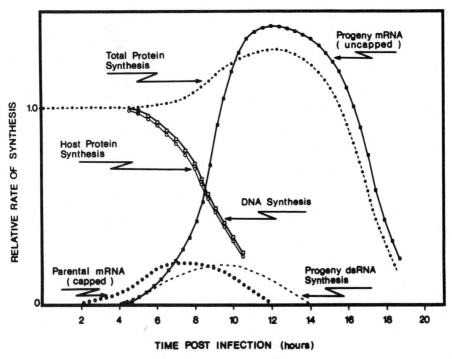

FIGURE 2. Macromolecular synthesis in L cells infected with reovirus. Adapted from Joklik (1974).

## III. EVENTS IN THE INFECTED CELL

The time course of reovirus replication in exponentially growing L cells is dependent on the temperature of incubation and the multiplicity of infection. The duration of the infection at 31°C is about twice that at 37°C. Correspondingly, the time of onset of each event in the infected cell is twice as long at 31°C. At higher multiplicities of infection, the time course of the infection is accelerated, but the final yield of virus is the same (Fig. 1). The time course of the infection is also accelerated if virions are treated with chymotrypsin in the presence of sodium ions to yield intermediate subviral particles. These particles are believed to enter the host cell by an alternate mechanism (Borsa *et al.*, 1981), which is discussed in Section III.B. The time course in the infected cell outlined here will be for cells infected with 5–10 PFU/cell and incubated at 37°C. The approximate time course of macromolecular syntheses in L cells infected with reovirus is schematically represented in Fig. 2. Adsorption of the virus to the cells is usually carried out at 4°C for 1–2 hr. The cells to which virus has adsorbed are then diluted and the temperature shifted to 37°C. This is defined as the start of the infection (zero hour). Penetration and uncoating of infecting virions followed by their escape from lysosomes is essentially complete in 2 hr. Shortly thereafter, capped parental transcripts begin to be synthesized, reaching a maximum at 6–8 hr and decreasing thereafter. Synthesis of viral polypeptides also begins with the onset of parental messenger RNA (mRNA) synthesis. During this time, the inhibition of both host DNA and host protein synthesis commences. Synthesis of progeny double-stranded (dsRNA) is detectable by 4–6 hr and reaches a maximum at 8–10 hr. Concomitant with the appearance of progeny dsRNA, there is an exponential increase in the synthesis of uncapped progeny transcripts and synthesis of viral proteins. Both attain a maximum rate at about 12 hr postinfection and decrease thereafter. Production of infectious progeny virions begins at about 6–8 hr and rises exponentially. Cell lysis and death commence at approximately 14 hr, and most cells are destroyed by 18 hr postinfection. Even though cell lysis does occur, the bulk of the progeny virions remain associated with cell debris. We will now proceed to analyze all the stages in reovirus multiplication in greater detail at the macromolecular level.

## A. Adsorption

Reovirus adsorbs efficiently to a wide range of cell types both in monolayer and in suspension culture. Attachment of an avian reovirus to nonpermissive mammalian cells has been reported (Spandidos and Graham, 1976b). Adsorption of reovirus to L cells occurs efficiently at temperatures between 4 and 37°C, with 60–80% of the viral inoculum being cell-associated by 1 hr (Silverstein and Dales, 1968). Inoculation of cells with reovirus is usually carried out at 4°C for two reasons. First, L cells can be maintained for 2 hr at concentrations in excess of $10^7$ cells/ml at 4°C without altering their viability, and adsorption of the virus is facilitated at these higher cell concentrations. Second, it has been shown that although reovirus adsorbs to cells at 4°C, penetration does not occur until the cells are returned to physiological temperatures (Silverstein and Dales, 1968). Infections initiated at 4°C are therefore highly synchronized, facilitating studies on the early events in the infectious cycle.

Joklik (1972) has calculated the rate of reovirus adsorption to L cells using labeled virus. He found that at a cell concentration of $10^7$ cells/ml in Puck's saline solution containing 0.02 M $Mg^{2+}$ and 1% fetal calf serum at 37°C, an adsorption rate constant of about $2.5 \times 10^{-9}$ $cm^3$/min per cell could be calculated (Joklik, 1972, 1974). Each cell can adsorb up to about 50,000 virus particles, which approaches close packing of the virus particles on the cell surface.

The interaction of reovirus particles with the surface of the host cell is a specific one. Adsorption occurs via specific receptors on the cell surface that bind to specific components of the virus (Lee et al., 1981b). These virus–cell interactions determine both the host range and the tissue tropism of reovirus. Specific virus–cell interactions have also been implicated in the agglutination of red blood cells by reovirus (Weiner *et al.*, 1978).

Little is known about the virus receptors found on the surface of host cells or erythrocytes. The virus receptors on different cell types, however, do display different degrees of specificity toward the different serotypes of reovirus. For example, Lee *et al.* (1981b) have shown that all three serotypes of human reovirus compete for the same receptors on the surface of L cells. On the other hand, Weiner *et al.* (1980a) have suggested that interactions of reovirus with different cell surface receptors are involved in the serotype-specific neurovirulence pattern observed in the central nervous system (CNS) of newborn mice. These investigators found that cell tropism is responsible for the selective infection of neuronal cells by reovirus type 3 (Margolis *et al.*, 1971) and the selective infection of ependymal cells by reovirus type 1 (Kilham and Margolis, 1969). Tardieu and Weiner (1982) have also shown that reovirus type 1 binds to the surface of isolated human and murine ciliated ependymal cells, while reovirus type 3 fails to show specific binding to these same cells.

Virus receptors on the surface of erythrocytes also display different degrees of specificity toward the serotypes of reovirus. Receptors on the surface of human erythrocytes react with all three reovirus serotypes. The resulting hemagglutination occurs with human eruthrocytes from all four ABO blood groups, although to differing degrees (Brubaker *et al.*, 1964). In contrast, receptors on the surface of bovine erythrocytes will interact exclusively with reovirus type 3 (Eggers *et al.*, 1962).

Information concerning the nature of the virus receptor on the surface of cells is limited. Studies on the erythrocyte virus receptors indicate that they are probably proteins, since pretreatment of erythrocytes with trypsin or chymotrypsin interferes with hemagglutination (Lerner and Miranda, 1968). Several studies have indicated that the virus receptors of human erythrocytes are not glycoproteins, since they are resistant to pretreatment with neuraminidase (Stanley, 1961a), and with a wide range of carbohydrases and sodium borohydride (Lerner and Miranda, 1968). In contrast, the virus receptors of bovine erythrocytes are probably glycoproteins, since they are inactivated by neuraminidase (Gomatos and Tamm, 1962). No information is available on the virus receptors of cells that support the replication of reovirus.

By comparison, more information is available, on the reovirus-encoded cell-attachment protein. Weiner *et al.* (1977, 1978, 1980a) used a genetic approach to determine the viral protein responsible for attachment of reovirus to the cell surface receptors. These investigators took advantage of the fact that reovirus type 1 and type 3 display different patterns of cell tropism in the CNS of newborn mice. As a result of cell tropism, reovirus type 3 infection results in the destruction of neuronal cells (Margolis *et al.*, 1971), while reovirus type 1 infection results in the destruction of ependymal cells (Kilham and Margolis, 1969). Weiner *et al.* (1977, 1980a) constructed recombinant clones containing nine genes from reovirus type 1 and the *S1* gene of type 3. In terms of cell tropism, this clone (1.HA3) behaved as type 3. Clone 1.HA3, like reovirus type 3, caused neuronal cell destruction in newborn mice inoculated intracerebrally (Weiner *et al.*, 1980a), but caused no damage to ependymal cells.

Clones that contained nine genes from reovirus type 3 and the *S1* gene from type 1 behaved like type 1. In terms of cell tropism, this clone (3.HA1), like reovirus type 1, caused the destruction of ependymal cells, but had no effect on neuronal cells. Furthermore, immunofluorescent studies showed no viral antigen in ependymal cells of mice infected with type 3 or clone 1.HA3 or neuronal cells of mice infected with type 1 or clone 3.HA1. It was concluded from these studies (Weiner *et al.*, 1980a) that reovirus virulence in the CNS relates to a specific interaction of ependymal or neuronal cell receptors with the viral protein $\sigma1$, which is encoded by the *S1* gene (McCrae and Joklik, 1978; Mustoe *et al.*, 1978b). These conclusions have been further substantiated by the work of Tardieu and Weiner (1982), using isolated human and murine ciliated ependymal cells. These workers have shown that reovirus type 1 and clone 3.HA1 bind specifically to ependymal cells, while reovirus type 3 and clone 1.HA3 fail to bind. Studies with similar recombinant clones showed that the ability of reovirus type 3, but not type 1, to agglutinate bovine erythrocytes was also a function of the $\sigma1$ protein (Weiner *et al.*, 1978).

Lee *et al.* (1981b) used a biochemical approach to confirm that the $\sigma1$ protein of reovirus is the cell-attachment protein. These investigators took advantage of the finding that infected cells contain significant amounts of unassembled reovirus-coded proteins (Huismans and Joklik, 1976). Lysates were prepared from [$^{35}$S]methionine-labeled cells infected with reovirus serotype 1, 2, or 3. Labeled lysates were subsequently added to monolayers of L cells, and the labeled proteins that bound to the cells were analyzed. In all cases, $\sigma1$ was the only protein bound to the surface of L cells. Furthermore, it was shown that the $\sigma1$ protein competes with reovirus particles for cell surface receptors. Lee *et al.* (1981b) also showed that monoclonal antibodies directed against $\sigma1$ (Lee *et al.*, 1981a) prevent adsorption of reovirus to L cells. These experiments indicated clearly that $\sigma1$ is the reovirus cell-attachment protein.

The cell-attachment protein $\sigma1$ is present on the surface of reovirus particles in very small quantities, comprising less than 1% of the total viral protein. Each virion contains approximately 24 molecules of $\sigma1$, which are probably arranged icosahedrally (Lee *et al.*, 1981b). It is therefore likely that the binding of reovirions to cell surface receptors involves no more than 2 molecules of $\sigma1$.

Several lines of evidence indicate that $\sigma1$ is a glycoprotein and that the carbohydrate moiety is involved in adsorption of reovirus to receptors on the cell surface. Gelb and Lerner (1965) found that the simple sugar *N*-acetyl-D-glucosamine specifically prevented agglutination of erythrocytes by reovirus, while over 20 other sugars tested had no effect. Lerner *et al.* (1966b) found that smaller aldoses and aldehydes also inhibited hemagglutination, while their polyhydroxy counterparts did not. Sugars, aldoses, and aldehydes that were active in inhibition of hemagglutination did not attach to purified preparations of virus, but did attach to the surface of erythrocytes. These data suggested that the virus–erythrocyte union may involve a free carbonyl group on oligosaccharide side chains of the virus.

Oxidation of reovirus with periodate (Tillotson and Lerner, 1966) and treatment with sodium borohydride (Lerner and Miranda, 1968) have also been shown to inhibit hemagglutination, infectivity, and production of homotypic antibodies. The latter has also been shown to be a function of the *S1* gene product, σ1 (Weiner and Fields, 1977). Treatment of reovirus with a series of carbohydrases was also found to have an inhibitory effect on hemagglutination (Gomatos and Tamm, 1962; Lerner *et al.*, 1966a; Lerner and Miranda, 1968) and a minor effect on infectivity (Nakashima *et al.*, 1979).

All these studies indicated that the σ1 protein of the reovirus outer capsid is glycosylated. However, all attempts to label reovirions with glucosamine or fucose have been unsuccessful (Hand and Tamm, 1971). More recently, Krystal *et al.* (1976) have detected the presence of glycosylated proteins in reovirions. However, these investigators did not find any evidence for glycosylation of the σ1 protein.

Treatment of reovirus with *p*-chloromercurobenzoate (Gomatos and Tamm, 1962) or *p*-hydroxymercurobenzoate (Lerner *et al.*, 1963) also inhibits hemagglutination, the inhibition being reversed by reduced glutathione. This suggested that free sulfhydryl groups are required for the interaction of σ1 with receptors on the surface of erythrocytes. A variety of nonspecific serum and phospholipid inhibitors of hemagglutination and adsorption have been reported (Schmidt *et al.*, 1964a,b).

## B. Penetration and Uncoating

The penetration of reovirions into the host cell has been studied primarily by electron microscopy (Silverstein and Dales, 1968; Dales *et al.*, 1965; Dales, 1965a; Anderson and Doane, 1966). Virions bound to cell surface receptors are taken into the cell by an active yet nonspecific process of phagocytosis, termed viropexis. Both single particles and clusters are seen in cytoplasmic inlets and in phagocytic vacuoles soon after infection. Loh *et al.* (1977) have shown that the penetration of reovirus into HeLa cells is inhibited by cationic polymers such as diethylaminoethyl (DEAE)–dextran, but the mechanism of this inhibition is not understood.

Phagocytic vacuoles containing virions then move to the interior of the cell and fuse with lysosomes. Viropexis and fusion of vacuoles with lysosomes occur rapidly (Silverstein and Dales, 1968), with 80% of the

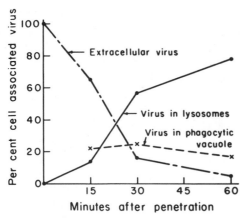

FIGURE 3. Distribution of reovirus in L cells by particle counts using the electron microscope. Reovirus (50 PFU/cell) was adsorbed at 4°C, and the virus–cell complexes were subsequently diluted into warm medium and incubated. At the designated intervals, aliquots were sampled for electron microscopy. At each time point, 100 cell-associated virus particles were counted, and their cellular location was noted. The results are plotted as percentage of virus in each cell compartment. Reprinted from Silverstein and Dales (1968), with permission.

infecting virions being within lysosomes by 60 min postinfection (Fig. 3). Once within the lysosomes, the particles are uncoated by hydrolytic enzymes (see below).

Recently, Borsa et al. (1979) have described a second mode of entry of reovirus particles into L cells. These investigators have shown that intermediate subviral particles (ISVPs) enter the host cell primarily by direct penetration of the plasma membrane. This was suggested on the basis of electron-microscopic observations and was confirmed by studies on the release of preloaded radioactive $^{51}$Cr from host cells following infection with ISVPs. These results indicated that ISVPs enter the host cell directly, without the involvement of phagocytic vacuoles.

These ISVPs can be produced in vitro by digestion of virions with chymotrypsin in the presence of sodium ions (Borsa et al., 1973a). Under these conditions, virions become partially uncoated, their density increasing from 1.36 to 1.38 g/cm$^3$ and their diameter decreasing from 73 to 64 nm (Borsa et al., 1973a). ISVPs differ from intact virions in that they have lost essentially, all the outer-capsid proteins, σ3, μ1C, and σ1. These particles have reduced amounts of μ1 and contain a significant amount of an additional protein that may represent a cleavage product of μ1C. These particles are refractory to further proteolytic digestion and retain an inactive transcriptase. However, on exposure to K$^+$ ions, these particles undergo further uncoating, to yield particles morphologically similar to cores and having an active transcriptase. This second stage of uncoating occurs both in vivo (Borsa et al., 1981) and in vitro (Borsa et al., 1973a, 1974a, 1974b) and is independent of further proteolysis. Therefore, when cells are infected with ISVPs, the particles penetrate the cell membrane and, on exposure to the K$^+$ environment of the cytoplasm, undergo the second stage of uncoating to give rise to particles that have an active RNA transcriptase. Such a mechanism allows ISVPs to become infectious without the involvement of lysosomal proteases. Such a bypass might explain the decreased eclipse phase (Borsa et al., 1981) and increased infectivity (Borsa et al., 1973b) of ISVPs relative to those observed for infection with intact virus. The production of partially uncoated particles that retain infectivity has also been reported by Shatkin and Lafiandra (1972) and by Joklik (1972). Although the latter investigators present results that differ somewhat from those described above, most of the discrepancies can be accounted for by different isolation conditions (Borsa et al., 1974b).

The production of particles similar to ISVPs may also explain the numerous reports to be found in the literature that suggest enhancement of reovirus infectivity by a range of proteolytic enzymes (Spendlove and Schaffer, 1965; Wallis et al., 1966; Spendlove et al., 1970; Cox and Clinkscales, 1976) or heating to 50°C in 2 M MgCl$_2$ (Wallis et al., 1964). However, since many of these investigators used unpurified virus preparations, one cannot rule out the possibility that the observed increases in infectivity were the result of virions being released from cell debris by proteolysis.

Particles similar to ISVPs may also be generated in vivo when infected cells are permitted to reach an advanced stage of cytopathic effect. This results in the exposure of released virus to cellular proteolytic enzymes in the Na$^+$ environment of the extracellular fluid. Such conditions are analogous to those used for the formation of ISVPs in vitro. The generation of ISVPs and the entry of such particles into host cells by direct penetration of the plasma membrane may be an important mechanism of reovirus infection in vivo. It is not known whether the interaction of ISVPs with target cells involves interaction of the σ1 protein with specific receptors on the cell surface. However, ISVPs generated in vitro do not contain detectable levels of σ1 (Borsa et al., 1973a). One must therefore assume that the interaction of ISVPs with the plasma membrane is nonspecific.

ISVPs therefore enter the host cell either by viropexis or by direct penetration of the cell membrane. The entry of intact virions, however, is restricted to a mechanism of viropexis and subsequent fusion of the virion-containing phagocytic vacuoles with lysosomes. On exposure to lysosomal hydrolases, the genome of reovirions is only partially uncoated. Very little if any naked double-stranded RNA (dsRNA) is found in infected cells (Silverstein and Dales, 1968; Chang and Zweerink, 1971). Lysosomal enzymes digested 40–50% of the viral proteins, but had no effect on the viral RNA (Silverstein et al., 1970; Chang and Zweerink, 1971; Silverstein et al., 1972). The genome RNA remained intact within parental subviral particles (SVPs), which are morphologically distinct from intact virions and chymotrypsin cores produced in vitro (Fig. 4). The inner capsid of parental SVPs remains intact and the genome RNA remains completely resistant to ribonucleases (Silverstein and Dales, 1968; Chang and Zweerink, 1971). Parental SVPs have a diameter of approximately 57 nm, which is smaller than that of virions (75 nm), and capsomers are clearly discernible on their surface. The density of parental SVPs in CsCl is 0.02 g/cm$^3$ greater than that of virions.

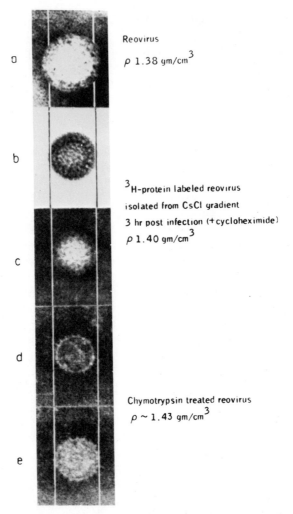

FIGURE 4. Electron micrographs of reovirus (a) and SVPs that were produced from reovirus *in vivo* (b,c) and *in vitro* (d,e). (b,c) Particles were obtained from the CsCl gradient fraction of density 1.40 g/cm$^3$; (d,e) particles were obtained by enzymatic digestion of reovirus with chymotrypsin as described by Skehel and Joklik (1969) and Banerjee and Shatkin (1970). × 192,000. Reprinted from Silverstein *et al.* (1972), with permission.

Uncoating of virions to parental SVPs begins 20–30 min after viral penetration of the host cell and is complete by 2–3 hr postinfection (Silverstein and Dales, 1968; Chang and Zweerink, 1971). The uncoating process is independent of the multiplicity of infection between 5 and 25 PFU/cell (Silverstein and Dales, 1968). Uncoating occurred at the same rate and to the same degree in the presence or absence of cycloheximide, the radiolabel lost from parental coat proteins becoming acid-soluble (Fig. 5). Uncoating also occurred in the presence of streptovitacin (Dales, 1965b). These results indicated that protein synthesis is not required for viral uncoating. The uncoating process is dependent on temperature. Hydrolysis of the outer-capsid proteins was not detected at 4°C, and the rate of hydrolysis was 3 times higher at 37 than at 20°C (Silverstein and Dales, 1968).

   The process of uncoating results in loss of almost all the proteins of the reovirus outer capsid. Analysis of virions uncoated to parental SVPs *in vivo* indicated a loss of the outer-capsid proteins σ3, μ1C, and σ1 (Chang and Zweerink, 1971; Silverstein *et al.*, 1972). The major capsid protein μ1C, however, was not completely removed from parental SVPs. Rather, it was cleaved to a polypeptide with an approximate molecular weight of 65,000, called δ, but the actual amount of the δ protein present in parental SVPs varied among preparations (Silverstein *et al.*, 1972). The detailed mechanism of uncoating *in vivo* has not yet been established. Chang and Zweerink (1971) have reported that the cleavage of σ3 and μ1C is coordinate, whereas Silverstein *et al.* (1972) have presented evidence for a sequential mechanism whereby σ3 is removed prior to cleavage of μ1C. It is of interest to note that uncoating of reovirions with chymotrypsin *in vitro* proceeds via a sequential mechanism (Joklik, 1972).

   Parental SVPs thus contain the reovirus core proteins λ1, λ2, μ1, and σ2 (Chang and Zweerink, 1971; Silverstein *et al.*, 1972) and must also contain λ3, since this protein has been shown to be the viral transcriptase (Dryna and Fields, 1982). In addition to the core proteins, parental SVPs also contain a variable amount of the μ1C cleavage product, δ. *In vivo*, no further proteolysis is observed, even after a prolonged intracellular residence in either the presence or the absence of cycloheximide or actinomycin D or both.

   Borsa *et al.* (1973a, 1974a,b) have shown that the uncoating of virions to cores *in vitro* proceeds via two distinct steps. The first step is dependent

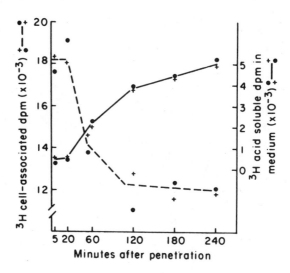

FIGURE 5. Hydrolysis of reovirus coat proteins. [³H]Protein-labeled reovirus–L cell complexes were formed in suspension at 4°C at a multiplicity of 100 PFU/cell. After adsorption, the cells were washed and placed in warm medium at a density of 8 × 10⁵ cells/ml. Half the cells received 20 μg/ml cycloheximide. At appropriate times, 3.2 × 10⁴ cells from both the cycloheximide-treated ( + ) and control cultures (●) were harvested by centrifugation. The radioactivity in the acid-insoluble precipitates (cells) and acid-soluble supernatants (medium) was assayed. Reprinted from Silverstein *et al.* (1970), with permission.

on proteolytis and results in the formation of ISVPs. The second step is independent of exogenous proteases and is actually inhibited by chymotrypsin. The conversion of ISVPs to cores is facilitated by $K^+$ ions or particle–particle interactions and is believed to be mediated via an endogenous reovirus protease. However, no direct evidence for the existence of such a protease is available.

Borsa et al. (1981) have suggested that uncoating in vivo may proceed via a pathway that is mechanistically identical to that elucidated for uncoating in vitro. These investigators suggest that parental SVPs isolated previously (Chang and Zweerink, 1971; Silverstein et al., 1972) represent the particles that have undergone the first stage of uncoating and, like ISVPs generated in vitro, have an inactive transcriptase. Borsa et al. (1981) maintain that isolation techniques used by previous investigators facilitated the second stage of uncoating and resulted in activation of the transcriptase. These investigators then go on to demonstrate the production in vivo of particles that resemble reovirus cores produced in vitro with chymotrypsin. These particles are produced whether cells are infected with ISVPs or with intact virions. These cores generated in vivo have a density of 1.44 $g/cm^3$ and a reduced diameter relative to parental SVPs and have lost the δ protein associated with parental SVPs or the δ2 protein associated with ISVPs. The authors concluded that there is a second step in the uncoating process in vivo that gives rise to particles with an activated transcriptase and that this second stage is similar to that observed in vitro (Borsa et al., 1973a, 1974a,b).

It must be noted, however, that the extraction of particles resembling cores from infected cells required the use of proteinase K digestion (Borsa et al., 1981). These authors have suggested that proteolysis releases the corelike particles from some intracellular matrix that cannot be disrupted by alternative extraction procedures. The use of proteolytis during the extraction, however, does raise the possibility that the corelike particles apparently formed in the infected cell could be artifactual. The authors attempted to control for such artifacts by adding purified ISVPs or virus prior to the enzymatic extraction and showed that none of these particles were shifted to an increased density. Such controls, however, do not preclude the possibility that the ISVPs or virions used to infect the cells are somehow changed during their residence within the cell and hence are different from those added prior to extraction. That is to say, the particles used in the infection may be modified in some way that activates the transcriptase, but also renders them sensitive to further proteolysis. In the absence of adequate controls, the use of protease digestion for extraction of such cores becomes suspect. Furthermore, even after allowance of 6 hr for uncoating in vivo, only a fraction of the input virus was extracted in the form of particles resembling cores. It is thus unlikely that such particles are generated from parental SVPs during the course of an infection. The uncoating of reovirions in vivo probably does not

proceed beyond the production of particles resembling parental SVPs previously isolated from infected cells (Chang and Zweerink, 1971; Silverstein *et al.*, 1972). Chang and Zweerink (1971) reported that at 2 hr postinfection, parental SVPs retained their full complement of A-rich
oligonucleotides. More recently, however, Carter and Lin (1979) reported
that concomitant with uncoating, the 5'-G-terminated oligonucleotides
undergo guanylylation and methylation, although it is not known
whether capping of the 5'-G-terminated oligonucleotides is effected by
viral or host enzymes. These capped oligonucleotides are subsequently
released from uncoated particles and are conserved in the host cell for at
least 5 hr after infection. It is also unclear whether capping and release
of oligonucleotides occurs within lysosomes or subsequent to escape of
parental SVPs to the cytoplasm. The function of these capped oligonucleotides has not been determined, although they may possibly play a
role in the inhibition of cap-dependent translation at late times postinfection (Skup and Millward, 1980a; Skup *et al.*, 1981).

The ability of reoviruses to use cellular lysosomes for uncoating is
unique among the animal viruses. Although a variety of animal viruses
penetrate their host cells by way of phagocytic vacuoles, these viruses or
their nucleoproteins invariably escape into the cytoplasm prior to fusion
of the phagocytic vacuoles with lysosomes (for a review, see Dales, 1973).
Most animal viruses are in fact destroyed by lysosomal hydrolases (Silverstein, 1975), and hence the lysosomal pathway in most cases defends
the cell against viral infection. Reoviruses, on the other hand, have
adapted a mechanism by which this defense mechanism is used against
the host cell.

The mechanism by which reoviruses initiate their genetic functions
is also unique. The genomic dsRNA has been shown to be conserved in
parental SVPs even after prolonged residence within the cell (Chang and
Zweerink, 1971; Silverstein *et al.*, 1970, 1972). Furthermore, dsRNA is
known to be inactive as a messenger. Therefore, reovirions contain a
dsRNA-dependent RNA polymerase that transcribes all ten dsRNA genome segments into messenger RNA (mRNA) (see Section D.5). The intracellular locus at which infecting virions synthesize mRNA, however,
remains equivocal. It has been clearly demonstrated that the uncoating
of virions to parental SVPs results in activation of the viral transcriptase
(Chang and Zweerink, 1971; Silverstein *et al.*, 1972). Cell-fractionation
and electron-microscopic experiments, on the other hand, suggested that
parental SVPs remain within the lysosomes throughout the infectious
cycle (Silverstein and Dales, 1968; Dales *et al.*, 1965). These results, coupled with the fact that parental SVPs have a diameter of 57 nm, suggested
the possibility that parental SVPs did not escape from lysosomes and
therefore synthesized mRNA within the lysosomes. Such a model, however, presents numerous problems. There is no known mechanism to
account for the presence within the vacuolar system of the nucleoside

triphosphates required for transcription. In fact, lysosomes contain acid phosphatases that degrade nucleoside phosphates. In addition, lysosomes contain ribonucleases that are capable of degrading viral mRNA. Therefore, if transcription were to occur within the lysosomes, one would have to evoke not only mechanisms for transport of nucleoside triphosphates into the lysosomes, but also mechanisms for transport of nascent transcripts out of parental SVPs and out of lysosomes (Silverstein *et al.*, 1976). The complexity of such models undermines the possibility that parental SVPs do not escape from lysosomes prior to commencing transcription.

Recent findings by Borsa *et al.* (1979) have provided a plausible model for the escape of parental SVPs from lysosomes. These investigators have found that ISVPs generated *in vitro* are capable of entering the cell by direct penetration of the cell membrane. These ISVPs closely resemble *in vivo* uncoated parental SVPs in both physical properties and protein composition (Borsa *et al.*, 1973a; Chang and Zweerink, 1971). It is therefore likely that parental SVPs are capable of interacting with and penetrating the lysosomal membrane by an analogous mechanism. The escape of parental SVPs from lysosomes, however, appears to be a relatively inefficient process, since some particles can be seen within lysosomes throughout the infectious cycle (Silverstein and Dales, 1968; Borsa *et al.*, 1979). Slow release of parental SVPs from lysosomes may also explain a curious lag in the onset of mRNA synthesis after infection. Although uncoating begins 20–30 min postinfection, no viral mRNA is detectable in the infected cell until 2 hr postinfection (Joklik, 1974).

Further evidence for the escape of parental SVPs from lysosomes after uncoating comes from the observation that at late times during infection, parental SVPs are reassembled into particles resembling virions (Chang and Zweerink, 1971; Silverstein *et al.*, 1972). These particles arise from the addition of *de novo*–synthesized σ3 proteins to the parental SVPs, and they differ from virions in that they do not contain any μ1C; rather, they contain the δ protein. The density of these reassembled parental SVPs is also slightly higher than that of virions. It is difficult to envisage a mechanism by which reassembly of uncoated particles could occur within lysosomes, where uncoating also occurs. Taken together, these observations strongly suggest that the escape of uncoated virions from lysosomes is an essential step in the infectious cycle. Although the exact site to which parental SVPs migrate and where they initiate mRNA synthesis has not been elucidated, the most likely site is the perinuclear region, where the assembly of progeny virions has been shown to occur (see Section C).

Several investigators have studied the effects of interferon on the early events during reovirus infection (Gupta *et al.*, 1974; Wiebe and Joklik, 1975; Galster and Lengyel, 1976). The results of these studies have indicated that early events in the infectious cycle, including adsorption, penetration, and uncoating of parental virions to SVPs, were unaltered in cells pretreated with interferon.

## C. Transcription of the Parental Genome

### 1. Activation of the Virion-Associated Transcriptase

The mechanism by which reoviruses initiate their genetic functions presented a problem to early investigators, since it was known that double-stranded RNA (dsRNA) could not function as messenger RNA (mRNA). The first viral mRNAs in the infected cell must therefore result from transcription of the parental genome or by separation of the two strands of the dsRNA genome. However, attempts to demonstrate strand separation in the infected cell proved unsuccessful (Gomatos, 1967; Silverstein and Dales, 1968; Shatkin and Sipe, 1968a), and the cellular polymerases tested could not utilize purified reovirus dsRNA as a template (Shatkin, 1965b). Furthermore, the synthesis of viral RNA *in vivo* was not inhibited by treatment of cells with actinomycin D at a concentration of 0.5 μg/ml (Shatkin, 1965a; Kudo and Graham, 1965; Loh and Soergel, 1966). At this concentration, actinomycin D inhibits L-cell RNA synthesis by 90%. Thus, it was unlikely that host enzymes were involved in the transcription of the reovirus genome. Watanabe *et al.* (1968b) subsequently reported the transcription of a limited number of the parental genome segments when infected cells were pretreated with cycloheximide. These authors suggested that the observed transcription was carried out by a preexisting polymerase that may be an integral part of the virion. Shortly thereafter, a dsRNA → single-stranded RNA (ssRNA) polymerase or transcriptase was found to be associated with purified reovirions. The transcriptase activity was masked or latent in intact virions, but could be activated either by heat shocking the virions (Borsa and Graham, 1968) or by treatment of virions with chymotrypsin (Shatkin and Sipe, 1968b). The latter treatment converts reovirions to reovirus core particles by virtue of the fact that proteolysis removes the outer capsid layer of proteins. Shortly thereafter, reovirus cores were shown to contain a nucleoside triphosphate phosphohydrolase activity (Kapuler *et al.*, 1970; Borsa *et al.*, 1970) and a guanosine-triphosphate-dependent pyrophosphate exchange activity (Wachsman *et al.*, 1970). Several years later, reovirus core particles were shown to contain at least one or possibly two RNA methylases (Shatkin, 1974; Faust and Millward, 1974). Purified reovirus cores were subsequently shown to contain a guanylyl transferase activity (Furuichi *et al.*, 1976; Furuichi and Shatkin, 1977). Together, these enzymes account for the ability of cores to transcribe all ten genome segments into ten mRNA species having the capped 5'-terminal structure $m^7GpppG(m)pC \ldots$ (Furuichi *et al.*, 1975; Faust *et al.*, 1975).

The enzymes involved in the synthesis of reovirus mRNA are not expressed in intact virus particles (Borsa and Graham, 1968; Shatkin and Sipe, 1968b), with the possible exception of the RNA methylases and the guanylyl transferase (Carter, 1977; Yamakawa *et al.*, 1982). The viral transcriptase, however, is completely inactive in mRNA synthesis in intact virions, although short abortive transcripts may be generated within the virion (Yamakawa *et al.*, 1981, 1982). The activity of the transcriptase of reovirions is regulated by proteins in the outer viral capsid. During uncoating *in vitro*, Joklik (1972) noted that removal of σ3 was not sufficient to activate the transcriptase and that the ability to transcribe mRNA appeared to coincide with cleavage or removal of the μ1C protein.

Shatkin and LaFiandra (1972) have reported the production of infectious subvital particles (SVPs) *in vitro* by digesting virions with chymotrypsin in the presence of sodium ions. Such particles were found to have an active transcriptase. These infectious SVPs have lost their σ3 proteins, but retain a polypeptide with a molecular weight of 65,000 that is derived from the cleavage of μ1C. Borsa *et al.* (1973a) have generated similar particles *in vitro*, but these particles had an inactive transcriptase. These results suggested that one or more of the outer-capsid proteins is involved in the regulation of the viral transcriptase *in vitro*.

Uncoating of virions to parental SVPs *in vivo* also results in the activation of the viral transcriptase (Chang and Zweerink, 1971; Silverstein *et al.*, 1972). Such parental SVPs have lost their σ3 protein, and their μ1C protein has been cleaved to a protein of molecular weight 65,000 protein called δ. Activation of the parental SVP transcriptase, however, was found to be reversible (Astell *et al.*, 1972). Incubation of parental SVPs with soluble proteins from infected cells resulted in formation of reassembled particles resembling virions. These particles arose by the reassociation of the σ3 protein with parental SVPs, possibly by interaction with the δ protein. This reassembly resulted in inactivation of the transcriptase activity. It was concluded from these observations that σ3 regulates the activity of the viral transcriptase.

More recently, Drayna and Fields (1982) studied the regulation of the viral transcriptase using a genetic approach. These authors took advantage of the observation that the different serotypes of reovirus require different conditions to activate their transcriptases with chymotrypsin. Using intertypic recombinants, they were able to demonstrate that the requirement for different conditions in the activation of the transcriptase was a function of the *M2* gene. This dsRNA segment has been shown to code for the μ1 protein (McCrae and Joklik, 1978; Mustoe *et al.*, 1978b), which in turn is cleaved to the μ1C protein (Zweerink and Joklik, 1970). These findings clearly demonstrated that μ1 or μ1C exerted a regulatory effect on the viral transcriptase, while no regulatory function was found for the σ3 protein. These findings agreed with previous results indicating that removal of σ3 was not sufficient for activation of the transcriptase and that cleavage of μ1C was required (Joklik, 1972). Similarly, all other SVPs with an active transcriptase (Chang and Zweerink, 1971; Silverstein *et al.*, 1972; Shatkin and LaFiandra, 1972) had at least a portion of their μ1C proteins cleaved to a polypeptide with a molecular weight of 60,000–65,000. The inactivation of the transcriptase in parental SVPs by reassociation with σ3 proteins (Astell *et al.*, 1972) must therefore be indirect. Inactivation may result from a conformational change in the δ protein on binding of σ3, which in turn inactivates the transcriptase. Alternatively, the spikes through which nascent transcripts are believed to be extruded (Gillies *et al.*, 1971) may be closed on binding of σ3, thus preventing mRNA synthesis. This is consistent with the finding that removal of the spikes with alkali (White and Zweerink, 1976) or perturbation with pyridoxal phosphate (Morgan and Kingsbury, 1980) also results in loss of transcriptase activity.

The regulatory function of the μ1 or μ1C proteins of reovirions, or of both, may be related to the numerous modifications seen in these proteins. These modifications include glycosylation (Krystal *et al.*, 1976) and phosphorylation (Krystal *et al.*, 1975), as well as polyadenylylation and ADP-ribosylation (Carter, 1979; Carter *et al.*, 1980).

Recently, Drayna and Fields (1982) have shown that the reovirus transcriptase activity resides in the λ3 protein. These investigators showed that the pH optima of the viral transcriptase varied among the three serotypes. Using intertypic recombinants and the characteristic pH optima of the different serotypes, it was shown that the transcriptase activity was associated with the *L1* gene, which encodes the λ3 protein (McCrae and Joklik, 1978). The exact location of the λ3 proteins in the viral core has not been determined, but they may be located on the inner surface of the cores, they may be associated with the genome, or they may occur free within the core (Joklik, 1981). Protein λ3 is a minor reovirus core protein present in 12 copies or less per virion, suggesting that each dsRNA genome segment is transcribed by one transcriptase molecule. Although the mechanism of transcription has not been determined, disruption of cores results in a complete loss of activity (White and Zweerink, 1976). It is therefore thought that enzymes are fixed components of the core shell, presumably at the base of the spikes, and that the dsRNA templates move past the transcriptase. Nascent transcripts may then be extruded through the 12 hollow cyclindrical spikes on the surface of the cores, where post transcriptional modifications can then occur (see Section C.2).

The transcriptase reaction catalyzed by the reovirus transcriptase is conservative. Neither strand of parental genome is found among the products of transcription (Skehel and Joklik, 1969; D.H. Levin *et al.*, 1970). The mechanism of nascent mRNA synthesis is therefore analogous to that of DNA-dependent RNA polymerases. The reaction catalyzed by the reovirus transcriptase is the asymmetric transcription of all ten dsRNA segments into ssRNA molecules; i.e., only one strand of each genome segment is transcribed (Skehel and Joklik, 1969; Banerjee and Shatkin, 1970; Joklik *et al.*, 1970; Shatkin and Banerjee, 1970). The transcripts are full-length copies of their dsRNA templates. This is indicated by the finding that the sedimentation coefficients of the transcripts are the same

TABLE I. Approximate Relative
Transcription Frequencies of Reovirus
Messenger RNA Species *in Vitro* and *in
Vivo*[a]

| Messenger RNA species | In vitro | In vivo |
|:---:|:---:|:---:|
| s4 | 1.0 | 1.0 |
| s3 | 1.0 | 1.0 |
| s2 | 0.88 | 0.5 |
| s1 | 0.88 | 0.5 |
| m3 | 0.5 | 0.5 |
| m2 | 0.5 | 0.3 |
| m1 | 0.5 | 0.15 |
| l3 | 0.275 | 0.05 |
| l2 | 0.275 | 0.05 |
| l1 | 0.275 | 0.05 |

[a] Reprinted from Zweerink and Joklik (1970), with permission.

as those of denatured dsRNA segments (Skehel and Joklik, 1969; Banerjee and Shatkin, 1970). Furthermore, it has been shown that dsRNA hybrids formed between transcripts and the corresponding minus strand of genome RNA possess exactly the same electrophoretic mobilities as genome RNA segments, indicating the absence of even short, unpaired, single-stranded tails (Skehel and Joklik, 1969; Ito and Joklik, 1972a,b). Recently, Hastings and Millward (1981) showed that the plus strands of RNA genome segments contained heterogeneous sequences at both the 5' termini and 3' termini that correspond to sequences at the 5' and 3' termini of *in vitro* transcripts. These results further suggested that the plus strand of genome dsRNA and *in vitro* transcripts are essentially identical sets of molecules.

Under optimal conditions, all ten gene segments are transcribed at the same rate (Skehel and Joklik, 1969). This means that at any given time, equal masses of all ten transcripts are formed. As a result, the number of molecules of any given transcript formed will be inversely proportional to its molecular weight (see Table I). The fact that equal numbers of each transcript are not made argues against the possibility that the ten genes are transcribed as a linked complex. Each segment must be transcribed independently, presumably by 1 molecule of λ3, of which there are approximately 10 per virion (Drayna and Fields, 1982). Under suboptimal conditions of *in vitro* transcription, however, marked changes are seen in the relative rates of transcription. For example, if the $Mg^{2+}$ concentration is lowered 10-fold, the ratio of the relative rate of synthesis of small ($S$) and large ($L$) gene transcripts is raised from 1 up to 5. Similarly, when the concentration of one of the nucleoside triphosphates is lowered to 2 μM, only the $S4$ gene segment is transcribed (Nichols *et al.*, 1972b).

The relative transcription frequencies observed *in vitro* are therefore dependent on the concentration of the triphosphate precursors and the concentration of $Mg^{2+}$.

The rate of nucleotide addition *in vitro* under optimum conditions has been estimated by determining the time required to form full-length transcripts. Skehel and Joklik (1969) have estimated a rate of 7–8 nucleotide residues/sec, while Banerjee and Shatkin (1970) estimated the rate to be 60 nucleotide residues/sec. Using electron microscopy and measuring the contour length of nascent transcripts extruded from reaction cores, Bartlett *et al.* (1974) have estimated the rate of synthesis to vary from 2 to 50 nucleotide residues/sec, the average rate being 18/sec. These estimated rates of transcription, however, appear to be overestimates. Shatkin and Banerjee (1970) found that the first *l* transcripts were completed within 1 min. The total mass of transcripts synthesized in 1 hr, however, corresponded to a rate of nucleotide addition that was 30 to 40-fold slower (Joklik, 1974). One way to explain this discrepancy would be to assume that initiation of transcription is a rapid process in relation to elongation of nascent transcripts. Although such a model presents conceptual difficulties, recent studies have suggested that elongation may indeed be the rate-limiting step in transcription (Zarbl *et al.*, 1980a; Yamakawa *et al.*, 1981). It has been found that even under optimum conditions for transcription, a molar excess of both capped and uncapped oligonucleotides was produced by the reovirus transcriptase in core particles. Furthermore, these oligonucleotides arose as a result of abortive transcription. These results are consistent with a model of transcription involving (1) synthesis of abortive transcripts by reiteration of RNA polymerase at "promoter" sites independently of transcription, (2) capping of some of the abortive transcripts, and (3) chain elongation and completion of a fraction of the capped oligonucleotides that escape the reiterative phase (Yamakawa *et al.*, 1981). Such a model is consistent with the hypothesis that elongation of transcription may be rate-limiting relative to initiation, possibly due to instability of the initiation complex (Zarbl *et al.*, 1980a).

*S*-Adenosyl-L-methionine (AdoMet) has been found to stimulate the transcriptase of orbiviruses 2-fold (Van Dijk and Huisman, 1980) and to stimulate the transcriptase of cytoplasmic polyhedrosis viruses by up to 70-fold (Furuichi, 1981). The mechanism of this stimulation is believed to be the result of a reduction in the $K_m$ of the initiating nucleotide at the promotor site. The transcriptase of reoviruses, however, is not stimulated by AdoMet, and its presence is not required for transcriptase activity.

## 2. Activation of the Reovirus Messenger RNA Capping Enzymes

Concomitant with activation of the viral transcriptase, one also observed activation of the enzymes involved in synthesis of 5'-terminal cap structures. The latter enzymes have been shown to function independently of the viral transcriptase (Furuichi et al., 1976; Furuichi and Shatkin, 1977; Nakashima et al., 1979; Yamakawa et al., 1982), although capping of nascent transcripts probably occurs shortly after initiation of transcription (Furuichi et al., 1976; Zarbl et al., 1980a; Yamakawa et al., 1981).

The mechanism of formation of reovirus mRNA 5'-terminal cap structures by reovirus cores has been studied in detail (Faust et al., 1975; Furuichi et al., 1976; Furuichi and Shatkin, 1976, 1977). The mechanism of formation is as follows:

$$\text{pppG} + \text{pppC} \xrightarrow{\text{RNA polymerase}} \text{pppGpC} + \text{PP}_i \tag{1}$$

$$\text{pppGpC} \xrightarrow{\text{nucleotide phosphohydrolase}} \text{ppGpC} + \text{P}_i \tag{2}$$

$$\overset{***}{\text{pppG}} + \text{ppGpC} \xleftrightarrow{\text{guanylyltransferase}} \overset{*}{\text{G}}\text{pppGpC} + \overset{**}{\text{PP}}_i \tag{3}$$

$$\text{GpppGpC} + \text{AdoMet} \xrightarrow{\text{methyltransferase 1}} \text{m}^7\text{GpppGpC} + \text{AdoHcy} \tag{4}$$

$$\text{M}^7\text{GpppGpC} + \text{AdoMet} \xrightarrow{\text{methyltransferase 2}} \text{m}^7\text{GpppGmpC} + \text{AdoHcy} \tag{5}$$

The removal of the γ-phosphate on nascent transcripts (reaction 2) is believed to occur rapidly, since mRNAs with triphosphorylated 5' termini are seldom synthesized (Furuichi and Shatkin, 1976). Hydrolysis of the γ-phosphate may therefore be coincident with the initiation of transcription, but the phosphohydrolase does function independently of transcription (Kapuler et al., 1970; Borsa et al., 1970).

The guanylyltransferase reaction (reaction 3) also appears to function independently of transcription and is believed to be reversible. Therefore, when mRNA synthesis was carried out in vitro in the presence of pyrophosphate, guanylylation occurred at a greatly reduced rate. In the presence of S-adenosyl-L-homocysteine (AdoHcy) used to inhibit methylation, Furuichi and Shatkin (1976) found that 93% of the mRNA molecules had uncapped 5' termini with the structure ppG. However, if the reaction also contained inorganic pyrophosphatase and a higher concentration of GTP, 93% of the transcripts had capped 5' termini with the structure GpppGp. The inorganic pyrophosphatase hydrolized pyrophosphate molecules generated by the polymerization of nucleoside triphosphates (reaction 1). In the absence of pyrophosphatase, pyrophosphate molecules accumulated in the reaction and resulted in pyrophosphorolysis of GpppG-terminated transcripts to yield uncapped ppG-terminated structures (reverse of reaction 3). In the presence of both AdoHcy and pyrophosphate, 97% of the mRNA molecules synthesized had uncapped, diphosphorylated 5' termini. Therefore, it was concluded that conditions of reduced GTP concentration in the presence of pyrophosphate and AdoHcy favor the production of uncapped, diphosphorylated 5' termini.

On the other hand, transcription at optimal GTP concentrations in the presence of AdoMet and inorganic pyrophosphatase favors the production of capped 5' termini. Under the latter conditions, 97% of the transcripts were found to have the 5'-terminal structure $m^7GpppGmpC$ . . . (Furuichi and Shatkin, 1976).

The reversal of the guanylyltransferase reaction (pyrophosphorolysis) is believed to account for the ribonucleoside-triphosphate-dependent pyrophosphate exchange activity observed in reovirus cores under conditions of mRNA synthesis (Wachsman et al., 1970). Wachsman and colleagues also found that reovirus cores possessed a GTP-dependent pyrophosphate exchange activity that was active in the absence of transcription. Thus, on incubation of cores with GTP, $Mg^{2+}$, and labeled $^{32}PP_i$, these investigators were able to observe an exchange reaction, although the mechanism for this exchange was not understood at that time. Recent results in our laboratory (D.R. Cleveland, H. Zarbl, and S. Millward, submitted) indicated that the reovirus guanylyltransferase reaction proceeds via a covalent enzyme–guanylate intermediate. Such a mechanism has been previously reported to be operative in the guanylyltransferase enzymes of vaccinia virus (Shuman and Hurwitz, 1981) and HeLa cells (Venkatesan and Moss, 1982).

We have found that incubation of reovirus cores with $[\alpha\text{-}^{32}P]$-GTP in the presence of $Mg^{2+}$ results in the formation of a covalent bond between the GMP from $[\alpha\text{-}^{32}P]$-GTP and the $\lambda$1 protein and a chymotrypsin cleavage product of $\lambda$1, denoted $\lambda$1C (Fig. 6). The GMP was found to be linked to the proteins via a phosphoramide bond. Formation of the enzyme–guanylate intermediate is not observed if pyrophosphate is included in the incubation mixture. Addition of inorganic pyrophosphatase, however, was found to markedly stimulate the formation of the covalent intermediate. After formation of the enzyme–guanylate intermediate, cores were purified free from unreacted $[\alpha\text{-}^{32}P]$-GTP and were shown to retain the enzyme–guanylate intermediate. The GMP moiety was then released with pyrophosphate, GDP, or ppG-terminated oligonucleotides, resulting in the formation of GTP, GpppG, and GpppGpC . . ., respectively. On the basis of these observations, we suggest that the reovirus guanylyltransferase reaction proceeds via a covalent enzyme–guanylate intermediate according to the following mechanism:

$$\lambda1–\lambda1C + pppG \leftrightarrow \lambda1–\lambda1C \sim GMP + PP_i \tag{6}$$

$$\lambda1–\lambda1C \sim GMP + ppG \leftrightarrow \lambda1–\lambda1C + GpppG \tag{7}$$

Such a mechanism provides an explanation for the GTP-dependent pyrophosphate exchange reaction (reaction 6) of reovirions (Wachsman et al., 1970) and the reversibility of the guanylyltransferase reaction (reaction 3) in the presence of pyrophosphate (Furuichi and Shatkin, 1976).

Little is known about the methyltransferase reactions catalyzed by

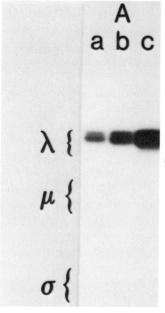

FIGURE 6. Identification of the reovirus guanylyltransferase. Purified reovirus cores (1 mg/ml) were incubated in 1 μM [α-$^{32}$P]-GTP, 4 mM MgCl$_2$, 100 mM Tris · HCl, pH 8.0, and 5 U/ml of inorganic pyrophosphatase for 3 hr at 20°C. The reaction was terminated by addition of EDTA to a final concentration of 10 mM. Labeled cores were then purified by isopycnic banding in CsCl gradients and dialyzed against 100 mM Tris · HCl, pH 8.0, 1 mM EDTA. Labeled cores were then dissolved in electrophoresis sample buffer, and sodium dodecyl sulfate–polyacrylamide gel electrophoresis was performed in a 10–18% gradient gel (Laemmli, 1970). Labeled proteins were detected by autoradiography. Lanes a–c contain increasing amounts of labeled core protein (D.R. Cleveland, H. Zarbl, and S. Millward, unpublished).

reovirus cores (reactions 4 and 5). Although the existence of two separate methylase enzymes has not been established, it seems unlikely that one enzyme could catalyze methylation at the N7 position of the guanine capping group as well as at the 2'-hydroxy group in the ribose moiety of the terminal guanosine. It is possible, however, that both activities reside in one polypeptide. The prime candidate is the λ2 protein of reovirus cores that makes up the 12 spikes found on the surface of reovirus cores (White and Zweerink, 1976). Morgan and Kingsbury (1980) have demonstrated that pyridoxal phosphate binds reversibly to a select population of the reovirus core proteins, λ1 and λ2. The interaction of pyridoxal phosphate with these proteins resulted in the inactivation of the viral transcriptase (Morgan and Kingsbury, 1980) as well as the enzymes involved in the posttranscriptional modification of viral mRNA (Morgan and Kingsbury, 1981). Results from our laboratory have indicated that λ1–λ1C is the reovirus guanylyltransferase (see above), and the work of Drayna and Fields (1982) showed that λ3 is the reovirus transcriptase. A process of elimination suggests that either the nucleotide phosphohydrolase or the methyltransferase activity or activities are associated with the λ2 protein. The nucleotide phosphohydrolase activity, however, appears to be closely coupled to initiation of transcription and hence should be closely associated with λ3 and the genomic RNA. The λ2 protein, on the other hand, is located on the outer aspect of viral cores. In fact, λ2 has been shown to be arranged in a pentameric configuration that makes up the 12 hollow cylindrical spikes found on the surface of reovirus cores (White and Zweerink, 1976; Ralph et al., 1980). Furthermore, both electron microscopy (Gillies et al., 1971; Bartlett et al., 1974) and laser light-scattering (Bellamy and Harvey, 1976) have indicated that the spikes, which are hollow and possess a 5-nm central channel (Luftig et al., 1972), may act as a conduit for the extrusion of mRNA. White and Zweerink (1976) have also indicated that λ1 is located beneath the reovirus spikes, since removal of spikes makes λ1 accessible to iodination. If nascent mRNA is indeed extruded through the spikes, it is likely that guanylylation occurs near the base of the spike, since λ1–λ1C is the guanylyltransferase. The possibility that methylation occurs on passage of transcripts through the spike is an alluring one. Taken together, these findings strongly suggest that λ2 possesses the reovirus methyltransferase activities. A definite assignment, however, awaits more direct evidence.

### 3. In Vivo Activation of Enzymes Associated with Reovirus

Activation of the virion-associated transcriptase following uncoating in vivo plays a crucial role in the reovirus life cycle, since genomic RNA is not completely uncoated. Transcripts from the parental genome are thus the sole carrier of genetic information from parent to progeny and, in fact, are the plus strands of the progeny genomic RNA (see Section D.1).

The mechanism of transcription in vivo by uncoated virions appears to be analogous to the mechanism of transcription determined in vitro using reovirus cores. Sensitive hybridization techniques have shown that transcripts formed in vivo are exactly the same length as the dsRNA genome segments from which they are transcribed, indicative of end-to-end transcription (Watanabe et al., 1967b; Bellamy and Joklik, 1967a; Prevec et al., 1968; Watanabe et al., 1968b; Zweerink and Joklik, 1970). Transcription in vivo is also asymmetric, since the transcripts do not self-anneal (Shatkin and Rada, 1967; Hay and Joklik, 1971). Competitive hybridization studies have clearly shown that the same strand of each genome segment is transcribed both in vitro and in vivo (Hay and Joklik, 1971).

Virions uncoated in vivo were also found to synthesize full-length transcripts in vitro (Change and Zweerink, 1971; Silverstein et al., 1972). Furthermore, transcripts synthesized by these purified parental SVPs have the same 5'-terminal cap structures as those synthesized by reovirus cores (Galster and Lengyel, 1976; Zarbl et al., 1980a). Parental transcripts isolated from infected cells also possess the same cap structures as those found at the 5' termini of in vitro transcripts (Desrosiers et al., 1976; Skup et al., 1981). Desrosiers et al. (1976) found that about 50% of the capped reovirus mRNA molecules found in the infected cell between 5 and 11 hr postinfection possessed type 2 cap structures ($m^7$GpppGmpCmp . . .) at their 5' termini. The existence of type 2 cap structures at the 5' end of reovirus mRNA from infected cells was also noted by Shatkin and Both (1976). Methylation of the cytidine residue penultimate to the cap is probably carried out by an enzyme of cellular origin (Moyer and Banerjee, 1976). These results indicate that the activation of the virion-associated capping enzymes is analogous to the activation *in vitro* with chymotrypsin. Transcripts synthesized *in vitro* by chymotrypsin cores have also been found to contain variable amounts of type 0 cap structures (m$^7$GpppGp . . .), in which the terminal guanosine is not 2-O-methylated (Furuichi *et al.*, 1976; Hastings and Millward, 1978). Type 0-cap structures are not found *in vivo* (Furuichi *et al.*, 1975; Desrosiers *et al.*, 1976; Skup *et al.*, 1981). This may indicate that methylation by cores *in vitro* is less efficient than methylation by particles uncoated *in vivo*.

The activity of the virion-associated transcriptase may also differ slightly in core particles as compared to particles uncoated *in vivo*. The transcriptase associated with reovirus cores has been shown to produce a molar excess of abortive transcripts, even under optimum conditions for transcription *in vitro* (Zarbl *et al.*, 1980a; Yamakawa *et al.*, 1981). These abortive transcripts presumably arise by multiple reiteration of the transcriptase at presumed promoter sites. Zarbl *et al.* (1980a) have suggested that such abortive transcripts may result from instability of the initiation complex. These investigators also examined transcription catalyzed by virions uncoated *in vivo* and found that capped, abortive transcripts were not formed by these particles. It is therefore plausible that the reiterative mechanism of transcription observed *in vitro* is an artifact. Reovirus cores lack all proteins of the outer capsid, while parental SVPs retain a fragment of molecular weight 65,000 denoted δ and derived by cleavage of the μ1C protein (Smith *et al.*, 1969; Chang and Zweerink, 1971; Silverstein *et al.*, 1972). Recent work by Drayna and Fields (1982) indicated that the μ1 or μ1C proteins or both regulate the activity of the virion-associated transcriptase. It is possible that the δ protein of parental SVPs stabilizes the initiation complex, thereby preventing or diminishing abortive transcription. This would indicate that initiation of transcription *in vivo* may be more efficient than that observed *in vitro*. Therefore, the overall rate of mRNA synthesis *in vivo* may exceed the rate observed *in vitro*.

Yamakawa *et al.* (1981), on the other hand, have suggested that an abortive mode of transcription and excess initiations relative to full-length mRNA production may be a general property of many transcriptases. It is known that the viral transcriptase produces abortive transcripts at the final stages of maturation (Stoltzfus and Banerjee, 1972; Bellamy *et al.*, 1972; Yamakawa *et al.*, 1981). These abortive transcripts probably result from the inactivation of the virion transcriptase as a result of conformational constraints associated with assembly of outer-coat proteins (see Section H.2). Abortive transcription also occurs independently of transcription in reovirus cores (Yamakawa *et al.*, 1981) and intact virions (Yamakawa *et al.*, 1982). Further studies will be required to determine whether abortive transcripts are produced during transcription *in vivo*.

## 4. *In Vivo* Regulation of Reovirus Transcription

Transcription of the reovirus genome *in vivo* can be divided into two stages, early and late. The two classes of mRNA are defined by their time of synthesis, by the type of particle synthesizing them, and by the structure of their 5' termini. Early mRNA is transcribed from the parental genome by the transcriptase associated with parental SVPs (Chang and Zweerink, 1971; Silverstein *et al.*, 1972) and is capped at its 5' termini (see Section III.C.3). Early mRNA synthesis is first detectable at about 2 hr postinfection (Zweerink and Joklik, 1970), and reaches a maximum between 6 and 8 hr postinfection at 37°C when cells are infected at 5–10 PFU per cell (Silverstein and Dales, 1968; Joklik, 1974). Synthesis of capped parental transcripts decreases thereafter, and by 12 hr postinfection, synthesis of capped transcripts has ceased (Fig. 7). Inactivation of the parental-SVP-associated transcriptase and capping enzymes probably results from reassembly of the outer capsomers as suggested by the data of Astell *et al.* (1972). The latter results suggest that capped parental transcripts are not synthesized to any significant extent at late times postinfection. This finding, coupled with the observation that only capped transcripts are used as templates for minus-strand synthesis (Furuichi *et al.*, 1975), is consistent with previous results (Schonberg *et al.*, 1971; Watanabe *et al.*, 1967a) that indicated that only transcripts synthesized at early times postinfection are used for replication (see Section D.1). This means that parental transcripts have a dual function, serving

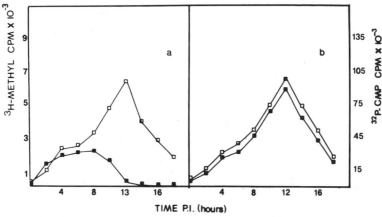

TIME P.I. (hours)

FIGURE 7. Time course of methylase and polymerase activities in SVP preparations. Cells were infected and samples were taken at the times indicated postinfection. SVPs ("slow-sedimenting fraction") were prepared from each sample and used to catalyze standard transcriptase reactions containing [methyl-$^3$H]-AdoMet and [$\alpha$-$^{32}$P]-GTP. Reactions were incubated for 60 min at 44°C. Samples were divided in half and precipitated with trichloroacetic acid (TCA) before and after RNase digestion. Data are presented as RNase-sensitive, TCA-precipitable radioactivity. (a) Progeny SVPs; (b) progeny SVPs treated with chymotrypsin. (■) Tritium; (□) $^{32}$P. Reprinted from Skup and Millward (1980b), with permission.

as mRNA for protein synthesis and as a template for minus-strand synthesis.

Late reovirus mRNA is defined as that mRNA transcribed from progeny genomes by the transcriptase associated with progeny SVPs. By definition, then, late mRNA expression is dependent on replication of the infecting viral genome, in keeping with the definition of late mRNA in other virus systems. Synthesis of mRNA by progeny SVPs begins 4–6 hr postinfection and reaches a maximum at about 12 hr postinfection, then declines throughout the remainder of the infectious cycle. Early studies showed that the addition to infected cells of inhibitors of protein synthesis such as cycloheximide or puromycin resulted in a cessation of dsRNA synthesis, i.e., minus-strand synthesis, while ssRNA synthesis continued unabated (Kudo and Graham, 1966; Watanabe et al., 1967a,b). Similar results were obtained by Ito and Joklik (1972a) using temperature-sensitive (ts) mutants, negative for dsRNA synthesis. At present, these data can be interpreted to mean that dsRNA synthesis is coupled to protein synthesis and that in the absence of new viral protein synthesis, very few, if any, new progeny SVPs are made. By the same token, the existing progeny SVPs cannot be assembled to form mature virus, so they continue to synthesize mRNA. From these kinds of experiments, one can estimate that 80–95% of the viral mRNA in infected cells is synthesized by progeny SVPs by about 12 hr postinfection. These same progeny SVPs have been shown to have masked guanylyltransferase and methyltransferase activities (Skup and Millward, 1980b) and, as a result, to synthesize uncapped mRNA (Zarbl et al., 1980b; Skup et al., 1981). The biological significance of these results lies in the observation that the protein-synthesizing apparatus of the host cell undergoes a transition from cap dependence to cap independence as a result of reovirus infection (Skup and Millward, 1980a). In effect, this obviates the necessity for viral mRNAs to compete with host mRNAs as seems to occur at early times in the infectious cycle when parental SVPs present the cell with capped mRNAs (Skup et al., 1981; Walden et al., 1981). Thus, in addition to early and late mRNAs' being synthesized by distinctively different and definable particles, they are also structurally distinct with respect to their 5' termini. This structural difference may explain why transcripts from progeny SVPs are not used as templates for minus-strand synthesis during replication, as is discussed in greater detail in Section D.1.

In vitro studies using reovirus cores (Skehel and Joklik, 1969; Shatkin and Banerjee, 1970), partially uncoated, infectious SVPs (Shatkin and LaFiandra, 1972), or isolated particles uncoated in vivo (Silverstein et al., 1972) indicated that there was no regulation of transcription in uncoated particles. In each case, all ten dsRNA genome segments were transcribed at the same rate. Studies concentrating on reovirus-specific transcription in vivo, however, have indicated that regulation of transcription does occur in the infected cell. Watanabe et al. (1968b), Millward and Graham (1974), and Nonoyama et al. (1974) pulse-labeled cells at various times postinfection with [$^3$H]uridine, extracted the labeled ssRNA, and hybridized the transcripts to $^{14}$C-labeled genome dsRNA. This method allowed the authors to quantitate the relative amounts of each species of ssRNA synthesized. The results of these experiments indicated that between 4 and 6 hr postinfection, which was the earliest time at which sufficient labeled ssRNA could be obtained for analysis, there appeared to be a preferential transcription of some genome segments. Those transcripts that are synthesized from a select number of genome segments very early in infection are defined in this review as "pre-early" mRNAs and are products of the parental-SVP-associated transcriptase. Because the onset of transcription in the infected cell is somewhat asynchronous and very little mRNA can be detected before 3 hr, it was probable that translation had already begun, obscuring the pre-early pattern of transcription prior to protein synthesis. To get around this problem, these investigators added cycloheximide at the time of infection. Cycloheximide had previously been shown not to affect adsorption or uncoating of the parental virions (Silverstein and Dales, 1968), but had been found to effect a total block in protein synthesis and, when added before 5 hr postinfection, to prevent the synthesis of virus-specific RNA (Watanabe et al., 1967a; Shatkin and Rada, 1967). Using more sensitive analytical techniques, it was found that a small amount of ssRNA was indeed synthesized in the presence of cycloheximide (Watanabe et al., 1968b; Nonoyama et al., 1974). Furthermore, in the absence of protein synthesis, it was found that only four of the ten viral genome segments, namely, L1, M3, S3, and S4, were transcribed. The largest segment (L1) and the two smallest segments (S3 and S4) were transcribed with approximately equal frequencies, while the medium-sized segment (M3) was transcribed at a frequency 5 times that of the other three (Fig. 8). This pre-early pattern of transcription was in sharp contrast to that seen between 10 and 13 hr postinfection, when all the genome segments were found to be transcribed with similar frequencies (Nonoyama et al., 1974). Analyses of transcripts pulse-labeled at times intermediate to the pre-early and late patterns of transcription were consistent with a transition from a state in which only four segments were transcribed to a state in which all ten genome segments were transcribed. This was true both in the presence and in the absence of cycloheximide, although the temporal sequence varied somewhat in the presence of the drug. These results were consistent with regulation of transcription at early times in vivo.

Further compelling evidence in support of the concept of regulated transcription came from the experiments of Lau et al. (1975). These investigators took advantage of the fact that inhibition of protein synthesis by cycloheximide is reversible. Cells were infected in the presence of this drug, and early transcripts were allowed to accumulate for a period of 17.5 hr at 31°C. Cycloheximide was then removed from the cells, and cordycepin was added to inhibit the formation of new viral transcripts. Cells were labeled with radioactive amino acids, and viral polypeptides

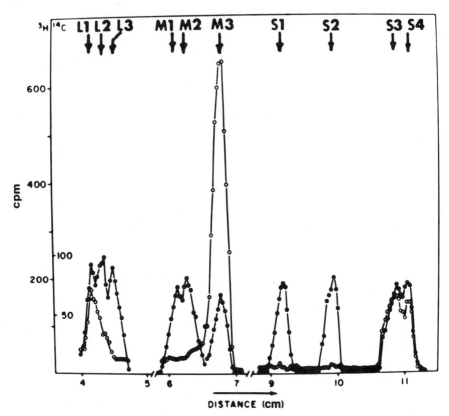

FIGURE 8. Polyacrylamide gel electrophoretic analysis of the hybrids formed between viral dsRNA and mRNA synthesized in infected cells when cycloheximide was present from the beginning of infection. (●) $^{14}$C-labeled genome hybrids; (○) $^{3}$H-labeled mRNA hybridized with genome dsRNA. Reprinted from Nonoyama et al. (1974), with permission.

were removed from cellular extracts by an immunoprecipitation technique. Analysis of the labeled viral polypeptides synthesized indicated that four viral polypeptides were the predominant products and included σ3, σNS, μNS, and λ3.* The σ3, σNS, μNS, and λ3 proteins are now known to be encoded by the *S4, S3, M3,* and *L1* genome segments. This is consistent with transcription of only four genome segments in cells infected in the presence of cycloheximide. Furthermore, the transcripts are active mRNA molecules that are subsequently translated into viral polypeptides.

The pattern of polypeptides synthesized in the absence of cordycepin was also analyzed. The polypeptides synthesized 30 min after the removal

---

* Lau *et al.* (1975) designated the λ protein synthesized as λ2. The sodium dodecyl sulfate–polyacrylamide gel system employed by the authors does not separate all three species of λ polypeptides. It was therefore assumed on the basis of gene coding assignments (McCrae and Joklik, 1978) that the largest protein synthesized was actually λ3, but that this fact was obscured by comigration of λ3 with the λ2 protein.

of cycloheximide consisted primarily of λ3, μNS, σNS, and σ3, consistent with translation of the four pre-early mRNA species that had accumulated in the cell. The viral polypeptides synthesized 3–4 hr after the removal of cycloheximide were the same as those synthesized in infected cells at 17.5 hr postinfection in the absence of the drug. These results are consistent with the transcription of only four of the ten genome segments in the absence of protein synthesis, followed by transcription of all genome segments after the onset of viral polypeptide synthesis.

Even though the transition from regulated transcription to a state wherein all genome segments are synthesized was dependent on the production of viral proteins, the transition did not require replication (Spandidos et al., 1976). This was demonstrated using defective virions that lack the L1 segment that encodes the viral polymerase (Drayna and Fields, 1982). These defective virions are uncoated in the same way as virions, and their genome is transcribed. The fact that these virions do not encode the viral polymerase (λ3) renders them unable to replicate their genome. Spandidos et al. (1976) examined the relative frequencies of transcription of the nine genome segments in defective virions. Using the quantitative procedure of Nonoyama et al. (1974), it was found that during the first 5 hr after infection, three genomic segments, M3, S3, and S4, were more frequently transcribed than the remaining six segments. During the succeeding 5 hr, there was a transition to a situation in which all nine segments were transcribed at the same relative frequencies. In addition to the defective virions, the authors also used class C ts mutants of reovirus, which had previously been shown to map in the L1 segment and to be defective in replication (Spandidos and Graham, 1975). Analysis of transcription in cells infected at a nonpermissive temperature indicated that at early times, four genome segments, L1, M3, S3, and S4, were the predominant transcripts, and later all ten segments were transcribed with the same relative frequencies.

Transcription of the defective genome and the class C mutant genome is therefore regulated in the same way as previously found for wild-type virus (Nonoyama et al., 1974). Both these mutant viruses, however, underwent the transition from a pre-early pattern to an early pattern in the absence of replication. Since no progeny virions were formed, it becomes evident that it is the transcriptase associated with parental particles that is regulated. By contrast, late transcription from progeny genomes does not appear to be regulated. Therefore, the early transcription associated with parental particles must be divided into pre-early and early parental transcription to avoid confusion with late transcription by progeny SVPs.

Two models can be evoked to explain the regulation of parental transcription. The first model supposes that in vivo–uncoated particles are capable of transcribing only four genome segments. It could then be postulated that one or more of the four early proteins recombines with parental SVPs and causes some form of change, conformational or other, that allows all ten segments to be transcribed. Such a model, however, is inconsistent with the observation that isolated particles uncoated *in vivo* synthesize all ten mRNA species *in vitro* (Silverstein *et al.*, 1972). The latter observation is more consistent with a second model for regulation, namely, one that proposes that a cellular repressor interacts with uncoated particles and prevents transcription of the six parental genome segments while allowing transcription of the four pre-early segments. One or more of the four early proteins might then interact with this cellular protein and derepress transcription of the remaining genome segments. The ability of isolated parental SVPs to transcribe all ten segments could then be explained by loss of the cellular repressor during the isolation procedure. Several other lines of evidence support this model of host-cell-induced repression of parental transcription.

Shatkin and LaFiandra (1972) obtained infectious SVPs by *in vitro* uncoating that closely resembled *in vivo*-uncoated SVPs. These SVPs were found to synthesize all ten mRNA species *in vitro*. Infection of cells with these particles in the presence of cycloheximide resulted in regulated transcription; i.e., the mRNA species synthesized were limited to *L1*, *M3*, *S3*, and *S4*. Whether the same pattern of early transcription occurred in the absence of cycloheximide was not determined. Nevertheless, this result strongly supports the notion that a cellular repressor is involved in the regulation of parental transcription (Nonoyama *et al.*, 1974). The most compelling evidence for host-induced repression of parental transcription comes from the work of Spandidos and Graham (1976b) using an avian reovirus. Although avian reovirus fails to replicate in mouse L cells, adsorption, penetration, and uncoating appear to proceed normally. During a nonpermissive infection of L cells with an avian reovirus, it was found that only four of the ten avian reovirus genome segments, *L1*, *M3*, *S3*, and *S4*, were transcribed. These are the same four pre-early transcripts found to be synthesized in permissive infections with reovirus type 3. Evidence indicating that translation of early avian transcripts did occur came from the observation that some mutant strains of avian reovirus will grow in L cells (Spandidos and Graham, 1976b). These results suggested that the avian reovirus polypeptides encoded by the early transcripts are incapable of inactivating the L-cell repressor. The inability of avian polypeptides to cause derepression was confirmed in experiments in which cells were coinfected with avian reovirus and the R2 strain of reovirus type 3, which was shown to replicate in these mixed infections. In mixed infections, all ten segments of the avian dsRNA genome are transcribed, although no replication of the avian genome occurs. These results are consistent with derepression of the transcriptase of avian parental particles by one of the early proteins synthesized by reovirus type 3. The mechanism that allows the four genome segments to escape repression is unknown. It is interesting to note that the viral proteins encoded by the four early transcripts include both the nonstructural proteins of reovirus, which presumably perform important functions in the cell. Among the four early proteins, λ3 and σNS have both been implicated in replication of the genome (Drayna and Fields, 1982; Gomatos et al., 1981). Another early protein, σ3, was shown to play a role in the inhibition of host protein synthesis (Sharpe and Fields, 1981, 1982). It is possible that the four early mRNAs code for functions required at early times to establish the infection and, as such, incur a selective advantage by escaping the host-induced repression (Millward and Graham, 1974). Further experiments will be required before definitive statements can be made.

Other investigators have also examined early transcription in vivo and have observed no regulation. Zweerink and Joklik (1970) reported that the in vivo transcription pattern remained constant over 2-hr periods from 2 to 8 hr postinfection. These investigators, however, used radioautography to measure the hybrids formed between labeled ssRNA extracted from the cells and genome dsRNA. Using this method, it was not possible to correct for the differences in efficiency of annealing among the genomic segments. Furthermore, Zweerink and Joklik (1970) did observe increased transcription of M3, S3, and S4 and have argued that these results support the suggestion that the four early mRNAs synthesized in the presence of cycloheximide may be artifactual (Joklik, 1974). The argument states that three of the four mRNAs synthesized in the presence of cycloheximide are also formed in the largest amounts in its absence. Thus, when the overall rate of transcription is very low at early times postinfection, these mRNAs (l1, m3, s3, and s4) might be preferentially extracted, while those formed only in small amounts may tend to be lost during mRNA isolation and hybridization. Such arguments, however, tend to be ruled out by the results of experiments of Lau et al. (1975), wherein transcripts synthesized in the presence of cycloheximide were allowed to accumulate for 17.5 hr. Under these conditions, the amounts of mRNAs synthesized were large enough to stimulate the synthesis of significant quantities of viral polypeptides.

It has also been argued that all ten genome segments are synthesized at nonpermissive temperatures by mutants defective in replication (Ito and Joklik, 1972a; Cross and Fields, 1972). Once again, these investigators did not correct for efficiency of annealing, nor did they use inhibitors of protein synthesis in their studies. Using both defective virions and ts mutants defective in replication, Spandidos et al. (1976) were able to demonstrate a pre-early parental transcription pattern. These results indicated clearly that synthesis of viral proteins and not replication was required for the transition from the pre-early parental to the early parental transcription pattern.

Silverstein et al. (1976) have argued that if inhibition of protein synthesis locks the parental transcription into the early stage, then only four transcripts should be made in cells treated with interferon. Wiebe and Joklik (1975) have not found this to be the case, but it should be noted that these authors did not find complete inhibition of parental transcription or translation in cells treated with interferon. One could argue that a sufficient number of the pre-early transcripts were translated to allow for derepression to the early transcription pattern. Cycloheximide, on the other hand, inhibits translation completely and therefore locks parental SVPs into the pre-early pattern. It is possible that the contradictions concerning regulation of transcription may be more apparent than real and may reflect diversity in the modes of analysis. Although the model involving a cellular repressor that is derepressed by viral proteins is alluring, confirmation of such a model awaits isolation and identification of the cellular repressor.

These data can be summarized in the form of a working model that describes the initial steps leading to reovirus replication in permissive host cells. Following uncoating of the infecting virus—a process that activates the transcriptase and capping enzymes—and release of the parental SVPs from lysosomes, four of the ten genome segments are expressed and can be translated by the existing host translational apparatus. We suppose that one or more of these pre-early transcripts codes for a protein involved in derepression of the remaining six genome dsRNA segments. As a result, all ten segments are now transcribed from the parental SVPs, and these constitute, by definition, the early mRNA pattern. It must be kept in mind that although we call these ten early transcripts mRNAs, they must also serve as templates for minus-strand synthesis leading to dsRNA and progeny SVP formation.

## D. Formation of Progeny Double-Stranded RNA

### 1. Replication of the Reovirus Genome Proceeds via a Conservative Mechanism

The fact that the reovirus genome is not uncoated completely *in vivo* (Silverstein and Dales, 1968) but is conserved within a parental subviral particle (SVP) (Silverstein *et al.*, 1970; Schonberg *et al.*, 1971) clearly indicated that replication of the reovirus double-stranded RNA (dsRNA) genome could not occur via a semiconservative mechanism analogous to that of dsDNA replication. The fact that both strands of each double-stranded genome segment are retained within the infecting particle suggested that the single-stranded transcripts must serve as the template on which dsRNA is formed. This indicated that parental transcripts must serve a dual role, being translated into viral polypeptides and serving as a template for minus-strand synthesis.

The validity of such a mechanism of replication has been verified by several lines of evidence. Silverstein *et al.* (1970) infected cells with reovirus that had been labeled with [³H]uridine and found that during uncoating, essentially all the labeled dsRNA was retained within parental SVPs. In the presence of cycloheximide, all the dsRNA remained associated with particles with a density of 1.41 g/cm$^3$ for up to 11 hr. In the absence of cycloheximide, however, the labeled parental genome was again found in particles that had the same density as mature virions. These results were interpreted as reflecting the reassembly of uncoated parental virions at a later stage in the infection. The appearance of the parental genome within particles with the same density as intact virions is consistent with either a semiconservative or a conservative mode of replication.

A more definitive approach was used by Schonberg et al. (1971), who infected cells with reovirus cores generated from [$^3$H]uridine-labeled virus with chymotrypsin. The fact that reovirus cores have a density of 1.43 g/cm$^3$, while progeny virions have a density of 1.36 g/cm$^3$, allowed these investigators to determine whether the two parental RNA strands remained in the original duplex or whether the complementary strands were displaced into progeny genomes. Cells were infected with [$^3$H]uridine-labeled cores, and the infection was allowed to proceed for 15 hr. The cells were then lysed and the cytoplasmic extracts analyzed by isopycnic centrifugation in CsCl gradients. It was found that the $^3$H-labeled parental RNA remained within cores (1.43 g/cm$^3$), and no label was found in progeny virions (1.36 g/cm$^3$) detected by plaque-forming ability. L cells infected with cores were also subjected to an infectious-center assay that indicated that every cell in the culture had been infected. This result confirmed that cells containing cores and those producing progeny virions were not mutually exclusive populations. Therefore, the absence of $^3$H-labeled RNA in the progeny genomes is indicative of a conservative mode of replication.

Further evidence suggesting that parental transcripts were used as templates for minus-strand synthesis was provided by the experiments of Schonberg et al. (1971), which indicated that the complementary strands of the genome were formed asynchronously. In these experiments, infected L cells were pulse-labeled with [$^3$H]uridine for 30 min during the midlogarithmic phase of viral replication. Total dsRNA was then extracted from the infected cell, denatured, and hybridized to an excess of unlabeled single-stranded RNA (ssRNA) made in vitro. This ssRNA has the same polarity as the plus strand of the genome segments. Hybridization of denatured dsRNA to an excess of plus strands allows for analysis of the distribution of the [$^3$H]uridine label between the plus and minus strands of dsRNA synthesized in vivo.

Schonberg et al. (1971) argued that analysis of the label distribution in hybrids between extracted dsRNA and excess cold ssRNA could provide insight into the mode of reovirus dsRNA replication. If replication occurred via a semiconservative mechanism, then the dsRNA extracted from pulse-labeled cells should be labeled equally in both strands. Hybridization of denatured dsRNA labeled in both strands to an excess of cold plus strands would give rise to hybrids that contain predominantly cold plus strands (Fig. 9c). Therefore, only 50% of the label associated

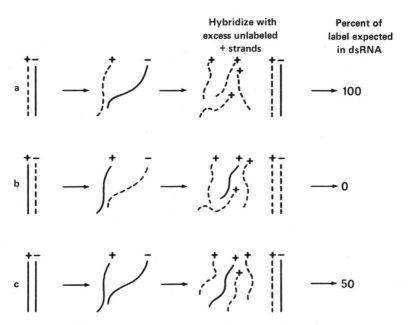

FIGURE 9. Models for the synthesis of reovirus dsRNA. Reprinted from Schonberg *et al.* (1971), with permission.

with the dsRNA would be expected to be conserved within the hybrids. If, on the other hand, minus strands were used as a template for plus-strand synthesis, then the label associated with extracted dsRNA is expected to be associated predominantly with plus strands. Under these conditions, annealing of denatured dsRNA to excess plus strands would result in hybrids from which all the label is chased (Fig. 9b). Conversely, if plus strands served as a template for minus-strand synthesis, then pulse-labeling should produce dsRNA molecules labeled predominantly in the minus strand. Hybridization of denatured dsRNA to an excess of cold plus strands would be expected to result in hybrids that retain 100% of the label (Fig. 9a).

To test these models, dsRNA was extracted from pulse-labeled cells and hybridized to excess plus strands and the resulting hybrids purified away from ssRNA. The results of these experiments indicated that all the label in the extracted dsRNA was indeed associated with the minus strands, since all label was conserved in the hybrids. These results were consistent with the notion that single-stranded transcripts are used as templates for the synthesis of minus strands. A similar analysis of dsRNA extracted from cells continuously labeled with [³H]uridine during the infection showed that the plus and the minus strand contained equal amounts of label. Together, these results were interpreted as being indicative of a lag between the time of plus-strand synthesis and the synthesis of minus strands on the plus-strand templates. Most of the plus strands used as templates for replication must therefore be formed at early times in the replicative cycle.

More recently, Morgan and Zweerink (1977) presented data that suggested that the synthesis of the complementary genomic strands is not completely asynchronous. Whereas Schonberg et al. (1971) analyzed total dsRNA extracted from pulse-labeled cells, Morgan and Zweerink (1977) studied the distribution of label between the plus and minus strands in replicase particles (see Sections D.2 and D.3) extracted from cells pulse-labeled for 30 min at 7½ hr after infection. The dsRNA associated with the replicase particles was then extracted and hybridized to an excess of cold plus strands. This technique is more sensitive than the one involving the use of total dsRNA and clearly demonstrated that plus strands synthesized between 7½ and 8 hr postinfection are capable of serving as templates for minus-strand synthesis. Nonetheless, 80% of the label in dsRNA extracted from replicase particles was associated with the minus strand. Morgan and Zweerink (1977) argued that this observed asynchrony is more apparent than real. When cells are labeled late in the infectious cycle, the presence of large pools of preformed plus strands may reduce the specific radioactivity of the labeled plus strands. Since pools of minus strands do not exist in the infected cell, this results in the asymmetric labeling of dsRNA.

Although the results of Morgan and Zweerink (1977) clearly indicate that plus strands synthesized at later times can still serve as templates for replication, they are also consistent with partial asynchrony in the synthesis of the complementary strands. The existence of a pool of plus-strand templates in the infected cell indicates that most of the templates must have been synthesized at earlier times. This contention is strengthened by the results of recombination experiments (Fields, 1971; Spandidos and Graham, 1976a) that indicated that only parental transcripts were used as templates for minus-strand synthesis. These investigators found that the frequency of recombination reaches a maximum at about 12 hr postinfection at 31°C and remains fairly constant thereafter (Fig. 10), indicating that the frequency of recombination does not increase during the synthesis of progeny transcripts. Since plus strands serve as templates for replication and recombination occurs at the single-strand level, the constant frequency of recombination strongly indicates that only parental transcripts are used for minus-strand synthesis. This means that while progeny transcripts account for 80–95% of the plus strands found in the infected cell, they are not used as templates for minus-strand synthesis. In light of these data, the arguments of Morgan and Zweerink (1977) that asynchrony in the synthesis of the complementary strands is merely apparent must be reexamined. The fact that the large quantities of progeny transcripts are not used for replication annuls the contention that the specific activity of pulse-labeled plus strands is diluted by the large pool of progeny transcripts found in the infected cell. The results of Morgan and Zweerink (1977) are therefore not contradictory to an asynchronous

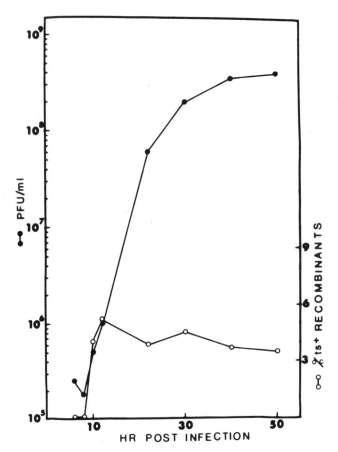

FIGURE 10. Viral growth curve and kinetics of recombination in a mixed infection with $R_2A$ and $R_2B$ mutants at 31°C. Multiplicity of infection was 10 PFU/cell for each mutant. Reprinted from Spandidos and Graham (1976a), with permission.

model for replication. In fact, they are in agreement with the results of Schonberg *et al.* (1971), which indicated that the bulk of the plus strands used for replication are synthesized early in the infection. The results of Morgan and Zweerink (1977) do indicate that parental transcripts synthesized at later times were used for replication, but accounted for only 20% of the label in the dsRNA. The parental transcripts synthesized at later times must therefore be diluted by a pool of parental transcripts synthesized at earlier times. The results of both investigators are therefore consistent with asynchronous synthesis of the complementary genome strands.

The inability of progeny transcripts to serve as templates for minus-strand synthesis was first suggested on the basis of constant genetic-recombination frequencies throughout the infectious cycle (Fields, 1971; Spandidos and Graham, 1976a). It was suggested that progeny transcripts served only as messenger RNA (mRNA) for protein synthesis. The mechanism that allows only parental transcripts to enter into replication complexes, however, remains obscure. Recent results from our laboratory have provided a tentative explanation for the exclusion of progeny mRNA from replication. It has been determined that progeny particles have inactive capping enzymes (Skup and Millward, 1980b) and, as a result, synthesize mRNAs the 5' termini of which are uncapped, having the structure pG . . . (Zarbl et al., 1980b; Skup et al., 1981). The 5' termini of mRNAs synthesized by parental SVPs, on the other hand, are capped, having the structure $m^7GpppGmp$. . . . Although no direct evidence is as yet available, the possibility that the absence of cap structures allows for the exclusion of progeny transcripts from replication is an attractive one. Such a model implies that the segregation of ten parental transcripts into replication complexes involves recognition of the 5'-terminal cap structures. This contention is reinforced by the fact that all the plus strands found in mature virion genome RNA have capped 5' termini (Furuichi et al., 1975). Since both reovirus and host capping enzyme act on uncapped ppG-terminated transcripts and do not cap monophosphorylated termini (Furuichi et al., 1976; Wei and Moss, 1977; Venkatesan and Moss, 1980), the possibility that progeny transcripts can be used in replication and are capped either before or after serving as templates for replication seems remote. All these observations are consistent with exclusion of progeny transcripts from replicative structures. Furthermore, the translational apparatus of L cells undergoes a transition from cap dependence to cap independence as a function of time postinfection (Skup and Millward, 1980a; Skup et al., 1981). The net effect of these events is the preferential translation of uncapped progeny transcripts. In effect, this liberates capped progeny transcripts from polysomes and obviates the potential dilemma of having two macromolecular events (minus-strand synthesis and translation) occurring from opposite ends of the same molecule. The role of the 5'-terminal cap structure in formation of replicase complexes has not been elucidated, but does merit further investigation.

## 2. Mechanism of Progeny Double-Stranded RNA Synthesis

The study of reovirus replication began with the observation that a large-particle fraction from infected cells catalyzed the synthesis of reovirus dsRNA in vitro (Watanabe et al., 1968a). Incubation of the large-particle fraction in the presence of four nucleoside triphosphates and $Mg^{2+}$ resulted in the synthesis of both ssRNA and dsRNA. The dsRNA synthesized by this fraction was found to be composed of the same size classes as the genome RNA segments and were nuclease-resistant. Sakuma and Watanabe (1971) then suggested that the transcriptase and replicase activities associated with the large-particle fraction were separate entities that could be separated on CsCl gradients. These investigators, however, treated the enzyme preparation with chymotrypsin prior to the analysis. There is no doubt that the transcriptase activity (density 1.43 g/cm$^3$) represented uncoating of immature and mature virions to cores. The replicase activity was found to be resistant to limited proteolysis and to be composed of complexes of varying densities. The latter property is also the result of chymotrypsin digestion (see p. 146). Nonetheless, the replicase preparation used provided useful information. Kinetic analysis indicated that minus-strand synthetic activity *in vitro* is short-lived and did not use the ssRNA molecules synthesized in the same fraction as templates for minus-strand synthesis. In addition, no free minus strands could be detected among the transcripts formed in the large-particle fraction. This suggested that replication *in vitro* proceeded by the synthesis of minus strands on a previously synthesized plus-strand template as predicted by Schonberg *et al.* (1971). Analysis of the dsRNA synthesized *in vitro* by the large-particle fraction confirmed that this was the case. When the *in vitro*–labeled dsRNA was denatured and reannealed to excess plus strands, essentially all the label was retained in the hybrids. The results indicated that only minus strands were synthesized by the replicase preparation. These results were subsequently corroborated by several investigators (Acs *et al.*, 1971; Sakuma and Watanabe, 1972a; Watanabe *et al.*, 1974). In further studies, synthesis of the minus strand was found to proceed in a 5'-to-3' direction (Sakuma and Watanabe, 1972a). In addition, the replicase enzyme could use both nucleoside triphosphates and diphosphates as a substrate (Schochetman and Millward, 1972). Sakuma and Watanabe (1972a) then isolated structures that contained intact plus strands and incomplete minus strands from *in vitro* replicase reactions. Such partial dsRNA intermediates were found for each of the three size classes of dsRNA. The existence of such intermediates was consistent with the contention that each plus strand has an individual site for initiating minus-strand synthesis.

Under optimal conditions for *in vitro* minus-strand synthesis, the reaction proceeds for a relatively short period of about 10 min (Sakuma and Watanabe, 1971; Zweerink *et al.*, 1972). The rate of the reaction could be slowed down by decreasing the concentration of one nucleoside triphosphate, but the total amount of minus strands synthesized never exceeded the amount of template (Zweerink *et al.*, 1972). These results are consistent with the notion that plus strands are transcribed into minus strands only once and that the newly synthesized minus strand remains associated with its template. In addition, these results suggested that no new structures capable of initiating minus-strand synthesis are formed *in vitro*. If this is the case, then minus-strand synthesis observed *in vitro* must represent the elongation and termination of minus strands the synthesis of which had already been initiated *in vivo*. This was confirmed by Watanabe *et al.* (1974), who studied the distribution of bromopyrimidines along the minus strands synthesized *in vitro* when heavy nucleoside triphosphates were used as substrates. It was concluded that all minus-strand synthesis observed *in vitro* represented completion of strands that had already been initiated *in vivo*. The inability of the large-particle fraction to initiate minus-strand synthesis *in vitro* may be due to the lack of capped parental transcripts, of additional factors required for initiation in this preparation, or of both. A more plausible explanation, however, may be that formation of replicase complexes is closely coupled to translation. This hypothesis is supported by the observation that replication *in vivo* is dependent on protein synthesis (Kudo and Graham, 1966; Shatkin and Rada, 1967; Watanabe *et al.*, 1967a). Inhibitors of protein synthesis cause a rapid inhibition of minus-strand synthesis *in vivo*. It appears that in the absence of translation, no further initiation of minus-strand synthesis can occur. Elongation of previously initiated minus strands, on the other hand, is independent of protein synthesis. These observations can be explained by a model in which single-stranded templates are incorporated into a replicase complex that contains the enzyme(s) responsible for minus-strand synthesis. Minus-strand synthesis would therefore occur only within the complex, and presumably the enzyme activity remains in the complex. The incorporation of ssRNA templates into particles containing viral polypeptides might therefore be tightly coupled to translation, suggesting that the proteins required for replicase complex formation do not accumulate in the infected cell. The assumption that the replicase activity remains template-associated may explain why each strand is replicated only once. It has therefore been postulated that the replicase activity and the viral transcriptase may be alternative activities of the same enzyme. Drayna and Fields (1982) have shown tht the viral transcriptase activity is associated with the core protein $\lambda 3$, which is encoded by the *L1* genome segment. Defective virions lacking the *L1* gene segment have been shown to be incapable of dsRNA synthesis, even though they transcribe normal mRNAs from the nine genome segments they contain (Spandidos *et al.*, 1976). These observations strongly imply that the *L1* gene codes for both viral RNA polymerases. If this is indeed the case, then particles containing the viral replicase activity would have to be precursors to mature virions. That this is indeed the case has been demonstrated by a number of investigators.

Acs *et al.* (1971) obtained results indicating that the replicase activity was sequestered into a provirion that also contains the plus-strand RNA templates. These investigators found that the replicase activity associated with the large-particle fraction was resistant to limited proteolysis and nonionic detergents. Pretreatment with ribonuclease, on the other hand, abolished the replicase activity. Once the *in vitro* replicase reaction was complete (10–15 min), however, the newly synthesized minus strand became resistant to ribonuclease. The resistance prevailed even under low-salt conditions that normally allow for nucleolytic digestion of dsRNA. These results demonstrated that minus-strand synthesis occurred in a provirion within which the template plus RNA is accessible to ribonuclease. Completion of minus-strand synthesis may then result in the condensation to a ribonuclease-impermeable particle. These findings led to the proposal that dsRNA is formed within the nascent core of developing progeny virions and that the dsRNAs remained within these particles. Such a model also explains why neither free minus strands nor free dsRNAs are detectable in the infected cell (Gomatos, 1967).

The synthesis of minus strands within a provirion was first demonstrated by Zweerink et al. (1972), who characterized the replicase activity associated with the large-particle fraction of infected cells. The replicase activity, more accurately called an ssRNA-dependent dsRNA polymerase (ss→dsRNA polymerase) was extracted using nonionic detergents and partially purified. When the ss→dsRNA polymerase was analyzed by sucrose velocity centrifugation, it was found to consist of a heterogeneous population of particles sedimenting between 300 and 600 S. Analysis on CsCl gradients showed that most of the particles with ss→dsRNA polymerase activity were rather uniform in terms of their density ($1.34$ g/cm$^3$), indicating a fairly constant RNA/protein ratio. Digestion of these particles with chymotrypsin led to an increase in the buoyant density of the ss→dsRNA polymerase from 1.34 to 1.43 g/cm$^3$. The density of the resulting replicase particles is the same as the density of viral cores generated by digestion of intact virions with chymotrypsin. Sakuma and Watanabe (1972b) subsequently demonstrated that the particles generated by digestion of the ss→dsRNA polymerase with chymotrypsin were also similar to cores in terms of: (1) particle size, (2) characteristics of elution from DEAE–Sephadex, (3) resistance to chymotryptic digestion, and (4) resistance of the dsRNA synthesized within the particles to nucleolytic digestion, in low-salt conditions. Electron-microscopic examination of particles with ss→dsRNA polymerase activity confirmed that the activity was indeed associated with particles resembling the viral core and having a diameter of approximately 40 nm (Morgan and Zweerink, 1975). The polypeptide composition of partially purified replicase particles was also found to be similar to that of the viral core. It can therefore be concluded that replication of the reovirus genome occurs within structures that resemble the inner nucleoprotein capsid of reovirus.

In all likelihood, these replicase particles are direct precursors of the progeny virions, and most of the available information is supportive of this contention. Zweerink (1974) has further characterized the particles with ss→dsRNA polymerase activity and found them to consist of a heterogeneous population with sedimentation coefficients ranging between 180 and 600 S (Fig. 11). Two peaks of activity were found reproducibly with corresponding sedimentation coefficients of 280 and 550 S. Accordingly, the ss→dsRNA activities in the different regions of the gradient were pooled as shown in Fig. 11. The dsRNA products synthesized by each class of polymerase were then analyzed by gel electrophoresis. The replicase particles with the lowest sedimentation coefficients (region 3 of Fig. 11) were found to synthesize predominantly dsRNA of the small-

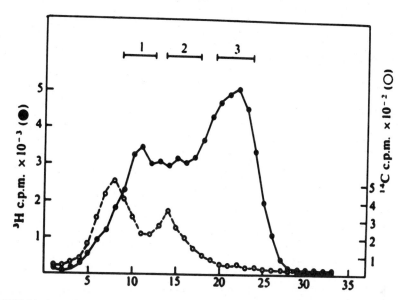

FIGURE 11. Distribution of ss→dsRNA polymerase activity. Cells were infected with reo-virus and incubated at 31°C for 18 hr. A large-particle fraction was prepared from the post-nuclear supernatant obtained by extraction of the cells with nonionic detergent. The pellet was resuspended and subjected to velocity sedimentation in a 20–40% glycerol gradient. Direction of sedimentation is from right to left. Fractions were collected and assayed for ss→dsRNA polymerase activity *in vitro*. (●) Incorporation of radioactivity from [$^3$H]-UTP into ribonuclease-resistant, trichloroacetic-acid-insoluble material; (○) sedimentation po-sitions of $^{14}$C-labeled virus (630 S) and top component (470 S). Reprinted from Zweerink (1974), with permission.

sized class (*S1–S4*). Those of intermediate sedimentation coefficients (re-gion 2 of Fig. 11) synthesized increased amounts of the middle-sized dsRNA segments (*M1–M3*). Correspondingly, the particles of highest se-dimentation rate (region 1 of Fig. 11) synthesized predominantly dsRNA segments of the large-sized class (*L1–L3*). However, once minus-strand synthesis was completed by the *in vitro* replicase reaction, the particles were found to sediment uniformly at about 550 S (Zweerink *et al.*, 1972). Furthermore, as the particles increased in rate of sedimentation, their proteins became increasingly more resistant to chymotryptic digestion. It was therefore proposed that the ten segments of ssRNA associate with viral proteins in the form of a provirion. Initially, these particles may replicate only a limited number of the ssRNA templates. Sequential al-terations in the structure of the replicase particles may then allow for sequential replication of the larger-size classes of genome segments. The changes in structure associated with the sequential replication would then result in the production of particles that more closely resemble vi-rions.

Further evidence that particles with ss→dsRNA polymerase activity are precursors to progeny virions is provided by experiments using temperature-sensitive (ts) mutants of reovirus. Matsuhisa and Joklik (1974) have characterized a member of the group C mutants (ts447) that fails to produce progeny dsRNA at the nonpermissive temperature. At elevated temperatures, this mutant fails to produce particles within which dsRNA is normally synthesized. The mutation was found to reside in the λ1 or σ2 protein, both of which are associated with replicase particles (Morgan and Zweerink, 1975; Zweerink et al., 1976). At nonpermissive temperatures, this mutant produces empty virions that also lack the inner capsid. These results leave little doubt that the corelike structures with ss→dsRNA polymerase activity constitute the inner ribonucleoprotein capsid of mature virions and hence are direct precursors to progeny virions.

### 3. Formation of Subviral Particles with Replicase Activity

Biochemical, structural, and genetic evidence (see Section D.2) clearly indicates that single-stranded parental transcripts are sequestered into some form of provirion. The fact that the individual transcripts are not synthesized with equal frequency by parental SVPs (Watanabe et al., 1968b; Nonoyama et al., 1974; Zweerink and Joklik, 1970) as well as the observed frequencies of genetic recombination (Fields, 1971; Spandidos and Graham, 1976a) strongly suggest that the parental transcripts are not linked. Each progeny virion, however, contains one of each of the ten dsRNA genome segments (Granboulan and Niveleau, 1967; Kavenoff et al., 1975; Spendlove et al., 1970). Silverstein et al. (1976) have estimated that the probability of forming a single particle containing one each of the ten ssRNA species is one in ten billion. It is obvious that some highly specific mechanism must exist for assembling ten discrete species of RNA into each virus particle. How this is accomplished remains one of the mysteries of reovirus replication, but two possible mechanisms may be envisaged. One possibility is that RNA–RNA interactions are involved. Plus strands may possess regions of complementary sequences that allow the appropriate strands to form complexes containing the correct number and species of templates. A more plausible model would involve interaction of the ssRNAs with proteins (structural or nonstructural), resulting in the formation of noncovalently linked complexes containing the ten appropriate ssRNA molecules. Segregation of ten ssRNA segments could then result from constraints imposed by the viral core or by viral-encoded ssRNA-binding protein(s) or by both. The existence of a viral-encoded polypeptide with affinity for ssRNA has been confirmed (Huismans and Joklik, 1976; Gomatos et al., 1981; Stomatos and Gomatos, 1982). The viral nonstructural protein σNS has been found to bind specifically to ssRNA with high affinity, and in fact this property allowed for its purification. It seemed likely that the σNS protein may also be complexed to reovirus ssRNA in vivo. Cells were thus labeled with [$^3$H]uridine and [$^{14}$C]amino acids between 16 and 20 hr postinfection at 31°C. A ribosomal fraction was prepared from these cells and analyzed by velocity sedimentation in sucrose. The profile revealed a broad band of newly synthesized protein and RNA in the region between 40 and 60 S. Analysis of the polypeptides in this region indicated that the only labeled protein present was σNS. The labeled RNA in this region of the gradient consisted primarily of the three size classes of reovirus mRNA. Further evidence for the existence of a σNS–RNA complex was provided by the finding that σNS was released from the complex by ribonuclease treatment or exposure to 0.5 M KCl. These results suggested that in the infected cell, σNS does occur extensively complexed with ssRNA, the molar ratio of protein to RNA being at least 20 and possibly higher. The significance of the affinity of the σNS protein for ssRNA is not clear. Two possible functions of this binding phenomenon are as follows: The σNS protein may function in viral morphogenesis by forming a primary complex with reovirus ssRNAs that are then sequestered into structures within which progeny dsRNAs are formed (Zweerink et al., 1972; Sakuma and Watanabe, 1972b; Zweerink, 1974). Alternatively, the σNS protein may play some as yet undefined role in the translation of viral mRNAs. The possibility that σNS may play a role in both these alternatives cannot be ruled out, since evidence is available to support either possibility.

The bulk of the evidence supports the idea that σNS is involved in the linking and sequestering of ssRNA into replicase particles. Group E ts mutants are dsRNA-negative at nonpermissive temperatures (Ito and Joklik, 1972a). Group E mutants have been shown to reside in the S3 gene segment (Ramig et al., 1978), which encodes the polypeptide σNS (McCrae and Joklik, 1978; Mustoe et al., 1978b). Furthermore, shifting group-E-mutant-infected cells to a nonpermissive temperature after dsRNA synthesis has begun does not inhibit further dsRNA synthesis. The σNS protein must therefore have a very early function, prior to the formation of replicase particles, and does not have to be expressed thereafter. Thus, σNS may be required in catalytic rather than stoichiometric amounts. This is consistent with the possibility that σNS has a function in linking ssRNAs to allow for their being packaged into replicase particles, while σNS itself is not incorporated into virions (Zweerink et al., 1971). The released σNS could presumably be recycled, accounting for the extremely early phase of temperature sensitivity displayed by group E mutants. One must then postulate that the mutation does not affect the ability of σNS to bind ssRNA; instead, the lesion may reside in some other property of the protein.

One possibility is that mutant σNS proteins fail to form particles of the type described by Gomatos et al. (1980, 1981). These particles are comprised almost exclusively of σNS, sediment between 13 and 19 S, have a diameter of 11–12 nm, and appear to be spherical or triangular in shape. Gomatos et al. (1981) showed that these particles bind to many different single-stranded nucleic acids with some degree of specificity (Stomatos and Gomatos, 1982). Different classes of σNS particles were found to bind either polycytidylate [poly(C)] or reovirus mRNAs, but not both at the same time. It was proposed that the specificity of binding was conferred by the configuration of the σNS proteins within different particles. The high capacity of these heterogeneous particles to bind different nucleic acids with some specificity has led to the proposal that they may function as condensing agents to bring together ten individual reovirus ssRNA templates in preparation for dsRNA synthesis. Another interesting aspect of these particles is that they possess a poly(C)-dependent RNA polymerase activity. Utilizing poly(C) as a template, these particles synthesize double-stranded poly(C):polyguanylate [poly(G)]. The polymerase catalyzes the synthesis of poly(G) on a poly(C) template, the size of the poly(G) synthesized being dependent on the size of the template. No other homopolymers could serve as a template for complementary strand synthesis. Synthesis is maximal in the presence of $Mn^{2+}$ ions and does not require a primer. In the presence of $Mg^{2+}$ ions, synthesis was greatly reduced, but was stimulated by the addition of the primer GpG. Analysis of the particles (13–19 S) by ion-exchange chromatography resulted in the separation of the particles into two forms of σNS. One form of σNS carried a greater net positive charge and was enzymatically active, while the second, less electropositive form was enzymatically inactive. Both forms were found to be associated with particles that have poly(C)-dependent RNA polymerase activity.

The function of this enzyme activity is not understood, but it may play a role in the replication of the viral genome. Particles with the poly(C)-dependent polymerase activity have been found to accumulate in the infected cell during the period of maximal reovirus genome replication at 30°C (Gomatos et al., 1980, 1981). Furthermore, at later times postinfection, when replication of viral genome segments has abated, the poly(C)-dependent RNA polymerase activity is found to be associated with SVPs (Gomatos and Kuechenthal, 1977). SVPs expressing the activity were found to sediment heterogeneously between 300 and 550 S. It appears that in the absence of minus-strand synthesis, the poly(C)-dependent polymerase activity accumulates within structures that normally synthesize reovirus dsRNA. This result, coupled to the ability of the enzyme to synthesize double-stranded poly(C):poly(G), has led to the hypothesis that the σNS protein may be the reovirus replicase enzyme (Gomatos and Kuechenthal, 1977; Gomatos et al., 1980). The bulk of available information suggests that λ3 is the reovirus transcriptase and ss→dsRNA polymerase (see Section III.D.2). Several additional observations also predict that σNS is probably not the ss→dsRNA polymerase. The poly(C)-dependent polymerase is maximally active in the presence of $Mn^{2+}$ ions (Gomatos and Kuechenthal, 1977), while the viral replicase is maximally active in the presence of $Mg^{2+}$ ions and is only slightly active with $Mn^{2+}$ ions (Watanabe et al., 1968a). In addition, analyses of the polypeptide composition of particles with ss→dsRNA polymerase activity have failed to detect the presence of σNS (Morgan and Zweerink, 1975; Zweerink et al., 1976). The σNS is associated with SVPs only when no further parental transcripts are available for incorporation into replicase particles (Gomatos and Kuechenthal, 1977). This finding, coupled to that of the ability of σNS to complex ssRNA (Huismans and Joklik, 1976; Gomatos et al., 1981; Stomatos and Gomatos, 1982), certainly implicates σNS in sequestering single-stranded templates into provirions. The σNS must, however, be displaced from these particles shortly thereafter, since the smallest particles with ss→dsRNA polymerase activity (180 S) already lack any detectable σNS (Morgan and Zweerink, 1975). It is therefore unlikely that σNS is the viral replicase enzyme.

If σNS is not the viral replicase enzyme, then what is the role of the poly(C)-dependent RNA polymerase activity associated with σNS particles? It is tempting to speculate that the poly(C)-dependent polymerase may represent a primase activity that synthesizes a short section of dsRNA at the 3' end of template RNA. Such a primer could then be elongated by the ss→dsRNA polymerase, which would displace the σNS from the template. As already mentioned, only poly(C) serves as a template for the activity associated with σNS (Gomatos et al., 1980, 1981). It is worth noting that all ten reovirus plus strands have a cytosine residue at their 3' termini (Banerjee et al., 1971). It is possible that the poly(C)-dependent polymerase activity can initiate complementary strand synthesis only at a 3'-terminal cytosine, and hence its apparent specificity in vitro. Even in the presence of $Mg^{2+}$ in vitro, the enzyme can initiate polymerization in the absence of a primer (Gomatos et al., 1980). Could the absence of σNS proteins from replicase particles explain their inability to initiate minus-strand synthesis in vitro? Clearly, a great deal of work will be required to clarify the role of the σNS in replication.

Experimental results have also implicated σNS in translation of viral mRNAs. Several studies have shown that reovirus mRNAs associated with polyribosomes are noncovalently linked (Ward and Shatkin, 1972; Christman et al., 1973; Kreft, 1980). The linked mRNA species have been found in complexes containing both host and viral proteins. These complexes varied in sedimentation rates, but most sedimented between 50 and 110 S. Particles of different sizes were shown to contain ten species of reovirus mRNA, and the complexes were protease-sensitive. The role of these complexes is unclear, but they may be involved in regulation of protein synthesis in infected cells via some as yet undetermined mechanism. Alternatively, such complexes may be precursors to particles in which minus-strand synthesis occurs. This would suggest that the formation of such provirions may be closely coupled to protein synthesis, accounting for the rapid inhibition of replication in the presence of protein inhibitors (Kudo and Graham, 1966; Shatkin and Rada, 1967; Watanabe et al., 1967a). Indeed, the association of virus like particles with polyribosomes has been observed in neuronal cells of newborn mice (Shestopalova *et al.*, 1977). Once again, much work has yet to be done before we can understand the role of these polysome-associated complexes in reovirus replication.

## 4. Replication and Morphogenesis

To further characterize the particles with ss→dsRNA polymerase activity, it was necessary to obtain such particles in a relatively pure form. Unfortunately, preparations of these particles were very heterogeneous and also contained particles with transcriptase activity (Watanabe *et al.*, 1968a; Sakuma and Watanabe, 1971; Zweerink *et al.*, 1972). It has therefore been assumed that the particles that possess ss→dsRNA polymerase also express the ds→ss polymerase (transcriptase) activities. Presumably, as minus-strand synthesis is completed, the resulting dsRNA can immediately serve as a template for mRNA synthesis. Whether or not this is a valid assumption remains a moot point.

More recently, Morgan and Zweerink (1975) demonstrated that the replicase and polymerase activities may actually be associated with discrete particles. Although this had been previously suggested by Sakuma and Watanabe (1971), the latter investigators had employed chymotryptic digestion in their isolation procedure, thereby complicating the interpretation. In the more recent studies, no chymotrypsin was used. By increasing the time of the glycerol gradient centrifugation, Morgan and Zweerink (1975) obtained fractions that contained ss→dsRNA polymerase but were relatively free of transcriptase activity. These replicase particles had a sedimentation coefficient of 180 S and banded in CsCl at a density of 1.35 g/cm$^3$. Analysis of the polypeptides by polyacrylamide gel electrophoresis (PAGE) showed that these particles contained the core proteins, λ1, λ2, μ1, and σ2. These particles almost certainly contain λ3, but the PAGE buffer system used did not allow for its detection. The λ2 protein was present in reduced amounts relative to λ1, consistent with the absence of spikes on the surface of these particles (White and Zweerink, 1976). The replicase particles also contained the outer-capsid proteins, μ1C, σ1, and σ3, but the relative amounts of these differed from those found in virions. The outer-capsid proteins are probably not essential for replication *per se*, since their removal with chymotrypsin did not impair the ss→dsRNA polymerase of the larger, more mature particles (Zweerink *et al.*, 1972; Zweerink, 1974). Morphologically, the replicase particles were approximately 40 nm in diameter, with no distinct features such as capsomers or spikes being discernible. The replicase particles were often surrounded by a diffuse layer that is presumably comprised of outer-capsid proteins associated with these particles.

As determined previously, the replicase particles synthesize dsRNA segments sequentially in particles with increasing sedimentation coefficients (Zweerink, 1974). The small-, medium-, and large-size classes of genome segments are synthesized in particles sedimenting at 300, 450, and 550 S, respectively. The 180 S replicase particles (Morgan and Zweerink, 1975) have also been shown to synthesize predominantly the small genome segments. Clearly, the observed increases in sedimentation rate cannot be accounted for by the dsRNA synthesized within these particles. In all probability, replication and morphogenesis occur concomitantly by addition of proteins and probably by rearrangements brought about by changes in the RNA–protein interactions within the particles (Acs et al., 1971; Zweerink et al., 1972).

Particles with transcriptase activity (220 S) had a density of about 1.42 g/cm$^3$ and contained all the dsRNA genome segments (Morgan and Zweerink, 1975). These particles contained the virus-specific polypeptides λ1, λ2 (in reduced amounts), μNS, and σ2, and must also contain λ3 (Drayna and Fields, 1982), although the latter was not detected. Morphologically, the transcriptase particles were also 40 nm in diameter and lacked spikes and capsomers. Particles with transcriptase activity are almost certainly derived from replicase particles. If this is the case, then the outer-capsid proteins associated with replicase particles must be displaced during the conversion to particles with transcriptase activity, possibly by addition of the μNS protein.

## 5. Intracellular Site of Replication and Morphogenesis

There have been numerous light- and electron-microscopic studies on the cytopathology of reovirus infection, and these have been extensively reviewed by Rosen (1968). The most characteristic feature of infected cells is the development of cytoplasmic inclusions or factories consisting of progeny virions. These factories can be visualized by a variety of techniques including the following: direct, light, and electron microscopy (Rhim et al., 1962; Gomatos et al., 1962; Dales, 1963; Spendlove et al., 1963a; Dales et al., 1965; Mayor and Jordan, 1965; Anderson and Doane, 1966; Fields et al., 1971), staining for RNA with acridine orange (Gomatos et al., 1962), staining with fluorescence-coupled antibody directed against dsRNA (Silverstein and Schur, 1970; Gavrilovskaya et al., 1974; MacDonald, 1980), staining for viral protein with fluorescein-coupled antibodies (Drouhet, 1960; Rhim et al., 1962; Spendlove et al., 1963a,b, 1964; Hassan et al., 1965; Oie et al., 1966; Fields et al., 1971), and autoradiography or staining with ferritin-conjugated antibodies coupled with electron microscopy (Dales et al., 1965). These perinuclear inclusions or factories appear at 6–8 hr after infection (37°C) and are first prevalent in the perinuclear region. Inclusions are found associated with the spindle fibers of the mitotic apparatus. At late times postinfection, these factories form cytoplasmic inclusions that are arrayed in a collar around the nucleus as seen by phase-contrast microscopy. Electron-microscopic observations indicate that the factories contain mature and immature, progeny virions on parallel arrays of microtubules (Fig. 12), and occasionally crystalline arrays of full and empty virions are evident.

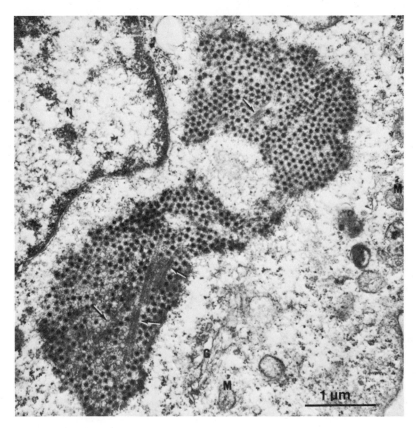

FIGURE 12. Large perinuclear inclusion of reovirus particles, most of which possess dense nucleoids. (→) Coated spindle tubules. ×25,000. Reprinted from Dales (1965a), with permission.

The microtubules, however, do not play an obligatory role in the replication of reovirus, since reovirus grows to normal yields in cells treated with colchicine (Dales, 1963) and other cytoskeleton poisons (Spendlove et al., 1963b). In the absence of colchicine, viral antigens spread throughout the cytoplasm in a reticular network. In the presence of this microtubule poison, antigens are found in large clumped masses. Although no polymerized microtubules are found in these treated cells, the progeny virions are still embedded in a reticulum of "kinky" filaments which consist of vimentin filaments that are disrupted as a result of infection (Sharpe et al., 1982). The characteristic perinuclear collar of progeny virions is not formed in treated cells.

Viral factories are distinct cytoplasmic entities that remain relatively intact if extracted from cells in the absence of detergents (Gomatos, 1967), but are not enclosed by any membrane structures. Both replication and transcription are catalyzed by particles contained within these factories (see above). The factories do not, however, contain any ribosomes, indicating that viral proteins are synthesized on polyribosomes outside the factories. Morphogenesis occurs within the factories, since incomplete virions accumulate within the factories of cells at nonpermissive temperatures with temperature-sensitive mutants defective in morphogenesis (Fields *et al.*, 1971). This means that viral proteins synthesized outside viral factories are transported into factories for assembly into virions. Conversely, mRNAs transcribed from progeny virions within the factories must be transported to the cytoplasm to be translated into viral proteins. No viral antigens were detected in nuclei of infected cells, indicating that reovirus replication is independent of host nuclear function. This notion is corroborated by the ability of enucleated cells to support reovirus replication (Follet *et al.*, 1975).

## E. Transcription from the Progeny Double-Stranded RNA Genome

### 1. Structure of Progeny Subviral Particles Active in Late Messenger RNA Synthesis

Late reovirus messenger RNAs (mRNAs) are defined as those transcribed from the progeny genome by the transcriptase associated with progeny subviral particles (SVPs). Synthesis of progeny transcripts begins 4–6 hr after infection, shortly after the onset of progeny double-stranded RNA (dsRNA) formation (Kudo and Graham, 1965). Cells in which the infection is allowed to proceed in the presence of cycloheximide, so as to prevent the formation of progeny SVPs (Watanabe *et al.*, 1967a,b), fail to produce progeny mRNA. Similarly, no progeny transcripts occur in cells infected at nonpermissive temperatures with temperature-sensitive mutants defective in dsRNA synthesis (Ito and Joklik, 1972a). Both these approaches demonstrated that 80–95% of the viral mRNA present in infected cells is synthesized by the transcriptase associated with immature progeny virions (progeny SVPs) and can thus be defined as late mRNA.

Synthesis of progeny transcripts reaches a maximum 10–12 hr postinfection, coinciding with the period of maximal dsRNA synthesis, and gradually abates thereafter (Kudo and Graham, 1965; Watanabe *et al.*, 1967a; 1968a). The enzyme that catalyzes the synthesis of late mRNAs is associated with a heterogeneously sedimenting population of particles that also expresses replicase (ss→dsRNA polymerase) activity (Watanabe *et al.*, 1968a; Sakuma and Watanabe, 1971; Zweerink *et al.*, 1972). It is therefore probable that transcription occurs within replicase particles soon after dsRNA synthesis is completed. More recent reports (Morgan and Zweerink, 1975, 1977; Zweerink *et al.*, 1976) have described the isolation of transcriptase particles that contain ten dsRNA segments, sediment at 220 and 400 S, and have buoyant densities of 1.42 and 1.44 g/cm$^3$, respectively, in CsCl. These particles were found to contain the viral core polypeptides $\lambda$1, $\lambda$2 (in reduced amounts), $\lambda$3 (although not detected by the gel system used), and $\sigma$2. In addition, these particles have associated with them variable amounts of the viral nonstructural protein $\mu$NS. Replicase particles, on the other hand, did not have any associated $\mu$NS proteins, but did contain variable amounts of the outer-capsid proteins $\mu$1C, $\sigma$1, and $\sigma$3 (Morgan and Zweerink, 1975). If replicase particles are direct precursors to transcriptase particles, then the $\mu$1C, $\sigma$1, and $\sigma$3 proteins associated with the former particles must be displaced on conversion to transcriptase particles. It is plausible that the proteins are displaced during the interaction of the transcriptase particles with the $\mu$NS protein. Electron-microscopic observations of the particles that have transcriptase activity indicate that they are about 40 nm in diameter and possess no distinctive features such as capsomers or spikes. Transcriptase particles isolated from the "slow-sedimenting fraction" described by Skup and Millward (1980b) have similar morphology and protein composition. The absence of spikes is consistent with the finding that these particles contained reduced amounts of the $\lambda$2 protein, which constitutes the spikes (White and Zweerink, 1976; Ralph *et al.*, 1980). The protein matrix of these particles appears to be loosely organized, as indicated by intensive staining with uranyl acetate. This loose assembly is believed to facilitate the release of newly formed transcripts in the absence of spikes, through which transcripts are usually extruded (Bartlett *et al.*, 1974). The $\mu$NS polypeptide associated with the transcriptionally active progeny particles may therefore play a role in maintaining the proper configuration within the particles to allow for mRNA extrusion (Morgan and Zweerink, 1975).

It is apparent from these results that many different classes of particles may participate in the synthesis of late mRNA. The various particles that express transcriptase activity are presumed to be intermediates along a morphogenetic pathway leading to mature-virus formation (see Section H.1). More detailed investigations will be necessary to clarify a role for the various progeny particles with transcriptase activities.

## 2. Enzyme Activities Associated with Progeny Subviral Particles

Recently, Skup and Millward (1980b) reported on the enzyme activities associated with partially purified preparations of progeny SVPs isolated from L cells at 12 hr postinfection (37°C), at which time late mRNA synthesis is maximal. The preparations were assayed for all the enzyme activities known to be associated with reovirus cores. In addition to an active transcriptase, progeny SVPs were shown to have an active nucleotide phosphohydrolase. This enzyme removes the $\gamma$-phosphate from nascent transcripts, producing a diphosphorylated 5' terminus. The phosphohydrolase normally generates a substrate for the viral guanylyltransferase activity leading to the formation of 5'-terminal cap structures (Furuichi *et al.*, 1976). However, when progeny SVPs were assayed for the viral guanylyltransferase activity, the enzyme was found to be latent within these particles. Similarly, the progeny SVPs were found to have masked methyltransferase activities. These RNA-modifying enzymes were not expressed by progeny SVPs unless these particles were first treated with chymotrypsin.

The appearance in infected cells of particles with active transcriptase enzymes and masked RNA capping enzymes was studied as a function of time postinfection. It was found that these particles first appeared 4–6 hr postinfection, attaining maximal levels at 12 hr after infection and declining gradually thereafter [see Fig. 7 (Section III.C.4)]. Parental particles with active capping enzymes were also present in these preparations at early times postinfection, but were virtually absent by 12 hr. Presumably, the parental particles are reassembled into virions by addition of outer-capsid proteins (Astell et al., 1972). The fact that progeny particles have latent mRNA capping enzymes suggested that the mRNA synthesized by these particles should be uncapped. Furthermore, since these particles accumulate in the infected cell, the bulk of the viral transcripts present in infected cells at late times should also be uncapped. Analysis of mRNA synthesized in vitro by progeny SVPs demonstrated that this was indeed the case (Zarbl et al., 1980b). Progeny transcripts failed to incorporate the β-phosphate from $[β-^{32}P]$-GTP into their 5' termini, indicating that the uncapped mRNA synthesized by these particles was different from uncapped ppG-terminated mRNA that can be synthesized in vitro by cores (Furuichi and Shatkin, 1976). The transcripts could, however, be labeled at their 5' termini with polynucleotide kinase, but only after treatment with phosphatase. Direct analysis established that the 5' termini of progeny transcripts had the structure pGpC. . . . The inability of progeny SVPs to cap nascent transcripts is not an artifact of the procedure used in extracting these particles, since viral mRNAs extracted from polysomes at late times after infection also had uncapped pGp-terminated 5' ends (Skup et al., 1981).

Desrosiers et al. (1976) had previously reported that viral mRNA synthesized in infected cells had modified 5' termini consisting of equal amounts of type 1 ($m^7$GpppGmpC . . .) and type 2 ($m^7$GpppGmpCmp . . .) cap structures. However, these investigators were specifically looking for capped mRNAs and hence labeled infected cells with [$^3$H-methyl]methionine. Furthermore, labeling of infected cells was carried out between 5 and 11 hr postinfection. During this period, parental SVPs are still actively transcribing capped mRNA (Skup and Millward, 1980b; Shup et al., 1981). Desrosiers et al. (1976) were therefore specifically labeling early mRNA (parental transcripts). Skup et al. (1981), on the other hand, labeled infected cells with [$^{32}$P]orthophosphoric acid to lable both capped and uncapped mRNAs. Whereas Desrosiers et al. (1976) had extracted total RNA from infected cells, Skup et al. (1981) specifically extracted viral mRNAs that were active in protein synthesis. When these criteria were applied, it was found that as the infection proceeded, uncapped mRNAs were preferentially translated, while capped parental transcripts were gradually excluded from polysomes. In addition, the translational apparatus of L cells has been shown to undergo a transition from cap dependence to cap independence, thereby allowing for preferential translation of uncapped progeny transcripts (Skup and Millward, 1980a). Therefore, the fact that progeny SVPs have masked capping enzymes, coupled to a viral-induced transition to cap-independent translation, accounts for the takeover of protein synthesis seen in L cells infected with reovirus (see Section F.3).

The fact that progeny transcripts have uncapped monophosphorylated termini indicated the presence of a pyrophosphatase activity in progeny SVPs. This enzyme hydrolyzes the β-phosphate from the diphosphorylated termini generated on nascent transcripts by the phosphohydrolase of progeny SVPs. It is not yet known whether this pyrophosphatase activity is associated with a viral protein or is a host enzyme inducted as a result of infection. The enzyme is present in progeny SVP preparations, since these particles synthesize pG-terminated mRNA. The pyrophosphatase activity has also been detected in lysates prepared from infected cells late in infection, while no activity has been detected in uninfected cells (Skup and Millward, 1980a; Zarbl and Millward, unpublished). The pyrophosphatase activity does not hydrolyze cap structures to yield monophosphorylated 5′ termini. Uncapped ppG-terminated mRNAs labeled in the β-phosphate were synthesized *in vitro* by reovirus cores and incubated in lysates prepared from infected cells (Skup and Millward, 1980a). The 5′-terminal structures of these uncapped mRNAs were determined before and after incubation in the infected lysates. The results indicated that the viral-induced pyrophosphatase activity found in the infected cell lysates specifically removed the β-phosphate from diphosphorylated 5′ ends. The mechanism for the formation of monophosphorylated 5′ termini of progeny transcripts is believed to be as follows:

$$pppG + pppC \xrightarrow[\text{(transcriptase)}]{\text{RNA polymerase}} pppGpC + PP_i \qquad (1)$$

$$pppGpC \xrightarrow{\text{nucleotide phosphohydrolase}} ppGpC + P_i \qquad (2)$$

$$ppGpC \xrightarrow{\text{polynucleotide pyrophosphatase}} pGpC + P_i \qquad (3)$$

Once the β-phosphate is removed, the transcripts are a substrate for neither the guanylyltransferase of reovirus (Furuichi *et al.*, 1976) nor the guanylyltransferase of the host cell (Venkatesan and Moss, 1980). It is therefore possible that the latency of capping enzymes in progeny SVPs is due to the presence of the pyrophosphatase activity. Treatment of progeny SVPs with chymotrypsin might remove or inactivate the protein with pyrophosphatase activity, thereby unmasking the guanylyltransferase and RNA methylases (Skup and Millward, 1980b). If this is indeed the case, the viral nonstructural protein μNS associated with immature progeny particles (Morgan and Zweerink, 1975; Zweerink et al., 1976) is the most likely candidate.

This nonstructural viral polypeptide is associated with progeny particles, which produce uncapped mRNA, but is not present in parental SVPs or cores, which synthesize capped mRNA. If the pyrophosphatase activity is indeed associated with the μNS protein, then nascent transcripts may have their β-phosphate removed prior to interaction with the guanylyltransferase. Such a mechanism alleviates the apparent contradiction of synthesizing uncapped mRNA in particles that appear to contain normal amounts of the λ1 protein, which is the reovirus guanylyltransferase (D.R. Cleveland, H. Zarbl, and S. Millward, in prep.).

An alternative explanation, however, can also be evoked. Progeny SVPs do not have on their surface the spikes through which nascent transcripts are believed to be extruded (Gillies et al., 1971; Bartlett et al., 1974). As discussed in Section C.2, it is possible that guanylylation and methylation of transcripts occur as they pass through these spikes. It is possible that the nascent transcripts synthesized by progeny SVPs are released via an alternative mechanism without being exposed to the capping enzymes and therefore remain uncapped. Such a model still requires that a pyrophosphatase activity be present in the progeny SVP preparations. The contention that μNS is the pyrophosphatase activity induced by reovirus is consistent with either explanation. Studies are currently in progress to determine whether the reovirus pyrophosphatase activity is indeed associated with the μNS protein. Experiments are also in progress to elucidate the mechanism that results in the masking of mRNA capping enzymes in progeny SVPs.

## F. Translation of Reovirus Messenger RNAs

A detailed analysis of the molecular mechanisms of reovirus mRNA translation is presented in Chapter 3. These mechanisms will be reviewed only briefly in this section, the main emphasis being on how they relate to the regulation of protein synthesis in the infected cell. The frequency at which any given protein is translated can be controlled at several levels, the first being at the level of transcription. The rates of transcription of the individual reovirus messenger RNAs (mRNAs) have already been discussed in previous sections. We will now consider the regulation of protein synthesis at the level of translation.

### 1. Reovirus-Encoded Polypeptides

Early studies on the viral polypeptides synthesized in infected cells indicated that only 8 primary translation products of reovirus mRNA were made (Zweerink and Joklik, 1970; Zweerink et al., 1971; Fields et al., 1972). Similar results were obtained during translation of endogenous viral mRNA in lysates prepared from reovirus-infected cells (McDowell and Joklik, 1971) or when cell-free lysates were primed with reovirus mRNA synthesized in vitro (Graziadei and Lengyel, 1972; McDowell et al., 1972; Graziadei et al., 1973). More recently, the viral proteins synthesized in infected cells and cell-free lysates have been reexamined using high-resolution discontinuous gel electrophoresis (Cross and Fields, 1976; Both et al., 1975a). When viral specific polypeptides were analyzed in this way, as many as 16 polypeptides were detected. Many of these probably arose via cleavage of primary translation products or may have been the result of other posttranslational modifications. It was thus necessary to determine which of the proteins detected were primary translation products and which double-stranded RNA (dsRNA) segments encoded their sequences.

Identification of the primary translation products of reovirus mRNAs required that each mRNA species be isolated and translated individually. The mRNA species transcribed from each of the ten dsRNA genome segments has been identified, since both types of RNA can be resolved on polyacrylamide gels and each mRNA species can be hybridized to the minus strand of its corresponding dsRNA genome segment (Floyd et al., 1974; Schuerch et al., 1975; Joklik, 1974). It proved difficult, however, to prepare sufficiently large amounts of each species of single-stranded RNA. McCrae and Joklik (1978) therefore utilized reovirus dsRNA segments as a source of reovirus mRNAs, since individual dsRNA segments can be isolated in bulk. It was found that under suitable conditions of denaturation, the plus strands of dsRNA could be translated into viral polypeptides in cell-free protein-synthesizing systems. By translating the individual, denatured gene segments and analyzing the products by gel electrophoresis, as well as comparing peptide maps of these products to those of virion proteins, the protein encoded by each dsRNA genome segment was unequivocally determined. At the same time, Mustoe et al. (1978b) used a genetic approach to arrive at the same gene coding assignments for the $\mu$ and $\sigma$ classes of viral polypeptides. These investigators made use of the observation that the three serotypes of reovirus show different mobilities in corresponding dsRNA segments and polypeptides (Ramig et al., 1977; Sharpe et al., 1978). Intertypic recombinants that consisted primarily of dsRNA segments of one serotype and a limited number of dsRNA segments of another serotype were isolated. Gene coding assignments could then be made on the basis of which substituted dsRNA segment resulted in the substitution of a corresponding polypeptide. The results of these experiments were in complete agreement with those of McCrae and Joklik (1978). Translation of purified mRNAs of the small-size class have also corroborated these assignments (K.H. Levin and Samuel, 1980). The results are summarized in Fig. 13. From these results,

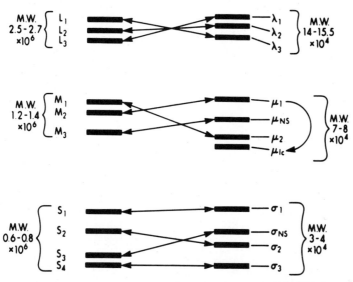

FIGURE 13. Summary of RNA coding assignments for reovirus type 3, strain Dearing. The positions of the dsRNA and polypeptide species are those assumed in the Loening's buffer E (0.036 M Tris, 0.03 M phosphate, pH 7.8) and Tris–glycine (Laemmli, 1970) systems, respectively. Reprinted from McCrae and Joklik (1978), with permission.

it is clear that all ten reovirus mRNA species are translated into unique viral polypeptides.

In addition to the ten primary proteins, a number of cleavage products are present either in virions or in infected cells. The $\mu$1 protein is extensively cleaved to $\mu$1C and in the cleaved form is a major outer-capsid component (Zweerink and Joklik, 1970; McCrae and Joklik, 1978). Antisera directed against individual viral polypeptides have indicated the existence of additional cleavage products including $\lambda$2C and $\mu$NSC (Joklik, 1980).

## 2. Relative Rates of Translation of Individual Messenger RNA Species

The relative rate at which any given viral protein is synthesized in the infected cell is to some extent regulated by the rate at which the corresponding mRNA species is transcribed (Zweerink and Joklik, 1970; Lau *et al.*, 1975). It is clear, however, that there is a profound control over the frequency with which each viral mRNA is translated. The mRNA species within a given size class are transcribed in almost equimolar quantities *in vitro* (Zweerink and Joklik, 1970; Both *et al.*, 1975a; Joklik, 1974) and are present in similar amounts in the infected cell (Zweerink and Joklik, 1970; Nonoyama *et al.*, 1974). The amount of each polypeptide synthesized during translation does not necessarily reflect the amount of each mRNA species present. Clearly, some mRNA species are translated more frequently than others. This is true whether translation is examined *in vivo* (Zweerink and Joklik, 1970; Zweerink *et al.*, 1971; Fields *et al.*, 1972; Skup *et al.*, 1981) or in cell-free lysates prepared from infected cells (McDowell and Joklik, 1971). Similar relative frequencies of translation are observed when reovirus mRNA synthesized *in vitro* is used to prime cell-free extracts from a wide variety of cell types (Graziadei and Lengyel, 1972; McDowell *et al.*, 1972; Graziadei *et al.*, 1973; Both *et al.*, 1975a; Samuel *et al.*, 1977; Skup and Millward, 1977, 1980a; K.H. Levin and Samuel, 1980) or when polysomal mRNA from infected cells is translated *in vitro* (Skup *et al.*, 1981). Although some differences may be apparent, it is clear that the relative efficiencies at which individual mRNAs are translated are inherent in the structure of the mRNA species. As will be discussed in Section III.F.3, the reovirus mRNAs translated at early times after infection are capped parental transcripts, while uncapped progeny transcripts are preferentially translated at later times. The frequency with which individual mRNA species are translated, however, shows no significant changes during the course of the infection (Zweerink and Joklik, 1970; Skup *et al.*, 1981). Therefore, the mRNA cap structures, while allowing for the discrimination against translation of host and parental subviral particle (SVP) transcripts, do not appear to affect the frequencies at which individual mRNAs are translated (Samuel *et al.*, 1977; Skup and Millward, 1980a; Skup *et al.*, 1981) in a cap-independent translation system. It therefore appears that the major, if not the sole, mechanism for regulating the translation frequencies of the reovirus mRNAs in a given environment resides in their unique sequence content or secondary structure.

A large number of studies have been carried out to elucidate the sequences found at the 5′ end of the reovirus mRNA species (Nichols *et al.*, 1972b; Kozak and Shatkin, 1976; 1977a,b, 1978; Kozak, 1977; Hastings and Millward, 1978; Darzynkiewicz and Shatkin, 1980; Kozak, 1982a,b; Antczak *et al.*, 1982; Gaillard *et al.*, 1982). The following features have been found in the 5′-terminal sequences of those six mRNAs that have been examined: (1) all begin with the sequences GCUA; (2) the noncoding sequences vary between 13 and 33 residues in length and have completely different sequences; (3) these regions are devoid of secondary structures such as hairpins; and (4) each mRNA has a sequence that is complementary to the 3′ end of 18 S ribosomal RNA. Recently, Li *et al.* (1980a,b) sequenced the 3′ ends of the *S1* and *S2* reovirus plus strands. These studies have revealed that there are sequences 6–10 residues long near the 3′ end of each reovirus mRNA that are complementary to sequences at the 5′ end of the same mRNA. These complementary sequences have been found for most reovirus mRNAs examined and have a relatively high free energy of association [−10 to −20 kcal (Joklik, 1981)]. In some cases, these complementary sequences overlap with the sequence near the 5′ end of the mRNA that is complementary to the 3′ end of 18 S ribosomal RNA (Shine and Dalgarno sequences), while in other cases there is no overlap. It is therefore possible that sequences immediately upstream from the initiation codon of reovirus mRNAs can associate either with complementary sequences near their own 3' end or with sequences at the 3' end of 18 S ribosomal RNA. Analysis of sequences at the 5' and 3' termini of all ten reovirus mRNAs failed to show a correlation between the stability of such interactions and the relative translational frequencies (Antczak et al., 1982). Kozak (1981; 1982a,b) has found the translational frequency to correlate with a consensus sequence flanking the initiation codon.

Recently, an in vitro assay was developed to compare the relative efficiencies at which mRNAs compete for limiting components in the translation-initiation machinery (Brendler et al., 1981a). Under certain conditions of $Mg^{2+}$ and $K^+$ concentrations, the relative efficiencies could mimic those found in vivo (Brendler et al., 1981a,b). It thus appears that competition among the reovirus mRNAs for components of the translation-initiation machinery plays a crucial role in determining their relative translation frequencies. Consistent with the observation that these relative rates are similar throughout the infectious cycle, it was also found that competition was independent of 5'-terminal cap structures (Brendler et al., 1981a). This indicates that the preferential translation of uncapped viral mRNAs over capped mRNAs is independent of competition.

## 3. Regulation of Protein Synthesis in L Cells Infected with Reovirus

Regulation of protein synthesis in the infected cell also occurs at the level of viral vs. host mRNA translation. When L cells are infected with reovirus, there is no rapid shutoff of host protein synthesis. Host protein synthesis continues relatively unabated at early times after infection and is gradually diminished so that by 10 hr, viral proteins comprise the majority of the translation products (Zweerink and Joklik, 1970; Joklik, 1974). This pattern of polypeptide synthesis is also observed when endogenous mRNA is translated in lysates prepared from infected cells (McDowell and Joklik, 1971; Skup and Millward, unpublished). The mechanism allowing for the preferential translation of viral mRNAs has been studied extensively and found to be operative at the level of initiation of translation. The host translational apparatus has been found to undergo a viral-induced transition from cap-dependent to cap-independent translation, leading to the preferential translation of uncapped progeny transcripts (Skup and Millward, 1980a; Skup et al., 1981).

Studies on the regulation of protein synthesis in infected cells were made possible with the development of nuclease-treated cell-free lysates from L cells that were highly efficient in their response to exogenous mRNA and exhibited negligible background translation (Skup and Millward, 1977). Lysates prepared from uninfected L cells translate mRNAs that have capped 5' termini. This is the expected result, since cap structures of the general form $m^7G(5')ppp(5')N(m)p$ have been found at the 5' ends of all cellular and most viral mRNAs. Furthermore, it has become clear that these cap structures play an essential role in the initiation step of translation. These topics have been reviewed recently (Shatkin and Both, 1976; Shatkin, 1976; Filipowicz, 1978; Kozak, 1978; Banerjee, 1980) and are dealt with in detail in Chapter 3. Briefly, however, cap structures at the 5' end of mRNAs have been found to be required for efficient translation by facilitating the binding of mRNAs to 40 S ribosomal sub-units during initiation of protein synthesis (Both *et al.*, 1975b,c; for a review, see Banerjee, 1980). This binding has been shown to be mediated by a polypeptide of molecular weight 24,000 present among host-cell initiation factors, called the cap-binding protein (Sonenberg and Shatkin, 1977; Sonenberg *et al.*, 1978; Sonenberg *et al.*, 1979a,b). This protein binds specifically to 5'-terminal cap structures, and its binding can be inhibited by cap analogues. A number of polypeptides related to this cap-binding protein have also been detected and have been shown to be related antigenically and by peptide mapping. The binding of these larger cap-binding proteins to the cap structure was found to be ATP/$Mg^{2+}$-dependent (Sonenberg, 1981). In addition, monoclonal antibodies directed against cap-binding proteins have been shown to preferentially inhibit initiation-complex formation with mRNAs having extensive secondary structure, while initiation-complex formation with mRNAs with relatively less secondary structure was only marginally inhibited (Sonenberg *et al.*, 1981a). It has thus been postulated that the mechanism by which the cap-binding proteins stimulate the formation of initiation complexes may involve unwinding of the secondary structure at the 5' end of eukaryotic mRNAs, thereby allowing attachment of the 40 S ribosomal subunit to the mRNA (Sonenberg, 1981; Sonenberg *et al.*, 1981a; Sonenberg *et al.*, 1982). Clearly, further studies will be required to elucidate the precise role of cap-binding proteins in the formation of initiation complexes. Nonetheless, it is evident that cap structures play a critical role in eukaryotic-cell protein synthesis.

Correspondingly, cell-free extracts from uninfected L cells have been shown to translate capped reovirus mRNAs with high efficiency, yielding normal viral polypeptides (Table II) (Skup and Millward, 1977, 1980a). Furthermore, translation of the capped mRNAs in these lysates is inhibited by cap analogues (Skup and Millward, 1980a) and by monoclonal antibodies directed against cap-binding proteins (Sonenberg *et al.*, 1981b). These same lysates failed to translate uncapped ppG-terminated reovirus mRNA prepared *in vitro* using reovirus cores. However, S-adenosyl-L-homocysteine (AdoHcy) must be added in the latter case to prevent capping, since ppG-terminated mRNA is a substrate for host capping enzymes (Venkatesan and Moss, 1980). Uncapped pG-terminated reovirus mRNAs synthesized by progeny SVPs *in vitro* (Zarbl *et al.*, 1980b) or isolated from polysomes of infected cells (Skup *et al.*, 1981) also fail to translate in uninfected cell lysates, even in the absence of AdoHcy.

Cell-free lysates were then prepared from L cells infected with reovirus at a time postinfection when synthesis of viral-specific proteins was maximal (Skup and Millward, 1980a). In contrast to lysates prepared from uninfected cells, infected-cell lysates translated capped reovirus mRNAs

TABLE II. Inhibition of Protein Synthesis by m$^7$G(5')ppp[a]

| Extract | Messenger RNA | m$^7$(G(5')ppp | Counts per minute | Inhibition (%) |
|---|---|---|---|---|
| Uninfected | — | – | 109 | — |
| | — | + | 129 | — |
| | Capped | – | 844,832 | — |
| | Capped | + | 78,632 | 91 |
| | Uncapped | – | 8,448 | — |
| | Uncapped | + | 8,571 | 0 |
| Infected | — | – | 175 | — |
| | — | + | 189 | — |
| | Capped | – | 331,190 | — |
| | Capped | + | 342,681 | 0 |
| | Uncapped | – | 798,371 | — |
| | Uncapped | + | 801,231 | 0 |

[a] Reprinted from Skup and Millward (1980a), with permission. All reaction mixtures contained 320 μM AdoHcy. The concentration of m$^7$G(5')ppp was 1 mM. Standard reactions were carried out for 90 min. Values represent [$^{35}$S]methionine counts incorporated into trichloroacetic-acid-precipitable counts.

at a reduced efficiency (25%). No viral polypeptides were synthesized, and the residual translation was insensitive to inhibition by cap analogues (Table II). On the other hand, uncapped ppG-terminated reovirus mRNAs translated efficiently in lysates from infected cells, yielding normal viral proteins. Predictably, the translation of uncapped mRNA was not sensitive to inhibition by cap analogues (Skup and Millward, 1980a) or antibodies directed against cap-binding proteins (Sonenberg et al., 1981b). Lysates from infected cells also translate with high efficiency uncapped pG-terminated reovirus mRNAs synthesized by progeny SVPs in vitro (Zarbl et al., 1980b) or isolated from polysomes of infected cells (Skup et al., 1981).

The ability of cell-free extracts from infected cells to translate uncapped mRNA preferentially over capped mRNA was found to increase as a function of time after infection (Fig. 14). The preferential translation of uncapped mRNAs by infected-cell lysates corresponded temporally with the period of maximal mRNA synthesis by progeny SVPs. The latter particles have been shown to have masked capping enzymes and hence synthesize uncapped mRNAs in vitro (Skup and Millward, 1980b; Zarbl et al., 1980b). It was therefore proposed that the mechanism by which reovirus usurps the translational machinery of the host cell involves a reovirus-induced transition from cap-dependent translation to cap-independent translation. Such a transition could then allow for preferential translation of uncapped progeny transcripts, while capped host transcripts would be excluded from polysomes. If such a mechanism is indeed operative, then viral mRNAs synthesized late in the infected cell should be uncapped. Total high-molecular-weight RNA was extracted from both normal and infected cells and translated in lysates prepared from normal

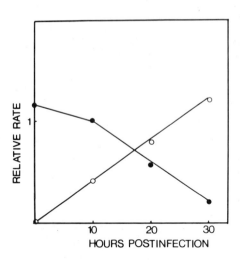

FIGURE 14. Transition from cap-dependent to cap-independent translation in infected L cells. L cells were infected with reovirus at 31°C. At the times indicated, samples of cells were removed and S-10 extracts were prepared. Kinetics of incorporation of [35S]methionine into hot acid-precipitable material were determined for each point. Initial rates of protein synthesis were determined under conditions of saturating mRNA concentrations. The initial rates of incorporation of [35S]methionine were $2.5 \times 10^4$ cpm/min with capped mRNA in S-10 extracts prepared at time zero after infection and $2.8 \times 10^4$ cpm/min with uncapped mRNA in S-10 extracts prepared at 30 hr after infection. (●) Translation of capped mRNA; (○) translation of uncapped mRNA. Reprinted from Skup and Millward (1980a), with permission.

and infected L cells (Table III and Fig. 15). The results of these experiments suggested that the bulk of the viral mRNAs present in the infected cell at late times were uncapped, since they failed to translate in lysates from uninfected cells. Host mRNAs, however, were still present in the infected cell and were translated in lysates from uninfected cells. Viral mRNAs from infected cells did, however, translate in lysates from infected cells, while little translation of host mRNAs was seen in these infected lysates. The translational properties of total high-molecular-weight RNA from infected cells were therefore consistent with the notion that late viral mRNAs are uncapped *in vivo*.

To show that a transition to cap-independent translation does indeed occur in the infected cell, further criteria needed to be established. This entailed the unequivocal demonstration that reovirus mRNA molecules

TABLE III. Translation of Total RNA from Uninfected and Infected L Cells[a]

| Source of RNA | Source of extract | AdoHcy[b] | [35S]-Met incorporated (cpm $\times 10^3$) |
|---|---|---|---|
| Uninfected | Uninfected | − | 36 |
| | Uninfected | + | 37 |
| | Infected | − | 12 |
| | Infected | + | 13 |
| Infected | Uninfected | − | 6 |
| | Uninfected | + | 7 |
| | Infected | − | 52 |
| | Infected | + | 53 |

[a] Reprinted from Skup *et al.* (1981), with permission.
[b] AdoHcy was added where indicated at a concentration of 320 μM.

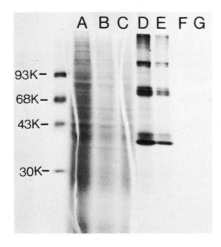

FIGURE 15. Autoradiogram of PAGE analysis of [35]S-labeled translation products of total high-molecular-weight RNA from normal and reovirus-infected L cells. (A) Lysate from uninfected cells primed with RNA from uninfected cells [5 × 10^4 cpm hot trichloroacetic acid (TCA)-precipitable material applied to gel]; (B) lysate from infected cells primed with RNA from uninfected cells (2.5 × 10^4 cpm hot TCA-precipitable material applied to gel); (C) lysate from uninfected cells primed with RNA from infected cells (2 × 10^4 cpm hot TCA-precipitable material applied to gel); (D) lysate from infected cells primed with RNA from infected cells (6 × 10^4 cpm hot TCA-precipitable material applied to gel); (E) polypeptides labeled *in vivo* in reovirus-infected cells with [35S]methionine; (F) lysate from uninfected cells, no mRNA; (G) lysate from infected cells, no mRNA. Electrophoresis was carried out in a 10–18% gradient gel. The lane far left contained radioactive polypeptide markers. Reprinted from Skup *et al.* (1981), with permission.

associated with polysomes at late times were uncapped. Polysomes were isolated from infected cells and the RNA extracted. The purification of viral mRNA associated with polysomes included the ability to form 80 S initiation complexes in both infected- and uninfected-cell lysates, thereby establishing their viability as mRNAs. The final step of purification involved annealing of the mRNA to an excess of reovirus dsRNA to select for viral sequences. Analysis of the 5′ termini of these viral mRNAs clearly demonstrated their uncapped nature. Therefore, it was concluded that viral mRNAs translated at late times in the infection were uncapped progeny transcripts.

Analysis of the 5′ termini of viral mRNAs associated with polysomes at different times postinfection indicated that the proportion of uncapped mRNA increased during the infection (Table IV). Concomitantly, the proportion of capped viral mRNAs being translated decreased. Once again, the transition corresponds temporally to synthesis of uncapped progeny transcripts and the ability of lysates prepared from infected cells to translate uncapped messages.

It was therefore concluded that reovirus induces a transition in the host translational apparatus. As a result, there is a gradual shift from cap-

TABLE IV. Distribution of Capped and
Uncapped 5′ Termini in Early, Intermediate,
and Late Messenger RNAs[a]

| Messenger RNA | Relative number of 5′ termini (%)[b] | |
| --- | --- | --- |
| | pGp | m[7]G(5′)ppp(5′)Gmpcp |
| Early | 29 | 71 |
| Intermediate | 75 | 25 |
| Late | 98 | 2 |

[a] Reprinted from Skup et al. (1981), with permission. Early, intermediate, and late refer to mRNA labeled between 3.5 and
6.5, 6.5 and 9.5, and 10 and 13 hr postinfection, respectively.
[b] Uniform labeling was assumed, and allowance was made for
the different number of P atoms.

dependent to cap-independent translation. This transition results in the
exclusion of capped viral and host mRNAs from polysomes as the infection progresses. At the same time, progeny SVPs are formed and begin to
synthesize large quantities of uncapped messages. As a consequence, the
relative proportion of viral to host proteins synthesized increases with
time postinfection, and eventually only viral proteins are made.

Recently, Walden et al. (1981) reported that competition among
mRNAs may play an important role in regulating the rate at which individual messages are translated in the infected cell. Any decrease in the
host translation rate was presumed to be due to competition with the
reovirus mRNAs for one or more limiting components of the host translational machinery. This interpretation was based on the observation that
low doses of cycloheximide specifically stimulated the translation of viral
mRNAs in SC-1 cells infected with reovirus. The differential effect of

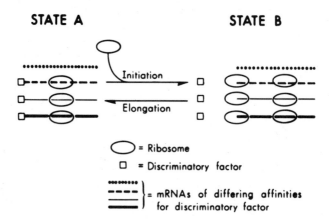

FIGURE 16. Scheme illustrating the proposed effects of cycloheximide treatment on translation in vivo. The site of cycloheximide action is the elongation ribosome, that is, the
transition from state B to state A. Reprinted from Walden et al. (1981), with permission.

cycloheximide on the translation of individual mRNA species was shown to result from a slowing down of the polypeptide elongation rate. Brendler *et al.* (1981a) showed that *in vitro*, individual mRNA species compete with differing efficiencies for limiting components of the translation-initiation machinery. A model for differential effect of cycloheximide on the translation of mRNAs with high or low affinity for the discriminatory factor is presented diagramatically in Fig. 16. It is assumed that the discriminatory factor can exist in one of two states, either bound to mRNA (state A) or free (state B). Slowing down the rate of elongation with cycloheximide could increase the average amount of time that the system spends in state B. This would have the effect of increasing the steady-state concentration of the unbound discriminatory factor and consequently increasing the probability that a low-affinity mRNA (dotted line in Fig. 16) will bind to the factor and be translated.

Results of the experiments by this group of investigators indicated that translation of several viral mRNAs was indeed stimulated by low levels of cycloheximide that inhibited translation of host mRNAs. It was also found that viral-specific polysomes were, on the average, smaller than those of comparable host mRNAs. These results suggested that viral mRNAs competed poorly with host mRNAs for limiting components of the translation system. The inability of viral mRNAs to compete efficiently with host mRNAs could then account for the unusually high amount of viral mRNA present at a time when viral protein synthesis is prominent. It was therefore concluded that in SC-1 cells infected with reovirus, competition of an excess of viral mRNAs with host mRNAs for one or more limiting components involved in the initiation of translation can account for the decline in host protein synthesis and the concomitant increase in viral protein synthesis. It was further suggested that a reovirus-induced transition from cap-dependent to cap-independent translation of the type reported in infected L cells (Skup and Millward, 1980a; Skup *et al.*, 1981) did not occur in SC-1 cells and that the majority of reovirus mRNAs present in infected SC-1 cells were capped (B.M. Detjen *et al.*, 1982).

These results appear to be in direct contradiction to those previously reported in L cells. Some caution must, however, be exercised in the comparison of results obtained in the two cell types. Walden *et al.* (1981) reported that the replication of reovirus was not "defective" in SC-1 cells, since yields of 1000–3000 PFU/infected cell were possible. The maximum burst of infectious virus, however, occurs between 24 and 48 hr postinfection at 37°C. In comparison to the viral life cycle in L cells, the cycle in SC-1 cells is considerably longer. In additiion, it was observed that the relative rates of host and viral protein synthesis were somewhat "variable." This variability was found to be related to the growth state of the cells and specifically to the capacity of the host translational machinery. Therefore, in an average infection, host protein synthesis was inhibited by only about 15% by 12 hr. Observations made at 12 hr postinfection

with reovirus in SC-1 cells probably correspond to an early time postinfection of L cells. The competition observed in SC-1 cells at 12 hr is
probably between capped mRNAs synthesized by parental SVPs and host
mRNAs. There is little doubt that such competition also occurs at early
times in infected L cells. Unfortunately, the studies of Walden et al. (1981)
were not extended beyond 12 hr. It remains to be determined whether
competition alone can account for discrimination between host and viral
mRNA at later times in SC-1 cells infected with reovirus.

It is clear that competition alone cannot account for the preferential
translation of viral mRNAs at late times in L cells infected with reovirus.
Translation of uncapped reovirus mRNA in lysates from infected cells
was found not to be affected by a 3-fold excess of capped mRNA (Skup
and Millward, 1980a). This result suggested that capped mRNAs compete
poorly with uncapped mRNAs in the infected lysate, even when capped
mRNAs are present in excess. It therefore seems unlikely that competition plays a significant role in the preferential translation of viral
mRNAs over that of host mRNAs in infected cells (see Addendum).

Competition of uncapped viral mRNAs for limiting factors of the
translational apparatus could, however, modulate the relative frequencies
with which the individual mRNA species are translated. The affinity of
different mRNA species for the discriminatory factor may be regulated
by the secondary or tertiary structures of the mRNAs (Brendler et al.,
1981a,b), as determined by their specific sequence. Although uncapped
reovirus mRNAs were found to be poor competitors in cap-dependent
cell-free lysates, the relative frequencies at which individual uncapped
mRNAs were translated were the same as those observed with capped
reovirus mRNAs. It is therefore clear that the mRNA discriminatory
component that is responsible for competition must be influenced by or
recognize features of the mRNAs that are unrelated to the 5' cap structures. This suggests that the cap-binding protein, which stimulates the
translation of capped mRNA, probably does not have a role in determining
the relative frequencies at which individual mRNAs are translated. Cap-
binding proteins may act subsequent to the interaction of the discriminatory factor with mRNAs, by allowing binding of the complex to 40 S
ribosomal subunits. Such a model is consistent with the observation that
capped and uncapped reovirus mRNAs are translated at the same relative
frequencies via cap-dependent or cap-independent mechanisms, respectively, both in vivo and in vitro.

The model for mRNA competition that is proposed to be operative
in SC-1 cells infected with reovirus may therefore account for the relative
translation frequencies of reovirus mRNAs observed in infected L cells.
It is unclear, however, whether competition alone can account for the
translation of viral mRNA in SC-1 cells infected with reovirus. The status
of the translational apparatus in SC-1 cells will have to be examined at
later times postinfection if comparisons are to be made to that of infected
L cells.

## 4. Molecular Mechanisms of the Reovirus-Induced Transition to Cap-Independent Translation

The molecular events involved in the reovirus-induced transition to cap-independent translation have not as yet been rigorously determined. The inhibition of host protein synthesis in HeLa cells infected with poliovirus, however, has been extensively studied. Poliovirus mRNAs are naturally uncapped, having a monophosphorylated 5' terminus with the structure pUp (Nomoto et al., 1976; Hewlett et al., 1976). Shortly after infection of HeLa cells with poliovirus, host protein synthesis is markedly inhibited, while uncapped viral mRNAs are preferentially translated. This inhibition results from a failure of capped host mRNAs to enter into initiation complexes (Leibowitz and Penman, 1971; Kaufman et al., 1976; Ehrenfeld and Manis, 1979), even though host-cell mRNA remains structurally intact and is translationally active in vitro (Fernandez-Munoz and Darnell, 1976; Kaufman et al., 1976; Ehrenfeld and Lund, 1977). This pattern of inhibition of host protein synthesis resembles that observed at late times postinfection in L cells infected with reovirus, indicating that a common mechanism may be operative. A brief review of the mechanisms involved in shutoff of host protein synthesis in poliovirus-infected HeLa cells will serve as a basis for presenting by comparison some of our preliminary findings in L cells infected with reovirus.

Several laboratories have found that after infection with poliovirus, some alteration occurs in the initiation factors of the host cell. As a result, crude initiation factors from infected cells stimulate the translation of poliovirus mRNA, but fail to stimulate translation of cellular mRNAs or capped viral mRNA such as those from vesicular stomatitis virus (VSV) (Kaufman et al., 1976; Helentjaris and Ehrenfeld, 1978; Rose et al., 1978; Helentjaris et al., 1979; Jen et al., 1980; Brown and Ehrenfeld, 1980). The failure of extracts from poliovirus-infected cells to translate capped mRNA from VSV can be overcome by initiation factors prepared from reticulocytes (Rose et al., 1978). The "restoring activity" has been attributed to the presence of cap-binding protein in these initiation-factor preparations (Trachsel et al., 1980). Cap-binding proteins can be covalently cross-linked to capped 5' ends of reovirus mRNA (Sonenberg and Shatkin, 1977) and have been shown to stimulate translation of capped mRNAs in vitro, while having no stimulatory effect on the translation of uncapped mRNAs (Sonenberg et al., 1980). As a result, it has been hypothesized that poliovirus infection results in the inactivation of cap-binding protein (Sonenberg et al., 1980; Trachsel et al., 1980). Accordingly, capped host mRNAs are not translated, while uncapped poliovirus mRNAs are translated. In support of this hypothesis, it has been shown that the ability of cap-binding proteins to cross-link with cap structures is reduced in poliovirus-infected cells as compared to mock-infected cells (Lee and Sonenberg, 1982). These investigators have also detected an activity residing in initiation-factor preparations from infected cells that rapidly inactivates cap-binding proteins and also impairs the restoring activity of mock-infected initiation-factor preparations. Hansen and Ehrenfeld (1981), on the other hand, have not detected a diminished ability of cap-binding proteins from poliovirus-infected cells to cross-link with cap structures. More recent studies indicate that poliovirus infection actually inhibits the ability of cap binding proteins to form active complexes with other initiation factors (Tahara et al., 1981; Brown et al., 1982; Etchison et al., 1982; Hansen et al., 1982a,b), thereby inhibiting the ability of capped mRNAs to enter into initiation complexes.

A similar mechanism for inhibition of host protein synthesis is also operative during infection of neuroblastoma cells with Semliki forest virus (Van Steeg et al., 1981). Late viral mRNAs, although capped, appear to be translated via a cap-independent mechanism. As a result of this transition, early viral and cellular mRNAs are not translated. It was found that the ability to translate host and early viral mRNAs could be restored by cap-binding protein preparations from reticulocytes, as had been shown for poliovirus-infected cell lysates. We were thus encouraged to determine whether a similar mechanism is involved in the shutoff of capped-mRNA synthesis in L cells infected with reovirus.

The ability of cap-binding proteins to restore the translation of capped reovirus mRNA in lysates from infected cells was examined (D. Skup, N. Sonenberg, and S. Millward, unpublished). Results of preliminary experiments have indicated that cap-binding protein does have some restoring activity in lysates from infected cells. It is therefore possible that inactivation of cap-binding proteins may also occur during reovirus infection. Further studies will be required before unequivocal conclusions can be drawn.

It is clear, nonetheless, that inactivation of cap-binding proteins alone cannot explain the ability of uncapped reovirus mRNAs to be translated in the infected cell. Unlike the naturally uncapped mRNAs, which are translated in lysates prepared from uninfected cells (Rose et al., 1978; for a review, see Banerjee, 1980), uncapped reovirus mRNAs cannot be translated in uninfected-cell lysates (Skup and Millward, 1980a). This is true even when cap-dependent translation is inhibited by addition of cap analogues or antibodies directed against cap-binding proteins. It was thus reasoned that some additional factor was required for the translation of uncapped reovirus mRNAs. Such a factor would presumably be reovirus-specific and must be present in lysates prepared from infected cells.

This possibility has been examined with the use of polyspecific antisera directed against all viral polypeptides of reovirus (D. Skup and S. Millward, unpublished). Standard in vitro translation reactions were carried out using lysates from infected and uninfected cells (Skup and Millward, 1980a). The effect of antisera and control sera on the translation of capped and uncapped reovirus mRNAs synthesized by cores in vitro was determined by measuring incorporation of [35S]methionine as trichloroacetic-acid-precipitable material. The results of these experiments are summarized in Table V.

TABLE V. Effect of Antireovirus Antisera on Translation of Capped and Uncapped Reovirus Messenger RNA's *in Vitro*[a]

| Extract | Messenger RNA | Antisera | Counts per minute | Inhibition (%) |
|---|---|---|---|---|
| Uninfection | Capped | — | 415,321 | — |
| | Capped | Control | 411,867 | — |
| | Capped | Antireovirus | 410,732 | — |
| | Uncapped | — | 1,037 | — |
| | Uncapped | Control | 1,174 | — |
| | Uncapped | Antireovirus | 1,004 | — |
| Infected | Capped | — | 103,761 | — |
| | Capped | Control | 113,843 | — |
| | Capped | Antireovirus | 105,932 | — |
| | Uncapped | — | 432,132 | — |
| | Uncapped | Control | 411,737 | — |
| | Uncapped | Antireovirus | 20,753 | 95 |
| | Progeny transcripts[b] | — | 443,721 | — |
| | Progeny transcripts[b] | Control | 422,538 | — |
| | Progeny transcripts[b] | Antireovirus | 45,753 | 90 |

[a] From Skup and Millward (unpublished).
[b] Represents uncapped pG-terminated reovirus mRNAs synthesized by progeny SVPs *in vitro* (Skup and Millward, 1980b; Zarbl *et al.*, 1980b).

The results indicate that antireovirus antibodies specifically inhibit the translation of uncapped mRNA in infected-cell-free lysates. This inhibition cannot be explained by precipitation of viral-specific polysomes, since the antisera had no effect on the translation of capped reovirus mRNA in uninfected-cell lysates. These findings therefore suggest that a viral protein may be required for the translation of uncapped viral mRNAs. Further experimentation, including purification of this factor, will determine whether or not this is indeed the case. The mechanisms by which reovirus induces a transition to cap-independent translation are clearly complex and merit further study (see Addendum).

## G. Effects on Host-Cell Functions

As discussed in the previous section, infection with reovirus causes a shutoff of host protein synthesis. The mechanism for this takeover of protein synthesis involves a transition of the host translational apparatus from cap dependence to cap independence. Since all host mRNAs are capped, the shutoff does not require that the virus also induce inhibition of host messenger RNA (mRNA) synthesis. Indeed, no inhibition of host transcription is observed during the viral replication cycle (Gomatos and Tamm, 1963; Kudo and Graham, 1965). Furthermore, host mRNAs are present in the infected cell at late times postinfection, even though they are not translated (Skup *et al.*, 1981), indicating stability of host transcripts in the infected cell.

Infection of L cells with reovirus does, however, result in the inhibition of host DNA synthesis (Ensminger and Tamm, 1969a). The extent to which host DNA synthesis is inhibited varies with cell-culture conditions, the serotype of reovirus used, and the multiplicity of infection. In monolayer cultures infected with reovirus type 3, inhibition of DNA synthesis begins by 12 hr postinfection. By 24 hr, the level of DNA synthesis in infected cells is 20–35% of that observed in mock-infected cells. Reovirus type 1, on the other hand, does not significantly decrease the rate of host-cell DNA synthesis.

In suspension cultures, inhibition of L-cell DNA synthesis begins at about 6–8 hr. At a multiplicity of infection of 5–10 PFU/cell, reovirus type 3 caused an 88% inhibition of host DNA synthesis by 12 hr postinfection. Under these conditions, reovirus type 1 inhibited L-cell DNA synthesis by only 25%. This variation was not due to differences in viral growth characteristics, since both serotypes grow at the same rate under these conditions, each cell producing about 1000 infectious virions (Sharpe and Fields, 1981).

Ensminger and Tamm (1970) studied reovirus-induced inhibition of DNA synthesis in synchronized cell cultures. They reported that DNA synthesis in the early S phase was relatively insensitive to inhibition. More recently, Cox and Shaw (1974) demonstrated that infection of cells with reovirus 8 hr before the onset of the S phase actually prevented DNA synthesis. Since inhibition of DNA synthesis begins about 6–8 hr postinfection, these results suggest that reovirus infection may actually prevent the initiation of the S phase of cell growth.

Numerous studies have demonstrated that the inhibition of DNA synthesis by reovirus occurs at the level of initiation by a reduction in the number of multifocal initiation sites without altering the rate of replication fork movement (Ensminger and Tamm, 1969b; Hand et al., 1971; Hand and Tamm, 1972, 1974; Hand and Kasupski, 1978). This mode of inhibition of DNA synthesis is analogous to that observed when cells are treated with inhibitors of protein synthesis. Such results suggest that the inhibition of DNA replication in infected cells may be nonspecific, resulting from the shutoff of host protein synthesis. This is corroborated by the observation that inhibition DNA synthesis and inhibition of host protein synthesis follow similar kinetics (Joklik, 1974).

Other lines of evidence indicate that alternative mechanisms may be involved in the inhibition of DNA synthesis. One possibility is that a viral-encoded protein synthesized at early times in the infected cell may directly or indirectly inhibit DNA synthesis. Shelton et al. (1981) have proposed that the σNS protein is a candidate for such an inhibitor. This conclusion was based on the ability of σNS to bind to double-stranded DNA or denatured DNA in vitro as well as the presence of this protein in crude nuclear extracts. Although suggestive, these data do not provide evidence for a direct role of σNS in DNA inhibition.

An equally plausible alternative is that one or more of the protein

TABLE VI. Functions of Viral Proteins

| dsRNA genome segment | Polypeptide species | Functional role | Ref. No.[a] |
|---|---|---|---|
| L1 | λ3 | Viral transcriptase activity | 1 |
| L2 | λ2 | Reovirus spikes (site of mRNA extrusion) | 2 |
| L3 | λ1/λ1C | Guanylyltransferase activity | 3 |
| M2 | μ1/μ1C | Regulation of viral transcriptase and resistance of virus to proteolytic digestion | 1 |
| S1 | σ1 | 1. Viral cell receptor and cell tropism | 4 |
|    |    | 2. Viral hemagglutinin | 5 |
|    |    | 3. Production of neutralizing antibodies | 6 |
|    |    | 4. Association of virion with microtubules | 7 |
|    |    | 5. Generation of host immune response | 8 |
|    |    | 6. Role in inhibition of host DNA synthesis | 9 |
| S3 | σNS | 1. ssRNA binding (viral mRNA linker?) | 10 |
|    |    | 2. Poly(C)-dependent RNA polymerase activity | 11 |
| S4 | σ3 | 1. Role in inhibition of host protein synthesis | 12 |
|    |    | 2. dsRNA-binding protein | 10 |

[a] References: (1) Drayna and Fields (1982); (2) White and Zweerink (1976); (3) D. R. Cleveland, H. Zarbl, and S. Millward (in prep.); (4) Weiner et al. (1977), Lee et al. (1981b); (5) Weiner et al. (1978); (6) Weiner and Fields (1977); (7) Babiss et al. (1979); (8) Finberg et al. (1979), Weiner et al. (1980b,c); (9) Sharpe and Fields (1981); (10) Huismans and Joklik (1976); (11) Gomatos et al. (1980); (12) Sharpe and Fields (1982).

components of the infecting virion is the inhibitory molecular species. This possibility has been investigated extensively. It was reasoned that if a capsid protein were involved, then top component (virion capsids devoid of genome RNA) should also inhibit DNA synthesis. Experiments indicated that this was not the case, since even high multiplicities (over 5000 particles/cell) of top component fail to inhibit DNA synthesis (M.-H.T. Lai et al., 1973; Hand and Tamm, 1973). These results suggested that a capsid protein was not involved in the reduction of DNA synthesis.

More recently, Sharpe and Fields (1981) presented evidence that the minor capsid protein σ1 is involved in the inhibition of DNA replication. These investigators exploited the variation among the serotypes of reovirus in their ability to inhibit host replication. Using the same intertypic recombinant strains described in Section A, they found that this difference mapped to the S1 gene, which encodes the viral cell receptor. Irradiation of these recombinant clones did not alter their effect on DNA synthesis, as had been shown to be the case for reovirus type 3 (M.-H.T. Lai et al., 1973; Hand and Tamm, 1973; Shaw and Cox, 1973). These results suggested that the σ1 protein associated with virions may exert the inhibitory effect directly. However, it was not determined whether irradiated virions transcribe limited amounts of s1 mRNA, giving rise to newly synthesized σ1 protein. The possibility that noncapsid σ1 molecules inhibit DNA synthesis is difficult to reconcile with the observation that reovirus cores, which lack σ1 but are infectious, do not cause inhibition of DNA replication (Cox and Shaw, 1976; Hand and Tamm,

1973). This suggests that σ1 molecules synthesized in the cytoplasm do not cause the inhibition, thereby implying that σ1 contained in the outer capsid is involved. The σ1 protein has been found to play a key role in virus–cell surface interactions (see Table VI). One interpretation is that inhibition of cellular DNA synthesis is mediated by an interaction of the virions with their cell surface receptors (Sharpe and Fields, 1981). The inability of top component to cause inhibition (M.-H.T. Lai *et al.*, 1973; Hand and Tamm, 1973) even though they contain σ1 (Smith *et al.*, 1969) is not consistent with this model. However, it has been found that the binding of top component to cells is aberrant, with only about 15% of the empty capsids binding under optimal conditions (Hand *et al.*, 1971). It is possible that these empty capsids do not have the proper configuration to interact with cell receptors. Further studies will be required to determine whether interaction of the σ1 protein with the cell surface receptor does indeed alter cellular DNA synthesis.

## H. Morphogenesis and Maturation

### 1. Assembly of Virions

Little is known about the way in which reovirions are assembled. As already discussed in a previous section, reovirus morphogenesis is coupled to replication of the genome. The most immature virions are those that contain single-stranded (ss) plus sense RNAs. The RNA in these complexes, composed of the viral polypeptides λ1, λ2, presumably λ3, μ1C, σ1, σ2, and σ3, is ribonuclease-sensitive (Morgan and Zweerink, 1975). These particles, which express the ss→double-stranded (ds)RNA polymerase, are a heterogeneous population of particles sedimenting between 180 and 550 S, although their density in CsCl is relatively uniform, averaging 1.34 g/cm$^3$ (Zweerink, 1974; Morgan and Zweerink, 1975). The particles sedimenting at about 250 S synthesize dsRNA primarily of the *S* class, those sedimenting at 450 S are enriched in the synthesis of dsRNA of the *M* class, and those that sediment at 550 S synthesize primarily dsRNA of the *L* class. The smaller particles thus give rise to the larger particles by sequential replication of genome segments and addition of viral polypeptides.

Once replication of the genome is complete, the particles appear to undergo a series of as yet uncharacterized rearrangements that lead to the formation of particles that resemble viral cores. The dsRNA within these particles is now resistant to ribonuclease even at low salt concentrations (Acs *et al.*, 1971). These particles have a density similar to that of reovirus core particles (1.42 g/cm$^3$), but once again sediment as a heterogeneous population (Morgan and Zweerink, 1975; Zweerink *et al.*, 1976). The particles have an active transcriptase and are composed of the polypeptides λ1, λ2, λ3 (although not detected), σ2, and μNS, and resemble cores morphologically. Several lines of evidence indicate that these particles are precursors to mature progeny virions. Zweerink *et al.* (1976)

have isolated several classes of subviral particles (SVPs) from infected cells. One class of particles sedimented at 400 S and contained the viral core polypeptides as well as the nonstructural polypeptides μNS. These particles are present in the infected cell at 4 hr postinfection, before mature virions are detectable, and are relatively scarce at later times. Furthermore, labeled uridine and amino acids are incorporated into these particles before they appear in virions. Pulse-labeling experiments with [$^3$H]uridine at 7 hr postinfection have demonstrated that after a 5-min labeling period, most of the label was incorporated into 400 S particles and occasionally 200 S particles. The latter class of particles probably corresponds to the 220 S transcriptase particles described by Morgan and Zweerink (1975). After more extended labeling periods, the relative amount of label sedimenting at 400 S decreased, while the amount of label sedimenting at about 600 S (immature virions) and 630 S (virions) increased. Particles sedimenting at 600 S were shown to contain all the viral capsid polypeptides, but still contained a small amount of μNS. They also lacked the A-rich oligonucleotides found in virions. These results strongly suggested that these 400 S particles were precursors to 600 S particles that in turn matured into virions.

This hypothesis was further corroborated by experiments in which infected cells were pulse-labeled [$^3$H]uridine and then incubated in 0.25M salt medium to prevent further dsRNA synthesis. During the 15-min pulse-labeling period, the bulk of the label was again found to be associated with 400 S particles. On incubation in the presence of 0.25 M NaCl for 75 min, the bulk of the radioactivity was associated with virion like particles, while few 400 S particles could be detected. These results demonstrated that 400 S particles have a relatively short half-life and are converted to virions.

Further evidence that 400 S particles are precursors to virions came from studies with temperature-sensitive (*ts*) mutants of reovirus (for a review, see Cross and Fields, 1977). Temperature-sensitive mutants of groups B and G are defective in morphogenesis and as a result accumulate particles resembling reovirus cores, at nonpermissive temperatures (Fields *et al.*, 1971; Morgan and Zweerink, 1974). The lesions of the B and G mutants have been shown to reside in the λ2 and σ3 polypeptides, respectively (Mustoe *et al.*, 1978a). Both these mutants synthesize dsRNA but fail to produce mature virions. Particles accumulated in cells infected with these mutants at 39°C (nonpermissive temperature) sediment at 400 S in sucrose density gradients, resemble core particles, and contain the polypeptides λ1, λ2, λ3, μ1, and σ2, and a small, variable amount of μNS, μ1C, and σ3. Similar particles but lacking μNS also accumulate in wild-type infections at 39°C. Particles sedimenting at 400 S also accumulate in cells infected at 37°C, but these do contain μNS and appear to be precursors to virions (Zweerink *et al.*, 1976), suggesting that μNS may be required in the conversion to whole virions.

Taken together, these results indicate that after replication is complete, particles undergo some form of rearrangement to yield particles that resemble cores and lose the outer-capsid proteins $\sigma1$, $\sigma3$, and $\mu1C$ associated with replicase particles. These particles still lack spikes and contain the nonstructural polypeptide $\mu NS$. It appears that these particles have a relatively short half-life. They are converted to particles sedimenting at 600 S by acquisition of the outer-capsid polypeptides and the 12 spikes composed of $\lambda2$. Displacement of the $\mu NS$ protein appears to be coincident with this process. The 600 S particles subsequently mature into virions.

Essentially nothing is known about the mechanisms of assembly. Formation of the two protein capsids of reovirus, however, appears to be independent of dsRNA formation, even though these occur simultaneously. This is demonstrated by the fact that empty capsids or top component are formed during the course of normal infections. These empty capsids appear to be equivalent to virions in their polypeptides composition and morphology (Smith et al., 1969). Formation of empty viral capsids was also observed during infection at nonpermissive temperatures with ts mutants defective in dsRNA synthesis.

Empty virions consisting of only the outer-capsid proteins have also been observed in mutants that are associated with one of the two inner-capsid proteins $\lambda1$ and $\sigma2$ (Matsuhisa and Joklik, 1974), both of which are located on the innermost surface of cores (Martin et al., 1973). These results suggest that assembly of virions is primarily dependent on protein–protein interactions. To this end, Lee et al. (1981a) have found that two of the outer-capsid polypeptides, $\mu1$–$\mu1C$ and $\sigma3$, exist as complexes in the infected cell. This was demonstrated by the formation of monoclonal antibody to this complex that reacted to neither protein individually. In addition, monoclonal antibodies directed against most of the viral structural and nonstructural polypeptides have been characterized. These monoclonal antibodies may prove extremely useful in the study of reovirus assembly and morphogenesis. These antisera could be used to isolate increasingly complex particles on the morphogenetic pathway in highly synchronized cells (Joklik, 1981). Such an approach may alleviate the problems encountered thus far in morphogenetic studies. The difficulties arise from the extremely heterogeneous population of SVPs present in the infected cell. Studies have been further hampered by the inability to perform classic pulse-chase experiments to demonstrate precursor–product relationships. This problem stems from the large pools of viral ssRNA and polypeptides present in the infected cell. Use of antisera directed against unique proteins may allow for isolation of discrete populations of SVPs, thereby enhancing the understanding of the stages involved in assembly.

## 2. Maturation and Production of Reovirus Oligonucleotides

In addition to the dsRNA genome segments, mature virions also contain several thousand molecules of small, adenine-rich, single-stranded oligonucleotides that comprise 15–20% of the total viral RNA (Bellamy and Joklik, 1967b; Shatkin and Sipe, 1968a). These oligonucleotides are found in all three serotypes of mammalian reoviruses (Shatkin and Sipe, 1968a) and have also been detected in avian reoviruses (Koide *et al.*, 1968). Oligonucleotides are found in virions regardless of the strain of host cell used in virus propagation (Bellamy and Joklik, 1967b). These oligonucleotides are contained within virus particles, where they are resistant to ribonuclease digestion (Shatkin and Sipe, 1968a), and are released from virions by proteolytic digestion of the outer capsid (Bellamy and Hole, 1970).

The reovirus oligonucleotides fall into two structurally distinct classes (Bellamy and Hole, 1970; Bellamy *et al.*, 1970, 1972; Stolzfus and Banerjee, 1972; Nichols *et al.*, 1972a) (see Table VII for a summary). The oligo(A) class consists of oligoadenylates ranging from 2 to 20 nucleotides in length. The 5'-G-terminated oligonucleotides consist of a series of molecules ranging from 2 to 9 nucleotides in length and having (p)(p)-pGpC . . . at their 5' termini. A small number of these oligonucleotides have also been shown to have guanylylated and methylated 5' termini (Carter, 1977). The nucleotide sequences of this latter class of oligomers closely resemble the sequences found at the 5' termini of reovirus messenger RNA synthesized *in vitro* (Nichols *et al.*, 1972b; Hastings and Millward, 1978). Furthermore, oligomers are not synthesized in the cytoplasm (Bellamy and Joklik, 1967b), but are synthesized within progeny virions during maturation. These results have led to the hypothesis that the 5'-G-terminated oligonucleotides are produced by multiple reinitiation events that occur because formation of long transcripts is prevented during the terminal stages of maturation. More than likely, abortive transcripts are produced after inactivation of the virion transcriptase by the addition of $\sigma 3$ proteins (Astell *et al.*, 1972).

The fact that the 5'-G-terminated oligomers are abortive transcripts has been further substantiated. It was found that reovirus cores abort

TABLE VII. Oligonucleotides Present in Reovirus Particle[a]

| Oligonucleotide(s) | Sequence | Chain length | Approximate number of molecules per virion |
|---|---|---|---|
| Oligoadenylates | $(p)(p)p(A)_{1-19}A_{OH}$ | 2–20 | 850 |
| | $(p)ppGC_{OH}$ | 2 | 50 |
| | $(p)ppGCU_{OH}$ | 3 | 900 |
| 5'-G-Terminated | $(p)(p)pGCUA_{OH}$ | 4 | 775 |
| oligonucleotides | $(p)ppGCU(A)_{1-3}A_{OH}$ | 5–9 | 130 |
| | $(p)ppGCUA(U)_{1-4}U_{OH}$ | 6–9 | 130 |
| | Others (probably capped) | 2–8 | 350 |

[a] From Nichols *et al.* (1972a).

50% of the initiated transcripts before they are more than a few nucleotides long (Zarbl et al., 1980a; Yamakawa et al., 1981). These abortive transcripts are either capped or uncapped and are the result of reiteration of the RNA polymerase at promoter sites, possibly due to instability of the initiation complex. Such a model is consistent with the observation that virions harvested from infections done at increasingly higher temperatures contain increasing amounts of oligomers (K.C. Lai and Bellamy, 1971). In addition, it has been found that abortive transcripts are synthesized within mature virions on incubation with the four ribonucleoside triphosphates and $Mn^{2+}$ ions (Yamakawa et al., 1981; 1982). The 5'-G-terminated oligonucleotides therefore represent abortive transcripts synthesized during the final stages of maturation that become trapped within the virions.

The origin of the reovirus oligoadenylates has also been extensively investigated. Carter et al. (1974) reported that the oligoadenylates are not essential for infectivity, nor are they required as a template/primer for their own synthesis. This was demonstrated by infecting cells with cores, prepared by digestion of virions with chymotrypsin, that have lost their oligomers. It was found that these infections yielded normal virions with a full complement of oligonucleotides. Stoltzfus et al. (1974) subsequently reported a poly(A) polymerase activity in highly purified reovirions. This enzyme synthesized oligoadenylates in the presence of ATP and $Mn^{2+}$ ions and was inactivated by chymotryptic digestion. These oligo(A) molecules were similar in size to those already present in virions. These results suggested that oligo(A) synthesis is mediated by the poly(A) polymerase during a late stage in morphogenesis. That this is indeed the case was corroborated by the experiments of Silverstein et al. (1974). These investigators showed that the poly(A) polymerase activity was associated with a large-particle (virus-factory-containing) cytoplasmic fraction of infected L cells. This fraction catalyzed the synthesis of oligoadenylates for about 1 hr in vitro. The products were resistant to ribonuclease, but synthesis was abruptly terminated on addition of chymotrypsin.

Addition of cycloheximide to infected cells during the midlogarithmic phase of viral replication results in rapid inhibition of viral protein synthesis and dsRNA synthesis. Synthesis of oligoadenylates, however, was found to continue for at least 1 hr under these conditions. Furthermore, all the label associated with oligo(A) was found to be localized within mature virions. These findings demonstrate that synthesis of oligoadenylates occurs within maturing virions and is independent of replication.

Further experiments (Silverstein et al., 1974) indicated that synthesis of oligo(A) was dependent on the presence of dsRNA. Infection of cells in the presence of 5 µg/ml of actinomycin D was found to inhibit the formation of both viral dsRNA and oligo(A). Synthesis of viral polypeptides was not affected qualitatively, although greatly reduced amounts of viral proteins were synthesized in the absence of replication (see Section E.1). Electron-microscopic examination of these cells revealed the presence of small viral factories containing a large proportion of empty virions, some containing ssRNA. These data strongly indicate that the synthesis of dsRNA precedes the formation of oligo(A). The notion that dsRNA is required for oligonucleotide synthesis by the oligo(A) polymerase is consistent with the observation that top component does not contain any oligonucleotides (Smith *et al.*, 1969). It therefore appears that dsRNA serves as a template for the synthesis of both classes of reovirus oligonucleotides and that this synthesis occurs within mature virions or SVPs during the last stages of morphogenesis. This contention is supported by the isolation of immature virions (600 S) that are morphologically very close to mature virions, but lack oligonucleotides (Zweerink *et al.*, 1976). Furthermore, these particles appear to be precursors to mature virions and acquire oligonucleotides during this final stage of morphogenesis.

On the basis of these observations, it has been hypothesized that the oligo(A) polymerase may actually represent an alernative activity of the virion-associated transcriptase and that it is regulated by outer-capsid proteins. This hypothesis is consistent with the observation that the poly(A) polymerase activity is abolished when the transcriptase is activated by chymotrypsin. Furthermore, the difference in particles with transcriptase activity compared to those with poly(A) polymerase activity resides in the cleavage pattern of the $\mu 1-\mu 1C$ polypeptides of the outer capsid (Carter, 1979). Conditions that resulted in cleavage of $\mu 1$ and $\mu 1C$ proteins to a protein with a molecular weight of 69,000 resulted in the formation of particles with an active transcriptase. Conditions that resulted in cleavage to polypeptide fragments with molecular weights of 64,000 and 60,000 resulted in the formation of particles with poly(A) polymerase activity. In view of the finding that the $\mu 1-\mu 1C$ proteins regulate the activity of the viral transcriptase (Drayna and Fields, 1982), these results corroborate the view that the poly(A) polymerase is an alternative mode of the viral transcriptase. This is further suggested by the fact that both activities have the same temperature optimum, but different cofactor requirements.

The oligoadenylates may arise by a mechanism similar to that involved in the synthesis of poly(A) by the *Escherichia coli* RNA polymerase (Bellamy *et al.*, 1972). Oligo(A) could arise via a "slippage mechanism," in which reiterative synthesis occurs along runs of uridine on a template. It has been shown that some of the revirus plus strands have a region of five consecutive uridylate residues close to their 5' termini (Nichols *et al.*, 1972b; Hastings and Millward, 1978). Such regions could provide a template for the synthesis of oligoadenylates.

It thus appears that both classes of reovirus oligonucleotides are synthesized by the virion-associated transcriptase. Synthesis of these oligomers occurs only after inactivation of the transcriptase during the last stages of maturation. Presumably, the activity of the transcriptase is modulated by addition of outer-capsid proteins, causing it to operate in alternate modes. Reovirus oligonucleotides appear to be by-products of maturation and are not formed in mutants that fail to produce mature infectious virus particles (Johnson *et al.*, 1976). Whether or not they serve a role early in the infectious cycle remains to be determined.

## IV. SUMMARY OF THE REOVIRUS LIFE CYCLE

The reovirus replication cycle as we now understand it is outlined in Fig. 17. Reovirus particles bind to the surface of host cells via an interaction of the σ1 proteins, located on the outer capsid, with specific receptor molecules found on the cell surface. After binding, the virions are taken up into the cell via an active process of phagocytosis. The phagocytic vesicles containing the virions then move into cytoplasm. Virus-containing vacuoles then fuse with lysosomes, exposing the virions to hydrolytic enzymes. The hydrolases partially digest the outer capsid of the virus, leaving the genome intact within a parental subviral particle (SVP). These particles are subsequently released into the cytoplasm. The uncoating process results in the activation of the viral transcriptase and capping enzymes. Therefore, when parental SVPs are released into the cytoplasm, they are competent to begin transcription of the genome to form viral messenger RNAs (mRNAs). At first, only four of the ten gen-

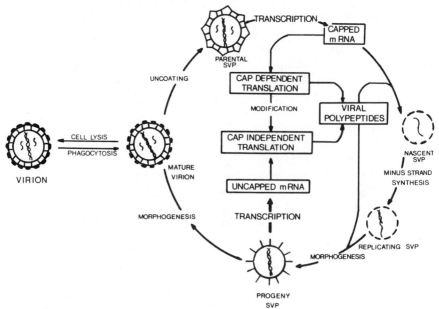

FIGURE 17. Model of the reovirus replicative cycle. Reprinted from Skup *et al.* (1981), with permission.

ome segments are transcribed, possibly due to repression by a cellular factor. The pre-early viral mRNAs transcribed are translated into viral polypeptides, one or more of which cause a derepression of the remaining six segments, thereby allowing for transcription of all ten genome segments. The parental transcripts have capped 5' termini and hence are able to compete with cellular messages for the host cap-dependent translational apparatus. All viral polypeptides are synthesized during this period. The proteins subsequently combine with capped parental transcripts by an obscure mechanism, which assures that one each of the ten mRNA species is incorporated into some form of nascent SVP. These nascent particles undergo morphogenesis, leading to the formation of replicating SVPs. Within these structures, each of the single-stranded mRNAs serves as a template for minus-strand synthesis only once, yielding the ten genomic segments of double-stranded RNA. Further morphogenesis leads to the formation of progeny SVPs that are active in mRNA synthesis and actually synthesize the bulk of the viral mRNAs found in the infected cell. These progeny SVPs, however, have latent mRNA capping enzymes and as a result synthesize uncapped transcripts. Uncapped transcripts are not compatible with the host cap-dependent translational machinery. During the course of the infection, however, the virus induces a transition in the translational apparatus of the host by an as yet poorly defined mechanism. As a result of this alteration, the translational machinery becomes cap-independent. This means that capped parental transcripts and capped cellular mRNAs are no longer translated. Uncapped progeny transcripts, on the other hand, are translated with high efficiency. These uncapped messages do not appear to be used in replication, indicating that their sole function is amplification of the pool of viral polypeptides in the infected cell. These progeny particles combine with viral protein via a poorly defined morphogenetic pathway, leading to the formation of mature virions. One of the final stages of maturation involves synthesis of the oligonucleotides found within virions. These are believed to arise by abortive transcription due to inactivation of the viral transcriptase during the final stage of morphogenesis. Mature virions accumulate in the infected cell and eventually cause cell death and lysis.

## REFERENCES

Acs, G., Klett, H., Schonberg, M., Christman, J., Levin, D.H., and Silverstein, S.C., 1971, Mechanism of reovirus double-stranded RNA synthesis *in vivo* and *in vitro*, *J. Virol.* 8(5):684–689.

Ahmed, R., Chakraborty, P.R., and Fields, B.N., 1980, Genetic variation during lytic reovirus infection: High passage stocks of wild-type reovirus contain temperature-sensitive mutants, *J. Virol.* 34(1):285–287.

Antczak, J.B., Chmelo, R., Pickup, D.J., and Joklik, W.K., 1982, Sequences at both termini of the 10 genes of reovirus serotype 3 (strain Dearing), *Virology* 121:307–319.

Anderson, N., and Doane, F.W., 1966, An electron microscope study of reovirus type 2 in L cells, *J. Pathol. Bacteriol.* 92:433–439.

Astell, C., Silverstein, S.C., Levin, D.H., and Acs, G., 1972, Regulation of the reovirus RNA transcriptase by a viral capsomere protein, *Virology* **48**:648–654.

Babiss, L.E., Luftig, R.B., Weatherbee, J.A., Weihing, R.R., Ray, U.R., and Fields, B.N., 1979, Reovirus serotypes 1 and 3 differ in their in vitro association with microtubules, *J. Virol.* **30**:863–874.

Banerjee, A.K., 1980, 5'-Terminal cap structure in eukaryotic messenger ribonucleic acids, *Microbiol. Rev.* **44**:175–205.

Banerjee, A.K., and Shatkin, A.J., 1970, Transcription in vitro by reovirus-associated RNA-dependent polymerase, *J. Virol.* **6**:1–11.

Banerjee, A.K., Ward, R., and Shatkin, A.J., 1971, Cytosine at the 3'-termini of reovirus genome and in vitro mRNA, *Nature (London) New Biol.* **232**:114–115.

Bartlett, N.M., Gillies, S.C., Bullivant, S., and Bellamy, A.R., 1974, An electron microscope study of reovirus reaction cores, *J. Virol.* **14**:315–326.

Bellamy, A.R., and Harvey, J.D., 1976, Biophysical studies of reovirus type 3. III. A laser-scattering study of the RNA transcriptase reaction, *Virology* **70**:28–36.

Bellamy, A.R., and Hole, L.V., 1970, Single-stranded oligonucleotides from reovirus type III, *Virology* **40**:808–819.

Bellamy, A.R., and Joklik, W.K., 1967a, Studies on reovirus RNA. II. Characterization of reovirus messenger RNA and of the genome RNA segments from which it is transcribed, *J. Mol. Biol.* **29**:19–26.

Bellamy, A.R., and Joklik, W.K., 1967b, Studies on the A-rich RNA of reovirus, *Proc. Natl. Acad. Sci. U.S.A.* **58**:1389–1395.

Bellamy, A.R., Hole, L.V., and Baguley, B.C., 1970, Isolation of the trinucleotide pppGpCpU from reovirus, *Virology* **42**:415–420.

Bellamy, A.R., Nichols, J.L., and Joklik, W.K., 1972, Nucleotide sequences of reovirus oligonucleotides: Evidence for abortive RNA synthesis during virus maturation, *Nature (London) New Biol.* **238**:49–51.

Borsa, J., and Graham, A.F., 1968, Reovirus: RNA polymerase activity in purified virions, *Biochem. Biophys. Res. Commun.* **33**:895–901.

Borsa, J., Grover, J., and Chapman, J.D., 1970, Presence of nucleoside triphosphate phosphohydrolase activity in purified virions of reovirus, *J. Virol.* **6**:295–302.

Borsa, J., Copps, T.P., Sargent, M.D., Long, D.G., and Chapman, J.D., 1973a, New intermediate subviral particles in the in vitro uncoating of reovirus virions by chymotrypsin, *J. Virol.* **11**(4):552–564.

Borsa, J., Sargent, M.D., Copps, T.P., Long, D.G., and Chapman, J.D., 1973b, Specific monovalent cation effects on modification of reovirus infectivity by chymotrypsin digestion in vitro, *J. Virol.* **11**:1017–1019.

Borsa, J., Long, D.G., Copps, T.P., Sargent, M.D., and Chapman, J.D., 1974a, Reovirus transcriptase activation in vitro: Further studies on the facilitation phenomenon, *Intervirology* **3**:15–35.

Borsa, J., Long, D.G., Sargent, M.D., Copps, T.P., and Chapman, J.D., 1974b, Reovirus transcriptase activation in vitro: Involvement of an endogenous uncoating activity in the second stage of the process, *Intervirology* **4**:171–188.

Borsa, J., Morash, B.D., Sargent, M.D., Copps, T.P., Lievaart, P.A., and Szekely, J.G., 1979, Two modes of entry of reovirus into L cells, *J. Gen. Virol.* **45**:161–170.

Borsa, J., Sargent, M.D., Lievaart, P.A., and Copps, T.P., 1981, Reovirus: Evidence for a second step in the intracellular uncoating and transcriptase activation process, *Virology* **111**:191–200.

Both, G.W., Lavi, S., and Shatkin, A.J., 1975a, Synthesis of all the gene products of the reovirus genome in vivo and in vitro, *Cell* **4**:173–180.

Both, G.W., Furuichi, Y., Muthukrishnan, S., and Shatkin, A.J., 1975b, Ribosome binding to reovirus mRNA in protein synthesis requires 5' terminal 7-methylguanosine, *Cell* **6**:185–195.

Both, G.W., Banerjee, A.K., and Shatkin, A.J., 1975c, Methylation-dependent translation of viral messenger RNAs in vitro, *Proc. Natl. Acad. Sci. U.S.A.* **72**:1189–1193.

Brendler, T., Godefroy-Colburn, T., Carlill, R.D., and Thach, R.E., 1981a, The role of mRNA competition in regulating translation. II. Development of a quantitative *in vitro* assay, *J. Biol. Chem.* **256**:11,747–11,754.

Brendler, T., Godefroy-Colburn, T., Yu, S., and Thach, R.E., 1981b, The role of mRNA competition in regulating translation. III. Comparison of *in vitro* and *in vivo* results, *J. Biol. Chem.* **256**:11,755–11,761.

Brown, B.A., and Ehrenfeld, E., 1980, Initiation factor preparations from poliovirus-infected cells restrict translation in reticulocyte lysates, *Virology* **103**:327–339.

Brown, D., Jones, C.L., Brown, B.A., Ehrenfeld, E., 1982, Translation of capped and uncapped VSV mRNAs in the presence of initiation factors from poliovirus-infected cells, *Virology* **123**:60–68.

Brubaker, M.M., West, B., and Ellis, R.J., 1964, Human blood group influence on reovirus hemagglutination titres, *Proc. Soc. Exp. Biol. Med.* **115**:1118–1120.

Carter, C.A., 1977, Methylation of reovirus oligonucleotides *in vivo* and *in vitro*, *Virology* **80**:249–259.

Carter, C.A., 1979, Activation of reovirion-associated poly(A) polymerase and oligomer methylase by cofactor-dependent cleavage of μ polypeptides, *Virology* **94**:417–429.

Carter, C.A., and Lin, B.Y., 1979, Conservation and modification of the pyrimidine-rich reovirus oligonucleotides after infection, *Virology* **93**:329–339.

Carter, C.A., Stoltzfus, C.M., Banerjee, A.K., and Shatkin, A.J., 1974, Origin of reovirus oligo(A), *J. Virol.* **13**:1331–1337.

Carter, C.A., Lin, B.Y., and Metlay, M., 1980, Polyadenylation of reovirus proteins, *J. Biol. Chem.* **255**:6479–6485.

Chang, C.-T., and Zweerink, H.J., 1971, Fate of parental reovirus in infected cell, *Virology* **46**(3):544–555.

Christman, J.K., Reiss, B., Kyner, D., Levin, D.H., Klett, H., and Acs, G., 1973, Characterization of a viral messenger ribonucleoprotein particle accumulated during inhibition of polypeptide chain initiation in reovirus infected L cells, *Biochim. Biophys. Acta* **294**(1):153–164.

Cox, D.C., and Clinkscales, C.W., 1976, Infectious reovirus subviral particles: Virus replication, cellular cytopathology and DNA synthesis, *Virology* **74**:259–261.

Cox, D.C., and Shaw, J.E., 1974, Inhibition of the initiation of cellular DNA synthesis after reovirus infection, *J. Virol.* **13**(3):760–761.

Cross, R.K., and Fields, B.N., 1972, Temperature-sensitive mutants of reovirus type 3: Studies on the synthesis of viral RNA, *Virology* **50**:799–809.

Cross, R.K., and Fields, B.N., 1976, Reovirus-specific polypeptides: Analysis using discontinuous gel electrophoresis, *J. Virol.* **19**:162–173.

Cross, R.K., and Fields, B.N., 1977, Genetics of reovirus, in: *Comprehensive Virology*, Vol. 9 (H. Fraenkel-Conrat and R.R. Wagner, eds.), pp. 291–340 Plenum Press, New York.

Dales, S., 1963, Association between the spindle apparatus and reovirus, *Proc. Natl. Acad. Sci. U.S.A.* **50**:268–275.

Dales, S., 1965a, Replication of animal viruses as studied by electron microscopy, *Am. J. Med.* **38**:699–715.

Dales, S., 1965b, Effects of Streptovitacin A on the initial events in the replication of vaccinia and reovirus, *Proc. Natl. Acad. Sci. U.S.A.* **54**:462–468.

Dales, S., 1973, Early events in cell–animal virus interactions, *Bacteriol. Rev.* **37**:103–145.

Dales, S., Gomatos, P.J., and Hsu, K.C., 1965, The uptake and development of reovirus in strain L cells followed with labeled viral ribonucleic acid and ferritin–antibody conjugates, *Virology* **25**:193–211.

Darzynkiewicz, E., and Shatkin, A.J., 1980, Assignment of reovirus mRNA ribosome binding sites to virion genome segments by nucleotide sequence analysis, *Nucleic Acids Res.* **8**:337–350.

Desrosiers, R.C., Sen, G.C., and Lengyel, P., 1976, Difference in 5′ terminal structure between the mRNA and the double-stranded virion RNA of reovirus, *Biochem. Biophys. Res. Commun.* **73**:32–39.

Detjen, B.M., Walden, W.E., and Thach, R.E., 1982, Transitional specificity in reovirus-infected mouse fibroblasts, *J. Biol. Chem.* **257**:9855–9860.

Drayna, D., and Fields, B.N., 1982, Activation and characterization of the reovirus transcriptase: Genetic analysis, *J. Virol.* **41**:110–118.

Drouhet, V., 1960, Lésions céllulaires provoquées par les réovirus (Virus echo 10): Anticorps fluorescents et étude cytochimique, *Ann. Inst. Pasteur* **98**:618–621.

Etchison, D., Milburn, S.C., Edery, I., Sonenberg, N., and Hershey, J.W.B., 1982, Inhibition of HeLa cell protein synthesis following poliovirus infection correlates with the proteolysis of a 220,000-dalton polypeptide associated with eukaryotic initiation factor 3 and a cap binding protein complex, *J. Biol. Chem.* **257**:14806–14810.

Eggers, H.J., Gomatos, P.J., and Tamm, I., 1962, Agglutination of bovine erythrocytes: A general characteristic of reovirus type 3, *Proc. Soc. Exp. Biol. Med.* **110**:879–881.

Ehrenfeld, E., and Lund, H., 1977, Untranslated vesicular stomatitis virus messenger RNA after poliovirus infection, *Virology* **80**:297–308.

Ehrenfeld, E., and Manis, S., 1979, Inhibition of 80S initiation complex formation by infection with poliovirus, *J. Gen. Virol.* **43**:441–445.

Ensminger, W.D., and Tamm, I., 1969a, Cellular DNA and protein synthesis in reovirus-infected L cells, *Virology* **39**:357–359.

Ensminger, W.D., and Tamm, I., 1969b, The step in cellular DNA synthesis blocked by reovirus infection, *Virology* **39**:935–938.

Ensminger, W.D., and Tamm, I., 1970, Inhibition of synchronized cellular deoxyribonucleic acid synthesis during Newcastle disease virus, mengovirus or reovirus infection, *J. Virol.* **5**:672–676.

Faust, M., and Millward, S., 1974, In vitro methylation of nascent reovirus mRNA by a virion-associated methyl transferase, *Nucleic Acids Res.* **1**:1739–1752.

Faust, M., Hastings, K.E.M., and Millward, S., 1975, $m^7G^{5'}$ ppp$^{5'}$ GmpCpUp at the 5' terminus of reovirus messenger RNA, *Nucleic Acids Res.* **2**:1329–1343.

Fernandez-Munoz, R., and Darnell, J.E., 1976, Structural differences between the 5' termini of viral and cellular mRNA in poliovirus-infected cells: Possible basis for the inhibition of host protein synthesis, *J. Virol.* **18**:719–726.

Fields, B.N., 1971, Temperature-sensitive mutants of reovirus type 3: Features of genetic recombination, *Virology* **46**:142–148.

Fields, B.N., and Eagle, H., 1973, The pH-dependence of reovirus synthesis, *Virology* **52**:581–583.

Fields, B.N., Raine, C.S., and Baum, S.G., 1971, Temperature-sentitive mutants of reovirus type 3: Defects in viral maturation as studied by immunofluorescence and electron microcopy, *Virology* **43**:569–578.

Fields, B.N., Laskov, R., and Scharff, M.D., 1972, Temperature-sensitive mutants of reovirus type 3: Studies on the synthesis of viral peptides, *Virology* **50**:209–215.

Filipowicz, W., 1978, Functions of the 5'-terminal $m^7G$ cap in eukaryotic mRNA, *FEBS Lett.* **96**:1–11.

Finberg, R., Weiner, H.L., Fields, B.N., Benacerraf, B., and Burakoff, S.J., 1979, Generation of cytolytic T lymphocytes after reovirus infection: Role of S1 gene, *Proc. Natl. Acad. Sci. U.S.A.* **76**:442–446.

Floyd, W.D., Stone, M.P., and Joklik, W.K., 1974, Separation of single-stranded ribonucleic acids by acrylamide–agarose–urea gel electrophoresis, *Anal. Biochem.* **59**:599–609.

Follet, E.A., Pringle, C.R., and Pennington, T.H., 1975, Virus development in enucleated cells: Echovirus, poliovirus, pseudorabies virus, reovirus, respiratory syncytial virus and Semliki forest virus, *J. Gen. Virol.* **26**:183–196.

Furuichi, Y., 1981, Allosteric stimulatory effect of S-adenosylmethionine on the RNA polymerase in cytoplasmic polyhedrosis virus: A model for the positive control of eukaryotic transcription, *J. Biol. Chem.* **256**:483–493.

Furuichi, Y., and Shatkin, A.J., 1976, Differential synthesis of blocked and unblocked 5'-termini in reovirus mRNA: Effect of pyrophosphate and pyrophosphatase, *Proc. Natl. Acad. Sci. U.S.A.* **73**:3448–3452.

Furuichi, Y., and Shatkin, A.J., 1977, 5'-Termini of reovirus mRNA: Ability of viral cores to form caps post-transcriptionally, *Virology* **77**:566–578.

Furuichi, Y., Muthukrishnan, S., and Shatkin, A.J., 1975, 5'-Terminal $m^7G(5')ppp(5')G^mp$

*in vivo*: Identification in reovirus genome RNA, *Proc. Natl. Acad. Sci. U.S.A.* **72**:742–745.

Furuichi, Y., Muthukrishnan, S., Tomasz, J., and Shatkin, A.J., 1976, Mechanism of formation of reovirus mRNA 5'-terminal blocked and methylated sequence, $m^7GpppG^mpC$, *J. Biol. Chem.* **251**:5043–5053.

Gaillard, R.K., Li, J.K.-K., Keene, J.D., and Joklik, W.K., 1982, The sequences at the termini of four genes of the three reovirus serotypes, *Virology*, **121**:320–326.

Galster, R.L., and Lengyel, P., 1976, Formation and characteristics of reovirus subviral particles in interferon-treated mouse L cells, *Nuclei Acids Res.* **3**:581–598.

Gavrilovskaya, I.N., Lavrova, I.K., Voroshilova, M.K., Chumakov, M.P., Poverenny, A.M., and Podgorodnichenko, V.K., 1974, Immunofluorescent demonstration of double-stranded RNA and virus antigen in RNA virus-infected cells, *Virology* **62**:276–279.

Gelb, L.D., and Lerner, A.M., 1965, Reovirus hemagglutination: Inhibition by *N*-acetyl-D-glucosamine, *Science* **147**:404–405.

Gillies, S., Bullivant, S., and Bellamy, A.R., 1971, Viral RNA polymerases: Electron microscopy of reovirus reaction cores, *Science* **174**:694–696.

Gomatos, P.J., 1967, RNA synthesis in reovirus-infected L929 mouse fibroblasts, *Proc. Natl. Acad. Sci. U.S.A.* **58**:1798–1805.

Gomatos, P.J., and Kuechenthal, I., 1977, Reovirus specific enzyme(s) associated with subviral particles responds *in vitro* to polyribocytidylate to yield double-stranded polyribocytidylate · polyriboguanylate, *J. Virol.* **23**(1):80–90.

Gomatos, P.J., and Tamm, I., 1962, Sites of reovirus type 3 and their interaction with receptor substances, *Virology* **17**:455–461.

Gomatos, P.J., and Tamm, I., 1963, Macromolecular synthesis in reovirus-infected L cells, *Biochim. Biophys. Acta* **72**:651–653.

Gomatos, P.J., Tamm, I., Dale, S., and Franklin, R.M., 1962, Reovirus type 3: Physical characteristic and interaction with L cells, *Virology* **17**:441–454.

Gomatos, P.J., Stamatos, N.M., and Sarkar, N.H., 1980, Small reovirus specific particle with polycytidylate-dependent RNA polymerase activity, *J. Virol.* **36**:556–565.

Gomatos, P.J., Prakash, O., and Stamatos, N.M., 1981, Small reovirus particles composed solely of sigma NS with specificity for binding different nuclei acids, *J. Virol.* **39**:115–124.

Granboulan, N., and Niveleau, A., 1967, Etude au microscope électronique du RNA de réovirus, *J. Microsc.* **6**:23–27.

Graziadei, W.D., III, and Lengyel, P., 1972, Translation of *in vitro* synthesized reovirus messenger RNA into proteins of the size of reovirus capsid proteins in a mouse L cell extract, *Biochem. Biophys. Res. Commun.* **46**(5):1816–1823.

Graziadei, W.D., III, Roy, D., Konigsberg, W., and Lengyel, P., 1973, Translation of reovirus RNA synthesized *in vitro* into reovirus proteins in a mouse L cell extract, *Arch. Biochem. Biophys.* **158**(1):266–275.

Gupta, S.L, Graziadei, W.D., III, Weideli, H., Sopori, M.L., and Lengyel, P., 1974, Selective inhibition of viral protein accumulation in interferon-treated cells: Nondiscriminate inhibition of the translation of added viral and cellular messenger RNAs in their extracts, *Virology* **57**:49–63.

Hand, R., and Kasupski, G.J., 1978, DNA and histone synthesis in reovirus infected cells, *J. Gen. Virol.* **39**(3):437–448.

Hand, R., and Tamm, I., 1971, Reovirus: Analysis of proteins from released and cell-associated virus, *J. Gen. Virol.* **12**:121–130.

Hand, R., and Tamm, I., 1972, Rate of DNA chain growth in mammalian cells infected with cytocidal RNA viruses, *Virology* **47**(2):331–337.

Hand, R., and Tamm, I., 1973, Reovirus: Effect of noninfective viral components on cellular DNA synthesis, *J. Virol.* **11**(2):223–231.

Hand, R., and Tamm, I., 1974, Initiation of DNA replication in mammalian cells and its inhibition by reovirus infection, *J. Mol. Biol.* **82**(2):175–183.

Hand, R., Ensminger, W.D., and Tamm, I., 1971, Cellular DNA replication in infections with cytocidal RNA viruses, *Virology* **44**:527–536.

Hansen, J., and Ehrenfeld, E., 1981, Presence of the cap-binding protein in initiation factor preparations from poliovirus-infected HeLa cells, *J. Virol.* **38:**438–445.

Hensen, R.K., Etchison, D., Hershey, J.W.R., and Ehrenfeld, E., 1982a, Association of cap-binding protein with eukaryotic initiation factor 3 in initiation factor preparations from uninfected and poliovirus-infected HeLa cells, *J. Virol.* **42:**200–207.

Hansen, J.L., Etchison, D.O., Hershey, J.W.B., and Ehrenfeld, E., 1982b, Localization of cap-binding protein in subcellular fractions of HeLa cells. *Mol. Cell. Biol.* **2:**1639–1643.

Hassan, S.A., Rabin, E.R., and Melnick, J.L., 1965, Reovirus myocarditis in mice: An electron microscopic, immunofluorescent, and virus assay study, *Exp. Mol. Pathol.* **4:**66–80.

Hastings, K.E.M., and Millward, S., 1981, Similar sets of terminal oligonucleotides from reovirus double-stranded RNA and viral messenger RNA synthesized *in vitro, lCan. J. Biochem.* **59:**151–157.

Hay, A.J., and Joklik, W.K., 1971, Demonstration that the same strand of reovirus genome RNA is transcribed *in vitro* and *in vivo, Virology* **44:**450–453.

Helentjaris, T., and Ehrenfeld, E., 1978, Control of protein synthesis in extracts from poliovirus-infected cells. I. mRNA discrimination by crude initiation factors, *J. Virol.* **26:**510–521.

Helentjaris, T., Ehrenfeld, E., Brown-Luedi, M.L., and Hershey, J.W.B., 1979, Alterations in initiation factor activity from poliovirus-infected HeLa cells, *J. Biol. Chem.* **254:**10,973–10,978

Hewlett, M.J., Rose, J.K., and Baltimore, D., 1976, 5'-Terminal structure of poliovirus polyribosomal RNA is pUp, *Proc. Natl. Acad. Sci. U.S.A.* **73:**327–330.

Huismans, H., and Joklik, W.K., 1976, Reovirus coded polypeptides in infected cells: Isolation of two native monomeric polypeptides with affinity for single-stranded and double-stranded RNA respectively, *Virology* **70**(2):411–424.

Ito, Y., and Joklik, W.K., 1972a, Temperature-sensitive mutants of reovirus. I. Patterns of gene expression by mutants of groups C, D and E, *Virology* **50:**189–201.

Ito, Y., and Joklik, W.K., 1972b, Temperature-sensitive mutants of reovirus. II. Anomalous electrophoretic migration behavior of certain hybrid RNA molecules composed of mutant plus strands and wild-type minus strands, *Virology* **50:**202–208.

Jen, G., Detjen, B.M., and Thach, R.E., 1980, Shutoff of HeLa cell protein synthesis by encephalomyocarditis virus and poliovirus: A comparative study, *J. Virol.* **35:**150–156.

Johnson, R.B., Jr., Soeiro, R., and Fields, B.N., 1976, The synthesis of A-rich RNA by temperature-sensitive mutants of reovirus, *Virology* **73**(1):173–180.

Joklik, W.K., 1972, Studies on the effect of chymotrypsin on reovirions, *Virology* **49:**700–715.

Joklik, W.K., 1974, Reproduction of Reoviridae, in: *Comprehensive Virology,* Vol. 2 (H. Fraenkel-Conrat and R.R. Wagner, eds.), pp. 231–334 Plenum Press, New York.

Joklik, W.K., 1980, The structure and function of the reovirus genome, *Ann. N. Y. Acad. Sci.* **354:**107–124.

Joklik, W.K., 1981, Structure and function of the reovirus genome, *Microbiol. Rev.* **45:**483–501.

Joklik, W.K., Skehel, J.J., and Zweerink, H.J., 1970, The transcription of the reovirus genome, *Cold Spring Harbor Symp. Quant. Biol.* **35:**791–801.

Kapuler, A.M., Mendelsohn, N., Klett, H., and Acs, G., 1970, Four base-specific nucleoside 5'-triphosphatases in the subviral core of reovirus, *Nature (London)* **225:**1209–1213.

Kaufman, Y., Goldstein, E., and Penman, S., 1976, Poliovirus-induced inhibition of polypeptide initiation *in vitro* on native polyribosomes, *Proc. Natl. Acad. Sci. U.S.A.* **73:**1834–1838.

Kavenoff, R., Talcove, D., and Mudd, J.A., 1975, Genome sized RNA from reovirus particles, *Proc. Natl. Acad. Sci. U.S.A.* **72:**4317–4321.

Kilham, L., and Margolis, G., 1969, Hydrocephalus in hamsters, ferrets, rats and mice following inoculations with reovirus type 1. II. Pathological studies, *Lab. Invest.* **21:**189–198.

Koide, F., Suzuka, I., and Sekiguchi, K., 1968, Some properties of an adenine-rich polynucleotide fragment from the avian reovirus, *Biochem. Biophys. Res. Commun.* **30**:95–99.

Kozak, M., 1977, Nucleotide sequences of 5'-terminal ribosome-protected initiation regions from two reovirus messages, *Nature (London)* **269**:390–394.

Kozak, M., 1978, How do eukaryotic ribosomes select initiation regions in messenger RNA? *Cell* **15**:1109–1123.

Kozak, M., 1981, Possible role of flanking nucleotides in recognition of the AUG initiator codon by eukaryotic ribosomes, *Nucleic Acids Res.* **9**:5233–5252.

Kozak, M., 1982a, Analysis of ribosome binding sites from the s1 message of reovirus. Initiation at the first and second AUG codons. *J. Mol. Biol.* **156**:807–820.

Kozak, M., 1982b, Sequences of ribosome binding sites from the large size class of reovirus mRNA, *J. Virol.* **42**:467–473.

Kozak, M., and Shatkin, A.J., 1976, Characterization of ribosome-protected fragments from reovirus messenger RNA, *J. Biol. Chem.* **251**:4259–4266.

Kozak, M., and Shatkin, A.J., 1977a, Sequences of two 5'-terminal ribosome-protected fragments from reovirus messenger RNA's, *J. Mol. Biol.* **112**:75–96.

Kozal, M., and Shatkin, A.J., 1977b, Sequences and properties of two ribosome binding sites from the small size class of reovirus messenger RNA, *J. Biol. Chem.* **252**:6895–6908.

Kozak, M., and Shatkin, A.J., 1978, Identification of features in 5'-terminal fragments from reovirus mRNA which are important for ribosome binding, *Cell* **13**:201–212.

Kreft, J., 1980, Reovirus-specific messenger ribonucleoprotein particles from HeLa cells, *Z. Naturforsch.* **35c**:1046–1051.

Krystal, G., Winn, P., Millward, S., and Sakuma, S., 1975, Evidence for phosphoproteins in reovirus. *Virology* **64**:505–512.

Krystal, G., Perrault, J., and Graham, A.F., 1976, Evidence for a glycoprotein in reovirus, *Virology* **72**:308–321.

Kudo, H., and Graham, A.F., 1965, Synthesis of reovirus ribonucleic acid in L cells, *J. Bacteriol.* **90**:936–945.

Kudo, H., and Graham, A.F., 1966, Selective inhibition of reovirus induced RNA in L cells, *Biochem. Biophys. Res. Commun.* **24**:150–155.

Laemmli, U.K., 1970, Cleavage of structural proteins during the assembly of the head of bacteriophage T4, *Nature (London)* **227**:680–685.

Lai, K.C., and Bellamy, A.R., 1971, Factors affecting the amount of oligonucleotides in reovirus particles, *Virology* **45**(3):821–823.

Lai, M.-H.T., Werenne, J.J., and Joklik, W.K., 1973, The preparation of reovirus top component and its effect on host DNA and protein synthesis, *Virology* **54**(1):237–244.

Lau, R.Y., Van Alstyne, D., Berckmans, R., and Graham, A.F., 1975, Synthesis of reovirus-specific polypeptides in cells pretreated with cycloheximide, *J. Virol.* **16**:470–478.

Lee, K.A.W., and Sonenberg, N., 1982, Inactivation of cap-binding proteins accompanies the shut-off of host protein systhesis by poliovirus, *Proc. Natl. Acad. Sci. U.S.A.* **79**:3447–3451.

Lee, P.W.K., Hayes, E.C., and Joklik, W.K., 1981a, Characterization of anti-reovirus immunoglobulins secreted by cloned hybridoma cell lines, *Virology* **108**:134–146.

Lee, P.W.K., Hayes, E.C., and Joklik, W.K., 1981b, Protein sigma-1 is the reovirus cell attachment protein, *Virology* **108**(1):156–163.

Leibowitz, R., and Penman, S., 1971, Regulation of protein synthesis in HeLa cells. III. Inhibition during poliovirus infection, *J. Virol.* **8**:661–668.

Lerner, A.M., and Miranda, Q.R., 1968, Cellular interactions of several enteroviruses and a reovirus after treatment with sodium borohydride or carbohydrases, *Virology* **36**:277–285.

Lerner, A.M., Cherry, J.D., and Finland, M., 1963, Hemagglutination with reovirus, *Virology* **19**:58–65.

Lerner, A.M., Bailey, E.J., and Tillotson, J.R., 1966a, Enterovirus hemagglutination: Inhibition by several enzymes and sugars, *J. Immunol.* **95**:1111–1115.

Lerner, A.M., Gelb, L.D., Tillotson, J.R., Carruthers, M.M., and Bailey, E.J., 1966b, Enterovirus hemagglutination: Inhibition by aldoses and a possible mechanism, *J. Immunol.* **96:**629–636.

Levin, D.H., Mendelsohn, N., Schonberg, M., Klett, H., Silverstein, S., Kapuler, A.M., and Acs, G., 1970, Properties of RNA transcriptase in reovirus subviral particles, *Proc. Natl. Acad. Sci. U.S.A.* **66:**890–897.

Levin, K.H., and Samuel, C.E., 1980, Biosynthesis of reovirus-specified polypeptides: Purification and characterization of the small-sized class mRNAs of reovirus type 3: Coding assignments and translational efficiencies, *Virology* **106:**1–13.

Li, J.K.-K., Keene, J.D., Scheible, P.P., and Joklik, W.K., 1980a, Nature of the 3′-terminal sequences of the plus and minus strands of the S1 gene of reovirus serotypes 1, 2, and 3, *Virology* **105:**41–51.

Li, J.K.-K., Scheible, P.P., Keene, J.D., and Joklik, W.K., 1980b, The plus strand of reovirus gene S2 is identical with its *in vitro* transcript, *Virology* **105:**282–286.

Loh, P.C., and Soergel, M., 1966, Growth characteristics of reovirus type 2: Actinomycin D and the preferential synthesis of viral RNA, *Proc. Soc. Exp. Biol. Med.* **122:**1248–1250.

Loh, P.C., Hashiro, G.M., and Yau, J.T., 1977, Effect of poly cations on the early stages of reovirus infection, *Microbios* **19**(77–78):213–230.

Luftig, R.B., Kilham, S.S., Hay, A.J., Zweerink, H.J., and Joklik, W.K., 1972, An ultrastructural study of virions and cores of reovirus type 3, *Virology* **48:**170–181.

MacDonald, R.D., 1980, Immuno fluorescent detection of double-stranded RNA in cells infected with reovirus, infectious pancreatic necrosis virus and infectious bursal disease virus, *Can. J. Microbiol.* **26**(2):256–261.

Margolis, G., Kilham, L., and Gonatos, N.K., 1971, Reovirus type 3 encephalitis: Observations of virus–cell interactions in neuronal tissues, *Lab. Invest.* **24:**101–109.

Martin, S.A., Pett, D.M., and Zweerink, H.J., 1973, Studies on the topography of reovirus and bluetongue virus capsid proteins, *J. Virology* **12:**194–198.

Matsuhisa, T., and Joklik, W.K., 1974, Temperature-sensitive mutants of reovirus. V. Studies on the nature of the temperature-sensitive lesion of the group C mutant ts 477, *Virology* **60:**380–389.

Mayor, H.D., and Jordan, L.E., 1965, Studies on reovirus. I. Morphologic observations on the development of reovirus in tissue culture, *Exp. Mol. Pathol.* **4:**40–50.

McCrae, M.A., and Joklik, W.K., 1978, The nature of the polypeptide encoded by each of the 10 double-stranded RNA segments of reovirus type 3, *Virology* **98:**578–593.

McDowell, M.J., and Joklik, W.K., 1971, An *in vitro* protein synthesizing system from mouse L fibroblasts infected with reovirus, *Virology* **45**(3):724–733.

McDowell, M.J., Joklik, W.K., Villa-Komaroff, L., and Lodish, H.F., 1972, Translation of reovirus messenger RNA's synthesized *in vitro* into reovirus polypeptides by several mammalian cell-free extracts, *Proc. Natl. Acad. Sci. U.S.A.* **69**(9):2649–2653.

Millward, S., and Graham, A.F., 1974, Reovirus: Early events (in the infected cell) and structure of the double-stranded RNA genome, in: *Viruses, Evolution and Cancer* (E. Kurstak and K. Maramorosch, eds.), pp. 651–675, Academic Press, New York.

Morgan, E.M., and Kingsbury, D.W., 1980, Pyridoxal phosphate as a probe of reovirus transcriptase, *Biochemistry* **19:**484–489.

Morgan, E.M., and Kingsbury, D.W., 1981, Reovirus enzymes that modify messenger RNA are inhibited by perturbation of the lambda proteins, *Virology* **113:**565–572.

Morgan, E.M., and Zweerink, H.J., 1974, Reovirus morphogenesis: Core-like particles in cells infected at 39° with wild-type reovirus and temperature-sensitive mutants of groups B and G, *Virology* **59:**556–565.

Morgan, E.M., and Zweerink, H.J., 1975, Characterization of transcriptase and replicase particles isolated from reovirus infected cells, *Virology* **68**(2):455–466.

Morgan, E.M., and Zweerink, H.J., 1977, Characterization of the double-stranded RNA in replicase particles in reovirus infected cells, *Virology* **77**(1):421–423.

Moyer, S.A., and Banerjee, A.K., 1976, *In vivo* methylation of vesicular stomatitis virus and its host-cell messenger RNA species, *Virology* **70:**339–351.

Mustoe, T.A., Ramig, R.F., Sharpe, A.H., and Fields, B.N., 1978a, A genetic map of reovirus.

III. Assignment of the double-stranded RNA-positive mutant groups A, B and G to genome segments, *Virology* **85:**545–556.

Mustoe, T.A., Ramig, R.F., Sharpe, A.H., and Fields, B.N., 1978b, Genetics of reovirus: Identification of the dsRNA segments encoding the polypeptides of the μ and σ size classes, *Virology* **89:**594–604.

Nakashima, K., LaFiandra, A.J., and Shatkin, A.J., 1979, Differential dependence of reovirus-associated enzyme activities on genome RNA as determined by psoralen photosensitivity, *J. Biol. Chem.* **254:**8007–8014.

Nichols, J.L., Bellamy, A.R., and Joklik, W.K., 1972a, Identification of the nucleotide sequences of the oligonucleotides present in reovirions, *Virology* **49:**562–572.

Nichols, J.L., Hay, A.J., and Joklik, W.K., 1972b, 5′-Terminal nucleotide sequence of reovirus mRNA synthesized *in vitro*, *Nature (London) New Biol.* **235:**105–107.

Nomoto, A., Lee, Y.F., and Wimmer, E., 1976, The 5′ end of poliovirus mRNA is not capped with $m^7G(5′)ppp(5′)Np$, *Proc. Natl. Acad. Sci. U.S.A.* **73:**375–380.

Nonoyama, M., and Graham, A.F., 1970, Appearance of defective virions in clones of reovirus, *J. Virol.* **6:**693–694.

Nonoyama, M., Watanabe, Y., and Graham, A.F., 1970, Defective virions of reovirus, *J. Virol.* **6:**226–236.

Nonoyama, M., Millward, S., and Graham, A.F., 1974, Control of transcription of the reovirus genome, *Nucleic Acids Res.* **1:**373–385.

Oie, H., Loh, P.C., and Soergel, M., 1966, Growth characteristics and immunocytochemical studies of reovirus type 2 in a line of human amnion cells, *Arch. Gesamte Virusforsch.* **18:**16–24.

Prevec, L., Watanabe, Y., Gauntt, C.J., and Graham, A.F., 1968, Transcription of the genomes of type 1 and type 3 reoviruses, *J. Virol.* **2:**289–297.

Ralph, S.J., Harvey, J.D., and Bellamy, A.R., 1980, Subunit structure of the reovirus spike, *J. Virol.* **36:**894–896.

Ramig, R.F., Cross, R.K., and Fields, B.N., 1977, Genome RNA's and polypeptides of reovirus serotypes 1, 2 and 3, *J. Virol.* **22:**726–733.

Ramig, R.F., Mustoe, T.A., Sharpe, A.H., and Fields, B.N., 1978, A genetic map of reovirus. II. Assignment of the double-stranded RNA-negative mutants groups C, D and E genome segments, *Virology* **85:**531–544.

Rhim, J.S., Jordan, L.E., and Mayor, H.D., 1962, Cytochemical, fluorescent-antibody and electron microscopic studies on the growth of reovirus (Echo 10) in tissue culture, *Virology* **17:**342–355.

Rose, J.K., Trachsel, H., Leong, K., and Baltimore, D., 1978, Inhibition of translation by poliovirus: Inactivation of a specific initation factor, *Proc. Natl. Acad. Sci. U.S.A.* **75:**2732–2736.

Rosen, L., 1962, Reovirus in animals other than man, *Ann. N.Y. Acad. Sci.* **101:**461–465.

Rosen, L., 1968, Reoviruses, in: *Virology Monographs*, Vol. 1 (S. Gard, C. Hallauer, and K.F. Meyer, eds.), pp. 73–107, Springer-Verlag, Vienna.

Sakuma, S., and Watanabe, Y., 1971, Unilateral synthesis of reovirus double-stranded ribonucleic acid by a cell free replicase system, *J. Virol.* **8:**190–196.

Sakuma, S., and Watanabe, Y., 1972a, Reovirus replicase-directed synthesis of double-stranded ribonucleic acid, *J. Virol.* **10:**628–638.

Sakuma, S., and Watanabe, Y., 1972b, Incorporation of *in vitro* synthesized reovirus double-stranded RNA into virus core-like particles, *J. Virol.* **10(5):**943–950.

Samuel, C.E., Farris, D.A., and Levin, K.H., 1977, Biosynthesis of virus specified polypeptides: System-dependent effect of reovirus mRNA methylation on translation *in vitro* catalyzed by ascites tumor and wheat embryo cell-free extracts, *Virology* **81:**476–481.

Schmidt, N.J., Dennis, J., Hoffman, M.N., and Lennette, E.H., 1964a, Inhibitors of echovirus and reovirus hemagglutination. I. Inhibitors in tissue culture fluids, *J. Immunol.* **93:**367–376.

Schmidt, N.J., Dennis, J., Hoffman, M.N., and Lennette, E.H., 1964b, Inhibitors of echovirus and reovirus hemagglutination. II. Serum and phospholipid inhibitors, *J. Immunol.* **93:**377–386.

Schochetman, G., and Millward, S., 1972, Ribonucleoside diphosphate precursors for *in vitro* reovirus RNA synthesis, *Nature (London) New Biol.* **239:**77–79.

Schonberg, M., Silverstein, S.C., Levin, D.H., and Acs, G., 1971, Asynchronous synthesis of the complementary strands of the reovirus genome, *Proc. Natl. Acad. Sci. U.S.A.* **68:**505–508.

Schuerch, A.R., Matsuhisa, T., and Joklik, W.K., 1974, Temperature-sensitive mutants of reovirus. VI. Mutants ts447 and ts556 particles that lack either one or two genome RNA segments, *Intervirology* **3:**36–46.

Schuerch, A.R., Mitchell, W.R., and Joklik, W.K., 1975, Isolation of intact individual species of single- and double-stranded RNA after fractionation by polyacrylamide gel electrophoresis, *Anal. Biochem.* **65:**331–345.

Sharpe, A.H., and Fields, B.N., 1981, Reovirus inhibition of cellular DNA synthesis: Role of the S1 gene, *J. Virol.* **38:**389–392.

Sharpe, A.H., and Fields, B.N., 1982, Reovirus inhibition of cellular RNA and protein synthesis: Role of the S4 gene, *Virology* **122:**381–391.

Sharpe, A.H., Chen, L.B., and Fields, B.N., 1982, The interaction of mammalian reoviruses with the cytoskeleton of monkey kidney CV-1 cells, *Virology* **120:**399–411.

Sharpe, A.H., Ramig, R.F., Mustoe, T.A., and Fields, B.N., 1978, A genetic map of reovirus. I. Correlation of genome RNA's between serotypes 1, 2 and 3, *Virology* **84:**63–74.

Shatkin, A.J., 1965a, Actinomycin and the differential synthesis of reovirus and L cell RNA, *Biochem. Biophys. Res. Commun.* **19:**506–510.

Shatkin, A.J., 1965b, Inactivity of purified reovirus RNA as a template for *E. coli* polymerases *in vitro*, *Proc. Natl. Acad. Sci. U.S.A.* **54:**1721–1728.

Shatkin, A.J., 1974, Methylated messenger RNA synthesis *in vitro* by purified reovirus, *Proc. Natl. Acad. Sci. U.S.A.* **71:**3204–3207.

Shatkin, A.J., 1976, Capping of eukaryotic mRNAs, *Cell* **9:**645–653.

Shatkin, A.J., and Banerjee, A.K., 1970, *In vitro* transcription of double-stranded RNA by reovirus-associated RNA polymerase, *Cold Spring Harbor Symp. Quant. Biol.* **35:**781.

Shatkin, A.J., and Both, G.W., 1976, Reovirus mRNA: Transcription and translation, *Cell* **7:**305–313.

Shatkin, A.J., and LaFiandra, A.J., 1972, Transcription by infectious subviral particles of reovirus, *J. Virol.* **10**(4):698–706.

Shatkin, A.J., and Rada, B., 1967, Reovirus-directed ribonucleic acid synthesis in infected L cells, *J. Virol.* **1:**24–35.

Shatkin, A.J., and Sipe, J.D., 1968a, Single-stranded, adenine-rich RNA from purified reoviruses, *Proc. Natl. Acad. Sci. U.S.A.* **59:**246–253.

Shatkin, A.J., and Sipe, J.D., 1968b, RNA polymerase activity in purified reoviruses, *Proc. Natl. Acad. Sci. U.S.A.* **61:**1462–1469.

Shaw, J.E., and Cox, D.C., 1973, Early inhibition of cellular DNA synthesis by high multiplicities of infectious and UV inactivated reovirus, *J. Virol.* **12**(4):704–710.

Shelton, I.H., Kasupski, G.J., Oblin, C., and Hand, R., 1981, DNA binding of a nonstructural reovirus protein, *Can. J. Biochem.* **59:**122–130.

Shestopalova, N.M., Reingold, V.N., Chumakova, A.B., and Voroshilova, M.K., 1977, A new complex between ribosomes and reovirus particles, *Biol. Cell.* **28:**9–12.

Shuman, S., and Hurwitz, J., 1981, Mechanism of mRNA capping by vaccinia virus guanylyltransferase: Characterization of an enzyme–guanylate intermediate, *Proc. Natl. Acad. Sci. U.S.A.* **78:**187–191.

Silverstein, S.C., 1975, The role of mononuclear phagocytes in viral immunity, in: *Mononuclear Phagocytes in Immunity, Infection and Pathology* (R. Van Furth, ed.), pp. 557–568, Blackwell, Oxford.

Silverstein, S.C., and Dales, S., 1968, The penetration of reovirus RNA and initiation of its genetic function in L-strain fibroblasts, *J. Cell Biol.* **36:**197–230.

Silverstein, S.C., and Schur, P.H., 1970, Immunofluorescent localization of double-stranded RNA in reovirus-infected cells, *Virology* **41:**564–566.

Silverstein, S.C., Schonberg, M., Levin, D.H., and Acs, G., 1970, The reovirus replicative cycle: Conservation of parental RNA and protein, *Proc. Natl. Acad. Sci. U.S.A.* **67:**275–281.

Silverstein, S.C., Astell, C., Levin, D.H., Schonberg, M., and Acs, G., 1972, The mechanisms of reovirus uncoating and gene activation *in vivo, Virology* **47**(3):797–806.

Silverstein, S.C., Astell, C., Christman, J., Klett, H., and Acs, G., 1974, Synthesis of reovirus oligoadenylic acid *in vivo* and *in vitro, J. Virol.* **13**(3):740–752.

Silverstein, S.C., Christman, J.K., and Acs, G., 1976, The reovirus replication cycle, *Annu. Rev. Biochem.* **45:**375–408.

Skehel, J.J. and Joklik, W.K., 1969, Studies on the *in vitro* transcription of reovirus RNA catalyzed by reovirus cores, *Virology* **39:**822–831.

Skup, D., and Millward, S., 1977, Highly efficient translation of messenger RNA in cell-free extracts prepared from L-cells, *Nucleic Acids Res.* **4:**3581–3587.

Skup, D., and Millward, S., 1980a, Reovirus induced modification of cap dependent translation in infected L cells, *Proc. Natl. Acad. Sci. U.S.A.* **77**(1):152–156.

Skup, D., and Millward, S., 1980b, mRNA capping enzymes are masked in reovirus progeny subviral particles, *J. Virol.* **34:**490–496.

Skup, D., Zarbl, H., and Millward, S., 1981, Regulation of translation in L-cells infected with reovirus, *J. Mol. Biol.* **151:**35–55.

Smith, R.E., Zweerink, H.J., and Joklik, W.K., 1969, Polypeptides components of virions, top component and cores of reovirus type 3, *Virology* **39:**791–810.

Sonenberg, N., 1981, ATP/Mg$^{++}$-dependent crosslinking of cap binding proteins to the 5′ end of eukaryotic mRNA, *Nucleic Acids Res.* **9:**1643–1656.

Sonenberg, N., and Shatkin, A.J., 1977, Reovirus mRNA can be covalently crosslinked via the 5′ cap to proteins in initiation complexes, *Proc. Natl. Acad. Sci. U.S.A.* **74:**4288–4292.

Sonenberg, N., Guertin, D., and Lee, K.A.W., 1982, Capped mRNAs with reduced secondary structure can function in extracts from poliovirus-infected cells, *Mol. Cell. Biology* **2:**1633–1638.

Sonenberg, N., Morgan, M.A., Merrick, W.C., and Shatkin, A.J., 1978, A polypeptide in eukaryotic initiation factors that crosslinks specifically to the 5′-terminal cap in mRNA, *Proc. Natl. Acad. Sci. U.S.A.* **75:**4843–4847.

Sonenberg, N., Morgan, M.A., Testa, D., Colonno, R.J., and Shatkin, A.J., 1979a, Interaction of a limited set of proteins with different mRNAs and protection of 5′-caps against pyrophosphatase digestion in initiation complexes, *Nucleic Acids Res.* **7:**15–29.

Sonenberg, N., Rupprecht, K.M., Hecht, S.M., and Shatkin, A.J., 1979b, Eukaryotic mRNA cap binding protein: Purification by affinity chromatography on Sepharose-coupled m$^7$GDP, *Proc. Natl. Acad. Sci. U.S.A.* **76:**4345–4349.

Sonenberg, N., Trachsel, H., Hecht, S., and Shatkin, A.J., 1980, Differential stimulation of capped mRNA translation *in vitro* by cap-binding protein, *Nature (London)* **285:**331–333.

Sonenberg, N., Guertin, D., Cleveland, D.R., and Trachsel, H., 1981a, Probing the function of the eukaryotic 5′ cap structure by using a monoclonal antibody directed against cap-binding proteins, *Cell* **27:**563–572.

Sonenberg, N., Skup, D., Trachsel, H., and Millward, S., 1981b, *In vitro* translation in reovirus- and poliovirus-infected cell extracts: Effects of anti-cap binding protein monoclonal antibody, *J. Biol. Chem.* **256:**4138–4141.

Spandidos, D.A., and Graham, A.F., 1975, Complementation between temperature-sensitive and deletion mutants of reovirus, *J. Virol.* **16:**1444–1452.

Spandidos, D.A., and Graham, A.F., 1976a, Recombination between temperature-sensitive and deletion mutants of reovirus, *J. Virol.* **18:**117–123.

Spandidos, D.A. and Graham, A.F., 1976b, Nonpermissive infection of L-cells by an avian reovirus: Restricted transcription of the viral genome, *J. Virol.* **19:**977–984.

Spandidos, D.A., Krystal, G., and Graham, A.F., 1976, Regulated transcription of the genomes of defective virions and temperature-sensitive mutants of reovirus, *J. Virol.* **18**(1):7–19.

Spendlove, R.S., and Schaffer, F.L., 1965, Enzymatic enhancement of infectivity of reovirus, *J. Bacteriol.* **89**:597–602.

Spendlove, R.S., Lennette, E.H., Knight, C.O. and Chin, J.N., 1963a, Development of viral antigen and infectious virus in HeLa cells infected with reovirus, *J. Immunol.* **90**:548–553.

Spendlove, R.S., Lennette, E.H., and John, A.C., 1963b, The role of the mitotic apparatus in the intracellular location of reovirus antigen, *J. Immunol.* **90**:554–560.

Spendlove, R.S., Lennette, E.H., Chin, J.N., and Knight, C.O., 1964, Effects of antimitotic agents on intracellular reovirus antigen, *Cancer Res.* **24**:1826–1833.

Spendlove, R.S., McClain, M.E., and Lennette, E.H., 1970, Enhancement of reovirus infectivity by extracellular removal or alteration of the virus capsid by proteolytic enzymes, *J. Gen. Virol.* **8**:83–94.

Stamatos, N.M., and Gamatos, P.J., 1982, Binding to selected regions of reovirus mRNAs by a nonstructural reovirus protein, *Proc. Natl. Acad. Sci. U.S.A.* **79**:3457–3461.

Stanley, N.F., 1961a, Reovirus—a ubiquitous orphan, *Med. J. Aust.* **2**:815–818.

Stanley, N.F., 1961b, Relationship of hepatoencephalomyelitis virus and reovirus, *Nature (London)* **189**:687.

Stoltzfus, C.M., and Banerjee, A.K., 1972, Two oligonucleotide classes of single-stranded ribopolymers in reovirus A-rich RNA, *Arch. Biochem. Biophys.* **152**:733–743.

Stoltzfus, C.M., Morgan, M., Banerjee, A.K., and Shatkin, A.J., 1974, Poly(A) polymerase activity in reovirus, *J. Virol.* **13**:1338–1345.

Tardieu, M., and Weiner, H.L., 1982, Viral receptors on isolated murine and human ependymal cells, *Science* **215**:419–421.

Tillotson, J.R., and Lerner, A.M., 1966, Effect of periodate oxidation on hemagglutinating and antibody-producing capacities of certain enteroviruses and reoviruses, *Proc. Natl. Acad. Sci. U.S.A.* **56**:1143–1150.

Trachsel, H., Sonenberg, N., Shatkin, A.J., Rose, J.K., Leong, K., Bergman, J.E., Gordon, J., and Baltimore, D., 1980, Purification of a factor that restores translation of VSV mRNA in extracts from poliovirus-infected HeLa cells, *Proc. Natl. Acad. Sci. U.S.A.* **77**:770–774.

Van Dijk, A.A., and Huismans, H., 1980, The *in vitro* activation and further characterization of the bluetongue virus-associated transcriptase, *Virology* **104**:347–356.

Van Steeg, H., Thomas, A., Verbeek, S., Kasperaitis, M., Voorma, H.O., and Benne, R., 1981, Shut-off of neuroblastoma cell protein synthesis by Semliki forest virus: Loss of ability of crude initiation factors to recognize early Semliki forest virus and host mRNA's, *J. Virol.* **38**:728–736.

Venkatesan, S., and Moss, B., 1980, Donor and acceptor specificities of HeLa cell mRNA guanylyltransferase, *J. Biol. Chem.* **255**: 2835–2842.

Venkatesan, S., and Moss, B., 1982, Eukaryotic mRNA capping enzyme—guanylate covalent intermediate, *Proc. Natl. Acad. Sci. U.S.A.* **79**:340–344.

Wachsman, J.T., Levin, D.H., and Acs, G., 1970, Ribonucleoside triphosphate-dependent pyrophosphate exchange of reovirus cores, *J. Virol.* **6**:563–565.

Walden, W.E., Godefroy-Colburn, T., and Thach, R.E., 1981, The role of mRNA competition in regulating translation. I. Demonstration of competition *in vivo*, *J. Biol. Chem.* **256**:11,739–11,746.

Wallis, C., Smith, K.O., and Melnick, J.L., 1964, Reovirus activation by heating and inactivation by cooling in MgCl₂ solutions, *Virology* **22**:608–619.

Wallis, C., Melnick, J.L., and Rapp, F., 1966, Effects of pancreatin on the growth of reovirus, *J. Bacteriol.* **92**:155–160.

Ward, R.L., and Shatkin, A.J., 1972, Association of reovirus messenger RNA with viral proteins: A possible mechanism for linking the genome segments, *Arch. Biochem. Biophys.* **152**(1):378–384.

Watanabe, Y., Kudo, H., and Graham, A.F., 1967a, Selective inhibition of reovirus ribonucleic acid synthesis by cycloheximide, *J. Virol.* **1**:36–44.

Watanabe, Y., Prevec, L., and Graham, A.F., 1967b, Specificity in transcription of the reovirus genome, *Proc. Natl. Acad. Sci. U.S.A.* **58**:1040–1046.

Watanabe, Y., Gauntt, C.J., and Graham, 1968a, Reovirus-induced ribonucleic acid polymerase, *J. Virol.* **2**:869–877.

Watanabe, Y., Millward, S., and Graham, A.F., 1968b, Regulation of transcription of the reovirus genome, *J. Mol. Biol.* **36**:107–123.

Watanabe, Y., Sakuma, S., and Shames, R., 1974, *In vitro* synthesis of reovirus genomic segments, *Jpn. J. Microbiol.* **18**:253–258.

Wei, C.-M., and Moss, B., 1977, 5′-Terminal capping of RNA by guanylyltransferase from HeLa cell nuclei, *Proc. Natl. Acad. Sci. U.S.A.* **74**:3758–3761.

Weiner, H.L., and Fields, B.N., 1977, Neutralization of reovirus: The gene responsible for the neutralization antigen, *J. Exp. Med.* **146**:1305–1310.

Weiner, H.L., Drayna, D., Averill, D.R., Jr., and Fields, B.N., 1977, Molecular basis of reovirus virulence: Role of the S-1 gene, *Proc. Natl. Acad. Sci. U.S.A.* **74**(12):5744–5748.

Weiner, H.L., Ramig, R.F., Mustoe, T.A., and Fields, B.N., 1978, Identification of the gene coding for the hemagglutinin of reovirus, *Virology* **86**(2):581–584.

Weiner, H.L., Powers, M.L., and Fields, B.N., 1980a, Absolute linkage of virulence and central nervous system cell tropism of reoviruses to viral hemagglutinin, *J. Infect. Dis.* **141**(5):609–616.

Weiner, H.L., Ault, K.A., and Fields, B.N., 1980b, Interaction of reovirus with cell receptors. I. Murine and human lymphocytes have a receptor for the hemagglutinin of reovirus type 3. *J. Immunol.* **124**:2143–2148.

Weiner, H.L., Greene, M.I., and Fields, B.N., 1980c, Delayed hypersensitivity in mice infected with reovirus. I. Identification of host and viral gene products responsible for the immune response, *J. Immunol.* **125**:278–282.

White, C.K., and Zweerink, H.J., 1976, Studies on the structure of reovirus cores: Selective removal of polypeptide λ2, *Virology* **70**:171–180.

Wiebe, M.E., and Joklik, W.K., 1975, The mechanism of inhibition of reovirus replication by interferon, *Virology* **66**:229–240.

Yamakawa, M., Furuichi, Y., and Shatkin, A.J., 1982, Reovirus transcriptase and capping enzymes are active in intact virions, *Virology* **118**:157–168.

Yamakawa, M., Furuichi, Y., Nakashima, K., LaFiandra, A.J., and Shatkin, A.J., 1981, Excess synthesis of viral mRNA 5′-terminal oligonucleotides by reovirus transcriptase, *J. Biol. Chem.* **256**:6507–6514.

Zarbl, H., Hastings, K.E.M., and Millward, S., 1980a, Reovirus core particles synthesize capped oligonucleotides as a result of abortive transcription, *Arch. Biochem. Biophys.* **202**:348–360.

Zarbl, H., Skup, S., and Millward, S., 1980b, Reovirus progeny subviral particles synthesize uncapped mRNA, *J. Virol.* **34**:497–505.

Zweerink, H., 1974, Multiple forms of ss→ds RNA polymerase activity in reovirus-infected cells, *Nature* (*London*) **247**:313–315.

Zweerink, H.J., and Joklik, W.K., 1970, Studies on the intracellular synthesis of reovirus-specified proteins, *Virology* **41**:501–518.

Zweerink, H.J., McDowell, M.J., and Joklik, W.K., 1971, Essential and nonessential noncapsid reovirus proteins, *Virology* **45**(3):716–723.

Zweerink, H.J., Ito, Y., and Matsuhisa, T., 1972, Synthesis of reovirus double-stranded RNA within virion-like particles, *Virology* **50**:349–358.

Zweerink, H.J., Morgan, E.M., and Skyler, J.S., 1976, Reovirus morphogenesis: Characterization of subviral particles in infected cells, *Virology* **73**:442–453.

# ADDENDUM

Detjen *et al.* (1982) found no transition to cap independent translation in reovirus-infected cells. This conclusion was based on the observations that the translation of endogenous reovirus mRNA in infected lysates was sensitive to $m^7GTP$, and that decapped globin mRNA or encephalomyocarditis (EMC) virus RNA was not translated more efficiently in infected lysates. Our results (R. Lemieux, H. Zarbl, and S. Millward, in prep.) also indicate that translation of late reovirus mRNA is sensitive to $m^7GTP$, although the reason for this is not clear. Furthermore, we have shown that decapped globin mRNA translates poorly in both infected and uninfected lysates, while EMC RNA is translated efficiently in both lysates. In addition, we have carried out competition experiments between L-cell mRNA and late reovirus mRNA. Results indicated that in uninfected lysates, late reovirus mRNA does compete, albeit poorly, with L-cell mRNA. In infected lysates, late reovirus mRNA is translated with high efficiency and does not compete with L-cell mRNA present in saturating amounts. We have also prepared fractionated translation systems from both uninfected and infected L cells, which consisted of an S-200 supernate, 0.5 M KCl washed ribosomes, and the ribosomal salt wash proteins, that retained their ability to discriminate between capped and uncapped reovirus mRNAs on reconstitution (R. Lemieux and S. Millward, in prep.). Each fraction was then tested for its ability to stimulate the translation of late reovirus mRNA in an unfractionated lysate prepared from uninfected cells. Results showed that the translation of late reovirus mRNA was greatly stimulated with any of the three fractions prepared from infected cells, but not by fractions from uninfected cells. In addition, fractions from infected cells failed to stimulate or inhibit the translation of globin mRNA, L-cell mRNA, tobacco mosaic virus RNA, and poliovirus mRNA. These results demonstrate the presence of a viral factor in infected cells which specifically stimulates the translation of late uncapped reovirus mRNAs. The bulk of the stimulatory activity was distributed equally between the S-200 supernate and the ribosomal salt wash. The specific activity of the viral factor was highest in the ribosomal salt wash, and preliminary experiments showed that this fraction was enriched in σ3, the major outer capsid protein of reovirus particles. This finding alluded to the possibility that σ3 may be the viral initiation factor which stimulates the translation of late viral mRNAs. We thus obtained a monospecific antibody directed against σ3 from Dr. P.W.K. Lee (University of Calgary, Calgary, Alberta). As already shown (Table V), polyspecific antisera against reovirus inhibited specifically the translation of uncapped reovirus mRNAs in infected lysates. Similar results have been obtained with the anti-σ3 antibody, further implicating σ3 in the translation of late viral mRNAs. This result has been reinforced by the genetic work of Sharpe and Fields (1982), who showed that inhibition of host protein synthesis maps in the S4 gene segment, which encodes the σ3 protein (Fig. 13).

CHAPTER 5

# Genetics of Reoviruses

Robert F. Ramig and Bernard N. Fields

## I. INTRODUCTION

The ultimate goal of genetic analysis of the mammalian reoviruses is to gain a detailed understanding of the structure and function of the genome and each of the viral polypeptide products. The early phases of genetic analysis were directed toward the isolation and physiological and genetic characterization of a collection of genetic markers, in this case, conditional lethal mutants of the temperature-sensitive type. More recent studies, particularly those since 1975, have emphasized the use of recombinant viruses generated during mixed infection of tissue-culture cells with pairs of viruses, most often of two different reovirus serotypes.

The segmented nature of the reovirus genome has certain unusual genetic implications. Most noteworthy is the observation that recombinant progeny are generated at either very high frequency or undetectably low frequency following mixed infection with two mutants. This "all-or-none" recombination has been shown to result from recombination via the mechanism of reassortment of genome segments. Thus, recombination in reovirus is more properly termed reassortment.

In this chapter, we will review briefly the structure of the reovirus genome and then consider in greater detail the types of mutants that have been identified in reovirus, the interactions that occur between pairs of mutants in mixed infection under permissive and nonpermissive conditions, the current status of the physical map of the reovirus genome, and the nature of the physiological lesions found in the different recombinationally defined mutant groups.

ROBERT F. RAMIG • Department of Virology and Epidemiology, Baylor College of Medicine, Houston, Texas 77030. BERNARD N. FIELDS • Department of Microbiology and Molecular Genetics, Harvard Medical School, Boston, Massachusetts 02115; and Department of Medicine, Brigham and Women's Hospital, Boston, Massachusetts 02115.

## II. NATURE OF THE REOVIRUS GENOME

The double-stranded nature of reovirus genome RNA (dsRNA) was first described by Gomatos *et al.* (1962) and has been confirmed in numerous subsequent studies. The dsRNA genome exists in the virion and is isolated as a series of unique pieces, or segments, rather than as a single molecule or chromosome. The virion dsRNA can be separated into three size classes by centrifugation or column chromatography. These size classes are designated *L* (large), *M* (medium), and *S* (small). The three size classes of genomic dsRNA can be further resolved into ten segments by polyacrylamide gel electrophoresis (Shatkin *et al.*, 1968; Prevec *et al.*, 1968; Skehel and Joklik, 1969). There are three electrophoretically resolved segments in the *L* size class (*L1–L3*, molecular weights 2.3–2.5

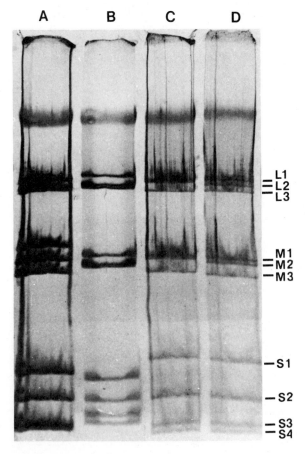

FIGURE 1. Double-stranded RNAs extracted from cells infected with the three reovirus serotypes (1, 2, and 3) and labeled with [$^{14}$C]uridine. Electrophoresis was carried out from top to bottom on a 10% polyacrylamide slab gel for 8 hr at 40 mA as described by Laemmli (1970). (A) Type 1 Lang; (B) type 2 Jones; (C) type 3 Abney; (D) type 3 Dearing. Adapted from Ramig *et al.* (1977b).

$\times 10^6$), three segments in the $M$ class ($M1$–$M3$, molecular weights 1.4–1.6 $\times 10^6$), and four segments in the $S$ class ($S1$–$S4$, molecular weights 0.6–0.9 $\times 10^6$). These segments are present in equimolar amounts in virion populations, indicating that each virion contains the full complement of ten segments. The molecular structure of the reovirus genome is considered in greater detail in Chapter 2.

A great deal of biochemical evidence supports the notion that segmentation of the reovirus genome is real and not an artifact of isolation procedures. Genetic studies support this concept by demonstrating that the genome functions in mixed infection as though it were segmented (see Section IV.B.).

Of considerable importance for the genetic analysis of reovirus was the finding that there is considerable electrophoretic heterogeneity among the genome segments of the three reovirus serotypes commonly used in the laboratory (Ramig et al., 1977b) (Fig. 1). This genomic heterogeneity is also observed among viruses freshly isolated from the field (Hrdy et al., 1979).

Finally, from the genetic viewpoint, it is important to note that the dsRNA genome is associated with viral particles throughout the infectious cycle. Genetic information is transferred from parent to progeny via a single-stranded RNA (ssRNA) intermediate form. Since both parental and progeny dsRNA genomes are always associated with viral ribonucleoprotein complexes, only the ssRNA is available for mixing and genetic interaction in the mixedly infected cell. The molecular events of the reovirus life cycle are considered in detail in Chapter 4.

## III. TYPES OF MUTANTS

### A. Temperature-Sensitive Mutants

Conditionally lethal, temperature-sensitive ($ts$) mutants of reovirus type 3 have been isolated in two laboratories following chemical mutagenesis (Ikegami and Gomatos, 1968; Fields and Joklik, 1969). These $ts$ mutants were the first used in studies of genetic interactions, morphogenesis, pathogenesis, and mutant physiology in reovirus. The conditional nature of the mutations made them well suited for genetic studies, since recombinants could be selected from among a virus population that was predominantly mutant simply by assay at the nonpermissive temperature (39°C). In many of these mutants, the $ts$ lesion was sufficiently "tight" and the reversion frequency was sufficiently low to allow their use in biochemical studies at nonpermissive temperatures. The genetic studies carried out with these mutants (see Section IV.B) showed that they defined seven of the ten expected reovirus recombination groups (Fields and Joklik, 1969; Cross and Fields, 1972). Thus, although the $ts$ mutants were quite useful, additional mutants were needed to define the missing three recombination groups. Additional mutant searches using

TABLE I. Prototype ts Mutants of the Ten Reovirus Mutant Groups[a]

| Recombination group | Prototype mutant | Typical EOP | Method of isolation |
|---|---|---|---|
| A | tsA(201) | 0.02 | Proflavin mutagenesis |
| B | tsB(352) | 0.0025 | Nitrous acid mutagenesis |
| C | tsC(447) | 0.0004 | Nitrosoguanidine mutagenesis |
| D | tsD(357) | 0.02 | Nitrous acid mutagenesis |
| E | tsE(320) | 0.3 | Nitrous acid mutagenesis |
| F | tsF(556) | 0.02 | Nitrosoguanidine mutagenesis |
| G | tsG(453) | 0.0002 | Nitrosoguanidine mutagenesis |
| H | tsH(26/8) | 0.002 | Serial high-MOI passage; rescue from suppressed pseudorevertant |
| I | tsI(138) | 0.001 | Rescue from suppressed pseudorevertant |
| J | tsJ(128) | 0.01 | Rescue from suppressed pseudorevertant |

[a] (EOP) Efficiency of plating; (MOI) multiplicity of infection.

chemical mutagenesis failed to reveal these missing mutant groups (Fields, Cross, and Chakraborty, unpublished data).

Temperature-sensitive mutants that define the remaining three recombination groups, as well as provide additional mutants in the previously identified mutant groups, have been isolated by several methods. Although the $ts^+$ phenotype of wild-type reovirus stocks passaged at low multiplicity is quite stable, Ahmed et al. (1980a) found that ts mutants could be isolated at fairly high frequency from wild-type stocks passaged serially at high multiplicity. At passage 11, they found that 15% of individual plaques isolated from type 3 had a ts phenotype. For type 1 and type 2, ts mutants could also be isolated at passage 11; however, the frequency was somewhat lower (about 5%). Most of the mutants isolated in this manner were members of previously identified mutant groups, but one mutant defined a new mutant group. When Ahmed et al. (1980b) examined the $ts^+$ virus shed by persistently infected L cells in which the infection was initiated with a ts mutant, they found that some of the $ts^+$ virus was extragenically suppressed (see Section V.B). Following backcross of the suppressed pseudorevertants to wild-type, ts mutants could be isolated. Most of the ts mutants rescued in this manner fell into previously identified mutant groups, but several defined a new mutant group. Ramig and Fields (1979) were also able to rescue ts mutants from suppressed pseudorevertants; in this case, the pseudorevertants were isolated as spontaneous revertants from each of the seven previously defined mutant groups (see Section V.B). As with the other unconventional methods of mutant isolation, many of these rescued ts mutants were in previously identified mutant groups. However, in this study, mutants that defined three new mutant groups were defined.

The general properties of the prototype ts mutants from each of the recombinationally defined mutant groups are summarized in Table I.

## B. Deletion Mutants

The equimolar distribution of genome segments isolated from reovirions has led to the conclusion that most virus particles contain the full complement of ten genome segments. However, on serial passage in tissue culture, deletion mutants of reovirus are generated (Nonoyama and Graham, 1970). These deletion mutants generally lack one genome segment, but can lack several genome segments. The *L1* genome segment is most frequently lost, but virus populations lacking segments *L3* and *M1* have also been detected. Deletions are generated within seven to eight passages for the Dearing strain of reovirus type 3. However, with certain of the *ts* mutants, deletions arise at earlier passage. In particular, *ts*C(447) yields particles that lack genome segments *L1*, *L3*, and *M1* within four passages (Schuerch *et al.*, 1974).

Although the most dramatic and easily demonstrated defect in deletion mutants is the deletion of one or more genome segments, other mutations are also found in deletion-mutant stocks. For example, Ahmed and Fields (1981) found that both *ts* and small-plaque/low-yield mutants could be rescued from deletion-mutant stocks. The *ts* mutants rescued from deletion mutants exhibited a very strong interference phenotype (see Section V.A), leading to the suggestion that the interference generally associated with deletion-mutant (defective interfering) stocks may, in fact, be due to the presence of interfering *ts* mutations in these stocks (Ahmed and Fields, 1981).

Deletion mutants are difficult to propagate because they require the presence of a helper virus. Thus, pure populations of deletion mutants are difficult to obtain. Spandidos and Graham (1975b) have found that relatively pure *L1* deletion stocks can be propagated by using the *ts*E(320) mutant as the helper virus. In this case, the complementation was asymmetric, so that the resulting virus stock was predominantly the *L1* deletion and contained virtually no helper virus. Deletion mutants are expected to be less dense than virus having a full complement of genome segments. However, Ahmed and Fields (1981) found that the denser fractions of a gradient were enriched for the deletion mutants. They were able to generate a relatively pure population of deletion mutants through several cycles of density-gradient centrifugation.

## C. Other Genetic Markers

A $\mu1-\mu1C$ polypeptide complex that showed aberrant electrophoretic migration was identified in several of the prototype *ts* mutants (Cross and Fields, 1976a). This aberrantly migrating polypeptide complex was shown to be independent of all *ts* mutations with which it was found except the *ts*A(201) lesion. The aberrant polypeptide complex provided a marker that could be used with *ts* mutants, yet not be selected for or against in the selection for *ts*[+] recombinants, and therefore provided a useful unselected outside marker in three-factor crosses (Cross and Fields, 1976b). The electrophoretic heterogeneity noted among the genome segments and polypeptides of the three reovirus serotypes (Ramig et al., 1977b) has also provided excellent genetic markers for mapping studies (see Section VI). Biological features of the virus such as type-specific neutralization and hemagglutination have also served as markers in genetic mapping studies (Weiner and Fields, 1977; Weiner et al., 1978).

## IV. GENETIC INTERACTIONS

Mixed infection of tissue-culture cells has provided the primary means of studying the genetic interactions between viral mutants. These studies have allowed us to determine the genetic grouping and the location of the various mutations and have provided a series of intertypic recombinant viruses that have been useful in studies of viral gene function.

### A. Complementation

Early studies of mixed infection with pairs of temperature-sensitive (ts) mutants at the nonpermissive temperature (39°C) showed little or no enhancement of yield as compared to the single infections with the mutants (Fields and Joklik, 1969). This result suggested that there was little or no complementation between reovirus ts mutants. The subsequent finding that many of the ts mutants have interfering phenotype at 39°C (see Section V.A) provided an explanation for the apparent lack of complementation (Chakraborty et al., 1979). It was found that the few noninterfering ts mutants exhibited efficient complementation in mixed infection at 39°C. However, the interfering phenotype of the interfering mutants was dominant, obscuring complementation in the majority of the mixed infections. As a result of the generally inefficient complementation observed between pairs of ts mutants, complementation has not been extensively evaluated or utilized in genetic studies and has not been used for genetic grouping of the ts mutants into functional groups.

Complementation between RNA-negative mutant pairs has been biochemically detected by measuring the synthesis of RNA in cells infected with mixed viruses at 39°C (Ito and Joklik, 1972b). As noted in Section III.B, complementation has also been observed between deletion mutants and ts mutants (Spandidos and Graham, 1975a,b). However, since reassortment occurs at 39°C (Chakraborty et al., 1979), appreciable numbers of $ts^+$ recombinants are clearly generated in mixed infections at 39°C. The increased yield of virus, or macromolecular synthesis, called complementation may represent the replication of $ts^+$ recombinant progeny formed during the complementation test. The yields of complementation tests have not been carefully examined to determine whether the increased yield represents growth of the mutants or growth of recombinants.

## B. Recombination (Reassortment)

The observation that the reovirus genome was segmented was made in the late 1960s (Shatkin *et al.*, 1968; Prevec *et al.*, 1968; Skehel and Joklik, 1969) and suggested that genetic interactions between reovirus mutants might have features unusual for animal viruses with RNA genomes. Specifically, the segmented nature of the genome predicted that recombination could occur by the mechanism of reassortment of genome segments. Furthermore, the absence of knowledge of enzymes (at that time) that could mediate intramolecular recombination between RNA molecules suggested that mutants residing on the same genome segment would be unable to recombine. Indeed, the earliest recombination tests performed with *ts* mutants of reovirus revealed that recombination frequencies were extraordinarily high or undetectably low, depending on the mutant pair (Fields and Joklik, 1969). Furthermore, a statistical analysis of a large number of recombination tests failed to provide any evidence for even weak linkage of markers (Fields, 1971). Thus, recombination analysis provided early evidence for the functional segmentation of the reovirus genome.

In this section, we will review the different types of crosses performed between pairs of reovirus mutants at both permissive and nonpermissive temperatures and discuss the conclusions that were reached on the basis of the observed recombination behavior.

### 1. Two-Factor Crosses

A detailed genetic analysis of mixed infection between pairs of *ts* mutants revealed that certain mutant pairs yielded a high proportion of $ts^+$ recombinants, while other pairs yielded no detectable $ts^+$ recombinants (Fields and Joklik, 1969) (Table II). Statistical analysis of the recombination frequencies shown by different mutant pairs indicated that they were not statistically different (Fields, 1971). Thus, the recombination observed was "all or none," suggesting that recombination occurred by the mechanism of reassortment of genome segments rather than intramolecular recombination between covalently closed molecules. This recombination behavior supported the notion that the reovirus genome is functionally segmented. The frequency of $ts^+$ recombinants obtained in these mixed infections was high, but did not reach the level of 25% $ts^+$ recombinants that would be expected if the reassortment mechanism were totally random. The lower-than-expected levels of recombination were not understood. The recent demonstration that

TABLE II. Presence and Absence of Genetic Recombination between Several *ts* Mutants at the Permissive Temperature (31°C)[a]

| Virus strain used in infection | Yield at 24 hr (PFU/ml) | | Efficiency of plating (39°C/31°C) | Presumed wild-type recombinants (% of 31°C titer) |
|---|---|---|---|---|
| | 39°C | 31°C | | |
| Wild-type | $3 \times 10^7$ | $3 \times 10^7$ | — | — |
| ts201 | $3 \times 10^4$ | $1 \times 10^8$ | $3 \times 10^{-4}$ | — |
| ts352 | $<1 \times 10^3$ | $1 \times 10^8$ | $1 \times 10^{-5}$ | — |
| ts447 | $2 \times 10^3$ | $5 \times 10^7$ | $4 \times 10^{-5}$ | — |
| ts234 | $<1 \times 10^3$ | $2 \times 10^8$ | $5 \times 10^{-6}$ | — |
| ts201 × 352 | $7 \times 10^6$ | $1 \times 10^8$ | $7 \times 10^{-2}$ | 7 |
| ts201 × 447 | $3 \times 10^6$ | $3 \times 10^7$ | $1 \times 10^{-1}$ | 10 |
| ts201 × 234 | $2 \times 10^4$ | $3 \times 10^8$ | $7 \times 10^{-5}$ | 0 |
| ts352 × 447 | $2 \times 10^7$ | $1 \times 10^8$ | $2 \times 10^{-1}$ | 20 |
| ts352 × 234 | $3 \times 10^6$ | $1 \times 10^8$ | $3 \times 10^{-2}$ | 3 |
| ts447 × 234 | $3 \times 10^6$ | $4 \times 10^7$ | $8 \times 10^{-2}$ | 8 |

[a] From Fields and Joklik (1969), with permission. (PFU) Plaque-forming units.

many *ts* mutants interfere with the growth of wild-type virus (Chakraborty *et al.*, 1979) may explain this discrepancy.

Since the recombination behavior of the *ts* mutants was consistent with a segmented genome, mutants that could recombine were assumed to reside on different reassorting genome segments. Likewise, mutant pairs unable to recombine were assumed to reside on the same genome segment. In this way, mutants were divided into groups between which recombination could occur and within which recombination did not occur. In this manner, five (Fields and Joklik, 1969), seven (Cross and Fields, 1972), and finally ten (Ramig and Fields, 1979) recombinationally defined mutant groups have been described. The assumption that each of these mutant groups represents mutations on a different genome segment has subsequently been verified (see Section VI).

The mixed infections discussed above were performed by standard genetic techniques; specifically, the crosses were performed at the permissive temperature, and recombinant progeny were detected in the yield by assay at the nonpermissive temperature. Analysis of mixed infections performed at the nonpermissive temperature revealed that recombination could occur at 39°C (Chakraborty *et al.*, 1979). However, the frequency of recombinants observed was affected by the interference phenotype of the mutants. Pairs of interfering mutants yielded significantly fewer recombinants than pairs of noninterfering mutants or mixed pairs. Thus, none of the mutants appeared to have a *ts* lesion that directly affected reassortment.

Additional experiments using two-factor crosses demonstrated several points relevant to recombination studies. Recombination frequencies were found to be maximal at the earliest times that recombinants could be detected, and these levels were maintained throughout the infectious cycle (Fields, 1971). This result suggested that each recombinant genome participated in only a single round of mating, a finding consistent with the simultaneous occurrence of recombination and the earliest stages of morphogenesis. The frequency of recombinants was found to depend on the multiplicity of infection (MOI), with the frequency of recombinants increasing in parallel with the MOI until 100% of the cells were mixedly infected (Fields, 1971). Since most virus stocks contain a high proportion of noninfectious particles, this result was interpreted to indicate that noninfectious particles do not play a significant role in genetic interactions. More recently, recombination has been shown to occur between noninfectious deletion mutants and wild-type (Ahmed and Fields, 1981). However, this recombination occurs at low frequency and is consistent with the general notion that noninfectious particles do not contribute significantly to observed recombination frequencies. Aggregates of complementing *ts* mutants and heterozygotes containing more than ten genome segments have also been shown to play no significant role in the recombination frequencies observed in two-factor crosses (Fields, 1973).

## 2. Three-Factor Crosses

The two-factor crosses described above strongly suggested that no linkage existed between markers on different genome segments. However, three-factor crosses, using an unselected outside marker, could provide a much more sensitive genetic test of linkage. The identification of $\mu1$ and $\mu1C$ polypeptides with aberrant electrophoretic migration in several of the prototype mutants (Cross and Fields, 1976a) provided a suitable third marker for use in three-factor crosses. The aberrant mobility of the $\mu1-\mu1C$ complex was a useful marker because it was genetically stable and, except for *ts*A(201), was independent of the *ts* lesion that conferred *ts* phenotype on the mutant (Cross and Fields, 1976b). This marker (designated $\mu^-$) could therefore be used as an unselected marker in crosses of the general type $tsX^- tsY^+ \mu^- \times tsX^+ tsY^- \mu^+$. Analysis of the segregation of the $\mu^-$ marker in the $tsX^+ tsY^+$ recombinants would provide a sensitive test for linkage of the $\mu^-$ marker to either of the *ts* markers. A number of three-factor crosses of this type were performed (Cross and Fields, 1976b). The $\mu^-$ marker segregated independently in the crosses *ts*B × *ts*D and *ts*C × *ts*D, but segregated with the *ts*A lesion in the cross *ts*A × *ts*D. This indicated that the $\mu^-$ marker was not linked to any of the *ts* mutants except the *ts*A mutant. We have subsequently been able to show that the *ts*A lesion lies on the genome segment that encodes the $\mu1-\mu1C$ polypeptide complex (see Section VI.A), explaining the linkage observed between *ts*A(201) and $\mu^-$ in three-factor crosses.

Thus, three-factor crosses were used to formally confirm that recombination in reovirus occurred via a mechanism of reassortment of genome segments. Furthermore, these experiments indicated that biochemical markers (such as the $\mu^-$ phenotype) would be useful for mapping the physical location of the ts lesions.

## 3. Intertypic Crosses

The finding that the genome RNAs and polypeptides of reovirus serotypes 1, 2, and 3 were electrophoretically heterogeneous (Ramig et al., 1977b) provided the biochemical markers necessary to physically map the entire reovirus genome. These biochemical markers allowed us to identify the parental origin of each genome segment and polypeptide in recombinants isolated following mixed infection with two different serotypes of virus. This type of cross has been performed in two ways. In one method, wild-type type 1 or type 2 was crossed with a ts mutant of type 3. Analysis of the $ts^+$ progeny by electrophoresis allowed rapid identification of recombinant progeny and $ts^+$ parental progeny. Since reassortment is efficient, high frequencies of $ts^+$ recombinants were isolated from these crosses. In these $ts^+$ recombinant progeny, one genome segment was always derived from the $ts^+$ parent, the segment depending on the ts mutant used in the cross (Ramig et al., 1978; Mustoe et al., 1978a). Alternatively, both the parental viruses, although of different serotype, could contain ts mutations (Ramig et al., 1983). This allowed easier selection of $ts^+$ recombinant progeny from among the yield of the cross. The $ts^+$ phenotype of the recombinants derived in this way demanded that two segments segregate nonrandomly, one always being derived from each parent. The use of intertypic recombination for mapping ts lesions is discussed in detail in Section VI.

## 4. Crosses with Deletion Mutants

In contrast to the ease with which recombinants are generated between ts mutants or wild-type viruses, it was initially reported that no recombinants were generated in mixed infection of L1 deletion mutants and ts mutants of reovirus type 3 (Spandidos and Graham, 1976). Since recombination is thought to occur at the level of single-stranded RNA and since L1 deletion mutants have normal transcriptase activity, the absence of recombination was somewhat surprising. The possibility that deletion mutants can recombine has recently been reexamined (Ahmed and Fields, 1981). Mixed infection of a ts mutant of type 1 and a deletion mutant derived from type 3 yielded $ts^+$ recombinants at low frequency. Electrophoretic analysis of genome RNA of the recombinants verified that $ts^+$ progeny from these crosses did indeed arise by recombination.

## V. NONGENETIC INTERACTIONS

Factors that affect the expression of temperature-sensitive (ts) lesions in animal viruses have not been extensively studied. Systematic studies of the genetics of reovirus have revealed two nongenetic phenomena that affect the expression of *ts* mutations. One of these phenomena, interference, has readily detectable effects on some standard genetic interactions. The other, extragenic suppression of *ts* phenotype, prevents the expression of *ts* phenotype by a virus the genotype of which includes a *ts* lesion.

## A. Interference

The failure of *ts* mutants in different recombination groups to complement in mixed infection at 39°C led Chakraborty *et al.* (1979) to systematically examine mutant interactions at the nonpermissive temperature. Among their observations was the demonstration that many of the reovirus *ts* mutants interfere strongly with the growth of wild-type virus in mixed infection. Interference was examined by infecting cells with mixtures of mutant and wild-type viruses, incubating the infected cells at the nonpermissive temperature (39°C), and quantitating the wild-type yield from the mixed infection by titration at 39°C. Interference was examined at several ratios of *ts* mutant to wild-type in the inoculum. A systematic examination of interference among the mutants of recombination groups A–G showed that in general, mutants with double-stranded (ds)RNA⁻ phenotype (groups D and E) did not interfere, whereas mutants from groups with dsRNA⁺ phenotype (groups A, B, and G) interfered strongly with the growth of wild-type virus (Table III). The exception to this rule was the group C mutant, which has a dsRNA⁻ phenotype, yet strongly interfered with the growth of wild-type. All the interfering mutants except *ts*A(201) reduced the yield of wild-type by at least 50% when the cells were infected at equal multiplicity of mutant and wild-type. When the interfering mutant predominated in the inoculum (ratio = 10 mutant/2 wild-type), the degree of interference increased significantly. Several different mutants in recombination group A were examined for interference with the result that some mutants interfered to a moderate degree, while one mutant did not interfere at any multiplicity ratio tested. Thus, interference was a property specific for each mutant and was not a property that could be generalized for all members of a recombination group. The mutant specificity of interference was subsequently confirmed (Ahmed *et al.*, 1980a). Interference with the growth of wild-type by the *ts* mutants was not observed at 31°C (Chakraborty *et al.*, 1979). However, some *ts* mutants isolated and examined in a subsequent study were found to interfere at both 31 and 39°C (Ahmed and Fields, 1981), indicating that the temperature at which the interference phenotype is displayed is also a mutant-specific property.

The interference properties of the *ts* mutants appeared to explain the pattern of genetic interactions observed in mixed infection at 39°C. For example, careful analysis showed that significant complementation occurred between group C, D, and E mutants, which do not generally in-

TABLE III. Interference of the Growth of Wild-Type Reovirus by *ts* Mutants at Nonpermissive Temperature[a]

| Virus used in infection | MOI[b] | Yield at 44 hr postinfection (PFU/ml) | | Inhibition of wild-type growth (%)[c] |
|---|---|---|---|---|
| | | 39°C | 31°C | |
| **Experiment I** | | | | |
| RV-3 WT | 12:0 | $2.4 \times 10^7$ | $6.9 \times 10^7$ | — |
| *ts*A(201) | 0:12 | $2.5 \times 10^3$ | $3.6 \times 10^6$ | — |
| RV-3 × *ts*A(201) | 6:6 | $2.7 \times 10^7$ | $6.8 \times 10^7$ | 0 |
| RV-3 × *ts*A(201) | 2:10 | $1.2 \times 10^7$ | $2.7 \times 10^7$ | 50.0 |
| *ts*B(352) | 0:12 | $<5.0 \times 10^2$ | $4.0 \times 10^6$ | — |
| RV-3 × *ts*B(352) | 6:6 | $9.0 \times 10^6$ | $3.2 \times 10^7$ | 62.0 |
| RV-3 × *ts*B(352) | 2:10 | $1.0 \times 10^6$ | $1.0 \times 10^7$ | 96.0 |
| *ts*C(447) | 0:12 | $<5.0 \times 10^2$ | $2.3 \times 10^6$ | — |
| RV-3 × *ts*C(447) | 6:6 | $8.7 \times 10^6$ | $1.2 \times 10^7$ | 64.0 |
| RV-3 × *ts*C(447) | 2:10 | $3.2 \times 10^5$ | $1.0 \times 10^6$ | 98.7 |
| *ts*D(357) | 0:12 | $3.5 \times 10^3$ | $1.8 \times 10^7$ | — |
| RV-3 × *ts*D(357) | 6:6 | $7.3 \times 10^7$ | $1.37 \times 10^8$ | 0 |
| RV-3 × *ts*D(357) | 2:10 | $2.8 \times 10^7$ | $1.0 \times 10^8$ | 0 |
| *ts*E(320) | 0:12 | $5.0 \times 10^4$ | $2.5 \times 10^8$ | — |
| RV-3 × *ts*E(320) | 6:6 | $7.0 \times 10^7$ | $1.37 \times 10^8$ | 0 |
| RV-3 × *ts*E(320) | 2:10 | $4.0 \times 10^7$ | $1.5 \times 10^8$ | 0 |
| *ts*G(453) | 0:12 | $<5.0 \times 10^2$ | $6.2 \times 10^6$ | — |
| RV-3 × *ts*G(453) | 6:6 | $7.0 \times 10^6$ | $2.7 \times 10^7$ | 71.0 |
| RV-3 × *ts*G(453) | 2:10 | $4.6 \times 10^5$ | $5.5 \times 10^6$ | 98.1 |
| *ts*G(12) | 0:12 | $<5.0 \times 10^2$ | $1.9 \times 10^6$ | — |
| RV-3 × *ts*G(12) | 6:6 | $4.7 \times 10^6$ | $2.8 \times 10^7$ | 80.0 |
| RV-3 × *ts*G(12) | 2:10 | $3.2 \times 10^5$ | $5.1 \times 10^6$ | 98.7 |
| **Experiment II** | | | | |
| RV-3 WT | 5 | $6.0 \times 10^7$ | $2.0 \times 10^8$ | — |
| *ts*A(340) | 10 | $7.0 \times 10^3$ | $8.0 \times 10^5$ | — |
| RV-3 × *ts*A(340) | 5:5 | $6.3 \times 10^7$ | $1.6 \times 10^8$ | 0 |
| RV-3 × *ts*A(340) | 2:10 | $6.4 \times 10^7$ | $1.8 \times 10^8$ | 0 |
| *ts*A(438) | 10 | $5.0 \times 10^2$ | $6.3 \times 10^6$ | — |
| RV-3 × *ts*A(438) | 5:5 | $5.0 \times 10^7$ | $1.5 \times 10^8$ | 17 |
| RV-3 × *ts*A(438) | 2:10 | $3.1 \times 10^7$ | $1.1 \times 10^8$ | 49 |
| *ts*B(271) | 10 | $5.0 \times 10^2$ | $6.3 \times 10^6$ | — |
| RV-3 × *ts*B(271) | 5:5 | $1.0 \times 10^7$ | $4.8 \times 10^7$ | 84 |
| RV-3 × *ts*B(271) | 2:10 | $2.1 \times 10^6$ | $1.2 \times 10^7$ | 97 |

[a] From Chakraborty *et al.* (1979). (PFU) Plaque-forming units; (MOI) multiplicity of infection.
[b] The first number corresponds to the wild-type MOI; the second, to the *ts* mutant MOI.
[c] Calculated from the 39°C titer.

terfere, while there was no detectable complementation observed between the interfering mutants of groups A, B, and G. Mixed infections of interfering and noninterfering mutants yielded no detectable complementation, indicating that the interfering phenotype was dominant in such infections (Chakraborty *et al.*, 1979). Studies of recombination at 39°C revealed that reassortment was very efficient between pairs of noninterfering mutants, an observation that correlated with complementation between these mutant pairs. Reassortment was also very efficient between pairs of interfering and noninterfering mutants, in contrast to the absence of complementation observed with such pairs. Reassortment between pairs of interfering mutants was observed, but was very inefficient and could be detected only in carefully performed experiments. The inefficient reassortment with interfering mutants correlated with the absence of complementation in these mixed infections (Chakraborty et al., 1979). These observations indicate that the interference exhibited by some of the ts mutants accounts for the lack of complementation observed for most mutant pairs. However, the dominance of the interfering phenotype observed in complementation is not observed in reassortment at 39°C. Only when both mutants interfere is reassortment affected.

Interference properties of viral mutants may account for other observations made in the reovirus system. High-passage virus stocks contain interfering virus, a property that has been traditionally assigned to the deletion mutants generally present in such stocks. Subcloning from high-passage stocks of wild-type revealed that significant numbers of ts mutants were present in these stocks. Furthermore, many of these ts mutants interfered strongly with the growth of wild-type virus (Ahmed et al., 1980a). This observation suggested that the interference characteristic of the high-passage stocks may result from the ts mutants contained in the stock, rather than from the deletion mutants as generally assumed. Ahmed and Fields (1981) have shown that interfering ts mutants could be rescued from a defective interfering [(DI) deletion] stock that contained virtually no standard virus. Thus, the deletion stock, in addition to lacking certain genome segments, had interfering ts mutations on other genome segments. The ts lesions of the ts interfering mutants rescued from the deletion mutant were members of recombination group G, a group previously associated with interfering phenotype. Ahmed and Fields (1981) proposed that the biological properties of reovirus DI particles (deletion mutants) were determined not by the genes deleted, but by mutations on genes that were present. Specifically, the interference property was due to expression of mutant genes and not to the absence of the deleted genes. In studies of persistently infected L cells, in which the persistent infection was initiated with a high-passage stock of the mutant tsC(447), Ahmed et al. (1980b) found that the phenotype of the virus shed from the cells changed to ts$^+$ with passage. Some of the ts$^+$ virus shed by the persistently infected cells was found to be pseudorevertant (see Section V.B). Temperature-sensitive mutants could be rescued from these pseudorevertants, and the rescued mutants had a strong interference phenotype. As a result of the isolation of interfering mutants from persistently infected cells, these investigators suggested that interference may play a role in the maintenance of the persistently infected state. However, this hypothesis must be considered with caution, since it has been demonstrated that in persistently infected cells, not only viral mutation and interference but also changes in the ability of the host cell to support viral replication occur (Ahmed et al., 1981). Thus, while interference does occur in persistent infection, the role of interference in the dynamics of these infections is not understood.

The mechanism of interference by reovirus mutants is not known. It has been proposed that interference is due to expression of mutant genes and incorporation of defective proteins into progeny virus particles (Ahmed and Fields, 1981). This hypothesis is supported by the observation that the interactions of viral RNA (reassortment) could occur in mixed infection in the presence of interference, whereas the interactions of the viral gene products (complementation) were inhibited in the presence of interfering mutants (Chakraborty et al., 1979). Further support for the notion that interference occurs at the level of protein synthesis or protein interactions comes from the work of Ikegami and Gomatos (1972). They found that when cells infected with wild-type virus were superinfected with a ts mutant, the synthesis of viral protein was inhibited. This suggested that the ts mutant interfered with the replication of wild-type. The superinfection was shown to have no effect on the synthesis of virus-specific RNA and the association of the RNAs with polyribosomes. Superinfection did, however, cause a significant decrease in protein synthesis and an 85% decrease in the association of virus-specific proteins with the polyribosomes. However, in these experiments, virus yields were not determined, so the inhibition of polysomal association of viral protein may be unrelated to interference.

The observations described here indicate that interference is an important property of many of the reovirus ts mutants. Furthermore, interference plays a role in the genetic interactions observed between ts mutants at nonpermissive temperature, may be responsible for the interfering properties associated with high-passage and deletion-containing virus stocks, and may have a role in the initiation and maintenance of persistent infection.

## B. Extragenic Suppression of ts Phenotype

The mechanisms of reversion have not been studied in detail in animal virus systems. We have characterized a number of revertants isolated from reovirus ts mutants and have found that extragenic suppression is the most common reversion pathway.

The group A mutant tsA(201) of reovirus type 3 was isolated following mutagenesis with proflavin (Fields and Joklik, 1969). This mutant synthesizes $\mu 1$ and $\mu 1 C$ polypeptides with altered electrophoretic mobility (Cross and Fields, 1976a). Reversion of tsA(201) to $ts^+$ phenotype is generally accompanied by a change in the electrophoretic mobility of $\mu 1$ and $\mu 1 C$. In one revertant clone, clone 101, the ts lesion and migration of $\mu 1$ and $\mu 1 C$ did not co-revert (Cross and Fields, 1976b). The absence of coreversion in this revertant clone suggested that it contained the tsA(201) lesion in suppressed form (Ramig et al., 1977a). To show that clone 101 contained a suppressed *ts* lesion, it was necessary to show that (1) the reversion event occurred outside the gene with the *ts* lesion and (2) the clone still contained the original *ts*A lesion.

Although the technique used to reveal suppressed lesions in pseudorevertants of reovirus was the simple back-cross to wild-type, the segmented genome that recombines by reassortment and the mechanisms of reversion placed certain limitations on the back-cross analysis (Ramig *et al.*, 1977a; Ramig and Fields, 1979). Reversion can occur through one of three pathways: (1) Reversion can occur by back-mutation at the nucleotide of the original mutation, restoring the *ts*$^+$ nucleotide sequence (true reversion). (2) A reversion event can occur at a second site in the same gene as the original mutation, for example, a frameshift mutation that restores reading frame (intragenic suppression). (3) Reversion can be mediated by a suppressor mutation in a gene different from the gene containing the original mutation (extragenic suppression). The effect of the suppressor mutation is to bypass the defect of the original mutation. Since suppression is revealed by back-cross of the revertant to wild-type and rescue of the original mutation by recombination, in reovirus we cannot distinguish between revertants that arise by true reversion and those that arise by intragenic suppression events. We can, however, distinguish between intragenic reversion and reversion by extragenic suppression because the two mutations can reassort during the back-cross. We reasoned that a suppressed *ts*A(201) lesion in clone 101 would be separated from its suppressor mutation by reassortment, if the *ts* lesion and the suppressor mutation were on different genome segments. Once separated from the suppressor mutation, the *ts* phenotype of the *ts*A(201) lesion would once again be expressed. Accordingly, clone 101 was back-crossed to wild-type, and clone 101, wild-type, and *ts*A(201) were self-crossed as controls (Ramig *et al.*, 1977a). Progeny plaques from the back-cross and control self-crosses were picked and passaged at permissive temperature. The temperature phenotype of each progeny clone was then quantitated as the efficiency of plating (EOP). All progeny clones analyzed from the clone 101 and wild-type controls had *ts*$^+$ phenotype, whereas all progeny from the *ts*A(201) had *ts* phenotype. The progeny from the back-cross had a bimodal distribution of temperature phenotypes (Fig. 2). One group of progeny clones had an EOP similar to that of the wild-type and clone 101 parents and was definitely *ts*$^+$. The other group had an EOP characteristic of *ts* virus and very similar to that of *ts*A(201). This rescue of phenotypically *ts* progeny from the back-cross provided unequivocal evidence that clone 101, although phenotypically *ts*$^+$, contained a suppressed *ts* lesion. Characterization of the rescued *ts* clones by recombination tests showed that all were in recombination group A. Thus, revertant clone 101 contained an extragenically suppressed *ts*A lesion. This type of revertant was more correctly called a pseudorevertant. This experiment identified extragenic suppression of *ts* phenotype as a mechanism by which mutations in reovirus could revert.

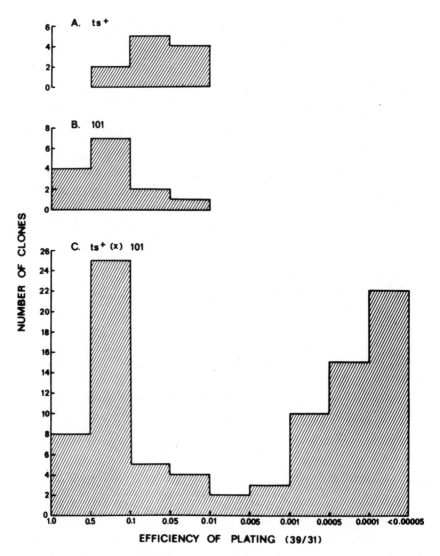

FIGURE 2. Distribution of EOP of progeny and control clones. A suspension culture of $10^7$ mouse L cells was mixedly infected with a multiplicity of infection (MOI) of 10 each with freshly cloned 101 and wild-type; 2 hr after infection, unadsorbed virus was removed by centrifuging the infected cells and resuspending them in fresh medium; 48 hr after infection, the cells were sonicated to release cell-associated virus and to disrupt viral aggregates. Appropriate dilutions were plated on L-cell monolayers and incubated for 13 days at 31°C. The culture plates were overlaid with neutral red agar, and after overnight incubation, plaques were picked. The plaques were passaged twice on L-cell monolayers at 31°C. The titer and EOP of second-passage virus were determined by plating on L-cell monolayers at 39 and 31°C. Plates at 39°C were overlaid and counted on day 5; plates at 31°C were overlaid and counted on day 13 after infection. Wild-type and clone 101 controls were the same, except that for single infection, an MOI of 20 was used. The EOP is the ratio of the titer at 39°C to that at 31°C. Adapted from Ramig et al. (1977a).

The methods used to reveal the presence of the parental $ts$A(201) lesion in pseudorevertant clone 101, while quite precise, were very time-consuming. As a result, we developed a much more rapid assay to detect pseudorevertants among revertants derived from leaky $ts$ mutants (Ramig and Fields, 1977). This method depends on the fact that certain of the reovirus $ts$ mutants are "leaky"; that is, the $ts$ lesion is not completely lethal at the nonpermissive temperature. A result of the incomplete lethality is that two populations of plaques are observed when a leaky $ts$ mutant is plated at the nonpermissive temperature. One population consists of large, clear, lytic plaques produced by revertants present in the mutant stock. The other population consists of small, faint plaques with diffuse edges that often have a clump of viable cells in their center. These plaques represent limited growth by the $ts$ mutant at the nonpermissive temperature. If the lytic-plaque morphology of a revertant is due to reversion by an intragenic event, a back-cross of the revertant to wild-type will yield only progeny with lytic morphology when plated at the nonpermissive temperature. However, if the lytic morphology of the revertant is due to reversion by extragenic suppression (i.e., a pseudorevertant), back-cross to wild-type will yield some progeny in which the parental $ts$ mutation has been separated from the suppressor mutation. These leaky, parental $ts$ progeny will make leak plaques when plated at the nonpermissive temperature. The leak plaques are easily distinguished from the background of lytic plaques representing wild-type and pseudorevertant progeny. This assay is a useful tool for rapidly screening revertants of leaky $ts$ mutants for extragenic suppression. The assay has a relatively high accuracy, providing a correct assessment of the pathway of reversion in 15 of 19 revertants that were subsequently confirmed by determination of the temperature phenotype of individual back-cross progeny (Ramig and Fields, 1979). Unfortunately, this assay is limited to use with revertants of leaky $ts$ mutants.

The finding that reversion of temperature phenotype from $ts$ to $ts^+$ could occur via a pathway of extragenic suppression led us to investigate the frequency with which this pathway was used in reovirus revertants (Ramig and Fields, 1979). A detailed study was undertaken in which 28 independently isolated spontanteous revertants, representing all reovirus recombination groups, were examined to determine whether reversion had occurred by intragenic events or extragenic suppression. The revertants were all examined by both the rapid leak assay and the phenotype determination for individual back-cross progeny. Of the 28 revertants examined, 25 were suppressed pseudorevertants, indicating that extragenic suppression was the major reversion pathway in reovirus $ts$ mutants. In all but one case, the parental $ts$ lesion was rescued from the pseudorevertant.

The high frequency of extragenic suppression of $ts$ phenotype found among revertants of reovirus $ts$ mutants suggested a possible general mechanism by which RNA viruses lacking DNA intermediates in their life cycles could overcome the effects of deleterious mutations (Ramig and Fields, 1979). Since these RNA viruses have no intramolecular recombination or recombination that occurs at extremely low rates (Cooper, 1968), they cannot easily generate viable combinations of genetic material from nonviable parental genomes. A high frequency of mutation to a suppressed pseudorevertant genotype would provide a means of bypassing the deleterious effects of mutations that accumulated in the absence of intramolecular recombination. In the segmented genome RNA viruses, suppressor mutations may be particularly advantageous, since the reassortment mechanism is highly efficient and could quickly spread suppressor mutations throughout the population.

Many of the suppressor mutations examined in reovirus have no temperature phenotype. This was demonstrated by failure to rescue nonparental *ts* lesions from pseudorevertants by back-cross. However, in about half the pseudorevertants examined, both the parental and nonparental *ts* mutations were rescued by back-cross to wild-type (Ramig and Fields, 1979). The nonparental *ts* mutations rescued from these pseudorevertants are candidates for suppressor mutations with *ts* phenotype. This possibility has not been examined. The nonparental *ts* mutations have, however, been placed into mutant groups by recombination tests with the seven previously identified recombination groups. About half the rescued nonparental *ts* mutants were in one of the previously identified recombination groups (A–G). However, ten of the nonparental *ts* mutants recombined with the prototype mutants of all the known mutant groups, indicating that they represented recombination groups not previously identified in reovirus. Further recombination analysis showed that the ten mutants fell into three new recombination groups that were designated H, I, and J. The frequent isolation of new mutants from suppressed pseudorevertants indicated that the selective pressures active in changes from *ts* to *ts*+ phenotype were different from those active in mutation from *ts*+ to *ts* genotype. This finding suggested that rescue of mutants from pseudorevertants may provide a generally applicable means of isolating mutants in previously undefined genes (Ramig and Fields, 1979).

Extragenic suppression of *ts* phenotype has been shown to be one of the events that occur during persistent infection with reovirus (Ahmed *et al.*, 1980b). When persistently infected L cells were established by infection with a highly defective population of *ts*C(447), the virus released from the cells gradually changed from *ts* to *ts*+ phenotype (Ahmed and Graham, 1977). Back-cross analysis of clones of the *ts*+ virus released from the persistently infected cells showed that two of seven clones examined contained extragenically suppressed *ts* lesions, while the remaining five clones had gained *ts*+ phenotype through intragenic events (Ahmed *et al.*, 1980b). These pseudorevertants were unusual in that recombination analysis showed that the parental *ts*C lesion could not be rescued. Furthermore, three different nonparental *ts* mutations were rescued from one of the pseudorevertants, indicating that multiple *ts* lesions could be present in the same pseudorevertant. Several of the *ts* mutations rescued were members of the newly identified mutant group H. These results indicated that although extragenic suppression of *ts* phenotype can occur, suppression alone did not account for the shift from *ts* to *ts*$^+$ phenotype observed during persistent infection.

The mechanism by which suppressor mutations overcome the defects of the mutations they mask is unknown. The hypotheses of others are attractive. The suppressor mutation may restore stoichiometric relationships altered by the original mutation (Floor, 1970), or it may produce compensating alterations in a second (suppressor) protein that is in physical contact with the parental (*ts*) protein (Jarvik and Botstein, 1975). In the absence of any evidence for stoichiometric alterations of protein synthesis in any of the reovirus *ts* mutants (Cross and Fields, 1976a), the second of these hypotheses must be favored. The possibility of translational suppression has been excluded by several lines of evidence that have been discussed in detail (Ramig and Fields, 1979). Two models are currently considered to be reasonable and testable hypotheses for the mechanism of suppression (Ramig, 1980): (1) Compensating protein interactions model. The *ts* lesion causes an alteration in the structure of a protein such that the protein interacts with a second protein in a manner that is thermolabile. In the suppressed pseudorevertant, the second protein is also altered so that its interaction with the *ts* protein is thermostable. This model predicts that a *ts* lesion could be suppressed by suppressor mutations in any protein with which the product of the *ts* gene interacts, either in the virion or during morphogenesis. (2) Mutator transcriptase model. This model shares most features with the compensating protein interactions model. However, in this case, the second suppressor mutation is the direct result of a virion transcriptase that reads with lower fidelity than the wild-type enzyme. Since the transcription products of reovirus function as both messenger RNA and template for the synthesis of progeny genomes, any mutations introduced by a mutator enzyme would be "locked" into the genome of progeny virions. Many of the mutations induced by the mutator transcriptase could be suppressor mutations. This model would account for both the high frequency of pseudorevertants among revertants of reovirus *ts* mutants and the high frequency of nonparental *ts* lesions that are rescued from pseudorevertants.

The biological significance of the extragenic suppression pathway of reversion is not understood in reovirus. However, results obtained with influenza virus, a virus that shares many properties with reovirus, indicate that reversion by this pathway can be biologically significant. An influenza vaccine that was attenuated by the presence of *ts* mutations was found to generate suppressed pseudorevertants during a vaccine trial

(Murphy *et al.*, 1980; Tolpin *et al.*, 1981). Subsequent analysis of the pseudorevertant revealed that it had regained its virulent character when tested in volunteers (Tolpin *et al.*, 1982).

The studies of extragenic suppression of *ts* phenotype in reovirus have proven to be quite fruitful. In addition to demonstration that extragenic suppression is the primary reversion pathway in reovirus, these studies have serious implications for the use of revertants as controls in experiments, since the revertant may still contain the *ts* lesion. The studies of suppression have suggested a way to isolate mutations in new genes and have shown that suppression is one of the complex series of interactions that occur during persistent infection. Extragenic suppression may represent a general mechanism by which viruses can escape the effects of deleterious mutations in the absence of intramolecular recombination.

## VI. PHYSICAL MAP OF THE REOVIRUS GENOME

Many early experiments strongly indicated that the groups of temperature-sensitive (*ts*) mutants defined by recombination tests represented lesions on different genome segments that reassorted during mixed infection to yield wild-type combinations of genome segments (Fields and Joklik, 1969; Fields, 1971). However, proof that recombination occurred by reassortment required markers for genome segments or gene products so that the parents could be differentiated and reassortment of parental genome segments or gene products demonstrated. An early attempt that used three-factor crosses between *ts* mutants and an electrophoretically aberrant μ1–μ1C protein complex as the unselected outside marker provided very strong genetic evidence for recombination by reassortment (Cross and Fields, 1976b). Physical proof of reassortment depended on identification of markers that could be used to identify the parental origin of each genome segment in putative reassortants. These markers were identifed when it was demonstrated that there was heterogeneity in electrophoretic migration rate (molecular weight?) among the genome segments of the three reovirus serotypes (Ramig *et al.*, 1977b). Since the parental origin of each genome segment in a recombinant could easily be determined by electrophoretic analysis, these markers were used to demonstrate that the mechanism of recombination in reovirus was indeed reassortment (Sharpe *et al.*, 1978). These electrophoretic markers for the genome segments also provided the tools necessary to map each of the *ts* mutant groups onto a specific genome segment. Another observation that greatly facilitated the mapping studies was the demonstration that the double-stranded RNA (dsRNA) genome segments could be resolved in gels as described by Laemmli (1970) for the resolution of proteins. The denaturing conditions in the gels allowed analysis of dsRNA without the extensive deproteinization generally required for electrophoresis of RNA

(Ramig et al., 1977b), facilitating the analysis of large numbers of recombinant clones.

The procedure used to map many of the ts lesions has been described in detail (Ramig et al., 1978). Briefly, crosses were made between ts mutants of reovirus type 3 (Dearing) and wild-type of type 1 (Lang) or type 2 (Jones). These crosses were expected to yield four progeny types: (1) the parental type 3 ts mutant, (2) the parental wild-type type 1 or type 2, (3) recombinant progeny containing the ts lesion from the type 3 parent, and (4) ts⁺ recombinant progeny resulting from replacement of the type 3 genome segment bearing the ts lesion by the corresponding genome segment derived from the ts⁺ type 1 or type 2 parent. Progeny types 1 and 3 were selected against by plating at nonpermissive temperature, leaving a mixture of ts⁺ recombinant and ts⁺ parental genotypes. Electrophoretic analysis of progeny clones allowed the classification of each clone as a ts⁺ parent or a ts⁺ recombinant. Only ts⁺ recombinant progeny clones (and occasional ts recombinant clones) were useful for mapping purposes. Later mapping experiments were performed with ts lesions present in both the parental viruses, allowing a much stronger selection for the desired ts⁺ recombinants, since progeny types 1, 2, and 3 could be selected against with temperature (Ramig et al., 1983). The use of temperature selection to enrich for ts⁺ recombinants placed a restriction on the reassortment of genome segments. Specifically, the genome segment of type 3 bearing the ts lesion was always replaced by the corresponding wild-type segment of the other parent. The remaining nine segments in the ts⁺ recombinant could be derived from either parent, since no selective pressures had been placed on the segregation of those segments. Therefore, if a number of ts⁺ recombinant progeny clones were analyzed, the parental origin of nine of the genome segments was seen to segregate randomly among the progeny. One segment segregated nonrandomly and was always derived from the ts⁺ parent (Table IV and Fig. 3). This segment replaced the segment bearing the ts lesion. In cases wherein both the parents contained ts lesions, the analysis was identical except that two genome segments (one from each parent) segregated nonrandomly. The occasional ts recombinant progeny clone that survived the selection procedure always had the segment from the ts parent that contained the ts lesion. This segment always corresponded to the segment derived from the ts⁺ parent in the ts⁺ recombinants. Thus, the pattern of segregation of genome segments in both ts⁺ and ts recombinant progeny was useful for mapping ts lesions.

The intertypic recombinant clones generated for mapping of ts lesions were also useful for assignment of the polypeptide gene products to the genome segments in which they are encoded. The parental origin of each genome segment in an intertypic recombinant could be determined by electrophoretic analysis. In addition to the markers that electrophoretic heterogeneity provided for the genome segments, similar markers for each polypeptide species were identified through the elec-

TABLE IV. Efficiency of Plating and Parental Origin of Genome Segments of Recombinant Clones Derived from Crosses between the Group G (ts453) Mutant of Reovirus Type 3 and Reovirus Type 1 or Type 2[a]

| Cross | Cross temperature | Clone No. | EOP | L1 | L2 | L3 | M1 | M2 | M3 | S1 | S2 | S3 | S4 |
|---|---|---|---|---|---|---|---|---|---|---|---|---|---|
| Type 1 × tsG(453) | 31.0 | BF51 | 0.86 | 1 | G | 1 | 1 | 1 | 1 | 1 | 1 | 1 | 1 |
| | | BF52 | 0.63 | 1 | G | 1 | 1 | 1 | 1 | 1 | 1 | 1 | 1 |
| | | 132 | 0.35 | 1 | 1 | 1 | 1 | 1 | 1 | 1 | G | G | 1 |
| | | 140 | 0.28 | 1 | 1 | 1 | 1 | G | G | 1 | 1 | G | 1 |
| Type 2 × tsG(453) | 31.0 | 250 | 0.37 | G | G | G | G | 2 | G | G | G | G | 2 |
| | | 251 | 0.33 | G | G | G | G | 2 | G | G | G | G | 2 |
| | | 252 | 0.86 | G | 2 | G | 2 | 2 | 2 | G | 2 | 2 | 2 |
| | | 253 | 0.33 | G | G | G | G | G | G | G | G | G | 2 |
| | | 254 | 0.25 | G | G | G | G | G | G | G | G | G | 2 |
| | | 255 | 0.33 | G | G | G | G | 2 | G | G | G | G | 2 |
| | | 256 | 0.20 | G | G | G | G | G | 2 | G | G | G | 2 |
| | | 257 | 0.50 | G | G | G | G | G | G | G | G | G | 2 |
| | | 258 | 0.40 | 2 | 2 | 2 | G | G | G | G | G | 2 | 2 |
| | | 259 | 0.54 | 2 | 2 | G | G | G | G | G | G | G | 2 |
| | | 260 | 0.40 | G | G | G | G | G | G | G | G | G | 2 |
| | | 261 | 0.45 | 2 | G | 2 | G | 2 | G | G | 2 | 2 | 2 |
| | | 262 | 0.55 | 2 | G | G | G | G | G | G | G | G | 2 |
| | | 263 | 0.50 | G | G | G | G | G | G | G | G | G | 2 |
| | | 264 | 0.90 | G | G | G | G | G | G | G | G | G | G |
| | | 246 | <0.10 | G | G | G | 2 | 2 | G | G | 2 | 2 | G |
| | | 247 | <0.015 | 2 | G | G | 2 | 2 | 2 | G | G | G | G |
| | | 248 | <0.08 | 2 | G | G | 2 | 2 | 2 | G | G | G | G |

[a] From Mustoe et al. (1978a). (EOP) Efficiency of plating.

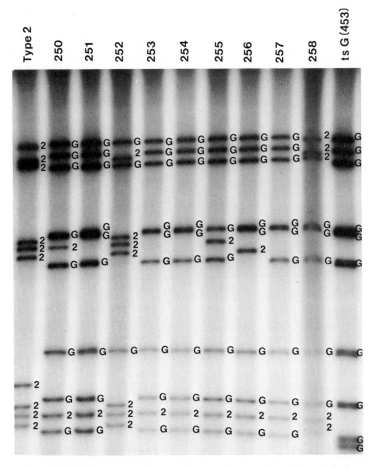

FIGURE 3. [32]P-labeled dsRNAs of $ts^+$ recombinant clones isolated from the cross $ts$G(453) × type 2. Electrophoresis was from top to bottom as described. All dsRNAs are labeled according to parent of origin. From Mustoe *et al.* (1978a).

trophoretic heterogeneity of the polypeptides of the three serotypes (Ramig *et al.*, 1977b). Thus, if the polypeptide species synthesized by an intertypic recombinant clone were examined by electrophoresis, the parent of origin for each species could be determined (Mustoe *et al.*, 1978b). Comparison of the parental origin of a specific polypeptide species with the parental origin of the genome segments allowed assignment of a polypeptide species to a specific genome segment when a number of clones segregating genome segments in different combinations were examined. In many cases, assignment could be made on the basis of a single recombinant clone that contained nine segments from one parent and only one segment from the other parent. Such a clone synthesized only one polypeptide species having the electrophoretic migration characteristic of a polypeptide from the parent contributing the single segment. This anal-

TABLE V. Genetic Map of the Reovirus Genome

| Genome segment | Polypeptide species[a] | ts Mutant group[b] | Biological function[c] |
|---|---|---|---|
| L1 | λ3 | tsD | Core |
| L2 | λ2 | tsB | Core |
| L3 | λ1 | tsI | Core |
| M1 | μ2 | tsH | Core |
| M2 | μ1 → μ1C | tsA | Outer capsid |
| M3 | μNS | (tsF)[d] | Nonstructural |
| S1 | σ1 | tsJ | Outer capsid (HA) |
| S2 | σ2 | tsC | Core |
| S3 | σNS | tsE | Nonstructural |
| S4 | σ3 | tsG | Outer capsid |

[a] From Mustoe et al. (1978b) and McCrae and Joklik (1978).
[b] From Ramig et al. (1978, 1983) and Mustoe et al. (1978a).
[c] From Smith et al. (1969), Cross and Fields (1976c), Mustoe et al. (1978b), and Weiner et al. (1978).
[d] The assignment of tsF to segment **M3** is tentative (see Section VI.F).

ysis was used to assign the polypeptides of the μ and σ size classes to genome segments. The electrophoretic heterogeneity among the λ polypeptide species of the serotypes was too slight to allow assignment of the λ species. Similar assignments have been made using the biochemical approach of translating denatured individual genome segments *in vitro* (McCrae and Joklik, 1978). The assignments made by both genetic and biochemical methods agree, lending confidence to the polypeptide-encoding assignments.

Analysis of genome-segment segregation in intertypic recombinants has been used to map nine of the ten mutant groups to genome segments (Ramig *et al.*, 1978, 1983; Mustoe *et al.*, 1978a). Each of the nine mutant groups mapped to a different genome segment. One group, F, has a very leaky prototype mutant that has proven difficult to map. However, preliminary results with this mutant suggest that it maps to the one segment to which a *ts* mutant has not been definitely assigned (Mustoe, Ramig, Ahmed, and Fields, unpublished data). Thus, the genome of reovirus appears to be saturated with mutations; i.e., each genome segment has been identified by a *ts* lesion. The genetic map of the reovirus genome is summarized in Table V. The location of each mutant group is discussed below. In addition, the relationship of assignments made with intertypic recombinants to assignments made by other methods is considered, as is the relationship of the assignment to the physiology of the mutant.

## A. tsA

Genetic analyses have been used to assign the tsA prototype mutant, tsA(201), to genome segment *M2* (Mustoe *et al.*, 1978a). The *M2* genome segment encodes protein μ1 and its cleavage product, μ1C (Mustoe *et al.*, 1978b), so *ts*A mutants synthesize *ts* forms of these proteins. This assignment conflicts with an earlier assignment of the *ts*A lesion to genome segment *L2* (Ito and Joklik, 1972a; Schuerch and Joklik, 1973). These investigators' analysis of the electrophoretic migration of hybrid dsRNAs composed of mutant (+) and wild-type (−) strands showed that 13 of 26 group A mutants had *L2* hybrid dsRNAs with retarded electrophoretic migration. Three revertants of *ts*A(340), a mutant that showed the retarded *L2* hybrid, did not have the anomalous hybrid *L2* dsRNA. On this basis, they assigned the *ts*A lesion to segment *L2*. We have suggested that the retarded migration of hybrid *L2* may have been due to a silent, second mutation in some of the mutant stocks (Mustoe *et al.*, 1978a). Indeed, several of the *ts* mutants and their revertants have been shown to contain silent mutations (Ramig and Fields, 1979; Rubin and Fields, 1980).

Support for the assignment of the *ts*A lesion to segment *M2* has come from other genetic analyses. Cross and Fields (1976b) showed with three-factor crosses that an anomalous electrophretic migration of the μ1–μ1C proteins found in several mutants segregated as though it were linked to the *ts*A lesion. Since μ1 and μ1C are encoded in genome segment *M2* (Mustoe *et al.*, 1978b), these data suggested that the *ts*A lesion was on *M2*. The genetic association of the *ts*A lesion with the μ1–μ1C proteins (and therefore segment *M2*) was further strengthened by the fact that in *ts*A mutants containing an aberrant μ1–μ1C, the ts phenotype and aberrant μ1–μ1C proteins generally co-reverted. The aberrant μ1–μ1C did not co-revert with the *ts* lesion in mutants of other groups that contained the aberrant proteins (Cross and Fields, 1976b).

The μ1C protein is a major outer-capsid protein of the reovirion (Smith *et al.*, 1969; Cross and Fields, 1976c; Mustoe *et al.*, 1978b). The synthesis of *ts* outer-capsid proteins by *ts*A mutants is consistent with the dsRNA$^+$ phenotype (Cross and Fields, 1972) and assembly of mature virions observed in *ts*A mutants (Fields *et al.*, 1971).

## B. *tsB*

The *ts*B prototype mutant, *ts*B(352), has been assigned to genome segment *L2* by analysis of intertypic recombinants (Mustoe *et al.*, 1978a). Mutants in group B thus produce a temperature-sensitive λ2 protein (McCrae and Joklik, 1978). No other genome segment assignments have been made for *ts*B. Since λ2, the *ts* protein of B mutants, is a component of the viral core (Smith *et al.*, 1969; Cross and Fields, 1976c; Mustoe *et al.*, 1978b), the observations that *ts*B mutants are dsRNA$^+$ (Cross and Fields, 1972) and assemble corelike particles at nonpermissive temperature (Fields *et al.*, 1971) suggest that the *ts* lesion does not prevent the assembly of core particles or the synthesis of dsRNA, but does affect the subsequent assembly of the outer capsid. Morgan and Zweerink (1974) have shown that the corelike particles synthesized by *ts*B at a nonpermissive temperature contained reduced amounts of λ2, suggesting that λ2 may have a role in determining the proper assembly of the outer capsid.

## C. *ts*C

Intertypic recombinants have been used to assign the group C prototype mutant, *ts*C(447), to genome segment *S2* (Ramig *et al.*, 1978). The group C mutants synthesize *ts* σ2 proteins (Mustoe *et al.*, 1978b). This assignment conflicts with the assignment of the *ts*C lesion to the *L1* genome segment (Spandidos and Graham, 1975a,b, 1976). The latter workers made this assignment on the basis of their observation that *ts*C was the only *ts* mutant that neither complemented *L1* deletion mutants nor was complemented by them, suggesting that *ts*C and the *L1* deletion were defective in the same function. This observation may be related to the fact that the genetic interactions of *ts*C with other mutants are restricted at a nonpermissive temperature (Chakraborty *et al.*, 1979). Matsuhisa and Joklik (1974) found that *ts*C mutants synthesize top-component particles that specifically lack the protein species λ1 and σ2. They therefore postulated that the *ts*C lesion was in either genome segment *L1* or *S2*. Their assignment to segment *S2* agrees with that made from the intertypic recombinants (Ramig *et al.*, 1978). The σ2 protein, synthesized in *ts* form by *ts*C mutants, is a component of the viral core (Smith *et al.*, 1969; Cross and Fields, 1976c; Mustoe *et al.*, 1978b). The synthesis of *ts* σ2 by the *ts*C mutants is consistent with the reduced synthesis of single-stranded RNA (ssRNA) and failure to synthesize dsRNA seen with *ts*C mutants (Cross and Fields, 1972; Ito and Joklik, 1972b). The location of the *ts*C lesion is also consistent with the accumulation of particles resembling empty outer capsids (Fields *et al.*, 1971) and an aberrant top-component particle (Matsuhisa and Joklik, 1974).

## D. *ts*D

The group D prototype mutant, *ts*D(357), has been assigned to genome segment *L1* by analysis of intertypic recombinants (Ramig *et al.*, 1978). The *L1* genome segment encodes protein λ3 (McCrae and Joklik, 1978). Other assignments have been made for the location of *ts*D; some of these assignments conflict and some agree with our assignment to *L1*. Controlled digestion of *ts*D virions by chymotrypsin suggested that mutants of this group had an altered μ1C protein (Ito and Joklik, 1972c), leading to the suggestion that the D lesion was on segment *M2*. However, aberrant μ1C proteins that are totally unrelated to the *ts*D lesion have been reported in the group D mutant used for these studies (Cross and Fields, 1976a,b), suggesting that the D lesion was unrelated to the aberrant digestion of μ1C. Analysis of hybrid dsRNA molecules composed of mutant (+) and wild-type (−) strands suggested that the group D lesion could be in either genome segment L1 or M2 (Ito and Joklik, 1972a; Schuerch and Joklik, 1973). However, the altered migration of the hybrid M2 may be related to the altered μ1C found in the group D mutant used for these studies (Cross and Fields, 1976a). Since the altered μ1C is neither temperature-sensitive nor linked to the tsD lesion (Cross and Fields, 1976b), the altered migration of the L1 hybrid dsRNA appears to represent the true location of the tsD lesion. The location of tsD on segment L1 by intertypic recombinants supports this hypothesis (Ramig et al., 1978). Cells infected with group D mutants accumulate a heterogeneous mixture of empty single and double capsid particles (Fields et al., 1971), consistent with the assignment of tsD to L1, a segment that encodes a core protein. This assignment is also consistent with the RNA$^-$ phenotype of the tsD mutants (Cross and Fields, 1972; Ito and Joklik, 1972b).

## E. tsE

Analysis of genome-segment segregation in intertypic recombinants has been used to assign the lesion of the tsE prototype mutant, tsE(320), to genome segment S3 (Ramig et al., 1978). Segment S3 encodes the nonstructural protein σNS (Mustoe et al., 1978b). This assignment agrees with the one other assignment made for tsE. Temperature-sensitive E mutants were found to yield hybrid dsRNAs in which the L2 and S3 species were retarded in gels (Ito and Joklik, 1972a; Schuerch and Joklik, 1973). However, revertants of the E mutant yielded only a retarded L2 dsRNA, suggesting that the retarded migration of the S3 dsRNA in the mutant was specifically associated with the tsE lesion. No viral structures are seen in infected cells (Fields et al., 1971), suggesting that the tsE lesion blocks a very early function. This early function is consistent with the production of a defective nonstructural protein if that protein acts very early in infection. σNS has been shown to have a binding capacity for ssRNA (Huismans and Joklik, 1976; Gomatos et al., 1981), a function expected to be required early in the reovirus infectious cycle. The ssRNA-binding property of σNS is also consistent with the RNA$^-$ phenotype of the E mutant (Cross and Fields, 1972; Ito and Joklik, 1972b).

## F. tsF

The tsF prototype mutant, tsF(556), is very leaky and has proven to be difficult to map using intertypic recombinants. However, preliminary mapping experiments with this mutant strongly suggest, but do not prove, that tsF resides on genome segment M3 (Mustoe, Ramig, Ahmed, and Fields, unpublished observations). The recent isolation of a group F mutant that is not leaky (Ahmed *et al.*, 1980a) should facilitate the absolute assignment of the F group to a genome segment. The tentative assignment of *ts*F to segment *M3* indicates that the nonstructural protein μNS is defective in group F mutants (Mustoe *et al.*, 1978b). The tentative assignment of the F lesion to a nonstructural protein (μNS) is consistent with the apparently normal morphogenesis (Fields *et al.*, 1971) and unimpaired synthesis of RNA (Cross and Fields, 1972) and protein (Cross and Fields, 1976a) seen with this mutant.

## G. *ts*G

The segregation patterns observed in intertypic recombinants have been used to assign the prototype *ts*G mutant, *ts*G(453), to genome segment *S4* (Mustoe *et al.*, 1978a). No other assignments of the G lesion have been made. Genome segment *S4* encodes the outer-capsid protein σ3 (Mustoe *et al.*, 1978b). This assignment is consistent with the observation that *ts*G is an RNA$^+$ mutant (Cross and Fields, 1972) that accumulates corelike particles at nonpermissive temperature (Fields *et al.*, 1971).

## H. *ts*H

The group H prototype mutant, *ts*H(26/8), has been assigned to genome segment *M1* on the basis of intertypic recombinants derived from two *ts* parents (Ramig *et al.*, 1983). Segment *M1* encodes protein μ2 (Mustoe *et al.*, 1978b). The *ts*H mutants were identified only recently and were isolated as nonparental *ts* mutations from extragenically suppressed pseudorevertants (Ramig and Fields, 1979; Ahmed *et al.*, 1980b). No physiological or morphogenetic studies have been done with the *ts*H mutants.

## I. *ts*I

The group I prototype mutant, *ts*I(138), has been assigned to genome segment *L3*, also on the basis of intertypic recombinants derived from two *ts* parents (Ramig *et al.*, 1983). Segment *L3* encodes protein λ1 (McCrae and Joklik, 1978). Like the group H mutants, the mutants of group I were identified only recently and were rescued from extragenically suppressed pseudorevertants (Ramig and Fields, 1979). No physiological or morphogenetic studies have been performed with the *ts*I mutants. It is interesting to note that in mapping the *ts*I lesion, suppression was noted in some of the recombinant progeny clones. This suppression could be detected because of violations of the expected segregation patterns in the suppressed progeny clones. The ability to demonstrate the presence of a class of phenotypically ts$^+$ recombinants that contained a suppressed parental *ts* lesion made unambiguous interpretation of the mapping data for *ts*I possible (Ramig *et al.*, 1983).


## J. *ts*J

Segregation analysis using intertypic recombinants derived from two *ts* parents was used to assign the *ts*J prototype mutant, *ts*J(128), to genome segment *S1* (Ramig *et al.*, 1983). Segment *S1* encodes the viral hemagglutinin, σ1 (Mustoe *et al.*, 1978b; Weiner *et al.*, 1978). The single mutant representing group J was rescued from a suppressed pseudorevertant (Ramig and Fields, 1979). No additional studies have been performed with this mutant. As with the *ts*I mutant, suppression was a complicating factor in the mapping of the *ts*J lesion (Ramig *et al.*, 1983).


## K. Summary

In summary, intertypic recombination has proven to be a very powerful technique for the mapping of reovirus *ts* lesions to genome segments. Nine of the ten mutant groups have been assigned to specific genome segments, with the tenth group tentatively assigned. These assignments are consistent in that only one mutant group has been assigned to each segment and a group has been assigned to every segment. In general, these assignments agree with the assignments of others and correlate well with the physiological and morphological properties of the mutant groups. Experiments with *ts* mutants can now be done with the site of the *ts* lesion and the protein species that is defective defined. This knowledge should simplify the interpretation of the results of experiments using the *ts* mutants.


## REFERENCES

Ahmed, R., and Graham, A.F., 1977, Persistent infection in L cells with temperature-sensitive mutants of reovirus, *J. Virol.* **23**:250–262.

Ahmed R., Chakraborty, P.R., and Fields, B.N., 1980a, Genetic variation during lytic reovirus infection: High passage stocks of wild type reovirus contain temperature-sensitive mutants, *J. Virol.* **34**:285–287.

Ahmed, R., Chakraborty, P.R., Graham, A.F., Ramig, R.F., and Fields, B.N., 1980b, Genetic variation during persistent reovirus infection: Presence of extragenically suppressed temperature-sensitive lesions in wild type virus isolated from persistently infection L cells, *J. Virol.* **34**:383–389.

Ahmed, R., and Fields, B.N., 1981, Reassortment of genome segments between reovirus defective interfering particles and infectious virus: Construction of temperature-sensitive and attenuated viruses by rescue of mutations from DI particles, *Virology* **111**:351–363.

Ahmed, R., Canning, W.M., Kauffman, R.S., Sharpe, A.H., Hallum, J.V., and Fields, B.N., 1981, Role of the host cell in persistent viral infection: Coevolution of L cells and reovirus during persistent infection, *Cell* **25**:325–332.

Chakraborty, P.R., Ahmed, R., and Fields, B.N., 1979, Genetics of reovirus: The relationship of interference to complementation and reassortment of temperature-sensitive mutants at nonpermissive temperature, *Virology* **94**:119–127.

Cooper, P.D., 1968, A genetic map of poliovirus temperature-sensitive mutants, *Virology* **35**:584–596.

Cross, R.K., and Fields, B.N., 1972, RNA synthesized in cells infected with reovirus type 3 mutants, *Virology* **50**:799–809.

Cross, R.K., and Fields, B.N., 1976a, Temperature-sensitive mutants of reovirus type 3: Evidence for aberrant μ1 and μ2 polypeptide species. *J. Virol.* **19**:174–179.

Cross, R.K., and Fields, B.N., 1976b, Use of an aberrant polypeptide as a marker in three-factor crosses: Further evidence for independent reassortment as the mechanism of recombination between temperature-sensitive mutants of reovirus type 3, *Virology* **74**:345–362.

Cross, R.K., and Fields, B.N., 1976c, Reovirus-specific polypeptides: Analysis using discontinuous gel electrophoresis, *J. Virol.* **19**:162–173.

Fields, B.N., 1971. Temperature-sensitive mutants of reovirus—features of genetic recombination, *Virology* **46**:142–148.

Fields, B.N., 1973, Genetic reassortment with reovirus mutants, in: *Virus Research* (C.F. Fox, ed.), p. 461, Academic Press, New York.

Fields, B.N., and Joklik, W.K., 1969, Isolation and preliminary genetic and biochemical characterization of temperature-sensitive mutants of reovirus, *Virology* **37**:335–342.

Fields, B.N., Raine, C.S., and Baum, S.G., 1971, Temperature-sensitive mutants of reovirus type 3: Defects in virus maturation as studied by immunofluorescence electron microscopy, *Virology* **43**:569–578.

Floor, E., 1970, Interaction of morphogenetic genes of bacteriophage T4, *J. Mol. Biol.* **47**:293–306.

Gomatos, P.J., Tamm, I., Dales, S., and Franklin, R.M., 1962, Reovirus type 3: Physical characteristics and interactions with L cells, *Virology* **17**:441.

Gomatos, P.J., Prakash, O., and Stamatos, N.M., 1981, Small reovirus particles composed solely of sigma NS with specificity for binding different nucleic acids. *J. Virol.* **39**:115–124.

Hrdy, D.B., Rosen, L., and Fields, B.N., 1979, Polymorphism of the migration of double stranded RNA genome segments of reovirus isolates from humans, cattle and mice, *J. Virol.* **31**:104–111.

Huismans, H., and Joklik, W.K., 1976, Reovirus-coded polypeptides in infected cells: Isolation of two native monomeric polypeptides for single stranded and double stranded RNA. *Virology* **70**:411–424.

Ikegami, N., and Gomatos, P.J., 1968, Temperature-sensitive conditional lethal mutants of reovirus 3. I. Isolation and characterization, *Virology* **36**:447.

Ikegami, N., and Gomatos, P.J., 1972, Inhibition of host and viral protein synthesis during infection at the nonpermissive temperature with ts mutants of reovirus 3, *Virology* **47**:306–312.

Ito, Y., and Joklik, W.K., 1972a, Temperature-sensitive mutants of reovirus. II. Anomalous electrophoretic migration behavior of certain molecules composed of mutant plus strands and wild type minus strands, *Virology* **50**:202–208.

Ito, Y., and Joklik, W.K., 1972b, Temperature-sensitive mutants of reovirus. I. Patterns of gene expression by mutants of groups C, D and E, *Virology* **50**:189–201.

Ito, Y., and Joklik, W.K., 1972c, Temperature-sensitive mutants of reovirus. III. Evidence that mutants of group C ("RNA negative") are structural polypeptide mutants, *Virology* **50**:282–286.

Jarvik, J., and Botstein, D., 1975, Conditional-lethal mutations that suppress defects in morphogenesis by altering structural proteins, *Proc. Natl. Acad. Sci. U.S.A.* **72**:2738–2742.

Laemmli, U.K., 1970, Cleavage of structural proteins during the assembly of the head of bacteriophage T4, *Nature (London)* **227**:680–685.

Matsuhisa, T., and Joklik, W.K., 1974, Temperature-sensitive mutants of reovirus. V. Studies on the nature of the temperature sensitive lesion of the C group mutant ts447, *Virology* **60**:380–389.

McCrae, M.A., and Joklik, W.K., 1978, The nature of the polypeptide encoded by each of the ten double-stranded RNA segments of reovirus type 3, *Virology* **89**:578–593.

Morgan, E.M., and Zweerink, H.J., 1974, Reovirus morphogenesis: Core-like particles in cells infected at 39 with wild type and temperature sensitive mutants of groups B and G, *Virology* **59**:556–565.

Murphy, B.R., Tolpin, M.D., Massicot, J.G., Kim, H.Y., Parrott, R.H., and Chanock, R.M., 1980, Escape of a highly defective influenza A virus mutant from its temperature sensitive phenotype by extragenic suppression and other types of mutation, *Ann. N.Y. Acad. Sci.* **354**:172–182.

Mustoe, T.A., Ramig, R.F., Sharpe, A.H., and Fields, B.N., 1978a, A genetic map of reovirus. III. Assignment of the double-stranded RNA positive mutant groups, A, B and G to genome segments, *Virology* **85**:545–556.

Mustoe, T.A., Ramig, R.F., Sharpe, A.H., and Fields, B.N., 1978b, Genetics of reovirus: Identification of the dsRNA segments encoding the polypeptides of the mu and sigma size classes, *Virology* **89**:594–604.

Nonoyama, M., and Graham, A.F., 1970, Appearance of defective virions in clones of reovirus, *J. Virol.* **6**:693.

Prevec, L., Watanabe, Y., Gauntt, C.J., and Graham, A.F., 1968, Transcription of the genomes of type 1 and type 3 reovirus, *J. Virol.* **2**:289.

Ramig, R.F., 1980, Suppression of temperature sensitive phenotype in reovirus: An alternative pathway from ts to ts$^+$ phenotype, in: *Animal Virus Genetics* (B.N. Fields, R. Jaenisch, and C.F. Fox, eds.), pp. 633–642, Academic Press, New York.

Ramig, R.F., and Fields, B.N., 1977, Method for rapidly screening revertants of reovirus temperatures sensitive mutants for extragenic suppression, *Virology* **81**:170–173.

Ramig, R.F., and Fields, B.N., 1979, Revertants of temperature sensitive mutants of reovirus: Evidence for frequent extragenic suppression, *Virology* **92**:155–167.

Ramig, R.F., White, R.M., and Fields, B.N., 1977a, Suppression of the temperature sensitive phenotype of a mutant of reovirus type 3, *Science* **195**:406–407.

Ramig, R.F., Cross, R.K., and Fields, B.N., 1977b, Genome RNAs and polypeptides of reovirus serotypes 1, 2, and 3, *J. Virol.* **22**:726–733.

Ramig, R.F., Mustoe, T.A., Sharpe, A.H., and Fields, B.N., 1978, A genetic map of reovirus. II. Assignment of the double stranded RNA negative mutant groups C, D and E to genome segments, *Virology* **85**:531–544.

Ramig, R.F., Ahmed, R., and Fields, B.N., 1983, A genetic map of reovirus: Assignment of the newly defined mutant groups H, I and J to genome segments (in press).

Rubin, D.H., and Fields, B.N., 1980, Molecular basis of reovirus virulence: Role of the M2 gene, *J. Exp. Med.* **152**:853–868.

Schuerch, A.R., and Joklik, W.K., 1973, Temperature-sensitive mutants of reovirus. IV. Evidence that anomalous electrophoretic migration behavior of certain double stranded RNA hybrid species is mutant group specific, *Virology* **56**:218–229.

Schuerch, A.R., Matsuhisa, T., and Joklik, W.K., 1974, Temperature-sensitive mutants of reovirus. VI. Mutant ts 447 and ts 556 particles that lack either one or two genome segments, *Intervirology* **3**:36–46.

Sharpe, A.H., Ramig, R.F., Mustoe, T.A., and Fields, B.N., 1978, A genetic map of reovirus. I. Correlation of genome RNAs between serotype 1, 2 and 3, *Virology* **84**:63–74.

Shatkin, A.J., Sipe, J.D., and Loh, P., 1968, Separation of 10 reovirus segments by polyacrylamide gel electrophoresis, *J. Virol.* **2**:968.

Skehel, J.J., and Joklik, W.K., 1969, Studies on the *in vitro* transcription of reovirus RNA catalyzed by reovirus cores, *Virology* **39**:822.

Smith, R.E., Zweerink, H.J., and Joklik, W.K., 1969, Polypeptide components of virions, top component and cores of reovirus type 3, *Virology* **39**:791–810.

Spandidos, D.A., and Graham, A.F., 1975a, Complementation of defective reovirus by ts mutants, *J. Virol.* **15**:954–963.

Spandidos, D.A., and Graham, A.F., 1975b, Complementation between temperature-sensitive and deletion mutants of reovirus, *J. Virol.* **16**:1444–1453.

Spandidos, D.A., and Graham, A.F., 1976, Recombination between temperature sensitive and deletion mutants of reovirus, *J. Virol.* **18**:117–123.

Tolpin, M.D., Massicot, J.G., Mullinix, M.G., Kim, H.W., Parrott, R.H., Chanock, R.M., and Murphy, B.R., 1981, Genetic factors associated with loss of the temperature sensitive phenotype of the influenza A/Alaska/77-ts-1A2 recombinant during growth *in vivo*, *Virology* **112**:505–517.

Tolpin, M.D., Clements, M.L., Levine, M.M., Black, R.E., Saah, A.J., Anthony, W.C., Cisneros, L., Chanock, R.M., and Murphy, B.R., 1982, Evaluation of a phenotypic revertant of the A/Alaska/77-ts-1A2 reassortment virus in hamsters and in seronegative adult volunteers: Further evidence that the temperature sensitive phenotype is responsible for attenuation of ts-1A2 reassortant viruses, *Infect. Immun.* **36**:645–650.

Weiner, H.L., and Fields, B.N., 1977, Neutralization of reovirus; The gene responsible for the neutralization antigen, *J. Exp. Med.* **146**:1303–1310.

Weiner, H.L., Ramig, R.F., Mustoe, T.A., and Fields, B.N., 1978, Identification of the gene coding for the hemagglutinin of reovirus, *Virology* **86**:581–584.

CHAPTER 6

# Pathogenesis of Reovirus Infection

ARLENE H. SHARPE AND BERNARD N. FIELDS

## I. INTRODUCTION

To successfully produce disease, a virus must enter its host, replicate within host cells, spread within the host and, in the case of systemic infection, overcome host immune defenses, and damage host tissues. At each stage of the infectious process, a number of viral and host factors determine the ultimate pathogenicity (capacity to produce disease) of the virus. Recent studies using the mammalian reoviruses as a model system have provided insights into the roles of viral and host components in the production of disease. The mammalian reoviruses offer an excellent system for the study of molecular aspects of viral pathogenesis because reovirus genetics has provided a means to identify specific viral components involved in the disease process.

This chapter will summarize our current understanding of reovirus pathogenesis. We will review the viral and host components involved in the interaction between reovirus and its host at each stage of the infectious process. Initially, we will summarize pertinent data concerning reovirus structure and genetics (for a more comprehensive presentation of reovirus structure and genetics, the reader is referred to Chapters 2 and 5, respectively).

ARLENE H. SHARPE • Department of Microbiology and Molecular Genetics, Harvard Medical School, Boston, Massachusetts 02115.    BERNARD N. FIELDS • Department of Microbiology and Molecular Genetics, Harvard Medical School, Boston, Massachusetts 02115; and Department of Medicine, Brigham and Women's Hospital, Boston, Massachusetts 02115.

## A. Reovirus Structure

The mammalian reoviruses are separable into three serotypes (1, 2, and 3) on the basis of hemagglutination-inhibition and antibody-neutralization tests (Rosen, 1960). The virions contain a segmented, double-stranded RNA (dsRNA) genome surrounded by two concentric protein shells (Gomatos et al., 1962). This double capsid consists of a core containing the genome and a closely applied inner capsid surrounded by a second outer capsid. The outer capsid shell is a relatively labile structure that is easily disrupted by heat or chymotrypsin, whereas the core is highly stable when treated with chymotrypsin or high concentrations of urea, guanidine, dimethylformamide, dimethylsulfoxide, or sodium dodecyl sulfate (Joklik, 1972, 1974; Smith et al., 1969).

The reovirus genome consists of ten segments of dsRNA that fall into three size classes, designated large ($L$), medium ($M$), and small ($S$) (Bellamy et al., 1967; Shatkin et al., 1968; Millward and Graham, 1970; Furuichi et al., 1975). There are three segments in the $L$ size class ($L1$, $L2$, and $L3$) with approximate molecular weights of $2.3–2.8 \times 10^6$, three $M$ segments ($M1$, $M2$, and $M3$) with molecular weights of $1.4–1.6 \times 10^6$, and four $S$ segments ($S1$, $S2$, $S3$, and $S4$) with molecular weights of approximately $0.6–0.9 \times 10^6$. Each of the dsRNA segments encodes a unique single-stranded RNA transcript of plus polarity that is translated into a single polypeptide (Shatkin and Rada, 1967; Both et al., 1975; Mustoe et al., 1978b; McCrae and Joklik, 1978). Like the genome segments that encode them, the viral proteins fall into three molecular-weight classes: large ($\lambda$), middle ($\mu$), and small ($\sigma$) proteins (Loh and Shatkin, 1968; Smith et al., 1969). In vitro studies utilizing the three size classes of dsRNA demonstrated that the $L$ segments encode the $\lambda$ polypeptides, the $M$ segments encode the $\mu$ polypeptides, and the $S$ segments encode the $\sigma$ polypeptides (Graziadei and Lengyel, 1972; McCrae and Joklik, 1978). There are ten structural proteins and three nonstructural proteins (Smith et al., 1969; Lee et al., 1981a). Two structural proteins, $\mu$1C and $\lambda$2C, and one nonstructural protein, $\mu$NSC, are derived by cleavage of $\mu$1, $\lambda$2, and $\mu$NS, respectively. Three of the structural proteins ($\mu$1C, $\sigma$1, and $\sigma$3) form the outer capsid of the virus (Smith et al., 1969; Mustoe et al., 1978b). The $\sigma$1 protein is the viral hemagglutinin (Weiner et al., 1978). Six of the structural proteins ($\lambda$1, $\lambda$2, $\lambda$3, $\mu$1, $\mu$2, and $\sigma$2) are contained within the viral core, which has transcriptase activity. Recent lactoperoxidase labeling studies have shown that the $\lambda$2 polypeptide, part of the viral core, is exposed on the outer surface of the virus (Hayes et al., 1981).

## B. Reovirus Genetics

Conditional, lethal temperature-sensitive (ts) mutants of reovirus type 3 were placed into groups (A–F) on the basis of their ability to yield wild-type ($ts^+$) recombinants in the progeny of pairwise crosses at the permissive temperature (Fields and Joklik, 1969; Cross and Fields, 1972, 1976a,b; see Chapter 5). Analysis of recombinants generated by coinfection of cells with reovirus of different serotypes demonstrated that recombinants arise by physical reassortment of genome segments between parental viruses (Sharpe et al., 1978). Such "intertypic" reassortants, which can be readily analyzed by polyacrylamide gel electrophoresis (PAGE) of dsRNA segments, have been used to map the location of the reovirus ts mutants; to correlate genome segments among serotypes 1, 2, and 3; and to identify the dsRNA segments that encode the viral polypeptides (Mustoe et al., 1978a,b; Ramig et al., 1977, 1978). In addition, these intertypic reassortants have enabled us to identify viral genome segments that are important in a number of stages of reovirus pathogenesis.

## C. General Approach for Genetic Analysis of Reovirus Virulence

Although structurally quite similar, the three serotypes of the mammalian reoviruses interact differently with mammalian hosts, producing very distinct patterns of disease (Kilham and Margolis, 1969; Margolis et al., 1971; Raine and Fields, 1973; Stanley and Joske, 1975a,b). Our general approach to the study of reovirus pathogenesis is to compare the pattern of infection produced by reovirus field isolates or laboratory strains of type 1 Lang, type 2 Jones, and type 3 Dearing and to perform a genetic analysis of any differences observed between two serotypes. A biological property that differs between two serotypes can be mapped to a single gene or a group of genes by examining the behavior of reassortant viruses that have dsRNA segments derived from both parental serotypes. Because each dsRNA segment can be isolated against the background of another serotype, the effect of one particular viral gene on a biological property can be studied unambiguously.

The first intertypic recombinants used in the study of reovirus pathogenesis contained segments derived from wild-type 1 or 2 and ts mutants of type 3 Dearing. These recombinants were useful in the identification of σ1 as the viral component involved in a number of virus–cell surface interactions (Weiner and Fields, 1977; Weiner et al., 1978; Sharpe and Fields, 1981). In other studies wherein μ1C and σ3 were found to be responsible for certain biological properties, this set of recombinants was not useful.

One of the potential problems in the use of ts mutants to construct intertypic recombinants is the presence of mutations in genome segments other than the one bearing the phenotype of temperature sensitivity. Such "silent" mutation could alter biological function of a ts mutant as compared to the type 3 Dearing wild-type strain. Specifically, when such ts mutants are used in genetic crosses with wild-type 1 or 2, $ts^+$ recombinants could be generated that contain mutagenized type 3 dsRNA segments that are not *ts*. These segments could produce aberrant polypeptides that may or may not be readily detectable by PAGE or two-dimensional gel systems, but differ from the wild-type 3 virus in their biological functions.

Several lines of evidence, in fact, have suggested that silent mutations are present in type 3 *ts* mutants and that these mutants produce aberrant biological behavior. Controlled digestion of group D virions by chymotrypsin suggested that the group D mutants have an altered $\mu$1C polypeptide in the virion (Ito and Joklik, 1972). However, mapping studies assigned the group D mutants to the *L1* dsRNA segment (Ramig *et al.*, 1978). Rubin and Fields (1980) found that recombinants derived from crosses involving the *ts* D and *ts* G mutants, both of which synthesize a $\mu$1C polypeptide of aberrant electrophoretic mobility, have aberrant behavior in sensitivity to digestion with chymotrypsin as compared to the wild-type 3 Dearing strain. In several other instances wherein the reovirus serotypes have been found to differ in a particular property, the recombinants also did not segregate the property in question to a particular segment. In these cases, it was not clear whether the property in question required several viral genes or whether silent mutations led to an inability to assign the property to a single dsRNA segment.

Such difficulties led to the construction of a second set of recombinants containing dsRNA segments derived from two wild-type parents. These recombinants have permitted the identification of one viral dsRNA segment as responsible for a particular property, whereas recombinants containing dsRNA segments from the type 3 *ts* mutants did not conclusively segregate the property in question to a particular dsRNA segment.

From these studies has emerged an important concept for the study of biological properties in viruses containing a segmented genome. Temperature-sensitive mutants derived from chemical mutagenesis should be used only with caution as parental viruses in the preparation of recombinants to be used in the study of biological properties. The presence of silent mutations in multiple genome segments can confuse the genetic analysis of biological properties. In addition to occurring in viral stocks subjected to chemical mutagenesis, mutation can also arise on multiple passages of virus. Ideally, nonmutagenized and low-passage stocks of parental viruses should be used to prepare recombinant viruses. The reassortment frequency of the mammalian reoviruses is sufficiently high to permit the isolation of a number of recombinant viruses arising spontaneously without a selection involving *ts* mutants. Such reassortant viruses will have a genetic background free of silent mutations and will be the optimal ones for use in the study of biological properties. It may be that the difficulties experienced by investigators studying determinants of pathogenesis in bunyavirus and influenza virus infections are related to the presence of silent mutations in viral stocks.

## II. OVERVIEW OF THE STAGES IN REOVIRUS PATHOGENESIS

Reovirus is thought to behave primarily as an enteric virus. Reoviruses thus initially enter the host through the gastrointestinal tract. Although the primary site of viral replication has not yet been determined, reovirus type 1 is known to penetrate the intestinal epithelium through M cells, a population of specialized epithelial cells that overlie Peyer's patches, and to appear rapidly in Peyer's patches (Wolf *et al.*, 1981). During the initial incubation period, reovirus multiplies at an undetermined site and begins to appear in more distal organs in increasing concentrations in 4–5 days. Eventually both type 1 and type 3 (but not type 2) appear in the nervous system, but in different cell types (Kilham and Margolis, 1969; Margolis *et al.*, 1971; Masters *et al.*, 1977; Raine and Fields, 1973). Type 1 appears primarily within ependymal cells, producing a nonlethal infection that frequently leads to hydrocephalus. Type 3, when inoculated into neonatal mice, produces a highly lethal encephalitis; the virus is localized in neurons in several parts of the brain. Both humoral and cellular immune responses appear following infection. Neutralizing and hemagglutination-inhibition antibodies are detected in serum after the first week. Serotype-specific cytolytic T cells, delayed-type hypersensitivity responses, and suppressor T cells are generated in response to reovirus infection.

After entering its host cells, reovirus can produce a lytic or a persistent infection. During lytic infections, reovirus alters host cell metabolism and cytoskeletal organization. During persistent infection, both the host cell and virus are altered: the virus changes and is less destructive to the host cell; the host cell changes and is capable of supporting viral replication without being lysed (see Section VII).

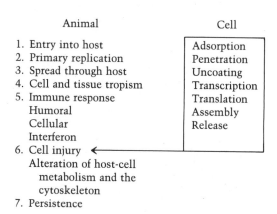

FIGURE 1. Stages in viral pathogenesis.

The following sections will present more details concerning the nature of reovirus interactions with the host at each of the stages summarized above (see Fig. 1).

## III. ENTRY

The upper alimentary tract is the natural portal of entry for reoviruses and other "enteric" viruses. Following introduction into the upper gastrointestinal tract, reoviruses come into direct contact with the contents of the intestinal lumen including intestinal fluids containing proteolytic enzymes and bile, as well as the fluids covering the surfaces of cells lining the gastrointestinal tract [which contain cellular proteases and secretory immunoglobulin A (IgA)]. The interactions of reovirus with the local environment within the intestinal lumen play an important role in determining virulence following entry through the upper gastrointestinal tract.

Studies of reovirus virulence following peroral inoculation have revealed that serotype 1 Lang and 3 Dearing differ in their capacities to grow in intestinal tissues and produce central nervous system (CNS) disease (Rubin and Fields, 1980). Following intracerebral inoculation of newborn mice, reovirus type 3 Dearing is highly neurovirulent, producing an acute, fatal encephalitis. In contrast, following introduction directly into the upper intestinal tract, type 3 Dearing is avirulent [$LD_{50} > 10^7$ plaque-forming units (PFU)]. Little or no virus reaches the CNS, and there is little viral growth in the brain. Type 1 Lang, on the other hand, produces a nonfatal ependymitis, regardless of whether it is inoculated intracerebrally or perorally. In fact, type 1 grows to similar titers in the CNS following peroral or intracerebral inoculation.

Comparison of the capabilities of reovirus types 1 and 3 to grow within the intestine revealed that type 1 grows well within the intestine, whereas type 3 does not grow but instead progressively loses titer. To determine the genetic basis for the differing capabilities of the two serotypes to successfully initiate local infection following entry through the gastrointestinal tract, a number of reassortants that contain genes derived from type 1 Lang and type 3 Dearing were examined for their ability to grow in intestinal tissue. Recombinants that possess the M2 double-stranded RNA (dsRNA) segment derived from type 1 behave like type 1 Lang; they grow in intestinal tissue. Recombinants that contain the type 3 M2 dsRNA behave like type 3 Dearing; they do not grow in intestinal tissue. Thus, the M2 dsRNA segment determines the capability of virus to grow in intestinal tissue.

The ability of virus to grow in intestinal tissue correlates with the efficient spread of virus from the gastrointestinal tract to the nervous system, as determined by viral growth within the CNS. Analysis of the capacity of reassortant viruses to spread from the gastrointestinal tract

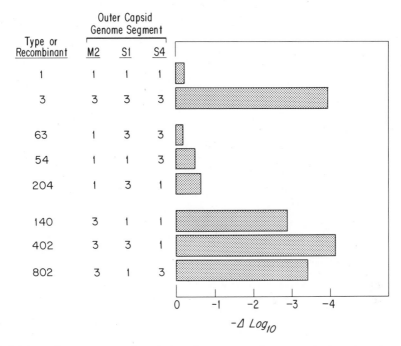

FIGURE 2. Reduction in titer ($\log_{10}$) after chymotrypsin digestion of selected viral stocks in the presence of $Cs^+$. The parental origins of the genome segments encoding the three outer-capsid polypeptides are listed. From Rubin and Fields (1980), by permission of the *Journal of Experimental Medicine*.

to the CNS indicates that the *M2* dsRNA segment also plays a role in determining the capability of virus to spread to the nervous system.

Because intestinal fluids and absorptive cells contain a variety of proteases, we reasoned that the differences in reovirus growth in the intestine might be related to the sensitivity of reoviruses to digestion by intestinal proteases. Chymotrypsin is found in large amounts and in high concentrations in the upper small intestine (Beck, 1973). It is also known to be capable of digesting the outer shell of the mammalian reoviruses *in vitro* (Joklik, 1972; Borsa *et al.*, 1973). For these reasons, the effect of chymotrypsin treatment on viral infectivity was studied *in vitro*. *In vitro* treatment of type 3 Dearing with chymotrypsin results in a marked loss of infectivity. In contrast, under comparable conditions, the infectivity of type 1 is either unaffected or slightly enhanced (Fig. 2). Genetic studies using viral reassortants revealed that the *M2* dsRNA segment also determines the differences in response to chymotrypsin treatment (Rubin and Fields, 1980).

That *in vitro* digestion of reovirus by chymotrypsin mimics the pattern of growth of reovirus serotypes and reassortants in the intestine as well as spread to the CNS suggests that proteases may play an important role in determining the capability of reoviruses to cause disease after

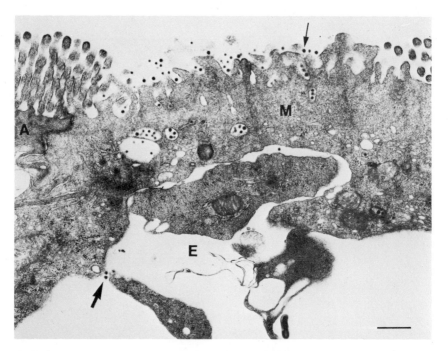

FIGURE 3. Interaction of reovirus type 1 with M cells. Events occurring 1 hr postinoculation with reovirus type 1 in a closed ileal loop made in a 10-day-old mouse. The electron micrograph shows a membranous epithelial cell (M) with reovirus type 1 (small arrow) adhering to its luminal surface, being transported within vesicles across the cell, and lying free (large arrow) within the extracellular space (E). The adjacent absorptive cell (A) has no adherent reovirus. Scale bar: 500 nm. Courtesy of Dr. Jacqueline Wolf.

introduction into the gastrointestinal tract. Proteases might affect viruses directly on passage through the gastrointestinal fluids or might affect the titer of virus released from primary sites of multiplication within the intestinal tract. Although the *in vivo* site of proteolysis of virus is not known, proteases within the intestinal lumen or at the surface of intestinal absorptive cells (Peter, 1973; Beck, 1973) could digest virus.

After a virus successfully traverses the harsh local environment in the intestinal lumen, it can either enter intestinal epithelial cells, leading to local inflammation and possibly diarrhea (significant local epithelial lesions may occur without overt symptoms), or enter the systemic circulation through the gastrointestinal tract, causing few local symptoms and producing disease at a distant site. Reovirus type 1 passes through the gastrointestinal tract of neonatal mice without causing apparent local pathology. During our studies on the early stages of type 1 reovirus infection in neonatal mice, we noted a striking enlargement of Peyer's patches. This enlargement was accompanied by rapid appearance of infectious virus in mesenteric lymph nodes, suggesting that virus was being transported to this site (Letvin *et al.*, 1981).

The site of anatomic localization of reovirus following direct inoculation into the gastrointestinal tract has now been determined using electron microscopy. Reovirus type 1 spares intestinal absorptive cells, but is found adherent to the surface of intestinal microfold (M) cells (Wolf et al., 1981). M cells are a population of specialized epithelial cells that overlie Peyer's patches (Owen and Jones, 1974). M cells have been shown to transport macromolecules such as horseradish peroxidase and ferritin from the intestinal lumen across epithelial cells to the intracellular space, permitting their uptake by mucosal mononuclear cells (Bockman and Cooper, 1973; Owen, 1977). Likewise, within 1 hr after entry into closed ileal loops of live suckling mice, type 1 reoviruses are seen in the M-cell cytoplasm within smooth-surfaced vesicles, within the intercellular space between the M cell and mucosal mononuclear cells, and on the surface of mononuclear cells (Fig. 3). These studies suggest that reovirus type 1 initially adheres to the surface of the M cell, then enters the cell by endocytosis, travels through the cell in vesicles, and finally exits the cell into the intercellular space beneath the M cell, where it interacts with mononuclear cells (Wolf et al., 1981). Thus, it appears that M cells can function to transport reovirus type 1 across the mucosal barrier, facilitating viral transport into Peyer's patches or more distant lymphoid tissue. Preliminary data suggest that reovirus type 3 may use a different pathway for entry (Wolf, unpublished). Whether this pathway represents a major route of entry for reovirus or a clearance pathway used by the host to eliminate virus remains to be seen.

Although reovirus is thought to behave primarily as an enteric virus, it can also act as a respiratory virus. Entry of reoviruses by the respiratory route is not well understood. The interaction between reoviruses and the local environment of the respiratory epithelium is under current investigation. Analogous to the gastrointestinal tract, the respiratory tract has a number of host components with which the virus must successfully interact to produce disease. It is tempting to speculate that the bronchus-associated lymphoid tissue (Bienenstock et al., 1976) may function in a role similar to that of the gut-associated lymphoid tissue in facilitating the entry of reovirus into the host and viral interaction with the immune system.

## IV. SPREAD

There are at least three routes by which viruses, after entering the mammalian host, can spread from their site of primary replication to distant sites. These include the circulatory system, the lymphatic system, and the nervous system. Preliminary experiments suggest that different reovirus serotypes may utilize distinct pathways for spread.

Studies of the clearance of reoviruses from the bloodstream following intravenous inoculation into adult mice have revealed that type 3 reovirus is cleared from the blood more rapidly than type 1 reovirus. In addition, type 3 virus appears to be poorly transported to a number of target organs in an infectious form following intravenous inoculation. Type 1 reovirus, on the other hand, appears to be transported efficiently through the bloodstream to the liver, spleen, and brain. The capacity of type 1 to spread efficiently through the bloodstream is a property of the viral hemagglutinin (Burstin and Fields, unpublished results). These data suggest that type 1 reovirus is spread at least in part by a vascular route. Preliminary studies suggest that type 3 reovirus may use a neural route of spread (unpublished). However, further experiments are needed to determine the precise pathways by which reovirus spreads from its site of entry to target organs.

## V. IMMUNE RESPONSE

The immune response to a viral infection is an important mechanism by which the host resolves an acute infection and protects itself from reinfection by the same virus. Both humoral and cellular immune responses are important in the host response to viral infections (Blanden, 1974; Ogra et al., 1975). At least two types of antibody are produced in response to reovirus infection: neutralizing and hemagglutination-inhibition (HI) antibodies. Three types of reovirus-specific T cells have been detected following inoculation of mice with reovirus: cytotoxic T lymphocytes ($T_{CTL}$), T cells that mediate delayed-type hypersensitivity ($T_{DTH}$), and suppressor T cells ($T_S$) (Table I).

### A. Humoral Immunity

Studies of the neutralization of reovirus infectivity by antibody led to the separation of the mammalian reoviruses into three serotypes on the basis of non-cross-reactivity in antibody-neutralization tests (Rosen, 1962). When neutralization tests were performed with intertypic reassortant viruses and type-specific antibody, the type as determined by neutralization testing correlated solely with the identity of the $S1$ double-stranded RNA (dsRNA) segment (Weiner and Fields, 1977). Therefore, the

TABLE I. Summary of Reovirus–Lymphocyte Interactions

| Interaction | References |
|---|---|
| Cytotoxic T lymphocytes ($T_{CTL}$) | Finberg et al. (1979, 1981), Letvin et al. (1981) |
| T-cell-dependent delayed-type hypersensitivity ($T_{DTH}$) | Weiner et al. (1980b), Greene and Weiner (1980), Rubin et al. (1981) |
| Suppressor T cells ($T_S$) and tolerance | Greene and Weiner (1980) |

polypeptide encoded by the *S1* dsRNA segment, the σ1 polypeptide, was identified as the viral protein responsible for the specific response to neutralizing antibody, thus determining type specificity. As with most viruses, the reovirus neutralization antigen is a capsid protein (Willcox and Mautner, 1976a,b; Webster and Laver, 1967).

Monoclonal antibodies have also been used to study neutralization of reovirus infectivity. Using monoclonal IgG antibodies prepared against proteins encoded by reovirus type 3 Dearing, Hayes *et al.* (1981) examined the ability of the monoclonal antibodies to neutralize infectivity of reovirus type 1 Lang, type 2 Jones, and type 3 Dearing. They found that antibodies directed against σ1, σ3, and λ2 could all neutralize infectivity, but only those directed against σ1 were type-specific. These studies confirm that the σ1 protein is the neutralization antigen that confers type-specificity and indicate that under certain circumstances antibodies to σ3 and λ2 can neutralize the infectivity of all three serotypes. The σ3 and λ2 proteins may represent antigens recognized by cross-reacting antisera. Whether one of them represents the reovirus group-specific complement-fixation antigen is unknown.

In a different study using monoclonal antibodies directed at the type 3 Dearing σ1 polypeptide, Burstin *et al.* (1982) demonstrated that not all antibodies that specifically immunoprecipitated the σ1 polypeptide neutralized type 3 infectivity. None of these monoclonal antibodies neutralized type 1 Lang or type 2 Jones. These data indicate that a specific region of the σ1 polypeptide is involved in neutralization of infectivity. Monoclonal antibodies that bind to it are good neutralizing antibodies, whereas monoclonal antibodies that bind to other regions are either poorly neutralizing or nonneutralizing.

Studies of HI have been performed with heteroimmune and monoclonal antibodies (Weiner *et al.*, 1978; Hayes *et al.*, 1981; Burstin *et al.*, 1982). Because hemagglutination is serotype-specific and the *S1* gene product is the viral hemagglutinin (Eggers *et al.*, 1962; Weiner *et al.*, 1978) (see also Section VI.A), it was anticipated that the *S1* dsRNA segment would be responsible for type specificity as tested by HI using type-specific antibody. HI tests using type 1 and type 3 heteroimmune antisera and intertypic reassortant viruses revealed that the *S1* gene product is indeed the type-specific antigen responsible for conferring specificity in HI tests (Weiner *et al.*, 1978).

In a detailed analysis of HI activity of monoclonal antibodies directed against the type 3 Dearing σ1 protein, Burstin *et al.* (1982) found that only certain of the monoclonal antibodies demonstrated HI activity against type 3. Interestingly, all but one of these monoclonal antibodies had little or no neutralizing activity. Similarly, monoclonal antibodies that very efficiently neutralized type 3 infectivity had no HI activity. Therefore, analysis of the behavior of monoclonal antibodies in HI and neutralization tests identified at least three antigenically distinct regions on the σ1 polypeptide: two monoclonal antibodies (A2 and G5) recognize a major site involved in neutralization of infectivity; a second group recognizes a site that blocks hemagglutination; a third antibody results in an interaction that leads to low levels of neutralization and HI; a fourth set has no detectable neutralization or HI activity. Taken together, these experiments imply that the σ1 polypeptide is divided into functionally distinct domains and that the major neutralization domain is distinct from the one involved in binding to erythrocytes.

In another analysis of HI using monoclonal antibodies directed against type 3 Dearing polypeptides, Hayes et al. (1981) found that antibodies to σ3 and λ2, as well as to σ1, exhibited type-specific HI activity. However, the HI activities of monoclonal antibodies directed against σ3 and λ2 were much less than those directed against σ1. Furthermore, similar to Burstin et al. (1982), they found that not all anti-σ1 monoclonal antibodies exhibited HI activity. Three monoclonal antibodies to σ1 differed in their relative capacity to inhibit hemagglutination, but all three could inhibit hemagglutination. The other antibody had a neutralization titer similar to the other three, but an HI of only 1% of those of the other monoclonal antibodies.

## B. Cellular Immunity

### 1. Cytotoxic T Lymphocytes

Cytotoxic T lymphocytes (CTL) are thought to play an important role in the immune response to viral infections, although their precise role is not clear (Zinkernagel and Welsh, 1976; Yap et al., 1978). CTL specific for viruses can be detected 3–5 days postinfection in mice and precede the antibody response (Doherty and Zinkernagel, 1976). These CTL show two types of specificity. CTL specific for one virus will not lyse target cells infected with unrelated viruses. Furthermore, for lytic interactions to occur, the virus-infected targets must express the same H-2K or H-2D glycoproteins of the major histocompatibility complex (MHC) as the CTL.

Following intraperitoneal inoculation of reoviruses into adult mice, reovirus-serotype-specific CTL are generated (Finberg et al., 1979). To demonstrate CTL activity, the spleen cells from an infected animal must be stimulated with reovirus-infected syngeneic cells in vitro prior to incubation with virus-infected target cells. Spleen cells from infected animals do not lyse infected target cells when assayed without restimulation with infected syngeneic cells. The degree of secondary cytotoxicity induced under these conditions is dependent on the primary dosage of reovirus given in vivo. As more virus is given in the primary immunization of adult mice, increasing cytotoxicity is observed. Furthermore, stimulatable secondary CTL persist for up to 8 weeks after primary infection.

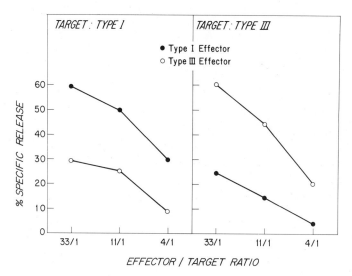

FIGURE 4. Serotype specificity of CTL generated in response to reovirus infection. C3H mice were inoculated with reovirus type 1 or 3 as described in the text. Spleen cells from infected animals were stimulated *in vitro* with syngeneic stimulator spleen cells infected with the same serotype used for *in vivo* infection. Effectors [(●) type 1; (○) type 3] were harvested after 4 days *in vitro* and assayed on type 1 (*left*) or type 3 (*right*) infected syngeneic macrophages. From Finberg *et al.* (1979), by permission of the *Proceedings of the National Academy of Sciences*.

The population of CTL generated under these conditions exhibits considerable serotype specificity (Fig. 4). Following infection of animals with reovirus type 1, there is significantly more lysis of type 1–infected peritoneal exudate cell (PEC) targets than of type 3–infected PEC targets at equivalent effector/target cell ratios. Likewise, secondary CTL derived from animals infected with type 3 lyse type 3–infected target cells better than type 1–infected target cells (Fig. 4).

Similar to other CTL, the reovirus-specific CTL exhibit specificity for *H-2K* or *H-2D* antigens. Maximal lysis occurs on syngeneic target cells infected with the appropriate virus.

Because reovirus-specific CTL exhibit serotype specificity, genetic experiments using reassortant viruses were performed to determine which viral protein is the antigen recognized by CTL. In such studies, mice were infected with type 1, type 3, or reassortant viruses. Spleen cells from the infected mice were restimulated with the virus used for *in vivo* infection. The secondary CTL that were generated were assayed on target cells infected with the viral serotype used for *in vivo* infection, with the other serotype, or with reassortant viruses. Maximal cell lysis occurred when the virus used to generate CTL and the virus used to infect target cells had the same *S1* dsRNA segment. Thus, the antigen encoded by the *S1* dsRNA segment, the σ1 polypeptide, not only determines serotype specificity in humoral immunity, but also determines serotype specificity of CTL.

Lower but significant levels of lysis occurred when the viruses used to infect target cells and to generate CTL did not have the same *S1* dsRNA segment. This degree of cross-reactivity could be explained in several ways: (1) A subpopulation of CTL could interact with the inappropriate σ1 protein through recognition of antigenic features common to the σ1 proteins of different serotypes. In a manner analogous to that described in the analysis of specificity of monoclonal antibodies directed toward specific regions of the σ1 protein, different subpopulations of CTL could recognize different regions of the σ1 protein. (2) A subpopulation of CTL may interact with viral antigens other than the σ1. (3) CTL may interact with the inappropriate σ1 protein weakly through antigenic features similar but not identical to those of the appropriate σ1. These weak interactions could sometimes lead to lysis. This cross-reactivity is susceptible to analysis by studies of populations of cloned reovirus-specific CTL.

The finding that the *S1* dsRNA segment is the predominant viral gene determining serotype specificity of CTL suggests that the viral protein encoded by this gene is recognized on the surface of target cells. For lysis to occur, CTL must recognize two molecular features on the surface of the target cell: a foreign antigen and a product of the MHC (Doherty and Zinkernagel, 1976). For enveloped viruses such as influenza and Sendai, one can easily envision how viral antigens can be recognized on the cell surface by CTL. Enveloped viruses incorporate large amounts of viral proteins into the infected cell membrane before the final stages of viral budding (Rifkin and Quigley, 1974). Lipid-containing viruses, such as Sendai, after irradiation, can "coat" the surface of cells with viral proteins (Schrader and Edelman, 1977). Reoviruses, however, are clearly different from enveloped viruses in that reoviruses are naked icosahedral viruses that do not bud from the cell surface and contain no lipid membrane.

How, then, do CTL recognize reovirus proteins on the cell surface? Immunofluorescence studies of infected target cells have revealed faint but unequivocal staining on the surface of infected cells (unpublished results). Whether these cell-surface antigens represent virus particles or proteins absorbed onto the cell surface or viral proteins synthesized within the infected cells and inserted into the cell membrane is not known. However, CTL recognition of virus particles absorbed onto a cell surface or of viral antigens expressed on an infected cell surface could be important in the control of viral infection. Through recognition of virions absorbed onto a cell surface, CTL could lyse cells before viruses have had a chance to replicate within the cell and produce progeny viruses, thereby halting infection. The killing of virally infected cells expressing viral antigens on their cell surfaces could also reduce the amount of virus produced.

To test the hypothesis that the σ1 protein is recognized by CTL on the surface of infected cells, the effect of serotype-specific antisera on CTL–target cell interactions was studied. The data indicated that serotype-specific antisera can block CTL–target interactions. These studies, however, cannot rule out the possibility that viral antigen could be capped and shed as a result of antibody cross-linking and thus be removed from target cells prior to their incubation with CTL.

Using recombinant viruses, the serotype-specific blocking of CTL by type-specific antisera was shown to depend on the type of the σ1 protein present in infected cells (Finberg *et al.*, 1981). Reovirus type 3 antiserum blocked the activity of type 3–specific CTL on target cells that had been infected with recombinants that had the type 3 *S1* dsRNA segment. Reovirus type 1 antiserum did not block the activity of type 3–specific CTL on target cells infected with recombinants that had the type 3 *S1* dsRNA segment. Similar results were obtained in the reciprocal experiments. Because the serotype specificity of reovirus antiserum is determined by the σ1 polypeptide and because the genetic studies in the antibody-blocking experiments utilized recombinants that differed only in the σ1 polypeptide, these experiments suggest that the σ1 protein is present on the surface of infected cells. These experiments, however, do not elucidate the nature of the σ1 antigen on the cell surface.

Further characterization of the cellular interactions that lead to the *in vitro* generation of secondary reovirus-specific CTL revealed that immune-associated antigen (Ia)-bearing adherent antigen-presenting cells (APC) stimulate the generation of secondary reovirus-specific CTL (Letvin *et al.*, 1981). When reovirus-specific splenic T cells are incubated with virus-infected PEC or splenic adherent cells (SAC) as stimulator cells, the generation of reovirus-specific CTL is much more efficient than when virus is presented by whole spleen cells. Depletion of the adherent cells from the APC population by the passage of whole spleen cells over a G10 column markedly reduces the efficiency with which such cells stimulate reovirus-specific CTL. Furthermore, when the stimulator SAC population is treated with anti-IA antigen and complement, there is a marked reduction in the efficiency with which the remaining cells stimulate antireovirus CTL. Similarly, the generation of reovirus-specific CTL by splenic T cells is dramatically reduced when the secondary stimulation *in vitro* occurs in the presence of monoclonal anti-Ia antibody of the same haplotype as the splenocytes.

To study the role of Ia-bearing adherent cells in the reovirus-infected mouse, various parameters of virus-specific T-cell function were compared in normal and UV-irradiated mice. Previous studies had demonstrated that daily exposure of a mouse to UV radiation produces a systemic depletion of APC (Greene *et al.*, 1979). UV irradiation of mice leads to a significant decrease in priming for the *in vitro* generation of secondary antireovirus CTL (Letvin *et al.*, 1981). To determine whether this depression reflects a selective APC defect, Letvin and co-workers examined the capacity of splenic T lymphocytes from irradiated mice to be primed for a reovirus-specific CTL function in the presence of normal APC. The results indicated that splenic T lymphocytes from UV-irradiated mice can be primed normally when presented antigen by cells from normal mice but not by cells from UV-irradiated mice. Furthermore, these studies confirm that Ia-bearing adherent cells are essential for the production of secondary CTL.

The studies described up to this point have addressed the cellular requirements and the specificity requirements for *in vitro* generation of secondary virus-specific CTL. These experiments have not explored the role of CTL or other immune cells in the resolution of a primary or secondary viral infection. In a study of the roles of immune cells in primary reovirus infection, the clearance of virus by normal mice, UV-irradiated mice, and "nude" athymic mice was compared (Letvin *et al.*, 1981). UV-irradiated mice could clear virus as well as normal mice whether virus was given by the intraperitoneal or the peroral route of infection. Viral titers in the spleen, mesenteric lymph nodes, and Peyer's patches of UV-irradiated and normal mice were equivalent 1 week postinfection. Since the defect in UV-irradiated mice is in APC, which have been shown to be necessary for the generation of secondary CTL, these data suggest that APC and CTL are not essential for clearance of a primary reovirus infection (Letvin *et al.*, 1980). Nude mice clear reovirus at least as well as, if not better than, normal mice. Defective in T-cell activity, nude mice exhibit increased natural killer (NK) cell activity compared to normal mice (Heberman *et al.*, 1975) and nonspecifically activated macrophages (Zinkernagel and Blanden, 1975). Whether phagocytic non-Ia-bearing cells or NK cells play a role in reovirus clearance, either through the production of a mediator such as interferon or by direct lysis of virally infected cells, remains to be seen.

## 2. Delayed-Type Hypersensitivity

Delayed-type hypersensitivity (DTH) is a cell-mediated immune reaction that can be elicited by subcutaneous injection of antigen, which results in a cellular infiltrate and edema that are maximal between 24 and 48 hr after antigen challenge. T cells that participate in DTH reactions ($T_{DTH}$) have been demonstrated in several viral systems by challenging infected animals with the virus to which they have been immunized (Zinkernagel, 1976; Lagrange *et al.*, 1978; Hudson *et al.*, 1979; Leung *et al.*, 1980).

DTH specific for reoviruses can be demonstrated in adult mice infected subcutaneously with reovirus and challenged in the footpad with reovirus (Weiner *et al.*, 1980b). The DTH response, assayed by maximal footpad swelling, occurs when $10^5–10^9$ viral particles are used as an immunizing dose and $5 \times 10^6$ particles are used as a challenge dose. In this system, footpad swelling is observed when animals are challenged at 5–9 days after subcutaneous injection of virus, with maximal swelling seen with challenge on day 7.

The DTH response to reovirus is serotype-specific. When mice are injected subcutaneously with type 1 reovirus and challenged with type

1 or type 3, a DTH response is seen only when animals are challenged with type 1. Likewise, when animals are immunized subcutaneously with type 3, a response is seen only when animals are challenged with type 3. To determine the viral protein responsible for eliciting a serotype-specific DTH response to reovirus, the DTH response of animals immunized with or challenged with viral recombinants was examined. A DTH response was elicitable when the parental serotype or recombinant virus used for challenge had the same S1 dsRNA segment as the parental serotype or recombinant virus used for subcutaneous immunization. All other viral dsRNA segments were ruled out as being major determinants of DTH serotype specificity. Thus, serotype-specific CTL and $T_{DTH}$ appear to have similar specificities for viral antigen. A low level of heterotypic response was observed, analogous to the cross-reactive lysis seen in serotype-specific CTL for reovirus.

The DTH response to reovirus can be transferred to naïve recipient mice by lymph-node cells obtained from a mouse 5 days after subcutaneous injection. The magnitude of the response is approximately 60% of that seen in the primary host. The immune cells responsible for the transfer of DTH are T cells, since the ability to transfer the DTH response is abrogated by treatment of donor lymph-node cells with anti-Thy 1.2 and complement *in vitro* prior to transfer into recipient mice. These immune cells transfer serotype-specific DTH; there is no response when animals that receive type 1 immune lymph-node cells are challenged with reovirus type 3. Similar to the adoptive transfer of DTH in other systems (Zinkernagel, 1976), the transfer of reovirus-specific DTH is restricted by the MHC. When mice of one inbred strain are infected subcutaneously with type 1 reovirus and the immune lymph-node cells are transferred to recipients of a variety of inbred strains, identity between *K*, *IA-IB*, and/or *D* regions of the *H-2* complex is found to be sufficient for the successful transfer of DTH.

The route of inoculation and viral infectivity also play a role in determining the ability of mice to generate immunocompetent T cells that are able to mediate *in vivo* reovirus-specific DTH (Greene and Weiner, 1980). In a series of experiments, the effects of inoculating both live and UV-inactivated virus by the subcutaneous, intraperitoneal, and intrave-

TABLE II. Summary of Effect of Route of Viral Administration on the Production of Delayed-Type Hypersensitivity

| Route of injection | Type 1 | | Type 3 | |
|---|---|---|---|---|
| | Live | UV-irradiated | Live | UV-irradiated |
| Intraperitoneal | DTH | DTH | DTH | DTH |
| Intravenous | DTH | Tolerance | DTH | Tolerance |
| Subcutaneous | DTH | DTH | DTH | DTH |
| Peroral | — | Tolerance | — | DTH |

nous routes were compared (Table II). In these experiments, $10^7$ particles of live or UV-inactivated virus were inoculated into mice by each of the aforementioned routes. All mice were challenged in the footpad 6 days later with live virus. Both live and UV-inactivated virus primed efficiently by the subcutaneous route. Both primed, though less efficiently, by the intraperitoneal route as compared to the subcutaneous route. However, live virus but not UV-inactivated virus primed by the intravenous route. The failure to induce DTH after priming with UV-inactivated virus could be due to active suppression of the DTH response or to the failure of intravenously injected UV-inactivated virus to induce $T_{DTH}$.

## 3. Immune Suppression

Immune suppression is a T cell–mediated activity in which multiple T-cell populations and soluble factors are required. Interactions among these T-cell subsets can result in the suppression of antibody synthesis by B cells or the inhibition of cellular immune reactions by effector T cells. For example, in extensively characterized systems, for the suppression of the immune response to sheep erythrocytes or haptens, at least three distinct T-cell subsets are needed (Eardley et al., 1979; Greene et al., 1982). Virus-specific $T_S$ have been shown to be generated during the immune response to viruses (Liew and Russell, 1980). The role of $T_S$ in virus–host interaction remains to be determined.

Reovirus-specific $T_S$ are generated following the intravenous administration of UV-inactivated reovirus (Greene and Weiner, 1980). The finding that a virus-specific DTH response failed to occur after intravenous inoculation of UV-inactivated reovirus suggested that intravenously administered UV-inactivated virus either did not induce DTH or elicited the generation of virus-specific $T_S$. To distinguish between these hypotheses, UV-inactivated reovirus type 1 was inoculated intravenously 1 hr after immunization with live virus. The intravenous administration of UV-inactivated virus completely abolished the DTH response to subcutaneously administered virus. The suppression of DTH occurred most reproducibly at a dose of $10^9$ UV-inactivated particles and could be induced by the administration of UV-inactivated virus as early as 14 days prior to subcutaneous inoculation with live virus. Giving UV-inactivated virus at the time of challenge did not prevent the DTH response.

The lack of DTH response (tolerance) following intravenous administration of UV-inactivated virus is serotype-specific. To determine the basis of this specificity, groups of mice were immunized subcutaneously with live reovirus type 1 or type 3, given either UV-inactivated reovirus type 1 or type 3 intravenously, and challenged with the viral serotype used for subcutaneous immunization. Animals immunized with reovirus type 1 were tolerized by the intravenous administration of UV-inactivated type 1 but not UV-inactivated type 3. The reciprocal specificity was also observed. A genetic analysis of this serotype-specific tolerance for DTH revealed that the viral gene responsible for serotype-specific unresponsiveness was the *S1* dsRNA segment. Animals immunized with type 1, given UV-inactivated type 1, type 3, or recombinant viruses intravenously, and challenged with type 1 failed to exhibit a DTH response when the intravenously administered UV-inactivated virus contained the type 1 *S1* dsRNA segment, regardless of the other nine dsRNA segments. Thus, the same viral determinant is responsible for serotype specificity in DTH, CTL, and neutralizing-antibody responses as well as tolerance.

Adoptive-transfer experiments demonstrated that the inability to develop a DTH response following intravenous administration of UV-inactivated virus was due in part to the generation of virus-specific T cells. When spleen cells from mice given UV-inactivated reovirus type 1 intravenously 5 days previously were transferred to syngeneic recipients that had been immunized subcutaneously with type 1, the DTH response was reduced following challenge with type 1 6 days later. Spleen cells from mice tolerized to reovirus type 1 did not affect the DTH response in animals immunized and challenged with type 3. Therefore, transfer of tolerance was serotype-specific. In addition, the cells responsible for the transfer of tolerance were T cells, since treatment of spleen cells from tolerized animals with anti-$\Theta$ and complement prevented the transfer of tolerance. Thus, several T-cell subsets (CTL, $T_{DTH}$, and $T_S$) have biological specificity for the $\sigma$1 polypeptide, suggesting that the binding site specific for $\sigma$1 is similar in these T-cell subsets.

In addition to suppressing DTH response, intravenous administration of UV-inactivated reovirus also impairs antibody response (Weiner and Greene, unpublished results). Intravenous administration of live reovirus produces a high-titer antibody response, whereas intravenous administration of UV-inactivated virus provokes a minimal response. Simultaneous administration of UV-inactivated virus intravenously and live virus subcutaneously results in a greater than 80% reduction of antibody production as compared to administration of live virus intravenously and subcutaneously. Similar studies on the generation of CTL in the presence of intravenously administered UV-inactivated virus suggest that intravenous administration of UV-inactivated virus inhibits the generation of CTL activity (Greene and Weiner, unpublished results).

Oral administration of UV-inactivated reovirus type 1, but not type 3, also suppresses the development of DTH responses when animals are primed with live virus and challenged 6 days later in the footpad (Rubin *et al.*, 1981). This suppression of the DTH response is serotype-specific, since oral feeding of UV-inactivated type 1 does not suppress the DTH response in animals immunized subcutaneously and challenged with type 3 reovirus. A genetic analysis of this serotype specificity using recombinant viruses has revealed that the *S1* dsRNA segment determines the serotype-specific suppression. The suppression of the DTH response is also dependent on the dose of UV-inactivated type 1 given orally. At doses of $10^3$ and $10^9$ particles, the DTH response is significantly reduced to 68 and 62% of control valves. At intermediate doses ($10^5$–$10^7$ particles), minimal suppression of DTH occurs (28%). Furthermore, live type 1 given orally does not suppress the DTH response.

The suppression of the development of DTH is transferable to naïve recipients by cells from either the spleen or the mesenteric lymph nodes of animals given UV-inactivated type 1 by the oral route. The cells responsible for the transfer of this suppression are T cells, since treatment of spleen cells or mesenteric lymph node cells with anti-Θ plus complement prior to transfer abolishes the transfer of suppression. Furthermore, transfer of this suppression is serotype-specific and a property of the *S1* dsRNA segment, as determined by studies of transfer of cells from tolerized mice to mice immunized subcutaneously and challenged with type 1 or recombinant viruses.

Oral administration of UV-inactivated type 3 does not suppress the DTH response (Rubin *et al.*, 1981). To determine the genetic basis for this inability to elicit suppression of DTH, mice were immunized subcutaneously with type 3 reovirus, given UV-inactivated type 1 or recombinant viruses orally, and then challenged with type 3 reovirus 6 days later. Oral administration of UV-inactivated type 3 or recombinants that had the type 3 *M2* dsRNA segment did not suppress DTH, whereas oral administration of recombinants that had the *M2* dsRNA segment from type 1 could elicit suppression. The presence of the *M2* dsRNA segment of type 3 abrogated the capacity of reovirus to suppress DTH responses after oral inoculation, regardless of the genotype of the *S1* dsRNA segment. Thus, at least two viral genes (*M2* and *S1*) are involved in producing a DTH response. It is possible that the sensitivity of the type 3 virus to degradation by proteases (which is a property of the *M2* dsRNA) is related to the inability of type 3 to induce suppression of DTH after oral administration. Virus could be degraded and not presented to the immune system.

The oral administration of antigens has been shown by a number of investigators to suppress the immune response (Hanson *et al.*, 1979; Mattingly and Waksman, 1978). The reason that oral administration of inactivated type 1 but not live type 1 can induce cellular immune hyporesponsiveness is not known. The finding that low concentrations of inactivated virus can produce striking inhibition of the DTH response suggests that reovirus may come into direct contact with the gut-associated lymphoid tissue. Whether the site of primary viral replication or systemic spread can affect the immune response to reovirus remains to be seen.

Fontana and Weiner (1980) have described a phenomenon whereby reovirus type 3 will inhibit the concanavalin A (Con A)–induced proliferation of spleen cells. This inhibition is dose-dependent and occurs with live and UV-inactivated type 3. Type 1 reovirus causes minimal inhibition. Genetic mapping studies have identified the type 3 *S1* dsRNA segment as the viral gene responsible for this inhibitory effect. This inhibition can be transferred by T cells. Fontana and Weiner interpret these results to mean that the inhibition is being mediated by $T_S$. However, such global inhibition of a mixed population of lymphocytes is not shown by other $T_S$. It seems likely that reovirus stimulates an undefined T-cell subset to produce a nonspecific inhibitor of proliferation.

## C. Interferon

Interferons are a family of proteins secreted by body cells in response to viruses as well as other stimuli. At least three distinct types of interferon can be produced, depending on the cell type and type of stimulus. These are leukocyte interferon ($\alpha$), fibroblast interferon ($\beta$), and immune interferon ($\gamma$) (Pestka and Baron, 1981). Interferons stimulate surrounding cells to produce other proteins that can regulate viral multiplication, the immune response, cell growth, and other functions. Interferon stimulates the maturation of NK cells, which may serve as a primary means of defense against viruses. Interferon is the most rapidly produced of the known body defenses against viruses. It is produced within hours of viral infection and continues to be produced throughout infection.

Several studies have examined the ability of reovirus to induce interferon production (Long and Burke, 1971; Oie and Loh, 1971). To gain insight into the mechanism involved in the induction of interferon, Lai and Joklik (1973) examined the ability of prototype mutants of six genetic groups to induce interferon production in mouse fibroblasts. The results of these studies indicated that each of the mutants can induce a quantity of interferon similar to wild-type 3 at the permissive temperature (31°C), but that they induce a markedly reduced amount at the nonpermissive temperature (39°C). There is no correlation between interferon induction and the ability of the mutants to synthesize viral dsRNA, single-stranded RNA, and protein. The concentration of interferon induced appears to be proportional only to the yield of infectious virus. Thus, this study failed to identify which viral function or functions are essential for interferon induction, but suggested that the induction process is initiated by a late event in the replication of reovirus or by the presence of mature progeny virions.

The ability of UV-inactivated reovirus to produce interferon has also been examined (Long and Burke, 1971; Lai and Joklik, 1973). The mechanism by which UV-irradiated reovirus induces interferon differs from that which occurs during a productive infection. UV irradiation decreases the interferon-inducing capacity of wild-type reovirus, but dramatically enhances the ability of a group C temperature-sensitive (ts) mutant, which is essentially unable to synthesize progeny dsRNA at 39°C, to induce interferon at 39°C, a temperature at which the wild-type virus is a poor inducer. Henderson and Joklik (1978) have investigated the mechanism by which tsC becomes such a good interferon inducer and concluded that UV irradiation labilizes the inner shell of the reovirus capsid and causes some of the dsRNA in parental virus particles to be liberated into the interior of the cell. These liberated dsRNA molecules are thought to induce interferon.

A comparison of the ability of different serotypes to induce fibroblast interferon has revealed that type 3 reovirus induces interferon in mouse L cells, whereas type 1 reovirus produces little or no interferon in these cells (Sharpe and Fields, unpublished results). A genetic analysis of the ability of reovirus to induce interferon has suggested that several viral genes are needed for reovirus to induce interferon (Brown, Sharpe, and Fields, unpublished results).

Reovirus is quite sensitive to the action of interferon (Wiebe and Joklik, 1975). To determine the stage in replication that is inhibited by interferon, a $ts$C mutant that is incapable of synthesizing progeny dsRNA and late messenger RNA (mRNA) at 39°C was used to study effects of interferon on early mRNA transcription and translation. These studies have indicated that transcription of early mRNA is only slightly sensitive to interferon, while translation of early mRNA is inhibited more dramatically. Late viral functions, such as the formation of progeny dsRNA, the transcription of late mRNA, and the formation of infectious virus are also suppressed. Because the $\lambda 1$ protein is a component of the dsRNA replication complex, on the formation of which the expression of late viral function depends, it has been suggested that the primary target for inhibition of replication of reovirus in interferon-treated cells is the translation of early mRNA and specifically the species that encodes the $\lambda 1$ polypeptide.

## D. Thymosin

A wide variety of immunological properties have been associated with thymosin, a family of polypeptides isolated from bovine calf thymus. Thymosin can restore T-cell immunity in thymectomized mice. It induces the expression of Thy1-positive spleen cells, anti-sheep red blood cell plaque-forming cells, T-cell mitogen responses, and increased graft vs. host activity in nude mouse thymocytes. Recent evidence suggests that thymosin can induce the appearance of Thy1-positive $T_S$ in nude mouse spleens (Marshall et al., 1981).

The finding that thymosin could reduce wasting disease and death in thymectomized mice afflicted with runting syndrome prompted Willey and Ushijima (1980, 1981) to question whether thymosin could alleviate the lethal runting syndrome that occurs in neonatal mice injected with reovirus type 3 (Walters et al., 1963). Previous studies had shown that the T cell plays a pivotal role in host-cell resistance to reovirus infection, since antithymocyte and antilymphocyte serum could restore the susceptibility of weanling mice to fatal infection with reovirus (Ida and Hinuma, 1971).

Willey and Ushijima found that thymosin administration to neonatal Swiss Webster mice that had been injected intraperitoneally with reovirus type 3 increased the survival rate of these mice. "Thymosin therapy" showed a circadian rhythm: if thymosin was given at 12:00 hr, no change in survival was observed, whereas if thymosin was given at 22:00 hr, an enhanced rate of survival was seen. Willey and Ushijima investigated the effect of thymosin on lymphocytes under the conditions of enhanced survival. They found that although thymosin administration increased T-cell maturation processes in normal mice, it did not increase T-cell responses, as determined by T-cell transformation assays with Con A or phytohemagglutinin P and DTH to oxazolone sensitization in the infected mice the survival of which had been enhanced. Further studies are needed to determine the mechanism by which thymosin enhances the survival of infected mice. It is possible that (1) the assays used thus far did not measure changes in T-cell subpopulations after thymosin injection; (2) thymosin stimulated a population of suppressor cells that inhibited T cells with other functions; (3) thymosin stimulated a population of immature T cells that would not respond in these assays; (4) thymosin stimulated the production of soluble factors; or (5) other lymphoid tissues (thymus, lymph nodes) are needed for the assay of thymosin effects on T-cell populations.

Willey and Ushijima observed that thymosin resulted in a decrease in the relatively high plasma corticosteroid levels present in infected mice. Corticosteroids have been shown to suppress immune response, to reduce the size of the thymus, and to depress the maturation sequence of thymocytes. Whether this decrease in corticosteroid levels is related to enhanced development and function of thymic lymphocytes remains to be determined.

## VI. VIRUS–HOST CELL INTERACTIONS

### A. Tissue Tropism: Virus–Receptor Interactions

The specific tropism of a particular virus for a certain host species and for different tissues in the host is an important factor determining viral virulence. As a consequence of differing tropisms, viruses produce differing patterns of disease. Tropism appears to be related to the interaction of viral surface structures with specific receptors on the surface of host cells. Such virus–receptor interactions result in the penetration of virus into the host cell and subsequent cell damage.

The mammalian reoviruses offer a fruitful system for the study of tissue tropism and virus–receptor interactions. The different serotypes interact differently with a variety of host cells, producing different patterns of disease. Furthermore, reovirus genetics provides a powerful approach to define the molecular basis for the differences in cell tropism and virulence. As a result of a number of genetic analyses, one viral outer capsid protein, the σ1 polypeptide, has been implicated as the determinant of cell and tissue tropism, through interactions with receptors on the surface of cells. Current studies focus on the precise nature of the interaction between the σ1 polypeptide and cell-surface receptors.

The first indication of the role of the σ1 polypeptide in cell-surface interactions came from the study of the interactions of the mammalian reoviruses with erythrocytes. A wide variety of animal viruses have the capability of binding to erythrocytes and causing them to agglutinate. This binding and agglutination, called hemagglutination, results from the interaction of a viral polypeptide with an erythrocyte surface receptor (Lerner *et al.*, 1963). With the various serotypes of reovirus, hemagglutination is type-specific. Reovirus type 1 agglutinates human erythrocytes, whereas type 3 agglutinates bovine erythrocytes (Eggers *et al.*, 1962). An examination of the hemagglutinating properties of recombinants derived from crosses between reovirus type 1 and type 3 revealed that the *S1* double-stranded RNA (dsRNA) segment determines the specificity of hemagglutination. The type of erythrocyte agglutinated by a recombinant virus was the same as that agglutinated by the parental serotype that contributed the *S1* segment to the recombinant (Weiner *et al.*, 1978). The parental origin of the other nine dsRNA segments did not affect the hemagglutination pattern. Thus, these studies indicated that the σ1 polypeptide is the viral hemagglutinin, the viral protein that interacts with an erythrocyte surface receptor.

Further evidence for the role of the σ1 polypeptide in virus–receptor interactions came from the study of reovirus tropism in cells of the CNS. Reovirus types 1 and 3 differ in their tropism for cells in the CNS and, presumably as a result of this difference, produce different patterns of virulence. Intracerebral inoculation of reovirus type 3 into newborn mice produces an acute, fatal encephalitis that is accompanied by the destruction of neuronal cells without damage to ependymal cells (Fig. 5). Neurons throughout the cerebral cortex are involved, but especially notable is massive involvement of the pyramidal layer of the hippocampus as well as an intermediate degree of severity in the basal ganglia and nuclei of the pons and medulla (Margolis *et al.*, 1971; Raine and Fields, 1973). When reovirus type 1 is inoculated intracerebrally, it produces a nonfatal infection involving ependymal cells with little or no effect on neurons (Kilham and Margolis, 1969). As a consequence of ependymal damage, hydrocephalus often develops.

Studies of the disease patterns in animals injected intracerebrally with recombinant viruses containing dsRNA segments from type 1 and type 3 revealed that the *S1* dsRNA segment is responsible for the differing neurotropisms of type 1 and type 3 (Weiner *et al.*, 1977). Reassortant

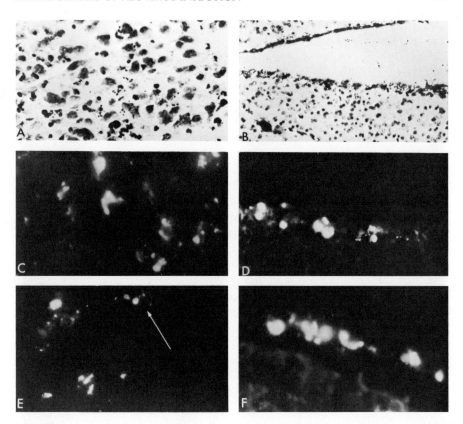

FIGURE 5. Brain tissue from mice injected with reovirus serotype 1 or 3 or single-segment recombinant clones 1.HA3 or 3.HA1, illustrating the CNS tropism of reoviruses. Clone 1.HA3 contains the *S1* dsRNA segment of type 3 and the other nine dsRNA segments of type 1. Clone 3.HA1 contains the *S1* dsRNA segment of type 1 and the other nine dsRNA segments of type 3. (A) Extensive neuronal necrosis in the temporal lobe, a pattern seen with both type 3 and clone 1.HA3. Hematoxylin–eosin. × 400. (B) Greatly enlarged ventricle with preservation of neuronal architecture, a pattern seen with both type 1 and clone 3.HA1. Hematoxylin–eosin. × 160. (C, E) Temporal lobe showing viral antigen in brain parenchyma, a pattern seen with both type 1 and clone 1.HA3. Immunofluorescence micrograph. × 630. The arrow in (E) identifies a neuronal cell body containing viral antigen in the axonal process. (D, F) Viral antigen in ependymal cells lining lateral ventricular wall cavities, a pattern seen with both type 1 and clone 3.HA1. Immunofluorescence micrograph. × 630.

viruses containing nine dsRNA segments from type 1 and the *S1* dsRNA segment from type 3 produced neuronal destruction and grew to high titers in the brain ($>5 \times 10^9$), similar to the type 3 parent (Weiner *et al.*, 1980c). Reassortants that had nine dsRNA segments from type 3 and the *S1* dsRNA segment from type 1 produced destruction of ependymal cells and grew to titers of $10^5–10^6$ in the brain, similar to the wild-type 1 parent. Immunofluorescent studies demonstrate that reovirus antigens are in ependymal cells of animals infected with reoviruses containing the type 1 *S1* dsRNA segment and in neurons of animals infected with reoviruses containing the type 3 *S1* dsRNA segment (Fig. 5). Thus, the *S1* dsRNA segment determines cell tropism within the CNS.

Further evidence for the critical role of the σ1 protein in neurovirulence came from the isolation and characterization of reovirus type 3 antigenic variants (Spriggs and Fields, unpublished results). Stable σ1 variants were isolated by selecting viral clones resistant to neutralization by a monoclonal antibody directed at the major neutralization site of σ1. When these variants were inoculated intracerebrally into suckling mice, all the variants were at least 4 $\log_{10}$ less neurovirulent than the type 3 Dearing strain from which they had been derived. Certain of the variants were found to be essentially avirulent ($LD_{50} > 10^7$ PFU) and impaired in their capacity to grow in the brains of mice. These data directly indicate that the type 3 σ1 protein is a major determinant of neurovirulence and further imply that particular regions of σ1 play a critical role in virulence. The neutralizing region of the σ1 protein appears to be important for σ1 binding to cells as well as to function as a target for immunological interactions. Hence, a specific domain of the σ1 protein has the specificity required to interact with cell receptors.

Further insight into the viral components that determine neurovirulence has come from the study of the relative virulence of field isolates of reovirus type 3 after intracerebral inoculation (Hrdy *et al.*, 1982). One isolate was found that had decreased neurovirulence; it has an $LD_{50}$ of greater than $10^4$ PFU/ml as opposed to the $LD_{50}$ of greater than $10^1$ for the Dearing strain. Although this isolate caused neuronal cell damage, similar to other type 3 isolates (and was thus "neurotropic"), it grew to only 1% of the level of type 3 Dearing in mouse brains. To determine the reason for the avirulence of this isolate, a series of recombinants were prepared from genetic crosses between the avirulent type 3 strain and T1 Lang or between the avirulent type 3 strain and T3 Dearing. An analysis of the capacity of these reassortants to grow in mouse brains and kill suckling mice revealed that the *M2* dsRNA was responsible for this property (Table III). Thus, differences in virulence between a normally neurovirulent type 3 virus and an avirulent isolate of the same serotype and CNS tropism resides not in the *S1* dsRNA segment, but in the *M2* dsRNA segment. Hence, two viral genes have been found that play distinct and different roles in neurovirulence. The *S1* dsRNA segment determines cell and tissue tropism and the *M2* dsRNA segment determines the relative capacity to grow in the CNS as well as at mucosal surfaces (such as the gastrointestinal tract). These observations illustrate that virulence is multigenic, with individual genes playing distinct and different roles.

Studies of Notkins and colleagues have confirmed the role of the *S1* gene in determining CNS tropism and have, in addition, demonstrated that the *S1* dsRNA segment also determines cell tropism within the pituitary (Onodera *et al.*, 1981). In their study of the capacity of the mammalian reoviruses to produce diabetes and polyendocrine disease in newborn mice, these workers observed that approximately 25–50% of mice

TABLE III. Genotypes and Intracranial Virulence in Suckling Mice of T3 (H/Ta), Type 1 (Lang), Type 3 (Dearing), and Reassortants[a]

| Virus | Genotype | | | | |
| | Outer capsid[b] | | | LD$_{50}$ (PFU) | Virulent? |
| | M2 | S1 | S4 | | |
|---|---|---|---|---|---|
| T1 (Lang) | 1 | 1 | 1 | $>10^6$ | No |
| T3 (Dearing) | 3 | 3 | 3 | $1 \times 10^1$ | Yes |
| T3 (H/Ta) | Ta | Ta | Ta | $3 \times 10^4$ | No |
| T3 (Dearing) × T3 (H/Ta) | | | | | |
| W14 | 3 | 3 | 3 | $5 \times 10^1$ | Yes |
| W38 | Ta | Ta | 3 | $3 \times 10^4$ | No |
| W52 | Ta | 3 | Ta | $1 \times 10^4$ | No |
| W62 | 3 | 3 | Ta | $<10^1$ | Yes |
| W73 | 3 | Ta | Ta | $<10^1$ | Yes |
| T1 (Lang) × T3 (H/Ta) | | | | | |
| Q1 | 1 | Ta | 1 | $<10^1$ | Yes |
| Q9 | Ta | Ta | Ta | $3 \times 10^3$ | No |
| Q15 | Ta | Ta | Ta | $1 \times 10^3$ | No |
| Q25 | Ta | Ta | Ta | $1 \times 10^3$ | No |
| Q33 | 1 | Ta | Ta | $1 \times 10^1$ | Yes |

[a] From Hrdy et al. (1982), by permission of the Proceedings of the National Academy of Sciences.
[b] (1) Genome segment originating from T1 (Lang); (3) genome segment originating from T3 (Dearing); (Ta) genome segment originating from T3 (H/Ta).

infected with type 1 Lang developed a runting syndrome consisting of retarded growth, oily hair, alopecia, and steatorrhea. The growth retardation suggested that virus might be damaging cells in the pituitary. Indeed, microscopic studies showed that type 1 reovirus antigens were present in the anterior lobe of the pituitary and that reovirus particles were present in the growth hormone (GH)–producing cells. Radioimmunoassays (RIAs) revealed that the GH concentration in the blood of infected, runted mice was reduced. In contrast to type 1, type 3 Dearing failed to produce the runting syndrome or infect the anterior pituitary cells of mice. Both serotypes, however, could infect pancreatic islet cells and produce diabetes.

A genetic analysis of the difference in the ability of type 1 and type 3 reovirus to infect the pituitary revealed that the S1 dsRNA segment is the viral gene that determines viral tropism for the anterior pituitary. Type 1 reovirus infection of the pituitary also induces the formation of antibodies to GH and insulin, whereas type 3 reovirus fails to induce these antibodies. Genetic analysis of this difference also implicated the S1 dsRNA segment in the induction of antibodies to GH. The relationship between viral infection of cells that produce GH and the induction of antibodies to GH is unclear. Furthermore, whether virus-induced runting is related to the viral destruction of GH-producing cells or to an effect of virally induced antibodies to GH is not known. Nonetheless, it is interesting to note that in a different experimental murine system with reovirus, the *S1* dsRNA segment is responsible for determining cell tropism. It is quite likely that the reovirus σ1 protein determines cell tropism in other systems as well.

In another murine model system, however, no specific determinants of viral tropism or virulence have yet been determined despite a genetic analysis. Intraperitoneal inoculation of weanling mice with type 3 produces a chronic obstructive jaundice, associated with viral replication in bile-duct mucosal cells (Phillips *et al.*, 1969; Stanley and Joske, 1975b). In contrast, reovirus type 1 does not cause hepatobiliary disease in weanling mice. When viral reassortants were examined for their capacity to produce hepatobiliary disease (hepatitis, extrahepatic cholangitis) in weanling (21-day-old) mice, no single viral gene was shown to be responsible for this property (Bangaru *et al.*, 1980). In fact, not a single reassortant virus, even one that had a predominantly type 3 background, produced hepatobiliary disease, though no reassortants that had all outer-capsid genes from type 3 were examined. These results are difficult to interpret, but suggest that the reovirus model of hepatobiliary disease is quite complex. Further definition of the experimental system may be necessary before a genetic analysis is possible. Study of the routes of infection that result in hepatobiliary disease with reoviruses, the replication of reoviruses in biliary duct or hepatic cells, and the effect of bile fluid on viral infectivity may prove useful. A study of this system may be especially important in light of the similarities between hepatobiliary lesions in human newborns with biliary atresia and those induced in weanling mice by reovirus type 3. Furthermore, recent serological data suggest that reovirus type 3 or an antigenically related virus may be the etiological agent of human biliary atresia (R. Morecki and M. Horwitz, personal communication).

All the aforementioned properties of the σ1 protein imply that this protein has a role in recognizing and binding to receptors on the surface of cells. Although there are as yet no direct data regarding the nature of cellular receptors for reovirus, studies have been performed that indicate that the σ1 protein has a role in attaching or binding to cells or cellular components. Monoclonal antibodies specific for the σ1 protein prevent absorption of reovirus particles to mouse L-cell fibroblasts (Lee *et al.*, 1981b; Nepom *et al.*, 1982b). The free σ1 protein present in lysates of infected cells is the only viral protein capable of absorbing to cells (Lee *et al.*, 1981b). In addition, the σ1 protein competes with virus particles for receptors on the surface of mouse L fibroblasts. These competition experiments indicate that virions of each of the three serotypes interfere with the attachment not only of the homologous σ1, but also of heterologous σ1, suggesting that virions of all three serotypes absorb to the same receptor on mouse L cells. Thus, the three σ1 proteins, which are quite different both biologically and chemically (Li *et al.*, 1980; Gentsch and Fields, 1981), may react with the same receptor on the plasma membrane of mouse L cells. This suggests that the σ1 protein interacts with the L-cell receptor via a domain that has been conserved among the three serotypes.

Using purified reovirus particles as a probe for cell-surface receptors, it is possible to demonstrate that reovirus type 3 but not type 1 binds to murine and human lymphocytes via the σ1 protein (Weiner et al., 1980a). An analysis of the binding of viral reassortants to lymphocytes using a fluorescence-activated cell sorter showed that the type 3 S1 dsRNA segment is responsible for the binding of reovirus type 3 to lymphocytes. The σ1 binding site is present on both T cells and B cells, though found on a greater percentage of B cells than T cells. This binding site does not appear to be a structure encoded by the major histocompatibility complex, since type 3 is able to bind to a cell line, R1.E, that is devoid of histocompatibility antigens. Analysis of reovirus binding to other cell types has confirmed that L cells bind both serotypes. In addition, type 1 reovirus was found to bind to rat ependymal cells better than type 3, a finding that parallels the in vivo tropism of reovirus type 1 (Kilham and Margolis, 1969; Weiner et al., 1980a).

Further characterization of the receptor for reovirus on the surface of lymphocytes has revealed that reovirus type 3 binds to a surface structure that can be independently modulated on the cell membrane, similar to other known surface structures (Epstein et al., 1981). Capping of reovirus receptors occurs after the binding of virions and antiviral antibody with kinetics identical to that seen for immunoglobulin capping. The viral receptor can be removed by treatment of cells with proteolytic enzymes. Protein synthesis is required for its reexpression. Cocapping studies indicate that the reovirus receptor is not associated with immunoglobulin, Ia, constant fragment (Fc), C3, Thy1.2, or H-2.

The binding of reovirus to ependymal cells has also been examined. Tardieu and Weiner (1982) prepared single-cell suspensions of viable ependymal cells from the CNS of mice according to the technique of Manthrope et al. (1977) and examined reovirus binding to ependymal cells by immunofluorescent staining techniques. When ependymal cells were incubated with reovirus type 1 followed by fluorescent labeling with a rabbit antiviral antibody and fluorescein-isothiocyanate-labeled goat anti-rabbit immunoglobulin, surface fluorescence was observed. No staining was seen when the cells were incubated with type 3 reovirus under identical conditions. This result parallels the in vivo experiments demonstrating tropism for type 1 to ependymal cells. Using recombinant viruses, this binding property was mapped to the σ1 protein, analogous to the mapping of type 1 reovirus tropism for ependyma (Weiner et al., 1978) to the σ1 protein. Ependymal cells of newborn and adult mice could bind reovirus type 1, suggesting that the age-dependent hydrocephalus observed in animals (Kilham and Margolis, 1969) is probably not due to the absence of viral receptors on the ependymal cell surface of older animals.

A novel immunological approach is also being employed to characterize reovirus receptors (Nepom *et al.*, 1982a,b). This approach is based on the network theory of Niels Jerne (1974), which proposes that antibodies themselves can act as antigens and provoke the synthesis of antiantibodies in the host animal. The features of an antibody molecule that make it immunogenic to the host are exclusively in the variable portions of the heavy and light chains of the antibody molecule. These features of antigenic uniqueness on the antibody molecule result from the structure in and around the antigen-combining site and are referred to as idiotypic determinants, or idiotypes. Antibodies that recognize such determinants are called antiidiotypic antibodies, or antiidiotypes.

The generation of an immune response to an antigen involves a complex series of events whereby a variety of cells communicate with one another. For antibody formation, for example, B and T lymphocytes interact. These cells recognize one another through surface receptors encoded by variable-region genes. Jerne proposed that the various functional cells of the immune system communicate with one another through the recognition of idiotypes and that the immune response regulates itself through a network of idiotypic and antiidiotypic responses. He suggested that an animal responds to an antigen by the proliferation of lymphocytes specific for that antigen. This leads to an increase in concentration of receptors (immunoglobulins in the case of B cells because circulating immunoglobulins carry the same variable region as the receptor antibody molecule on the B cell) specific for antigen. The increase in receptor concentration makes the receptor sufficiently immunogenic so that it stimulates the proliferation of other lymphocytes with receptors specific for idiotypic determinants on the first set of receptors. Thus, an animal makes antibodies to antigen and antibodies to the idiotype. These antiidiotypes can stimulate other lymphocytes with receptors specific for antiidiotype to proliferate. Jerne visualized this process occurring in successive groups of lymphocytes, eventually encompassing the entire immune system.

Central to this theory, then, is the concept that an idiotype has a dual function, acting as antigen and antibody. The idiotype-containing receptor for antigen on a cell reacts with antigen in its capacity as antibody, but is also seen as an idiotype-containing antigen by other antibodies. By analogy, the $\sigma 1$-combining site of an antibody for the $\sigma 1$ protein not only reacts with $\sigma 1$ as antibody but also acts as an antigen, stimulating the production of an antiidiotypic antibody that recognizes the $\sigma 1$-binding site. It has been reasoned that if two proteins have a binding site for $\sigma 1$, then antibody made against the binding site of one of the proteins should also recognize the binding site of the other protein. Thus, antiidiotypic antibody to $\sigma 1$ might recognize not only the $\sigma 1$-binding site on antibody but also the $\sigma 1$-binding site on a cell-surface receptor. An investigation of the antigen-binding site on an antibody might therefore provide insight into the nature of the binding site of the cell-surface receptor for σ1. This hypothesis is supported by the finding that antibody to antiinsulin antibody can mimic insulin in binding to the insulin target receptor (Flier *et al.*, 1976). Thus, this approach to virus–receptor interactions is an effort to determine whether viral receptors share idiotypic determinants with antibodies to virus.

To prepare an antiidiotypic antiserum with specificity for the type 3 σ1 protein, antireovirus antibodies that selectively bound the σ1 protein were prepared from heteroimmune mouse antireovirus serum (Nepom *et al.*, 1982a). These σ1-specific fractions were used as immunogens to make antiidiotypic antibodies in a rabbit. This rabbit antiserum was passed over a column containing normal mouse immunoglobulin to remove anti-Fc immunoglobulins and was then characterized by competitive binding RIAs. These assays evaluated the ability of various antireovirus immunoglobulins to inhibit the binding of $^{125}$I-labeled anti-type 3 σ1 to the rabbit antisera. Both antireovirus type 1 and 3 immunoglobulins could inhibit this interaction to the same extent, suggesting that a large number of antireovirus specificities were present that cross-reacted in this type of inhibition assay. To identify σ1-specific binding structures, a series of monoclonal antibodies specific for the σ1 protein were examined for their ability to inhibit this assay. It was reasoned that inhibition would occur if a significant amount of the rabbit antiserum recognized a major determinant shared between the monoclonal antibody and the anti-σ1 antibody. One monoclonal antibody, G5, was found to greatly inhibit the binding of the rabbit antiserum to the anti-type 3 immunoglobulin. This G5 monoclonal antibody, which was described in Section V, binds to the domain of the σ1 protein that is the viral neutralization site. This G5 immunoglobulin appears to have a high proportion of the idiotypic determinants recognized by the rabbit antiserum.

Since the G5 monoclonal antibody is directed against the σ1 protein and the rabbit immunoglobulin used in the inhibition assays was selected for σ1-binding activity, it was postulated that a shared σ1-binding-site determinant with specificity for the type 3 σ1 had been identified. This G5 monoclonal antibody, then, could be used to assay for the presence of this shared determinant, designated Id3, in a more sensitive competitive RIA, since it contains the σ1-binding site (idiotype). A solid-phase RIA was designed using the G5 antibody and type 3 σ1 protein that had been purified according to the method of Gentsch (unpublished results). This assay evaluated the ability of antiidiotypic antibodies to inhibit the binding of G5 antibody to radiolabeled, purified type 3 σ1. Rabbit antiserum to antireovirus type 3 σ1 caused marked inhibition, whereas normal rabbit serum or rabbit antiserum raised to antibody specific for the type 1 σ1 protein failed to do so even at high concentrations. (Antibodies of the same subclass and allotype as G5 were used as controls.) These findings suggest that the idiotypic determinant, Id3, is indeed contained within the antigen-combining site of the G5 antibody.

To determine whether the Id3 determinant identifies a σ1-binding idiotype structure on cells, the antiidiotypic antiserum was used to probe cells that bind reovirus type 3. In some cases, the antiidiotypic antiserum used for these studies was enriched for Id3-specific reactivity by affinity chromatography on an immunoadsorbent containing the G5 monoclonal (idiotype-bearing) antibody. Indirect immunofluorescence studies demonstrated the ability of the antiidiotypic antiserum to bind to the R1.1 cell line, a T-cell line that binds type 3 but not type 1 reovirus, and to CNS cells that type 3 reovirus could infect. The YAC lymphoma, which binds neither type 1 or type 3, was negative for antiidiotypic antiserum binding in these immunofluorescent studies. Similarly, a correspondence between reovirus binding and antiidiotype staining was seen with B- and T-lymphocyte populations, suggesting a close association or identity between the presence of a cellular viral receptor and the Id3 marker.

Support for the recognition of both a T-cell determinant and G5 idiotype-containing antibody by the antiidiotypic sera came from studies using a monoclonal antibody prepared to the Id3 specificity (Noseworthy, unpublished). This monoclonal antibody, made by syngeneic immunization to the G5 antibody, binds to the G5 idiotype but has only one specificity, since it comes from a cloned hybridoma cell line. Absorption of this monoclonal antibody on R1.1 cells removed 97% of the G5-binding antibody, as measured by RIA.

These studies, then, suggest that the antiidiotypic antibody can identify a cell-surface determinant on lymphocytes that is shared by anti-$\sigma$1 immunoglobulins. If this determinant is linked to the viral receptor, then the effect of the antiidiotypic antibody on such cells might mimic virus itself. Indeed, similar to reovirus type 3, the antiidiotypic antiserum was observed to inhibit the conconavalin A–dependent proliferation of lymphocytes. Normal rabbit antiserum and rabbit antiserum raised to anti-reovirus type 1 could not inhibit this stimulation. Hence, these studies suggest that antiidiotypic antisera may provide a useful tool for the analysis of virus–cell receptor interactions.

## B. Viral Modification of Host Cells: Alteration of Cellular Metabolism

Despite extensive knowledge of the molecular biology of viral replication, little is known about how cytocidal viruses alter the host cell during infection and produce cellular injury. The role of individual viral components in the production of cellular damage is poorly understood. The mammalian reoviruses offer an excellent system for the study of viral functions that modify and damage host cells because reovirus genetics provides a means to identify the viral genes involved in these processes.

Studies with reoviruses type 2 and 3 have shown that these viruses alter host-cell macromolecular synthesis. Type 1 reovirus has not been reported to significantly alter host-cell metabolism.

Infection of mouse L cells with reovirus type 3 Dearing results in an inhibition of cellular DNA synthesis. This inhibition begins at 8–10 hr postinfection, before the onset of cytopathic effects (Gomatos and Tamm, 1963). It reflects a true decrease in DNA synthesis and not an alteration of DNA precursor pools (Ensminger and Tamm, 1970; Shaw and Cox, 1973). This inhibition occurs without detectable degradation of cellular DNA or alteration of the activity of DNA polymerase or enzymes involved in the conversion of thymidine triphosphate (Ensminger and Tamm, 1969b). Autoradiographic techniques and sedimentation analyses in alkaline sucrose density gradients show that nascent DNA chains grow with normal kinetics under conditions wherein DNA synthesis is inhibited overall (Ensminger and Tamm, 1969b; Hand et al., 1971, 1972). DNA synthesis appears to be inhibited because reovirus blocks the multifocal initiation of new DNA-chain synthesis on replication units (Ensminger and Tamm, 1969b; Hand et al., 1971; Hand and Tamm, 1974). Studies using synchronized L cells have shown that whereas DNA synthesis in the early part of the S phase is relatively unaffected by infection with reovirus, in the late S phase it is inhibited strongly. Infected cells are able to enter the S phase early during infection, but later the cells are blocked in $G_1$ (Ensminger and Tamm, 1970; Cox and Shaw, 1974). Electron-microscopic examination of the ultrastructural changes in the nuclei has revealed that the initial decrease in DNA synthesis is accompanied by decompaction of condensed chromatin within nuclei. Several hours later, nuclei show margination and clumping of heterochromatin (Chaly et al., 1980).

Not all cells are equally sensitive to inhibition of cellular DNA synthesis by type 3 reovirus. Shaw and his co-workers found that there is a differential sensitivity of normal and transformed human cells to the effect; inhibition occurs in SV40-transformed WI38 cells but not in normal WI38 cells (Duncan et al., 1978). This differential inhibition correlates with a difference in cytopathology. Infection of transformed cells terminated in cell lysis by 96 hr postinfection, whereas normal cells were productively infected and continued to produce virus for days after infection without detectable cytopathology.

Studies performed by several groups of investigators have suggested that reovirus infection does not produce a concomitant decrease in cellular protein synthesis, suggesting that the action of reovirus on DNA replication is direct and specific (Ensminger and Tamm, 1969a; Gomatos and Tamm, 1963). However, other investigators have suggested that type 3 reovirus inhibits host-cell DNA synthesis as a consequence of preventing synthesis of essential host proteins (Joklik, 1974).

The time of the beginning of inhibition of DNA synthesis is related to the number of infectious particles infecting a cell, suggesting that the inhibitory process is related to a virus-specific function. Studies of the effects of noninfective reovirus components on cellular DNA synthesis show that reovirus inactivated by UV light, so as to abolish its infectivity but not its transcriptase activity, inhibits DNA synthesis. However, reovirus cores, which have low infectivity but possess transcriptase activity, and reovirus empty capsids do not inhibit DNA replication. Reovirus oligoadenylates that have been adsorbed to cells in the presence of diethylaminoethyl–dextran inhibit DNA replication weakly (Hand and Tamm, 1973; Shaw and Cox, 1973; Lai *et al.*, 1973). Taken together, these results have been interpreted to suggest that none of the polypeptides of the parental virus causes the inhibition, since neither empty capsid nor cores produced the inhibition. It has been suggested that a nonstructural viral protein could mediate the inhibition.

Hand and co-workers examined the affinity of reovirus proteins for DNA, postulating that such affinity could provide a mechanism for inhibition (Shelton *et al.*, 1981). They examined cytoplasmic and nuclear extracts of infected cells by DNA affinity chromatography and found that two classes of reovirus-encoded proteins, σ and μ, had affinity for native and denatured DNA. The σNS had the strongest affinity for DNA, but the σ3 protein also had marked affinity. In addition, they did observe that a small fraction of purified virus had DNA affinity and therefore could not rule out the possibility that other structural proteins could be involved in inhibition of DNA synthesis. These results contrast with the studies of Huismans and Joklik (1976) that tested both σ3 and σNS for binding to native DNA and obtained negative results. A possible explanation for the differences in the findings is that the DNA affinity chromatography procedures were performed with crude infected-cell extracts, whereas the studies of Huismans and Joklik were performed with partially purified forms of the proteins. The affinity of DNA-binding proteins can be altered by the presence of other proteins in solution.

Using a genetic approach, Sharpe and Fields (1981) also investigated reovirus inhibition of cellular DNA synthesis. They found that while type 3 reovirus inhibits DNA synthesis in L cells, type 1 reovirus exerts little or no effect on L-cell DNA synthesis. Through the study of the effects of viral reassortants on L-cell DNA synthesis, the type 3 *S1* dsRNA segment was found to be responsible for the capacity of reovirus to inhibit L-cell DNA synthesis. As described extensively in this review, the σ1 protein, encoded by the *S1* dsRNA segment, is involved in a number of virus–receptor interactions. Whether inhibition of cellular DNA synthesis is mediated through an interaction at the cell surface is not known.

Studies on the effects of type 3 Dearing on L-cell protein synthesis have produced conflicting results that relate to the cell culture technique used for study. No inhibition of cellular protein synthesis was found in reovirus-infected L cells grown as monolayer cultures for up to 16 hr postinfection (Gomatos and Tamm, 1963). In contrast, an inhibition of protein synthesis was observed in reovirus-infected L cells grown in suspension culture, beginning 10 hr postinfection (Kudo and Graham, 1965). Ensminger and Tamm (1969a) examined this controversy and attributed the inhibitory effect observed in suspension cultures to cell damage. At times when protein-synthesis inhibition was seen, the cells no longer excluded trypan blue. However, these studies examined total rather than host-specific protein synthesis. When Joklik and his co-workers examined host-specific protein synthesis, they found that host-specific protein synthesis is inhibited 3–6 hr earlier than total protein synthesis in suspension-culture cells (Zweerink and Joklik, 1970). This protein-synthesis inhibition is dependent on the multiplicity of infection and on the temperature of incubation. The inhibition correlates closely with the onset of viral messenger RNA (mRNA) synthesis, suggesting that host protein synthesis may be inhibited by direct competition between host and viral RNA for some component of the protein-synthesizing machinery.

Recent *in vitro* studies have suggested that the L-cell translation machinery becomes modified as a result of type 3 virus infection (Skup and Millward, 1980). Cell-free extracts prepared from infected L cells at late times postinfection preferentially translate uncapped reovirus mRNA and translate capped mRNA at reduced efficiency. This contrasts with the translational preference of extracts prepared from uninfected cells, which translate capped reovirus mRNA at high efficiency and uncapped mRNA at lower efficiency. This transition from cap dependence to cap independence suggests a possible mechanism for viral takeover of host protein synthesis. Parental subviral particles synthesize single-stranded RNA (ssRNA) copies of the dsRNA genome. These ssRNA transcripts are capped (Furuichi *et al.*, 1975) and can be translated at early times by the normal cap-dependent translational machinery of the host cell. At later times, the primary producers of viral mRNA are progeny subviral particles (SVPs) that contain an active RNA polymerase and an inactive methyltransferase activity (Skup and Millward, 1980). These progeny SVPs produce uncapped mRNA (Zarbl *et al.*, 1980), which is preferentially translated by a virally modified translational apparatus. This transition in the host-cell translational machinery may be analogous to the modification of the translational machinery by poliovirus. In a cell-free synthesizing system, the ribosomal salt wash from poliovirus-infected cells does not stimulate translation of host mRNAs (capped mRNAs), but does stimulate the translation of poliovirus mRNA, which is uncapped. Poliovirus appears to inhibit cellular protein synthesis by inactivation of some crucial property of the cap-binding protein (Trachsel *et al.*, 1980).

Only a few studies have examined the effects of type 2 reovirus on host-cell macromolecular synthesis. Loh and Soergel have shown that type 2 reovirus inhibits protein and DNA synthesis without affecting the synthesis of RNA (Loh and Soergel, 1965) in RA (human amnion) cells (Loh and Soergel, 1967). Inhibition of protein synthesis occurs prior to the inhibition of DNA synthesis in these cells. Loh and co-workers have also examined the cytotoxic effects produced by type 2 virus that had been extensively irradiated ($10^5$ erg/sec per cm$^2$). After irradiation for 2–10 min, type 2 reovirus acquires the ability to produce accelerated cy-

topathic effects in HeLa cells without producing virus or viral components (Loh and Oie, 1969). The cytotoxic effects are preceded by marked
inhibition of host cellular RNA, DNA, and protein synthesis. This ability
is lost on prolonged irradiation for 60 min, which produces an alteration
of a number of viral properties. Experiments studying the effects of empty
capsids and urea-degraded virus indicate that protein components of the
outer capsid structure are involved in the production of the cytotoxicity
phenomenon (Subasinghe and Loh, 1972). The relationship between these
findings and the effects of infectious type 2 reovirus on host-cell macromolecular synthesis is uncertain.

     In a genetic study of the effects of the mammalian reoviruses on host-
cell protein synthesis, Sharpe and Fields (unpublished results) found that
type 2 reovirus inhibits protein synthesis in L-cell monolayers more rapidly and efficiently than does type 3 reovirus. Inactivation of type 2 reovirus by UV irradiation destroys the ability of type 2 reovirus to inhibit
protein synthesis, suggesting that viral replication is required to mediate
the inhibition. Through an analysis of the capacities of viral reassortants
to inhibit L-cell protein synthesis, the type 2 *S4* dsRNA segment was
shown to be responsible for this property. Reovirus types 2 and 3 were
also found to differ in their capacities to inhibit L-cell RNA synthesis,
type 2 causing a more efficient inhibition than type 3. The type 2 *S4*
dsRNA segment was also found to be responsible for this property. Studies
of Huismans and Joklik (1976) had shown that the σ3 protein specifically
binds to dsRNA. It remains to be determined whether this property and
the effect on protein and RNA synthesis are related.

     The identification of the *S4* gene as the dsRNA segment responsible
for the capacity of reovirus type 2 to inhibit L-cell protein and RNA
synthesis, however, does demonstrate that there is a difference in the
way that the reoviruses mediate inhibition of DNA, RNA, and protein
synthesis. The finding that two different genes mediate the inhibition of
DNA and protein synthesis lends support to the idea that inhibition of
DNA synthesis is not secondary to inhibition of protein synthesis.


## C. Viral Modification of Host Cells: Cytoskeletal Organization

     Viruses may exert their pathological effects not only through inter-
ference with cellular metabolism, but also through disruption of the host-
cell cytoskeleton. Microscopic studies of reovirus-infected cells indicate
that cytological changes accompany reovirus replication and morphogenesis (Rhim *et al.*, 1962; Spendlove *et al.*, 1963).

     Reoviruses enter cells within phagocytic vacuoles and remain within
these vacuoles until the vacuoles fuse with primary and secondary lysosomes (Dales *et al.*, 1965). Within lysosomes, reoviruses are uncoated
and the viral transcriptase is activated. Nascent virions are synthesized
subsequently in regions of the cytoplasm known as viral factories. The

intracytoplasmic route by which subviral particles move to these sites of replication is unknown. In addition, the mechanism of viral-factory formation is not clear.

Factories first appear as phase-dense granular material scattered throughout the cytoplasm. As infection progresses, these discrete granules coalesce and move toward the nucleus, eventually forming phase-dense perinuclear inclusions. As the number of progeny viruses increases, so do the size and extent of these factories. At late stages in infection (48–72 hr), these inclusions form a network that gradually spreads throughout the entire cytoplasm. The temporal development of this pattern of inclusion formation depends on the temperature of infection as well as the cell type. The stages of inclusion formation are delayed more at 31°C than at higher temperatures (Fields *et al.*, 1971). The stages appear to occur sooner in mouse L cells than in monkey kidney cells at the same temperature (Rhim *et al.*, 1962; Fields *et al.*, 1971).

The cellular protein and structural elements that are present within viral factories remain to be clarified. Although the factory is a discrete cytoplasmic entity, it is not enclosed by a membrane. Electron-microscopic studies have revealed that ribosomes are not present within factories. Viral proteins thus must be synthesized on polysomes outside the factory and transported back into the factory prior to assembly onto virus. Conversely, for viral mRNAs to be translated, the mRNAs must be synthesized within the factory and transported to adjacent polysomes within the cytoplasm.

Electron-microscopic studies have also indicated that several types of filaments are located within viral factories (Dales, 1963; Dales *et al.*, 1965). Viral particles are occasionally aligned on parallel arrays of microtubules that are thought to be covered with viral protein, possibly the viral hemagglutinin (Babiss *et al.*, 1979). Between microtubules is a complex filament network consisting of masses of densely twisted or kinky filaments with a diameter of 50–80 Å. The density of these filaments is not known, but their size is similar to that of intermediate filaments. These filaments are observed to be in intimate contact with virus particles and the viral proteins coating the microtubules. Such filaments occur regularly in wavy bundles in the cytoplasm of uninfected mouse L cells. Dales *et al.* (1965) have suggested that these filaments change from the wavy to the kinky type in areas where virus is being formed and that such filaments may provide a surface for the attachment of viral mRNAs or for viral morphogenesis.

The availability of specific probes to components of the cytoskeleton offers another approach to the study of reovirus interactions with cytoplasmic filaments. Although electron-microscopic studies could describe filament structure and distribution, the extent of cytoplasmic-filament organization was not fully appreciated until indirect-immunofluorescence techniques were developed (Weber *et al.*, 1975; Brinkley *et al.*, 1978; Lazarides, 1975). These immunofluorescence studies revealed that

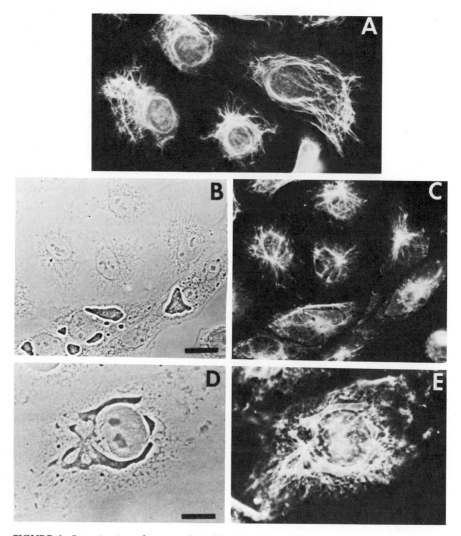

FIGURE 6. Organization of intermediate filaments in mock-infected and reovirus type 3–infected CV-1 cells at 48 hr postinfection. (A) Fluorescence micrograph showing intermediate-filament organization in mock-infected CV-1 cells. (B, D) Phase-contrast micrographs of type 3–infected CV-1 cells showing phase-dense inclusion bodies. (C, E) Fluorescence micrographs of the same cells as in (B) and (D), respectively, showing intermediate-filament organization in reovirus-infected CV-1 cells. Note that intermediate-filament organization is affected by reovirus infection. Cells were subjected to indirect immunofluorescence microscopy using antibody against vimentin. Scale bars: (A–C) 40 μm; (D, E) 25 μm. From Sharpe et al. (1982), by permission of Virology.

cytoplasmic filaments form networks within the cell cytoplasm. Electron-microscopic analyses of reovirus-infected cells have described reovirus interactions with filaments in regions of the cytoplasm, but have not permitted an analysis of the overall effects of reovirus on cytoskeletal organization.

Sharpe *et al.* (1982) used immunocytochemical techniques to further define the composition of viral factories and to examine the effects of reovirus infection on the three major filamentous systems in the cell cytoplasm: microtubules, microfilaments, and intermediate filaments. They found that reovirus infection produces a major disruption of vimentin filament (a type of intermediate filament) organization without producing a discernible disorganization of microtubules or microfilament bundles in African green monkey kidney CV-1 cells. By 12 hr postinfection, before intracytoplasmic inclusions were visible, the perinuclear organization sites of vimentin filaments disappeared. As infection progressed, there was further disruption of these filaments such that wavy filaments with no apparent organization were observed in the cytoplasm (Fig. 6). In addition to disrupting vimentin-filament organization, reovirus infection also appeared to cause a reorganization of vimentin filaments. Vimentin protein appears to be present in filamentous structures within viral inclusions. These filamentous structures may in fact be the complex filamentous network previously visualized by the electron-microscopic studies of viral factories carried out by Dales *et al.* (1965).

The presence of vimentin filaments in the viral factories suggests that these filaments may be participating in the formation of unique viral structures that are distinct from vimentin-filament organization in the cytoplasm of the noninfected cells. It is possible that reovirus is using the vimentin-filament system to create a discrete cytoplasmic entity, the viral factory, which functions as the site of viral replication and assembly. It has been proposed that vimentin filaments function in the coordination of the organization of cytoplasmic contents (Lazarides, 1980). With such a function, intermediate filaments may be expected to play a crucial role in the regulation of cell shape and to be influenced by the cell's metabolic state and response to injury. That vimentin filaments are found within viral factories suggests that they may play a critical role in the organization of the viral replication process. Possibly, intermediate filaments are one of the major cellular targets involved in producing the cytopathic effects that result from reovirus infection (e.g., rounding of cells, inclusion formation). It remains to be determined whether intermediate filament disorganization (or reorganization or both) is a primary and central step in viral infection, leading ultimately to cell death.

The disruption of vimentin filament organization by reovirus is especially interesting in view of the finding that agents that disrupt protein synthesis, such as diphtheria toxin and cycloheximide, specifically disrupt the organization of vimentin filaments without affecting microtubule or microfilament organization (Sharpe *et al.*, 1980). It is possible that the disruption of intermediate filaments by reovirus is related to the capacity of reovirus to inhibit host-cell protein synthesis.

Viral factories do not appear to contain actin or tubulin proteins, since antibodies to these proteins did not stain viral inclusions. However, antibodies to tubulin visualized microtubules that coursed through regions of the cytoplasm containing viral factories without interruption or distortion. These findings are consistent with the electron microscopic studies that show that reoviruses are aligned on parallel arrays of micro-tubules within viral factories (Dales *et al.*, 1965) and can bind to micro-tubules *in vitro* (Babiss *et al.*, 1979).

It is uncertain whether microtubules play a role in viral growth, since colchicine treatment of infected cells does not reduce viral yield (Spend-love *et al.*, 1964). Disruption of microtubules by colchicine treatment, however, alters the morphology of viral inclusions. Following the treat-ment of virally infected cells with colchicine, only small inclusions lo-cated at the cell periphery are observed, and the large perinuclear inclu-sions usually observed in cells infected with reovirus are absent. It is thus possible that microtubules play a role in the coalescence of viral inclu-sions and in their migration toward the nucleus.

In addition to altering vimentin filament organization, reovirus in-fection also leads to a disorganization of mitochondrial distribution in living cells. As visualized with the fluorescent probe Rhodamine 123 (Walsh *et al.*, 1979), mitochondria have a characteristic discontinuous distribution within the CV-1 cell cytoplasm. Following reovirus infec-tion, mitochondria are aggregated around the nucleus and occasionally present at the cell periphery. Mitochondria are excluded from viral in-clusions. Although reovirus infection alters mitochondrial distribution in cultured cells, it is unclear whether reovirus infection alters mito-chondrial function. The accumulation of Rhodamine 123 by mitochon-dria reflects the transmembrane potential (Johnson *et al.*, 1981). Because the accumulation of Rhodamine 123 in cells infected with reovirus is similar to that of uninfected cells, it is likely that the respiratory activity of infected and uninfected cells is similar.

## VII. PERSISTENT INFECTION

In addition to causing acute infections, viruses can produce infections in which they persist for months or years. These persistent infections are associated with a variety of pathogenetic mechanisms and clinical man-ifestations. Fenner *et al.* (1974) have divided persistence into three cat-egories: (1) latent infections—persistent infections with intermittent acute episodes of disease between which virus is usually not seen; (2) chronic infections—persistent infections in which virus is always de-monstrable and often shed, but disease is either absent or associated with an immunological disturbance; and (3) slow infections—persistent in-fections with a long incubation period followed by slowly progressive disease that is usually lethal.

Persistence at the cellular level is necessary for persistent infection of the whole animal. For this reason, a number of studies on persistent viral infections have focused on the establishment and maintenance of persistent infections in tissue-culture cells. These studies have suggested that for a lytic virus to establish a persistent infection in a susceptible cell, either the virus, the cell, or both have to mutate (for reviews, see Holland et al., 1980; Younger and Preble, 1980).

Reovirus is ordinarily a lytic virus, since infection of mouse L cells with low-passage stocks of wild-type reovirus causes cell damage and subsequent cell death. However, reovirus can also produce persistent infections in cell cultures. Bell and Ross (1966) demonstrated that reovirus can establish persistent infection in human embryonic fibroblasts, resulting in production and release of progeny virions without modification of cell growth. Taber et al. (1976) described the establishment of persistent reovirus infection in Chinese hamster ovary cells, though they were unable to determine whether viral or host mutants were involved in the establishment and maintenance of persistence. Ahmed and Graham (1977) showed that persistent infections could be established by infecting mouse L cells with reovirus stocks that had been serially passaged at high multiplicity of infection (MOI). Reovirus genetics has permitted an analysis of viruses used to establish these persistent infections and of viruses produced in such persistent infections. These studies have identified viral genes that play critical roles in the persistence of reovirus in L cells (Ahmed et al., 1980a,b; Ahmed and Fields, 1981, 1982).

Persistent infections can be established with much greater ease when the cells are infected with virus stocks that have been serially passaged at high MOI (Holland and Levine, 1978). It is known that when animal viruses are passaged at high MOI, deletion mutants are generated (Huang and Baltimore, 1977). Analysis of high-passage stocks of reovirus that have been used to establish persistent reovirus infections has revealed that such stocks contain not only defective interfering (DI) particles (deletion mutants) (Nonoyama et al., 1970; Schuerch et al., 1974; Ahmed and Graham, 1977), but also temperature-sensitive (ts) mutants not present in the original wild-type stock (Ahmed et al., 1980a,b). A significant proportion of these ts mutants have a ts lesion in the S4 double-stranded RNA (dsRNA) segment, the segment that encodes the σ3 polypeptide. The finding of ts mutants in high-passage stocks is particularly significant, because such high-passage stocks are often used to establish persistent infections, and in many cases ts mutants are later recovered from persistently infected cell cultures. It has been generally assumed that these ts mutants were generated during the course of persistent infection. The results of Ahmed and Fields, however, suggest the possibility that ts mutants present in high-passage stocks used to establish persistent infections are selected during the course of persistent infection.

Serially passaged stocks of reovirus also contain a large proportion of DI particles lacking the L1 and L3 dsRNA segments (Ahmed et al., 1980a,b). Because such stocks are needed for the establishment of persistent infections in L cells, Ahmed and Fields investigated the genetic properties of reovirus DI particles. They found that reovirus DI particles could reassort with either wild-type or ts virus. Earlier studies, based on infectious-center assays, had reported that no recombination occurred when defective virus was crossed with *ts* mutants (Spandidos and Graham, 1976). When Ahmed and Fields (1981) repeated this experiment, they found that although the percentage of cells making plaques was low, the actual number of plaques in the crosses between type 3 DI and *ts* mutants was significantly higher than in single infection. Studies on viral interference have clarified this situation (Ahmed and Fields, 1981; Chakraborty *et al.*, 1979). When certain *ts* mutants of reovirus are grown with $ts^+$ virus at the nonpermissive temperature, they interfere with the growth of the $ts^+$ virus. Similarly, the lack of detectable complementation and low frequency of recombination between pairs of interfering *ts* mutants at the nonpermissive temperature is most likely due to viral interference. Likewise, Ahmed and Fields (1981) showed that interference by DI particles occurs and may explain the low frequency of recombination between DI particles and infectious virus.

Having established that DI particles can reassort genome segments, Ahmed *et al.* (1980a,b) performed genetic crosses between a DI stock derived from a $ts^+$ reovirus type 3 and a cloned $ts^+$ reovirus type 1 and analyzed the progeny reassortants. They found that *ts* and small-plaque/low-yield mutants could be rescued from the reovirus DI particles. Both the *ts* and the small-plaque/low-yield mutations were located on the *S4* dsRNA segment. Some of the *ts* viruses made by the rescue of mutations from DI particles could interfere with the growth of the wild-type reovirus at both the permissive and the nonpermissive temperature. Thus, reovirus DI particles, in addition to lacking certain dsRNA segments, contain mutations in other genes. Furthermore, it is possible that such mutations, rather than deleted genes, are responsible for the biological properties of DI particles, such as viral interference. Specifically, interference by DI stocks of reovirus may be due to mutation in the σ3 protein. Hence, it is likely that DI particles provide a source of mutant genes and contribute to the emergence of genetic variants within persistent infections.

These genetic experiments have demonstrated that high-passage stocks of reovirus are comprised of a genetically heterogeneous population and contain many types of viral mutants, *ts* mutants, and growth-attenuated mutants. These studies suggest that the "low virulence" of high-passage stocks is due at least in part to mutations in the *S4* dsRNA segment, because the majority of *ts* mutants in such stocks are in the *S4* dsRNA segment and the DI particles contain *ts* and small-plaque/low-yield mutation on the *S4* dsRNA segment.

More direct evidence for the role of the *S4* dsRNA segment in the establishment of persistent reovirus infection came from the following experiment: Mouse L cells were coinfected with wild-type 2 reovirus and type 3 DI under conditions wherein persistent infections were established. The genotype of virus selected during the establishment of persistent infection was examined to determine whether particular genes were always derived from the type 3 DI parent. Presumably, genes essential for the establishment of a persistent infection would always be derived from a defective virus. Ahmed and Fields (1982) found that the S4 dsRNA segment was always derived from the type 3 DI virus in three persistently infected cell lines established independently, using different stocks of DI virus. Hence, these studies indicate that the S4 gene plays an important role in the establishment of persistent reovirus infection. In lytic infections, the S4 gene is responsible for inhibiting host-cell RNA and protein synthesis as well as regulating the transcriptase activity of subviral particles (Sharpe and Fields, unpublished results; Astell et al., 1972). The selection of a defective S4 dsRNA segment could relate to an alteration in the S4 dsRNA segment such that it no longer inhibits host-cell RNA and protein synthesis or that it interferes with viral replication.

A mutant S1 dsRNA segment with altered mobility on PAGE was also selected during the establishment of persistent infection. The significance of the mutant S1 dsRNA segment is not known. Possibly, a mutation in the S1 dsRNA segment such that it no longer produces a σl protein capable of inhibiting host-cell DNA synthesis would be advantageous. The significance of a mutant S1 gene might also relate to the selection of mutant L cells that have an altered receptor for the σl protein.

In addition to analyzing reovirus stocks used to establish persistent infection, Ahmed et al. (1980b) studied the evolution of reovirus during persistent infection using a genetic approach. An examination of three persistently infected cell lines maintained over a 1-year period revealed that reovirus undergoes extensive mutation during persistent infection, although no consistent pattern of virus evolution was observed. In one cell line, initiated by infection with a high-passage stock of the tsC mutant, there was a rapid selection of the ts$^+$ phenotype. Genetic analysis of ts$^+$ clones revealed that although these clones expressed a ts$^+$ phenotype, several contained multiple ts lesions in a suppressed state. Cold-sensitive (cs) mutants were isolated from another persistently infected cell line, whereas in the third line there was a selection of ts mutants that contained multiple ts lesions. Neither the cs nor the ts mutants expressed their defects at 37°C, the temperature at which the persistently infected cells were maintained. Thus, it appears that both cs and ts phenotypes were unselected markers and did not play a crucial role in the maintenance of persistent infections. These studies thus illustrate the complexity of genetic changes that occur during the course of persistent infection. Furthermore, they question the significance of mutants isolated from persistently infected cells. The isolation of such mutants indicates that virus has changed during the course of the persistent infection, but does not necessarily imply that a particular mutation plays a critical role in the maintenance of persistent infection, especially since many mutants do not exhibit their phenotypes at the temperature at which the persistently infected cell cultures have been maintained.

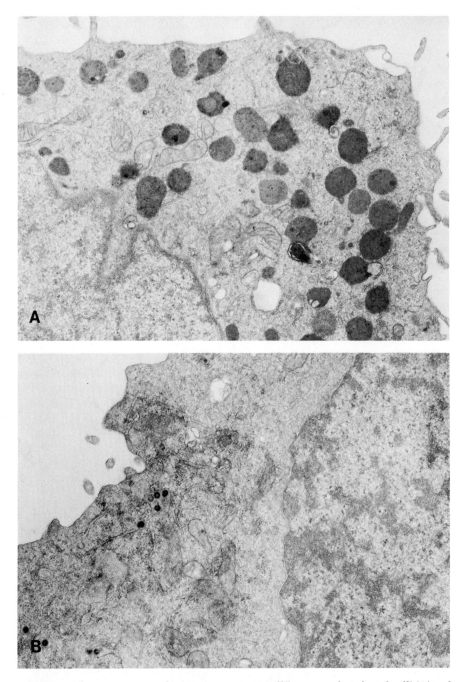

FIGURE 7. Electron micrographs comparing an LR cell (persistently infected cell) (A) and an uninfected L cell (B). (A) Note the presence of multiple dense vesicles in the cytoplasm of the LR cell. The vesicles appear to be lysosomes by acid phosphatase staining. (B) The same structures are present in fewer numbers in uninfected L cells. × 10,000. Courtesy of Michael Canning and Dr. John Robinson.

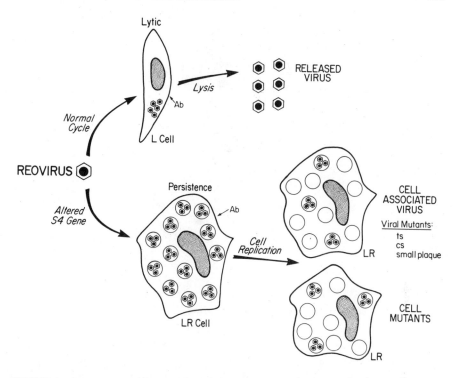

FIGURE 8. Comparison of lytic and persistent reovirus infections. The figure illustrates the different interactions of reoviruses with host cells during lytic and persistent infection. Details are in the text.

A better understanding of such viral mutants came from studies that examined concurrent changes in the host cell in persistently infected cell cultures. Ahmed *et al.* (1981) found that during the course of persistent infection, there was a coselection of both mutant L cells and mutant reovirus. Mutant L cells were isolated after "curing" a persistently infected cell line with antireovirus serum. These mutant cells, designated LR cells, were shown to be free of reovirus. No virus was detected by infectious-center assay, plaque assay, or biochemical assays designed to detect viral protein or dsRNA. Persistent infections in these mutant cells could be established by infecting with either wild-type virus or reovirus isolated from persistently infected cell cultures (p.i. virus). Infection of mutant cells with wild-type reovirus resulted in a low-level, persistent infection with inefficient viral replication, whereas infection with virus isolated from persistently infected cell cultures led to a persistent infection accompanied by efficient viral replication. Infection of the original L cells with either wild-type or the p.i. virus resulted in a lytic infection with no surviving cells. Thus, these studies demonstrated that the host cell plays an important role in the maintenance of persistent infection. It appears that the mutant cells arose during the course of the persistent

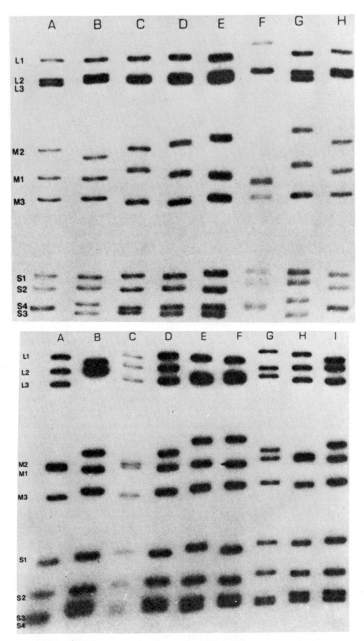

FIGURE 9. ³²P-labeled dsRNA from type 1 and type 3 reovirus isolates from humans, cattle, and mice, showing the extensive variability in patterns of migration of the ten dsRNA genome segments. Electrophoresis is from top to bottom in a single 10% acrylamide gel containing the Laemmli Tris–glycine buffer system (Laemmli, 1970). *Top:* Type 1 reovirus isolates: (A) Type 1 (Lang); (B) human type HT1a, TM; (C) human type HT1b, WA; (D) human type HT1c, WA; (E) bovine type BT1a, CP; (F) bovine type BT1b, CP; (G) bovine type BT1c, JE; (H) type 1 (Lang). The RNA genome segments of type 1 (Lang) are identified.

infection, although there are no conclusive data to exclude the possibility that the L-cell mutants were present at low levels in the original L-cell population.

The nature of the mutations in the LR cells is under current investigation. Most of the LR cells are morphologically distinct from parental L cells (Fig. 7). They are larger and less fibroblastlike than L cells. Preliminary results indicate that LR cells may have an altered cytoskeleton. These cells contain a large number of lysosomelike structures in the cytoplasm (Canning, unpublished results). Whether these changes are related to the ability of LR cells to escape viral injury remains to be determined.

In summary, genetic analyses of the establishment and maintenance of reovirus persistent infection have documented the strong selective pressures exerted on both the virus and the host cell. Alterations in both the virus and the host cell play critical roles (Fig. 8). Mutation in the *S4* dsRNA segment plays an important role in the establishment of persistent infection. Mutation in the L-cell host, such that it supports viral replication without lysis, appears to be an important factor in the maintenance of persistently infected cell lines. The evolution of mutant L cells appears to be followed by the selection of viral mutants that can replicate more efficiently in the mutant cells.

## VIII. MOLECULAR EPIDEMIOLOGY OF MAMMALIAN REOVIRUSES

The mammalian reoviruses are widely distributed in humans and in a number of animal species. During the late 1950s and early 1960s, Leon Rosen and co-workers collected a large number of reovirus isolates during field investigations in several parts of the world (Rosen and Abinanti, 1960; Rosen *et al.*, 1963). They grouped these isolates according to the traditional method of serological analysis (Rosen, 1960). The availability of a more precise technique for the analysis of double-stranded RNA (dsRNA) viruses has permitted a more detailed study of the molecular epidemiology of the mammalian reoviruses (Hrdy *et al.*, 1979). Through the examination of the electrophoretic migration patterns of the dsRNA segments of these field isolates, Hrdy, Rosen, and Fields were able to study variability within a series of isolates from a single serotype, variation within a serotype among isolates originating from different species, and variation within a single outbreak.

Capital-letter abbreviations denote sites of isolation: (TM) Toluca, Mexico; (WA) Washington, D.C.; (CP) College Park herd; (JE) Jessup herd. *Bottom:* Type 3 reovirus isolates: (A) Type 3 (Dearing); (B) human type HT3a, TA; (C) human type HT3b, WA; (D) human type HT3c, WA; (E) bovine BT3a, CP; (F) bovine type BT3b, CP; (G) murine type MT3a, France; (H) Type 3 (Dearing); (I) bovine type BT3c, JE. RNA genome segments of type 3 (Dearing) are identified. (TA) Tahiti. From Hrdy *et al.* (1979), by permission of the *Journal of Virology*.

A series of 94 isolates of reovirus from humans, cattle, and mice exhibited extensive variability in the migration pattern of the ten dsRNA segments. This variation was observed in all three serotypes and involved all ten segments, including the segment responsible for serotype specificity (Fig. 9). Thus, this study revealed that no single dsRNA pattern distinguishes a particular serotype. This variability in migration of the dsRNA segments was seen among several isolates within a serotype, among the three serotypes, and among isolates derived from different species or geographic areas. Thus, the RNA pattern of an isolate does not identify the serotype, the host of origin, or the geographic locale.

Analysis of isolates collected during a single outbreak revealed that multiple genetic variants of a single serotype may be present during a single outbreak. Certain isolates emerged and disappeared during a 2-year period.

The migration of the slowest $S$ segment was of special interest, since it is this segment that encodes the protein responsible for serological specificity in the laboratory strains. Preliminary determination of serotype could be made by analysis of the migration of the slowest $S$ segment. In many instances for the type 1 and type 3 strains, the slowest $S$ segment had a relatively constant migration pattern (Fig. 9). In contrast, the type 2 isolates exhibited striking variations in the migration of the corresponding $S$ segment. This observation is consistent with the previous finding of serological subtypes among type 2 strains. Such variability may suggest that type 2 is an evolutionarily unstable serotype.

While it had long been postulated and recently shown that genome-segment exchange takes place between human and nonhuman species infected with influenza virus (Laver and Webster, 1973, 1982), it is not known whether such interspecies exchange occurs with reoviruses. Such genetic exchange would permit rapid evolution by exchange of genetic material and would assist in the maintenance of a nonpersistent virus in host populations. Finally, it is interesting to note that the family Reoviridae includes insect-borne orbiviruses, the phytoreoviruses of plants, and the cytoplasmic polyhedrosis viruses of insects. It is unknown whether exchange of genetic material can occur between the mammalian reoviruses and any of these other dsRNA viruses. Such an exchange would have wide-ranging implications for the evolution of viruses and the diseases that they could produce.

## IX. CONCLUDING REMARKS

The mammalian reoviruses offer an excellent system for the study of viral pathogenesis because the genetics of reovirus provides a means to identify the role of individual viral components in the production of viral virulence, making the study of virus–host interactions more amenable to direct experimentation. Studies on reovirus pathogenesis have

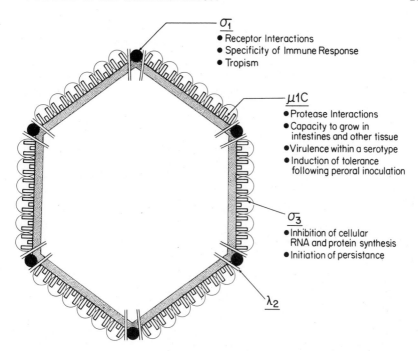

FIGURE 10. Summary of the role of the reovirus outer capsid in viral pathogenesis. This schematic diagram of the outer capsid of the mammalian reoviruses shows the organization of capsomers in the icosahedral virion. The drawing shows how the three outer-capsid proteins, σ1, σ3, and μ1C, are organized: σ1 is located at the vertices of the icosahedron; μ1C and σ3 are associated with each other (1 μ1C2 σ3) and are located on the flat surface of the icosahedron. The two core proteins protrude from the core to the outer surface. The roles of the outer-capsid proteins in pathogenesis are summarized.

revealed that the outer-capsid proteins of reovirus play a critical role in viral virulence. The σ1 protein is responsible for determining cell and tissue tropism and for inhibiting cellular DNA synthesis. It is also the major antigen determining specificity in humoral and cellular immune responses. The μ1C protein plays a role in determining the capacity of reoviruses to interact with cellular proteases and to initiate infections within the gastrointestinal tract. It is also involved in determining the nature of the immune response to viral infections. The σ3 protein is responsible for inhibiting cellular RNA and protein synthesis, is a site of frequent mutations that result in decreased ability of reoviruses to lyse cells, and plays a role in the establishment of persistent infection. Thus, virulence is multigenic, with a number of viral components playing critical and distinct roles (for a summary, see Fig. 10).

It is quite likely that the findings concerning the mechanisms of reovirus virulence are relevant to other viruses. The interaction of reoviruses with the host at each stage of pathogenesis serves as a useful model for other viruses and microbial agents. A central theme emerging from these studies is that although the outer-capsid proteins clearly need to work together, each protein has a distinct role in viral pathogenesis. Understanding how each of the capsid proteins functions at each stage of pathogenesis should continue to provide important concepts for understanding the molecular basis of viral disease. An increased understanding of viral pathogenesis should provide new approaches for the prevention and treatment of viral diseases.

ACKNOWLEDGMENTS. We gratefully acknowledge our many outstanding colleagues in the Departments of Microbiology and Molecular Genetics, Pathology, and Neuroscience at Harvard Medical School. We are especially grateful to Dr. Gordon Freeman and Dr. Mark Greene for helpful comments regarding the immunological section of this review. We also thank Janis Segal and Linda Plaza for their outstanding assistance in the preparation of this review. This research was supported by grants from the National Institutes of Health (National Institute of Allergy and Infectious Diseases) and the Institute of Neurologic Diseases and Blindness.

# REFERENCES

Ahmed, R., and Fields, B.N., 1981, Reassortment of genome segments between reovirus defective interfering particles and infectious virus: Construction of temperature-sensitive and attenuated viruses by rescue of mutations from DI particles, *Virology* **111**:351.

Ahmed, R., and Fields, B.N., 1982, Role of the S4 gene in the establishment of persistent reovirus infection in L cells, *Cell* **28**:605.

Ahmed, R., and Graham, A.F., 1977, Persistent infections in L cells with temperature-sensitive mutants of reovirus, *J. Virol.* **23**:250.

Ahmed, R., Chakraborty, P.R., and Fields, B.N., 1980a, Genetic variation during lytic virus infection: High passage stocks of wild type reovirus contain temperature-sensitive mutants, *J. Virol.* **34**:285.

Ahmed, R., Chakraborty, P.R., Graham, A.F., Ramig, R.F., and Fields, B.N., 1980b, Genetic variation during persistent reovirus infection: Presence of extragenically suppressed ts lesions in wild type virus isolated from persistently infected cells, *J. Virol.* **34**:383.

Ahmed, R., Canning, W.M., Kauffman, R.S., Sharpe, A.H., Hallum, J.V., and Fields, B.N., 1981, Role of the host cell in persistent viral infection: Co-evolution of L cells and reovirus during persistent infection, *Cell* **25**:325.

Astell, C., Silverstein, S.C., Levin, D.H., and Acs, G., 1972, Regulation of the reovirus RNA transcriptase by a viral capsomere protein, *Virology* **48**:648.

Babiss, L.E., Luftig, R.B., Weatherbee, J.A., Weihing, R.R., Ray, U.R., and Fields, B.N., 1979, Reovirus serotypes 1 and 3 differ in their *in vitro* association with microtubules, *J. Virol.* **30**:863.

Bangaru, B., Morecki, R., Glaser, J.H., Gartwer, L.M., and Horwitz, M.S., 1980, Comparative studies of biliary atresia in the human newborn and reovirus-induced cholangitis in weanling mice, *Lab. Invest.* **43**:456.

Beck, I., 1973, The role of pancreatic enzymes in digestion, *Am. J. Clin. Nutr.* **26**:311.

Bell, T.M., and Ross, M.G.R., 1966, Persistent latent infection of human embryonic cells with reovirus type 3, *Nature (London)* **212**:412.

Bellamy, A.R., Shapiro, L., August, J.T., and Joklik, W.K., 1967, Studies on reovirus RNA. I. Characterization of reovirus genome RNA, *J. Mol. Biol.* **29**:1.

Bienenstock, J., Clancy, R.L., and Perey, D.Y.E., 1976, Bronchus associated lymphoid tissue (BALT): Its relationship to mucosal immunity, in: *Immunologic and Infectious Re-*

*actions in the Lung* (C.H. Kirkpatrick and H.Y. Reynolds, eds.), pp. 29–58, Marcel Dekker, New York and Basel.

Blanden, R.V., 1974, T cell response to viral and bacterial infection, *Transplant. Rev.* **19**:56.

Bockman, D.E., and Cooper, M.D., 1973, Pinocytosis by epithelium associated with lymphoid follicles in the bursa of Fabricius, appendix, and Peyer's patches: An electron microscopic study, *Am. J. Anat.* **136**:455.

Borsa, J., Sargent, M.D., Copps, T.P., Long, D.G., and Chapman, J.D., 1973, Specific monovalent cation effects on modification of reovirus infectivity by chymotrypsin digestion in vitro, *J. Virol.* **11**:1017.

Both, G.W., Banerjee, A.K., and Shatkin, A.J., 1975, Methylation-dependent translation of viral messenger RNAs in vitro, *Proc. Natl. Acad. Sci. U.S.A.* **72**:1189.

Brinkley, B.R., Fuller, G.M., and Highfield, D.P., 1978, Cytoplasmic microtubules in normal and transformed cells in culture: Analysis of tubulin antibody immunofluorescence, *Proc. Natl. Acad. Sci. U.S.A.* **72**:4981.

Burstin, S.J., Spriggs, D.R., and Fields, B.N., 1982, Evidence for functional domains on the reovirus type 3 hemagglutinin, *Virology* **117**:146.

Chakraborty, P.R., Ahmed, R., and Fields, B.N., 1979, Genetics of reovirus: The relationship of interference to complementation and reassortment of ts mutants at non-permissive temperature, *Virology* **94**:119.

Chaly, N., Johnstone, M., and Hand, R., 1980, Alterations in nuclear structure and function in reovirus-infected cells, *Clin. Invest. Med.* **2**:141.

Cox, D.C., and Shaw, J.E., 1974, Inhibition of the initiation of cellular DNA synthesis after reovirus infection, *J. Virol.* **13**:760.

Cross, R.K., and Fields, B.N., 1972, RNA synthesized in cells infected with reovirus type 3 mutants, *Virology* **50**:799.

Cross, R.K., and Fields, B.N., 1976a, Reovirus specific polypeptides: Analysis using discontinuous gel electrophoresis, *J. Virol.* **19**:162.

Cross, R.K., and Fields, B.N., 1976b, Use of an aberrant polypeptide as a marker in three-factor crosses: Further evidence for independent reassortment as the mechanism of recombination between temperature-sensitive mutants of reovirus type 3, *Virology* **74**:345.

Dales, S., 1963, Association between the spindle apparatus and reovirus, *Proc. Natl. Acad. Sci. U.S.A.* **50**:268.

Dales, S., Gomatos, P.J., and Hsu, K.C., 1965, The uptake and development of reovirus in strain L cells followed with labelled viral ribonucleic acid and ferritin–antibody conjugates, *Virology* **25**:193.

Doherty, P.C., and Zinkernagel, R.M., 1976, Specific immune lysis of paramyxovirus-infected cells by H-2-compatible thymus derived lymphocytes, *Immunology* **31**:27.

Duncan, M.R., Stanish, S.M., and Cox, D.C., 1978, Differential sensitivity of normal and transformed human cells to reovirus infection, *J. Virol.* **28**:444.

Eardley, D.D., Shen, F.W., Cantor, H., and Gershon, R.K., 1979, Genetic control of immunoregulatory circuits: Genes linked to the Ig locus govern communication between regulatory T cell sets, *J. Exp. Med.* **150**:44.

Eggers, H.J., Gomatos, P.J., and Tamm, I., 1962, Agglutination of bovine erythrocytes: A general characteristic of reovirus type 3, *Proc. Soc. Exp. Biol. Med.* **110**:879.

Ensminger, W.D., and Tamm, I., 1969a, Cellular DNA and protein synthesis in reovirus-infected cells, *Virology* **39**:357.

Ensminger, W.D., and Tamm, I., 1969b, The step in cellular DNA synthesis blocked by reovirus infection, *Virology* **39**:935.

Ensminger, W.D., and Tamm, I., 1970, Inhibition of synchronized cellular deoxyribonucleic acid synthesis during Newcastle disease virus, mengovirus or reovirus infection, *J. Virol.* **5**:672.

Epstein, R.L., Powers, M.L., and Weiner, H.L., 1981, Interaction of reovirus with cell surface receptors. III. Reovirus type 3 induces capping of viral receptors on murine lymphocytes, *J. Immunol.* **127**:1800.

Fenner, F., McAuslan, B.R., Mims, C.A., Sambrook, J., and White, D.O., 1974, Persistent infections, in: *The Biology of Animal Viruses* (F. Fenner, B.R., McAuslan, C.A. Mims, J. Sambrook, and D.O. White, eds.), pp. 452–476, Academic Press, New York and London.

Fields, B.N., and Joklik, W.K., 1969, Isolation and preliminary genetic and biochemical characterization of temperature-sensitive mutants of reovirus, *Virology* **37**:335.

Fields, B.N., Raine, C.S., and Baum, S.G., 1971, Temperature-sensitive mutants of reovirus type 3: Defects in viral maturation as studied by immunofluorescence and electron microscopy, *Virology* **43**:569.

Finberg, R., Weiner, H.L., Fields, B.N., Benacerraf, B., and Burakoff, S.J., 1979, Generation of cytolytic T lymphocytes after reovirus infection: Role of S1 gene, *Proc. Natl. Acad. Sci. U.S.A.* **76**:442.

Finberg, R., Weiner, H.L., Burakoff, S.J., and Fields, B.N., 1981, Type-specific reovirus antiserum blocks the cytotoxic T cell–target cell interaction: Evidence for the association of the viral hemagglutinin of a nonenveloped virus with the cell surface, *Infect. Immun.* **31**:646.

Flier, J.S., Kahn, C.R., Jarrett, D.B., and Roth, J., 1976, Characterization of antibodies to the insulin receptor: A cause of insulin-resistant diabetes in man, *J. Clin. Invest.* **58**:1442.

Fontana, A., and Weiner, H.L., 1980, Interaction of reovirus with cell surface receptors. II. Generation of Suppressor T Cells by the Hemagglutinin of Reovirus Type 3, *J. Immunol.* **125**:2660.

Furuichi, Y., Muthukrishnan, S., and Shatkin, A.J., 1975, 5′-Terminal m(7)G(5′)ppp(5′)G(m)p *in vivo*: Identification in reovirus genome RNA, *Proc. Natl. Acad. Sci. U.S.A.* **72**:742.

Gentsch, J.R., and Fields, B.N., 1981, Tryptic peptide analysis of the outer capsid polypeptides of mammalian reovirus serotypes 1, 2, and 3, *J. Virol.* **38**:208.

Gomatos, P.J., and Tamm, I., 1963, Macromolecular synthesis in reovirus-infected cells, *Biochim. Biophys. Acta* **72**:651.

Gomatos, P.J., Tamm, I., Dales, S., and Franklin, R.M., 1962, Reovirus type 3: Physical characteristics and interaction with L cells, *Virology* **17**:441.

Graziadei, W.D., and Lengyel, P., 1972, Translation of *in vitro* synthesized reovirus mRNAs into proteins of the size of reovirus capsid proteins in a mouse L cell extract, *Biochem. Biophys. Res. Comm.* **46**:1816.

Greene, M.I., and Weiner, H.L., 1980, Delayed hypersensitivity in mice infected with reovirus. II. Induction of tolerance and suppressor T cells to viral specific gene products, *J. Immunol.* **125**:283.

Greene, M.I., Sy, M.S., Kripke, M., and Benacerraf, B., 1979, Impairment of antigen presenting cell function by ultraviolet radiation, *Proc. Natl. Acad. Sci. U.S.A.* **76**:6591.

Greene, M.I., Nelles, M., Sy, M.S., and Nisonoff, A., 1982, The regulatory influence of ligand and idiotypes, *Adv. Immunol.* **32**:253.

Hand, R., and Tamm, I., 1972, Rate of DNA chain growth in mammalian cells infected with cytocidal RNA viruses, *Virology* **47**:331.

Hand, R., and Tamm, I., 1973, Reovirus: Effect of noninfective viral components on cellular deoxyribonucleic acid synthesis, *J. Virol.* **11**:223.

Hand, R., and Tamm, I., 1974, Initiation of DNA synthesis in mammalian cells and its inhibition by reovirus infection, *J. Mol. Biol.* **82**:175.

Hand, R., Ensminger, W.D., and Tamm, I., 1971, Cellular DNA replication in infections with cytocidal RNA viruses, *Virology* **44**:527.

Hanson, D.G., Vaz, N.M., Rawlings, L.A., and Lynch, J.M., 1979, Inhibition of specific immune responses by feeding protein antigens. II. Effects of prior passive and active immunization, *J. Immunol.* **122**:2261.

Hayes, E.C., Lee, P.W.K., Miller, S.E., and Joklik, W.K., 1981, The interaction of a series of hybridoma IgGs with reovirus particles, *Virology* **108**:147.

Heberman, R.B., Nunn, M.E., and Lavrin, D.H., 1975, Natural cytotoxic reactivity of mouse lymphoid cells against syngeneic and allogeneic tumors. I. Distribution of reactivity and specificity, *Int. J. Cancer* **16**:216.

Henderson, D.R., and Joklik, W.K., 1978, The mechanism of interferon induction by UV-irradiated reovirus, *Virology* **91**:389.

Holland, J.J., and Levine, A., 1978, Mechanisms of viral persistance, *Cell* **14**:447.

Holland, J.J., Kennedy, S.I.T., Semler, B.L., Jones, C.L., Roux, L., and Grabau, E.A., 1980, Defective interfering RNA viruses and the host-cell response, in: *Comprehensive Virology*, Vol. 16 (H. Fraenkel-Conrat and R. Wagner, eds.), pp. 137–192, Plenum Press, New York.

Hrdy, D.B., Rosen, L., and Fields, B.N., 1979, Polymorphism of the migration of double-stranded RNA genome segments of reovirus isolates from humans, cattle, and mice, *J. Virol.* **31**:104.

Hrdy, D.B., Rubin, D.H., and Fields, B.N., 1982, Molecular basis of reovirus neurovirulence: Role of the M2 gene in avirulence, *Proc. Natl. Acad. Sci. U.S.A.* **79**:1298.

Huang, A.S., and Baltimore, D., 1977, Defective interfering animal viruses, in: *Comprehensive Virology*, Vol. 10 (H. Fraenkel-Conrat and R.R. Wagner, eds.), pp. 73–116, Plenum Press, New York.

Hudson, B.W., Wolff, K., and DeMartini, J.C., 1979, Delayed-type hypersensitivity responses in mice infected with St. Louis encephalitis virus: Kinetics of the response and effects of immunoregulatory agents, *Infect. Immun.* **24**:71.

Huismans, H., and Joklik, W.K., 1976, Reovirus-coded polypeptides in infected cells: Isolation of two native monomeric polypeptides with affinity for single-stranded and double-stranded RNA, respectively, *Virology* **70**:441.

Ida, S., and Hinuma, Y., 1971, Effect of antilymphocyte serum on reovirus infection of mice, *Infect. Immun.* **3**:304.

Ito, Y., and Joklik, W.K., 1972, Temperature-sensitive mutants of reovirus. III. Evidence that mutants of group D ("RNA-negative") are structural polypeptide mutants, *Virology* **50**:282.

Jerne, N.K., 1974, Towards a network theory of the immune system, *Ann. Immunol.* (*Inst. Pasteur*) **125C**:373.

Johnson, L.V., Walsh, M., and Chen, L., 1980, Localization of mitochondria in living cells with rhodamine 123, *Proc. Natl. Acad. Sci. U.S.A.* **77**:990.

Johnson, L.V., Walsh, M.L., Bockus, B.J., and Chen, L.B., 1981, Monitoring of relative mitochondrial membrane potential in living cells by fluorescence microscopy, *J. Cell Biol.* **88**:526.

Joklik, W.K., 1972, Studies on the effect of chymotrypsin on reovirions, *Virology* **49**:700.

Joklik, W.K., 1974, Reproduction of Reoviridae, in: *Comprehensive Virology*, Vol. 2 (H. Fraenkel-Conrat and R. Wagner, eds.), pp. 297–334, Plenum Press, New York.

Kilham, L., and Margolis, G., 1969, Hydrocephalus in hamsters, ferrets, rats and mice following inoculations with reovirus type 1, *Lab. Invest.* **21**:183.

Kudo, H., and Graham, A.F., 1965, Synthesis of reovirus ribonucleic acid in L cells, *J. Bacteriol.* **90**:936.

Laemmli, U.K., 1970, Cleavage of structural proteins during the assembly of the head of bacteriophage T4, *Nature* (*London*) **227**:680.

Lagrange, P.H., Tsiand, H., and Hurtrel, B., 1978, Delayed-type hypersensitivity to rabies virus in mice: Assay of active or passive sensitization by the footpad test, *Infect. Immun.* **21**:931.

Lai, M.-H.T., and Joklik, W.K., 1973, The induction of interferon by temperature-sensitive mutants of reovirus, UV-irradiated reovirus, and subviral reovirus particles, *Virology* **51**:191.

Lai, M.T., Werenne, J.J., and Joklik, W.K., 1973, The preparation of reovirus top component and its effect on host DNA and protein synthesis, *Virology* **54**:237.

Laver, W.G., and Webster, R.G., 1973, Studies on the origin of pandemic influenza. III. Evidence implicating duck and equine influenza viruses as possible progenitors of the Hong Kong strain of human influenza, *Virology* **51**:383.

Laver, W.G., and Webster, R.G., 1982, Molecular mechanisms of variation in influenza viruses, *Nature* (*London*) **296**:115.

Lazarides, E., 1975, Tropomyosin antibody: The specific localization of tropomyosin in nonmuscle cells, *J. Cell Biol.* **65**:549.

Lazarides, E., 1980, Intermediate filaments as mechanical integrators of cellular space, *Nature (London)* **283**:249.

Lee, P.W.K., Hayes, E.C., and Joklik, W.K., 1981a, Characterization of anti-reovirus immunoglobulins secreted by cloned hybridoma cell lines, *Virology* **108**:134.

Lee, P.W.K., Hayes, E.C., and Joklik, W.K., 1981b, Protein σ1 is the reovirus cell attachment protein, *Virology* **108**:156.

Lerner, A.M., Cherry, J.D., and Finland, M., 1963, Hemagglutination with reoviruses, *Virology* **19**:58.

Letvin, N.L., Greene, M.I., Benacerraf, B., and Germain, R.N., 1980, Immunologic effects of whole body ultraviolet (UV) irradiation: Selective defect in splenic adherent cell function *in vitro, Proc. Natl. Acad. Sci. U.S.A.* **77**:2881.

Letvin, N.L., Kauffman, R.S., and Finberg, R., 1981, T lymphocyte immunity to reovirus: Cellular requirements for generation and role in clearance of primary infections, *J. Immunol.* **127**:2334.

Leung, K.N., Ada, G.L., and McKenzie, J.F.C., 1980, Specificity, Ly phenotype and H-2 compatibility requirements of effector cells in delayed-type hypersensitivity responses to murine influenza virus infection, *J. Exp. Med.* **151**:815.

Li, J. K.-K., Keene, J.D., Scheible, P.P., and Joklik, W.K., 1980, Nature of the 3′ terminal sequence of the plus and minus strands of the S1 gene of reovirus serotypes 1, 2 and 3, *Virology* **105**:41.

Liew, F.Y., and Russell, S.M., 1980, Delayed-type hypersensitivity to influenza virus: Induction of antigen-specific suppressor T cells for delayed-type hypersensitivity to hemagglutinin during influenza virus infection in mice, *J. Exp. Med.* **151**:799.

Loh, P.C., and Oie, H.K., 1969, Growth characteristics of reovirus type 2: Ultraviolet light inactivated virion preparations and cell death, *Arch. Gesamte Virusforsch.* **26**:197.

Loh, P.C., and Shatkin, A.J., 1968, Structural proteins of reoviruses, *J. Virol.* **2**:1353.

Loh, P.C., and Soergel, M., 1965, Growth characteristics of reovirus type 2: Actinomycin D and the synthesis of viral RNA, *Proc. Natl. Acad. Sci. U.S.A.* **54**:857.

Loh, P.C., and Soergel, M., 1967, Macromolecular synthesis in cells infected with reovirus type 2 and the effect of ARA-C, *Nature (London)* **214**:622.

Long, W.F., and Burke, D.C., 1971, A comparison of interferon induction by reovirus RNA and synthetic double-stranded polypeptides, *J. Gen. Virol.* **12**:1.

Manthrope, C.M., Wilkin, G.P., and Wilson, G.E., 1977, Purification of viable ciliated cuboidal ependymal cells from rat brain, *Brain Res.* **134**:407.

Margolis, G., Kilham, L., and Gonatos, N., 1971, Reovirus type III encephalitis: Observations of virus–cell interactions in neural tissues. I. Light microscopy studies, *Lab Invest.* **24**:91.

Marshall, G.D., Thurman, G.B., Rossio, J.L., and Goldstein, A., 1981, *In vivo* generation of suppressor T cells by thymosin in congenitally athymic nude mice, *J. Immunol.* **126**:741.

Masters, C., Alpers, M., and Kakulas, B., 1977, Pathogenesis of reovirus type 1 hydrocephalus in mice: Significance of aqueductal changes, *Arch. Neurol.* **34**:18.

Mattingly, J.A., and Waksman, B.H., 1978, Immunologic suppression after oral administration of antigen. I. Specific suppressor cells formed in rat Peyer's patches after oral administration of sheep erythrocytes and their systemic migration, *J. Immunol.* **25**:941.

McCrae, M.A., and Joklik, W.K., 1978, The nature of the polypeptide encoded by each of the 10 double-stranded RNA segments of reovirus type 3, *Virology* **89**:578.

Millward, S., and Graham, A.F., 1970, Structural studies on reovirus: Discontinuities in the genome, *Proc. Natl. Acad. Sci. U.S.A.* **65**:422.

Mustoe, T.A., Ramig, R.F., Sharpe, A.H., and Fields, B.N., 1978a, A genetic map of reovirus. III. Assignment of dsRNA positive mutant groups A, B, and G to genome segments, *Virology* **85**:545.

Mustoe, T.A., Ramig, R.F., Sharpe, A.H., and Fields, B.N., 1978b, Genetics of reovirus: Identification of the dsRNA segments encoding the polypeptides of the μ and σ size classes, *Virology* **89**:594.

Nepom, J.T., Weiner, H.L., Dichter, M.A., Spriggs, D., Graham, C.F., Powers, M.L., Benacerraf, B., Fields, B.N., and Greene, M.I., 1982a, Identification of a hemagglutinin-specific idiotype associated with reovirus recognition shared by lymphoid and neural cells, *J. Exp. Med.* **155**:155.

Nepom, J.T., Tardieu, M., Epstein, R.L., Weiner, H.L., Gentsch, J., Fields, B.N., and Greene, M.I., 1982b, Antidiotype to a reovirus binding site related idiotype interacts with peripheral lymphocyte subsets, *Surv. Immunol. Res.* **1**:255.

Nonoyama, M., Watanabe, Y., and Graham, A.F., 1970, Defective virions of reovirus, *J. Virol.* **6**:226.

Ogra, P.L., Morag, A., and Tiku, M.L., 1975, Humoral immune response to viral infections, in: *Viral Immunology and Immunopathology* (A. Notkins, ed.), pp. 57–78, Academic Press, New York.

Oie, H.K., and Loh, P.C., 1971, Reovirus type 2: Induction of viral resistance and interferon production in fathead minnow cell, *Proc. Soc. Exp. Biol. Med.* **136**:369.

Onodera, T., Toniolo, A., Ray, U.R., Jensen, A.B., Knazek, R.A., and Notkins, A.L., 1981, Virus-induced diabetes mellitus. XX. Polyendocrinopathy and autoimmunity, *J. Exp. Med.* **153**:1457.

Owen, R.L., 1977, Sequential uptake of horseradish peroxidase by lymphoid follicle epithelium of Peyer's patches in the normal unobstructed mouse intestine: An ultrastructural study, *Gastroenterology* **72**:440.

Owen, R.L., and Jones, A.L., 1974, Epithelial cell specialization within human Peyer's patches: An ultrastructural study of intestinal lymphoid follicles, *Gastroenterology* **66**:189.

Pestka, S., and Baron, S., 1981, Definition and classification of the interferons, in: *Methods in Enzymology*, Vol. 78, Part A (S. Pestka, ed.), pp. 3–14, Academic Press, New York.

Peter, T.J., 1973, The hydrolysis of glycine oligopeptides by guinea pig intestinal mucosa and by isolated brush borders, *Clin. Sci. Mol. Med.* **45**:803.

Phillips, P.A., Keast, D., Papadimitriou, J.M., Walters, M.N., and Stanley, N.F., 1969, Chronic obstructive jaundice induced by reovirus type 3 in weanling mice, *Pathology* **1**:193.

Raine, C.S., and Fields, B.N., 1973, Ultrastructural features of reovirus type 3 encephalitis, *J. Neuropathol. Exp. Neurol.* **32**:19.

Ramig, R.F., Cross, R.K., and Fields, B.N., 1977, Genome RNAs and polypeptides of reovirus serotypes 1, 2, and 3, *J. Virol.* **22**:726.

Ramig, R.F., Mustoe, T.A., Sharpe, A.H., and Fields, B.N., 1978, A genetic map of reovirus. II. Assignments of dsRNA negative mutant groups C, D, and E to genome segments, *Virology* **85**:531.

Rhim, J.S., Jordan, L.E., and Mayor, H.D., 1962, Cytochemical, fluorescent-antibody and electron microscopic studies on the growth of reovirus (ECHO 10) in tissue culture, *Virology* **17**:342.

Rifkin, D.B., and Quigley, J.P., 1974, Virus-induced modification of cellular membranes related to viral structure, *Annu. Rev. Microbiol.* **28**:325.

Rosen, L., 1960, Serologic grouping of reoviruses by hemaglutination-inhibition, *Am. J. Hyg.* **71**:243.

Rosen, L., 1962, Reoviruses in animals other than man, *Ann. N. Y. Acad. Sci.* **101**:461.

Rosen, L., and Abinanti, F.R., 1960, Natural and experimental infection of cattle with human types of reovirus, *Am. J. Hyg.* **71**:424.

Rosen, L., Abinanti, F.R., and Hovin, J.F., 1963, Further observations on the natural infections of cattle with reoviruses, *Am. J. Hyg.* **77**:38.

Rubin, D.H., and Fields, B.N., 1980, Molecular basis of reovirus virulence: Role of the M2 gene, *J. Exp. Med.* **152**:853.

Rubin, D., Weiner, H.L., Fields, B.N., and Greene, M.I., 1981, Immunologic tolerance after oral administration of reovirus: Requirement for two viral gene products for tolerance induction, *J. Immunol.* **127**:1697.

Schrader, J.W., and Edelman, G.M., 1977, Joint recognition by cytotoxic T cells of inactivated Sendai virus and products of the major histocompatibility complex, *J. Exp. Med.* **145**:523.

Schuerch, A.R., Matshuhisa, I., and Joklik, W.K., 1974, Temperature-sensitive mutants of reovirus. VI. Mutants ts447 and ts556 particles that lack one or two L genome RNA segments, *Intervirology* **3**:36.

Sharpe, A.H., and Fields, B.N., 1981, Reovirus inhibition of cellular DNA synthesis: Role of the S1 gene, *J. Virol.* **38**:389.

Sharpe, A.H., Ramig, R.F., Mustoe, T.A., and Fields, B.N., 1978, A genetic map of reovirus. I. Correlation of genome RNAs between serotypes 1, 2 and 3, *Virology* **84**:63.

Sharpe, A.H., Chen, L.B., Murphy, J.R., and Fields, B.N., 1980, Specific disruption of vimentin filament organization in monkey kidney CV-1 cells by diphtheria toxin, exotoxin A, and cycloheximide, *Proc. Natl. Acad. Sci. U.S.A.* **77**:7267.

Sharpe, A.H., Chen, L.B., and Fields, B.N., 1982, The interaction of mammalian reoviruses with the cytoskeleton of monkey kidney CV-1 cells, *Virology* **120**:399.

Shatkin, A.J., and Rada, B., 1967, Reovirus-directed ribonucleic acid synthesis in infected L cells, *J. Virol.* **1**:24.

Shatkin, A., Sipe, J.D., and Loh, P., 1968, Separation of ten reovirus segments by polyacrylamide gel electrophoresis, *J. Virol.* **2**:986.

Shaw, J.E., and Cox, D.C., 1973, Early inhibition of cellular DNA synthesis by high multiplicities of infectious and UV-irradiated reovirus. *J. Virol.* **12**:704.

Shelton, I.H., Kasupski, G.J., Oblin, C., and Hand, R., 1981, DNA binding of a nonstructural reovirus protein, *Can. J. Biochem.* **59**:122.

Skup, D., and Millward, S., 1980, Reovirus-induced modification of cap-dependent translation in infected L cells, *Proc. Natl. Acad. Sci. U.S.A.* **77**:152.

Smith, R.E., Zweerink, H.J., and Joklik, W.K., 1969, Polypeptide components of virions, top component, and cores of reovirus type 3, *Virology* **39**:791.

Spandidos, D.A., and Graham, A.F., 1976, Recombination between temperature-sensitive and deletion mutants of reovirus, *J. Virol.* **18**:117.

Spendlove, R.S., Lennette, E.H., Knight, C.O., and Chin, J.H., 1963, Development of viral antigen and infectious virus in HeLa cells infected with reovirus, *J. Immunol.* **90**:548.

Spendlove, R.S., Lennette, E.H., Chin, J.N., and Knight, C.O., 1964, Effect of antimitotic agents on intracellular reovirus antigen, *Cancer Res.* **24**:1826.

Stanley, N.F., and Joske, R.A., 1975a, Animal model: Chronic murine hepatitis induced by reovirus type 3, *Am. J. Pathol.* **80**:181.

Stanley, N.F., and Joske, R.A., 1975b, Animal model: Chronic biliary obstruction caused by reovirus type 3, *Am. J. Pathol.* **80**:185.

Subasinghe, H.A., and Loh, P.C., 1972, Reovirus cytotoxicity: Some properties of the UV-irradiated reovirus and its capsid proteins, *Arch. Gesamte Virusforsch.* **39**:172.

Taber, R., Alexander, V., and Whitford, W., 1976, Persistent reovirus infection of CHO cells resulting in virus resistance, *J. Virol.* **17**:513.

Tardieu, M., and Weiner, H., 1982, Viral receptors on isolated murine and human ependymal cells, *Science* **215**:419.

Trachsel, H., Sonenberg, N., Shatkin, A.J., Rose, J.K., Leong, K., Bergmann, J.E., Gordon, J., and Baltimore, D., 1980, Purification of a factor that restores translation of VSV mRNA in extracts from poliovirus-infected HeLa cells, *Proc. Natl. Acad. Sci. U.S.A.* **77**:770.

Walsh, M.L., Jen, J., and Chen, L.B., 1979, Transport of serum components into structures similar to mitochondria, in: *Cell Proliferation*, Vol. 6, pp. 513–520, Cold Spring Harbor Press, New York.

Walters, M., Leaks, N.I., Joske, P.I., and Stanley, N.F., 1963, Murine infection with reovirus: Pathology of the acute phase, *Br. J. Exp. Pathol.* **444**:27.

Weber, K., Pollack, R., and Bibring, T., 1975, Antibody against tubulin: The specific visualization of cytoplasmic microtubules in tissue culture cells, *Proc. Natl. Acad. Sci. U.S.A.* **72**:459.

Webster, R.G., and Laver, W.G., 1967, Preparation and properties of antibody directed specifically against the neuraminidase of influenza virus, *J. Immunol.* **99**:49.

Weiner, H.L., and Fields, B.N., 1977, Neutralization of reovirus: The gene responsible for the neutralization antigen, *J. Exp. Med.* **146**:1305.

Weiner, H.L., Drayna, D., Averill, D.R., Jr., and Fields, B.N., 1977, Molecular basis of reovirus: Role of the S1 gene, *Proc. Natl. Acad. Sci. U.S.A.* **74**:5744.

Weiner, H.L., Ramig, R.F., Mustoe, T.A., and Fields, B.N., 1978, Identification of the gene coding for the hemagglutinin of reovirus, *Virology* **86**:581.

Weiner, H.L., Ault, K.A., and Fields, B.N., 1980a, Interaction of reovirus with cell surface receptors. I. Murine and human lymphocytes have a receptor for the hemagglutinin of reovirus type 3, *J. Immunol.* **124**:2143.

Weiner, H.L., Greene, M.I., and Fields, B.N., 1980b, Delayed hypersensitivity in mice infected with reovirus. I. Identification of host and viral gene products responsible for the immune response, *J. Immunol.* **125**:278.

Weiner, H., Powers, M.L., and Fields, B.N., 1980c, Absolute linkage of virulence with the central nervous system cell tropism of reovirus to hemagglutinin, *J. Infect. Dis.* **141**:609.

Wiebe, M.E., and Joklik, W.K., 1975, The mechanism of inhibition of reovirus replication by interferon, *Virology* **66**:229.

Willcox, N., and Mautner, V., 1976a, Antigenic determinants of adenovirus capsids. I. Measurement of antibody cross-reactivity, *J. Immunol.* **116**:19.

Willcox, N., and Mautner, V., 1976b, Antigenic determinants of adenovirus capsids. II. Homogeneity of hexons, and assessibility of their determinants in the virion, *J. Immunol.* **116**:25.

Willey, D.E., and Ushijima, R.N., 1980, The circadian rhythm of thymosin therapy during acute reovirus type 3 infection of neonatal mice, *Clin. Immunol. Immunopathol.* **16**:72.

Willey, D.E., and Ushijima, R.N., 1981, The effect of thymosin on murine lymphocyte responses and corticosteroid levels during acute reovirus type 3 infection of neonatal mice, *Clin. Immunol. Immunopathol.* **19**:35.

Wolf, J.L., Rubin, D.H., Finberg, R., Kaufman, R.S., Sharpe, A.H., Trier, J.S., and Fields, B.N., 1981, Intestinal M cells: A pathway for entry of reovirus into the host, *Science* **212**:471.

Yap, K.L., Ada, F.L., and McKenzie, I.F.C., 1978, Transfer of specific cytotoxic T lymphocytes protects mice inoculated with influenza virus, *Nature (London)* **273**:238.

Younger, J.S., and Preble, O.T., 1980, Viral persistence: Evolution of viral populations, in: *Comprehensive Virology*, Vol. 16 (H. Fraenkel-Conrat and R. Wagner, eds.), pp. 73–135, Plenum Press, New York.

Zarbl, H., Skup, D., and Millward, S., 1980, Reovirus progeny subviral particles synthesize uncapped mRNA, *J. Virol.* **34**:497.

Zinkernagel, R.M., 1976, H-2 restriction of virus-specific T cell-mediated effector functions *in vivo*. II. Adoptive transfer of delayed-type hypersensitivity to murine lymphocytic choriomeningitis virus is restricted by the K and D region of H-2, *J. Exp. Med.* **144**:776.

Zinkernagel, R.M., and Blanden, R.V., 1975, Macrophage activation in mice lacking thymus-derived (T) cells, *Experientia* **31**:591.

Zinkernagel, R.M., and Welsh, R.M., 1976, H-2 compatibility requirement for virus-specific T cell-mediated effector functions *in vivo*. I. Specificity of T cells conferring antiviral protection against lymphocytic choriomeningitis virus is associated with H-2K and H-2D, *J. Immunol.* **117**:1495.

Zweerink, H.J., and Joklik, W.K., 1970, Studies on the intracellular synthesis of reovirus-specified proteins, *Virology* **44**:501.

## CHAPTER 7

# Orbiviruses

BARRY M. GORMAN, JILL TAYLOR,
AND PETER J. WALKER

## I. INTRODUCTION

The name *Orbivirus* was proposed by Borden *et al.* (1971) to describe a
number of arthropod-borne viruses that by morphological and physicochemical criteria formed a distinct group. The derivation from the Latin
"orbis" ("ring" or "circle") was appropriate, since negatively stained virus
particles when examined in the electron microscope have large doughnut
shaped capsomers. Borden *et al.* (1971) distinguished ten serological
groups of orbiviruses including the viruses of African horsesickness and
bluetongue of sheep. Although it had long been known that these viruses
were arthropod-borne, they differed from most of the recognized arboviruses in their relative resistance to inactivation by solvents and detergents. Since most of the arboviruses known at that time contained lipid
envelopes, sensitivity to solvents and detergents became one criterion
for the classification of arboviruses (Theiler, 1957).

Verwoerd (1969) purified bluetongue virus and showed that its genome consisted of double-stranded RNA (dsRNA). He was the first to recognize the need to establish a new taxonomic group to include the known
dsRNA viruses and those arboviruses with physicochemical and morphological properties similar to those of bluetongue virus. Although not
all the viruses he listed had been shown to contain dsRNA, Verwoerd
(1970a) proposed the name diplornaviruses for the new taxonomic group.

The International Committee on Taxonomy of Viruses decided that
the bluetonguelike viruses would be defined by the genus *Orbivirus*

BARRY M. GORMAN, JILL TAYLOR, AND PETER J. WALKER • Queensland Institute
of Medical Research, Bramston Terrace, Brisbane 4006, Australia.

within the family Reoviridae (Fenner, 1976). The reoviruses and orbiviruses differ in ecology, and they have distinctive structures with concomitant differences in the stability of virus particles, but the genome strategies of viruses of the two genera have many features in common.

Verwoerd *et al.* (1970) resolved the bluetongue virus genome into 10 segments by polyacrylamide gel electrophoresis (PAGE). The patterns of separation of the genome segments of a reovirus and a bluetongue virus were different, and the molecular weight of the genome of bluetongue virus was estimated to be $12 \times 10^6$ compared to $15 \times 10^6$ for the reovirus (Verwoerd *et al.*, 1970). Representative viruses from each of the recognized serogroups have been found to contain genomes similar to that of bluetongue virus (for a review, see Gorman, 1979) except for viruses of the Colorado tick fever serogroup. Recent evidence that the genomes of Colorado tick fever viruses consist of 12 segments of dsRNA (Knudson, 1981) and have a molecular weight of approximately $18 \times 10^6$ indicates that these viruses will become part of a new taxonomic group. However, for historical reasons, the Colorado tick fever viruses will be discussed in this review of orbiviruses.

The orbiviruses have been reviewed recently in this series (Verwoerd *et al.*, 1979), and the viruses of African horsesickness and bluetongue have been reviewed extensively because of their importance in veterinary medicine and the livestock industries (Bowne, 1971; Cox, 1954; Howell, 1963a,b; Howell and Verwoerd, 1971). Since these reviews include descriptions of the pathology of the diseases caused by the viruses, this aspect will not be considered herein.

In this chapter, we will review the biological and biochemical properties of orbiviruses, particularly as they relate to the segmented nature of the virus genomes.

## A. Classification of Orbiviruses

The described orbiviruses are differentiated into 12 serological groups (Table I). No common generic antigen has been found, but viruses within each group share antigens detectable in complement-fixation (CF) tests, immunodiffusion tests, and fluorescent-antibody tests. The serotypes within each group are distinguishable by specific reactions in serum-neutralization tests.

The emphasis placed by different workers on the significance of cross-reactions in serological tests has created some confusion in the classification of orbiviruses. Within most of the serogroups, the serotypes cannot be distinguished by their reactions in the group-specific tests; in others, the serotypes can be clustered within the serogroup by their cross-reactions in CF tests. The serotypes of the Kemerovo serogroup can be clustered into four subgroups [see Table V (Section I.C.7)]. The viruses Warrego and Mitchell River are recognized as serotypes of the Warrego

TABLE I. Orbivirus Serological Groups

| Serogroup | Number of serotypes |
|---|---|
| African horsesickness | 9 |
| Bluetongue | 20 |
| Epizootic hemorrhagic disease | 7 |
| Eubenangee | 3 |
| Colorado tick fever[a] | 2 |
| Palyam | 6 |
| Changuinola | 7 |
| Corriparta | 3 |
| Kemerovo | 20 |
| Warrego | 2 |
| Wallal | 2 |
| Equine encephalosis | 5 |

[a] Recent evidence (Knudson, 1981) suggests that these viruses should be excluded from the genus *Orbivirus*.

serogroup, although the viruses are distantly related and give low-level cross-reactions in CF tests (Borden *et al.*, 1971; Doherty *et al.*, 1973).

The problem is further complicated by reports of low-level cross-reactions among viruses that are normally considered as members of distinct serogroups. There has been reluctance to include viruses of the Eubenangee and epizootic hemorrhagic disease (EHD) of deer serogroups in the bluetongue virus serogroup despite many reports of distant serological relationships among viruses in these groups (Moore and Lee, 1972; Moore, 1974; St. George *et al.*, 1978; Gorman and Taylor, 1978; Della-Porta *et al.*, 1979). The bluetongue viruses are important pathogens of sheep, and it appears difficult to justify the inclusion of viruses of the Eubenangee serogroup, which are not known to cause disease in animals, in an extended bluetongue virus serogroup. Gorman (1979) suggested that the recognized serogroups of orbiviruses be clustered into higher taxonomic units to allow a more meaningful comparison of the relationships among viruses. The exclusive use of immunological reactions in defining orbivirus groups underestimates the degree of genetic similarity among viruses.

The division of some serogroups into numbered serotypes and the formation of other serogroups with viruses of different names has also led to confusion in defining orbiviruses. Historically, the bluetongue and horsesickness groups were established by workers isolating viruses that were related to known serotypes but gave distinct reactions in cross-protection tests in animals or in serum-neutralization tests. Most of the other orbiviruses were isolated as by-products in programs to isolate and identify viruses as the causative agents of arthropod-transmitted disease. The identification of arboviruses as distinct serotypes is made at the Yale Arbovirus Research Unit, New Haven, Connecticut. New viruses are given names and are registered in the *International Catalogue of Arboviruses* (Berge, 1975). In this way, most of the orbivirus serogroups are named from the first member isolated and consist of viruses with different names. The use of names to describe serotypes in some groups and numbers to describe viruses of the horsesickness and bluetongue serogroups obscures the fact that the serogroups are composed of viruses related by the use of the same serological tests. It is unlikely that future isolates of bluetongue viruses would be given trivial names or that named viruses of, for example, the Kemerovo group would be allocated numbers. A compromise system of numbering and designation of strain was suggested by Gorman (1979). In the discussion of relationships among orbiviruses, we will refer to serotype numbers or names to avoid reference to a coding system and to avoid unnecessary confusion.

The difficulty in classification of orbiviruses may be due to the fact that the viruses contain segmented genomes and that gene reassortment occurs at least between serologically related viruses (for a review, see Gorman, 1979). The extent of variation possible through genetic reassortment is not known. The phenomenon has been described only for serotypes of the Wallal serogroup (Gorman *et al.*, 1978), the bluetongue serogroup (Gorman *et al.*, 1982), and the Eubenangee serogroup (Taylor, unpublished results), and has not yet been shown to occur between viruses in different serogroups. Within the genus *Orbivirus*, the recognized serological groups are considered species of viruses (Fenner, 1976; Matthews, 1979). The viruses of the bluetongue, Eubenangee, and Wallal serogroups are designated as distinct species so that the available evidence suggests that gene reassortment occurs within defined species of viruses.

Although evidence for gene reassortment between viruses in different serogroups has not yet been found, the possibility of genetic reassortment between serologically unrelated viruses leads to a consideration of genetically interacting groups of viruses. If the process of independent assortment of genes between two viruses is considered analagous to sexual reproduction in higher organisms, then genetic diversity within and among orbivirus species can be explained by reference to modern concepts of the structure of natural populations of organisms.

Aspects of speciation in orbiviruses are discussed in Section VI. However, low-level cross-reactions in serological tests between viruses normally considered as members of distinct serogroups could result from rare reassortment events between viruses in different serogroups. The introduction of novel genes into the available gene pool of an *Orbivirus* species (serogroup) could lead to greater diversity within that group, but need not represent a significant shift of the major antigenic determinants in the group. It is equally possible that the shared minor antigens reflect common ancestry of virus species that have diverged significantly in evolution. The shared antigen may indicate true evolutionary resemblance between distinct populations of organisms.

## B. Disease Relationships

Bluetongue virus causes a degenerative and sometimes fatal disease of sheep that has been described as being "characterized by congestion of the buccal and nasal mucosa and the coronary tissue of the hooves, and stiffness due to muscle degeneration" (Robertson, 1976). Signs of the disease can range from subclinical to severe depending on the susceptibility of the infected host and on the pathogenicity of the infecting strain of virus (Bowne, 1971). In enzootic areas, infection of cattle is mostly mild or inapparent (Owen et al., 1965; Bowne et al., 1968), but outbreaks of clinical bluetongue in cattle have been reported (Komarov and Goldsmit, 1951). The virus is also thought to cause hydranencephaly in calves and lambs (McKercher et al., 1970; Cordy and Schultz, 1961; Griner et al., 1964; Osburn et al., 1971a,b; Luedke et al., 1970; Barnard and Pienaar, 1976). Mortality due to bluetongue virus has also been observed in a wide range of naturally and experimentally infected nondomesticated ruminants (Hoff and Trainer, 1978).

EHD is similar in many clinical and pathogenic signs to experimental bluetongue in deer (Karstad and Trainer, 1967). The major differences are the absence of extensive buccal erosions and foot lesions and a greater tendency toward thrombosis. Naturally occurring mortality due to EHD has been reported for white-tailed deer, mule deer, and pronghorn antelope (Shope et al., 1960; Hoff et al., 1973). Serological evidence of EHD infection of sheep and cattle has been reported, but experimental infection of sheep resulted in no overt symptoms of disease (Verwoerd et al., 1979). Included in the EHD serogroup is Ibaraki virus, which causes an acute febrile illness in cattle (but not sheep) with clinical and pathological signs similar to those of bluetongue disease (Omori et al., 1969; Inaba et al., 1966; Inaba, 1975).

Two serogroups of orbiviruses are associated with equine disease. African horsesickness viruses cause an acute febrile illness in horses, mules, and donkeys that may often be fatal (Henning, 1956). Clinical signs may manifest as a pulmonary edema or as a cardiovascular disease with localized subcutaneous edema of the head, and cardiac hemorrhage, edema, necrosis, and myocarditis (Maurer and McCully, 1963). Erasmus (1972) also described a mild form with no clinical symptoms other than a remittent febrile reaction and an acute form with both pulmonary and cardiac involvement. Subclinical infection has been reported in zebra, which may act as a reservoir host (Erasmus et al., 1976b). A relationship between the severity of disease and the immunological status of the susceptible host has been suggested (Erasmus, 1972). Virus strains may contain mixed populations of particles with different tissue tropisms. Virus replication occurs in cells of the lymph nodes, spleen, thymus, and pharyngeal mucosa (Erasmus, 1972).

Equine encephalosis virus was isolated from sporadic cases of peracute death of horses in South Africa in 1967. Autopsy examination revealed general venous congestion, fatty liver degeneration, brain edema, and areas of catarrhal enteritis (Erasmus *et al.*, 1970). Since that time, three other serologically related viruses have been isolated from horses with febrile illness of varying degrees of severity and were found in high concentration in the organs of aborted fetuses (Erasmus *et al.*, 1976a). Although the incidence of clinical disease is low, serological studies have revealed a high incidence of natural infection.

Several tick-borne orbiviruses have been identified as human pathogens. Colorado tick fever virus causes an acute febrile illness that is rarely fatal. Several hundred cases of the disease occur annually in the northwestern mountainous regions of the United States, where the virus is endemic (Burgdorfer, 1977). A serologically related virus (Eyach) that was isolated in the Federal Republic of Germany was not associated with human disease (Rehse-Kupper *et al.*, 1976). Tick-borne orbiviruses of the Kemerovo serogroup have been associated with aseptic meningitis in humans in central and eastern Europe (Libikova *et al.*, 1970; Libikova and Casals, 1971). The number of recorded clinical cases is small, partly due to difficulties in eliminating other viruses as potential causes of the disease. However, serological evidence suggests a relatively high rate of antibody conversion in endemic areas (Libikova *et al.*, 1964, 1970, 1978; Gresikova *et al.*, 1965). Antibodies to Tribec (Kemerovo group) and Eyach (Colorado tick fever group) were found in patients with various neuropathies (Málková *et al.*, 1980).

## C. Isolation of Orbiviruses and the Relationships among Them

### 1. African Horsesickness

African horsesickness has probably been recognized as a clinical entity since the occupation of the Cape of Good Hope by the Dutch East India Company (Henning, 1956). M'Fadyean (1900) showed that the agent was filterable, using samples of infected horse blood taken to London by the English veterinarian W. Robertson (Waterson and Wilkinson, 1978). The agent had been under study in South Africa from 1887, and according to Howell (1962), the significance of the antigenic plurality of strains was realized by Theiler as early as 1908. In numerous experiments over many years, Theiler showed that immunity in horses and mules after challenge with homologous strains was solid, but when heterologous strains were used, a percentage of animals contracted the disease in varying degrees of severity (Theiler, 1921). The finding by Alexander (1933) that the white mouse was susceptible to an intracerebral inoculation of virus led to the development of neutralization tests in mice. McIntosh (1958) showed that the strains could be grouped into seven immunological types, and Howell (1962) added two more types. Davies and Lund (1974) reported the isolation in Kenya of a strain of virus (G-75) that was not neutralized by antiserum to any of the nine known serotypes and must therefore be considered a new serotype (Davies, 1976).

McIntosh (1958) found seven serotypes of horsesickness virus indistinguishable by CF tests. Pavri (1961) prepared hemagglutinins to horse erythrocytes from horsesickness-virus-infected mouse brain suspensions and found that the hemagglutination-inhibition (HT) types corresponded to the neutralization types (Pavri and Anderson, 1963).

Ozawa and Hafez (1973) used hyperimmune rabbit antisera prepared against horsesickness, bluetongue, and Ibaraki (EHD group) viruses to detect relationships in agar gel diffusion tests. An antiserum to horsesickness virus type 9 cross-reacted with bluetongue type 16, but there was no cross-reaction between horsesickness virus and antiserum to bluetongue virus. The significance of this report is difficult to assess, since no other published evidence for relationships between the two virus serogroups exists. The extensive research on viruses in both serogroups at the Veterinary Research Institute at Onderstepoort for more than 80 years suggests no serological relationships between the two groups of viruses. Verwoerd and Huismans (1969) compared the genomes of bluetongue type 10 and horsesickness type 4 by hybridization of RNA of one virus with that of the other. They found a slight but variable amount of hybridization (about 4%) between the RNAs of the viruses, compared to extensive hybridization (64%) between the RNAs of two bluetongue serotypes. The analysis was based on comparisons of total genome RNA of the viruses. Since each of the four small genome segments of both viruses constitutes less than 5% of the total molecular weight of the genomes, the possibility cannot be overlooked that regions of sequence homology in one of the genome segments could account for the small and variable amount of hybridization found. The difficulties in interpreting data obtained from hybridization of genomes of orbiviruses are discussed in Section V.B. The available evidence does not preclude the possibility of genetic relationships between viruses of the horsesickness and bluetongue serological groups.

Horsesickness is widely distributed through Africa, Egypt, Iran, West Pakistan, Afghanistan, and India. Transmission is mainly by nocturnal biting insects, with *Culicoides pallidipennis* being the main vector (Mellor *et al.*, 1975).

## 2. Bluetongue Complex of Orbivirus Serogroups

Viruses in the bluetongue, EHD, and Eubenangee serogroups cross-react in some serological tests and are considered a cluster of viruses (Gorman, 1979). The EHD and Eubenangee serogroups were established before their relationships with bluetongue viruses were recognized, and serotypes within each group are listed in Table II. There is reluctance to type viruses as bluetongue unless reactions of virtual identity are found in the group-specific tests [CF, agar–gel precipitin (AGP) and fluorescent-antibody. The problems in classification of bluetongue-related viruses are discussed in Section I.A, but the relationships among the viruses must be taken into account in any taxonomic arrangement and discussion of the evolution of viruses. Viruses in each serogroup will be discussed separately and the relationships among them indicated.

### a. Bluetongue

The literature on the isolation and identification of bluetongue viruses has been reviewed extensively (Bowne, 1971; Howell, 1963b; Howell and Verwoerd, 1971).

The first descriptions of malarial catarrhal fever in sheep were given by Hutcheon (1902), while Theiler and Robertson (quoted in Spreull, 1905) proved that the agent was filterable. Theiler (1908) immunized sheep by injection of a mild strain of virus (now designated as type 4) that had been serially passaged in sheep. The early literature suggested that reinfection of recovered sheep was not unusual (Spreull, 1905; Theiler, 1906). Despite evidence that the vaccine was not safe and the resultant immunity was inadequate, the monovalent vaccine was used for more than 40 years in South Africa. The vaccine consisted of Theiler's original strain from 1907 until 1928, when it was replaced by the Veglia strain. In 1948, Theiler's strain was reintroduced.

Although distinct antigenic types of horsesickness virus were recognized by Theiler as early as 1908 (see Howell, 1962), Neitz (1948) was the first to recognize different antigenic types of bluetongue virus. In a series of cross-protection tests in sheep using ten strains of bluetongue virus isolated over a period of 40 years, Neitz concluded that each strain produced solid immunity against itself but a variable degree of protection to challenge by heterologous strains. It was not possible to group strains on the basis of common antigenic structure (Neitz, 1948).

The first polyvalent vaccine was prepared from the strains Theiler (type 4), Mimosa Park (type 3), Cyprus (type 3), and Estantia (type 12). From time to time, the composition of the vaccine was altered to include new strains empirically selected on the basis of their occurrence in the field (Howell, 1969). Kipps (1956) demonstrated a common antigen in 6 bluetongue virus strains using CF tests. Howell (1960) defined 12 antigenic types of viruses by serum-neutralization tests in primary lamb kidney cells and subsequently classified 244 naturally occurring isolates into 16 antigenic groups (Howell, 1970). The significance of immunologically distinct serotypes in cross-protection immunity is difficult to assess. Neitz (1948) demonstrated reciprocal cross-immunity between the Bekker and Theiler strains, which Howell classified as type 4. Similarly, the Cyprus and Mimosa Park strains cross-protect (Neitz, 1948) and were classified as type 3 by Howell (1969). No extensive studies using 20 recognized serotypes (Verwoerd *et al.*, 1979) in cross-protection tests in animals have been reported.

TABLE II. Bluetongue Complex of Orbivirus Serogroups

| Serotype | Prototype strain | Isolation | | Ref. No[a] |
| | | Year | Country | |
|---|---|---|---|---|
| Bluetongue | | | | |
| 1 | Biggarsberg | 1958 | South Africa | 1 |
| 2 | Ermelo 22/59 | 1959 | South Africa | 1 |
| 3 | Cyprus sample B | 1943 | Cyprus | 1 |
| 4 | Vaccine batch 603 | 1900 | South Africa | 1 |
| 5 | Mossop | 1953 | South Africa | 1 |
| 6 | Strathene | 1958 | South Africa | 1 |
| 7 | Utrecht | 1955 | South Africa | 1 |
| 8 | Ermelo 89/59 | 1959 | South Africa | 1 |
| 9 | University Farm | 1942 | South Africa | 1 |
| 10 | Ermelo 91/59 | 1959 | South Africa | 1 |
| 11 | Nelspoort | 1944 | South Africa | 1 |
| 12 | Bynespoort | 1941 | South Africa | 1 |
| 13 | Mt. Currie | 1959 | South Africa | 1 |
| 14 | Ermelo 87/59 | 1959 | South Africa | 1 |
| 15 | Onderstepoort 133/60 | 1960 | South Africa | 1 |
| 16 | Hazara | 1960 | Pakistan | 1 |
| 17 | Wyoming 2790 | 1962 | United States | 2 |
| 18 | — | — | South Africa | 1 |
| 19 | — | — | South Africa | 1 |
| 20 | CSIRO 19 | 1975 | Australia | 3 |
| Epizootic hemorrhagic disease | | | | |
| 1 | New Jersey | 1955 | United States | 4 |
| 2 | Alberta | 1962 | Canada | 5 |
| 3 | Ib Ar 22619 | 1967 | Nigeria | 6 |
| 4 | Ib Ar 33853 | 1968 | Nigeria | 6 |
| 5 | Ib Ar 49630 | 1970 | Nigeria | 6 |
| 6 | XBM/67 | 1967 | South Africa | 7 |
| 7 | Ibaraki-2 | 1959 | Japan | 4 |
| Eubenangee | | | | |
| 1 | Eubenangee In 1074 | 1963 | Australia | 4 |
| 2 | Pata Dak Ar 1327 | 1968 | Central African Republic | 4 |
| 3 | Tilligerry NB 7080 | 1971 | Australia | 4 |

[a] References: (1) Howell and Verwoerd (1971); (2) Barber (1979); (3) St. George et al. (1978); (4) Berge (1975); (5) Barber and Jochim (1975); (6) V. H. Lee et al. (1974); (7) Lecatsas and Erasmus (1973).

The validity of the classification of bluetongue viruses into distinct serotypes may also be questioned. In early work, Howell (1960) used sheep convalescent sera and found some cross-neutralization between heterologous strains of bluetongue virus. These cross-reactions were virtually eliminated in subsequent tests by using hyperimmune guinea pig sera (Howell and Verwoerd, 1971). Thomas and Trainer (1971) used convalescent and hyperimmune sera of calves to compare seven bluetongue virus strains in plaque-reduction tests. Convalescent sera cross-reacted with heterologous virus strains, but the reactions were less than those observed using hyperimmune sera. Barber and Jochim (1973) studied the serological characters of ten strains of bluetongue virus isolated in the United States by plaque-reduction tests. Using hyperimmune sheep sera, the strains were classified into four serotypic groups, but cross-reactions among all ten strains were detected using hyperimmune rabbit sera.

The classification of viruses into neat discrete groups seems improbable given the virtually continuous nature of biological variation. Thomas *et al.* (1979) found extensive cross-reactions among isolates of bluetongue virus in North America and suggested that an antigenic continuum of virus isolates existed instead of clearly defined antigenic groups. The serological comparison of the Australian isolate CSIRO 19 with reference bluetongue viruses provides further evidence of the need to reassess the serum-neutralization tests used to identify bluetongue virus serotypes. Although the strain had been designated as a new serotype (type 20) by the World Reference Centre, The Veterinary Research Institute, Onderstepoort (Verwoerd *et al.*, 1979), Della-Porta *et al.* (1981a) found the isolate indistinguishable from type 4 using plaque-reduction, plaque-inhibition, or quantal microtiter neutralization tests. The viruses could be distinguished only by analysis of linear-regression curves of plaque-reduction assay results (Westaway, 1965). Della-Porta *et al.* (1981a) commented that the isolate CSIRO 19 could be considered as a subtype of type 4 and questioned its designation as a new serotype. Since animals inoculated with CSIRO 19 were protected against challenge with virulent type 4 or type 17 viruses, the practical value of a classification system based on *in vitro* reactions using carefully selected antisera must be questioned.

Polyvalent live vaccines are in current use in South Africa, where animals are inoculated with pentavalent vaccines three times at weekly intervals (B.J. Erasmus, personal communication). Thus, each animal is exposed to 15 bluetongue virus serotypes. Bluetongue viruses contain segmented genomes (Section II.B), and recombinant viruses deriving genome segments from 3 bluetongue virus serotypes have been isolated in cell culture (Section IV.B.2) and detected in nature (Section V.C.1). The problems associated with antigenic diversity in bluetongue viruses may be increased by the use of polyvalent live vaccines.

A wide variety of immunological tests have been used in attempts to differentiate bluetongue viruses (for a review, see Verwoerd *et al.*, 1979). Recent applications include enzyme-linked immunosorbent assays (ELISAs) (Manning and Chen, 1980; Hubschle *et al.*, 1981) and a renewed interest in HI tests (Hubschle, 1980; Van der Walt, 1980).

A passive hemagglutination (PHA) test for detection of bluetongue virus antibodies had been described (Blue *et al.*, 1974) in which partially purified virus was used to sensitize tannic-acid-treated equine erythrocytes and the erythrocytes were used in a PHA test. Although some type specificity was found, the test has not been widely used.

Despite many attempts, no hemagglutinin had been detected for bluetongue virus (Howell and Verwoerd, 1971; Verwoerd et al., 1979). Recently, Hubschle (1980) and Van der Walt (1980) reported that partially purified bluetongue viruses agglutinated erythrocytes of various species and that the reaction was independent of pH between 6 and 9 and of temperature between 4 and 37°C. Van der Walt (1980) found that an avirulent strain of type 10 agglutinated erythrocytes of sheep, goose, rabbit, and man. Hubschle (1980) reported that the BT8 strain from the United States (bluetongue type 10) agglutinated sheep erythrocytes only and that bluetongue viruses of serotypes 3, 8, and 10 from South Africa agglutinated erythrocytes of sheep, guinea pig, mouse, and chicken. Van der Walt (1980) found the hemagglutination test to be serotype-specific.

Manning and Chen (1980), using peroxidase-conjugated anti-species immunoglobulin G, found the ELISA test group reactive for bluetongue viruses. A similar group-reactive test was described by Hubschle et al. (1981).

Jochim and Jones (1980) developed a hemolysis in gel (HIG) test for bluetongue viruses. Like the CF test, the HIG test was group-reactive, but bluetongue and EHD viruses were differentiated in the test.

The isolation of strains of bluetongue virus is shown in Table II. The designation of prototype strains for bluetongue serotypes is based on Howell and Verwoerd (1971). No details of the isolation of types 18 and 19 have been published, but the serotypes are referred to in Verwoerd et al. (1979).

Bluetongue viruses have a wide geographic distribution. The disease is endemic on the African continent, where most serotypes have been recorded. Periodic epizootics have occurred in Israel, the Middle East, Cyprus, Spain, and Portugal. The introduction of bluetongue viruses into some of these regions by windblown infected Culicoides has been suggested by Sellers et al. (1979). From data obtained in surveys of animal sera for antibodies to bluetongue and related viruses, it is apparent that infection with these viruses is widespread on each of the continents.

Epizootics of bluetongue disease were first reported in the United States in the late 1940s and early 1950s in California and Texas, but the disease may have occurred in Texas in 1923 (Robinson et al., 1967). Serotypes 10, 11, 13, and 17 have been active in the United States, and at least one serotype has been isolated in each of 29 states. All four serotypes are active in California, Montana, and Colorado (Barber, 1979).

Three serotypes of bluetongue virus have been isolated in Australia (St. George et al., 1978, 1980). The relationship between serotype 20 and other serotypes has been discussed previously. Della-Porta et al. (1981b) have designated the isolate CSIRO 156 as serotype 1 and the isolate CSIRO 154 as related to but distinguishable from type 6. Although the viruses are able to produce clinical signs of bluetongue disease in experimentally inoculated animals (St. George and McCaughan, 1979; St. George et al., 1980), there is no evidence of disease in Australia.

In a major study of arthropod-borne viruses in Northern Australia, the CSIRO Division of Animal Health laboratories have isolated a number of orbiviruses that have been identified as members of the bluetongue, Eubenangee, EHD, Palyam, Corriparta, Warrego, and Wallal serogroups. Detailed descriptions of the isolations of all these viruses have not been published, but the large number of isolates in each of the serogroups provides valuable material for research. In particular, isolates of the bluetongue, Eubenangee, EHD, and Palyam serogroups have revealed the complexity of the interrelationships among viruses in these groups. A summary of the earlier programs of isolation and identification of viruses was published in the proceedings of a symposium on arbovirus research in Australia (St. George and French, 1979).

## b. Epizootic Hemorrhagic Disease

Outbreaks of EHD have occurred since 1890 in various parts of the southeastern United States. The New Jersey strain was isolated in 1955 and compared with the South Dakota strain isolated in 1956 (Shope et al., 1960). In cross-protection tests, deer that had recovered from infection with either strain resisted heterologous challenge. In neutralization tests, the viruses appeared to differ (Mettler et al., 1962), but effective comparison of the two strains was hampered by the lack of susceptible laboratory animals or cell culture for both viruses. The New Jersey strain was more lethal in deer than the South Dakota strain (Shope et al., 1960). The original isolate of the South Dakota strain has apparently been lost (R.E. Shope, personal communication).

The virus has since been isolated from white-tailed deer, mule deer, or antelope in South Dakota, North Dakota, Michigan, Kentucky, Wyoming, Washington, North Carolina, and Indiana, in Alberta, Canada, and from cattle in Colorado (Hoff and Trainer, 1978; Foster et al., 1980). The Alberta strain is designated as type 2 (Barber and Jochim, 1975).

Three isolates of viruses from *Culicoides* spp. and one from *C. schultzei* collected at Ibadan, Nigeria (Lee et al., 1974; Lee, 1979) are probably distinct serotypes. The virus XBM/67 isolated in South Africa is referred to as a serotype (Verwoerd et al., 1979), but no detailed serological comparison with other EHD strains has been reported.

Ibaraki virus, originally isolated from the blood of a cow in Japan, is related to but distinct from the Alberta strain of serotype 2 (Campbell et al., 1978). The virus is described as producing a bluetonguelike illness in cattle (Inaba, 1975). Several viruses related to Ibaraki virus in group-specific tests have been isolated from healthy cattle and from *C. brevitarsis* in Australia (St. George et al., 1979; St. George, personal communication).

Huismans et al. (1979) found low but reproducible homology in cross-hybridization of the genomes of EHD-New Jersey and bluetongue type 10 viruses, but were unable to locate the region of homology on any genome segment. Huismans and Erasmus (1981), in cross-precipitation of virus-induced proteins with antisera to both viruses, located the shared antigenic determinants on a protein ($P_7$) in the nucleocapsids of the viruses. The result confirms the serological relationship between the EHD and bluetongue serogroups.

### c. Eubenangee Serogroup

Eubenangee virus was isolated from a mixed pool of 11 species of mosquitoes near Innisfail in north Queensland (Doherty et al., 1968). The virus shares antigenic determinants detected in CF tests with Pata virus isolated from Aedes palpalis in the Central African Republic (Borden et al., 1971). The third recognized serotype, Tilligerry virus, was isolated from Anopheles annulipes at Nelson Bay, New South Wales (Gard et al., 1973), and in CF tests is more closely related to Eubenangee virus than to Pata virus (Marshall et al., 1980). Although the weak but definite CF relationship is usually regarded as sufficient evidence to include Pata virus in the Eubenangee serogroup, similar cross-reactions between Pata virus and EHD-New Jersey (Borden et al., 1971) and between Eubenangee and bluetongue viruses are usually regarded as insufficient evidence for inclusion of the Eubenangee group viruses in either the bluetongue or the EHD serogroup. The practical value of such divisions is obvious, but viruses of the three serogroups must be classified in a single taxon.

Most of the isolations of Eubenangee group viruses have been made from pools of mosquitoes. In an extensive study involving isolation of viruses from arthropods collected in the Northern Territory of Australia, four isolations of Eubenangee-related viruses were made from Culex annulirostris, one from Anopheles farauti, and only one from the biting midge Culicoides marksi. This contrasts with the isolations of bluetongue and EHD viruses consistently from species of Culicoides.

The exact serological relationships among the six isolates and other serotypes of the Eubenagee group have not been established. The isolate from A. farauti (CSIRO 23) is interesting in that antiserum prepared against the isolate neutralized infectivity of bluetongue type 1 in plaque-inhibition tests, although CSIRO 23 was not neutralized by antiserum prepared against bluetongue type 1 (Della-Porta et al., 1979). The isolation of a bluetongue type 1 virus (CSIRO 156) in Australia (St. George et al., 1980; Della-Porta et al., 1981b) suggests that there may be a genetic relationship between the Australian isolate of the Eubenangee group (CSIRO 23) and bluetongue virus (CSIRO 156).

The genomes of Tilligerry virus and bluetongue type 20 were found to be related to the extent of less than 4% as determined by RNA–RNA hybridization assays, but no homologous regions in any of the genome segments could be detected by PAGE of the reaction mixtures (Gorman et al., 1981). The significance of low levels of homology is difficult to assess. The molecular weight of the smallest genome segment of Tilligerry virus has been estimated at $2.5 \times 10^5$ in a genome of $11.9 \times 10^6$ (Gorman and Taylor, 1978). Complete homology between the smallest genome segments of two orbiviruses would be expressed as 2% homology in the whole genome. The experimental error in estimates of whole-genome homology makes such estimates meaningless in determining genetic relationships between orbiviruses.

No homologous regions could be detected in PAGE of RNA–RNA hybridization assay mixtures of bluetongue type 20 and the isolate CSIRO 23 (Gorman et al., 1981) or between bluetongue type 1 (CSIRO 154) and the isolate CSIRO 23 [see Fig. 5 (Section V.B)].

Inhibitors in serum presumed to be antibody to Eubenangee viruses have been found in cattle, kangaroos, and wallabies (Doherty, 1972), and the arthropods from which the viruses have been isolated in Australia feed preferentially on cattle (Kay et al., 1978; 1979). However, in an extensive study of arbovirus infections in sentinel herds of cattle in Australia and Papua New Guinea, no antibodies to Eubenangee virus were detected (St. George, 1980).

The isolate CSIRO 23 multiplied to high titer in both *Culicoides variipennis* and *C. nebeculosus* after intrathoracic inoculation, but in *C. variipennis* only after oral ingestion (Mellor and Jennings, 1980). The infection rates of 7.9% for the Eubenangee group virus compared with 30–35% infection rates for bluetongue virus (Jones and Foster, 1974) and for horsesickness virus (Mellor et al., 1975) in the same strain of *C. variipennis*.

## 3. Colorado Tick Fever

Colorado tick fever was recognized as an acute febrile illness of man by Becker (1930). Florio et al. (1946, 1950a,b) proved that the virus was tick-borne.

A CF test was developed by De Boer et al. (1947), and there appeared to be a single serotype involved in infection in man. Because of the ease of isolation of virus from patients and ticks, the geographic distribution has been determined as essentially that of the major vector, *Dermacentor andersoni*, in mountainous northwestern United States and Canada. Vertebrate reservoirs include rodents, ground squirrels, pine squirrels, chipmunks, meadow voles, and porcupines (Burgdorfer, 1977). In 1972, a second serotype (Eyach) was isolated from forest-dwelling *Ixodes ricinus* near Tubingen, West Germany (Rehse-Kupper et al., 1976). The report suggested that no antibodies to the virus were found in man, but antibodies to Eyach virus have been detected in patients with various neuropathies (Málková et al., 1980).

Colorado tick fever persists in association with erythrocytes in human patients and in experimentally infected animals (Emmons et al., 1972; Hughes et al., 1974). Oshiro et al. (1978) found virus inclusions within erythroblasts, reticulocytes, and erythrocytes from mice infected *in utero* or when newborn.

In an electron-microscopic study, Murphy *et al.* (1968) commented that Colorado tick fever virus resembled reovirus in structure and in its mode and site of maturation in cells. A double-shelled structure for Colorado tick fever virus has not yet been demonstrated. Recent evidence (Knudson, 1981) that the genome consists of 12 segments of dsRNA indicates that the virus is distinct from other orbiviruses.

## 4. Palyam Serogroups

Viruses of the Palyam serogroup have been isolated in India, Australia, and Africa. Six distinct serotypes have been described (Table III), but it is likely that others will be found in the large number of Palyam-related viruses described.

Lee (1979) reported the isolation of 44 viruses related to Abadina virus from species of *Culicoides* and one from *Aedes fowleri* collected in Nigeria between 1967 and 1970. St. George *et al.* (1979) described 35 isolations of D'Aguilar-related viruses from blood taken from sentinel cattle in the Northern Territory of Australia. They suggest that the isolates could be divided into three distinct serotypes. In six instances, two serologically distinct viruses were isolated from particular cows on two separate occasions. St. George and Dimmock (1976) had previously described the isolation of a Palyam group virus from a healthy cow.

Swanepoel and Blackburn (1976) isolated Nyabira virus from the aborted fetus of a cow. The virus reacted in CF tests with Palyam group viruses, but appeared to be more closely related to D'Aguilar virus. The virus was distinguishable from D'Aguilar virus in neutralization tests.

Cross-reactions in AGP tests between Abadina virus and EHD and bluetongue virus have been reported (Moore, 1974). The relationships between viruses of the Palyam serogroup and those of the bluetongue complex of serogroups have not been examined systematically. Della-Porta *et al.* (1979) described an interesting situation in which serum from a cow experimentally infected with a Eubenangee virus (CSIRO 23) neutralized bluetongue type 1 but not bluetongue type 20 or Ibaraki virus (EHD serogroup). The animal was subsequently challenged with a Palyam

TABLE III. Isolation of Palyam Serogroup Viruses

| Serotype | Prototype strain | Source | Country of isolation |
|---|---|---|---|
| 1 | Palyam IG 5287 | *Culex vishnui* | India[a] |
| 2 | Kasba IG 15534 | *Culex vishnui* | India[a] |
| 3 | Vellore IG 68886 | *Culex* spp. | India[a] |
| 4 | D'Aguilar B 8112 | *C. brevitarsis* | Australia[a] |
| 5 | Abadina Ib Ar 22388 | *Culicoides* spp. | Nigeria[b] |
| 6 | Nyabira | Cattle | Zimbabwe[c] |

[a] Berge (1975).   [b] Lee *et al.* (1974).   [c] Swanepoel and Blackburn (1976).

TABLE IV. Changuinola Serogroup[a]

| Serotype | Prototype strain | | Isolation |
|---|---|---|---|
| 1 | Changuinola | BT436 | Phlebotomines, 1960, Panama |
| 2 | Irituia | Be Ar 28873 | *Oryzomys* sp., 1961, Brazil |
| 3 | Gurupi | Be Ar 35646 | *Phlebotomus* sp., 1961, Brazil |
| 4 | Ourem | Be Ar 41067 | *Phlebotomus* sp., 1962, Brazil |
| 5 | Caninde | Be Ar 54342 | *Phlebotomus* sp., 1963, Brazil |
| 6 | Jamanxi | Be Ar 243090 | *Phlebotomus* sp., 1973, Brazil |
| 7 | Altamira | Be Ar 264277 | *Phlebotomus* sp., 1974, Brazil |

[a] From the *International Catalogue of Arboviruses* (Berge, 1975; Karabatsos, 1978; and additions).

group virus (CSIRO 58) and developed more cross-reactive antibodies that neutralized both bluetongue serotypes 1 and 20 and Ibaraki virus to a lesser extent. The result illustrates the difficulty in interpreting serological tests on animals that may be exposed to infection with different orbiviruses.

## 5. Changuinola Serogroup

The Changuinola group viruses have been isolated from phlebotomines, mosquitoes, and *Bradypus* sloths in Central and South America. There are seven distinct serotypes (Table IV) recorded in the *International Catalogue of Arboviruses* (Berge, 1975; Karabatsos, 1978; and additions). De Oliva and Knudson (1980) list more than 270 isolates of viruses that are included in the serogroup by CF test and suggest that there may be many distinct serotypes.

Changuinola virus has been isolated from the blood of a human patient presenting with a mild febrile illness.

## 6. Corriparta Serogroup

Corriparta virus was isolated from *Culex annulirostris* at Kowanyama in north Queensland (Doherty *et al.*, 1963) and was shown to have physical and morphological properties similar to those of orbiviruses (Carley and Standfast, 1969). The isolation of the virus from a black-fronted dotterel (*Charadrius melanops*) at Kowanyama in 1965 was reported by Whitehead *et al.* (1968). Subsequent isolations in Australia of serologically related viruses were made from *Aedomyia catastica* in the Ord region of Western Australia (Liehne *et al.*, 1976) and from *C. annulirostris* at Beatrice Hill, Northern Territory (T.D. St. George, personal communication).

A second serotype (Acado) was isolated from a mixed pool of *Culex antennatus* and *C. univittatus* in the Ilubabor region of Ethiopia (Berge, 1975).

Two isolates of the Corriparta serogroup from *C. melanonion* and *C. declarator* collected in Brazil have not been compared with the recognized serotypes in serum-neutralization tests (Knudson, 1980).

Hemagglutinin for goose cells has been prepared from mouse brain infected with Acado virus (Berge, 1975), but repeated attempts to produce hemagglutinin from Corriparta virus–infected cells have failed. Limited surveys for antibody to Corriparta viruses suggest that they infect man, cattle, horses, marsupials, and birds (Doherty, 1967, Doherty *et al.*, 1970).

## 7. Kemerovo Serogroup

Viruses of the Kemerovo serogroup are tick-borne, with birds as the predominant vertebrate host. The serogroup can be divided into four antigenic complexes (Table V) that often reflect the origin of the isolates (Casals, 1971; Libikova and Casals, 1971; Main *et al.*, 1976).

Libikova and Casals (1971) showed minor antigenic differences between strains of Kemerovo group viruses, but designated two separable serotypes. The viruses are maintained in the Palearctic region among small mammals and birds by two similar species of hard ticks: Kemerovo by *Ixodes persulcatus* in western Siberia and Tribec by *I. ricinus* in Slovakia. The viruses have been associated with human disease (Section I.B).

Viruses of the Great Island complex are maintained among colonies of sea birds by *I. uriae* in widely scattered areas of the northern and southern hemispheres. By CF tests, the viruses are closely related to one another and to a lesser extent to the Kemerovo antigenic complex (Main *et al.*, 1976).

The Chenuda complex includes five different viruses from soft ticks of the genera *Argas* and *Ornithodorus* parasitizing pigeons, swallows, gulls, and cormorants in Africa, Asia, and North and South America (Hoogstraal, 1972).

Wad Medani virus has been isolated from sheep in the Sudan and from hard ticks of the genera *Boophilus*, *Hyalomma*, *Ambylomma*, and *Rhipicephalus* in the Sudan, West Pakistan, India, the U.S.S.R., and Jamaica. The closely related virus Seletar was isolated from *B. microplus* in Singapore and Malaysia. These viruses are distantly related serologically to the Kemerovo serogroup viruses in other antigenic complexes (Main *et al.*, 1980).

The Kemerovo serogroup of orbiviruses provides interesting models in the study of virus evolution. Main *et al.* (1973) found two distinct but closely related serotypes, Great Island and Bauline, among puffins and petrels on Great Island, with little or no indication of other serotypes. They suggested that the antigenic identity of serotypes may be maintained by the isolation of the primary hosts of the vector and that each serotype developed in separate demes on the island. Yunker (1975) sug-

TABLE V. Kemerovo Serogroup

| Type | Prototype virus | Isolation of prototype virus | | |
|---|---|---|---|---|
| | | Tick | Location | Year |
| Kemerovo complex | | | | |
| 1 | Kemerovo | *Ixodes persulcatus* | Siberia | 1962 |
| 2 | Tribec | *I. ricinus* | Czechoslovakia | 1963 |
| 3 | Lipovnik | *I. ricinus* | Czechoslovakia | 1963 |
| Great Island complex | | | | |
| 1 | Great Island | *I. uriae* | Newfoundland | 1971 |
| 2 | Bauline | *I. uriae* | Newfoundland | 1971 |
| 3 | Cape Wrath | *I. uriae* | Scotland | 1973 |
| 4 | Yaquina Head | *I. uriae* | Oregon | 1970 |
| 5 | Ohotskiy | *I. uriae* | Tyuleniy Island | 1969 |
| 6 | Nugget | *I. uriae* | Macquarie Island | 1972 |
| 7 | Poovoot | *I. uriae* | Alaska | — |
| 8 | Kenai | *I. signatus* | Alaska | — |
| 9 | Tindholmur | *I. uriae* | Faroe Islands | 1974 |
| 10 | Mykines | *I. uriae* | Faroe Islands | 1974 |
| Chenuda complex | | | | |
| 1 | Chenuda | *Argas reflexus hermanni* | Egypt | 1954 |
| 2 | Mono Lake | *A. cooleyi* | California | 1966 |
| 3 | Huacho | *Ornithodorus amblus* | Peru | 1967 |
| 4 | Baku | *O. capensis* | Caspian Sea | 1970 |
| 5 | Sixgun City | *Argas cooleyi* | Texas | 1969 |
| Wad Medani complex | | | | |
| 1 | Wad Medani | *Rhipicephalus sanguineus* | Egypt | 1952 |
| 2 | Seletar | *Boophilus microplus* | Singapore | 1961 |

gested that each virus within defined complexes has been influenced by common patterns of geography, ecology, and behavior. These included a discontinuous distribution, the nidiculous activity of the ticks that maintained them, and the homing–colonial instinct of the bird-hosts. Isolation produces a large number of strong insular variants, which in some cases develop into separate species. Thus, each serotype attained its antigenic identity on separate islands before the serotypes were brought together (Yunker, 1975). Main (1978) suggested that the antigenic composition of the virus population was constantly changing in response to the immunity levels in the avian populations.

## 8. Warrego Serogroup

Viruses of the Warrego serogroup have been found only in Australia. Two serotypes are recognized, and the viruses can be distinguished easily in CF tests (Borden et al., 1971; Doherty et al., 1973).

The prototype strain of Warrego virus and 7 isolates indistinguishable from it by CF test were isolated from species of Culicoides and from Culex annulirostris collected in 1969 in southwest Queensland (Doherty et al., 1973). A further 18 strains indistinguishable from Warrego virus in CF tests were isolated from Culicoides spp. at Beatrice Hill in the Northern Territory of Australia (T.D. St. George, personal communication). A virus isolate from Anopheles meraukensis collected at Kowanyama, north Queensland, was closely related by CF to the prototype Warrego virus (Doherty et al., 1979). Detailed comparisons of the 27 Warrego-related viruses by serum-neutralization tests have not been made, and it is possible that many distinct serotypes exist.

The other serotype, Mitchell River virus, was isolated from a mixed pool of 13 Culicoides collected in 1969 at Kowanyama. Reisolation of the virus from the original insect pool was not possible, but the virus was distantly related to Warrego and unrelated to other known arthropod-borne viruses. Although many Warrego-related viruses have been isolated in Australia, no other isolates of Mitchell River–related viruses have been made.

Neutralizing antibodies to both Warrego and Mitchell River viruses have been found in cattle in Australia (Doherty et al., 1973) and to Mitchell River virus in cattle in Japan (Miura et al., 1980). In Australia, neutralizing antibodies to both serotypes have also been found in wallabies and kangaroos (Doherty et al., 1973).

## 9. Wallal Serogroup

Viruses of the Wallal serogroup have been isolated only in Australia. The prototype strain of Wallal virus and 6 strains indistinguishable from it in CF tests were isolated from Culicoides spp. in southwest Queensland (Doherty et al., 1973). An isolate from C. marksi collected in the Northern

Territory was found to be antigenically distinguishable from Wallal virus and was designated Mudjinbarry virus (Doherty et al., 1978). Subsequently, 24 isolates from *C. marksi*, 1 from *C. dycei* collected in northern Australia (T.D. St George, personal communication), and 1 from *Culex* spp. collected in New South Wales (I.D. Marshall, personal communication) have been related to Wallal virus by CF, but detailed serological classification of the viruses has not been undertaken. Attempts to characterize the isolates by analysis of genome segments of the viruses revealed extensive heterogeneity within the group (Gorman et al., 1977b). The results are discussed in more detail in Section V.

Surveys for antibodies suggest that vertebrate hosts for Wallal viruses are restricted to kangaroos and wallabies (Doherty et al., 1973).

## 10. Equine Encephalosis Serogroup

Equine encephalosis virus was isolated in South Africa in 1967 from horses that died from a previously unknown peracute illness (Erasmus et al., 1970). A number of isolates were obtained from other fatal cases as well as from the blood of febrile horses from different parts of the country. The isolates were indistinguishable by serum-neutralization tests, and the prototype strain, Cascara, was shown by its physicochemical properties and by electron microscopy (Lecatsas et al., 1973) to be an orbivirus. The results of serological investigations indicated that there was widespread incidence of the disease in the early months of 1967 and that it had not occurred in South Africa during the previous ten years.

Gamil virus (M 9/71) was isolated from blood collected in March 1971 from a horse that exhibited a febrile reaction, general inappetence, and slight depression (Erasmus et al., 1976a). Electron microscopy of negatively stained preparations revealed particles typical of orbiviruses and virus nucleocapsids (Lecatsas and Gorman, 1972).

During December 1973, a number of fatal cases caused by type 4 horsesickness virus were reported in northeastern Transvaal. A monovalent vaccine was developed by plaque selection from the vaccine strain 32/62 of type 4 (Erasmus, 1976). In January 1974, 30 horses were inoculated with a candidate vaccine strain as part of a field trial. None of the animals showed any untoward reactions, but 17 days after infection one mare developed a severe conjunctivitis and swelling of the eyelids followed by a febrile reaction. Kaalplaas virus was isolated from the horse and reisolated after passage of infected blood in a second horse.

During April and May 1976, viruses were isolated from three aborted equine fetuses, and these were found to be identical to Bryanston virus that was isolated in June from fatal cases of acute cardiac failure in horses (Erasmus et al., 1976a).

The four groups of viruses were indistinguishable by CF tests, but were not related to horsesickness, bluetongue, EHD, Palyam, Corriparta, Warrego, or Mitchell River viruses. No cross-reactions in serum-neutralization tests were detected between Cascara, Gamil, Kaalplaas, and Bryanston viruses (Erasmus *et al.*, 1976a).

Detailed comparisons of the nucleic acids and proteins of viruses of the encephalosis group and other orbiviruses have yet to be made. The genetic relationship between the encephalosis group viruses and other orbiviruses could provide interesting information on the evolution of or- biviruses. A virus that had been unrecognized for at least the ten years prior to 1967 appeared on four separate occasions over the next nine years as four distinct serotypes of encephalosis virus.

## 11. Serologically Ungrouped Orbiviruses

A number of orbiviruses have been isolated that cannot be related to viruses in any of the recognized serogroups by standard serological tests (Table VI). The best-characterized of these is Orungo virus isolated in Africa, which causes a mild febrile illness in man (Tomori *et al.*, 1976). Antibodies to the virus have also been found in sheep, cattle, and monkeys, and Tomori and Fabiyi (1976) suggest that *Aedes* spp. may be involved in natural transmission of the virus between man and domestic and wild animals. Lebombo virus has been isolated from human plasma (Berge, 1975).

## D. Physicochemical Properties Used in Classification of Orbiviruses

The orbiviruses can be readily distinguished from other Reoviridae by their characteristic appearance in negative-contrast electron micros-

TABLE VI. Serologically Ungrouped Orbiviruses[a]

| Virus | Prototype strain | Isolation | | Vertebrate hosts |
| | | Arthropod | Locality | |
|---|---|---|---|---|
| Ife | Ib Ar 57245 | — | Nigeria, 1971 | *Eidolon helvium* (bat) |
| Japanaut | Mk 6357 | Culicine mosquitoes | Papua New Guinea, 1965 | Bats |
| Lebombo | Ar 136 | *Aedes circumluteolus* | South Africa, 1956; Nigeria | Man, rodents |
| Llano Seco | BFN 3112 | *Culex tarsalis* | California, 1971 | — |
| Umatilla[b] | 69V 2161 | *C. pipiens* | Oregon, 1969 | — |
| Orungo | UgMp 359 | *Anopheles funestus* | Uganda, 1959; Nigeria | Man, sheep, cattle, monkeys |
| Paroo River | GG 868 | *C. annulirostris* | New South Wales, Australia, 1973 | |

[a] From the *International Catalogue of Arboviruses* (Berge, 1975; Karabatsos, 1978; and additions).
[b] Umatilla is related by CF and indirect fluorescent-antibody tests to Llano Seco virus.

copy (Palmer *et al.*, 1977). A useful property that distinguishes them from reoviruses is the stability of reoviruses at low pH. The genus was formed to include relatively solvent-resistant arboviruses (Borden *et al.*, 1971), and resistance to lipid solvents and detergents and loss of infectivity at low pH remain useful properties in preliminary characterization of viruses isolated from arthropods.

## 1. pH Stability

Hamparian *et al.* (1963) used acid lability as one criterion in the classification of viruses. Loss of infectivity at low pH appears to be characteristic of orbiviruses and contrasts with the stability of reoviruses at low pH.

Inactivation of bluetongue virus infectivity occurs below pH 6.5 (Owen, 1964). Verwoerd (1969) observed a marked sensitivity of the virus to high salt concentrations and concluded from inactivation curves that the virus was most stable at pH 9. Other reports of acid lability of orbiviruses include EHD (Trent and Scott, 1966), horsesickness (Ozawa, 1968), Colorado tick fever (Murphy *et al.*, 1968), Corriparta (Carley and Standfast, 1969), Changuinola, Eubenangee, Kemerovo, Lebombo, and Palyam (Borden *et al.*, 1971), and Wallal, Warrego, and Mitchell River (Gorman, 1978).

Verwoerd *et al.* (1972) showed that the surface polypeptides of bluetongue virus were lost in solutions of cesium chloride at low pH. In a study of the activation of bluetongue virus transcriptase, Van Dijk and Huismans (1980) found that removal of the smaller of the surface polypeptides by chymotrypsin required the addition of a high concentration of divalent cations. Dissociation of the polypeptide from the core particle at high monovalent-cation concentration could be achieved only by reducing the pH of the solution below 7. Under these conditions, dissociation of some of the core polypeptides also occurred.

The mechanism of inactivation of orbivirus infectivity at low pH is not known. From the work of Verwoerd *et al.* (1972) and Van Dijk and Huismans (1980), it may be inferred that lowering pH reduces the stability of the virus particles and under appropriate conditions leads to the loss of surface polypeptides and hence a loss in virus infectivity.

## 2. Effect of Solvents and Detergents

Franklin (1962) classified animal viruses according to the presence or absence of structural lipids. The lipid envelopes of most arboviruses made them sensitive to extraction with lipid solvents and detergents, and the subsequent loss of infectivity became one criterion in the classification of arboviruses (Theiler, 1957). Since most of the orbiviruses were initially isolated from arthropods and standard tests applied to them, they were usually described as relatively resistant to the action of diethyl ether and sodium deoxycholate. In practice, this meant that using standard tests that led to inactivation of lipid-enveloped viruses, the orbiviruses showed a slight but reproducible loss in infectivity. This suggested two possibilities: (1) that the orbiviruses contained some lipid that had a role in virus infectivity or (2) that the standard tests were unsuitable when applied to these viruses, perhaps because of the denaturing effects of these reagents.

Bluetongue virus was found to be resistant to ether and deoxycholate (Howell, 1963b; Studdert, 1965; Svehag et al., 1966) and to chloroform (Bowne, 1971). Verwoerd (1969) reported that after purification, the stability of bluetongue virus was less apparent. Howell (1962) found horsesickness virus resistant to ether and deoxycholate. Borden et al. (1971) demonstrated resistance of EHD, Palyam, Orungo, and Lebombo viruses to ether, chloroform, and deoxycholate. They found Irituia virus of the Changuinola serogroup and Wad Medani virus of the Kemerovo group relatively sensitive to these reagents. The Florio strain of Colorado tick fever virus at the 75th mouse-brain passage was inactivated by ether, chloroform, and deoxycholate, but low-passage isolates of the same strain and the Condon strain were more resistant (Borden et al., 1971). Corriparta, Eubenangee, D'Aguilar, Warrego, and Mitchell River viruses resisted treatment with ether, but infectivity of Wallal virus was significantly reduced, and of these only Corriparta virus resisted treatment with chloroform (Gorman, 1978).

The observation in thin-section electron microscopy of enveloped forms of bluetongue (Bowne and Jones, 1966; Bowne and Ritchie, 1970; Foster and Alders, 1979), horsesickness (Breese et al., 1969), Eubenangee (Schnagl et al., 1969), and Colorado tick fever viruses (Murphy et al., 1968) led to conflicting interpretations of the role of lipid in orbivirus structure. Els and Verwoerd (1969) interpreted surrounding membranes in bluetongue virus particles as pseudoenvelopes, since they occurred in a very small proportion of particles.

No reliable information on the lipid composition of orbiviruses is available. A small amount of phospholipid and cholesterol in purified bluetongue viruses was thought to be of cellular origin (Howell and Verwoerd, 1971). Purified bluetongue virus consists of seven polypeptides, five in the nucleocapsid and two forming the layer on the surface of the virus (Verwoerd et al., 1972; Martin and Zweerink, 1972). The earlier work on the lipid composition of bluetongue virus (Verwoerd, 1969) had been done using purified nucleocapsids rather than purified virions, and no chemical analyses have been made of structures including surface polypeptides.

The failure to inactivate orbiviruses using phospholipase (Gorman, 1978) together with their general resistance to ether suggests that lipid is not essential for virus infectivity. Reports of inactivation of orbiviruses with chloroform should be viewed with caution, since even reoviruses lose infectivity in the presence of chloroform (Engler and Broome, 1969).

The effect of chloroform more likely represents denaturation of protein, rather than extraction of lipid. The loss of infectivity by treatment of an orbivirus with solvents and detergents probably reflects the instability of the virus and the loss of surface polypeptides. The general method of release of virus from infected cells by cell lysis (Section III) also suggests that the orbiviruses, like the reoviruses, contain no essential lipid.

## II. STRUCTURAL COMPONENTS AND THEIR FUNCTIONS

### A. Morphology of Orbiviruses

Studdert *et al.* (1966) examined bluetongue virus by electron microscopy. Negatively stained virus particles were characteristically 53 nm in diameter, with no evidence of an enveloped particle. These investigator's suggested that the virus was icosahedral, with 92 capsomers, and was similar to reoviruses. Owen and Munz (1966) described icosahedral particles with 92 capsomers 60 nm in diameter and fewer enveloped particles 100 nm in diameter. Evidence for an outer envelope was also presented by Bowne and Jones (1966) in a study of virus particles in the salivary glands of *Culicoides variipennis*. Els and Verwoerd (1969) suggested that what other workers had represented as enveloped particles were cellular membranes occasionally wrapped around one or more particles. These structures, called pseudoenvelopes, were removed by treatment with Tween-80 and ether without loss of infectivity. Els and Verwoerd (1969) described a virus particle 55 nm in diameter with 32 capsomers, in contrast to the 92 suggested by other workers. No evidence was found of a double-layered capsid, as is present in reoviruses.

Bowne and Ritchie (1970) indicated that bluetongue virus had a complicated morphology. The most commonly observed particle had a diameter of 60–70 nm in which the reoviruslike symmetry was obscured by fine hairlike structures extending from the capsomers. The particle with typical reoviruslike symmetry of diameter 55–60 nm was not often observed. Still another nonenveloped form was seen in which the reoviruslike pattern was covered by a thick feltwork overlay of fibrillar material that increased the diameter to 70–80 nm. In material extracted with trifluorotrichloroethane, the smaller particle predominated, but infectivity was lost, suggesting that bare capsid particles were not as stable as the particles in which the capsomers were covered with a fibrillar network. A fourth particle was described in which core particles were surrounded by a membrane.

Verwoerd *et al.* (1972) showed that two polypeptides form a diffuse protein layer surrounding the nucleocapsid and obscure the arrangement of structural units in the nucleocapsid. This observation solved the discrepancy seen in reports on the structure and size of the virion. Obviously, the particle 55 nm in diameter with 32 clearly discernible capsomers observed by Els and Verwoerd (1969) represented a subviral particle, while the larger ill-defined particles represented the virion possessing the outer layer (Fig. 1). The structure of the bluetongue virus capsid (Verwoerd *et al.*, 1972) was closer than previously thought to that of reovirus, both in size of the complete particles and in the possession of a double layer. However, the assembly of the viruses is different. In reovirus, the outer layer consists of structural units arranged in a regular way and an inner capsid that is less structured. In bluetongue virus, the outer layer is diffuse and unstructured, whereas the inner layer consists of structural units clustered as conventional pentamer–hexamer morphological units or capsomers. Martin and Zweerink (1972) confirmed that bluetongue virus consisted of a clear nucleocapsid structure with 32 capsomers (diameter 63 nm) surrounded by two polypeptides that gave the complete virion a diameter of 69 nm.

The prominent ring-shaped capsomers on the surface of the inner capsid are thought to be elongated hollow tubes radiating from a central core (Els and Verwoerd, 1969). Each capsomer consists of a number of smaller subunits (Els and Verwoerd, 1969; Oellermann *et al.*, 1970; Murphy *et al.*, 1971; Palmer *et al.*, 1977). Penetration of stain into the inner capsid of incomplete or "empty" virus particles reveals a cavity of icosahedral structure (Studdert *et al.*, 1966; Murphy *et al.*, 1968; Els and Verwoerd, 1969; Oellermann *et al.*, 1970; Lecatsas and Erasmus, 1973; Palmer *et al.*, 1977). Analysis of ultrathin sections of cells infected with XBM/67 virus (EHD serogroup) suggested that in complete particles, this cavity may contain a substructure in the form of 12 spherical units (Lecatsas and Erasmus, 1973).

Failure to appreciate that the orbivirus capsid consists of two protein shells appears to have contributed to confusion in the early reports of orbivirus particle size and morphology. Larger particles of diffuse appearance were often considered to be contaminated with cellular debris. The relative ease with which the outer shell dissociates has caused many observers to mistake nucleocapsids for complete virions. Lecatsas and Gorman (1972) demonstrated a diffuse outer layer on five Australian orbiviruses for which nucleocapsids had previously been described as complete virions (Schnagl and Holmes, 1971). Verwoerd *et al.* (1979) have suggested that the absence of a fixation step in the negative-stain technique may have caused uncontrolled swelling of particles, resulting in discrepancies in size estimations.

The reports on the structure of other orbiviruses should then be interpreted to take into account the more recent studies on the morphology and structure of bluetongue virus.

Carley and Standfast (1969) reported that Corriparta virus had a diameter of 57 nm and suggested that the virion contained 92 capsomers. Schnagl *et al.* (1969) reported that Eubenangee virus had a diameter of 62 nm. The precise arrangement of capsomers was not discernible, but they suggested that it was consistent with a 92-capsomer icosahedral

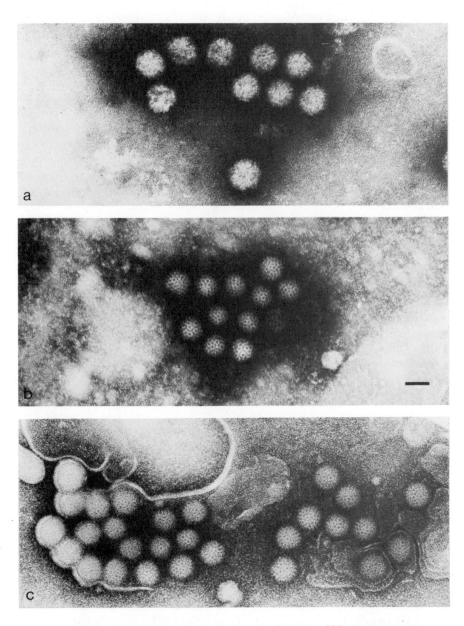

FIGURE 1. Electron micrographs of negatively stained virions of bluetongue type 20 virus
(a), nucleocapsids of bluetongue type 20 virus (b), and various particles from cells infected
with Corriparta virus (c). Note that virions have obscured structure and capsomers become
discernible after removal of surface proteins. In (c), a membranelike structure covers the
particles. × 127,500. Scale bar: 50 nm. Unpublished micrographs of A.J. Melzer.

structure. Less frequently, viruslike particles without any capsomer structure and with diameters of 67–86 nm were observed. Schnagl and Holmes (1971) found that D'Aguilar, Warrego, and Mitchell River viruses were all similar, appearing polygonal with obvious capsomer structure, and having diameters of 70, 71, and 71 nm, respectively. Particles with a closely fitting envelope were not observed, but occasionally particles surrounded by pseudoenvelopes were seen. Lecatsas and Gorman (1972) found that the nucleocapsids of Corriparta, Eubenangee, D'Aguilar, Warrego, and Mitchell River viruses measured 61, 65, 63, 67, and 59 nm, respectively, whereas the complete virions had diameters of 66, 77, 68, 76, and 65 nm, respectively.

Polson and Deeks (1963) precipitated horsesickness virus with polyethylene glycol and found virus particles 70–80 nm in diameter with clear structure and an estimated 92 capsomers. Oellermann et al. (1970) described horsesickness virus as morphologically similar to bluetongue virus, since both possessed icosahedral shape and measured 55 nm in diameter with only one layer of 32 structural units. Bremer (1976) found horsesickness and bluetongue viruses similar in structure, with minor differences in the arrangement of proteins in the outer and inner shells.

Thomas and Miller (1971) suggested that EHD virus was similar to bluetongue virus in shape and general capsid arrangement. By direct measurement, the diameter of both capsids was 53 nm. Tsai and Karstad (1970) claimed a mean diameter of 62.3 nm for EHD virus particles by negative staining. Although a clear-cut capsomer number could not be given, many virus particles appeared to be composed of 42 morphological units. Ito et al. (1973) reported that Ibaraki virus (EHD serogroup) measured 55 nm in diameter and had no envelope. The capsid of the virus consisted of a single layer of 32 capsomeres arranged in 5 : 3 : 2 symmetry. Virions enclosed in pseudoenvelopes were observed occasionally. Suzuki et al. (1978) reported similar double-layered structures for Ibaraki and bluetongue viruses.

Several groups have described mature orbivirus particles enclosed in an envelope structure (Owen and Munz, 1966; Murphy et al., 1968; Oellermann et al., 1970; Bowne and Ritchie, 1970). Enveloped particles appear to form when a small proportion of mature particles escape the infected cell by membrane "budding" prior to cell lysis (Bowne and Jones, 1966; Lecatsas and Erasmus, 1967; Murphy et al., 1968; Schnagl et al., 1969; Bowne and Ritchie, 1970; Schnagl and Holmes, 1971; Ito et al., 1973). The significance of the "budding" process in the orbivirus replication cycle is not known.

Murphy et al. (1968) compared Colorado tick fever virus with reovirus type 2 by electron microscopy. Identical virion diameters (80 nm) were obtained for both viruses. However a smaller inner-capsid diameter was found for reovirus (45 nm). A consistent difference between the amount of structural detail visible in the two viruses was evident. The capsomer structure of Colorado tick fever virus appeared "ragged," resulting in an inability to define surface capsomer patterns. Murphy *et al.* (1971) reexamined Colorado tick fever virus together with eight other viruses including bluetongue virus. They reported that the capsid architecture of all the viruses examined was strongly indicative of a common icosahedral construction. The arrangement of structural proteins in Colorado tick fever virus has not been studied.

## B. Orbivirus Genome

The replication of orbiviruses is unaffected by inhibitors of DNA synthesis, particularly mitomycin C and halogenated deoxyuridine derivatives (Ozawa, 1967; Verwoerd, 1969; Carley and Standfast, 1969; Tsai and Karstad, 1973; Schnagl and Holmes, 1975; Gorman *et al.*, 1977a). Cells stained with the fluorochrome dyes acridine orange and coriphosphine-o have been shown to develop inclusions in the perinuclear region early in infection (Livingstone and Moore, 1962; Green, 1969; Schnagl and Holmes, 1971). The ribonuclease resistance and alkali solubility of these cytoplasmic inclusions suggested that they contained double-stranded RNA (dsRNA) (Green, 1969). Nucleic acid extracted from purified EHD virus produced a positive orcinol reaction (RNA indicator), but no reaction with the DNA indicator diphenylamine (Kontor and Welch, 1976). Verwoerd (1969) demonstrated the double-strandedness of the bluetongue virus genome by the thermal denaturation curve of the RNA, its resistance to ribonuclease degradation, and its base composition. Verwoerd *et al.* (1970) fractionated the RNA into five components by sucrose-gradient sedimentation analysis and into ten components by PAGE. The size of the components varied from $0.5 \times 10^6$ to $2.8 \times 10^6$, with a total molecular-weight estimate of $1.5 \times 10^7$ for the viral nucleic acid.

Verwoerd (1969) found that the proportion of $G + C$ base pairs in bluetongue virus RNA was 42.4% compared with estimates of 43.8% (Bellamy *et al.*, 1967) and 48.1% (Shatkin and Sipe, 1968) for RNA of the Dearing strain of reovirus type 3. The $G + C$ content has been reported as 45.2% for Ibaraki virus RNA (Suzuki *et al.*, 1977) and as 38.2% for Wallal virus RNA (Walker, 1981). The base composition of individual genome segments of Mudjinbarry virus varied from 36.5 to 42.8% (Table VII). On average, the larger genome segments had a lower $G + C$ content than smaller segments (Walker, 1981).

Verwoerd *et al.* (1970) demonstrated bluetongue virus RNA to be relatively resistant to ribonuclease but sensitive to alkali and to produce on heating a sharp hyperchromic shift at a temperature characteristic of the breakdown of a double-helical structure.

The denaturation temperature $(T_m)$ of double-stranded polynucleotides is dependent on the ionic strength of the solvent and for dsDNA is proportional to the $G + C$ content under standard conditions (Marmur and Doty, 1962). In 0.015 M sodium chloride, 0.0015 M sodium citrate, pH

TABLE VII. Base Compositions of Individual Genome Segments of
Mudjinbarry Virus

| Segment | Composition (%) | | | | | Ratio |
| | Cytosine | Adenine | Guanine | Uracil | G + C | A + G/C + U |
|---|---|---|---|---|---|---|
| 1 | 18.8 | 31.6 | 18.5 | 31.1 | 37.3 | 1.00 |
| 2 | 19.0 | 31.6 | 17.5 | 31.9 | 36.5 | 0.96 |
| 3 | 19.1 | 30.8 | 19.6 | 30.5 | 38.7 | 1.02 |
| 4 | 20.4 | 28.8 | 20.0 | 30.8 | 40.4 | 0.95 |
| 5 | 19.5 | 30.1 | 19.3 | 31.1 | 38.8 | 0.98 |
| 6 | 21.8 | 28.5 | 20.3 | 29.4 | 42.1 | 0.95 |
| 7 | 23.0 | 28.3 | 19.8 | 28.9 | 42.8 | 0.93 |
| 8 | 21.6 | 29.0 | 18.4 | 31.0 | 40.0 | 0.90 |
| 9 | 20.6 | 29.4 | 20.4 | 29.6 | 41.0 | 0.99 |
| 10 | 20.9 | 29.4 | 19.0 | 30.7 | 39.9 | 0.94 |
| Weighted average[a] | 19.4 | 31.2 | 17.8 | 31.6 | 37.2 | 0.96 |
| Wallal virus | 19.8 | 31.6 | 18.4 | 30.2 | 38.2 | 1.00 |

[a] Determined on the basis of molecular-weight estimates of Gorman *et al.* (1978).

7.4 [0.1 × standard saline–citrate (SSC)], the $T_m$ of RNA extracted from
bluetongue virus type 10 was found to be 80°C (Verwoerd *et al.*, 1970).
Under similar conditions, the $T_m$ of Ibaraki virus RNA was 74°C (Suzuki
*et al.*, 1977); of Warrego virus RNA, 78.5°C; of Mitchell River virus RNA,
75.5°C (Gorman *et al.*, 1977a); of Wallal virus RNA, 77°C (Walker, 1981);
and of RNA of the Dearing strain of reovirus type 3, 85°C (Bellamy *et al.*,
1967). Widely different $T_m$ values (86 and 94°C in 0.1 × SSC) have been
reported for closely related strains of EHD virus (Tsai and Karstad, 1973;
Kontor and Welch, 1976). Since this would represent a substantial difference in base composition, it appears that at least one of these estimates
is in error. Both values are significantly greater than estimates obtained
for RNA from other orbiviruses.

The hyperchromic shifts reported for the RNA genomes of Ibaraki
virus (43%), Warrego virus (45%), Mitchell River virus (42%), and Wallal
virus (46%) are higher than the variable 30–36% reported for bluetongue
virus RNA (Verwoerd *et al.*, 1970). An approximate inverse correlation
has been reported between the degree of hyperchromicity and the G + C
content of DNA (Marmur and Doty, 1959). The values reported for bluetongue virus RNA appear low.

Verwoerd *et al.* (1970) demonstrated that the genome of bluetongue
virus could be fractionated into five overlapping peaks by sedimentation
through sucrose density gradients. A low-molecular-weight (2–4 S) component similar to that observed for reovirus (Bellamy *et al.*, 1967; Shatkin
and Sipe, 1968) was reported in early experiments (Verwoerd, 1969). Although this component was later shown to be absent from highly purified
virus, retrospective consideration suggests that these particles were in-

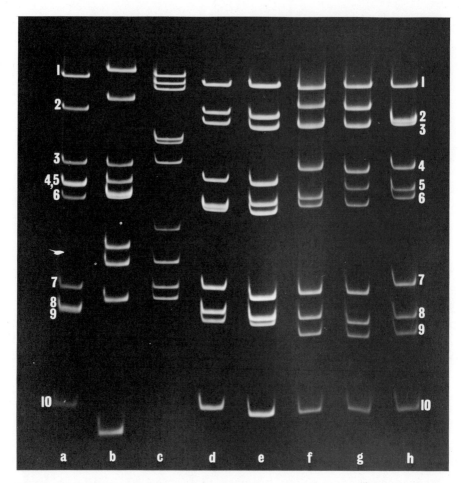

FIGURE 2. PAGE of dsRNA genome segments of Corriparta (MRMI) (a), Nugget (MI14847)—Kemerovo serogroup (b), Reovirus (Dearing strain) (c), bluetongue type 4 (isolate 922, South Africa) (d), bluetongue type 1 (isolate 494, South Africa) (e), bluetongue type 1 (isolate CSIRO 156, Australia) (f), bluetongue type 6 (isolate CSIRO 154, Australia) (g), and bluetongue type 20 (isolate CSIRO 19, Australia) (h). Electrophoresis was performed in a 7.5% polyacrylamide gel (Gorman *et al.*, 1977a) and stained with ethidium bromide.

complete virions (Verwoerd *et al.*, 1970). The low-molecular-weight RNA of reovirus is lost when the outer capsid layer is removed (Joklik, 1972). Hence, the question of a slow-sedimenting RNA in bluetongue virus remains uncertain. However, Bremer (1976) has clearly shown complete particles of horsesickness virus to contain only high-molecular-weight RNA.

Verwoerd *et al.* (1970) fractionated bluetongue virus RNA into ten components by PAGE. By coelectrophoresis of the RNA of bluetongue type 10 and that of the Dearing strain of reovirus type 3, Verwoerd *et al.* (1972) estimated that the molecular weight of the genome segments of bluetongue virus type 10 ranged from $2.5 \times 10^6$ for the slowest-migrating segment (designated as segment 1) to $0.3 \times 10^6$ for the fastest-migrating segment (designated as segment 10). The molecular-weight estimate for the genome of bluetongue virus type 10 of $11.8 \times 10^6$ was considerably smaller than the accepted genome molecular weight of $15 \times 10^6$ for reoviruses.

The genomes of viruses representing each of the recognized serogroups of orbiviruses except for the Colorado tick fever viruses have been separated into ten segments by PAGE (Fig. 2 and Table VIII).

Green (1970) isolated RNA from Colorado Tick fever virus and centrifuged the RNA and ribonuclease-treated RNA in sucrose density gradients. Two peaks were observed (14 and 4 S) in untreated RNA and a 14 S peak in the ribonuclease-treated RNA. The 4 S peak was assumed to be cellular RNA released from infected cells, not removed on purification of the virus. The 14 S ribonuclease-resistant peak was assumed to be the double-stranded genome of Colorado tick fever virus. The genomes of Colorado tick fever viruses were separated into 12 segments by PAGE (Knudson, 1981), and the patterns of separation differed markedly from those of other orbiviruses. The total molecular weight of the genome was estimated at $18 \times 10^6$ and is larger than any RNA genome of viruses infecting animals within the family Reoviridae. It appears necessary to reevaluate the taxonomic positon of Colorado tick fever virus.

Heterogeneity in the patterns of electrophoretic separation of genome segments of serologically related orbiviruses was first reported for sero-

TABLE VIII. Separation of Double-Stranded RNA Genomes of Orbiviruses by Polyacrylamide Gel Electrophoresis

| Serogroup | Virus | Reference |
| --- | --- | --- |
| African horsesickness | Serotype 3 | Bremer (1976) |
| Bluetongue | Serotype 10 | Verwoerd et al. (1972) |
|  | 10 | Martin and Zweerink (1972) |
|  | 4, 10, 20 | Huismans and Bremer (1981) |
|  | 10, 11, 13, 17 | Sugiyama et al. (1982) |
|  | 1, 4, 10, | Gorman et al. (1981) |
|  | 15, 17, 20 |  |
| EHD | New Jersey | Huismans et al. (1979) |
|  | Ibaraki | Suzuki et al. (1978) |
| Eubenangee | Eubenangee | Schnagl and Holmes (1975) |
|  | Tilligerry | Gorman and Taylor (1978) |
| Palyam | D'Aguilar | Schnagl and Holmes (1975) |
| Changuinola | Changuinola | Knudson (1981) |
| Corriparta | Corriparta | Gorman (1978) |
| Kemerovo | Nugget | Gorman et al. (1983) |
| Warrego | Warrego | Gorman et al. (1977a) |
|  | Mitchell River | Gorman et al. (1977a) |
| Wallal | Wallal | Gorman et al. (1978) |
|  | Mudjinbarry | Gorman et al. (1978) |

types of the Warrego serogroup (Gorman et al., 1977a). Coelectrophoresis of the RNAs of the two viruses revealed 18 bands and suggested that the viruses were distantly related. Only low-level cross-reactions in complement-fixation (CF) tests had been described for these viruses (Borden et al., 1971; Doherty et al., 1973), and no cross-neutralization can be detected between the viruses. The genome segments of two serotypes of the Eubenangee serogroup were easily distinguished by PAGE (Gorman and Taylor, 1978), as were serotypes of the Wallal serogroup (Gorman et al., 1978) and the bluetongue serogroup (Gorman et al., 1981). The basis for the differences in separation of genome segments of related viruses is not known. Using a technique of oligonucleotide mapping of each of the genome segments of the two serotypes of the Wallal serogroup, Walker et al. (1980) found that differences in base sequence were not confined to segments with different electrophoretic mobilities. Although only six of the ten genome segments could be distinguished in the gel electrophoresis system used (Gorman et al., 1978), each genome segment in each virus produced a distinct oligonucleotide map (Walker et al., 1980).

Although examination of orbivirus RNA by PAGE or gradient sedimentation has invariably shown the genome to exist as discrete segments, some workers have suggested that within the virion, the segments are weakly linked to form a continuous genome. Tsai and Karstad (1973) observed the release of dsRNA from EHD virions by treatment with urea or sodium perchlorate. Particles treated with urea liberated filaments that frequently formed from three to five loops without discontinuities. For 80% of the particles, the total length of the looped filaments was 2.5–4.5 μm. The longest looped filament was 4.7 μm in length, which corresponds to a molecular weight of approximately $10.8 \times 10^6$ (Langridge and Gomatos, 1963). Sodium perchlorate treatment released extended linear filaments up to 5.8 μm in length (molecular weight approximately $13.3 \times 10^6$). Urea treatment of bluetongue virus also generates looped structures (Els, 1973). Foster et al. (1978) have reported acid–phenol-purified bluetongue virus RNA to form ten loops, each twice the size of a genome segment, suggesting a diploid genome. Several other viruses containing segmented dsRNA genomes have been reported to have a continuous genome in situ (Dunnebacke and Kleinschmidt, 1967; Vasques and Kleinschmidt, 1968; Nishimura and Hosaka, 1969; Kavenoff et al., 1973, 1975). However, this proposition appears in direct conflict with biochemical evidence of genome structure and replication.

## C. Gene Products

### 1. Polypeptides Specified during Virus Infection

Ten virus-specified polypeptides have been detected in cells infected with bluetongue type 20 (Gorman et al., 1981) and with other viruses of the bluetongue, Eubenangee, and Wallal serogroups (unpublished observations of the authors). In pulse-chase experiments involving radiolabeling of bluetongue virus–induced polypeptides, no secondary cleavage of the primary gene products was observed (Huismans, 1979; Gorman et al., 1981). This contrasts with reovirus in which, under similar experimental conditions, secondary cleavage of polypeptide $\mu l$ to the capsid polypeptide, $\mu 1C$, can be detected (Joklik, 1974). Bluetongue virus type 20 depressed host-cell protein synthesis so that from 12 hr after infection, [$^{35}$S]methionine was incorporated mainly into virus protein and the virus-induced polypeptides separated by PAGE were labeled in order of increasing mobility in the gel system used (Fig. 3). The numbering does not necessarily correspond with that used by other workers to designate virus-

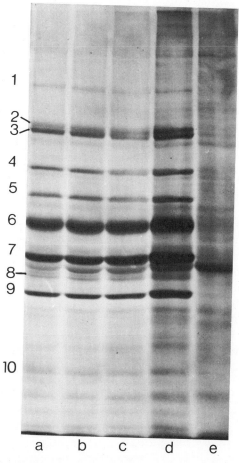

FIGURE 3. PAGE of proteins induced by bluetongue virus type 20 in BHK 21 cells. (a) Virus-infected cells treated with methionine-free Eagle's Minimal Essential Medium (EMEM) plus [$^{35}$S]methionine for 10 min at 12 hr after infection. (b–d) As for (a) with removal of [$^{35}$S]methionine and incubation in MEM plus excess unlabeled methionine for 15 min (b), 40 min (c), or 75 min (d). Uninfected cells labeled for 10 min with [$^{35}$S]methionine. Electrophoresis was performed in an 8% polyacrylamide gel by the method of Laemmli (1970).

specific polypeptides, nor does it infer relationships with numbered segments in the electrophoretic separation of RNA (Fig. 2).

### a. Virion Polypeptides

Seven polypeptides were found to be present in the capsid of bluetongue virus type 10 (Verwoerd *et al.*, 1972), four of which were major components and three minor components. The number and size distribution were similar to those of reovirus type 3 analyzed in the same gel electrophoresis system. Bluetongue virus was found to be unstable in cesium chloride gradients. Utilizing this degrading effect, Verwoerd *et al.* (1972) were able to demonstrate that two of the bluetongue virus polypeptides were present as a diffuse layer surrounding the nucleocapsid. An RNA-dependent RNA polymerase was activated by removal of these two polypeptides. Loss of either one or both of the outer polypeptides resulted in a loss of infectivity. Particle counts per plaque-forming unit (PFU) were estimated at 5 for virus isolated from sucrose gradients; 6 in cesium chloride gradients at pH 8.0 (particle density 1.38 g/cm$^3$), in which the virus remained intact; 10$^5$ in cesium chloride gradients at pH 6.0 (1.39 g/cm$^3$), in which one polypeptide (P$_2$) was lost; and 4.7 × 10$^5$ in cesium chloride gradients at pH 7.0 (1.43 g/cm$^3$), in which both polypeptides P$_2$ and P$_5$ were lost. Martin and Zweerink (1972) also found two virus particles on purification of the station 8 strain of bluetongue virus type 10. The light particles (1.36 g/cm$^3$ in cesium chloride gradients) contained seven polypeptides and had a specific infectivity of 5 × 10$^8$ PFU/A260 unit. The dense particles (1.38 g/cm$^3$) contained five polypeptides and had a specific infectivity 10- to 100-fold less than that of light particles. The inner capsid consists of five polypeptides, P$_1$, P$_3$, P$_4$, P$_6$, and P$_7$, the smallest of which (P$_7$) is inaccessible to enzymatic labeling with iodine (Martin *et al.*, 1973) and appears to be less exposed on the surface of the nucleocapsid than are the other polypeptides.

In subsequent work, Huismans (1979) reported that additional extractions with fluorocarbon were necessary to dissociate bluetongue virus type 10 from cellular material. In that study, the virus was considered purified when no more than seven polypeptides could be detected in a virus sample (Huismans, 1979). Huismans and Bremer (1981) used the same purification techniques in a comparative study of bluetongue virus types 4, 10, and 20. Samples that were resolved into seven bands by electrophoresis in continuous 7.5% polyacrylamide gel slabs in the presence of urea in phosphate buffer (Stone *et al.*, 1974) could be resolved into eight components for types 4 and 20 when electrophoresed in discontinuous gel slabs of 12.5% polyacrylamide in Tris–glycine (Laemmli, 1970). The minor nucleocapsid polypeptide P$_6$ was resolved into two bands for types 4 and 20, but remained as a single band for type 10.

In their purification of Ibaraki virus (EHD serogroup), Suzuki *et al.*, (1978) released virus from infected cells by fluorocarbon extraction procedures similar to those used by Verwoerd *et al.* (1972) and Martin and Zweerink (1972) in the purification of bluetongue virus. Suzuki *et al.* (1978) found that repeated extraction of Ibaraki virus resulted in a loss of infectivity, and after velocity sedimentation through sucrose and cesium chloride equilibrium gradients, they described three virus particles. Virions (density 1.36 g/cm$^3$) contained seven polypeptides after electrophoresis in 7.5% polyacrylamide gels in a continuous phosphate buffer system, subviral particles (density 1.38 g/cm$^3$) contained reduced amounts of polypeptides $P_2$ and $P_5$ and had a specific infectivity about 200-fold less than that of virions, and nucleocapsids (1.40 g/cm$^3$) contained no $P_2$ and $P_5$ and had a specific infectivity about 200,000-fold lower than that of complete viruses. Huismans *et al.* (1979) found EHD virus, New Jersey strain, difficult to purify because of a strong tendency of the virus to adhere to material presumed to be of cellular origin. After repeated extraction with fluorocarbon, the yield of virus was low, and purified virus, which in contrast to bluetongue virus was stable on cesium chloride gradients of pH 7.0 and 8.0, had a density of 1.36 g/cm$^3$. Eight polypeptides were detected after electrophoresis in a continuous phosphate buffer system (Stone *et al.*, 1974). The additional polypeptide of molecular weight 92,000 (labeled $P_3a$) was apparently located on the surface of the virion, since purified EHD broke down on storage to nucleocapsid particles with density 1.40 g/cm$^3$ consisting of polypeptides $P_1$, $P_3$, $P_4$, $P_6$, and $P_7$.

An eighth structural polypeptide of molecular weight 73,000 (labeled $P_4a$) has also been described for horsesickness virus (Bremer, 1976) using the same electrophoresis conditions used by Huismans *et al.* (1979) in the analysis of EHD virus. The stability of horsesickness virus on cesium chloride gradients differed from that of bluetongue virus (Verwoerd *et al.*, 1972) and EHD virus (Huismans *et al.*, 1979) in that horsesickness virus banded on cesium chloride gradients at pH 8 showed some loss of polypeptide $P_3$ and at pH 6 complete loss of $P_3$ and loss of most of polypeptide $P_5$. Polypeptide $P_2$ of horsesickness virus had a molecular weight identical to that of polypeptide $P_3$ of bluetongue virus. The minor component $P_4a$ was associated with nucleocapsids banding at a density of 1.40 g/cm$^3$ in cesium chloride at pH 6.0 (Bremer, 1976).

### b. Nonvirion Polypeptides

Huismans (1979) studied protein synthesis in baby hamster kidney cells and mouse L fibroblasts infected with bluetongue virus type 10. The study deals with nine of the virus-induced polypeptides, since no low-molecular-weight polypeptide derived from the smallest genome segment (10) was detected in infected cells. Genome segment 10 is translated into a low-molecular-weight polypeptide that can be detected in infected cells (Fig. 3) (Gorman *et al.*, 1981), and isolated segment 10 can be translated after denaturation with methylmercuric hydroxide in an *in vitro* system

using rabbit reticulocyte lysate (Sangar *et al.*, 1981). Huismans identified
the seven structural polypeptides $P_1–P_7$ described by Verwoerd *et al.*
(1972) and two nonstructural polypeptides, one of molecular weight
54,000, which he labeled $P_5a$ (since its electrophoretic mobility was intermediate to that of structural polypeptides $P_5$ and $P_6$), and the other of
molecular weight 40,000, which he labeled $P_6a$. It is likely that these
polypeptides are equivalent to the nonstructural polypeptides of medium
molecular weight ($\mu$NS) and low molecular weight ($\delta$NS) of reoviruses
(Joklik, 1974). The molecular weight of $P_6a$ approximates that of cellular
actin, and in radiolabeling experiments, the polypeptide is often difficult
to identify against a background of cellular actin.

Huismans (1979) found that synthesis of viral polypeptides was obscured by host-cell synthesis until 8 hr after infection, but by 11 hr, all
seven capsid polypeptides were detected as well as $P_5a$ and $P_6a$, which
were synthesized in large amounts. Cellular components obscured the
synthesis of $P_6a$ even late in the infectious cycle, but its synthesis was
determined by precipitation of virus-induced polypeptides, with immune
serum, from extracts of infected cells.

Polypeptide $P_6a$ was reported to be transferred to the particulate fraction of cytoplasmic extracts less rapidly than $P_5a$ but before most viral
structural proteins. Of the viral capsid components, polypeptides $P_3$ and
$P_7$ were most rapidly incorporated into particulate structures. These proteins are major components of the viral inner capsid (Verwoerd *et al.*,
1972; Martin and Zweerink, 1972). The precipitation of $P_6a$ by antiserum
to purified virus led Huismans (1979) to suggest an antigenic relationship
between a viral capsid protein and this noncapsid polypeptide and to
postulate that $P_6a$ might be a modified capsid polypeptide. The observation that nucleocapsid polypeptides $P_3$ and $P_7$ and the nonstructural
polypeptide $P_6a$ are incorporated into particulate structures more rapidly
than other polypeptides suggests that they may have a role in the early
events in assembly of virus particles. Polypeptides of Wallal virus equivalent to $P_6a$ and $P_7$ have a high affinity for single-stranded RNA (ssRNA)
as judged by their ability to bind to RNA templates. Chromatography of
a soluble fraction of cytoplasmic extracts from Wallal virus–infected cells
on columns of polyadenylate–Sepharose showed that polypeptides $P_6a$
and $P_7$ were bound in low-ionic-strength buffers and eluted in high-salt
fractions (Gorman, unpublished observations). The nonstructural polypeptide of similar size in reovirus $\delta$NS binds strongly to ssRNA (Huismans and Joklik, 1976). Small reovirus-specific particles with polycytidylate-dependent RNA polymerase activity purified from virus-infected
cells were composed entirely of $\delta$NS (Gomatos *et al.*, 1980). The high
capacity of $\delta$NS to bind many different nucleic acids and select the nucleic
acid bound to small particles suggests that it may act as a condensing
protein to bring together ten ssRNA templates in preparation for dsRNA
synthesis (Gomatos *et al.*, 1981).

## D. Comments on the Comparative Structures of Reoviruses and Orbiviruses

Verwoerd *et al.* (1972) purified bluetongue virus type 10 and compared the virion polypeptides with those of reovirus type 3 by PAGE in a continuous phosphate-based system. They described seven polypeptides in each virus, but subsequent analysis of reovirus type 3 in PAGE systems using Tris–glycine buffers resolved a third λ polypeptide (Cross and Fields, 1976; Ramig *et al.*, 1977; McCrae and Joklik, 1978). Two nonstructural polypeptides were detected in cells infected with reoviruses (Zweerink *et al.*, 1971), and nonstructural polypeptides of similar molecular weight have been described in cells infected with bluetongue virus (Huismans, 1979).

The polypeptides encoded by each of the genome segments of reovirus type 3 have been determined by translation of denatured dsRNA molecules *in vitro* in wheat germ systems (McCrae and Joklik, 1978) and by genetic mapping methods using reovirus recombinants (Mustoe *et al.*, 1978a,b). The genetic map clearly defines the eight structural and two nonstructural polypeptides of reoviruses.

The seven structural and two nonstructural polypeptides of bluetongue virus range in molecular weight from approximately 30,000 to 150,000 (Verwoerd *et al.*, 1972; Martin and Zweerink, 1972; Huismans, 1979), and there is sufficient coding capacity in one or other of the largest nine genome segments to code for these polypeptides. Genome segment 10, of molecular weight approximately $3 \times 10^5$, could theoretically code for a polypeptide of molecular weight 15,000. A small polypeptide (10) has been detected in cells infected with bluetongue virus type 20 (Fig. 3) (Gorman *et al.*, 1981), and denatured genome segment 10 can be translated into a corresponding polypeptide in an *in vitro* system using rabbit reticulocyte lysates (Sangar *et al.*, 1981). Polypeptides of similar size have been detected in cells infected with other orbiviruses (unpublished observations of the authors), and the role of the small polypeptide in orbivirus replication requires further investigation.

Genome segment 10 of reovirus type 3 codes for the smallest polypeptide, δ3, which is located on the surface of the virion (Chapter 2). The attachment of δ3 to subviral particles appears to be the last stage in reovirus morphogenesis, and fully infectious subviral particles that lack δ3 have been described (Joklik, 1972). Under appropriate conditions, digestion of virions with chymotrypsin produced particles that were similar to virions in their physical, biological, and enzymatic properties but lacked about one third of the capsid polypeptide complement. The particles lacked δ3 and a 12,000-dalton fragment of the other major surface polypeptide, μIC, did not have transcriptase activity, and were referred to as "paravirions" (Joklik, 1972). Other subviral particles were generated that lacked δ3 and an 8000–dalton fragment of μIC, and these resembled infectious subviral particles (SVPs) found *in vivo* in reovirus-infected cells (Shatkin and LaFiandra, 1972). Recent evidence suggests that most of the $\mu$1C that exists in infected cells is complexed with $\delta$3, while the uncleaved form, the precursor $\mu$1, exists in free form (P.W.K. Lee *et al.*, 1981). These results suggest a close structural association of the two major surface polypeptides of reoviruses.

Two major polypeptides ($\delta$3 and $\mu$1C) and one minor polypeptide ($\delta$1) are located on the surface of reoviruses, while only two major polypeptides constitute the "fuzzy layer" of orbiviruses. It may be reasonable to speculate that the gene product of segment 10 of orbiviruses is complexed with one of the surface polypeptides $P_2$ or $P_5$ or, alternatively, that infectious orbivirus particles with seven polypeptides are assembled without the addition of a polypeptide equivalent to $\delta$3 of reoviruses. Purified bluetongue virus particles consisting of seven polypeptides could be regarded as structures analogous to "paravirions" or SVPs of reoviruses. Like infectious SVPs of reoviruses found *in vivo*, bluetongue virus particles possess no oligonucleotides, but unlike reovirus SVPs, the transcriptase of bluetongue virus remains inactive until both surface polypeptides $P_2$ and $P_5$ are removed (Van Dijk and Huismans, 1980).

The purification procedures based on extraction of infected cells with fluorocarbon and subsequent treatment with diethyl ether and detergents lead to inconsistencies in reported structures of orbiviruses (Section II.C.1.a). It is possible that the procedures that were based on techniques for purifying reoviruses are not satisfactory when applied to the more labile orbiviruses.

## E. Polypeptides Involved in Immunological Reactions

Attempts have been made to correlate antigenic differences among viruses with molecular structure. Huismans and Howell (1973) hybridized RNAs of bluetongue virus strains and found relatively large differences among the serotypes, but they suggested that the immunological specificity was determined mainly by genome segment 2 and to a lesser extent by segment 6. Cross-hybridization experiments between virulent and attenuated strains of the same serotypes suggested that the process of attenuation involved changes in genome segments 2 and 6. Thus, the same genome segments were involved in the determination of both the immunological specificity and the virulence of the virus. Direct comparison of the electrophoretic mobilities of RNA genome segments and virion polypeptides led to the assumption that genome segments 2 and 6 might code for the surface polypeptides $P_2$ and $P_5$ (Verwoerd *et al.*, 1972; Martin and Zweerink, 1972), and the hybridization studies were assumed to support the proposition that the antigens on the polypeptides were involved in the type-specific reactions between bluetongue viruses (Huismans and Howell, 1973).

De Villiers (1974) compared the molecular sizes of the capsid polypeptides of different serotypes of bluetongue virus with those of bluetongue virus type 10A and found significant differences among serotypes. The molecular weight of $P_2$ differed in 9 of the 11 serotypes. The molecular weight of $P_5$ differed in 4 of the 11 serotypes, and 2 of these were strains for which a significant difference in molecular weight was not detected in $P_2$. Although changes in electrophoretic mobilities of proteins do not necessarily reflect antigenic alterations, the results were interpreted as further evidence that the major determinants of serotype specificity are located on the surface proteins of bluetongue viruses.

Oellermann and Carter (1977) observed that complete virions, and virus particles lacking the surface polypeptide $P_2$, could adsorb to sheep erythrocytes, but that core particles lacking both $P_2$ and $P_5$ could not. They suggested that $P_5$ was involved in the adsorption of bluetongue virus to cells.

A more direct approach to the identification of serotype-specific and group-specific antigens of bluetongue virus was reported by Huismans and Erasmus (1981). They prepared soluble fractions from cytoplasmic extracts of cells infected with bluetongue viruses in the presence of radiolabeled protein hydrolysates. Immune precipitations were carried out with guinea pig, sheep, or rabbit sera that had been selected on the basis of their activities in serological tests. The guinea pig sera selected were those routinely used for serotyping bluetongue viruses and showed little or no cross-reactivity in serum-neutralization tests. In homologous reactions, the sera precipitated the surface polypeptide $P_2$ (but not the other surface polypeptide, $P_5$), the nonstructural polypeptide $P_6a$, and the major nucleocapsid polypeptides $P_3$ and $P_7$. Precipitation of $P_3$, $P_6a$, and $P_7$ occurred with heterologous antisera and was therefore independent of serotype. Heterologous antisera were less effective than homologous antiserum in precipitation of $P_2$. Rabbit antisera also contained antibodies to polypeptide $P_5$. The use of heterologous rabbit antisera led to reduced precipitation of $P_2$, but precipitation of $P_5$ did not consistently correlate with serotype. In some cases, $P_5$ was precipitated strongly with heterologous sera, and in others it was precipitated in small amounts.

The results were interpreted as indicating that the serotype-specific antigen in bluetongue viruses is located on the surface polypeptide $P_2$. Consistent results were obtained in immune precipitation of polypeptide $P_2$ of the New Jersey strain of EHD with homologous guinea pig antiserum, but not with antiserum to the Alberta strain (serotype 2).

The group-specific antigens of bluetongue viruses were identified using ascitic fluids from bluetongue virus–infected mice. The fluids, which contained high CF-antibody titers and low neutralizing-antibody titers, precipitated $P_7$ and $P_6a$ (Huismans and Erasmus, 1981). Antiserum to EHD-New Jersey virus precipitated polypeptides $P_7$, $P_6a$, and $P_3$ of bluetongue virus, while antiserum to bluetongue virus precipitated polypeptides $P_7$ and $P_3$ of EHD virus (Huismans et al., 1979). The group-specific antigens appear to be located on polypeptides in the nucleocapsids of orbiviruses.

## III. REPLICATION

### A. Cell Culture

The orbiviruses replicate in a wide variety of cultures of mammalian and insect origin. The viruses were most often isolated by intracerebral inoculation of newborn mice, but primary isolations have also been made in cultures of lamb kidney cells and chick embryo fibroblasts and in continuous cell lines (BHK 21, Vero, LLC-MK2, MS, PS). Embryonated chicken eggs are commonly used in the study of orbiviruses. The viruses also replicate in mosquito cell lines, often without production of cytopathic effect (Buckley, 1969, 1972).

The replication of orbiviruses has been studied mainly in continuous cell lines (BHK 21, MDBK, Vero), but studies of the morphogenesis of bluetongue virus have also been made in *Culicoides variipennis* (Bowne and Jones, 1966).

### B. Thin-Section Electron Microscopy

Lecatsas (1968a) conducted a sequential study of the formation of bluetongue virus in baby hamster kidney cells. Absorbed virus particles were seen on cell surfaces within minutes, followed by the appearance of phagocytic vesicles containing virus particles. Between 30 min and 2 hr after infection, virus nucleoids or complete virus particles were seen within liposomes near the nucleus. At 2 hr, the rough endoplasmic reticulum (RER) began to show progressive swelling to form vesicles lined with ribosomes. Cromack *et al.* (1971) described similar events occurring in bovine kidney cells (MDBK) infected with bluetongue virus.

Dense fine-granular inclusions near the nucleus were first observed about 8 hr after infection. Lecatsas (1968a) also observed dense inclusion bodies apparently associated with the cristae of mitochondria, but these were not seen by Cromack *et al.* (1971). At 8 hr, the particles adjacent to the nucleus were vaguely defined, and some appeared hollow with fine filaments attached to a central core. Complete virions were also present in the inclusions. Incomplete particles were not observed elsewhere in the cytoplasm (Cromack *et al.*, 1971). Bowne and Jones (1966) noted granular inclusions containing complete particles and also incomplete particles, consisting of either a nucleoid or an empty capsid, within salivary-gland cells of infected *C. variipennis*. The granular inclusions appear to be the matrix for progeny virus replication. Bowne and Jochim (1967) found that the first sign of abnormal activity in bluetongue virus–infected cells was the formation of pinocytotic vesicles in the plasma membrane. These vesicles were mostly empty and apparently moved to the interior of the cell, where they eventually clustered around the nucleus. Not all the cytoplasmic vesicles originated from the plasma membrane, since all membranous elements of the cell became vesiculated. One type of inclusion body originated from the external nuclear membrane and contained a nonspecific granular matrix. Granular masses with morphological structure similar to that of nucleoli appeared in the nucleus. These structures stained positively for DNA. However, similar structures that stained positively for RNA subsequently appeared in the cytoplasm. These inclusion bodies were non-membrane-bound and pleomorphic. Another constant feature observed in conjunction with infected cells was the large array of parallel tubular masses slightly larger in diameter than the core particle. "Kinky" filaments were described admixed with the tubules.

Lecatsas (1968a) described the appearance of bundles of parallel-oriented tubular elements from about 8 hr after infection with bluetongue virus that were similar to those reported in reovirus replication (Gomatos et al., 1962; Anderson and Doane, 1966; Lecatsas, 1968b). These were generally described as about the same diameter as virus particles. Cromack et al. (1971) measured the diameter of the tubular structures in bluetongue virus–infected cells as 47 nm and commented that their electron-staining properties were not similar to those of nucleic acid. They suggested that the tubules were composed of soluble antigen or another virus protein.

The association of orbiviruses with tubular structures is a constant feature of descriptions of virus morphogenesis. Tsai and Karstad (1970) showed that replication of epizootic hemorrhagic disease (EHD) virus was closely related to tubular structures. Aggregates of virus particles were observed within membrane-bound vesicles. The development of virus particles was associated with intracytoplasmic viral matrices consisting of moderately electron-dense granules. Their observations in ultrathin sections indicated that EHD virus possessed two discernible coats, but the finding was discounted, since the observations from negative-contrast electron microscopy did not support the presence of an inner shell structure. Thomas and Miller (1971) observed tubules and fibrils in cells infected with EHD virus similar to those observed in cells infected with bluetongue virus. The tubules were demonstrated in association with viruses, their diameter corresponding roughly to that of the virion. In both EHD- and bluetongue virus–infected cells, masses of free virions with some poorly defined material adjacent to the capsomers were seen in degenerating cells.

Ozawa et al. (1965) found dense oval bodies in horsesickness virus–infected cells. They suggested that replication occurred in the nucleus. However, Lecatsas and Erasmus (1967) found no mature or immature horsesickness virus particles in the nucleus of infected cells. Infection by horsesickness virus was characterized by swelling of the RER, an increased number of fine filaments, the appearance of thick filaments in the cytoplasmic matrix, the appearance of inclusion bodies in the mitochondria, and production of particles associated with the Golgi complex.

Murphy *et al.* (1968) observed Colorado tick fever virus particles with electron-dense cores associated with intracytoplasmic granular matrices, arrays of intracytoplasmic filaments, and fine kinky threads. Occasionally, virus particles were enveloped by membranes of cytoplasmic organelles. Intranuclear filaments in dense arrays were frequently found in infected cells. No nuclear change other than margination of chromatin in some cells late in infection occurred. Oshiro and Emmons (1968) also reported numerous intracytoplasmic fibrillar structures in cells infected with Colorado tick fever virus. Intranuclear bundles of fibers were also observed.

Lecatsas *et al.* (1969) found Corriparta virus morphologically similar to bluetongue virus in thin sections. The presence of dense granular cytoplasmic inclusions in infected cells was also considered a specific factor distinguishing Corriparta, bluetongue, and horsesickness viruses from reovirus. Schnagl and Holmes (1971) found that D'Aguilar, Warrego, and Mitchell River viruses appeared within and around discrete fibrillogranular intracytoplasmic inclusions, but several varieties of associated fibrillar structures were listed. Intranuclear filaments similar to those found with Colorado tick fever virus (Murphy *et al.*, 1968; Oshiro and Emmons, 1968) were observed with D'Aguilar virus.

Murphy *et al.* (1971) demonstrated a similar morphology and mode of morphogenesis with Tribec, Chenuda, Wad Medani, Irituia, Lebombo, Palyam, Colorado tick fever, EHD, and bluetongue viruses. The viruses developed from a granular or reticular cytoplasmic matrix and with rare exception were unenveloped. Filamentous and tubular structures were characteristically associated with their sites of maturation entirely within the cytoplasm. The tubules were usually similar in diameter to virus particles, and it was suggested that they may represent virus subunits assembled in an anomalous manner.

Recent evidence indicates that the tubular structures consist of a virus-coded nonstructural polypeptide (Huismans and Els, 1979). In a study of protein synthesis in cells infected with bluetongue virus, Huismans (1979) detected a polypeptide of molecular weight 54,000 that was synthesized at a rate faster than any other virus-induced polypeptide and was transferred very rapidly from a soluble fraction to a particulate fraction of infected cells. The polypeptide, which he designated $P_5a$, was found to be the main component of a cytoplasmic complex with a sedimentation coefficient of 400 S. When examined by negative-contrast electron microscopy, the complexes consisted of highly purified tubular structures of diameter 68 nm. Similar complexes isolated from EHD virus sedimented at about 400 S and were 54 nm in diameter. By contrast, the complex isolated from horsesickness-infected cells sedimented around 300 S and had tubules with a diameter of 18 nm. In each case, a non-structural polypeptide of similar size to bluetongue virus $P_5a$ was reported to be the only major component. The tubules contained no RNA and appeared to be empty. Bluetongue virus tubules first appeared 2–4 hr after infection and later were seen to be associated with virus particles. Immunoprecipitation studies demonstrated an antigenic reaction between tubules of bluetongue virus and EHD virus, confirming a serological relationship first detected by complement-fixation tests (Borden et al., 1971; Moore and Lee, 1972; Barber and Jochim, 1973). The role of the tubular structures in replication is not understood, and the relationship between tubular structures seen in the replication of reoviruses and orbiviruses has not been investigated.

The predisposition of reovirus to replicate near microtubules was first reported by Dales (1963), who noticed that virions were often assembled close to coated tubules derived from the mitotic spindle. Pretreatment of cells with colchicine, which disassembles cellular microtubules, severely altered the pattern of virus formation, but did not prevent viral synthesis or block viral assembly. Subsequent papers on reoviruses and orbiviruses refer to spindle tubules and microtubules, when in some cases the reports obviously deal with larger structures.

The larger tubules described in cells infected with reoviruses and orbiviruses are distinct from classic cellular microtubules, but they do have features in common and appear to be constructed from proteins of similar size. Microtubules found in virtually every eukaryotic cell are hollow cylindrical polymers of 25-nm diameter. The basic building block is the highly conserved protein tubulin, which is a 6 S dimeric protein composed of two nonidentical protein chains designated $\alpha$ and $\beta$ of molecular weight 50,000. The tubular structures isolated from cells infected with orbiviruses consist entirely of the virus-specified polypeptide $P_5a$ of approximate molecular weight 54,000. The diameters of the purified tubules from cells infected with EHD and bluetongue virus were consistent with those of tubular structures observed in thin sections of virus-infected cells and were about twice as great as that of cellular microtubules. By contrast, tubular structures isolated from cells infected with horsesickness virus were smaller than normal cellular microtubules.

Tubular structures have not been isolated from reovirus-infected cells, but a polypeptide of molecular weight similar to that of $P_5a$ has been described (Zweerink et al., 1971). Recent evidence suggests that the polypeptide exists in two forms in infected cells. The proteins are designated $\mu NS$ and $\mu NSC$, since $\mu NSC$ is about 5000 daltons smaller than $\mu NS$ and is regarded as a cleavage product of the nonstructural polypeptide (Lee et al., 1981).

Mature virions are released from the cell by either cell lysis or extrusion. Lecatsas (1968a) noted complete virus particles extracellularly near small discontinuities in the plasma membrane and suggested that the virus was liberated through these discontinuities. He found no evidence of budding from cell membranes. Bowne and Jochim (1967) reported that virus particles were released singly or en masse at the time of cell lysis. There are numerous reports of orbiviruses budding from the cell membranes, but the general observation is of release of virus at the time of cell lysis. At cytopathic effect, most of the infectious virus remains associated with cellular material (Howell and Verwoerd, 1971; Verwoerd *et al.*, 1979), but Cromack *et al.* (1971) found that bluetongue virus did not accumulate in bovine kidney cells but was released as it was made. The finding of the latter authors that only a small proportion of the virus remained cell-associated conflicts with most reports on the release of orbiviruses from infected cells.

## C. RNA Synthesis

Synthesis of virus-specific single-stranded RNA (ssRNA) commences shortly after infection of mouse L fibroblasts with bluetongue virus. The rate of synthesis continues at a relatively low level until after the appearance of newly synthesized double-stranded RNA (dsRNA) and progeny virus particles (Huismans, 1970a). The viral ssRNA is heterogeneous in size distribution and hybridizes completely with each of the ten segments of the viral genome (Huismans, 1970a; Verwoerd, 1970b; Verwoerd and Huismans, 1972). The ten single-stranded species are found in association with the polysomal fraction of infected cells in the same relative proportions as they appear in total cytoplasmic extracts, suggesting that they are of positive polarity and function as messenger RNA (mRNA). Each RNA is a complete copy of one strand of each genome segment (Verwoerd and Huismans, 1972; Huismans and Verwoerd, 1973). The relative proportions of the different RNA species are not equimolar but remain constant for many hours after infection, indicating some form of transcriptional control (Huismans and Verwoerd, 1973). There is no evidence that ssRNA complementary to polysomal RNA is present in infected cells, suggesting that, as for reovirus, dsRNA is the only available template for synthesis of (+) RNA.

Verwoerd (1970b) demonstrated that the early synthesis of bluetongue virus mRNA is unaffected by cycloheximide-induced inhibition of protein synthesis, suggesting that unless a suitable RNA polymerase is present in normal cells, the enzyme must be virion-associated. It was subsequently shown that removal of the two outer-capsid proteins ($P_2$ and $P_5$ from the virus particle induces an RNA-dependent RNA polymerase (transcriptase) activity in the bluetongue virus inner capsid (Verwoerd *et al.*, 1972; Martin and Zweerink, 1972; Verwoerd and Huismans, 1972). All ten genome segments are transcribed *in vitro*. The relative proportions of transcriptase products are not significantly different from those of the viral mRNA species found in infected cells. This indicates that transcriptional control is a function solely of the inner capsid of the virion (Huismans and Verwoerd, 1973). In this respect, the bluetongue virus polymerase differs from that of reovirus, for which the distributions of products synthesized *in vitro* and *in vivo* are different (Skehel and Joklik, 1969; Banerjee and Shatkin, 1970; Zweerink and Joklik, 1970).

The mechanism of transcriptional control for bluetongue virus is not known. Huismans and Verwoerd (1973) observed that to a broad approximation (error 40%), the transcription rates for most segments were proportional to length except that segment 5 was transcribed at twice the anticipated rate and the product of segment 10 was produced in half the expected quantities. Transcription rates, particularly those observed *in vitro*, may provide useful information about the arrangement of template segments within the virion.

Little is known of the process by which dsRNA is synthesized in infected cells. Bluetongue virus dsRNA first appears several hours after infection, and the maximum rate of synthesis coincides with the accumulation of progeny virus particles (Huismans, 1970a). The mechanism by which ten genome segments are packaged into virions is now known. For reovirus, the synthesis of dsRNA occurs on ten (+) RNA templates packaged in subviral particles (Sakuma and Watanabe, 1971; Zweerink *et al.*, 1972; Zweerink, 1974). Further work is required to determine whether such particles exist in bluetongue virus–infected cells. No information is yet available on the nature of RNA synthesis in cells infected with other orbiviruses.

## D. Protein Synthesis

Infection of cells with bluetongue virus leads to rapid inhibition of cellular protein and DNA synthesis (Huismans, 1970a,b). The rate of inhibition was dependent on multiplicity of infection, but appeared not to depend on viral replication. Inhibition of macromolecular synthesis occurred in cells infected with UV-irradiated virus or in cells infected in the presence of interferon. Huismans (1971) found that polyribosomes were degraded in virus-infected cells, and he suggested some inactivating effect of virion polypeptides.

Early viral proteins were detected 2–4 hr after infection with bluetongue virus (Huismans, 1979). There appears to be little evidence of translational control during infection. The ten mRNA species are associated with polyribosomes in a ratio corresponding to that in which they are synthesized (Huismans and Verwoerd, 1973). Using arbitrary coding assignments based on relative electrophoretic mobilities of genome segments and viral proteins (Verwoerd *et al.*, 1972; Martin and Zweerink, 1972), Huismans (1979) found that the ratio of transcription to translation frequency deviated by less than 2-fold. A possible exception occurred with polypeptide $P_1$. Assuming that the polypeptide was translated from the largest of the mRNAs, the relative abundance of mRNA species 1 was nearly 4 times greater than predicted from the amounts of $P_1$ detected.

These results should be viewed with caution, since there is no direct evidence for coding assignments of genome segments to proteins for bluetongue virus type 10. An assignment based on electrophoretic mobilities in arbitrarily chosen PAGE systems led to misinterpretation of coding assignments of genome segments and proteins of reoviruses (McCrae and Joklik, 1978). The relative mobilities of dsRNA molecules and of proteins vary depending on the electrophoretic systems used. Individual genome segments of bluetongue virus type 4 isolated from polyacrylamide gels (Gorman *et al.*, 1977b) were translated in an *in vitro* rabbit reticulocyte system (Sangar *et al.*, 1981). The relative electrophoretic mobilities of the gene products did not correlate with the relative mobilities of genome segments.

## IV. GENETICS

The genetic interactions among orbiviruses have been studied mainly for viruses of the Wallal serogroup and for bluetongue-group viruses. The studies have been limited to interactions between temperature-sensitive (*ts*) conditional lethal mutants or between wild-type viruses. No cold-sensitive, host-range, or other mutants have been described.

### A. Temperature-Sensitive Mutants of Orbiviruses

#### 1. Isolation of Temperature-Sensitive Mutants

Shipman and De la Rey (1976) reported the isolation of 19 *ts* mutants of bluetongue virus type 10. Purified virus was treated with *N*-methyl-*N'*-nitro-*N*-nitrosoguanidine (MNNG) or nitrous acid, or virus was grown in culture in the presence of proflavine or 5-fluorouracil (5-FU). The mutagen-treated virus stocks were assayed for plaques in petri dishes under agar at a permissive temperature of 28°C for 24 hr. The outline of each plaque was scored, and the dishes were transferred to 38°C for further growth. Nonenlarging plaques were selected as potentially temperature-sensitive and plaque-cloned twice. Of 829 nonenlarging plaques, only 19 stable mutants with an efficiency of plating (38°C/28°C) of $10^{-3}$ or less were found.

The mutagens differed in efficiency. After treatment of bluetongue virus with MNNG, only 1% of the original infectivity was retained, and the frequency of induction of *ts* mutants was 5%. Only 0.01% of infective virus could be recovered after treatment with nitrous acid, and the frequency of induction of mutants was 2%. The infectivity of bluetongue virus grown in the presence of 2 mg 5-FU/ml culture medium was reduced to about 20% of that grown in normal medium, while growth in the presence of 10 μg proflavine/ml culture medium reduced infectivity to 0.2% of control values. The frequency of induction of mutants by 5-FU and proflavine was 3 and 0.5%, respectively.

Treatment of Wallal virus with nitrous acid under conditions similar to those used for bluetongue virus resulted in complete inactivation of virus infectivity (Taylor, unpublished observations). Incorporation of 0.5 mg 5-FU/ml culture medium during growth of Wallal and Mudjinbarry (serotype 2 of the Wallal serogroup) viruses reduced infectivity by 98 and 92%, respectively. At temperatures above 37°C, the plaque size for each virus was reduced, and no plaques could be detected above 39°C. Using a permissive temperature of 32°C and a nonpermissive temperature of 37°C, 13 stable mutants of Wallal virus were isolated from 125 nonenlarging plaques and 4 stable mutants of Mudjinbarry virus from 51 nonenlarging plaques (Taylor, unpublished observations). Although the selection methods used in the two studies were not identical, the behavior of Wallal-serogroup viruses appeared to contrast with that of bluetongue type 10 in the study of Shipham and De la Rey (1976). Growth of Warrego and Mitchell River viruses in the presence of 0.5 mg 5-FU/ml growth medium reduced virus yields by 97% for Mitchell River virus and by more than 99% for Warrego virus. Under the same conditions, the yield of Tilligerry virus (Eubenangee serogropu) was reduced by 85% in the presence of 5-FU.

## 2. Characterization of Temperature-Sensitive Mutants

Shipham and De la Rey (1979) characterized *ts* mutants of bluetongue virus representative of each of six recombination groups.

A shift to the nonpermissive temperature (38°C) after incubation for 24 hr at the permissive temperature led to a marked reduction in virus yields for mutants in recombination groups II, III, V, and VI. For these mutants, the critical temperature-sensitive stage occurred late in the infectious cycle. For mutants in recombinant groups I and IV, an incubation period of 24 hr at the nonpermissive temperature prior to shiftdown to the permissive temperature led to a reduction in virus yield, suggesting that the temperature-sensitive stage occurred early in the replication cycle.

Representative mutants of groups II and III were able to synthesize RNA at both 28 and 38°C, but a mutant representative of group VI was unable to synthesize RNA at the higher temperature. Shipham and De la Rey (1979) suggested that the mutant was defective in synthesis of single-stranded RNA. The defective protein or mutant gene has not been identified in any of the bluetongue virus mutants.

Shipham (1979) compared the viral-induced polypeptides in the soluble and particulate fractions of cells infected with a group II mutant (F207). The virion outer-shell polypeptides $P_2$ and $P_5$ remained in the soluble fraction of cytoplasmic extracts and failed to bind to corelike particles in infected cells. The author suggested that the temperature-sensitive defect could be in one or both of the surface polypeptides or be due to an inability of the core particle to bind the surface polypeptide. The mutant gene was not identified.

No physiological characterization of *ts* mutants of other orbiviruses has been undertaken.

## B. Gene Reassortment

### 1. High-Frequency Recombination

Since the orbiviruses contain segmented genomes, wild-type ($ts^+$) progeny from cells infected with two different *ts* mutants could be generated by reassortment of RNA genome segments or by intramolecular recombination within a genome segment. Temperature-sensitive mutants bearing lesions in different RNA segments should recombine at high frequency, but mutants defective in the same segment would not recombine to produce $ts^+$ progeny unless intramolecular recombination occurred. Recombination by reassortment of gene segments leads to the proposition that there could be ten recombination groups for orbiviruses.

To perform two-factor ($ts \times ts$) genetic crosses between orbiviruses, cells are simultaneously infected with two different *ts* mutants at the permissive temperature, and the virus yields are titrated at both the permissive and the nonpermissive temperature. The recombination frequency is determined as the ratio of the yield of virus plaques at the nonpermissive temperature in the mixed culture and the yield at the permissive temperature. The yield of virus at the nonpermissive temperature is adjusted for yield of each mutant individually at that temperature, and the frequency is expressed as a percentage. The theoretical frequency is assumed to be twice the calculated frequency of $ts^+$ recombinants, since reciprocal double mutants cannot be detected.

In a recombination test using two mutants A and B at a permissive temperature of 28°C and a nonpermissive temperature of 38°C, the recombination frequency is calculated from the following formula.

$$\% = \frac{\text{yield (mixed culture) 38°C} - \text{yield (A) 38°C} - \text{yield (B) 38°C}}{\text{yield (mixed culture) 28°C}} \times 100$$

Shipham and De la Rey (1976) classified 19 *ts* mutants of bluetongue virus type 10 into six recombination groups. The frequency of recombination was variable from experiment to experiment using the same pairs of mutants in comparable experiments, but the observed recombination frequencies of 0.1–19% between selected mutants were interpreted as the result of gene reassortment between mutants. Other pairs of mutants consistently failed to recombine in mixed infections, suggesting that the *ts* lesion occurred in the same segment for these mutant pairs.

The "all-or-none" type of genetic recombination is consistent with that described for reoviruses, orthomyxoviruses, and bunyaviruses (for reviews, see Cross and Fields, 1977; Hightower and Bratt, 1977; Bishop and Shope, 1979).

Gorman *et al.* (1978) demonstrated genetic recombination between serotypes of the Wallal group of orbiviruses. Fifteen mutants of Wallal and Mudjinbarry viruses were divided into two groups within which no recombination occurred. Recombination frequencies between mutants in different groups ranged from 1 to 30% and varied from test to test despite efforts to standardize conditions. The mutants were divided into the same recombination groups at both high (20 PFU/cell) and low (0.02 PFU/cell) input multiplicities of infection. High-frequency recombination at low-input multiplicity has also been described for influenza A viruses (Hirst, 1973) and was taken to indicate the involvement of noninfectious particles in the recombination process. This interpretation may not be valid for orbiviruses. The experimental procedure described by Gorman *et al.* (1978) allowed for multiple rounds of infection during the low-multiplicity test. The recombination frequency may have been amplified during high-multiplicity infections, which would have occurred after the first round of infection. For bluetongue virus mutants, the frequency of recombination increased during the second half of the replication cycle, suggesting that recombination may be a late event associated with the maturation of progeny virions (Shipham and De la Rey, 1976).

## 2. Isolation of Recombinants

Reassortment of genome segments between serotypes of an orbivirus serogroup has been confirmed by isolation of recombinant viruses with genome segments derived from each parental virus.

The differences in migration rates in PAGE of RNA segments of serologically related viruses (Section II.B) can be exploited to isolate recombinants between viruses. Wild-type plaques ($ts^+$) appearing at the nonpermissive temperature after crossing *ts* mutants of two serotypes of Wallal virus were analyzed for RNA content by PAGE (Gorman *et al.*, 1978). The electrophoretic patterns of the parent viruses, Wallal and Mudjinbarry, were distinct in segments 2, 3, 4, 5, 6, and 10. By comparison of the electrophoretic profiles of the RNA segments of $ts^+$ plaques with parental virus RNA, 3 plaque clones of 60 selected were found to have genome segments derived from each of the parent viruses. The isolation of recombinants verified that recombination had taken place by physical exchange of genome segments, but a consistent relationship between segment reassortment and loss of temperature sensitivity could not be demonstrated by PAGE of RNA genome segments. The parental origins of

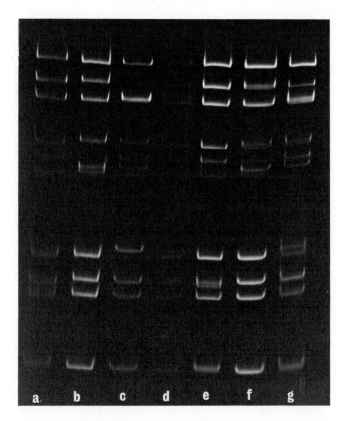

FIGURE 4. PAGE of dsRNA genome segments of bluetongue virus type 6 (isolate CSIRO 154) (a), bluetongue virus type 1 (isolate CSIRO 156) (b), bluetongue virus type 20 (isolate CSIRO 19) (c), recombinant virus (d), coelectrophoresis of RNA of bluetongue virus type 6 and recombinant virus (e), coelectrophoresis of RNA of bluetongue virus type 1 and recombinant virus (f), and coelectrophoresis of RNA of bluetongue virus type 20 and recombinant virus (g). Electrophoresis was performed in a 7.5% polyacrylamide gel (Gorman *et al.*, 1977a) and stained with ethidium bromide.

segments 1, 7, 8, and 9 of recombinants could not be determined, since the corresponding RNA segments of the parental viruses had identical electrophoretic mobilities.

Walker *et al.* (1980) denatured the double-stranded RNA (dsRNA) genome segments of parental and recombinant viruses and treated the RNA with ribonuclease $T_1$. After electrophoresis of the oligonucleotides in a two-dimensional PAGE system, unique fingerprints were obtained for each of the genome segments of Wallal and Mudjinbarry viruses (see Section V.C). Analysis of the fingerprints of genome segments of recombinant viruses established that the replacement of segment 1 of Wallal virus by the corresponding segment of Mudjinbarry virus occurred in each of the $ts^+$ plaques examined.

A combination of comparative PAGE of RNA genome segments and of oligonucleotides of each of the genome segments has been an effective method for determining the parental origin of each genome segment in recombinant orbiviruses. Two recombination groups had been established for mutants of the Wallal serogroup (Gorman et al., 1978). The ts mutations were located in segment 1 for one recombination group and in segment 2 for the other (Walker et al., 1980).

Recombinant viruses have been isolated from cultures simultaneously infected with three distinct serotypes of bluetongue virus (Gorman et al., 1982). The pattern of separation of the RNA genome segments of the three serotypes was distinct after PAGE (Fig. 4). Plaque clones of each of the three serotypes were inoculated simultaneously into cultures of pig kidney cells, and at cytopathic effect, the cultures were assayed for PFU. Individual plaques were selected and their RNA analyzed by PAGE. Although the parental origin of each segment has not been determined uniquely, recombinant viruses deriving genome segments from each of the parent viruses were detected. The origin of genome segments as determined by PAGE of dsRNA for recombinant is shown in Table IX.

## C. Mapping Gene Function

The techniques of PAGE of genome RNA segments and of oligonucleotides derived from them can be used to define the genetic composition of recombinant viruses and determine the parental origin of each gene. Defined recombinant viruses can be used to assign function to each gene. For example, genetic markers for virulence can be located by analysis of virulent and avirulent viruses and recombinant viruses with genes derived from each of them. The value of defined recombinant viruses has

TABLE IX. Origin of Genome Segments of Bluetongue Virus Recombinant

| Segment | Serotype origin of segment | | |
|---------|:---:|:---:|:---:|
| 1  | 1 | 6 | 20[a] |
| 2  | — | 6 | — |
| 3  | — | 6 | — |
| 4  | — | 6 | — |
| 5  | — | 6 | — |
| 6  | — | — | 20 |
| 7  | 1 | 6 | —[a] |
| 8  | 1 | — | 20[a] |
| 9  | 1 | 6 | 20[a] |
| 10 | 1 | — | — |

[a] Segment origin not specifically determined.

been demonstrated by the work of Fields and his colleagues using recombinant reoviruses (see Chapter 5).

The analysis of orbivirus recombinants has been limited to the study of the proteins involved in the serum-neutralization reactions between serotypes of the Wallal virus group (Walker, 1981). Recombinant viruses of known genetic composition (Walker *et al.*, 1980) were compared with Wallal and Mudjinbarry viruses in plaque-reduction neutralization tests. The patterns of segregation of genome segments in the recombinant viruses suggested that genome segment 2 and possibly genome segment 5 code for polypeptides involved in the type-specific serum-neutralization reaction. The electrophoretic patterns of proteins synthesized in cells infected with Wallal and Mudjinbarry viruses were distinct in four of the ten virus-induced polypeptides. Analysis of the PAGE separation patterns of the viruses and selected recombinants suggested that polypeptides PII, PIV, and PVIII were encoded in, respectively, genome segments 3, 6, and 8 (Walker, 1981). The product of genome segment 5 was identified as polypeptide PV, but the type-specific polypeptide encoded in genome segment 2 was not identified.

## V. GENETIC RELATIONSHIPS AMONG ORBIVIRUSES

The establishment of serological groups has been important in assessment of relationships among viruses and in the development of concepts of evolution of viruses. Serological tests cannot be used to estimate the relatedness of each of the genome segments of orbiviruses. The detection of shared antigens in complement-fixation (CF) tests has application in defining groups of orbiviruses, but is restricted to comparing antigenic sites on a limited number of virus-specified proteins. Serum-neutralization tests measure the relatedness of antigenic determinants carried by one protein, the genetic information for which is contained in one of the ten genome segments (Section II.E).

Comparison of the genomes of orbiviruses by techniques such as RNA–RNA hybridization and oligonucleotide mapping provide estimates of the relatedness of each of the genome segments and reveal extensive genetic diversity within orbiviruses (see Section II.E).

### A. Comparison of Genome Segments of Orbiviruses

Attempts have been made to compare related orbiviruses by PAGE of their genome segments. Electrophoretic separation of the genome segments of Warrego and Mitchell River viruses showed considerable differences in the RNA patterns (Gorman *et al.*, 1977a). Coelectrophoresis of the RNAs of the two viruses revealed 18 bands, suggesting that the viruses are distantly related. Only low-level cross-reactions are found in

CF tests, and no cross-neutralization can be detected between the viruses. However, the serotypes of the Wallal serological group, which are indistinguishable by CF tests, also showed considerable differences in the electrophoretic separation of their RNAs (Gorman et al., 1978). Variations in the electrophoretic mobilities of genome segments of serotypes of the Eubenangee (Gorman and Taylor, 1978) and bluetongue (Gorman et al., 1981; Huismans and Bremer, 1981; Sugiyama et al., 1982) serogroups have also been described.

It has been suggested that cytoplasmic polyhedrosis viruses (CPVs) (Payne et al., 1977) and rotaviruses (Rodger and Holmes, 1979) could be classified by the electrophoretic separation patterns of their double-stranded RNA (dsRNA) genomes. Payne et al. (1977) proposed that the electrophoretic profiles were more useful than serological methods in grouping CPVs. Analysis of oligonucleotide fingerprints of each of the RNA genome segments of serotypes of Wallal virus has shown substantial sequence heterology not restricted to segments with different electrophoretic mobilities (Walker et al., 1980). The patterns of separation of genome segments cannot be correlated with differences in sequence between corresponding segments in viruses being compared. The differences in electrophoretic mobilities of RNA segments can be enhanced by selection of conditions for PAGE. Little difference can be detected in the profiles of genome segments of Wallal and Mudjinbarry viruses in gels of 3% polyacrylamide. Distinct differences can be observed only for segments that migrate at position 6. Optimal resolution of seven of the ten dsRNA segments can be obtained in gels of 10% polyacrylamide (Walker, 1981).

Analysis of 17 Wallal serogroup isolates by PAGE showed that the dsRNA profiles could not be used to classify the viruses, nor was there any consistency in pattern for viruses isolated within days of each other at a single site (Gorman et al., 1977b). Knudson (1980) analyzed the genomes of a large number of isolates in different serogroups and found extensive variation in PAGE profiles. He implied that viruses originating from specific areas were similar, but extensive variability in the dsRNA profiles was observed for isolates from different regions.

## B. Comparison of Orbivirus Genomes by RNA–RNA Hybridization

Orbiviruses synthesize single-stranded RNA (ssRNA) molecules from the dsRNA genome templates using a virion-associated RNA polymerase (transcriptase). In bluetongue virus, the loss of surface polypeptides activates the transcriptase in vivo and in vitro (Verwoerd and Huismans, 1972; Van Dijk and Huismans, 1980). The ssRNA isolated from infected cells or from an in vitro transcription system will hybridize uniquely with isolated denatured dsRNA, but not with itself (Verwoerd and Huismans, 1969). The electrophoretic mobilities of the renatured hybrid double-stranded molecules were identical to the corresponding native dsRNA molecules (Verwoerd and Huismans, 1972).

Huismans and Howell (1973) studied the relationships between serotypes of bluetongue virus by cross-hybridization of messenger RNA (mRNA)[+] with the minus strand of genome RNA. The technique involved isolating [3]H-labeled mRNA from cells infected with one virus and hybridizing it with saturating amounts of denatured [32]P-labeled dsRNA from the virus to be compared. After removal of ssRNA, the double-stranded molecules were analyzed by PAGE. Since the bulk of denatured dsRNA reassociated to form the original ten genome segments, [32]P activity in the gel showed the location of the homologous genome segments. A small proportion of the denatured dsRNA hybridized with the [3]H-labeled mRNA to form a duplex, and the position of the [3]H label in the gel allowed an assessment to be made of the degree of hybridization between the heterologous strands. A change in the migration position of heterologous duplex molecules compared with the renatured genome segment was attributed to regions of mismatching in the hybrid molecules (Ito and Joklik, 1972).

Cross-hybridization of the RNAs of three strains of bluetongue virus type 4 isolated in South African in 1901 and 1958 and in Cyprus in 1971 revealed altered electrophoretic mobilities in each of the heterologous molecules compared with the homologous dsRNA segments (Huismans and Howell, 1973). In some reactions, no hybrid was formed between segment 10 and the heterologous virus. The interpretation that altered electrophoretic mobilities are caused by mismatched duplex molecules suggests that base-sequence divergence had occurred in corresponding genes in each of the viruses. Assuming common ancestry, the results suggest mutational change in each genome segment of viruses separated geographically and temporally.

Huismans and Bremer (1981) compared RNA from bluetongue type 20 isolated from Australia with types 4 and 10 from South Africa. Despite the close serological relationship between types 4 and 20 (Della-Porta *et al.*, 1981a), the RNA homology between the Australian and South African viruses was less than 30% compared with more than 70% between South African serotypes. Cross-hybridization of individual genome segments of type 4 with RNA of type 20 detected regions of homology in segments 7 and 10, but the alterations in electrophoretic mobilities of the hybrids, compared with the original segments, suggested extensive regions of mismatched base pairs. Gorman *et al.* (1981) found no significant reassociation of mRNA[+] derived from bluetongue virus type 20 with strands of opposite polarity of bluetongue virus types 1, 4, and 10 isolated in South Africa and type 17 from the United States. Since the smallest genome segments each represent less than 5% of the total nucleic acid, the mixtures were analyzed by electrophoresis in 7.5% polyacrylamide gels. Single-stranded RNA does not enter these gels, and no double-stranded duplexes could be detected in the mixtures. Short regions of homology could have escaped detection by the methods used, but the results indicated a considerable sequence divergence between the virus isolated in Australia and bluetongue virus serotypes isolated in other geographic regions.

From six to eight duplex molecules were detected by PAGE after cross-hybridization of RNA from bluetongue virus serotypes isolated in South Africa (Huismans and Howell, 1973). In many cases, the electrophoretic mobilities of the hybrids were altered significantly, and it was difficult to determine which genome segment the hybrids were derived from. However, some generalizations are possible. Invariably, only two hybrids were detected in the region of genome segments 1, 2, and 3, and with one exception in ten cross-hybridizations performed, only two hybrids were detected in the region of genome segments 4, 5, and 6. Since the two surface polypeptides on bluetongue viruses are encoded in one genome segment from each region (Section II.E), the results can be interpreted as indicating that extensive sequence divergence has occurred between segments coding for the surface polypeptides of the South African serotypes of bluetongue virus.

Similar results were obtained in cross-hybridization of RNA from three bluetongue virus serotypes isolated in Australia. Hybridization of ssRNA from bluetongue type 1 (CSIRO 156) with type 20 (CSIRO 19) and type 6 (CSIRO 154) resulted in eight hybrids detectable by PAGE (Fig. 5). No hybrids could be detected between bluetongue type 1 and a Eubenangee serogroup virus (CSIRO 23). In similar tests, no hybrids could be detected between bluetongue virus type 20 and selected South African serotypes (Gorman et al., 1981), while Huismans and Bremer (1981) found limited homology between Australian and South African viruses.

Assuming common ancestry for bluetongue viruses, the observations are consistent with evolution of viruses in isolation on the two continents, leading to significant differences in the gene pools of the two virus populations. It may also be significant that RNA of bluetongue virus type 16 isolated in Pakistan hybridized to eight of the genome segments of type 3 from South Africa (Huismans and Howell, 1973). There appeared to be no cross-hybridization of segments 2 and 10.

## C. Oligonucleotide Fingerprinting

Small changes in nucleotide sequences of RNA viruses can be detected by digesting the RNA with specific enzymes and separating the oligonucleotides produced by PAGE to produce a fingerprint for each virus (De Wachter and Fiers, 1972; Billeter et al., 1974). The technique has revealed extensive genetic diversity in viruses that are indistinguishable by serological methods.

Oligonucleotide fingerprints can be obtained for each of the genome segments of orbiviruses after separation of the dsRNA segments by PAGE,

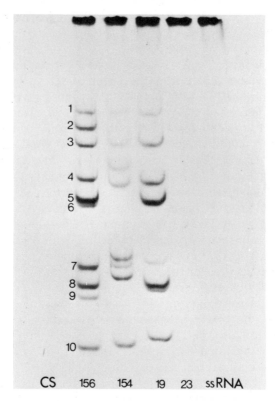

FIGURE 5. PAGE in 7.5% polyacrylamide of RNA–RNA hybridization products after melting 5 μg dsRNA of bluetongue virus serotypes 1 (CSIRO 156), 6 (CSIRO 154), 20 (CSIRO 19), and the Eubenangee serogroup virus CSIRO 23 and reannealing in the presence of radiolabeled ssRNA isolated from cells infected with bluetongue virus type 1. Hybrid molecules were detected by fluorography (Gorman *et al.*, 1981).

denaturation of each segment to ssRNA, and treatment with ribonuclease $T_1$ (Walker *et al.*, 1980; Sugiyama *et al.*, 1982). The fingerprints are of limited complexity, since a relatively large sample of oligonucleotides is obtained from genome segments, these oligonucleotides ranging in molecular weight from approximately $2.5 \times 10^5$ to $2.5 \times 10^6$ (Section II.B). The use of oligonucleotide fingerprinting in determining the origin of genome segments in recombinant viruses was discussed in Section IV.B, but the technique can also be used to examine genetic diversity in natural populations of orbiviruses.

## 1. Segment Reassortment among Orbiviruses

Sugiyama *et al.* (1981) compared oligonucleotide fingerprints of two strains of bluetongue virus type 11 and one of type 10 isolated in the United States. The strains of type 11 shared at least 50% of large oligonucleotides in all but one genome segment. Fingerprints of segment 3 were dissimilar, but one was related to the fingerprint of segment 3 of the type 10 virus. Of the 37 large oligonucleotides of segment 3 of the two viruses, 32 were common. The result suggests that the three bluetongue viruses contain segments derived from a common gene pool and represent naturally occurring reassortant viruses.

The PAGE patterns of genome segments and oligonucleotide fingerprints of isolates of Wallal virus obtained from a single pool of 100 biting midges suggested that the isolates were recombinants of two serotypes in the pool. Two distinct oligonucleotide maps were detected for most of the genome segments, and the segments were arranged in different constellations in different isolates (Walker *et al.*, in prep.).

## 2. Heterogeneity in the Gene Pool of Orbiviruses

Oligonucleotide fingerprints of corresponding genome segments of serotypes of Wallal virus are distinct (Walker *et al.*, 1980). Similar fingerprints were obtained for segment 7 of the viruses in which 11 of the 29 large oligonucleotides detected were common (Fig. 6). The extent of conservation of sequence in each genome segment has not been determined, but it is unlikely to exceed 85–90% (Walker, 1981). The oligonucleotides common to each segment 7 appear to represent a core of sequences not significantly altered in the evolution of the viruses.

Sugiyama *et al.* (1982) have detected close genetic relationships between serotypes of bluetongue virus isolated in the United States. The CA8 strain of serotype 10 and the Station strain of serotype 11 share at least 50% of the large oligonucleotides in six genome segments. Four

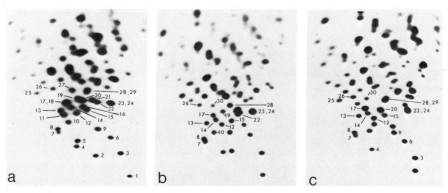

FIGURE 6. Comparison of ribonuclease $T_1$ fingerprints of genome segment 7 of serotypes of the Wallal serogroup. (a) Strain Ch 12048 R [serotype 3 (Walker, 1981)]; (b) Strain Ch 12048 [serotype 1 (Wallal virus)]; (c) Strain NT 14952 [serotype 2 (Mudjinbarry virus)]. Of the largest oligonucleotides in (a), 29 were assigned the numbers 1–29, and corresponding oligonucleotides in (b) and (c) were assigned the same numbers. Oligonucleotide 30 was present only in (b) and (c). The serotypes share 11 oligonucleotides; (a) and (b) share 14, (b) and (c) share 12, and (a) and (c) share 18. The methods used in fingerprinting are described by Walker *et al.* (1980).

segments produced dissimilar fingerprints. The most closely related fingerprints were obtained from segment 8, in which only 3 of 26 large oligonucleotides were not shared. Oligonucleotide fingerprints obtained from genome segments of the Australian serotype 20 (Gorman *et al.*, 1981) show no relationship to those obtained for the United States serotypes (Sugiyama *et al.*, 1982), but direct comparisons of viruses from the two countries have not been made. No cross-hybridization of the RNA of type 20 with type 17 from the United States could be detected (Gorman *et al.*, 1981), and it is possible that the gene pools of bluetongue viruses in the United States and in Australia are distinct.

Sugiyama *et al.* (1982) examined genome segment fingerprints of bluetongue viruses of serotype 11 isolated in Colorado in 1963, 1970, 1974, and 1975 and in Texas, Idaho, California, and Oregon in 1973. The segments of each virus were analyzed as composite fingerprints of segments 1–3, 4–6, and 7–10 and compared with characteristic oligonucleotides of the prototype strain of type 11 isolated in Texas in 1962. Each strain was related to but distinguishable from the prototype strain, with alterations in each of the genome segments. Some fingerprints were distinct, suggesting shifts in genetic structure possibly due to genome segment reassortment. In general, the results indicated substantial sequence divergence between isolates of the one serotype over 12 years. It may be significant, however, that a strain isolated in Colorado in 1970 (70–48 S) differed from the prototype strain from Texas in 1962 in only 2 of 101 oligonucleotides analyzed, and these changes were detected in segment 10. This suggests that the rate of genetic drift may be slow for individual genome segments and that the virus population consists of a distribution of many different sequences.

## VI. SPECIATION IN ORBIVIRUSES

The genetic diversity within orbivirus serogroups can be explained by reference to modern concepts of the structure of natural populations of organisms (Mayr, 1980). The nature of genetic variation was not fully understood while species were seen as uniform types. Mutation was thought of as affecting *the* wild-type, and the classic view that the wild-type gene was almost always superior to other alleles led to misunderstanding of the role of selection in the evolution of populations. Species and populations were not seen as highly variable aggregates consisting of genetically unique individuals, but as uniform types.

The concept of serotype inhibits appreciation of virus populations as variable aggregates. For orbiviruses, serotype specificity is conferred by antigenic determinants encoded in one of the genome segments, but studies on the genomes of the viruses show extensive variation in each segment of closely related viruses. In influenza viruses, mutations occur not only in the immunologically important hemagglutinin and neuraminidase genes, but also in other RNA segments (Young *et al.*, 1979; Ortin *et al.*, 1980). Antigenic variability in foot-and-mouth disease virus was shown to be a consequence of variability at many sites on the genome (Domingo *et al.*, 1980).

Although the wild-type sequence is often regarded as a standard copy of an optimally adapted phenotype, analysis of the wild-type distribution of Qβ showed that only a small fraction of sequences was identical with that assigned to wild-type (Domingo *et al.*, 1978). An average sequence defines the wild-type. Eigen and Schuster (1979) used the term "quasi-species" to represent an organized distribution of sequences and proposed that the quasi-species was the target for selection.

A species in the phylogenetic sense is the largest aggregate of individual organisms that evolves as a unit. The biological species concept stresses the community gene pool and reproductive isolation. Estimates of the degree of similarity among viruses can be obtained by the use of serological reactions including immunoprecipitation of virus-specified proteins, by techniques such as RNA hybridization and oligonucleotide mapping, and recently by comparing sequences of nucleic acids and proteins. The techniques estimate similarity and are of value in grouping viruses into taxa. Viruses with segmented genomes exchange genetic information through processes analogous to sexual reproduction, and it is important to define the gene pool of genetically interacting groups of viruses.

Viruses of the bluetongue serogroup are closely related by the use of any of the aforementioned techniques and are able to reassort genetic information (Gorman *et al.*, 1982). Bluetongue viruses isolated in South Africa and Australia are more distantly related by the use of RNA–RNA hybridization techniques (Gorman *et al.*, 1981; Huismans and Bremer, 1981) and share close serological relationships (Della-Porta *et al.*, 1981a), and the proteins on which the antigenic sites are located can be detected by immunoprecipitation techniques (Huismans and Bremer, 1981). While the geographic isolation of the viruses has produced significant differences in the nucleotide sequences of the two groups, they are able to interact genetically (Gorman *et al.*, 1982). The available evidence suggests that while viruses of the bluetongue and epizootic hemorrhagic disease (EHD) serogroups share common antigens on proteins in their nucleocapsids, the genetic homology between a bluetongue virus isolated in South Africa and an EHD virus isolated in the United States is low and homologous regions could not be located on any genome segment (Huismans *et al.*, 1979). There is as yet no evidence that the viruses can interact genetically, but the distant relationships between them suggest common evolutionary ancestors for both groups of viruses.

## VII. CONCLUSION

The genus *Orbivirus* consists of viruses with common morphological features, and the present classification into serogroups is confusing because of the cross-reactions between viruses in some serogroups. Eventually, it will be possible to divide the genus into genetically interacting groups of viruses with defined relationships between them indicating the evolutionary links between species of orbiviruses.

ACKNOWLEDGMENTS. Orbivirus research at the Queensland Institute of Medical Research is supported by grants from the National Health and Medical Research Council of Australia and the Australian Meat Research Committee. We thank Mr. A.J. Melzer and Mr. D.A. Smith for providing photographs and Miss R. McDowell and Miss C.R. Hawkins for typing the manuscript.

# REFERENCES

Alexander, R.A., 1933, Preliminary note on the infection of white mice and guinea-pigs with the virus of horsesickness, *J. S. Afr. Vet. Med. Assoc.* **4**:1.

Anderson, N., and Doane, F.W., 1966, An electron-microscope study of reovirus type 2 in L cells, *J. Pathol. Bacteriol.* **92**:433.

Banerjee, A.K., and Shatkin, A.J., 1970, Transcription *in vitro* by reovirus-associated ribonucleic acid-dependent polymerase, *J. Virol.* **6**:1.

Barber, T.L., 1979, Temporal appearance, geographic distribution and species of origin of bluetongue virus serotypes in the United States, *Am. J. Vet. Res.* **40**:1654.

Barber, T.L., and Jochim, M.M., 1973, Serological characterization of selected bluetongue virus strains from the United States, *Proc. 77th Annu. Meet. U.S. Anim. Health Assoc.*, pp. 352–359.

Barber, T.L., and Jochim, M.M., 1975, Serotyping bluetongue and epizootic hemorrhagic disease virus strains, *Proc. 18th Annu. Meet. Am. Assoc. Vet. Lab. Diagn.*, Portland, Oregon, pp. 149–162.

Barnard, B.J. H., and Pienaar, J.G., 1976, Bluetongue virus as a cause of hydranencephaly in cattle, *Onderstepoort J. Vet. Res.* **43**:155.

Becker, F.E., 1930, Tick-borne infections in Colorado. II. A survey of the occurrence of infections transmitted by the wood tick, *Colo. Med.* **27**:87.

Bellamy, A.R., Shapiro, L., August, J.T., and Joklik, W.K., 1967, Studies on reovirus RNA. 1. Characterization of reovirus genome RNA, *J. Mol. Biol.* **29**:1.

Berge, T.O. (ed.), 1975, *International Catalogue of Arboviruses*, 2nd ed., DHEW Publication No. (CDC) 75-8301, Washington, D.C.

Billeter, M.A., Parsons, J.T., and Coffin, J.M., 1974, The nucleotide sequence complexity of avian tumor virus RNA, *Proc. Natl. Acad. Sci. U.S.A.* **71**:3560.

Bishop, D.H.L., and Shope, R.E., 1979, Bunyaviridae, in: *Comprehensive Virology*, Vol. 14 (H. Fraenkel-Conrat and R.R. Wagner, eds.), pp. 1–156, Plenum Press, New York.

Blue, J.L., Dawe, D.L., and Gratzek, J.B., 1974, The use of passive hemagglutination for the detection of bluetongue viral antibodies, *Am. J. Vet. Res.* **35**:139.

Bordon, E.C., Shope, R.E., and Murphy, F.A., 1971, Physicochemical and morphological relationships of some arthropod-borne viruses to bluetongue virus—a new taxonomic group: Physicochemical and serological studies, *J. Gen. Virol.* **13**:261.

Bowne, J.G., 1971, Bluetongue disease, *Adv. Vet. Sci. Comp. Med.* **15**:1.

Bowne, J.G., and Jochim, M.M., 1967, Cytopathologic changes and development of inclusion bodies in cultured cells infected with bluetongue virus, *Am. J. Vet. Res.* **28**:1091.

Bowne, J.G., and Jones, R.H., 1966, Observations on bluetongue virus in the salivary glands of an insect vector, *Culicoides variipennis*, *Virology* **30**:127.

Bowne, J.G., and Ritchie, A.E., 1970, Some morphological features of bluetongue virus, *Virology* **40**:903.

Bowne, J.G., Luedke, A.J., Jochim, M.M., and Metcalf, H.E., 1968, Bluetongue disease in cattle, *J. Am. Vet. Med. Assoc.* **153**:662.

Breese, S.S., Jr., Ozawa, Y., and Dardiri, A.H., 1969, Electron microscopic characterization of African horse-sickness viruses, *J. Am. Vet. Med. Assoc.* **155**:391.

Bremer, C. W., 1976, A gel electrophoretic study of the protein and nucleic acid components of African horsesickness virus, *Onderstepoort J. Vet. Res.* **43**:193.

Buckley, S. M., 1969, Susceptibility of *Aedes albopictus* and *A. aegypti* cell lines to infection with arboviruses, *Proc. Soc. Exp. Biol. Med.* **131**:625.

Buckley, S.M., 1972, Propagation of 3 relatively solvent-resistant arboviruses in Singh's *Aedes albopictus* and *A. aegypti* cell lines, *J. Med. Entomol.* **9**:168.

Burgdorfer, W., 1977, Tick-borne diseases in the United States: Rocky Mountain spotted fever and Colorado tick fever, *Acta Trop.* **34**:103.

Campbell, C.H., Barber, T.L., and Jochim, M.M., 1978, Antigenic relationship of Ibaraki, bluetongue, and epizootic hemorrhagic disease viruses, *Vet. Microbiol.* **3**:15.

Carley, J.G., and Standfast, H.A., 1969, Corriparta virus: Properties and multiplication in experimentally inoculated mosquitoes, *Am. J. Epidemiol.* **89**:583.

Casals, J., 1971, Classification of arboviruses transmitted by ticks by means of classical serology and according to physiochemical properties, in: *International Symposium on Tick-Borne Arboviruses (Excluding Group B)*, (M. Gresikova, ed.), pp. 13–20, Slovak Academy of Sciences, Bratislava.

Cordy, D.R., and Shultz, G., 1961, Congenital subcortical encephalopathies in lambs, *J. Neuropathol. Exp. Neurol.* **20**:554.

Cox, H.R., 1954, Bluetongue, *Bacteriol. Rev.* **18**:239.

Cromack, A.S., Blue, J.L., and Gratzek, J.B., 1971, A quantitative ultrastructural study of the development of bluetongue virus in Madin–Darby bovine kidney cells, *J. Gen. Virol.* **13**:229.

Cross, R.K., and Fields, B.N., 1976, Reovirus-specific polypeptides: Analysis using discontinuous gel electrophoresis, *J. Virol.* **19**:162.

Cross, R.K., and Fields, B.N., 1977, Genetics of Reoviruses, in: *Comprehensive Virology*, Vol. 9 (H. Fraenkel-Conrat and R.R. Wagner, eds.), pp. 291–340, Plenum Press, New York.

Dales, S., 1963, Association between the spindle apparatus and reovirus, *Proc. Natl. Acad. Sci. U.S.A.* **50**:268.

Davies, F.G., 1976, AHS vaccine, *Vet. Rec.* **98**:36.

Davies, F.G., and Lund, L.J., 1974, The application of fluorescent antibody techniques to the virus of African horsesickness, *Res. Vet. Sci.* **17**:128.

De Boer, C.J., Kunz, L.J., Koprowski, H., and Cox, H.R., 1947, Specific complement-fixing diagnostic antigens for Colorado tick fever, *Proc. Soc. Exp. Biol. Med.* **64**:202.

Della-Porta, A.J., McPhee, D.A., and Snowdon, W.A., 1979, The serological relationships of orbiviruses, in: *Arbovirus Research in Australia, Proceedings of the Second Symposium 1979* (T.D. St. George and E.L. French, eds.), pp. 64–71, CSIRO-QIMR, Brisbane.

Della-Porta, A.J., Herniman, K.A.J., and Sellers, R.F., 1981a, A serological comparison of the Australian isolate of bluetongue virus type 20 (CSIRO 19) with bluetongue group viruses, *Vet. Microbiol.* **6**:9.

Della-Porta, A.J., McPhee, D.A., Wark, M.C., St. George, T.D., and Cybinski, D.H., 1981b, Serological studies of two additional Australian bluetongue virus isolates, CSIRO 154 and CSIRO 156, *Vet. Microbiol.* **6**:233.

De Oliva, O.F.P., and Knudson, D.L., 1980, DsRNA analyses of Changuinola (CGL) serogroup viruses, *Annu. Rep. Yale Arbovirus Res. Unit*, New Haven, pp. 53–56.

De Villiers, E.-M., 1974, Comparison of the capsid polypeptides of various bluetongue virus serotypes, *Intervirology* **3**:47.

De Wachter, R., and Fiers, W., 1972, Preparative two-dimensional poly-acrylamide gel electrophoresis of $^{32}$P-labeled RNA, *Anal. Biochem.* **49**:184.

Doherty, R.L., 1967, Studies of aborigines at Aurukun and Weipa missions, north Queensland. 2. Other laboratory studies, *Aust. Paediatr. J.* **3**:213.

Doherty, R.L., 1972, Arboviruses of Australia, *Aust. Vet. J.* **48**:172.

Doherty, R.L., Carley, J.G., Mackerras, M.J., and Marks, E.N., 1963, Studies of arthropod-borne virus infections in Queensland. III. Isolation and characterization of virus strains from wild-caught mosquitoes in north Queensland, *Aust. J. Exp. Biol. Med. Sci.* **41**:17.

Doherty, R.L., Standfast, H.A., Wetters, E.J., Whitehead, R.H., Barrow, G.J., and Gorman, B.M., 1968, Virus isolation and serological studies of arthropod-borne virus infections in a high rainfall area of north Queensland, *Trans. R. Soc. Trop. Med. Hyg.* **62**:862.

Doherty, R.L., Wetters, E.J., Gorman, B.M., and Whitehead, R.H., 1970, Arbovirus infection in western Queensland: Serological Studies, 1963–1969, *Trans. R. Soc. Trop. Med. Hyg.* **64**:740.

Doherty, R.L., Carley, J.G., Standfast, H.A., Dyce, A.L., Kay, B.H., and Snowdon, W.A., 1973, Isolation of arboviruses from mosquitoes, biting midges, sandflies and vertebrates collected in Queensland, 1969 and 1970, *Trans. R. Soc. Trop. Med. Hyg.* **67**:536.

Doherty, R.L., Standfast, H.A., Dyce, A.L., Carley, J.G., Gorman, B.M., Filippich, C., and Kay, B.H., 1978, Mudjinbarry virus, an orbivirus related to Wallal virus isolated from midges from the Northern Territory of Australia, *Aust. J. Biol. Sci.* **31**:97.

Doherty, R.L., Carley, J.G., Kay, B.H., Filippich, C., Marks, E.N., and Frazier, C.A., 1979, Isolation of virus strains from mosquitoes collected in Queensland, 1972–1976, *Aust. J. Exp. Biol. Med. Sci.* **57**:509.

Domingo, E., Sabo, D., Taniguchi, T., and Weissmann, C., 1978, Nucleotide sequence heterogeneity of an RNA phage population, *Cell* **13**:735.

Domingo, E., Davila, M., and Ortin, J., 1980, Nucleotide sequence heterogeneity of the RNA from a natural population of foot-and-mouth-disease virus, *Gene* **11**:333.

Dunnebacke, T.H., and Kleinschmidt, A.K., 1967, Ribonucleic acid from reovirus as seen in protein monolayers by electron-microscopy, *Z. Naturforsch.* **22b**:159.

Eigen, M., and Schuster, P., 1979, *The Hypercycle*, Springer-Verlag, Berlin.

Els, H.J., 1973, Electron microscopy of bluetongue virus RNA, *Onderstepoort J. Vet. Res.* **40**:73.

Els, H.J., and Verwoerd, D.W., 1969, Morphology of bluetongue virus, *Virology* **38**:213.

Emmons, R.W., Oshiro, L.S., Johnson, H.N., and Lennette, E.H., 1972, Intraerythrocytic location of Colorado tick fever, *J. Gen. Virol.* **17**:185.

Engler, R., and Broome, C., 1969, Inactivation of reovirus type 2 by a combination of chloroform and moderate temperature, *Appl. Microbiol.* **18**:940.

Erasmus, B.J., 1972, The pathogenesis of African horsesickness, in: *Proc. 3rd Int. Conf. Equine Infect. Dis.*, Paris, pp. 1–11, S. Karger, Basel.

Erasmus, B.J., 1976, A new approach to polyvalent immunization against African horsesickness, in: *Proc. 4th Int. Conf. Equine Infect. Dis.*, Lyon, pp. 401–403, Veterinary Publications, Princeton, New Jersey.

Erasmus, B.J., Adelaar, T.F., Smit, J.D., Lecatsas, G., and Toms, T., 1970, The isolation and characterization of equine encephalosis virus, *Bull. Off. Int. Epizootol.* **74**:781.

Erasmus, B.J., Boshoff, S.T., and Pieterse, L.M., 1976a, the isolation and characterization of equine encephalosis and serologically related orbiviruses from horses, in: *Proc. 4th Int. Conf. Equine Infect. Dis.*, Lyon, pp. 447–450, Veterinary Publications, Princeton, New Jersey.

Erasmus, B.J., Young, E., Pieterse, L.M., and Boshoff, S.T., 1976b, The susceptibility of zebra and elephants to African horsesickness virus, in: *Proc. 4th Int. Conf. Equine Infect. Dis.*, Lyon, pp. 409–413, Veterinary Publications, Princeton, New Jersey.

Fenner, F., 1976, The classification and nomenclature of viruses, *Intervirology* **6**:1–12.

Florio, L., Mugrage, E.R., and Stewart, M.O., 1946, Colorado tick fever, *Ann. Intern. Med.* **25**:466.

Florio, L., Miller, M.S., and Mugrage, E.R., 1950a, Colorado tick fever: Isolation of virus from *Dermacentor andersoni* in nature and a laboratory study of the transmission of the virus in the tick, *J. Immunol.* **64:**257.

Florio, L., Miller, M.S., and Mugrage, E.R., 1950b, Colorado tick fever: Isolations of virus from *Dermacentor variabilis* obtained from Long Island, New York with immunological comparisons between eastern and western strains, *J. Immunol.* **64:**265.

Foster, N.M., and Alders, M.A., 1979, Bluetongue virus: A membraned structure, in: *37th Ann. Proc. Electron Microscopy Soc. Amer.* (G.W. Bailey, ed.), pp. 48–49.

Foster, N.M., Alders, M.A., and Walton, T.E., 1978, Continuity of the dsRNA genome of bluetongue virus, *Curr. Microbiol* **1:**171.

Foster, N.M., Metcalf, H.E., Barber, T.L., Jones, R.H., and Luedke, A.J., 1980, Bluetongue and epizootic hemorrhagic disease virus isolation from vertebrate and invertebrate hosts at a common geographic site, *J. Am. Vet. Med. Assoc.* **176:**126.

Franklin, R.M., 1962, The significance of lipids in animal viruses: An essay on virus multiplication, *Prog. Med. Virol.* **4:**1.

Gard, G., Marshall, I.D., and Woodroofe, G.M., 1973, Annually recurrent epidemic polyarthritis and Ross River virus activity in a coastal area of New South Wales. II. Mosquitoes, viruses and wildlife, *Am. J. Trop. Med. Hyg.* **22:**551.

Gomatos, P.J., Tamm, I., Dales, S., and Franklin, R.M., 1962, Reovirus type 3: Physical characteristics and interaction with L cells, *Virology* **17:**441.

Gomatos, P.J., Stamatos, N.M., and Sarker, N.H., 1980, Small reovirus-specific particle with polycytidylate-dependent RNA polymerase activity, *J. Virol.* **36:**556.

Gomatos, P.J., Prakash, O., and Stamatos, N.M., 1981, Small reovirus particles composed solely of sigma NS with specificity for binding different nucleic acids, *J. Virol.* **39:**115.

Gorman, B.M., 1978, Susceptibility of orbiviruses to low pH and to organic solvents, *Aust. J. Exp. Biol. Med. Sci.* **56:**359.

Gorman, B.M., 1979, Variation in orbiviruses, *J. Gen. Virol.* **44:**1.

Gorman, B.M., and Lecatsas, G., 1972, Formation of Wallal virus in cell culture, *Onderstepoort J. Vet. Res.* **39:**229.

Gorman, B.M., and Taylor, J., 1978, The RNA genome of Tilligerry virus, *Aust. J. Exp. Biol. Med. Sci.* **56:**369.

Gorman, B.M., Walker, P.J., and Taylor, J., 1977a, Electrophoretic separation of double-stranded RNA genome segments from Warrego and Mitchell River viruses, *Arch. Virol.* **54:**153.

Gorman, B.M., Taylor, J., Brown, K., and Melzer, A.J., 1977b, Analysis of ribonucleic acid of orbiviruses, *Annu. Rep. Queensland Inst. Med. Res.* **31:**15.

Gorman, B.M., Taylor, J., Walker, P.J., and Young, P.R., 1978, The isolation of recombinants between related orbiviruses, *J. Gen. Virol.* **41:**333.

Gorman, B.M., Taylor, J., Walker, P.J., Davidson, W.L., and Brown, F., 1981, Comparison of bluetongue type 20 with certain viruses of the bluetongue and Eubenangee serological groups of orbiviruses, *J. Gen. Virol.* **56:**251.

Gorman, B.M., Taylor, J., Finnimore, P.M., Bryant, J.A., Sangar, D.V., and Brown, F., 1982, A comparison of bluetongue viruses isolated in Australia with exotic bluetongue virus serotypes, in: *Arbovirus Research in Australia, Proceedings of the Third Symposium, 1982* (T.D. St. George and B.H. Kay, eds.), pp. 101–109, CSIRO-QIMR, Brisbane.

Gorman, B.M., Taylor, J., Morton, H.C., Melzer, A.J., and Young, P.R., 1983, Characterization of Nugget virus, a serotype of the Kemerovo group of orbiviruses, *Aust. J. Exp. Biol. Med. Sci.* (in press).

Green, I.J., 1969, Histochemical and immunofluorescent studies of the growth of Colorado tick fever virus in Vero monkey kidney cells, *Bacteriol. Proc.* **69:**185.

Green, I.J., 1970, Evidence for the double-stranded nature of the RNA of Colorado tick fever virus, an ungrouped arbovirus, *Virology* **49:**878.

Gresikova, M., Nosek, J., Kozuch, O., Ernek, E., and Lichard, M., 1965, Study on the ecology of Tribec virus, *Acta Virol.* **9:**83.

Griner, L.A., McCrory, B.R., Foster, N.M., and Meyer, H., 1964, Bluetongue associated with abnormalities in newborn lambs, *J. Am. Vet. Med. Assoc.* **145:**1013.

Hamparian, V.V., Hilleman, M.R., and Ketler, A., 1963, Contributions to characterization and classification of animal viruses, *Proc. Soc. Exp. Biol. Med.* **112:**1040.

Henning, M.W., 1956, *Animal Diseases in South Africa*, 3rd ed., Central News Agency, South Africa.

Hightower, L.E., and Bratt, M.A., 1977, Genetics of Orthomyxoviruses, in: *Comprehensive Virology*, Vol. 9 (H. Fraenkel-Conrat and R.R. Wagner, eds.), pp. 535–598, Plenum Press, New York.

Hirst, G.K., 1973, Mechanism of influenza recombination. I. Factors influencing recombination rates between temperature-sensitive mutants of strain WSN and the classification of mutants into complementation–recombination groups, *Virology* **55:**81.

Hoff, G.L., and Trainer, D.O., 1978, Bluetongue and epizootic hemorrhagic disease viruses: Their relationship to wildlife species, *Adv. Vet. Sci. Comp. Med.* **22:**111.

Hoff, G.L., Richards, S.H., and Trainer, D.O., 1973, Epizootic of hemorrhagic disease in North Dakota deer, *J. Wildl. Manage.* **37:**331.

Hoogstraal, H., 1972, Birds as tick hosts and as reservoirs and disseminators of tick-borne infectious agents, *Wiad. Parazytol.* **18:**703.

Howell, P.G., 1960, A preliminary antigenic classification of strains of bluetongue virus, *Onderstepoort J. Vet. Res.* **28:**357.

Howell, P.G., 1962, The isolation and identification of further antigenic types of African horsesickness virus, *Onderstepoort J. Vet. Res.* **29:**139.

Howell, P.G., 1963a, African horse-sickness, in: *Emerging Diseases of Animals*, pp. 73–108, FAO Agricultural Studies No. 61, Rome.

Howell, P.G., 1963b, Bluetongue, in: *Emerging Diseases of Animals*, pp. 111–153, FAO Agricultural Studies No. 61, Rome.

Howell, P.G., 1969, The antigenic classification of strains of bluetongue virus, their significance and use in prophylatic immunization, D.V.Sc. thesis, University of Pretoria.

Howell, P.G., 1970, The antigenic classification and distribution of naturally occurring strains of bluetongue virus, *J. S. Afr. Vet. Med. Assoc.* **41:**215.

Howell, P.G., and Verwoerd, D.W., 1971, Bluetongue virus, in: *Virology Monographs* (S. Gard, C. Hallauer, and K.F. Meyers, eds.), pp. 35–74, Springer-Verlag, New York.

Hubschle, O.J.B., 1980, Bluetongue virus hemagglutination and its inhibition by specific sera, *Arch. Virol.* **64:**133.

Hubschle, O.J., Lorenz, R.J., and Matheka, H.-D., 1981, Enzyme linked immuno sorbent assay for detection of bluetongue virus antibodies, *Am. J. Vet. Res.* **42:**61.

Hughes, L.E., Casper, E.A., and Clifford, C.M., 1974, Persistence of Colorado tick fever in red blood cells, *Am. J. Trop. Med. Hyg.* **23:**530.

Huismans, H., 1970a, Macromolecular synthesis in bluetongue virus-infected cells. I. Virus-specific ribonucleic acid synthesis, *Onderstepoort J. Vet. Res.* **37:**191.

Huismans, H., 1970b, Macromolecular synthesis in bluetongue virus-infected cells, II. Host cell metabolism, *Onderstepoort J. Vet. Res.* **37:**199.

Huismans, H., 1971, Host cell protein synthesis after infection with bluetongue virus and reovirus, *Virology* **46:**500.

Huismans, H., 1979, Protein synthesis in bluetongue virus-infected cells, *Virology* **92:**385.

Huismans, H., and Bremer, C.W., 1981, A comparison of an Australian bluetongue virus isolate (CSIRO 19) with other bluetongue virus serotypes by cross-hybridization and cross-immune precipitation, *Onderstepoort J. Vet. Res.* **48:**59.

Huismans, H., and Els, H.J., 1979, Characterization of the microtubules associated with the replication of three different orbiviruses, *Virology* **92:**397.

Huismans, H., and Erasmus, B.J., 1981, Identification of the serotype-specific and group-specific antigens of bluetongue virus, *Onderstepoort J. Vet. Res.* **48:**51.

Huismans, H., and Howell, P.G., 1973, Molecular hybridization studies on the relationships between different serotypes of bluetongue virus and on the difference between virulent and attenuated strains of the same serotype, *Onderstepoort J. Vet. Res.* **40:**93.

Huismans, H., and Joklik, W.K., 1976, Reovirus-coded polypeptides in infected cells: Isolation of two native monomeric polypeptides with affinity for single-stranded and double-stranded RNA, respectively, *Virology* **70**:411.

Huismans, H., and Verwoerd, D.W., 1973, Control of transcription during the expression of the bluetongue virus genome, *Virology* **52**:81.

Huismans, H., Bremer, C.W., and Barber, T.L., 1979, The nucleic acid and proteins of epizootic hemorrhagic disease virus, *Onderstepoort J. Vet. Res.* **46**:95.

Hutcheon, D., 1902, Malarial catarrhal fever of sheep, *Vet. Rec.* **14**:629.

Inaba, Y., 1975, Ibaraki disease and its relationship to bluetongue, *Aust. Vet. J.* **51**:178.

Inaba, Y., Ishii, S., and Omori, T., 1966, Bluetongue-like disease in Japan, *Bull. Off. Int. Epizoot.* **66**:329.

Ito, Y., and Joklik, W.K., 1972, Temperature-sensitive mutants of reovirus. II. Anomalous electrophoretic migration behaviour of certain hybrid RNA molecules composed of mutant plus strands and wild-type minus strands, *Virology* **50**:202.

Ito, Y., Tanaka, Y., Inaba, Y., and Omori, T., 1973, Electron microscopy of Ibaraki virus, *Arch. Gesamte Virusforsch.* **40**:29.

Jochim, M.M., and Jones, S.C., 1980, Evaluation of a hemolysis-in-gel test for detection and quantitation of antibodies to bluetongue virus, *Am. J. Vet. Res.* **41**:595.

Joklik, W.K., 1972, Studies on the effect of chymotrypsin on reovirions, *Virology* **49**:700.

Joklik, W.K., 1974, Reproduction of Reoviridae, in: *Comprehensive Virology*, Vol. 2 (H. Fraenkel-Conrat and R.R. Wagner, eds.), pp. 231–334, Plenum Press, New York.

Jones, R.H., and Foster, N.M., 1974, Oral infection of *Culicoides variipennis* with bluetongue virus: Development of susceptible and resistant lines from a colony population, *J. Med. Entomol.* **11**:316.

Karabatsos, N. (ed.), 1978, Supplement to International Catalogue of Arboviruses including certain other viruses of vertebrates, *Am. J. Trop. Med. Hyg.* **27**:372.

Karstad, L., and Trainer, D.O., 1967, Histopathology of experimental bluetongue disease of white-tailed deer, *Can. Vet. J.* **8**:247.

Kavenoff, R., Klotz, L.C., and Zimm, B.H., 1973, On the nature of chromosome-sized DNA molecules, *Cold Spring Harbor Symp. Quant. Biol.* **38**:1.

Kavenoff, R., Talcove, D., and Mudd, J.A., 1975, Genome-sized RNA from reovirus particles, *Proc. Natl. Acad. Sci. U.S.A.* **72**:4317.

Kay, B.H., Boreham, P.F.L., Dyce, A.L., and Standfast, H.A., 1978, Blood feeding of biting midges (Diptera: Ceratopogonidae) at Cape York Peninsula, north Queensland, *J. Aust. Entomol. Soc.* **17**:145.

Kay, B.H., Boreham, P.F.L. and Williams, G.M., 1979, Host preferences and feeding patterns of mosquitoes (Diptera: Culicidae) at Kowanyama, Cape York Peninsula, northern Queensland, *Bull. Entomol. Res.* **69**:441.

Kipps, A., 1956, Complement fixation with antigens prepared from bluetongue virus-infected mouse brains, *J. Hyg.* **54**:79.

Knudson, D.L., 1980, DsRNA analyses of other orbiviruses, *Annu. Rep. Yale Arbovirus Res. Unit*, New Haven, p. 63.

Knudson, D.L., 1981, Genome of Colorado tick fever virus, *Virology* **112**:381.

Komarov, A., and Goldsmit, L., 1951, A bluetongue-like disease of cattle and sheep in Israel, *Refu. Vet.* **8**:96.

Kontor, E.J., and Welch, A.B., 1976, Characterization of an epizootic haemorrhagic disease virus, *Res. Vet. Sci.* **21**:190.

Laemmli, U.K., 1970, Cleavage of structural proteins during the assembly of the head of bacteriophage T4, *Nature (London)* **227**:680.

Langridge, R., and Gomatos, P.J., 1963, The structure of RNA: Reovirus RNA and transfer RNA have similar three-dimensional structures, which differ from DNA, *Science* **141**:694.

Lecatsas, G., 1968a, Electron microscopic study of the formation of bluetongue virus, *Onderstepoort J. Vet. Res.* **35**:139.

Lecatsas, G., 1968b, Electron microscopic studies on reovirus type 1 in BHK21 cells, *Onderstepoort J. Vet. Res.* **35**:151.

Lecatsas, G., and Erasmus, B.J., 1967, Electron microscopic study of the formation of African horsesickness virus, *Arch. Gesamte Virusforsch.* **22**:442.

Lecatsas, G., and Erasmus, B.J., 1973, Core structure in a new virus, XBM/67, *Arch. Gesamte Virusforsch.* **42**:264.

Lecatsas, G., and Gorman, B.M., 1972, Visualization of the extra capsid coat in certain bluetongue-type viruses, *Onderstepoort J. Vet. Res.* **39**:193.

Lecatsas, G., Erasmus, B.J., and Els, H.J., 1969, Electron microscopic studies on Corriparta virus, *Onderstepoort J. Vet. Res.* **36**:321.

Lecatsas, G., Erasmus, B.J., and Els, H.J., 1973, Electron microscopic studies on equine encephalosis virus, *Onderstepoort J. Vet. Res.* **40**:53.

Lee, P.W.K., Hayes, E.C., and Joklik, W.K., 1981, Characterization of anti-reovirus immunoglobulins secreted by cloned hybridoma cell lines, *Virology* **108**:134.

Lee, V.H., 1979, Isolation of viruses from field populations of *Culicoides* (Diptera: Ceratopogonidae) in Nigeria, *J. Med. Entomol.* **16**:76.

Lee, V.H., Causey, O.R., and Moore, D.L., 1974, Bluetongue and related viruses in Ibadan, Nigeria: Isolation and preliminary identification of viruses, *Am. J. Vet. Res.* **35**:1105.

Libikova, H., and Casals, J., 1971, Serological characterization of Eurasian Kemerovo group viruses. I. Cross complement fixation tests, *Acta Virol.* **15**:65.

Libikova, H., Mayer, V., Kozuch, O., Rehacek, J., Ernek, E., and Albrecht, P., 1964, Isolation from *Ixodes persulcatus* ticks of cytopathic agents (Kemerovo virus) differing from tick-borne encephalitis virus and some of their properties, *Acta Virol.* **8**:289.

Libikova, H., Tesarova, J., and Rajcani, J., 1970, Experimental infection of monkeys with Kemerovo virus, *Acta Virol.* **14**:64.

Libikova, H., Heinz, F., Ujhazyova, D., and Stunzner, D., 1978, Orbiviruses of the Kemerovo complex and neurological diseases, *Med. Microbiol. Immunol.* **166**:255.

Liehne, C.G., Leivers, S., Stanley, N.F., Alpers, M.P., Paul, S., Liehne, P.F.S., and Chan, K.H., 1976, Ord River arboviruses—isolations from mosquitoes, *Aust. J. Exp. Biol. Med. Sci.* **53**:499.

Livingston, C.W., and Moore, R.W., 1962, Cytochemical changes of bluetongue virus in tissue culture, *Am. J. Vet. Res.* **23**:701.

Luedke, A.J., Jochim, M.M., Bowne, J.G., and Jones, R.H., 1970, Observations on latent bluetongue virus infection in cattle, *J. Am. Vet. Med. Assoc.* **156**:1871.

Main, A.J., 1978, Tindholmur and Mykines: Two new Kemerovo group orbiviruses from the Faroe Islands, *J. Med. Entomol.* **15**:11.

Main, A.J., Downs, W.G., Shope, R.E., and Wallis, R.C., 1973, Great Island and Bauline: Two new Kemerovo group orbiviruses from *Ixodes uriae* in eastern Canada, *J. Med. Entomol.* **10**:229.

Main, A.J., Shope, R.E. and Wallis, R.C., 1976, Cape Wrath: A new Kemerovo group orbivirus from *Ixodes uridae* (Acari: Ixodidae) in Scotland, *J. Med. Entomol.* **13**:304.

Main, A.J., Kloter, K.O., Camicas, J.-L., Robin, Y., and Sarr, M., 1980, Wad Medani and Soldado viruses from ticks (Ixodoidea) in West Africa, *J. Med. Entomol.* **17**:380.

Málková, D., Holubová, J., Kolman, J.M., Marhoul, Z., Hanzal, F., Kulková, H., Markvat, K., and Simková, L., 1980, Antibodies against some arboviruses in persons with various neuropathies, *Acta Virol.* **24**:298.

Manning, J.S., and Chen, M.F., 1980, Bluetongue virus: Detection of antiviral immunoglobulin G by means of enzyme-linked immunosorbent assay, *Curr. Microbiol.* **4**:381.

Marmur, J., and Doty, P., 1959, Heterogeneity in deoxyribonucleic acids. I. Dependence on composition of the configurational stability of deoxyribonucleic acids, *Nature (London)* **183**:1427.

Marmur, J., and Doty, P., 1962, Determination of the base composition of deoxyribonucleic acid from its thermal denaturation temperature, *J. Mol. Biol.* **5**:109.

Marshall, I.D., Woodroofe, G.M., and Gard, G.P., 1980, Arboviruses of coastal south-eastern Australia, *Aust. J. Exp. Biol. Med. Sci.* **58**:91.

Martin, S.A., and Zweerink, H.J., 1972, Isolation and characterization of two types of blue-tongue virus particles, *Virology* **50**:495.

Martin, S.A., Pett, D.M., and Zweerink, H.J., 1973, Studies on the topography of reovirus and bluetongue virus capsid polypeptides, *J. Virol.* **12**:194.

Matthews, R.E.F., 1979, Classification and nomenclature of viruses: Third report of the International Committee on Taxonomy of Viruses, *Intervirology* **12**:200.

Maurer, F.D., and McCully, R.M., 1963, African horsesickness—with emphasis on pathology, *Am. J. Vet. Res.* **24**:235.

Mayr, E., 1980, Some thoughts on the history of the evolutionary synthesis, in: *The Evolutionary Synthesis* (E. Mayr and W.B., Provine, eds.), pp. 1–48, Harvard University Press, Cambridge.

McCrae, M.A., and Joklik, W.K., 1978, The nature of the polypeptide encoded by each of the 10 double-stranded RNA segments of reovirus type 3, *Virology* **89**:578.

McIntosh, B.M., 1958, Immunological types of horsesickness virus and their significance in immunization, *Onderstepoort J. Vet. Res.* **27**:465.

McKercher, D.G., Saito, J.K., and Singh, K.V., 1970, Serologic evidence of an etiologic role for bluetongue virus in hydranencephaly in calves, *J. Am. Vet. Med. Assoc.* **156**:1044.

Mellor, P.S., and Jennings, M., 1980, Replication of Eubenangee virus in *Culicoides nebeculosis* (Mg.) and *Culicoides variipennis* (Coq.), *Arch. Virol.* **63**:203.

Mellor, P.S., Boorman, J., and Jennings, M., 1975, The multiplication of African horsesickness virus in two species of *Culicoides* (Diptera, Ceratopogonidae), *Arch. Virol.* **47**:351.

Mettler, N.E., MacNamara, L.G., and Shope, R.E., 1962, The propagation of the virus of epizootic hemorrhagic disease of deer in newborn mice and HeLa cells, *J. Exp. Med.* **116**:665.

M'Fadyean, J., 1900, African horsesickness, *J. Comp. Pathol.* **13**:1.

Miura, Y., Inaba, Y., Hayashi, S., Takahashi, E., and Matumoto, M., 1980, A survey of antibodies to arthropod-borne viruses in Japanese cattle, *Vet. Microbiol.* **5**:277.

Moore, D.L., 1974, Bluetongue and related viruses in Ibadan, Nigeria: Serologic comparison of bluetongue, epizootic hemorrhagic disease of deer and Abadina (Palyam) viral isolates, *Am. J. Vet. Res.* **35**:1109.

Moore, D.L., and Lee, V.H., 1972, Antigenic relationship between the virus of epizootic haemorrhagic disease of deer and bluetongue virus, *Arch. Gesamte Virusforsch.* **37**:282.

Murphy, F.A., Coleman, P.H., Hansen, A.K., and Gray, G.W., 1968, Colorado tick fever virus, an electron microscopic study, *Virology* **35**:28.

Murphy, F.A., Borden, E.C., Shope, R.E., and Harrison, A., 1971, Physicochemical and morphological relationships of some arthropod-borne viruses to bluetongue virus—a new taxonomic group: Electron microscopic studies, *J. Gen. Virol.* **13**:273.

Mustoe, T.A., Ramig, R.F., Sharpe, A.H., and Fields, B.N. 1978a, A genetic map of reovirus. III. Assignment of the double-stranded RNA-positive mutant groups A, B, and G to genome segments, *Virology* **85**:545.

Mustoe, T.A., Ramig, R.F., Sharpe, A.H., and Fields, B.N., 1978b, Genetics of reovirus: Identification of the dsRNA segments encoding the polypeptides of the μ and σ size classes, *Virology* **89**:594.

Neitz, W.O., 1948, Immunological studies on bluetongue in sheep, *Onderstepoort J. Vet. Sci. Anim. Ind.* **23**:93.

Nishimura, A., and Hosaka, Y., 1969, Electron microscopic study of RNA of cytoplasmic polyhedrosis virus of the silkworm, *Virology* **38**:550.

Oellermann, R.A., and Carter, P., 1977, The immunological response to intact and dissociated bluetongue virus in mice, *Onderstepoort J. Vet. Res.* **44**:201.

Oellermann, R.A., Els, H.J., and Erasmus, B.J., 1970, Characterization of African horsesickness virus, *Arch. Gesamte Virusforsch.* **29**:163.

Omori, T., Inaba, Y., Morimoto, T., Tanaka, Y., Ishitani, R., Kurogi, H., Munakata, K., Matsuda, K., and Matumoto, M., 1969, Ibaraki virus, an agent of epizootic disease of cattle resembling bluetongue. I. Epidemiologic, clinical and pathologic observations and experimental transmission to calves, *Jpn. J. Microbiol.* **13**:139.

Ortin, J., Najera, R., Lopez, C., Davila, M., and Domingo, E., 1980, Genetic variability of Hong Kong (H3N2) influenza viruses: Spontaneous mutations and their location in the viral genome, *Gene* **11**:319.

Osburn, B.I., Silverstein, A.M., Prendergast, R.A., Johnson, R.T., and Parshall, C.J., 1971a, Experimental viral-induced congenital encephalopathies. I. Pathology of hydranencephaly and porencephaly caused by bluetongue vaccine virus, *Lab. Invest.* **25**:197.

Osburn, B.I., Johnson, R.T., Silverstein, A.M., Prendergast, R.A., Jochim, M.M., and Levy, S.E., 1971b, Experimental viral-induced congenital encephalopathies. II. The pathogenesis of bluetongue vaccine virus infection in fetal lambs, *Lab. Invest.* **25**:206.

Oshiro, L.S., and Emmons, R.W., 1968, Electron microscopic observations of Colorado tick fever virus in BHK-21 and KB cells, *J. Gen. Virol.* **3**:279.

Oshiro, L.S., Dondero, D.V., Emmons, R.W., and Lennette, E.H., 1978, The development of Colorado tick fever virus within cells of the haemopoietic system, *J. Gen. Virol.* **39**:73.

Owen, N.C., 1964, Investigation into the pH stability of bluetongue virus and its survival in mutton and beef, *Onderstepoort J. Vet. Res.* **31**:109.

Owen, N.C., and Munz, E.K., 1966, Observations on a strain of bluetongue virus by electron microscopy, *Onderstepoort J. Vet. Res.* **33**:9.

Owen, N.C., Du Toit, R.M., and Howell, P.G., 1965, Bluetongue in cattle: Typing of viruses isolated from cattle exposed to natural infections, *Onderstepoort J. Vet. Res.* **32**:3.

Ozawa, Y., 1967, Studies on the replication of African horsesickness virus in two different cell line cultures, *Arch. Gesamte Virusforsch.* **21**:155.

Ozawa, Y., 1968, Studies on the properties of African horsesickness virus, *Jpn. J. Med. Sci. Biol.* **21**:27.

Ozawa, Y., and Hafez, S.M., 1973, Antigenic relationship between African horsesickness and bluetongue viruses, *Proc. 3rd Int. Conf. Equine Infect. Dis.*, Paris, pp. 31–37, S. Karger, Basel.

Ozawa, Y., Hopkins, I.G., Hazrati, A., Modjitabai, A., and Kaveh, P., 1965, Cytology of monkey kidney cells infected with African horsesickness virus, *Nature (London)* **26**:1321.

Palmer, E.L., Martin, M.L., and Murphy, F.A., 1977, Morphology and stability of infantile gastroenteritis virus: Comparison with reovirus and bluetongue virus, *J. Gen. Virol.* **35**:403.

Pavri, K.M., 1961, Haemagglutination and haemagglutination-inhibition with African horsesickness virus, *Nature (London)* **189**:249.

Pavri, K.M., and Anderson, C.R., 1963, Haemagglutination-inhibition tests with different types of African horsesickness virus, *Indian J. Vet. Sci.* **33**:113.

Payne, C.C., Piasecka-Serafin, M., and Pilley, B., 1977, The properties of two recent isolates of cytoplasmic polyhedrosis viruses, *Intervirology* **8**:155.

Polson, A., and Deeks, D., 1963, Electron microscopy of neurotropic African horsesickness virus, *J. Hyg.* **61**:149.

Ramig, R.F., Cross, R.K., and Fields, B.N., 1977, Genome RNAs and polypeptides of reovirus serotypes 1, 2, and 3, *J. Virol.* **22**:726.

Rehse-Kupper, B., Casals, J., Rehse, E., and Ackermann, R., 1976, Eyach—an arthropod-borne virus related to Colorado tick fever virus in the Federal Republic of Germany, *Acta Virol.* **20**:339.

Robertson, A., 1976, *Handbook on Animal Diseases in the Tropics*, 3rd ed., P. Burgess and Son (Abingdon), Abingdon, Oxfordshire.

Robinson, R.M., Hailey, T.L., Livingston, C.W., and Thomas, J.W., 1967, Bluetongue in desert bighorn sheep, *J. Wildl. Manage.* **31**:165.

Rodger, S.M., and Holmes, I.H., 1979, Comparison of the genomes of simian, bovine, and human rotaviruses by gel electrophoresis and detection of genomic variation among bovine isolates, *J. Virol.* **30**:839.

Sakuma, A., and Watanabe, Y., 1971, Unilateral synthesis of reovirus double-stranded ribonucleic acid by a cell-free replicase system, *J. Virol.* **8**:190.

Sangar, D.V., Taylor, J., and Gorman, B.M., 1981, The identification of genome segments coding for bluetongue virus polypeptides, in: *Abstracts of the Fifth International Congress for Virology*, p. 428, Strasbourg, France.

Schnagl, R.D., and Holmes, I.H., 1971, A study of Australian arboviruses resembling bluetongue virus, *Aust. J. Biol. Sci.* **24**:1151.

Schnagl, R.D., and Holmes, I.H., 1975, Polyacrylamide gel electrophoresis of the genomes of two orbiviruses: D'Aguilar and Eubenangee, *Intervirology* **5**:300.

Schnagl, R.D., Holmes, I.H., and Doherty, R.L., 1969, An electron microscope study of Eubenangee, an Australian arbovirus, *Virology* **38**:347.

Sellers, R.F., Gibbs, E.P., Herniman, K.A., Pedgley, D.E., and Tucker, M.R., 1979, Possible origin of the bluetongue epidemic in Cyprus, August 1977, *J. Hyg.* **83**:547.

Shatkin, A.J. and LaFiandra, A.J., 1972, Transcription by infectious subviral particles of reovirus, *J. Virol.* **10**:698.

Shatkin, A.J., and Sipe, J.D. 1968, Single-stranded adenine-rich RNA from purified reoviruses, *Proc. Natl. Acad. Sci. U.S.A.* **59**:246.

Shipham, S., 1979, Further characterization of the ts mutant F207 of bluetongue virus, *Onderstepoort J. Vet. Res.* **46**:207.

Shipham, S.O., and De la Rey, M., 1976, The isolation and preliminary genetic classification of temperature-sensitive mutants of bluetongue virus, *Onderstepoort J. Vet. Res.* **43**:189.

Shipham, S.O., and De le Rey, M., 1979, Temperature-sensitive mutants of bluetongue virus: Genetic and physiological characterization, *Onderstepoort J. Vet. Res.* **46**:87.

Shope, R.E., MacNamara, L.G., and Mangold, R., 1960, A virus-induced epizootic hemorrhagic disease of the Virginia white-tailed deer (*Odocoileus virginianus*), *J. Exp. Med.* **111**:155.

Skehel, J.J., and Joklik, W.K., 1969, Studies on the *in vitro* transcription of reovirus RNA catalyzed by reovirus cores, *Virology* **39**:822.

Spreull, J., 1905, Malarial catarrhal fever (bluetongue) of sheep in South Africa, *J. Comp. Pathol.* **18**:321.

St. George, T.D., 1980, A sentinel herd system for the study of arbovirus infections in Australia and Papua New Guinea, *Vet. Sci. Commun.* **4**:39.

St. George, T.D., and Dimmock, C.K., 1976, The isolation of D'Aguilar virus from a cow, *Aust. Vet. J.* **52**:598.

St. George, T.D., and French, E.L. (eds.), 1979, *Arbovirus Research in Australia, Proceedings of the Second Symposium, 1979*, CSIRO-QIMR, Brisbane.

St. George, T.D., and McCaughan, C.I., 1979, The transmission of the CSIRO19 strain of bluetongue virus type 20 to sheep and cattle, *Aust. Vet. J.* **55**:198.

St. George, T.D., Standfast, H.A., Cybinski, D.H., Dyce, A.L., Muller, M.J., Doherty, R.L., Carley, J.G., Filippich, C. and Frazier, C.L., 1978, The isolation of a bluetongue virus from *Culicoides* collected in the Northern Territory of Australia, *Aust. Vet. J.* **54**:153.

St. George, T.D., Cybinski, D.H., Bainbridge, M.H., and Scanlan, W.A., 1979, The use of sentinel cattle and sheep for the isolation of arboviruses in the Northern Territory of Australia in 1979, in: *Arbovirus Research in Australia, Proceedings of the Second Symposium, 1979* (T.D. St. George and E.L. French, eds.), pp. 84–88, CSIRO-QIMR, Brisbane.

St. George, T.D., Cybinski, D.H., Della-Porta, A.J., McPhee, D.A., Wark, M.C., and Bainbridge, M.H., 1980, The isolation of two bluetongue viruses from healthy cattle in Australia, *Aust. Vet. J.* **56**:562.

Stone, K.R., Smith, R.E., and Joklik, W.K., 1974, Changes in membrane polypeptides that occur when chick embryo fibroblasts and NRK cells are transformed with avian sarcoma viruses, *Virology* **58**:86.

Studdert, M.J., 1965, Sensitivity of bluetongue virus to ether and sodium deoxycholate, *Proc. Soc. Exp. Biol. Med.* **118**:1006.

Studdert, M.J., Pangborn, J., and Addison, R.B., 1966, Bluetongue virus structure, *Virology* **29**:509.

Sugiyama, K., Bishop, D.H.L., and Roy, P., 1981, Analyses of the genomes of bluetongue viruses recovered in the United States. I. Oligonucleotide fingerprint studies that indicate the existence of naturally occurring reassortant BTV isolates, *Virology* **114**:210.

Sugiyama, K., Bishop, D.H.L., and Roy, P., 1982, Analyses of the genomes of bluetongue virus isolates recovered from different states of the United States and at different times, *Am. J. Epidemiol.* **115**:332.

Suzuki, Y., Saito, Y., and Nakagawa, S., 1977, Double-stranded RNA of Ibaraki virus, *Virology* **76**:670.

Suzuki, Y., Nakagawa, S., Namiki, M., and Saito, Y., 1978, RNA and protein of Ibaraki virus, *Kitasato Arch. Exp. Med.* **51**:61.

Svehag, S.E., Leendertsen, L., and Gorham, J.R., 1966, Sensitivity of bluetongue virus to lipid solvents, trypsin and pH changes and its serological relationship to arboviruses, *J. Hyg.* **64**:339.

Swanepoel, R., and Blackburn, N.K., 1976, A new number of the Palyam serogroup of orbiviruses, *Vet. Rec.* **99**:360.

Theiler, A., 1906, Bluetongue in sheep, *Annu. Rep. Dir. Agric. Transvaal, 1904–5*, pp. 110–121.

Theiler, A., 1908, Inoculation of sheep against bluetongue and results in practice, *Vet. J.* **64**:600.

Theiler, A., 1921, African horsesickness (pestis equorum), *Union S. Afr. Dep. Agric. Sci. Bull.*, No. 19.

Theiler, A., 1957, Action of sodium deoxycholate on arthropod-borne viruses, *Proc. Soc. Exp. Biol. Med.* **96**:380.

Thomas, F.C., and Miller, J., 1971, A comparison of bluetongue virus and EHD virus: Electronmicroscopy and serology, *Can. J. Comp. Med.* **35**:22.

Thomas, F.C., and Trainer, D.O., 1971, Bluetongue virus: Some relationships among North American isolates and further comparisons with EHD virus, *Can. J. Comp. Med.* **35**:187.

Thomas, F.C., Morse, P.M., and Seawright, G.L., 1979, Comparisons of some bluetongue virus isolates by plaque neutralization and relatedness tests, *Arch. Virol.* **62**:189.

Tomori, O., and Fabiyi, A., 1976, Neutralizing antibodies to Orungo virus in man and animals in Nigeria, *Trop. Geogr. Med.* **28**:233.

Tomori, O., El-Bayoumi, M.S.M., and Fabiyi, A., 1976, Virological and serological studies of a suspected yellow fever virus outbreak in Mabudi area of Benne Plateau State of Nigeria, *Niger. Med. J.* **6**:135.

Trent, D. W., and Scott, L.V., 1966, Colorado tick fever in cell culture. II. Physical and chemical properties, *J. Bacteriol.* **91**:1282.

Tsai, K.-S., and Karstad, L., 1970, Epizootic hemorrhagic disease virus of deer: An electron microscopic study, *Can. J. Microbiol.* **16**:427.

Tsai, K.-S., and Karstad, L., 1973, Ultrastructural characterization of the genome of epizootic hemorrhagic disease virus, *Infect. Immun.* **8**:463.

Van der Walt, N.T., 1980, A haemagglutination and haemagglutination-inhibition test for bluetongue virus, *Onderstepoort J. Vet. Res.* **47**:113.

Van Dijk, A.A., and Huismans, H., 1980, The *in vitro* activation and further characterization of the bluetongue virus associated transcriptase, *Virology* **104**:347.

Vasques, C., and Kleinschmidt, A.K., 1968, Electron microscopy of RNA strands released from individual reovirus particles, *J. Mol. Biol.* **34**:137.

Verwoerd, D.W., 1969, Purification and characterization of bluetongue virus, *Virology* **38**:203.

Verwoerd, D.W., 1970a, Diplornaviruses: A newly recognized group of double-stranded RNA viruses, *Prog. Med. Virol.* **12**:192.

Verwoerd, D.W., 1970b, Failure to demonstrate *in vitro* as opposed to *in vivo* transcription of the bluetongue virus genome, *Onderstepoort J. Vet. Res.* **37**:225.

Verwoerd, D.W., and Huismans, H., 1969, On the relationship between bluetongue, African horsesickness, and reoviruses: Hybridization studies, *Onderstepoort J. Vet. Res.* **36**:175.

Verwoerd, D.W., and Huismans, H., 1972, Studies on the *in vitro* and *in vivo* transcription of the bluetongue virus genome. *Onderstepoort J. Vet. Res.* **39**:185.

Verwoerd, D.W., Louw, H., and Oellermann, R.A., 1970, Characterization of bluetongue virus ribonucleic acid, *J. Virol.* **5**:1.

Verwoerd, D.W., Els, H.J., de Villiers, E.-M., and Huismans, H., 1972, Structure of the bluetongue virus capsid, *J. Virol.* **10**:783.

Verwoerd, D.W., Huismans, H., and Erasmus, B.J., 1979, Orbiviruses, in: *Comprehensive Virology*, Vol. 14 (H. Fraenkel-Conrat and R.R. Wagner, eds.), pp. 285–345, Plenum Press, New York.

Walker, P.J., 1981, A study of structural characteristics and genetic interactions of closely related orbiviruses, Ph.D. thesis, University of Queensland.

Walker, P.J., Mansbridge, J.N., and Gorman, B.M., 1980, Genetic analysis of orbiviruses by using RNase $T_1$ oligonucleotide fingerprints, *J. Virol.* **34**:583.

Waterson, A.P., and Wilkinson, L., 1978, *An Introduction to the History of Virology*, Cambridge University, Cambridge.

Westaway, E.G., 1965, The neutralization of arboviruses. II. Neutralization in heterologous virus–serum mixtures with four group B arboviruses, *Virology* **26**:528.

Whitehead, R.D., Doherty, R.L., Domrow, R., Standfast, H.A. and Wetters, E.J., 1968, Studies of the epidemiology of arthropod-borne virus infections at Mitchell River Mission, Cape York Peninsula, north Queensland. III. Virus studies of wild birds, 1964–1967, *Trans. R. Soc. Trop. Med. Hyg.* **62**:439.

Young, J.F., Desselberger, V., and Palese, P., 1979, Evolution of human influenza A viruses in nature: Sequential mutations in the genomes of new $H_1N_1$ isolates, *Cell* **18**:73.

Yunker, C.E., 1975, Tick-borne viruses associated with seabirds in North America and related islands, *Med. Biol.* **53**:302.

Zweerink, H.J., 1974, Multiple forms of SS → DS RNA polymerase activity in reovirus-infected cells, *Nature (London)* **247**:313.

Zweerink, H.J., and Joklik, W.K., 1970, Studies on the intracellular synthesis of reovirus-specified proteins, *Virology* **41**:501.

Zweerink, H.J., McDowell, M.J., and Joklik, W.K., 1971, Essential and nonessential noncapsid reovirus proteins, *Virology* **45**:716.

Zweerink, H.J., Ito, Y., and Matsuhisa, T., 1972, Synthesis of reovirus double-stranded RNA within virionlike particles, *Virology* **50**:349.

CHAPTER 8

# Rotaviruses

## Ian H. Holmes

## I. INTRODUCTION

Electron microscopy played an important role in the recent recognition of a large group of viruses associated with diarrheal disease in the young of humans, of many kinds of animals, and of birds, so it is fitting that the name chosen for them should be based on their particle morphology. Flewett *et al.* (1974a) derived the name *Rotavirus* from the Latin "rota" ("wheel") because in negatively stained preparations, the inner capsid subunits suggest the appearance of spokes supporting a smooth rim of outer capsid. In early publications, members of the group were referred to as "orbivirus-like" (Bishop *et al.*, 1973; Middleton *et al.*, 1974), "infantile gastroenteritis virus (orbivirus group)" (Petric *et al.*, 1975), and "reovirus-like agents" (Fernelius *et al.*, 1972; Kapikian *et al.*, 1974), and the term "Duovirus" was also proposed (Davidson *et al.*, 1975a), but *Rotavirus* came into more common use and was adopted officially for the genus (Matthews, 1979).

Historical aspects of the discovery of rotaviruses and of their role in diarrheal diseases have been reviewed by Flewett (1977), Flewett and Woode (1978), and Holmes (1979), so the stories will not be retold here, but the indebtedness of later investigators to the pioneering studies of J.S. Light and H.S. Hodes, L.M. Kraft, and C.A. Mebus will become evident as we discuss the basic properties of the viruses and diseases.

*Characteristics and Classification.* Early studies of the morphology and morphogenesis of the viruses of epizootic diarrhea of infant mice (EDIM) and of Nebraska calf scours suggested that they belonged with the reoviruses, and human infantile enteritis virus was also immediately

IAN H. HOLMES • Department of Microbiology, University of Melbourne, Parkville, Victoria 3052, Australia.

recognized as orbivirus- or reoviruslike (Adams and Kraft, 1967; Banfield *et al.*, 1968; Bishop *et al.*, 1973; Flewett *et al.*, 1973; Middleton *et al.*, 1974).

Rotavirus particles are icosahedral, 65–75 nm in diameter, have two concentric layers of capsid, and contain a segmented genome of double-stranded RNA. The 11 segments range in molecular weight from about 0.2 to $2 \times 10^6$. Their growth is normally restricted to enterocytes of the small intestine of mammalian or avian species, but they can also be cultivated in kidney cells in the presence of trypsin. Rotaviruses cause diarrheal disease in all the species that they infect, especially in the young, but adult infections are also common. They are classified as a genus in the family Reoviridae (Matthews, 1979), and the type species is listed as "human rotavirus."

Since rotaviruses are not strictly host-specific (see Section VI.B) and more than one human serotype exists (Section VI.D.2), the type species will have to be more precisely designated. The WHO–FAO Committee for Comparative Virology suggests a bovine rotavirus [Nebraska calf diarrhea virus (NCDV)] as the type on the grounds that it had been studied in more detail (Derbyshire and Woode, 1978), but at present the choice on this criterion would be the simian rotavirus SA 11 (Malherbe and Strickland-Cholmley, 1967). Since SA 11 virus is related to a human rotavirus serotype (see Section VI.D.2) and is widely available throughout the world, it may well be the best eventual choice.

Until recently, all rotaviruses were thought to share common inner-capsid antigens that were detectable by immunofluorescence or complement fixation (Flewett and Woode, 1978), but serologically unrelated viruses that nevertheless seem to have all other rotavirus characteristics have now been found, though rather rarely, in pigs, turkeys, and humans. These "pararotaviruses" (Bohl *et al.*, 1982) are discussed further in Sections II.A, II.C, II.F, and VI.D.2.

## II. VIRIONS

### A. Morphology

Rotavirus particles are isometric and resemble those of reovirus, but by negative staining, the outer rim of complete particles appears smooth (Fig. 1A), whereas the periphery of reovirus particles is less sharply defined (Flewett *et al.*, 1974a; Holmes *et al.*, 1975). Rotavirus preparations usually also contain a significant proportion of particles that lack the outer capsid layer, and these single-shelled particles are often described as rough because their periphery shows projecting subunits of the inner capsid (Fig. 1B). Such particles closely resemble orbivirus inner capsids (Bishop *et al.*, 1974; Els and Lecatsas, 1972; Middleton *et al.*, 1974).

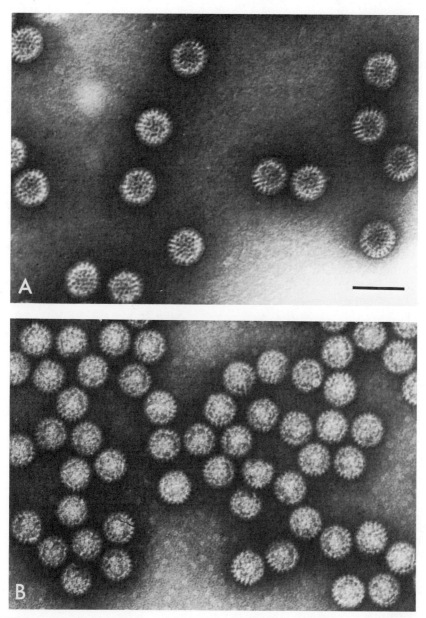

FIGURE 1. Particles of simian rotavirus SA 11, negatively stained with ammonium molybdate. (A) Complete or double-shelled particles; (B) innercapsid or single-shelled particles. Scale bar: 100 nm.

Differences in negative-staining technique probably account for most of the variation in reported size estimates; for example, Woode *et al.* (1976b) showed that particles on formvar–carbon films appeared 10% larger than those on pure carbon supports, when negatively stained with phosphotungstate. According to figures published on murine, bovine, human, porcine, simian, and avian rotaviruses, the diameter of the inner capsid is 55–65 nm and that of the outer capsid 65–75 nm (McNulty, 1979; Woode *et al.*, 1976b). Particles penetrated by negative stain reveal an inner core (38 nm), and isolated, thin-walled cores are sometimes seen in virus preparations treated with EDTA and trypsin (Palmer *et al.*, 1977) or in human stools (Payne *et al.*, 1981). They can also be prepared by treatment of rotaviruses with sodium thiocyanate (Almeida *et al.*, 1979).

Tubules of diameter 50–100 nm and consisting of hexagonally packed subunits (spacing 10 nm) are found in association with some but not all strains of murine, human, bovine, and equine rotaviruses (Banfield *et al.*, 1968; Flewett *et al.*, 1974c; Holmes *et al.*, 1975; Kimura, 1981; Suzuki and Konno, 1975; Woode *et al.*, 1976b).

In thin sections, rotavirus particles have an electron-dense core (25–30 nm) surrounded by a moderately electron-dense capsid layer 50–70 nm in diameter (McNulty, 1979). Enveloped particles (diameter 70–90 nm) are also seen in cisternae of the endoplasmic reticulum (see Section IV.F), and the envelope has sometimes been mistaken for the outer capsid layer, but by fixing and embedding purified preparations of double- and single-shelled particles of simian rotavirus SA 11, Petrie *et al.* (1981) have clarified the situation. They showed that double-shelled particles appeared as evenly electron-dense, smooth-edged ovals with dense nucleoids, whereas single-shelled particles had a more granular appearance, with ragged edges and protruding threads. Enveloped particles were clearly larger (Petrie *et al.*, 1981).

The obvious subunit structure, especially visible in the inner capsid of rotaviruses, has led to a number of attempts to define the capsid structure and symmetry. Various interpretations have been suggested; all agree on icosahedral symmetry, but propose triangulation numbers $T = 3, 4, 9$, or 16 (Esparza and Gil, 1978; Kogasaka *et al.*, 1979; Palmer *et al.*, 1977; Stannard and Schoub, 1977). The problem is the old one of double-sided images produced by negative staining, as was shown by Woode *et al.* (1976b), who were able to make large ring-shaped "capsomers" appear and disappear simply by tilting their specimens on a goniometer stage in the electron microscope.

By freeze–drying and shadowing particles of bovine rotavirus, Roseto *et al.* (1979) were able to obtain clear single-sided images and showed that the structure of the inner capsid is a skewed one with $T = 13$ (Fig. 2). Although they calculated that this structure should have 132 capsomers, in fact if a model is built, it is found that a $T = 13$ structure cannot be divided into nonoverlapping hexamers but only described in terms of 260 trimeric subunits (Holmes, 1982). The basic trimers had already been

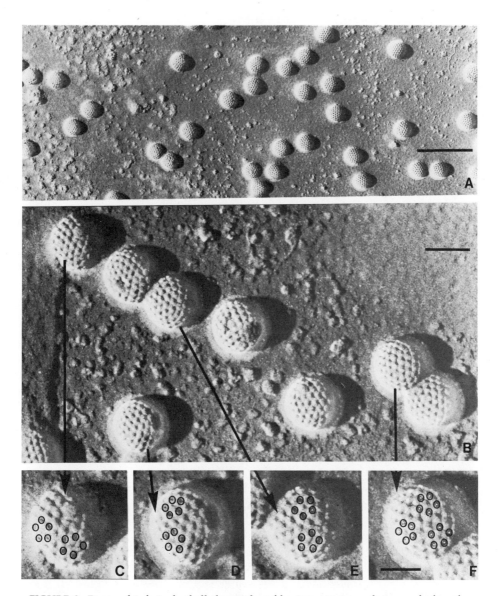

FIGURE 2. Freeze–dried single-shelled particles of bovine rotavirus, platinum-shadowed.
(A) Lower-magnification field illustrating the golfball aspect. Scale bar: 200 nm. (B) High-
magnification field showing the arrangement of surface subunits. Scale bar: 50 nm. (C–F)
Four particles from (B) enlarged further, with sets of five hollows surrounding the five-fold
axis hollows indicated. Note the skewed icosahedral arrangement. Scale bar: 25 nm. Kindly
provided by A. Roseto, slightly modified from Roseto et al. (1979), and reproduced with the
permission of Academic Press, Inc.

demonstrated by negative staining of disintegrating particles (Esparza and Gil, 1978; Martin et al., 1975).

The outer capsid has generally been assumed to follow the same symmetry as the inner one, and this conclusion was supported by the freeze–drying study, which showed a smooth surface perforated by small holes, but no actual subunits could be seen (Roseto et al., 1979).

Morphologically, the rotaviruses that lack the common inner-capsid antigen appear to be indistinguishable from conventional rotaviruses (Bridger et al., 1982; Bohl et al., 1982; McNulty et al., 1981).

## B. Physicochemical Properties

Centrifugation in cesium chloride density gradients separates complete (double-shelled) rotavirus particles (density 1.36 g/cm$^3$) from inner-capsid (single-shelled) particles (density 1.38 g/cm$^3$) (Newman et al., 1975; Rodger et al., 1975b; Tam et al., 1976). Infectivity depends on the presence of the outer capsid (Bridger and Woode, 1976; Elias, 1977) and thus is lost when intact particles are converted to the single-shelled form by treatment with either of the calcium-chelating agents EDTA and ethyleneglycol bis($\beta$-aminoethyl ether)-$N,N'$-tetraacetic acid (EGTA) (Cohen, 1977; Estes et al., 1979b).

For bovine rotavirus in sucrose gradients, sedimentation coefficient estimates are 500 and 450–478 S (Newman et al., 1975; Liebermann et al., 1979). For human rotavirus particles first separated on the basis of density in cesium chloride, complete particles of density 1.36 g/cm$^3$ sedimented at 520–530 S, while single-shelled particles of density 1.38 g/cm$^3$ gave a value of 380–400 S (Tam et al., 1976).

Rotavirus infectivity and particle integrity are resistant to fluorocarbon extraction and exposure to ether, chloroform, or deoxycholate (Fernelius et al., 1972; Much and Zajac, 1972; Welch and Thompson, 1973; Tam et al., 1976; Estes et al., 1979b), but chloroform is reported to destroy the hemagglutinating ability of Nebraska bovine rotavirus, which is also a property of double-shelled (smooth) particles (Bishai et al., 1978). Sodium dodecyl sulfate (SDS) at 0.1% inactivates simian rotavirus SA 11, but exposure to certain nonionic detergents actually enhances infectivity, probably by dispersing aggregates (Ward and Ashley, 1980).

Bovine rotavirus infectivity is stable within the pH range 3–9, but SA 11 rotavirus is slightly more acid-labile and is inactivated below pH 4 (Malherbe and Strickland-Cholmley, 1967; Welch and Thompson, 1973; Palmer et al., 1977; Estes et al., 1979b). All rotaviruses are stable at −70°C, and bovine and human samples have retained infectivity for months at 4 or even at 20°C, when stabilized by 1.5 mM CaCl$_2$ (Shirley et al., 1981). Bovine and simian rotaviruses are relatively stable even at 45–50°C, losing 10–90% of infectivity per hour, but their stability varies with the diluent; for example, they are less stable in 1 M MgCl$_2$ or in

Tris- or phosphate-buffered saline than in maintenance medium or even water (Welch and Thompson, 1973; Estes *et al.*, 1979b). Murine rotavirus is reported to be much more labile, even at 4°C (Much and Zajac, 1972). The hemagglutinating ability of bovine rotavirus is lost very rapidly at 45°C, and both bovine and SA 11 hemagglutinins are destroyed by repeated freezing and thawing (Bishai *et al.*, 1978; Bastardo and Holmes, 1980).

Disinfectants active against rotaviruses include a chlorinated phenolic compound, cresols, an iodophore, and formalin (although 4–10% formaldehyde is required for rapid action), but both SA 11 and ovine rotaviruses are highly resistant to hypochlorite (Snodgrass and Herring, 1977; Tan and Schnagl, 1981). Ethanol, 95%, was the most effective disinfectant of all, and the hemagglutinins and particle integrity of bovine and simian rotaviruses are also destroyed by ethanol at 25 or 50% (vol./vol.) or methanol at 50% (Bishai *et al.*, 1978; Bastardo and Holmes, 1980; Tan and Schnagl, 1981).

## C. Genomes

In early studies employing orcinol and diphenylamine reactions and thermal denaturation, it was shown that bovine and murine rotaviruses contained double-stranded RNA (dsRNA) (Welch, 1971; Much and Zajac, 1972; Welch and Thompson, 1973; Rodger *et al.*, 1975b), and by gel electrophoresis the bovine rotavirus genome was found to consist of 11 segments ranging in molecular weight from about 2.2 to 0.2 $\times$ $10^6$ (Newman *et al.*, 1975; Rodger *et al.*, 1975b). Similar results were soon obtained for rotaviruses of human, ovine [or bovine ("O" agent)], simian, and porcine origins (Schnagl and Holmes, 1976; Kalica *et al.*, 1976; Todd and Mc-Nulty, 1976, 1977).

Although minor differences were evident in the electrophoretic band patterns, the grouping of four large segments, then two of medium size, a closely running triplet that could not always be resolved, and finally two small segments was immediately recognizable as a "rotavirus" pattern (Fig. 3) and was very easy to distinguish from those of reoviruses and orbiviruses (Schnagl and Holmes, 1976; Todd and McNulty, 1976). Many molecular-weight estimates have been published, but all have been obtained by electrophoresis of dsRNA and using reoviruses as standards. Since the last five segments or rotaviral RNA are all smaller than reovirus segment 10 and a completely denaturing system is probably essential for precise determinations, the absolute accuracy of the molecular-weight estimates is uncertain. As a representative set of values, those obtained for molecular weights of genome segments of bovine (Nebraska) rotavirus by Barnett *et al.* (1978) are 2.18, 1.73, 1.64, 1.48, 0.94, 0.77, 0.50, 0.50, 0.50, 0.29, and 0.22 $\times$ $10^6$, giving a total molecular weight of 10.75 $\times$ $10^6$. The most recent (unpublished) estimates obtained by comparing

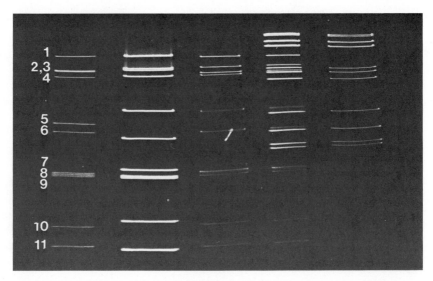

FIGURE 3. Electrophoretic fractionation patterns of the dsRNA genomes of human rotavirus (Hu/W. Australia/76) (A), bovine rotavirus (Bo/Australia/3/75) (B), simian rotavirus SA 11 (C), co-run of SA 11 plus reovirus 3 (D), and reovirus type 3 (Abney) (E) on a Laemmli 10% polyacrylamide gel. Courtesy of S.M. Rodger.

glyoxal-denatured RNA or cDNA copies with DNA standards suggest that the published molecular weights are about 10% too low (M.L. Dyall-Smith, A.A. Azad, and G.W. Both, A.R. Bellamy, J.E. Street and L.J. Siegman, personal communications). For the U.K. strain of bovine rotavirus, the current estimates of base pairs per segment are 3300, 2600, 2550, 2370, 1550, 1340, 1050, 1050, 1050, 760, 680 giving a total of 18,300 nucleotide pairs and a total molecular weight of $12.08 \times 10^6$ (M.L. Dyall-Smith, personal communication).

Since rotaviral serology has until recently remained technically difficult, gel electrophoresis of genome RNA was rapidly recognized as a very useful method for distinguishing among different isolates. Where segments of different strains had similar electrophoretic mobilities, it was necessary to co-run mixtures of the RNAs to establish small differences, and the change over to slab gels was a technical advance that greatly facilitated such comparisons. Segment-pattern differences among rotaviruses obtained from single species, e.g., human, bovine, or equine, were then established (Verly and Cohen, 1977; Kalica et al., 1978b; Rodger and Holmes, 1979; Rodger et al., 1980), and further applications of this approach ("electropherotyping") for epidemiological studies are discussed in Section VI.C.3.

The most noticeable deviations from the "average" rotaviral RNA electrophoretic patterns (see Fig. 4) have been reported for murine rotavirus, the segment 11 of which runs close to segment 10 (M. Smith and Tzipori, 1979); avian rotaviruses, with segment 5 migrating close to segment 4 and segments 10 and 11 co-running (Todd *et al.*, 1980); human rotaviruses, with segment 11 displaced, giving a "short" pattern (Dyall-Smith and Holmes, 1981a) see Sections V.C and VI.C.3); and finally the "pararotaviruses" of pigs, chickens, and humans, which are all quite distinct and show marked displacements of segment 5, 7, or 9 (Bohl *et al.*, 1982; Bridger *et al.*, 1982; McNulty *et al.*, 1981; S.M. Rodger, personal communication).

Comparison of rotaviral genomes by hybridization studies will undoubtedly be important in the future. In the first report of this kind, Matsuno and Nakajima (1982) showed that a Japanese human rotavirus isolate (strain TK20) was closely related to the human Wa strain isolated in the United States (see Sections III.B and VI.D), whereas Wa strain RNA had very little sequence homology with bovine (Nebraska) or simian (SA 11) strains. Using a Northern blot hybridization technique which makes it possible to look for different degrees of homology, Street *et al.* (1982) and Schroeder *et al.* (1982) studied a number of human and animal rotavirus strains and found a low order of sequence relatedness was general for corresponding segments, but hybridization under more stringent conditions showed considerable sequence diversity even among apparently similar human strains. The first actual sequence comparisons can be found in Appendix 1.

Suitable methods for extraction of genome RNA from feces or crude rotaviral preparations have been described by Rodger and Holmes (1979), Clarke and McCrae (1981), Croxson and Bellamy (1981), and Theil *et al.* (1981). For gels to be stained with ethidium bromide, the sharpest resolution of segments is generally obtained by using the discontinuous buffer electrophoresis system of Laemmli (1970). A silver staining technique recently described by Herring *et al.* (1982) increases both the sensitivity of detection and the resolution of rotaviral genome segments and is highly recommended.

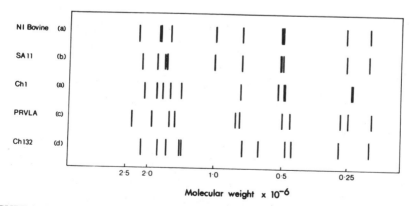

FIGURE 4. Diagrammatic representation of the migration of the dsRNA genome segments of various rotaviruses on polyacrylamide gels, based on data from Todd *et al.* (1980) (a), S.M. Rodger, personal communication (b), Bridger *et al.* (1982) (c), and McNulty *et al.* (1981) (d).

## D. Capsid Proteins

Analyses of capsid proteins were first reported for bovine rotaviruses (Newman *et al.*, 1975; Rodger *et al.*, 1975b; Bridger and Woode, 1976; Matsuno and Mukoyama, 1979), but comparisons with human, simian, ovine, porcine, equine, murine, and lapine rotaviruses soon followed (Rodger *et al.*, 1977; Todd and McNulty, 1977; Thouless, 1979). Surprisingly, the major component of the outer capsid was shown to be glycosylated (Rodger *et al.*, 1977; Cohen *et al.*, 1979; Matsuno and Mukoyama, 1979).

Thouless (1979) introduced a simple and easily remembered terminology for the polypeptides indicating whether they belonged to the inner capsid (I) or outer capsid (O) or were non-structural (NS), but unfortunately further studies have led to one amendment ($I_4$ had to become $O_1$) and considerable disagreement about the total number of polypeptides that are structural. Since we are not yet concerned with nonstructural proteins, a full comparison of nomenclature is left for Section IV.D (see Table I), and the proteins will be designated here by the prefix "p" followed by numbers denoting their molecular weights ($\times 10^3$), using revised estimates for simian rotavirus SA 11 (Dyall-Smith and Holmes, 1981a). The molecular weights of corresponding proteins of different rotaviruses do vary, but few precise comparisons have been made (Thouless, 1979; Espejo *et al.*, 1980b; Dyall-Smith and Holmes, 1981b).

There is general agreement on one minor and two major proteins of the rotaviral inner capsid, p116, p96, and p42 (Espejo *et al.*, 1981; Estes *et al.*, 1981; Novo and Esparza, 1981), but Dyall-Smith and Holmes

TABLE I. Molecular-Weight Estimates and Nomenclature of Rotavirus Polypeptides Found in Infected Cells

| SA 11[a] | SA 11[b] | Various[c] | UK bovine[d] | SA 11[e] |
|---|---|---|---|---|
| p113 | 125 VP1 | $I_1$ | VP1 | VP1 |
| p96 | 94 VP2 | $I_2$ | VP2 | VP2 |
| p91 | — | $I_{3a}$ | VP3 | NCVP1 |
| p84 | 88 VP3 | $I_{3b}$ | VP4 | VP3 |
| p57 | 53 | $(O_1)$ | VP5 | NCVP2 |
| p42 | 41 VP6 | $(I_4)$ | VP6 | VP6 |
| gp34 | 38 VP7a | $(O_2)$ | VP7 | VP7 |
| | 36 VP7 | — | VP7c | — |
| p33 | 34 | $NS_1$ | VP8 | NCVP3 |
| p31 | 32 | $NS_2$ | VP9 | NCVP4 |
| p26 | 28 VP8 | $O_3$ | VP10 | — |
| gp25 | — | $O_4$ | VP10c | NCVP5 |
| | 27 VP9 | — | VP11, 11c | NCVP6 |
| p21 | 20 | $NS_3$ | VP12 | pNCVP5 |

[a] Dyall-Smith and Holmes (1981a).    [b] Mason *et al.* (1980).
[c] Thouless (1979).    [d] McCrae and Faulkner-Valle (1981).
[e] Arias *et al.* (1982).

(1981b) would add a minor protein p91 and McCrae and Faulkner-Valle (1981) include the bovine rotavirus protein corresponding to p91 and also the one corresponding to p33; this latter protein is believed to be non-structural by all other workers. Estes *et al.* (1981) have shown that pro-teolytic breakdown products of p96 can be found in the region of p91 and below, but Dyall-Smith and Holmes (1981b) confirmed that their p91 is a primary gene product (see Section V.C) and not related to p96.

In the virus particle and in infected cells, p42 is found to be aggregated via disulfide bonds (Bastardo *et al.*, 1981). Novo and Esparza (1981) have shown that the corresponding p45 in bovine rotavirus is found in the form of trimers, and since it comprises about 80% of the inner capsid, it must be the main component of the visible morphological units.

Analysis following limited proteolysis of corresponding proteins of simian (SA 11) and human (Wa) rotaviruses showed marked similarities in the patterns of digestion of inner-capsid polypeptides (Dyall-Smith and Holmes, 1981b), but no studies on sequences have yet been reported.

With regard to the outer capsid, Espejo *et al.* (1981) and Estes *et al.* (1981) showed that the primary outer-shell polypeptide p84 (which they estimate as p88) is cleaved specifically by trypsin after the assembled virion has been released from its host cell, to form p62 and p28. This has clarified some apparent discrepancies; for example, in bovine (Nebraska) virus, Novo and Esparza (1981) found only the uncleaved p84 in their preparations. Dyall-Smith and Holmes (1981b) have confirmed that their p84 is related to p62, but were not able to detect the expected p28 cleavage product. In the nomenclature of Thouless (1979), this means that the assignment of $I_4$ to the inner capsid was incorrect, but it is in fact the precursor of p62, which she called $O_1$.

The major outer-shell component is glycoprotein 34 (gp34), which can occur in multiple forms (Dyall-Smith and Holmes, 1981b; Estes *et al.*, 1981). The apparent molecular weights and proportions of (usually) two versions vary from one rotavirus strain to the next, and probably even among different current stocks of SA 11 rotavirus, but two glyco-sylated bands are particularly evident in some reports on bovine rotavi-ruses (Cohen *et al.*, 1979; Novo and Esparza, 1981).

In addition to the cleavage product p28 mentioned above, Estes *et al.* (1981) reported another minor outer-capsid component, p27. Both Bas-tardo *et al.* (1981) and McCrae and Faulkner-Valle (1981) considered that p58 and gp25 could be capsid components, but most other investigators believe that they are nonstructural. It appears that we are at the limits of the techniques of purification and analysis now being employed.

## E. Enzymes

Cohen (1977) showed that when bovine rotavirus particles were treated with the chelating agent EDTA or subjected to heat shock, the outer capsid dissociated and an RNA-dependent RNA polymerase associated with the inner-capsid particles was activated. The activity was measured by incorporation of UTP into an acid-insoluble product and required the presence of all four ribonucleoside triphosphates. The optimum conditions for the reaction were pH 8, 50°C, and a concentration of $Mg^{2+}$ around 8–10 mM. The product was sensitive to ribonuclease, showing that it was single-stranded RNA, but became ribonuclease-resistant following hybridization with an excess of genome RNA.

These findings were confirmed and extended by Hruska *et al.* (1978), who obtained similar results with another bovine rotavirus (Nebraska strain) and a number of human rotavirus samples, partially purified from stools.

Cohen *et al.* (1979) used the specific chelating agent EGTA to demonstrate that the concentration of free calcium ions was the crucial factor in activation of the polymerase, which occurred even in the presence of magnesium. By hybridization of [$^3$H]uridine-labeled products of the polymerase with unlabeled, denatured dsRNA from the virus, followed by ribonuclease treatment and polyacrylamide gel electrophoresis (PAGE), Cohen and Dobos (1979) proved that complete transcripts of all the genome segments were produced, but the *in vitro* reaction appeared to produce an excess of transcripts of the segments of low molecular weight.

RNA-dependent RNA polymerase in simian rotavirus SA 11 was activated similarly by Mason *et al.* (1980), but they found it advantageous to add bentonite to the reaction mixtures. Even so, the higher-molecular-weight transcripts that were demonstrated by gel electrophoresis following a 2-hr reaction at 40°C appeared to be degraded if the incubation was continued to 18 hr.

The porcine rotaviruslike agent that lacks the common group antigen was also converted to the single-capsid form when treated with EDTA, but no studies of polymerase in these particles have yet been reported (Bridger *et al.*, 1982).

A polyadenylate [poly(A)] polymerase activity was reported to be associated with complete, double-shelled human rotavirus particles purified from stools, but could not be detected in cell-culture-grown bovine rotavirus (Gorziglia and Esparza, 1981). The reaction was carried out under conditions established for poly(A) polymerase activity of reovirus (Stoltzfus *et al.*, 1974).

## F. Antigenic Determinants

A common (group) antigen shared by bovine and human rotaviruses was originally detected by immunoelectron microscopy (IEM) and immunofluorescence (IF) (Flewett *et al.*, 1974a). It is also detectable by complement fixation (CF), immune adherence hemagglutination assay (IAHA), and enzyme-linked immunosorbent assays (ELISA), and has also been found in rotaviruses of porcine, ovine, lapine, equine, murine, simian, and avian origin (Kapikian *et al.*, 1976a; Woode *et al.*, 1976b; Matsuno *et al.*, 1977b; Thouless *et al.*, 1977b; Scherrer and Bernard, 1977; Yolken *et al.*, 1977; McNulty *et al.*, 1979). The only rotaviruses that lack this antigen are the "pararotaviruses" from pigs, chickens, and humans (Bohl *et al.*, 1982; Bridger *et al.*, 1982, McNulty *et al.*, 1981; S. Rodger and R. Bishop, personal communication).

Studies by IEM on reactions of various rotaviruses with homologous and heterologous antirotaviral sera showed that the group antigen was located on the inner capsid and was thus masked in intact, double-shelled particles (Woode *et al.*, 1976b; Bridger, 1978). Peak CF reactivity was also associated with particles lacking the outer capsid when bovine rotavirus was purified in cesium chloride gradients (Fauvel *et al.*, 1978). In immunodiffusion reactions between crude (fecal) bovine and human rotavirus preparations and homologous and heterologous sera, electron microscopy of the precipitated "line of identity" by Mathan *et al.* (1977) showed antibody-cross-linked aggregates of intact and broken inner capsids and significant amounts of amorphous material interpreted as unassembled capsid subunits from infected cells.

Some discrimination between certain human and animal rotavirus strains on the basis of inner-capsid antigens has been achieved using selected hyperimmune guinea pig or gnotobiotic calf infection sera, by CF and more recently by ELISA or IAHA (Zissis and Lambert, 1978, 1980; Yolken *et al.*, 1978b; Zissis *et al.*, 1981; Kapikian *et al.*, 1981). This subgrouping is discussed in more detail in Section VI.D, but must be mentioned here because the work of Kapikian *et al.* (1981) and Kalica *et al.* (1981b) on a range of reassortants between a human and a bovine rotavirus showed that the subgroup specificity was always associated with genome segment 6 of either parental virus. Since it is known that this segment codes for the major inner-capsid polypeptide p42 or p45 (M.L. Smith *et al.*, 1980; McCrae and McCorquodale, 1982), this protein must carry the subgroup determinants. It also carries the common group determinant(s) because it is immunoprecipitated by the most commonly encountered class of antirotaviral monoclonal antibodies, which react with a range of rotaviruses (H. Greenberg, personal communication; S. Sonza and A. Breschkin, personal communication).

A greater degree of discrimination between rotavirus strains can be obtained using neutralization (N) or hemagglutination-inhibition tests, and the finding that immune sera could be specific by N test yet widely cross-reactive by IF, CF, or IAHA led to the conclusion that distinct antigens were involved (Woode *et al.*, 1976b; Matsuno *et al.*, 1977b; Thouless *et al.*, 1977b). This was confirmed when it was found that hemagglutination, like infectivity, was a property of intact, double-shelled virions, and that the neutralizing specificities of sera were reflected by their IEM reactions with complete but not single-shelled particles (Fauvel *et al.*, 1978; Bridger, 1978). Thus, it was clear that the antigen involved in neutralization and hence in serotype specificity must be located in the outer capsid.

Serological analysis of human–bovine rotavirus reassortants and determination of the parental origin of their genome RNA segments by electrophoretic comparisons enabled Kalica et al. (1981b) to show that in bovine (UK) and human (Wa) rotaviruses, RNA segment 9 codes for the neutralization antigen. By testing antisera prepared by immunization of rabbits with individual polypeptides of simian rotavirus SA 11 that were separated by PAGE, Bastardo et al. (1981) demonstrated directly that the major outer-shell glycoprotein (gp34) elicits the best neutralizing response. The antiserum thus produced was type-specific. The antigenicity of gp34 was sensitive to reduction with mercaptoethanol, so the electrophoresis had to be carried out under nonreducing conditions, but it survived at least partial denaturation with SDS.

Monoclonal antibodies that neutralize SA 11 rotavirus have been prepared by A. Breschkin and S. Sonza (personal communication) and shown to immunoprecipitate gp34. Most are specific for SA 11, but some also neutralize other rotaviruses, so there must be both serotype-specific and some shared antigenic determinants on this glycoprotein. This observation may explain why many antisera show a degree of cross-reactivity in rotavirus N tests. Most of the neutralizing monoclonal antibodies also inhibit hemagglutination. One monoclonal antibody that neutralized SA 11 virus appears to react not with gp34 but with a polypeptide of somewhat lower molecular weight, which has not yet been identified. Bastardo et al. (1981) also obtained a lower but significant level of neutralization with a serum prepared against trace amounts of an unreduced protein identified as p26, which could be either a cleavage fragment of p84 or possibly the product of RNA segment 10 or 11. The suggestion that segment 11 might code for a minor neutralization antigen is supported by the finding of Matsuno et al. (1980) that a reassortant between SA 11 and bovine rotavirus that derived its major neutralization antigen and RNA segments 4 and 10 from SA 11 virus but segment 11 from the bovine parent showed an increased cross-reaction with antibovine rotavirus serum.

## III. CELL SPECIFICITY AND CULTIVATION

### A. Differentiated Nature of Host Cells

In infected animals, the main or usually the only sites of replication of rotaviruses are the columnar epithelial cells lining the villi of the small intestine. This has been shown by immunofluorescence (IF) staining for rotaviral antigens during histopathological studies of naturally or experimentally infected mice, calves, piglets, and lambs (Wilsnack et al., 1969; Mebus et al., 1971b; Hall et al., 1976; Pearson and McNulty, 1977; Snodgrass *et al.*, 1977a). Necessarily limited studies in humans by electron microscopy or IF suggest the same conclusion (Bishop *et al.*, 1973; Davidson *et al.*, 1975b). In mice and lambs, rotaviral infection of the colonic epithelium has also been found (Banfield *et al.*, 1968; Snodgrass *et al.*, 1977a).

The villous epithelium consists of a continuously differentiating population of cells. Cell division occurs in the crypts, and the cells mature as they move up the sides of the villi, elongating from a cuboidal to a columnar shape and developing a brush border with numerous new enzyme activities en route. Their lifetime as differentiated cells is limited to a few days, from the time they reach one-third or halfway along the villi until they fall off at the tips.

Both electron-microscopic and IF studies suggest that only the differentiated epithelial cells are susceptible to rotaviruses, since no particles or antigens are found in the proliferating cells in the crypts, but both appear in the enterocytes on the apical halves of the villi (Adams and Kraft, 1967; Mebus *et al.*, 1971b; Stair *et al.*, 1973; Snodgrass *et al.*, 1977a). Following the accelerated loss of infected cells from villous tips, the villi become covered with immature cuboidal epithelial cells that appear resistant to rotavirus infection, and Mebus *et al.* (1971b) suggested that this may limit the duration of the infection. The immature cells may lack receptors for the virus (Holmes *et al.*, 1976).

Even when grown in organ culture, pieces of small intestine do not retain their differentiated epithelial cells for long, and attempts to propagate rotaviruses in such organ cultures have met with only limited success (Rubenstein *et al.*, 1971; Wyatt *et al.*, 1974).

## B. Cell-Culture Adaptation and the Role of Trypsin

Ironically, the first rotaviruses adapted to cell-culture growth were the simian rotavirus SA 11 isolated from a healthy monkey and the "O" agent from abattoir waste, long before they were known to have anything to do with diarrheal disease (Malherbe and Strickland-Cholmley, 1967). It was only after a great deal of patient and painstaking work that cell-culture adaptation of the Nebraska strain of bovine rotavirus was achieved (Mebus *et al.*, 1971a; Fernelius *et al.*, 1972) and other bovine strains were isolated in the United Kingdom, France, and Northern Ireland (Bridger and Woode, 1975; L'Haridon and Scherrer, 1977; McNulty *et al.*, 1976a). In each case, cultures of kidney epithelial cells were the most successful for rotavirus growth. Infected cells were monitored by IF staining, because cytopathic effects were variable.

There were many attempts to grow human rotaviruses in cell culture, but they proved to be more difficult. Banatvala *et al.* (1975) found that they could infect moderate numbers of cells if they centrifuged human rotavirus inocula onto cell monolayers, and this became the basis of a successful method for diagnosis and serotyping of human rotaviruses (Bryden *et al.*, 1977; Thouless *et al.*, 1978), but on further passaging, the percentage of infected cells diminished rapidly, and very few particles were produced (Albrey and Murphy, 1976; Wyatt *et al.*, 1976a).

Although they grew bovine rotavirus satisfactorily, Welch and Twiehaus (1973) had little success with plaque formation, so Matsuno *et al.* (1977a) made a major advance when they were able to develop a plaque assay using the sensitive rhesus monkey kidney cell line MA-104 and both trypsin and DEAE–dextran in the overlay. Other workers soon found that bovine, porcine, and avian rotaviruses could be serially passaged in the presence of low concentrations of trypsin (Babiuk *et al.*, 1977; Theil *et al.*, 1977; Almeida *et al.*, 1978; McNulty *et al.*, 1979, 1981).

By ingenious experiments with trypsin inhibitors and the enzyme immobilized on agarose beads, Barnett *et al.* (1979) proved that trypsin was acting directly on the virus, not on the host cell. Various other proteases were found to have an enhancing effect on rotavirus growth, but Ramia and Sattar (1980) showed that with all except elastase and subtilisin, the effect was abolished by soybean trypsin inhibitor and was presumably due to traces of trypsin contaminating the other enzymes. It was also realized that the inhibitory effect of antibody-free fetal calf serum on rotavirus growth that had been noted in earlier studies (Welch and Twiehaus, 1973; Barnett *et al.*, 1975) was due to its trypsin-inhibitory capacity (Graham and Estes, 1980).

As discussed in Section II.D, the essential effect of trypsin is specific cleavage of one of the outer-capsid proteins (Espejo *et al.*, 1981; Estes *et al.*, 1981). Clark *et al.* (1981) showed that this cleavage does not affect the efficiency or rate of attachment of bovine rotavirus to cells, but activates a previously noninfectious fraction of particles by facilitating its uncoating in the cell. Trypsin enhances replication of even those rotaviruses that had been culture-adapted in its absence, such as simian rotavirus SA 11, but the effect is much more marked with cells in which it multiplies poorly, such as the Vero line of monkey kidney, than in MA-104 cells, which give quite good virus yields in the absence of trypsin (Graham and Estes, 1980). Presumably, cultures of cells like those of the MA-104 line contain sufficient protease of trypsinlike specificity to activate SA 11 virus, but not enough for many other rotaviruses. In the intestine, of course, trypsin would always be present, and probably bound to the surface of the enterocytes (Goldberg *et al.*, 1968).

For unknown reasons, human rotaviruses proved to be the most difficult to cultivate *in vitro*. The first success was obtained by Wyatt *et al.* (1980) with the Wa strain; after 11 passages through gnotobiotic piglets, the virus was trypsin-treated and could be serially passaged thereafter in primary African green monkey kidney (AGMK) cells. In the presence of trypsin Wa strain replicates moderately well also in MA-104 and CV-1 monkey kidney cell lines. Drozdov *et al.* (1979) have reported a similar isolation, without piglets.

Very recently, almost routine isolation of human rotaviruses of different serotypes has been achieved by a number of researchers in Japan (Sato *et al.*, 1981; Urasawa *et al.*, 1981; R. Kono, personal communication; T. Konno, personal communication). Trypsin treatment, MA-104 or CV-1 cells, and rolling of the cultures during the initial passages appear to be the critical factors, and several passages are needed before the isolate can be regarded as established. Isolation of a range of human rotaviruses will greatly facilitate identification of serotypes (see Section VI.D), and the aforementioned achievement is an excellent example of how scientists gradually build on earlier results.

## C. Adaptation by Reassortment

The fact that viruses with segmented genomes are frequently capable of genetic reassortment during mixed infections led Greenberg *et al.* (1981) to try to rescue genes of uncultivable human rotaviruses by this means. The nonadapted human rotavirus strains Wa and DS-1, which could normally be passaged only once or twice in AGMK cells, and temperature-sensitive mutants of the cultivable bovine rotavirus (UK strain) were chosen. After mixed infection of the AGMK cells at 34°C, reassortants were selected from plaques produced at 39°C and/or in the presence of antiserum capable of neutralizing the bovine parent. A number of reassortants that grew well in cell culture but carried the neutralization antigens of either Wa or DS-1 strains were isolated. A more extensive series of rescues of 33 noncultivable human rotavirus strains including 12 which are serologically distinct from either Wa or DS-1 has been reported since (Greenberg *et al.*, 1982). Such reassortants have been important in the identification of antigenic polypeptides, as was discussed in Section II.F. Genome segment 4 of the bovine rotavirus appeared to be essential for cultivability under the conditions employed (Greenberg *et al.*, 1982).

## D. Assay Systems

Because of the difficulty of conventional isolation of rotaviruses despite their importance in human and animal disease, a great deal of effort has been put into the development of alternatives to the usual infectivity assays, mainly for diagnostic purposes. Thus, it seemed opportune to briefly mention these methods at this point, but readers who are interested primarily in diagnostic methods should refer to reviews such as those by Ellens *et al.* (1978), Kapikian *et al.* (1979), and de Leeuw and Guinee (1981) for details and comparative evaluation of the available methods.

1. Methods That Do Not Rely on Infectivity

At first, rotavirus detection and quantitation depended almost entirely on electron microscopy using negative-staining methods. Preliminary concentration (and sometimes partial purification) has mostly involved differential centrifugation (Fernelius et al., 1972; Flewett et al., 1973, 1974b; Bishop et al., 1974), but ammonium sulfate precipitation and concentration using dried polyacrylamide gel (lyphogel) also have their exponents (Caul et al., 1978; Rogers et al., 1981). Immunoelectron microscopy is not necessary for rotavirus recognition, but has often been employed in differential diagnosis of the various viruses involved in diarrheal disease (Kapikian et al., 1974; Flewett and Boxall, 1976). The use of supporting films coated with specific antibodies may increase the sensitivity of rotavirus detection (Nicolaieff et al., 1980; Obert et al., 1981), but on the whole, electron-microscopic methods are not ideal for quantitation.

Precipitation methods include the fluorescent virus precipitin test of L.G. Foster et al. (1975), in which aggregates obtained after mixing virus suspensions with fluorescein-conjugated rotavirus antibodies are scored by UV microscopy, and (counter)immunoelectrophoresis, which is convenient but not very sensitive or quantifiable (Grauballe et al., 1977; Middleton et al., 1976; Spence et al., 1977; Tufvesson and Johnsson, 1976).

Radioimmunoassays (RIAs) and enzyme-linked immunosorbent assays (ELISAs) are most popular for rotavirus antigen detection and quantitation, because they can be readily applied to large numbers of samples (Kalica et al., 1977; Middleton et al., 1977a; Cukor et al., 1978; Yolken et al., 1977; Scherrer and Bernard, 1977; Ellens et al., 1978; Kapikian et al., 1979; de Leeuw and Guinee, 1981). As a method for rotavirus detection, ELISA is about 100 times more sensitive than immunoelectrophoresis (Grauballe et al., 1981b) and equivalent to RIA, if antisera of equivalent titers are used (Sarkkinen et al., 1980)

Bradburne et al. (1979) used erythrocytes coupled to rotaviral antibodies as the detection system in an assay otherwise similar to an RIA or ELISA, which they christened the solid-phase aggregation of coated erythrocytes (SPACE) test. It offers speed and simplicity, but at the cost of lower sensitivity than ELISA. Another rapid test that looks very promising for field use also employs fixed erythrocytes coupled to purified rotaviral antibodies; it is the reverse passive hemagglutination test of Sanekata et al. (1979). Alternatively, rotaviral antibodies can be adsorbed on to protein A–carrying Staphylococcus aureus cells, and these are agglutinated by samples containing rotavirus antigen (Hebert et al., 1981).

Finally, hemagglutination of human or sheep erythrocytes can be employed for assay of rotaviruses such as SA 11, but not all bovine and very few human rotavirus strains appear to hemagglutinate (Spence et al., 1976, 1978; Inaba et al., 1977; Kalica et al., 1978a; Shinozaki, 1978).

As mentioned in Section II.F, hemagglutination is a property of particles with an intact outer capsid (Fauvel *et al.*, 1978).

## 2. Infectivity Assays

Assays of rotaviral infectivity in infant animals are possible, but rotaviruses are notorious for contaminating the environment and are extremely contagious, so susceptible animals have to be maintained in strict isolation. In her pioneering studies of murine rotavirus [epizootic diarrhea of infant mice (EDIM)], Kraft (1958) developed the use of filter-topped cages for her litters of mice, but although gnotobiotic calves and piglets have been used extensively in rotavirus research, they are too expensive for use in titrations.

Even with rotaviruses that have not been cell-culture-adapted, infectivity titers can be estimated by fluorescent focus assays. Monolayers of bovine kidney, LLC-MK2, or MA-104 cells are exposed to dilutions of the rotavirus to be assayed, with centrifugation during the adsorption period for human strains; then, after incubation for about 16 hr to complete the first cycle of replication, infected cells are stained and counted by immunofluorescence (Barnett *et al.*, 1975; Bryden *et al.*, 1977; Moosai *et al.*, 1979).

Plaque assays are generally the method of choice for infectivity titrations, since in quantal assays rotaviral cytopathic effects can be variable and difficult to read. Following the lead of Matsuno *et al.* (1977a), similar plaque assays for other rotaviruses have been developed (Ramia and Sattar, 1979; E.M. Smith *et al.*, 1979). Inclusion of trypsin in the overlay is usually essential, but plaquing of bovine rotavirus in AGMK cells not only without trypsin or DEAE–dextran but even with fetal calf serum in the overlay has been described (Kapikian *et al.*, 1979).

## E. Growth Kinetics

Not many growth curves for rotaviruses have been published. Welch and Twiehaus (1973) infected primary bovine embryonic kidney cells with an estimated multiplicity of bovine rotavirus (Nebraska strain) of 10 $ID_{50}$ per cell. An increase in titer was detected by 4 hr, and there was a steady rise in titer until 18 hr, with most of the virus cell-associated. Similar results were obtained for Nebraska bovine rotavirus by Kurogi *et al.* (1976) and Matsuno and Mukoyama (1979), for Northern Ireland bovine rotavirus by McNulty *et al.* (1977), and for simian rotavirus SA 11 in MA-104 cells by Estes *et al.* (1979a), but lower multiplicities of infection were used and the duration of a single cycle of growth was difficult to estimate.

The most convincing single-cycle growth curves so far published are for bovine rotaviruses: the Nebraska strain in Madin-Darby bovine kidney cells and the UK strain of bovine rotavirus in the BSC-1 line of AGMK cells, infected at high multiplicity (20 plaque-forming units per cell) (Clark *et al.*, 1979; McCrae and Faulkner-Valle, 1981). Synthesis of new viral double-stranded RNA was detected by 3 hr postinfection, and maximum yields of progeny virus were obtained by 10–12 hr.

## F. Effects on Host Cells

Cytopathic effects in rotavirus-infected cultures have been reported for SA 11 virus and the "O" agent by Malherbe and Strickland-Cholmley (1967) and Estes *et al.* (1979a) and for bovine rotavirus by Welch and Twiehaus (1973) and McNulty *et al.* (1977). The main visible changes are cytoplasmic vacuolization and the appearance of small, eosinophilic intracytoplasmic inclusions. A very characteristic feature of rotavirus-infected cultures that was noted by C.A. Mebus is "flagging" (Holmes, 1979). Infected cells tend to remain attached to the monolayer by a single process, so they wave about when the medium moves. Cytopathic effects are more notable in rolled than in stationary cultures, probably because there is more cell detachment, but replication may also be enhanced (Kurogi *et al.*, 1976; McNulty *et al.*, 1977).

At the biochemical level, a shutdown of host-cell protein synthesis is detected if the infecting multiplicity is sufficiently high (i.e., if a sufficient proportion of cells are infected) (Thouless, 1979). McCrae and Faulkner-Valle (1981) noted that this shutdown occurred by 4 hr postinfection, but found that the overall levels of DNA, RNA, and protein synthesis in BSC-1 cells infected with UK bovine rotavirus were not markedly affected. On the other hand, Carpio *et al.* (1981) reported margination of chromatin and a decreased synthesis of DNA, RNA, and proteins in BSC-1 cells infected with bovine rotavirus strain C486, with apparent breakdown of cellular DNA.

Welch and Twiehaus (1973) showed that interferon was produced by bovine embryonic kidney cells exposed to UV-inactivated bovine rotavirus, whereas unirradiated virus was a less effective inducer. La Bonnardiere *et al.* (1980) found bovine rotavirus a poor inducer, and only weakly susceptible to interferon. In MA-104 cells pretreated ("primed") with low concentrations of interferon, however, infectious simian rotavirus SA 11 can induce significant quantities of interferon (McKimm-Breschkin and Holmes, 1982), and the replication of SA 11 virus is sensitive to interferon if the infected cells are simultaneously exposed to UV-treated SA 11 particles. This suggests that the rotavirus normally fails to activate the antiviral enzymes produced in cells exposed to interferon, but if the enzymes are activated by UV-treated particles, they effectively block rotavirus multiplication. It thus seems possible that interferon may have a more significant role in limiting rotavirus replication *in vivo* than the *in vitro* experiments at first suggested, since in the intestine other agents capable of inducing interferon and of activating the antiviral enzymes would be present at the same time as the rotavirus.

## IV. REPLICATION

### A. Adsorption

Whether the high degree of host-cell specificity of rotaviruses depends only on specificity of adsorption or on subsequent intracellular events does not appear to have been investigated, but the former is generally assumed. On the basis of the apparent development of rotavirus sensitivity during differentiation of intestinal epithelial cells (Section III.A) and the reported decrease in susceptibility of mice to murine rotavirus (EDIM) at about the time of weaning, Holmes et al. (1976) suggested that the cell-membrane receptor for rotaviruses might be lactase. Lactase activity is found on the kidney cells in which rotaviruses are often propagated, but other brush-border components may have a similar distribution.

Treatment of human group O and sheep erythrocytes with neuraminidase rendered them inagglutinable by simian rotavirus SA 11 (Bastardo and Holmes, 1980), and similar treatment of MA-104 cells makes them refractory to SA 11 virus infection (G. Raghu, personal communication), so in each case the presence of sialic acids on the receptors must be essential.

By electron microscopy, adsorption of rotavirus particles to host-cell plasma membranes has been demonstrated in vitro (Petrie et al., 1981) and in vivo (Coelho et al., 1981). In the latter case, murine rotavirus particles were mostly associated with the tips of microvilli, but some were situated between the microvilli, and endocytosis probably occurs at the base of the microvilli.

### B. Penetration and Uncoating

Although rotavirus particles have been seen in lysosomelike bodies within infected cells (Lecatsas, 1972; Petrie et al., 1981), uptake under single-cycle conditions has not yet been studied. On the other hand, as Cohen et al. (1979) point out, their findings on "uncoating" or disassembly of the rotaviral outer capsid at low calcium ion concentrations are very relevant to the penetration step in rotaviral infection. As soon as the incoming particle left the extracellular zone, in which the calcium concentration is relatively high, and passed through the plasma or vesicular membrane of the cell into the cytoplasmic compartment, where the $Ca^{2+}$ concentration is maintained at less than 1 $\mu$M (Carafoli and Crompton, 1978), its outer capsid would be dissociated and its endogenous RNA polymerase would be activated. Clark *et al.* (1981) presented evidence that the process of infection by particles with outer-capsid protein p84 that has not been cleaved by trypsin is probably blocked at this point.

## C. Transcription

Following uncoating of the incoming rotavirus particle, transcription would presumably begin immediately in the cytoplasm. Some properties of the RNA-dependent RNA polymerase (Cohen, 1977) have already been mentioned in Section II.E. The transcripts do not self-hybridize, so they are transcribed fom only one strand of the genomic double-stranded RNA (dsRNA) (Cohen, 1977), but their behavior during electrophoresis following hybridization with excess genomic RNA shows that they are full-length copies of each of the 11 segments (Cohen and Dobos, 1979; Bernstein and Hruska, 1981). *In vitro,* there seems to be a selective degradation of the higher-molecular-weight transcripts during long-term incubations even in the presence of bentonite (Cohen and Dobos, 1979; Mason *et al.*, 1980), and the same appears to be true in SA 11 virus–infected MA-104 cells (Dyall-Smith and Holmes, 1981a; J.L. McKimm-Breschkin, personal communication). The transcripts do not appear to contain poly(A) tracts, but function as messenger RNA, directing synthesis of viral polypeptides in cell-free translation systems (Cohen and Dobos, 1979; Mason *et al.*, 1980).

## D. Biosynthesis of Polypeptides

If cells are infected at a sufficiently high multiplicity to cause a rapid shutdown of cellular protein synthesis, newly synthesized rotaviral proteins can be detected by 4 hr postinfection (McCrae and Faulkner-Valle, 1981). At least 12 virus-specific polypeptides can be identified; molecular-weight estimates and the relationship among the systems of nomenclature used by different research groups are shown in Table I. In addition to the structural proteins discussed in Section II.D., there are between one and six nonstructural proteins, with most workers agreeing on four. These are the primary gene product p57, which was originally confused with the outer-capsid tryptic-cleavage product of p84 (Thouless, 1979; Smith *et al.*, 1980), p33 and p31, and gp25. McCrae and Faulkner-Valle (1981), Arias *et al.* (1982), and Ericson *et al.* (1982) have shown that the smallest polypeptide, p21, or $NS_3$ of Thouless (1979), is in fact the precursor of gp25. This glycoprotein is doubly unusual in that detectable amounts of its precursor are generally seen in infected cells in the absence of tunicamycin treatment (Thouless, 1979; Dyall-Smith and Holmes,

1981a; Arias *et al.*, 1982), and no other example of a nonstructural glycoprotein comes to mind.

Multiple forms of the structural glycoprotein gp34 can be found in infected cells, but the unglycosylated precursor is seen only in cells treated with tunicamycin (McCrae and Faulkner-Valle, 1981; Dyall-Smith and Holmes, 1981b; Arias *et al.*, 1982; Ericson *et al.*, 1982). The carbohydrate portions of both gp34 and gp25 can be hydrolyzed by endoglycosidase H, which indicates that they are simple mannose-rich oligosaccharides rather than more complex Golgi-processed ones (Arias *et al.*, 1982).

Ultrastructural studies suggest that both transcription and biosynthesis of viral polypeptides must occur within or around the moderately electron-dense, granular or finely fibrillar areas designated "viroplasm" or viral inclusions. These resemble the corresponding "virus factories" in reovirus- or orbivirus-infected cells and have been observed in cells infected with murine, simian, human, bovine, ovine, and porcine rotaviruses (Banfield *et al.*, 1968; Lecatsas, 1972; Holmes *et al.*, 1975; McNulty *et al.*, 1976c; Snodgrass *et al.*, 1977a; Saif *et al.*, 1978; Pearson and McNulty, 1979). Their protein composition is unknown, but no doubt includes some of the nonstructural proteins.

The tubules mentioned in Section II.A have also been seen within infected cells, where they appear as thin-walled structures 50–60 nm in diameter and may run from the cytoplasm into the nucleus. They are commonly associated with murine and porcine rotaviruses and have once been reported in human rotavirus–infected cells (Banfield *et al.*, 1968; Saif *et al.*, 1978; Pearson and McNulty, 1979; Suzuki and Konno, 1975), but are not produced by simian rotavirus SA 11. Their antigenic relationship to inner capsids (Holmes *et al.*, 1975; Kimura, 1981) suggests that they are probably composed of structural protein(s), but no purification or analysis has been published. Similar tubules are common in orbivirus-infected cells (Murphy *et al.*, 1971). Huismans and Els (1979) have shown that the orbivirus-associated tubules consist of a nonstructural protein, even though they appear to be precipitated by antisera prepared against purified particles. Thus, the rotavirus tubules could also turn out to be nonstructural.

Laminar structures that were described by Lecatsas (1972) as "membranous elements" and by Altenburg *et al.* (1980) as "tubular structures" are found in cells infected with SA 11 rotavirus. They occur both in nuclei and in cytoplasm, but differ from the tubules mentioned above. They are 15–20 nm thick, and fine cross-striations can be seen at high magnification (F.A. Murphy, personal communication). They closely resemble the cross-striated filaments found in cells infected with Colorado tick fever virus (Murphy *et al.*, 1968). With some antisera, they can be detected by immunofluorescence, and appear to be ribbon-shaped and up to 5 µm long (M. Dyall-Smith, personal communication). Neither their function nor their composition is known, but they are probably paracrystalline aggregates of a nonstructural protein.

## E. Viral RNA Synthesis

Biochemical studies on bovine rotavirus–infected cells have shown that newly synthesized genomic dsRNA can be detected by 2–3 hr postinfection, and all the dsRNA segments appear to be made at the same time (McCrae and Faulkner-Valle, 1981). It is not known whether the second strands are synthesized on plus-strand templates or whether the process occurs in viral precursor particles as has been shown for reoviruses (Joklik, 1974).

In studies of rotavirus morphogenesis, electron-dense (after staining) cores appear within particles while they are still associated with the viroplasm matrix (Banfield *et al.*, 1968; Holmes *et al.*, 1975; Pearson and McNulty, 1979) or during budding into rough endoplasmic reticulum (RER) vesicles (Fig. 5). Large numbers of "coreless" particles were produced in the porcine kidney PK-15 cell line infected with a high multiplicity of lamb rotavirus that was not adapted to serial passage in PK-15 cells (McNulty *et al.*, 1978). It would be very interesting to know whether such particles, the cores of which fail to stain during processing for thin-

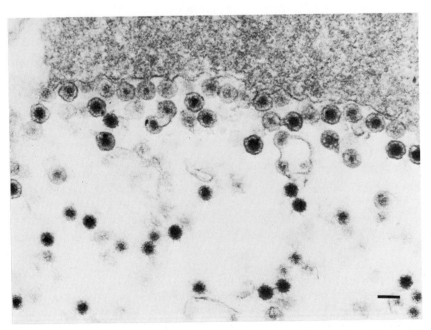

FIGURE 5. Simian rotavirus SA 11 in CV-1 cells, showing viroplasm and particles budding into the lumen of the RER. Note the increase in staining of the cores during budding. Scale bar: 100 nm. Kindly provided by F.A. Murphy and A.K. Harrison.

section electron microscopy contain single-stranded RNA, or any RNA at all, but no sequel to this study has appeared. As the authors suggested, the coreless particles could be defective interfering particles produced as a result of high-multiplicity passages.

## F. Assembly

Electron-microscopic studies on the morphogenesis of rotavirus particles provide the necessary background for a discussion of the assembly stage of replication, but for an understanding of the process, we also need to consider recent biochemical findings. In the following account, I have cited published experimental results wherever possible, but parts of the hypothesis, which is based on discussions between the author and M.L. Dyall-Smith, have yet to be proven.

The most distinctive feature of rotavirus morphogenesis, in which rotaviruses differ from reoviruses or other genera in the family, is the budding process whereby particles accumulate in vesicles of the RER. This has been noted by almost all workers who have studied rotavirus-infected cells, and when the plane of the section is favorable, it can be seen that the particles bud directly from the edge of the viroplasm through adjacent areas of RER membrane (Adams and Kraft, 1967; Banfield et al., 1968; Lecatsas, 1972; McNulty et al., 1976c; Saif et al., 1978; Pearson and McNulty, 1979; McNulty, 1979; Altenburg et al., 1980). It has also been remarked that most of the particles in RER vesicles are unenveloped, but as illustrated in Fig. 6 (Holmes et al., 1975), enveloped particles are usually located near the RER membrane. By the criteria of Petrie et al. (1981), most of the unenveloped particles in such vesicles are double-shelled. At present, it is technically impossible to determine whether the particles within the envelopes are single- or double-shelled, but we believe them to be single-shelled because of the considerations discussed below.

If the idea of automatic uncoating of rotaviruses in the low-calcium environment of the cytoplasm (Section IV.B) is accepted, it is hard to imagine how the outer capsids of progeny particles could assemble under the same conditions, even if single-shelled particles and the outer-capsid polypeptides were all present near the viroplasm. On the other hand, from what is known of the distribution of calcium in various kinds of cells, the $Ca^{2+}$ concentration within the endoplasmic reticulum is more likely to be sufficiently high for complete particle assembly (Carafoli and Crompton, 1978). The problem then becomes one of transfer of single-shelled particles and the outer-capsid components to the lumen of the endoplasmic reticulum.

We suggest that the observed budding process indicates how the inner capsid is transferred, and predict that the outer-capsid polypeptides, or at least the major structural glycoprotein (gp34 in the case of SA 11 virus) and possibly the second viral glycoprotein gp25, will be synthesized on membrane-bound ribosomes and secreted directly into the RER. A preliminary study of bovine rotavirus–infected cells employing immunoperoxidase staining (Chasey, 1980) directly supports this suggestion, since the reaction indicating viral antigens occurred not only on budding particles but also along the RER membrane nearby. Arias *et al.* (1982) have shown by endoglycosidase H digestion that the glycoproteins of SA 11 virus seem to contain only high-mannose oligosaccharide residues, which suggests that the glycosylation occurs at the RER membrane and that further processing (Robbins *et al.*, 1977) does not occur; they also suggest that maturation of particles occurs in the lumen of the endoplasmic reticulum, and Korolev *et al.* (1981) came to the same conclusion on morphological grounds.

One problem still remains: following budding, the inner capsid within what we shall now call the pseudoenvelope would still be separated from the outer-capsid components and $Ca^{2+}$ ions in the lumen of the RER, so the pseudoenvelope must be removed. Pseudoenvelope removal depends on a glycosylated product synthesized in infected cells during rotavirus multiplication, and it is tempting to assume that it is

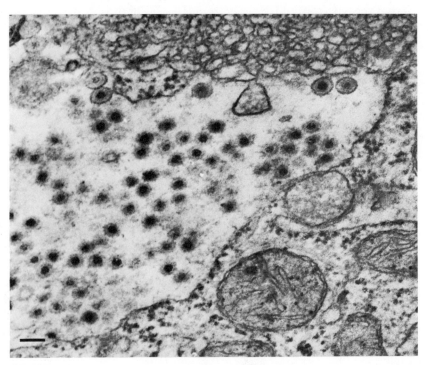

FIGURE 6. Portion of the cytoplasm of a human rotavirus–infected duodenal epithelial cell showing a mass of convoluted smooth membrane beside a cisterna of the RER, which contains enveloped and unenveloped virus particles and amorphous material. Scale bar: 100 nm. From Holmes *et al.* (1975), by permission of the American Society for Microbiology.

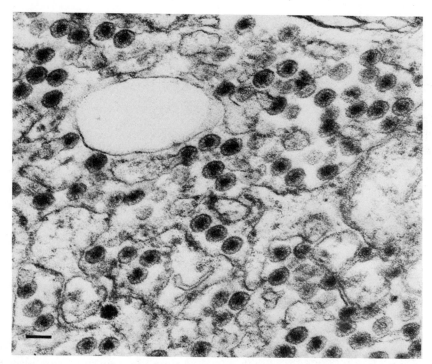

FIGURE 7. SA 11 rotavirus morphogenesis in tunicamycin-treated MA-104 cell. Note that all the particles in the endoplasmic reticulum vesicles retain their "pseudoenvelopes." Scale bar: 100 nm. Kindly provided by M.L. Dyall-Smith.

one or another of the rotavirus glycoproteins; at any rate, as Fig. 7 shows, if the infected cells are tunicamycin-treated, the pseudoenvelopes remain around all the particles within RER vesicles (M.L. Dyall-Smith, personal communication), and following lysis of the cells, only single-capsid particles are found. That tunicamycin-treated cells fail to produce complete double-shelled particles (but not the mechanism of the effect) has been noted independently by N. Ikegami, M. Petric, and R.T. Espejo (personal communications).

As the micrographs (Fig. 6 and 7) show, the RER vesicles in which rotaviral particles are found also contain considerable amounts of amorphous electron-dense material that could be viral outer-capsid components, and the eventual fate of the psuedoenvelope membranes may be to form the masses of convoluted smooth membrane (Fig. 6) that are another previously unexplained feature of rotavirus morphogenesis. They adjoin or even protrude into rotavirus-containing vesicles (Saif *et al.*, 1978) and could be produced by a membrane-fusion or -destabilization process.

In the normal (i.e., not tunicamycin-treated) rotavirus-infected cell, it is thus postulated that inner capsids bud into RER vesicles, where their pseudoenvelopes are removed, so that the outer-capsid components, which have been secreted separately through the RER membrane, can assemble onto them. Thus complete double-shelled particles are formed free in the lumen of the RER where the calcium ion concentration is sufficient to allow the assembly to occur.

## V. GENETICS

### A. Isolation of Temperature-Sensitive Mutants

Temperature-sensitive (*ts*) mutants of the UK strain of bovine rotavirus have been isolated by Greenberg *et al.* (1981) following growth of the virus in AGMK cells in the presence of 200 or 400 µg per ml of 5-azacytidine. Clones were selected that grew and produced cytopathic effect at 34°C, but not at 39°C. Mutants were assigned to one of four recombination groups (A–D) on the basis of experiments in which pairs of mutants were grown together in AGMK or CV-1 cells for 36 hr at 34°C; yields were then assayed at 39°C and compared with the yields from similar cultures infected singly with each mutant. Recombination indices ranged from about 20 to greater than 1000. In all, seven mutants were obtained, with efficiencies of plating (EOPs) (39°C/34°C) between 0.002 and less than 0.00006. No biochemical characterization of the *ts* lesions in these mutants has yet been reported, but they have been used most successfully in the production of reassortment viruses, which are discussed in Section V.B.

Faulkner-Valle *et al.* (1982) have also isolated a number of *ts* mutants of UK bovine rotavirus which could be classified into five recombination groups. Representatives of two of the groups were unable to synthesize dsRNA or polypeptides at 39.5°C, but the others appeared to synthesize both RNA and polypeptides at the restrictive temperature. Ramig (1982) has reported the isolation of ten stable *ts* mutants of simian rotavirus SA 11 following mutagenesis with nitrous acid or hydroxylamine. They also fell into five recombination groups but the nature of the *ts* lesions has not yet been investigated.

M.E. Begin and S.M. Rodger (personal communication) have also isolated *ts* mutants of the Northern Ireland strain of bovine rotavirus and of simian rotavirus SA 11 following treatment of inocula with nitrous acid (to 10% survival) or growth in the presence of 200 µg/ml 5-azacytidine, in which case yields were about 1% of the normal. Both mutagens gave similar results; frequencies of isolation of *ts* mutants were low, only about 0.1% of clones tested had EOPs (39°C/34°C) of less than $10^{-3}$, and were thus candidates for biochemical characterization. Four recombination groups have been identified, and for this purpose interstrain recombination appeared to be as efficient as intrastrain recombination (C.P.

Hum, personal communication). Studies of *ts* functions are currently in progress.

## B. Genome Reassortment

By coinfection of MA-104 cells with bovine rotavirus (Lincoln strain) and UV-irradiated SA 11 virus, followed 24 hr later by plaque selection under an overlay containing anti-bovine rotavirus serum, Matsuno *et al.* (1980) obtained the first rotavirus reassortant clone. Electrophoretic comparisons of its double-stranded RNA (dsRNA) genome segments with those of the parental viruses showed that the reassortant had derived segments 4, 5, and 10 from SA 11 and segments 1, 2, 3, 6, and 11 from the bovine rotavirus, but it was not possible to determine the origin of segments 7, 8, and 9. By neutralization and hemagglutination-inhibition tests, the reassortant resembled SA 11 virus.

Greenberg *et al.* (1981, 1982) were able to obtain cultivable (i.e., cell-culture-adapted) reassortants that behaved in neutralization tests like human rotavirus strains, by mixed infection of AGMK cells with *ts* mutants of cultivable bovine rotavirus and noncultivable human rotaviruses. Following growth at 34°C, reassortants were selected by plaquing at 39°C under bovine rotavirus antiserum. The human strains thus "rescued" were Wa and DS-1, which represent different serotypes, so the reassortants have been most useful in studies of human rotavirus serotyping. These and additional reassortants produced by similar methods were employed in determining the genes of human and bovine rotaviruses that code for neutralization and subgroup antigens by Kalica *et al.* (1981b). This study was mentioned in Section II.F and is further discussed in Section V.C.

## C. Gene Coding Assignments

Following the method of *in vitro* translocation of separated, denatured dsRNA segments that had been applied to reovirus by McCrae and Joklik (1978), M.L. Smith *et al.* (1980) identified the polypeptides encoded by dsRNA segments 1–6 of SA 11 rotavirus, and later the coding assignments of segments 10 and 11 were added (Dyall-Smith and Holmes, 1981a). Assignments for the remaining three segments have recently been completed (P. Kantharidis and M.L. Dyall-Smith, personal communication; G.W. Both and A.R. Bellamy, personal communication). The correlation between SA 11 genome segments and proteins is illustrated in Figure 8.

Arias *et al.* (1982) tackled the same problem using similar techniques, except that *in vitro* protein synthesis was carried out with a rabbit re-

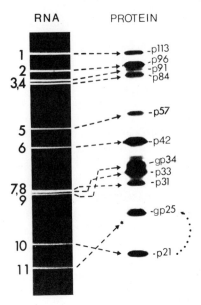

FIGURE 8. Genome coding assignments of simian rotavirus SA 11, showing the relationships between dsRNA segments as separated by electrophoresis in polyacrylamide gel and polypeptides detected in SA 11–infected cell cultures by [$^{35}$S]methionine labeling at 7.5 hr postinfection, gel electrophoresis, and fluorography. Kindly provided by M.L. Dyall-Smith.

ticulocyte lysate instead of a wheat germ preparation. There was complete agreement regarding the assignments for segments 1, 2, 5, 6, and 10, and although partial separation of segments 7–9 was achieved, it remained impossible to assign their three protein products individually. Arias *et al.* (1982) considered that the outer capsid protein that is cleaved by trypsin (Espejo *et al.*, 1981) (see Section III.B) was the product of genome segment 3, whereas it clearly appeared to be produced by segment 4 in the earlier study (M.L. Smith *et al.*, 1980). This disagreement may arise from a strain difference that has arisen during passaging of the SA 11 rotavirus stocks. Although they originally came from the same source, minor differences in the electrophoretic mobilities of certain dsRNA segments and of gp34 do seem to have evolved.

Although Arias *et al.* (1982) confirmed the finding of Dyall-Smith and Holmes (1981a) that segment 10 of SA 11 codes for what they respectively called pNCVP5 and NS$_3$, they further proved that it is the precursor of the glycoprotein gp25 (NCVP5 or O$_4$). Thus, the conclusion of Dyall-Smith and Holmes (1981a) that the segment 11 product was converted to gp25 must be incorrect. Instead, it must be a separate protein (molecular weight 24,000) that was seen as an *in vitro* translation product but was not detected in infected cells by Dyall-Smith and Holmes (1981a). For unknown reasons, Arias *et al.* (1982) were unable to translate segment 11 *in vitro*, so for SA 11 virus some uncertainties remain, but the situation is clearer for bovine rotavirus.

By translation of denatured dsRNA segments of the UK strain of bovine rotavirus, McCrae and McCorquodale (1982) showed that segments 1–6 coded for the proteins VP1–6 (see Table I for equivalents), as in SA 11 virus. In bovine as in the simian rotavirus, segment 10 codes for the precursor (VP12) of the glycoprotein VP10, and in this case the product of segment 11, VP11, was detected in infected cells. In the bovine rotavirus also, segments 7, 8, and 9 were very difficult to resolve, but McCrae and McCorquodale (1982) consider that they encode VP8, the precursor of VP7, and VP9, respectively.

A peculiarity of the RNA electrophoretic patterns of human rotaviruses is that the smallest segments 10 and 11 of some strains appear to move more slowly than those of others that resemble simian, bovine, and other mammalian rotaviruses (Espejo et al., 1979; Kalica et al., 1981a). Segment 11 of strains with "short" electrophoretic patterns lines up approximately with segment 10 of strains with "long" patterns, and in fact both code for the same kind of protein, corresponding to p21 of SA 11 virus (Dyall-Smith and Holmes, 1981a). Similarly, segment 10 of a strain with a "short" pattern is analogous to segment 11 of a "long" pattern, and it is the considerable variation in mobility of this segment that gives rise to the different patterns.

As mentioned in Section II.F, studies of the neutralization type specificities and of the parental origins of genome segments of a series of reassortants between human (Wa) and bovine (UK) rotaviruses enabled Kalica et al. (1981b) to identify human rotavirus RNA segment 9 as the carrier of neutralization specificity for this human strain. Since in SA 11 rotavirus neutralizing monoclonal antibodies react with the outer-capsid glycoprotein gp34 (A. Breschkin and S. Sonza, personal communication), it is reasonable to assume that in the human (Wa) rotavirus, RNA segment 9 thus codes for the corresponding glycoprotein, gp35 (Dyall-Smith and Holmes, 1981a). In the case of the human rotavirus strain DS-1, it appears to be segment 8 that encodes the neutralization antigen (H. Greenberg et al., personal communication), so it is apparent that the relative positions of RNA segments 7–9 can vary among rotavirus strains. Corresponding dsRNA segments in different rotavirus strains can now be identified by Northern blot hybridization carried out under conditions of low stringency, using DNA copies of segments 7–9 of UK bovine rotavirus cloned in the plasmid PBR 322 (M.L. Dyall-Smith, A.A. Azad and I.H. Holmes, personal communication).

## VI. BIOLOGY OF ROTAVIRUSES

### A. Diseases and Pathogenesis

Rotaviruses typically cause acute diarrheal disease in infants or the young of many kinds of animals (For review see Flewett, 1977; Flewett and Woode, 1978; McNulty, 1978). The first interspecies transmission of a rotavirus was probably achieved by Light and Hodes (1949) in the course of an investigation of severe infantile gastroenteritis in Baltimore in 1941–1942.

"Scours" (e.g., white scours, milk scours, etc.) is an acute infectious diarrheal disease that very commonly affects calves and piglets and sometimes foals and intensively-reared lambs. Rotaviruses appear to be the most common causative agents, although similar symptoms are also produced by coronaviruses and certain bacteria, and the most severe disease may be caused by multiple infections (Tzipori et al., 1981). In calves and pigs particularly, rotaviral enteritis causes severe economic losses both through deaths and through retardation of weight gain in animals that recover.

## 1. Epizootic Diarrhea of Infant Mice

Epizootic diarrhea of infant mice (EDIM) or murine rotavirus infection was described by Cheever and Mueller (1947) and Kraft (1958). The disease can be a major problem in mouse-breeding units, since it is highly contagious and hard to eradicate. The incubation period is about 2 days. The symptoms are yellow, watery diarrhea and dehydration in sucklings 3–16 days old. Susceptibility to symptomatic disease depends markedly on age (Kraft, 1957; Wolf et al., 1981), and adult mice get asymptomatic infections. Since all individuals in the litter are affected, the task of cleaning up gets beyond the mother mouse, and the sucklings frequently end up stuck together by their dried excreta in rather bizarre rosettes. Losses can be high, although in endemically infected colonies, most infant mice recover within 4–5 days.

## 2. Calf Scours

Bovine rotaviral enteritis or calf scours has been described by Mebus et al. (1969), Mebus (1976), and Woode and Bridger (1975). The incubation period in experimental infections with large doses of virus can be as short as 13 hr. In order of appearance, the symptoms are depression, anorexia, and diarrhea with white or yellow feces for 6–12 hr. The calves may die as a result or dehydration or secondary bacterial infection, or recover in 3–4 days. Natural rotavirus infections appear to be most common in the 2nd week of life (Acres and Babiuk, 1978).

## 3. Piglet Enteritis

Rotaviral diarrhea in pigs has been reviewed recently by Bohl (1979). The symptoms resemble those caused by transmissible gastroenteritis of swine (a coronavirus), but are generally less severe, although fatal dehydration can occur in very young piglets. They include anorexia, diarrhea, and vomiting, after a 1 to 2-day incubation period (Lecce et al., 1976; McNulty et al., 1976d; Woode et al., 1976a). Illness occurs in piglets

1–8 weeks of age, often soon after weaning, especially if it is early or abrupt (Lecce and King, 1978). Weight gains are severely depressed.

## 4. Disease in Other Domestic Animals and Birds

Equine rotaviruses are associated with diarrhea in foals that was previously attributed to hormonal changes in the mares that suckled them (Flewett et al., 1975). A number of fatal infections of very valuable animals have been reported (Tzipori and Walker, 1978; Conner and Darlington, 1980). In lambs, symptomatic infections seem to occur only under conditions of intensive rearing under cover (McNulty et al., 1976b; Snodgrass et al., 1976), although antibodies in sheep are widespread in countries such as Australia where the disease has not been noticed. In chickens and turkeys, at least three serological types of rotaviruses have been associated with symptoms of diarrhea, poor or abnormal appetite, dehydration, and increased mortality (Jones et al., 1979; McNulty et al., 1979, 1980). The chicken pararotavirus (McNulty et al., 1981) does not appear to cause severe disease.

## 5. Human Infantile Enteritis

### a. Clinical Features

The clinical features of human rotaviral enteritis have been reported in considerable detail (Davidson et al., 1975a; Flewett, 1977; Shepherd et al., 1975; Hamilton et al., 1976; Rodriguez et al., 1977; Delage et al., 1978). After an incubation period of 1–2 days, vomiting and then diarrhea occurs, and there is generally a mild fever (38–39°C). The diarrhea lasts for 4–5 days, and moderate dehydration is common. Severe illness is most common in the 6- to 18- month age group. Deaths due to rapid, severe dehydration occur even in centers where fluid–electrolyte therapy is available (Carlson et al., 1978), and the disease is believed to be a major cause of infant mortality in many parts of the world (Kapikian et al., 1980).

On the other hand, it must be remembered that the clinical descriptions have almost always been based on patients admitted to hospital, and the bulk of rotavirus infections in children are milder. Asymptomatic infections in neonates seem to be common (Cameron et al., 1975; Madeley and Cosgrove, 1975; Murphy et al., 1977; Chrystie et al., 1978). An association of upper respiratory tract symptoms with rotaviral enteritis has been suggested (Carr et al., 1976; Lewis et al., 1979), but others have found an equally high incidence of respiratory symptoms in age-matched children with diarrhea due to other causes (Rodriguez et al., 1977; Mäki, 1981). Like adenoviruses, rotaviruses may be a cause of intussusception in infants and young children (Konno et al., 1978b), though this is still under investigation. Long-term diarrhea with rotavirus excretion for more than 6 weeks and rotaviral antigens detectable in the serum has been reported in immunodeficient children (Saulsbury et al., 1980). In many places, multiple infections with rotavirus and other enteric pathogens can be quite common (Evans et al., 1977), but it is not yet known whether they are significantly more severe than single infections.

### b. Pathogenesis and Pathology

Only limited investigations of the pathogenesis and pathology of rotavirus disease in humans have been possible, but many detailed studies of natural and experimental rotaviral infections in other animals have been reported, and all show very similar changes, so it is possible to generalize (McNulty, 1978; Mebus et al., 1971b; Mebus and Newman, 1977; Pearson and McNulty, 1977; Snodgrass et al., 1977a; Theil et al., 1978). Rotaviruses infect epithelial cells lining the apical halves of villi of the small intestine (see Section III.A), causing vacuolation and premature shedding, so the villi are shortened and become covered with immature, cuboidal epithelial cells from the crypts, which appear to be insusceptible. The infection may progress from the upper to the lower small intestine. By immunofluorescence (IF), infection of epithelial cells in the colon has also been found in mice, humans, and lambs, but not in calves and only rarely in pigs (Banfield et al. 1968; Hamilton et al., 1976; Snodgrass et al., 1977a; Theil et al., 1978). Histopathological changes in the lungs of pigs, lambs, and calves have also been noted, but no infected cells have been demonstrated outside the intestine (Mebus, 1976; McNulty, 1978).

Similar pathology was found in experimental infections of pigs with bovine rotavirus or human rotavirus and of calves with human rotavirus (Hall et al., 1976; Davidson et al., 1977; Mebus et al., 1977). The immature epithelial cells that replace those destroyed by the virus have reduced levels of disaccharidases, including lactase, and glucose-coupled sodium transport is also impaired (Davidson et al., 1977; Davidson and Barnes, 1979). Undigested lactose in milk exerts an osmotic effect, inflammation and bacterial degradation of the lactose in the large bowel compound this, and diarrhea is the end result (Flewett, 1977).

The villi return to normal within 3–4 weeks. Two stages in recovery were demonstrated by Crouch and Woode (1978) in a most interesting study in which porcine rotavirus was titrated and histopathology noted at five points along the small intestine of piglets, at various intervals after experimental infection. It was clear that after a peak of infectivity and of epithelial cells showing IF at about 20 hr postinfection, the infectivity titer dropped sharply from about $10^8$ median tissue-culture infectious doses ($TCID_{50}$)/ml to about $10^5$ $TCID_{50}$/ml, and the percentage of fluorescing cells dropped from 75 to 5% by about 40 hr postinfection. The lower levels were maintained until about 4 days postinfection, when the infectivity titer dropped to zero. It was suggested that the first phase of recovery was nonimmune in nature, and likely to be due to the loss of susceptible cells and possibly the production of interferon. Final elimination of the virus in the second phase probably followed the production of local antibodies.

## B. Geographic and Host Range

It was once interesting to record the countries from which rotaviruses had been reported, but now the lists are so long, for both human and bovine rotaviruses especially, that one can only say the distribution is worldwide. Even the most isolated human populations have antibodies as evidence of past infections, and occasional epidemics occur (S.O. Foster et al., 1980; Linhares et al., 1981).

Similarly, it appears likely that rotaviruses are to be found in all species of mammals surveyed and in chickens and turkeys, the only two avian species that have been studied. Curiously, although rotaviruses in mice were among the first to be found (Cheever and Mueller, 1947), they have not been reported in rats. Discoveries of rotaviruses in calves, pigs, lambs, and foals have been mentioned previously (Mebus et al., 1969; Rodger et al., 1975a; Lecce et al., 1976; McNulty et al., 1976b,d; Woode et al., 1967a; Snodgrass et al., 1976; Flewett et al., 1975).

Rotaviruses appear to be common in rabbits (Bryden et al., 1976; Petric et al., 1978), but not in guinea pigs. They have been found in apes and monkeys (Ashley et al., 1978; Soike et al., 1980; Stuker et al., 1980), cats (Chrystie et al., 1979; Snodgrass et al., 1979), dogs (Eugster and Sidwa, 1979), antelopes, deer, and other ruminants in a zoo (Reed et al., 1976; Tzipori and Caple, 1976), and goats (Scott et al., 1978). Finally, a variety of chicken and turkey rotaviruses have been found (Jones et al., 1979; McNulty et al., 1979).

A number of experimental, cross-species transmissions have been reported. No attempts to infect mice with rotaviruses from other species have been successful, but piglets are susceptible to bovine, equine, simian, and human strains (Woode and Bridger, 1975; Hall et al., 1976; Woode et al., 1976b; Tzipori and Williams, 1978; Bridger et al., 1975; Middleton et al., 1975; Torres-Medina et al., 1976). Human rotaviruses have also been transmitted to calves (Light and Hodes, 1949; Mebus et al., 1976), lambs (Snodgrass et al., 1977b), monkeys (Wyatt et al., 1976b), and dogs (Tzipori, 1976).

The latter transmission may be particularly significant from the human point of view, since the puppies appeared to be highly susceptible to the human virus, opportunities for virus exchange between dogs and children are obvious in almost all parts of the world, and a recent random sampling revealed a surprisingly high prevalence of rotavirus excretion by dogs (Roseto et al., 1980). Naturally enough, familiarity with the epidemiology of influenza leads one to an interest in the possibility of zoonotic rotavirus infections of humans. Although no quantitative comparisons have yet been made of minimal infective doses or relative susceptibilities, the current general impression is that rotaviruses are moderately, though certainly not absolutely, host-specific in nature.

## C. Epidemiology

### 1. Mode of Transmission

Since rotaviruses replicate in the cells lining the intestinal tract, feces are the source of subsequent infections, and they contain very large numbers of rotavirus particles. In humans, estimates of peak excretion soon after the onset of diarrhea range from about $10^9$ to more than $10^{10}$ particles/g feces (Davidson et al., 1975a; Chrystie et al., 1978), and a sample of piglet feces contained at least $3 \times 10^7$ infectious doses/ml (Woode et al., 1976a). In addition, the studies described in Section II.B showed that rotaviruses are rather stable, and their infectivity would be preserved in the feces of young animals because of the high calcium content (Shirley et al., 1981). As a result, there is more than one possible mode of transmission, and it is difficult to determine which of them, if any, is most important overall.

Airborne spread of dust or droplets carrying virus was postulated by Kraft (1957) to explain the efficient spread of murine rotavirus from cage to cage within rooms despite all precautions to avoid cross-contamination. The addition of filter tops to the cages prevented the accidental transmission and confirmed her hypothesis (Kraft, 1958). Mebus et al. (1969) experimentally infected a calf with bovine rotavirus via an aerosol spray, but it was clear that at least part of the inoculum would be swallowed, and they later showed that there was no evidence of infection of the upper respiratory tract (see Section VI.A.5.b). Airborne spread of human rotaviruses in hospitals, perhaps associated with diaper-changing, has been suggested by Chrystie et al. (1975) and Campbell and Lang (1979). The high attack rate and speed of spread of a human rotavirus epidemic in the Truk Islands also led S.O. Foster et al. (1980) to consider airborne transmission, although they favored the idea of multiplication in the respiratory tract.

Circumstantial evidence of transmission of human rotavirus infections via water has been obtained by Lycke et al. (1978), Vollet et al. (1979), Linhares et al. (1981), Zamotin et al. (1981), and Sutmoller et al. (1982). In the case of an epidemic in a small Swedish town that affected over 3000 people of all ages, it appeared that the freshwater supply was contaminated with sewage effluent (Lycke et al., 1978). It is known that infectious rotaviruses can survive for considerable periods in water (Hurst and Gerba, 1980), and they are highly resistant to chlorination (Snodgrass and Herring, 1977; Tan and Schnagl, 1981). They also seem to be adsorbed less by soil than enteroviruses, and thus can circulate via groundwater and be found in wells (Keswick and Gerba, 1980; Murphy et al., 1983).

Transfer of rotaviruses by direct contact with infected individuals or a contaminated environment is the idea favored by most workers dealing with domestic animals, and it also has its supporters among those investigating nosocomial infections in hospitals, where rotaviral enteritis is a considerable and costly problem (Middleton et al., 1977b; Ryder et al., 1977). In a study of diarrhea and rotavirus infection associated with differing regimens for postnatal care of newborn babies, Bishop et al. (1979) found that the incidence of diarrhea was far higher in babies in communal nurseries than in those "rooming in" with their mothers, and stated that the most important factors influencing incidence of diarrhea were proximity to other newborn babies and frequency of handling by unrelated adults. In the same vein, Lecce et al. (1978) described an artificial rearing regimen for piglets involving early weaning in which a buildup of rotavirus occurred repeatably over several weeks, resulting in earlier and more severe infections in the later arrivals.

The efficacy of increased hand-washing in reducing incidence of diarrhea in circumstances where rotavirus infections are common lends support to the direct-contact hypothesis (Ryder et al., 1977; Koopman, 1978; Halvorsrud and Orstavik, 1980; Black et al., 1981a).

## 2. Incidence of Infections

In general, symptomatic rotaviral infections occur mainly in neonatal or very young animals, but repeated infections occur at intervals throughout life, often without symptoms (McNulty, 1978; Holmes, 1979). In contrast to the situation in other animals, and for unknown reasons, human neonatal infections are frequently mild or asymptomatic (Cameron et al., 1978; Chrystie et al., 1978; Jesudoss et al., 1979), whereas the most serious infections occur in the 6-month to 2-year age group (Davidson et al., 1975a; Bryden et al., 1975; Kapikian et al., 1976b). The actual incidence of rotavirus gastroenteritis requiring hospitalization in a defined population in the United States was found to be 2–3/1000 children aged 0–2 years per year (Rodriguez et al., 1980). In a prospective study in which a cohort of 104 infants and their families were followed for an average of 16 months, Gurwith et al. (1981) found that rotavirus was by far the most common enteropathogen. It was detected 82 times in 72 children. Altogether, 237 episodes of diarrhea were investigated in the 104 infants and 62 siblings.

Symptomatic infections are less common in older children, but epidemics occur in schoolchildren (Hara et al., 1978). Adult diarrheal infections tend to be vastly underreported, but both symptomatic and asymptomatic rotavirus infections with quite high attack rates have been documented in parents, hospital personnel, and other adults in contact with pediatric patients (Kim et al., 1977; Kapikian et al., 1976b; von

Bonsdorff et al., 1976; Rodriguez et al., 1979; Wenman et al., 1979), and epidemics have occurred even in the elderly (Cubitt and Holzel, 1980; Harvorsrud and Orstavik, 1980). Numerous diarrheal adult infections were a feature of epidemics in isolated populations (S.O. Foster et al., 1980; Linhares et al., 1981), and rotaviruses are a significant cause of diarrhea in adult travelers (Bolivar et al., 1978; Vollet et al., 1979; Ryder et al., 1981; Sheridan et al., 1981).

Seasonal and age factors are compounded in domestic animals with controlled breeding seasons, but in temperate climates there is a winter peak of rotavirus infections in both mice and humans (Cheever, 1956; Middleton et al., 1974; Davidson et al., 1975a; Kapikian et al., 1976b; Konno et al., 1978a). The reason for this is unknown, and in the tropics there is no consistent pattern; infections frequently occur throughout the year. Reported variations in incidence in dry and wet seasons are not consistent from country to country and may not be consistent in a single country from year to year (Hieber et al., 1978; Maiya et al., 1977; Soenarto et al., 1981; V.I. Mathan, personal communication).

The level of immunity, especially passive immunity in young animals, will obviously affect the incidence of infection. This has been noticed in mice, where successive litters of the same mother show milder symptoms (Cheever, 1956), and in humans, where there is a reduced incidence of infections in breast-fed as opposed to bottle-fed babies, but discussion of this will be left for Section VI.E. Weanling diarrhea was recognized as a problem in humans long before rotaviruses were known (Gordon et al., 1963), and in both humans and pigs, the shock of weaning and increased possibility of contamination of food may be as important as the lack of mother's milk in the observed increase of disease (Bishop et al., 1979; Lecce and King, 1978).

### 3. Strain Variation

A further complication of rotavirus epidemiology, which is gradually being taken into account as techniques improve, is the existence of different serotypes of rotaviruses. Strains and serotypes of rotaviruses are discussed in the next section, but it is necessary to mention here publications on the epidemiology of human rotavirus subgroups and the use being made of double-stranded RNA (dsRNA) electrophoretic patterns in epidemiological studies. Since the distinction between rotavirus subgroups and serotypes was not drawn until recently (Kapikian et al., 1981) and the methods available for serotype identification were formerly too difficult, the techniques applied in these studies [mainly enzyme-linked immunosorbent assay (ELISA)] were those that recognize subgroups. This must be kept in mind, since at the time the authors considered they were dealing with serotypes, and probably they frequently were, since there is at least a partial correlation between serotype and subgroup among human rotaviruses (see Section VI.D).

Almost as soon as it had been shown in longitudinal studies that individual children frequently experienced more than one rotavirus episode within their first couple of years (Wyatt *et al.*, 1979b), it was found that sequential infections were often caused by viruses of different subgroups (Fonteyne *et al.*, 1978; Rodriguez *et al.*, 1978). Surveys conducted over 4 years in Washington, D.C., showed that although antibodies to viruses of subgroups 1 and 2 were acquired at equal rates, subgroup 2 rotaviruses were associated with 75% of the illnesses that required hospital treatment (Yolken *et al.*, 1978b; Brandt *et al.*, 1979). Thus, it was suggested that the subgroup 2 viruses were more virulent than those of subgroup 1, but the difference may not be so great as it appeared, since strains belonging to a third subgroup may have been included with subgroup 2 (Zissis *et al.*, 1981).

As mentioned in Section II.C, gel electrophoresis of genome RNA was rapidly recognized as a very useful method for distinguishing different rotavirus strains for epidemiological purposes. It is more discriminating than serology, and will remain invaluable for studies of the evolution and spread of new epidemic strains even now that serotyping is becoming easier, as has been found in studies of influenza (Hinshaw *et al.*, 1978).

The method was first applied with enthusiasm by Espejo *et al.* (1977, 1979, 1980a,c), and a number of similar studies have now been published (Kalica *et al.*, 1978b; Croxson and Bellamy, 1979; Rodger *et al.*, 1981; Lourenco *et al.*, 1981; Schnagl *et al.*, 1981). Figures 9 and 10 show the

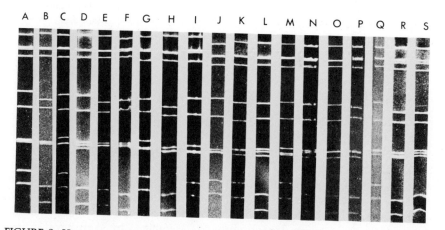

FIGURE 9. Human rotavirus "electropherotypes" as defined by differing dsRNA segment patterns produced by electrophoresis of rotaviral RNA extracted from a total of 335 fecal samples from children and neonates from various Melbourne hospitals over a 6-year period. Small differences cannot be judged by comparison with a composite picture such as this containing results from a number of gels, which is intended only to convey a general idea of the variations; samples producing similar patterns must be coelectrophoresed to establish identity or distinctness. From Rodger *et al.* (1981), by permission of the American Society for Microbiology.

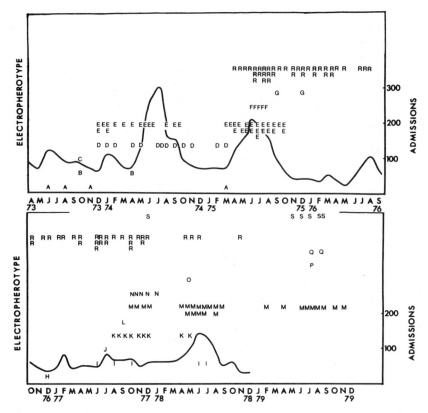

FIGURE 10. Occurrence of human rotavirus electropherotypes obtained from children in Melbourne hospitals between May 1973 and November 1979. (———) Total number of children admitted to these hopsitals with acute gastroenteritis (all causes). From Rodger *et al.* (1981), by permission of the American Society for Microbiology.

range of "electropherotypes" found in Melbourne, Australia, and their temporal distribution over 6 years (Rodger *et al.*, 1981). In the same study, it was noted, and subsequently confirmed in more detail by Kalica *et al.* (1981a), that rotaviruses with genomes that gave short electrophoretic patterns (such as those labeled A, C, G, I, J, K, M, and Q in Fig. 9) are found to belong to subgroup 1, while the long patterns correlate with subgroup 2 [or the putative subgroup 3 (G. Zissis and R. Bishop, personal communication)]. The correlation may not always hold because it is indirect and may be coincidental, since, as discussed in Sections II.F and V.C, the subgroup antigens depend on segment 6 rather than segment 11. Since so many fine variations of mobility occur among segments 7–9, it is not possible to read off serotype from the RNA electrophoretic pattern.

It is clear from Fig. 10 that a number of different rotaviruses can be present simultaneously in a large population [or even in a much smaller one, as was found by Schnagl *et al.* (1981)], that some types are much more common in hospitalized patients than others, which were rarely seen, and that predominant types can persist through two or three annual epidemic periods.

## D. Strains and Serotypes

### 1. Strains

The origins and nomenclature of the first cultivable rotavirus strains are listed in Table II. Now that cultivation is no longer a novelty, the number of true isolates is rapidly increasing, and a standardized system of nomenclature is urgently needed. For this reason, I have included cryptograms similar to those used to identify influenza isolates, as suggested by Rodger and Holmes (1979). It is not yet possible to incorporate neutralization serotype or subgroup designations, but these can readily be added in future as they are determined.

### 2. Serotypes

Shortly after it became possible to carry out infectivity assays on bovine rotaviruses, it was noted that some sera from humans and pigs,

TABLE II. Prototype Cultivable Rotaviruses

| Origin | Strain designation | Cryptogram | Ref. No. |
|---|---|---|---|
| Simian (vervet) | SA 11 | Si/S.Africa/H96/58 | 1 |
| Bovine (or ovine) | "O" agent | Bo/S.Africa/9/65 | 1 |
| Bovine | NCDV Lincoln | Bo/USA/23/69 | 2 |
| | NCDV Cody | Bo/USA/2/67 | 2 |
| | U.K. | Bo/England/27/73 | 3 |
| | N.I. | Bo/N.Ireland/447/75 | 4 |
| Porcine | OSU | Po/USA/ /75 | 5 |
| Human | DS-1 | Hu/USA/G621/76 | 6 |
| | Wa | Hu/USA/G187/74 | 7 |
| | M | Hu/USA/G529/76 | 8 |
| Simian (rhesus) | RRV-2 | Si/USA/2/79 | 9 |
| Canine | K-9 | Cn/USA/36/79 | 10 |
| Feline | Taka | Fe/USA/11/80 | 11 |
| Turkey | Ty1 | Ty/N.Ireland/415/79 | 12 |
| Chicken | Ch1 | Ch/N.Ireland/363/78 | 12 |
| Turkey | Ty3 | Ty/N.Ireland/78/79 | 13 |
| | 132 | Ty/N.Ireland/132/80 | 14 |

[a] References: (1) Malherbe and Strickland-Cholmley (1967); (2) Mebus et al. (1971a); (3) Bridger and Woode (1975); (4) McNulty et al. (1976a); (5) Theil et al. (1977); (6) Rodriguez et al. (1978); (7) Wyatt et al. (1980); (8) Kalica et al. (1978b); (9) Stuker et al. (1980); (10) Fulton et al. (1981); (11) Hoshino et al. (1981); (12) McNulty et al. (1979); (13) McNulty et al. (1980); (14) McNulty et al. (1981).

as well as bovine sera, contained neutralizing antibodies against them (Woode et al., 1976b; Matsuno et al., 1977a). The first series of cross-neutralization tests involving human, bovine, porcine, equine, ovine, murine, and lapine rotaviruses and corresponding convalescent antisera were performed by Thouless et al. (1977b). They were able to distinguish rotaviruses of different species origin, although the differences between neutralization titers of homologous and heterologous sera were often only 4- to 8-fold. Murine rotavirus was neutralized to a significant degree by almost all the sera, and it was conjectured that since only single-capsid particles were seen in highly infectious preparations, perhaps in the case of murine rotavirus such particles are infectious and can be neutralized by antibodies to the common inner-capsid antigens.

It has since been realized that in contrast to observations made on other viral genera, convalescent sera show a broader cross-reactivity by neutralization than hyperimmune sera prepared by repeated injections of particular rotaviruses into guinea pigs or rabbits, provided the animals do not have preexisting antibodies to other rotaviruses (Wyatt et al., 1980; T.H. Flewett, personal communication; G.N. Woode, personal communication). Preexisting antibodies were noted in the rabbits immunized by Malherbe and Strickland-Cholmley (1967) against simian rotavirus SA 11 and the "O" agent. These sera showed a high degree of cross-neutralization, whereas similar sera prepared in rotavirus antibody–free rabbits by Lecatsas (1972) showed no cross-reaction at all. Similarly, Linhares et al. (1981) noted a boost of anti-type 3 antibodies in Tiriyo Indians following an outbreak of Birmingham type 1 human rotavirus (see below). As yet, these observations have not been followed up, but they need to be borne in mind by anybody undertaking the production of sera for specific identification of rotaviruses.

The original cultivable bovine rotaviruses are closely related by neutralization tests, and no information is available on porcine isolates, but two serotypes of turkey rotavirus have been identified (McNulty et al., 1980).

Human rotavirus serotyping turned out to be more complex than was realized at first, and it was especially difficult since it has only recently become possible to grow stocks of prototype strains. By neutralization of fluorescent foci produced by nonadapted human rotavirus samples, Thouless et al. (1978) in Birmingham, England, were able to distinguish two or three serotypes, and they later confirmed the third and suggested that a fourth or even fifth may exist (Flewett et al., 1978; Beards et al., 1980).

Another approach was originated by Zissis and Lambert (1978), who found that hyperimmune guinea pig antisera prepared against some human rotaviruses reacted poorly against others by complement fixation (CF). Using sera prepared by Zissis against his "types" 1 and 2, Yolken et al. (1978b) were able to make the same discrimination by ELISA, and Zissis and Lambert (1980) devised their own ELISA system that provided a slightly clearer distinction. Recently, Zissis *et al.* (1981) produced a third antiserum against a human rotavirus isolate that could not be typed by their previous sera, and have thus identified a third group of strains, some of which were previously considered to be "type" 2.

As mentioned earlier (Sections II.F and VI.C.3), experiments on a range of human and animal rotaviruses and on cultivable reassortants carrying human rotavirus serological specifities recently showed that the antigens detected by neutralization were distinct from those identified by ELISA or immune adherence hemagglutination assay (IAHA) (Kapikian *et al.*, 1981). Accordingly the "type" specificities distinguished by the current ELISA or IAHA tests will henceforth be defined as subgroups 1, 2, and 3, while the antigens distinguishable by neutralization will define the serotypes. It is too early to attempt a tabulation of strains according to this new nomenclature, since insufficient information has been published, but it appears that human rotavirus strains with the "Wa" or Birmingham No.1 serotype usually also have the subgroup 2 antigen, whereas the "DS-1" serotype (Birmingham 2 or 4) is associated with subgroup 1 antigen (Wyatt *et al.*, 1980; Greenberg *et al.*, 1981; Kapikian *et al.*, 1981; T.H. Flewett, personal communication). A third human serotype, "M" or Birmingham No. 3, is particularly interesting in that it seems to be closely related to simian rotaviruses SA 11 and RRV-2 and to some other rotaviruses of animal origin (Wyatt *et al.*, 1982; A.Z. Kapikian, T.H. Flewett, and G.N. Woode, personal communications). Whether it is associated in human strains with the subgroup 3 antigen of Zissis *et al.* (1981) is not yet known. At least one recent Japanese isolate of human rotavirus appears to differ from strain Wa by neutralization (Sato *et al.*, 1981).

Although research on the "pararotaviruses" is still in its early stages, it appears so far that they differ in both serotype and subgroup specificities from other rotaviruses, and no cross-protection was observed between the C or S strains of porcine pararotavirus and the classic porcine rotavirus OSU (Bohl *et al.*, 1982).

## E. Immunity

It has been shown that in young animals, colostrally derived serum antibodies do not prevent rotavirus infections (Woode *et al.*, 1975; Snodgrass and Wells, 1976; McNulty, 1978). Similarly, in human neonates, transplacentally acquired (maternal) serum antibodies are not protective *per se*, and older children and adults can be susceptible to rotavirus infection despite the presence of serum antibodies that indicate past infections (Kapikian *et al.*, 1974; McLean and Holmes, 1981).

On the other hand, there is abundant evidence of protection against rotavirus diarrhea mediated by neutralizing antibodies in the lumen of the intestine. Maternal colostral antibodies have been shown to be protective in calves, lambs, piglets, and humans (Woode et al., 1975; Snod-grass and Wells, 1976; Bohl, 1979; McLean and Holmes, 1981), and even heterologous colostral or serum antibodies added to the feed are effective, provided that they are capable of neutralizing the infecting rotavirus (Lecce et al., 1976; Snodgrass and Wells, 1978; Bartz et al., 1980).

In studies of immunity to rotaviruses, it is important to distinguish the immunoglobulin classes to which the antibodies belong, and this has been greatly facilitated by the development of ELISAs or radioimmu-noassays for measuring anti-rotaviral immunoglobulin A (IgA), IgG, or IgM of humans, pigs, calves, or other animals (Yolken et al., 1978a; Cukor et al., 1979; Grauballe and Vestergaard, 1980; McLean et al., 1980; Sheridan et al., 1981; Corthier and Franz, 1981; Acres and Babiuk, 1978). In early bovine colostrum, antirotaviral antibodies are found in both IgA and $IgG_1$ fractions, but by 14–28 days after calving, they are predominantly of $IgG_1$type (Snodgrass et al., 1980). In pigs, the total amount of antirotaviral IgG in colostrum is greater than that of IgA or IgM, but specific activity is highest in IgA, and in milk the antibody is mainly IgA (Corthier and Franz, 1981; Hess and Bachmann, 1981). In humans, only IgG antibody is transferred across the placenta, but almost all the rotavirus-specific immunoglobulin in colostrum and milk is secretory IgA (ScIgA) (Yolken et al., 1978a; Cukor et al., 1979; McLean and Holmes, 1980).

Both in animals and in man, the antibody titers in colostrum are maximal soon after parturition, but decrease sharply over the next few days (Woode et al., 1975; Thouless et al., 1977a; McLean and Holmes, 1980). Thus, rotavirus infections in suckling animals or humans tend to be postponed rather than completely prevented, although very often such infections under partial antibody protection are mild or asymptomatic; it is clear that the protective effect of colostrum depends on both its antibody titer and the volume ingested (Snodgrass and Wells, 1978; McNulty, 1978; Acres and Babiuk, 1978; McLean and Holmes, 1981). In humans, higher than average levels of colostral and milk antibodies against rotavirus almost certainly result from recent maternal infections, and boosts in lacteal antibody titers have been observed during lactation when mothers have experienced diarrheal episodes (Yolken et al., 1978a; Cukor et al., 1979; McLean and Holmes, 1981).

Although it has not yet been possible to carry out similar measurements in the rotavirus-infected human intestine, it is believed that earlier work on the effects of secretory antibodies on poliovirus infections of the nasopharynx are highly relevant. Ogra and Karzon (1969) found that preexisting poliovirus antibodies in nasal secretions had a quantitative rather than an all-or-none effect, so that in the presence of higher levels of secretory antibody, both the titers excreted and the duration of excretion of vaccine poliovirus were decreased. Almost certainly, a similar "dampening" of rotavirus production in the gut could result in asymptomatic infection instead of diarrheal illness.

Remarkably little is known about the development of active local immunity following human intestinal infection with rotaviruses. Even though the susceptibility of infants to serious illness extends up to about 2 years of age, whereas most animals are vulnerable only for a much shorter time, the ready availability of experimental animals has permitted much more work to be done on domestic livestock and laboratory animals, although unfortunately none of them is a particularly good model for humans (Shearman et al., 1972; Husband and Watson, 1978).

It has been reported that no immune response was detectable by CF in neonates following rotavirus infections in the first week of life (Crewe and Murphy, 1980), but this may remain an open question until a more sensitive test for IgA is employed. Keller et al. (1969) demonstrated an intestinal IgA response to poliovirus in neonates, but they were somewhat older (1–2 weeks) when immunized. In four slightly older children and one adult following a family outbreak of rotavirus enteritis, Sonza and Holmes (1980) detected specific IgA, IgM, and IgG coproantibodies with peak titers between 10 and 30 days after onset of symptoms. The response was transient and coproantibodies were undetectable by about 2 months postinfection, but Grauballe et al. (1981a) have since claimed that they could detect IgA coproantibodies 4–7 months after rotavirus infections in a larger series of patients, so further work is required to settle this question also.

Grauballe et al. (1981a) suggested that the presence of antirotaviral ScIgA in serum was diagnostic of a recent rotavirus infection, but this would not apply in late pregnancy, when it is normally found in serum (McLean and Holmes, 1980). Nonsecretory IgA is more commonly present in adult sera, but diagnostic increases in antirotaviral IgA were demonstrated in a study of traveler's diarrhea (Sheridan et al., 1981). Serum IgM responses are detectable in young children, but apparently not in adults (Konno et al., 1975; Ørstavik and Haug, 1976; Ørstavik et al., 1976; Sarkkinen et al, 1979; Sheridan et al., 1981).

Practically no investigations have been reported on the specificity of antibody responses following sequential infections with different rotavirus serotypes. Similarly, nothing is known about possible cellular immune responses to rotavirus infections, but the observation by Saulsbury et al. (1980) that immunodeficient children who became chronic excretors of rotavirus could be successfully treated with milk containing specific antibodies suggests that these are the most important.

## F. Prospects for Control

Current experience suggests that no foreseeable improvements in hygiene on farms, in homes, or even in hospitals are likely to control rotavirus infections completely, although this does not mean that the incidence of disease could not be lowered by changes in management procedures and improvements in water supplies and general living standards. Encouragement of feeding of colostrum is obviously sensible, but in both animals and humans, such passive protection is effective for only a short time (Section VI.E). Consequently, most interest has been shown in the possibility of developing vaccines to stimulate local enteric immuhity (W.H.O. Scientific Working Group, 1980; Kapikian et al., 1980).

A live, attenuated bovine vaccine was developed from the NCDV Lincoln rotavirus strain and has been widely used. Although reports on early trials of its efficacy were enthusiastic (Mebus et al., 1973), trials conducted under different conditions have failed to show protection (Acres and Radostits, 1976; de Leeuw et al., 1980). In the latter case, it was suggested that the vaccine was probably neutralized by colostral antibodies, and of course that problem is well known in relation to human poliovirus immunization.

At least under experimental conditions, the homologous vaccine approach does work, but results obtained with rotaviruses of heterologous origin are more variable. Woode et al. (1978) found that the NCDV vaccine protected three gnotobiotic calves against challenge with the more virulent UK bovine rotavirus, but cross-protection by foal or human rotaviruses was effective in only one of three calves tested with each. Wyatt et al. (1979a) showed that preimmunization (in utero) with bovine rotavirus protected gnotobiotic calves from symptomatic infection with human rotavirus, but Lecce and King (1979) found that bovine vaccine did not protect piglets against challenge with a field strain of porcine rotavirus. The idea of a heterologous animal rotavirus vaccine for human use has lost most of its appeal now that cell-culture growth of a range of human strains is possible (Section III.B).

Because of the current lack of a suitable experimental animal model for human rotavirus infections, the development and testing of attenuated strains is a major problem, even though recent advances in understanding of rotaviral genetics and antigenic structure resulting from basic virological research are most encouraging. Nevertheless, we are only at the beginning of the molecular genetic phase of rotavirus research, and there is a long way to go if the sophisticated strategies being explored for the production of a live influenza virus vaccine are to be followed (Chanock, 1981). We do not even know how many rotavirus serotypes are involved in human or animal infections.

An alternative possibility might be immunization of young animals or infants with wild-type rotaviruses under partial (lacteal) antibody protection. Experimentally, this approach has succeeded in piglets (Bridger and Brown, 1981), and it probably approximates the natural mode of acquisition of primary immunity in man and many other species, but its practical application would require very careful contol of the antibody level. If necessary, lacteal antibody levels could be increased by maternal immunization, as has been shown by the observation of natural human infections mentioned in Section VI.E and by experimental vaccinations of cattle (Snodgrass et al., 1980). Note that for this purpose, parenteral immunization with inactivated virus or perhaps single polypeptides produced by expression of cloned genes would be applicable, since parenteral immunization can boost SIgA responses provided the subject has previously experienced an immunizing intestinal infection (Svennerholm *et al.*, 1977; Snodgrass *et al.*, 1980).

At least in humans, control of dehydration and thereby of infant mortality due to diarrheal disease of any cause is seen as an achievable goal and is being pursued actively by the World Health Organization (W.H.O., 1979). A suitably formulated glucose–electrolyte solution for oral rehydration has been developed and shown to be effective, and the task has become one of organizing production so that the solution can be made widely available wherever it is needed and of educating mothers and primary-health workers in its use (W.H.O., 1976; Mahalanabis *et al.*, 1981).

## VII. CONCLUDING REMARKS

The magnitude of the disease problems produced by rotaviruses in humans and domestic animals is so great that it is very difficult to obtain accurate statistics, but the best available estimates are striking (Acres and Babiuk, 1978; Kapikian *et al.*, 1980). It already appears that rotavirus is a major cause of diarrheal disease in most parts of the world, that human gastroenteritis associated with rotavirus is more severe than that caused by other agents, and that the estimated human death toll in Africa, Asia, and Latin America due to diarrheal diseases exceeds the total of all the other major infectious diseases put together (Kapikian *et al.*, 1980). Consequently, it has been suggested that to save lives, the concentration of effort should be on development of vaccines for enterotoxigenic *Escherichia coli* and rotaviruses (Black *et al.*, 1981b).

Currently, investigations on rotaviruses are popular because the potential of rotaviruses for producing disease is equaled by their potential as a field for new basic research with immediate practical applications. In this review, I have necessarily omitted all but the merest mention of a large body of work on seroepidemiology and on other agents that cause diarrheal disease, in order to cover as much as possible of the basic virology. I have attempted to draw attention to areas in which much more work is obviously needed, such as immunity, immunization strategies, definition and comparisons of serotypes, and genetics. New and improved methods are needed for studies of molecular epidemiology so that it can be determined whether genetic exchanges are occurring between rotavirus strains in different animal and human populations. Given the number of different rotaviruses that appear to exist, the frequency of rotavirus infections, and the fact that there does not seem to be a strict species barrier for rotaviruses, mixed infections that could lead to the creation and selection of new reassortant strains seem very probable. Not only are these interesting questions, but also they must be answered if control of rotaviral disease is to become a reality.

## APPENDIX

   Comparison of the first available sequences for a rotavirus gene, encoding a non-structural protein known as VP8 of UK bovine rotavirus and as p33 or NCVP3 of simian rotavirus SA 11 (cf. Table 1). The sequence of the coding strand of a cloned DNA copy of segment 7 of UK bovine

```
                              SA11 SiRV Segment 8                                    C              46
                              UK BoRV Segment 7   5'-GGCTTTTAAAGCGTCTCAGTCGCCGTTTGAGCCTTGCGGTGTAGCC


                                                           G          A      GT       A121
ATG GCT GAG CTA GCT TGC TTT TGT TAT CCC CAT TTG GAG AAC GAT AGC TAT AAA TTT ATT CCG TTT AAC AAC TTG
Met Ala Glu Leu Ala Cys Phe Cys Tyr Pro His Leu Glu Asn Asp Ser Tyr Lys Phe Ile Pro Phe Asn Asn Leu25
                                                         Arg                              Ser

      G   T               A           G     A           A              T . C        G
GCT ATA AAA TGC ATG TTG ACG GCA AAA GTA GAC AGA AAA GAT CAG GAT AAA TTC TAT AAC TCA ATA ATT TAT GGT196
Ala Ile Lys Cys Met Leu Thr Ala Lys Val Asp Arg Lys Asp Gln Asp Lys Phe Tyr Asn Ser Ile Ile Tyr Gly50
                                        Lys

                       T   C     C   G G   T                                               A271
ATT GCA CCA CCG CCA CAA TTC AAA AAA CGC TAT AAT ACA AAT GAC AAT TCA AGA GGA ATG AAC TAT GAA ACT TCG
Ile Ala Pro Pro Pro Gln Phe Lys Lys Arg Tyr Asn Thr Asn Asp Asn Ser Arg Gly Met Asn Tyr Glu Thr Ser75
                                                   Ser

   T               CG T                   C           T                          A  G 346
ATG TTC AAT AAA GTG GCG GTA CTA ATT TGT GAA GCA TTG AAT TCA ATT AAA GTT ACA CAA TCT GAT GTT GCG AAT
Met Phe Asn Lys Val Ala Val Leu Ile Cys Glu Ala Leu Asn Ser Ile Lys Val Thr Gln Ser Asp Val Ala Asn100
                   Ala                                                                      Ser

   A   T      A A             G     T A  G   C           A  A       G          G        A421
GTG CTC TCA AGA GTA GTT TCT GTA AGA CAT CTG GAA AAT TTG GTG CTG AGG AGA GAA AAT CAT CAA GAC GTG CTT
Val Leu Ser Arg Val Val Ser Val Arg His Leu Glu Asn Leu Val Leu Arg Arg Glu Asn His Gln Asp Val Leu125
                  Lys Ile

   C   T   A      G T G T  C G          A C             T              C  G   A 496
TTT CAC TCG AAA GAA CTA CTA TTA AAA TCA GTG TTA ATA GCT ATT GGT CAC TCA AAA GAA ATT GAA ACG ACT GCC
Phe His Ser Lys Glu Leu Leu Leu Lys Ser Val Leu Ile Ala Ile Gly His Ser Lys Glu Ile Glu Thr Thr Ala150

              A   G                          A   T    A              A  C C G      T 571
ACT GCT GAA GGG GGA GAA ATT GTT TTT CAA AAT GCT GCG TTT ACG ATG TGG AAA TTG ACG TAT TTA GAA CAC AAA
Thr Ala Glu Gly Gly Glu Ile Val Phe Gln Asn Ala Ala Phe Thr Met Trp Lys Leu Thr Tyr Leu Glu His Lys175

C                           T     A    A A   C                          G 646
TTA ATG CCA ATT TTG GAT CAA AAT TTC ATT GAG TAT AAG ATT ACA GTG AAT GAA GAT AAA CCA ATT TCA GAA TCA
Leu Met Pro Ile Leu Asp Gln Asn Phe Ile Glu Tyr Lys Ile Thr Val Asn Glu Asp Lys Pro Ile Ser Glu Ser200
                                                       Leu

   C  C      G           A C   G    A C T A     A      T 721
CAT GTG AAA GAA CTC ATT GCT GAG TTG CGA TGG CAG TAT AAC AAG TTT GCG GTA ATA ACA CAT GGT AAA GGT CAC
His Val Lys Glu Leu Ile Ala Glu Leu Arg Trp Gln Tyr Asn Lys Phe Ala Val Ile Thr His Gly Lys Gly His225

            A  A                               C              C 796
TAC AGA GTT GTC AAG TAT TCA TCA GTT GCG AAT CAT GCA GAT AGA GTT TAT GCT ACT TTC AAG AGT AAT AAT AAA
Tyr Arg Val Val Lys Tyr Ser Ser Val Ala Asn His Ala Asp Arg Val Tyr Ala Thr Phe Lys Ser Asn Asn Lys250

          A G A A         C A T T     A      A      G T             C 871
AAT GGA AAT GTA CTG GAA TTT AAT TTG GAC CAA AGG ATA ATT TGG CAA AAC TGG TAT GCG TTT ACG TCT TCA
Asn Gly Asn Val Leu Glu Phe Asn Leu Leu Asp Gln Arg Ile Ile Trp Gln Asn Trp Tyr Ala Phe Thr Ser Ser275
             Met Ile

          C                A            G   G            G C     C    G 946
ATG AAA CAA GGT AAT ACT CTT GAC ATA TGT AAG AAA CTA CTC TTC CAG AAG ATG AAA AGA GAA AGT AAT CCA TTT
Met Lys Gln Gly Asn Thr Leu Asp Ile Cys Lys Lys Leu Leu Phe Gln Lys Met Lys Arg Glu Ser Asn Pro Phe300
                              Glu

   G  G                                                          A     GAA        A1028
AAA GAA CTG TCA ACT GAT AGA AAG ATG GAT GAA GTT TCT CAA ATA GGA ATT TAATTCGTTATCGGTTTGAAGGTGGGTATGG
Lys Gly Leu Ser Thr Asp Arg Lys Met Asp Glu Val Ser Gln Ile Gly Ile317


   A  T
CAGAGCAAGAATAGAAAGCGCTTATGTGACC-3'1059
```

rotavirus was determined by T.C. Elleman, P.A. Hoyne, M.L. Dyall-Smith and A.A. Azad and is shown in full, with the deduced amino acid sequence immediately below it. The corresponding sequence of segment 8 of simian rotavirus SA 11 was obtained by G.W. Both, A.R. Bellamy, J.E. Street and L.J. Siegman, who we thank for providing their data prior to publication. The SA11 sequence is identical except where base changes are indicated above the bovine rotavirus sequence, and where these have produced amino acid changes, these are indicated below the bovine amino acid sequence. A high degree of conservation of amino acid sequence is evident. The 121 nucleotide differences result in only 12 amino acid changes, i.e. over 85% of the mutational differences are silent, and the alternative amino acids are very similar in 11 of the 12 sites of variation.

# REFERENCES

Acres, S.D., and Babiuk, L.A., 1978, Studies on rotaviral antibody in bovine serum and lacteal secretions, using radioimmunoassay, *J. Am. Vet. Med. Assoc.* **173**:555.

Acres, S.D., and Radostits, O.M., 1976, The efficacy of a modified live reo-like virus vaccine and an *E. coli* bacterin for prevention of acute undifferentiated neonatal diarrhea of beef calves, *Can. Vet. J.* **17**:197.

Adams, W.R., and Kraft, L.M., 1967, Electron microscopic study of the intestinal epithelium of mice infected with the agent of epizootic diarrhea of infant mice (EDIM virus), *Am. J. Pathol.* **51**:39.

Albrey, M.B., and Murphy, A.M., 1976, Rotavirus growth in bovine monolayers, *Lancet* **1**:753.

Almeida, J.D., Hall, T., Banatvala, J.E., Totterdell, B.M., and Chrystie, I.L., 1978, The effect of trypsin on the growth of rotavirus, *J. Gen. Virol.* **40**:213.

Almeida, J.D., Bradburne, A.F., and Wreghitt, T.G., 1979, The effect of sodium thiocyanate on virus structure, *J. Med. Virol.* **4**:269.

Altenburg, B.C., Graham, D.Y., and Estes, M.K., 1980, Ultrastructural study of rotavirus replication in cultured cells, *J. Gen. Virol.* **46**:75.

Arias, C.F., Lopez, S., and Espejo, R.T., 1982, Gene protein products of SA 11 simian rotavirus genome. *J. Virol.* **41**:42.

Ashley, C.R., Caul, E.O., Clarke, S.K.R., Corner, B.D., and Dunn, S., 1978, Rotavirus infections of apes, *Lancet* **2**:477.

Babiuk, L.A., Mohammed, K., Spence, L., Fauvel, M., and Petro, R., 1977, Rotavirus isolation and cultivation in the presence of trypsin, *J. Clin. Microbiol.* **6**:610.

Banatvala, J.E., Totterdell, B., Chrystie, I.L., and Woode, G.N., 1975, *In-vitro* detection of human rotaviruses, *Lancet* **2**:821.

Banfield, W.G., Kasnic, G., and Blackwell, J.H., 1968, Further observations on the virus of epizootic diarrhoea of infant mice: An electron microscopic study, *Virology* **36**:411.

Barnett, B.B., Spendlove, R.S., Peterson, M.W., Hsu, L.Y., Lasalle, V.A., and Egbert, L. N., 1975, Immunofluorescent cell assay of neonatal calf diarrhea virus, *Can. J. Comp. Med.* **39**:462.

Barnett, B.B., Egbert, L.N., and Spendlove, R.S., 1978, Characteristics of neonatal calf diarrhea virus ribonucleic acid, *Can. J. Comp. Med.* **42**:46.

Barnett, B.B., Spendlove, R.S., and Clark, M.L., 1979, Effect of enzymes on rotavirus infectivity, *J. Clin. Microbiol.* **10**:111.

Bartz, C.R., Conklin, R.H., Tunstall, C.B., and Steele, J.H., 1980, Prevention of murine rotavirus infection with chicken egg yolk immunoglobulins, *J. Infect. Dis.* **142**:439.

Bastardo, J.W., and Holmes, I.H., 1980, Attachment of SA 11 rotavirus to erythrocyte receptors, *Infect. Immun.* **29**:1134.

Bastardo, J.W., McKimm-Breschkin, J.L., Sonza, S., Mercer, L.D., and Holmes, I.H., 1981, Preparation and characterization of antisera to electrophoretically purified SA 11 virus polypeptides, *Infect. Immun.* **34**:641.

Beards, G.M., Pilfold, J.N., Thouless, M.E., and Flewett, T.H., 1980, Rotavirus serotypes by serum neutralization, *J. Med. Virol.* **5**:231.

Bernstein, J.M., and Hruska, J.F., 1981, Characterization of RNA polymerase products of Nebraska calf diarrhea virus and SA 11 rotavirus, *J. Virol.* **37**:1071.

Bishai, F.R., Blaskovic, P., and Goodwin, D., 1978, Physicochemical properties of Nebraska calf diarrhea virus hemagglutinin, *Can. J. Microbiol.* **24**:1425.

Bishop, R.F., Davidson, G.P., Holmes, I.H. and Ruck, B.J., 1973, Virus particles in epithelial cells of duodenal mucosa from children with acute non-bacterial gastroenteritis, *Lancet* **2**:1281.

Bishop, R.F., Davidson, G.P., Holmes, I.H. and Ruck, B.J., 1974, Detection of a new virus by electron microscopy of faecal extracts from children with acute gastroenteritis, *Lancet* **1**:149.

Bishop, R.F., Cameron, D.J.S., Veenstra, A.A., and Barnes, G.L., 1979, Diarrhea and rotavirus infection associated with differing regimens for postnatal care of newborn babies, *J. Clin. Microbiol.* **9**:525.

Black, R.E., Dykes, A.C., Anderson, K.E., Wells, J.G., Sinclair, S.P., Gary, G.W., Jr., Hatch, M.H., and Gangarosa, E.J., 1981a, Handwashing to prevent diarrhea in day-care centers, *Am. J. Epidemiol.* **113**:445.

Black, R.E., Merson, M.H., Huq, I., Alim, A.R., and Yunus, M., 1981b, Incidence and severity of rotavirus and *Escherichia coli* diarrhoea in rural Bangladesh: Implications for vaccine development, *Lancet* **1**:141.

Bohl, E.H., 1979, Rotaviral diarrhoea in pigs: Brief review, *J. Am. Vet. Med. Assoc.* **174**:613.

Bohl, E.H., Saif, L.J., Theil, K.W., Agnes, A.G., and Cross, R.F., 1982, Porcine pararotavirus: Detection, differentiation from rotavirus, and pathogenesis in gnotobiotic pigs, *J. Clin. Microbiol.* **15**:312.

Bolivar, R., Conklin, R.H., Vollet, J.J., Pickering, L.K., Du Pont, H.L., Walters, D.L., and Kohl, S., 1978, Rotavirus in traveller's diarrhea: Study of an adult student population in Mexico, *J. Infect. Dis.* **137**:324.

Bradburne, A.F., Almeida, J.D., Gardner, P.S., Moosai, R.B., Nash, A.A., and Coombs, R.R.A., 1979, A solid-phase system (SPACE) for the detection and quantification of rotavirus in faeces, *J. Gen. Virol.* **44**:615.

Brandt, C.D., Kim, H.W., Yolken, R.H., Kapikian, A.Z., Arrobio, J.O., Rodriguez, W.J., Wyatt, R.G., Chanock, R.M., and Parrott, R.H., 1979, Comparative epidemiology of two rotavirus serotypes and other viral agents associated with pediatric gastroenteritis, *Am. J. Epidemiol.* **110**;243.

Bridger, J.C., 1978, Location of type-specific antigens in calf rotaviruses, *J. Clin. Microbiol.* **8**:625.

Bridger, J.C., and Brown, J.F., 1981, Development of immunity to porcine rotavirus in piglets protected from disease by bovine colostrum, *Infect. Immun.* **31**:906.

Bridger, J.C., and Woode, G.N., 1975, Neonatal calf diarrhoea: Identification of a reovirus-like (rotavirus) agent in faeces by immunofluorescence and immune electron microscopy, *Br. Vet. J.* **131**:528.

Bridger, J.C., and Woode, G.N., 1976, Characterization of two particle types of calf rotavirus, *J. Gen. Virol.* **31**:245.

Bridger, J.C., Woode, G.N., Jones, J.M., Flewett, T.H., Bryden, A.S., and Davies, H., 1975, Transmission of human rotaviruses to gnotobiotic piglets, *J. Med. Microbiol.* **8**:565.

Bridger, J.C., Clarke, I.N., and McCrae, M.A., 1982, Characterization of an antigenically distinct porcine rotavirus, *Infect. Immun.* **35**:1058.

Bryden, A.S., Davies, H.A., Hadley, R.E., Flewett, T.H., Morris, C.A., and Oliver, P., 1975, Rotavirus enteritis in the West Midlands during 1974, *Lancet* **2**:241.

Bryden, A.S., Thouless, M.E., and Flewett, T.H., 1976, Rotavirus in rabbits, *Vet. Rec.* **99**:323.

Bryden, A.S., Davies, H.A., Thouless, M.E., and Flewett, T.H., 1977, Diagnosis of rotavirus infection by cell culture, *J. Med. Microbiol.* **10**:121.

Cameron, D.J.S., Bishop, R.F., Davidson, G.P., Townley, R.R.W., Holmes, I.H., and Ruck, B.J., 1975, Rotavirus infections in obstetric hospitals, Lancet 2:124.

Cameron, D.S., Bishop, R.F., Veenstra, A.A., and Barnes, G.L., 1978, Noncultivable viruses and neonatal diarrhea: Fifteen-month survey in newborn special care nursery, J. Clin. Microbiol. 8:93.

Campbell, C., and Lang, W.R., 1979, An epidemic of gastroenteritis in Auckland 1978, N. Z. Med. J. 90:233.

Carafoli, E., and Crompton, M., 1978, The regulation of intracellular calcium, in: Current Topics in Membranes and Transport, Vol. 10 (F. Bronner and A. Kleinzeller, eds.), pp. 151–216, Academic Press, New York.

Carlson, J.A.K., Middleton, P.J., Szymanski, M.T., Huber, J., and Petric, M., 1978, Fatal rotavirus gastroenteritis—analysis of 21 cases, Am. J. Dis. Child. 132:477.

Carpio, M.M., Babiuk, L.A., Misra, V., and Blumenthal, R.M., 1981, Bovine rotavirus–cell interactions: Effect of virus infection on cellular integrity and macromolecular synthesis, Virology 114:86.

Carr, M.E., McKendrick, D.W., and Spyridakis, T., 1976, The clinical features of infantile gastroenteritis due to rotavirus, Scand. J. Infect. Dis. 8:241.

Caul, E.O., Ashley, C.R., and Egglestone, S., 1978, An improved method for the routine identification of faecel viruses using ammonium sulphate precipitation, FEMS Microbiol. Lett. 4:1.

Chanock, R.M., 1981, Strategy for development of respiratory and gastrointestinal tract viral vaccines in the 1980s, J. Infect. Dis. 143:364.

Chasey, D., 1980, Investigation of immunoperoxidase-labelled rotavirus in tissue culture by light and electron microscopy, J. Gen. Virol. 50:195.

Cheever, F.S., 1956, Epidemic diarrheal disease of suckling mice, Ann. N. Y., Acad. Sci. 66:196.

Cheever, F.S., and Mueller, J.H., 1947, Epidemic diarrheal disease of suckling mice. I. Manifestations, epidemiology and attempts to transmit the disease, J. Exp. Med. 85:405.

Chrystie, I.L., Totterdell, B., Baker, M.J., Scopes, J.W., and Banatvala, J.E., 1975, Rotavirus infections in a maternity unit, Lancet 2:79.

Chrystie, I.L., Totterdell, B.M., and Banatvala, J.E., 1978, Asymptomatic endemic rotavirus infections in the newborn, Lancet 1:1176.

Chrystie, I.L., Goldwater, P.N., and Banatvala, J.E., 1979, Rotavirus infection in a domestic cat, Vet. Rec. 105:404.

Clark, S.M., Barnett, B.B., and Spendlove, R.S., 1979, Production of high-titre bovine rotavirus with trypsin, J. Clin. Microbiol. 9:413.

Clark, S.M., Roth, J.R., Clark, M.L., Barnett, B.B., and Spendlove, R.S., 1981, Trypsin enhancement of rotavirus infectivity: Mechanism of enhancement, J. Virol. 39:816.

Clarke, I.N., and McCrae, M.A., 1981, A rapid and sensitive method for analysing the genome profiles of field isolates of rotavirus, J. Virol. Methods 2:203.

Coelho, K.I.R., Bryden, A.S., Hall, C., and Flewett, T.H., 1981, Pathology of rotavirus infection in suckling mice: A study by conventional histology, immunofluorescence, ultrathin sections, and scanning electron microscopy, Ultrastruct. Pathol. 2:59.

Cohen, J., 1977, Ribonucleic acid polymerase activity associated with purified calf rotavirus, J. Gen. Virol. 36:395.

Cohen, J., and Dobos, P., 1979, Cell free transcription and translation of rotavirus RNA, Biochem. Biophys. Res. Commun. 88:791.

Cohen, J., Laporte, J., Charpilienne, A., and Scherrer, R., 1979, Activation of rotavirus RNA polymerase by calcium chelation, Arch. Virol. 60:177.

Conner, M.E., and Darlington, R.W., 1980, Rotavirus infection in foals, Am. J. Vet. Res. 41:1699.

Corthier, G., and Franz, J., 1981, Detection of antirotavirus immunoglogulins A, G and M in swine colostrum, milk and feces by enzyme-linked immunosorbent assay, Infec. Immun. 31:833.

Crewe, E., and Murphy, A. M., 1980, Further studies on neonatal rotavirus infections, Med. J. Aust. 1:61.

Crouch, C.F., and Woode, G.N., 1978, Serial studies of virus multiplication and intestinal damage in gnotobiotic piglets infected with rotavirus, *J. Med. Microbiol.* **11:**325.

Croxson, M.C., and Bellamy, A.R., 1979, Two strains of human rotavirus in Auckland, *N. Z. Med. J.* **90:**235.

Croxson, M.C., and Bellamy, A.R., 1981, Extraction of rotaviruses from faeces by treatment with lithium dodecyl sulfate, *Appl. Envir. Microbiol.* **41:**255.

Cubitt, W.D., and Holzel, H., 1980, An outbreak of rotavirus infection in a long-stay ward of a geriatric hospital, *J. Clin. Pathol.* **33:**306.

Cukor, G., Berry, M.K., and Blacklow, N.R., 1978, Simplified radioimmunoassay for detection of human rotavirus in stools, *J. Infect. Dis.* **138:**906.

Cukor, G., Blacklow, N.R., Capossa, F.E., Panjvani, Z.F.K., and Bednarek, F., 1979, Persistence of antibodies to rotavirus in human milk, *J. Clin. Microbiol.* **9:**93.

Davidson, G.P., and Barnes, G.L., 1979, Structural and functional abnormalities of the small intestine in infants and young children with rotavirus enteritis, *Acta Pediatr. Scand.* **68:**181.

Davidson, G.P., Bishop, R.F., Townley, R.R.W., Holmes, I.H., and Ruck, B.J., 1975a, Importance of a new virus in acute sporadic enteritis in children, *Lancet* **1:**242.

Davidson, G.P., Goller, I., Bishop, R.F., Townley, R.R.W., Holmes, I.H., and Ruck, B.J., 1975b, Immunofluorescence in duodenal mucosa of children with acute enteritis due to a new virus, *J. Clin. Pathol.* **28:**263.

Davidson, G.P., Gall, D.G., Petric, M., Butler D.G. and Hamilton, J.R., 1977, Human rotavirus enteritis induced in conventional piglets: Intestinal structure and transport, *J. Clin. Invest.* **60:**1402.

Delage, G., McLaughlin, B., and Berthiaume, L., 1978, A clinical study of rotavirus gastroenteritis, *J. Pediatr.* **93:**455.

De Leeuw, P.W., and Guinee, P.A.M., 1981, Laboratory diagnosis in neonatal calf and pig diarrhoea, in: *Current Topics in Veterinary Medicine and Animal Science*, Vol. 13, Martinus Nijhoff, The Hague.

De Leeuw, P.W., Ellens, D.J., Talmon, F.P., Zimmer, G.N., and Kommerij, R., 1980, Rotavirus infections in calves: Efficacy of oral vaccination in endemically infected herds, *Res. Vet. Sci.* **29:**142.

Derbyshire, J.B., and Woode, G.N., 1978, Classification of rotaviruses: Report from the World Health Organization Food and Agriculture Organization comparative virology program, *J. Am. Vet. Med. Assoc.* **173:**519.

Drozdov, S.G., Shekoyan, L.A., Korolev, M.B., and Andzhaparidze, A.G., 1979, Human rotavirus in cell culture: Isolation and passaging, *Vopr. Virusol.* **4:**389.

Dyall-Smith, M.L., and Holmes, I.H., 1981a, Gene-coding assignments of rotavirus double-stranded RNA segments 10 and 11, *J. Virol.* **38:**1099.

Dyall-Smith, M.L., and Holmes, I.H., 1981b, Comparisons of rotavirus polypeptides by limited proteolysis: Close similarity of certain polypeptides of different strains, *J. Virol.* **40:**720.

Elias, M.M., 1977, Separation and infectivity of two particle types of human rotavirus, *J. Gen. Virol.* **37:**191.

Ellens, D.J., de Leeuw, P.W., Straver, P.J., and van Balken, J.A.M., 1978, Comparison of five diagnostic methods for the detection of rotavirus antigens in calf faeces, *Med. Microbiol. Immunol.* **166:**157.

Els, H.J., and Lecatsas, G., 1972, Morphological studies on simian virus SA 11 and the "related" O agent, *J. Gen. Virol.* **17:**129.

Ericson, B.L., Graham, D.Y., Mason, B.B., and Estes, M.K., 1982, Identification, synthesis, and modifications of simian rotavirus SA 11 polypeptides in infected cells, *J. Virol.* **42:**825.

Esparza, J., and Gil, F., 1978, A study on the ultrastructure of human rotavirus, *Virology* **91:**141.

Espejo, R.T., Calderon, E., and Gonzalez, N., 1977, Distinct reovirus-like agents associated with acute infantile gastroenteritis, *J. Clin. Microbiol.* **6:**502.

Espejo, R.T., Calderon, E., Gonzalez, N., Salomon, A., Martuscelli, A., and Romero, P., 1979, Presence of two distinct types of rotavirus in infants and young children hospitalized with acute gastroenteritis in Mexico City, 1977, *J. Infect. Dis.* **139:**474.

Espejo, R.T., Avendano, L.F., Munoz, O., Romero, P., Eternod, J.G., Lopez, S., and Moncaya, J., 1980a, Comparison of human rotaviruses isolated in Mexico City and in Santiago, Chile, by electrophoretic migration of their double-stranded ribonucleic acid genome segments, *Infect. Immun.* **30:**342.

Espejo, R., Martinez, E., Lopez, S., and Munoz, O., 1980b, Different polypeptide composition of two human rotavirus types, *Infect. Immun.* **28:**230.

Espejo, R.T., Munoz, O., Serafin, F., and Romero, P., 1980c, Shift in the prevalent human rotavirus detected by ribonucleic acid segment differences, *Infect. Immun.* **27:**351.

Espejo, R.T., Lopez, S., and Arias, C., 1981, Structural polypeptides of simian rotavirus SA 11 and the effect of trypsin, *J. Virol.* **37:**156.

Estes, M.K., Graham, D.Y., Gerba, C.P., and Smith, E.M., 1979a, Simian rotavirus SA 11 replication in cell cultures, *J. Virol.* **31:**810.

Estes, M.K., Graham, D.Y., Smith, E.M., and Gerba, C.P., 1979b, Rotavirus stability and inactivation, *J. Gen. Virol.* **43:**403.

Estes, M.K., Graham, D.Y., and Mason, B.B., 1981, Proteolytic enhancement of rotavirus infectivity: Molecular mechanisms, *J. Virol.* **39:**879.

Eugster, A.K., and Sidwa, T., 1979, Rotaviruses in diarrheic feces of a dog, *Vet. Med. Small Anim. Clinician* **74:**817.

Evans, D.G., Olarte, J., Du Pont, H.L., Evans, D.J., Jr., Galindo, E., Portnoy, B.L., and Conklin, R.H., 1977, Enteropathogens associated with pediatric diarrhea in Mexico City, *J. Pediatr.* **91:**65.

Faulkner-Valle, G.P., Clayton, A.V., and McCrae, M.A., 1982, Molecular biology of rotaviruses, III. Isolation and characterization of temperature-sensitive mutants of bovine rotavirus, *J. Virol.* **42:**669.

Fauvel, M., Spence, L., Babiuk, L.A., Petro, R., and Bloch, S., 1978, Hemagglutination and hemagglutination-inhibition studies with a strain of Nebraska calf diarrhea virus (bovine rotavirus), *Intervirology* **9:**95.

Fernelius, A.L., Ritchie, A.E., Classick, L.G, Norman, J.O., and Mebus, C.A., 1972, Cell culture adaptation and propagation of a reovirus-like agent of calf diarrhea from a field outbreak in Nebraska, *Arch. Gesamte Virusforsch.* **37:**114.

Flewett, T.H., 1977, Acute non-bacterial infectious gastroenteritis—an essay in comparative virology, in: *Recent Advances in Clinical Virology*, No. 1 (A.P. Waterson, ed.), pp. 151–169, Churchill Livingstone, Edinburgh.

Flewett, T.H., and Boxall, E., 1976, The hunt for viruses in infections of the alimentary system: An immuno-electron-microscopical approach, *Clin. Gastroenterol.* **5:**359.

Flewett, T.H., and Woode, G.N., 1978, The rotaviruses, *Arch. Virol.* **57:**1.

Flewett, T.H., Bryden, A.S., and Davies, H., 1973, Virus particles in gastroenteritis, *Lancet* **2:**1497.

Flewett, T.H., Bryden, A.S., Davies, H., Woode, G.N., Bridger, J.C., and Derrick, J.M., 1974a, Relation between viruses from acute gastroenteritis of children and newborn calves, *Lancet* **2:**61.

Flewett, T.H., Bryden, A.S., and Davies, H., 1974b, Diagnostic electron microscopy of faeces. I. The viral flora of the faeces as seen by electron microscopy, *J. Clin. Pathol.* **27:**603.

Flewett, T.H., Davies, H., Bryden, A.S., and Robertson, M.J., 1974c, Diagnostic electron microscopy of faeces. II. Acute gastroenteritis associated with reovirus-like particles, *J. Clin. Pathol.* **27:**608.

Flewett, T.H., Bryden, A.S., and Davies, H., 1975, Virus diarrhoea in foals and other animals, *Vet. Rec.* **96:**477.

Flewett, T.H., Thouless, M.E., Pilfold, J.N., Bryden, A.S., and Candeias, J.A.N., 1978, *Lancet* **2:**632.

Fonteyne, J., Zissis, G., and Lambert, J.P., 1978, Recurrent rotavirus gastroenteritis, *Lancet* **1:**983.

Foster, L.G., Peterson, M.W., and Spendlove, R.S., 1975, Fluorescent virus precipitin test, *Proc. Soc. Exp. Biol. Med.* **150**:155.

Foster, S.O., Palmer, E.L., Gary, G.W., Jr., Martin, M.L., Herrmann, K.L., Beasley, P., and Sampson, J., 1980, Gastroenteritis due to rotavirus in an isolated Pacific Island group: An epidemic of 3439 cases, *J. Infect. Dis.* **141**:32.

Fulton, R.W., Johnson, C.A., Pearson, N.J., and Woode, G.N., 1981, Isolation of a rotavirus from a newborn dog with diarrhea, *Am. J. Vet. Res.* **42**:841.

Goldberg, D.M., Campbell, R., and Roy, A.D., 1968, Binding of trypsin and chymotrypsin by small intestine, *Biochim. Biophys. Acta* **167**:613.

Gordon, J.E., Chitkara, I.D., and Wyon, J.B., 1963, Weanling diarrhea. *Am. J. Med. Sci.* **245**:345.

Gorziglia, M., and Esparza, J., 1981, Poly (A) polymerase activity in human rotavirus, *J. Gen. Virol.* **53**:357.

Graham, D.Y., and Estes, M.K., 1980, Proteolytic enhancement of rotavirus infectivity: Biologic mechanisms, *Virology* **101**:432.

Grauballe, P.C., and Vestergaard, B.F., 1980, ELISA for rotavirus: Optimation of the reaction for the measurement of IgG and IgA antibodies against human rotavirus in human sera and secretions, *Dev. Clin. Biochem.* **1**:143.

Grauballe, P.C., Genner, J., Meyling, A., and Hornsleth, A., 1977, Rapid diagnosis of rotavirus infections: Comparison of electron microscopy and immuno-electro-osmophoresis for the detection of rotavirus in human infantile gastroenteritis, *J. Gen. Virol.* **35**:203.

Grauballe, P.C., Hjelt, K., Krasilnikoff, P.A., and Schiøtz, P.O., 1981a, ELISA for rotavirus-specific secretory IgA in human sera, *Lancet* **2**:588.

Grauballe, P.G., Vestergaard, B.F., Meyling, A., and Genner, J., 1981b, Optimized enzyme-linked immunosorbent assay for determination of human and bovine rotavirus in stools: Comparison with electron microscopy, immunoelectro-osmophoresis and fluorescent antibody techniques, *J. Med. Virol.* **7**:29.

Greenberg, H.B., Kalica, A.R., Wyatt, R.G., Jones, R.W., Kapikian, A.Z., and Chanock, R.M., 1981, Rescue of noncultivatable human rotavirus by gene reassortment during mixed infection with ts mutants of a cultivable bovine rotavirus, *Proc. Natl. Acad. Sci. U.S.A.* **78**:420.

Greenberg, H.B., Wyatt, R.G., Kapikian, A.Z., Kalica, A.R., Flores, J., and Jones, R., 1982, Rescue and serotypic characterization of noncultivable human rotavirus by gene reassortment, *Infect. Immun.* **37**:104.

Gurwith, M., Wenman, W. Hinde, D., Feltham, S., and Greenberg, H., 1981, A prospective study of rotavirus infection in infants and young children, *J. Infect. Dis.* **144**:218.

Hall, G.A., Bridger, J.C., Chandler, R.L., and Woode, G.N., 1976, Gnotobiotic piglets experimentally inoculated with neonatal calf diarrhoea reovirus-like agent (rotavirus), *Vet. Pathol.* **13**:197.

Halvorsrud, J., and Orstavik, I., 1980, An epidemic of rotavirus-associated gastroenteritis in a nursing home for the elderly, *Scand. J. Infect. Dis.* **12**:161.

Hamilton, J.R., Gall, D.G., Butler, D.G., and Middleton, P.J., 1976, Viral gastroenteritis: Recent progress, remaining problems, in: *Acute Diarrhoea in Childhood, Ciba Foundation Symposium*, No. 42, (K. Elliott and J. Knight, eds.), pp. 209–219, Elsevier/Excerpta Medica/North-Holland, Amsterdam.

Hara, M., Mukoyama, J., Tsuruhara, T., Ashiwara, Y., Saito, Y., and Tagaya, I., 1978, Acute gastroenteritis among schoolchildren associated with reovirus-like agent, *Am. J. Epidemiol.* **107**:161.

Hebert, J.P., Caillet, R., Hacquard, B., and Fortier, B., 1981, Use of *Staphylococcus aureus* protein "A" to detect rotavirus in stools, *Pathol. Biol. (Paris)* **29**:101.

Herring, A.J., Inglis, N.F., Ojeh, C.K., Snodgrass, D.R., and Menzies, J.D., 1982, Rapid diagnosis of rotavirus infection by direct detection of viral nucleic acid in silver-stained polyacrylamide gels, *J. Clin. Microbiol.* **16**:473.

Hess, R.G., and Bachmann, P.A., 1981, Distribution of antibodies to rotavirus in serum and lacteal secretions of naturally infected swine and their suckling pigs, *Am. J. Vet. Res.* **42**:1149.

Hieber, J.P., Shelton, S., Nelson, J.D., Leon, J., and Mohs, E., 1978, Comparison of human rotavirus disease in tropical and temperate settings, *Am. J. Dis. Child.* **132**:853.

Hinshaw, V.S., Bean, W.J., Jr., Webster, R.G., and Easterday, B.C., 1978, The prevalence of influenza viruses in swine and the antigenic and genetic relatedness of influenza viruses from man and swine, *Virology* **84**:51.

Holmes, I.H., 1979, Viral gastroenteritis, *Prog. Med. Virol.* **25**:1.

Holmes, I.H., 1982, Basic virology of rotaviruses, in: *Viral Infections of the Gastrointestinal Tract* (D.A.J. Tyrrell and A.Z. Kapikian, eds.), pp. 111–124, Marcel Dekker, New York.

Holmes, I.H., Ruck, B.J., Bishop, R.F., and Davidson, G.P., 1975, Infantile enteritis viruses: Morphogenesis and morphology, *J. Virol.* **16**:937.

Holmes, I.H., Rodger, S.M., Schnagl, R.D., Ruck, B.J., Gust, I.D., Bishop, R.F., and Barnes, G.L., 1976, Is lactase the receptor and uncoating enzyme for infantile enteritis (rota) viruses? *Lancet* **1**:1387.

Hoshino, Y., Baldwin, C.A., and Scott, F.W., 1981, Isolation and characterization of feline rotavirus, *J. Gen. Virol.* **54**:313.

Hruska, J.F., Notter, M.F.D., Menegus, M.A., and Steinhoff, M.C., 1978, RNA polymerase associated with human rotaviruses in diarrhea stools, *J. Virol.* **26**:544.

Huismans, H., and Els, H.J., 1979, Characterization of the tubules associated with the replication of three different orbiviruses, *Virology* **92**:397.

Hurst, C.J., and Gerba, C.P., 1980, Stability of simian rotavirus in fresh and estuarine water, *Appl. Envir. Microbiol.* **39**:1.

Husband, A.J., and Watson, D.L., 1978, Immunity in the intestine, *Vet. Bull.* **48**:911.

Inaba, Y., Sato, K., Takahashi, E., Kurogi, H., Satoda, K., Omori, T., and Matumoto, M., 1977, Hemagglutination with Nebraska calf diarrhea virus, *Microbiol. Immunol.* **21**:531.

Jesudoss, E.S., John, T.J., Maiya, P.P., Jadhav, M., and Spence, L., 1979, Prevalence of rotavirus infection in neonates, *Indian J. Med. Res.* **70**:863.

Joklik, W.K., 1974, Reproduction of Reoviridae, in: *Comprehensive Virology*, Vol. 2 (H. Fraenkel-Conrat and R.R. Wagner, eds.), pp. 231–334, Plenum Press, New York.

Jones, R.C., Hughes, C.S., and Henry, R.R., 1979, Rotavirus infection in commercial laying hens, *Vet. Rec.* **104**:22.

Kalica, A.R., Garon, C.F., Wyatt, R.G., Mebus, C.A., van Kirk, D.H., Chanock, R.M., and Kapikian, A.Z., 1976, Differentiation of human and calf reoviruslike agents associated with diarrhea using polyacrylamide gel electrophoresis of RNA, *Virology* **74**:86.

Kalica, A.R., Purcell, R.H., Sereno, M.M., Wyatt, R.G., Kim, H.W., Chanock, R.M., and Kapikian, A.Z., 1977, A microtitre solid-phase radioimmunoassay for detection of the human reovirus-like agent in stools, *J. Immunol.* **118**:1275.

Kalica, A.R., James, H.D., Jr., and Kapikian, A.Z., 1978a, Hemagglutination by simian rotavirus, *J. Clin. Microbiol.* **7**:314.

Kalica, A.R., Sereno, M.M., Wyatt, R.G., Mebus, C.A., Chanock, R.M., and Kapikian, A.Z., 1978b, Comparison of human and animal rotavirus strains of gel electrophoresis of viral RNA, *Virology* **87**:247.

Kalica, A.R., Greenberg, H.B., Espejo, R.T., Flores, J., Wyatt, R.G., Kapikian, A.Z., and Chanock, R.M., 1981a, Distinctive ribonucleic acid patterns of human rotavirus subgroups 1 and 2, *Infect. Immun.* **33**:958.

Kalica A.R., Greenberg, H.B., Wyatt, R.G., Flores, J., Sereno, M.M., Kapikian, A.Z., and Chanock, R.M., 1981b, Genes of human (strain Wa) and bovine (strain UK) rotaviruses that code for neutralization and subgroup antigens, *Virology* **112**:385.

Kapikian, A.Z., Kim, H.W., Wyatt, R.G., Rodriguez, W.J., Ross, S., Cline, W.L., Parrott, R.H., and Chanock, R.M., 1974, Reoviruslike agent in stools: Association with infantile diarrhea and development of serologic tests, *Science* **185**:1049.

Kapikian, A.Z., Cline, W.L., Kim, H.W., Kalica, A.R., Wyatt, R.G., van Kirk, D.H., Chanock, R.M., James, H.D., Jr., and Vaughn, A.L., 1976a, Antigenic relations among five reoviruslike (RVL) agents by complement fixation (CF) and development of new substitute CF antigens for the human RVL agent of infantile gastroenteritis, *Proc. Soc. Exp. Biol. Med.* **152**:535.

Kapikian, A.Z., Kim, H.W., Wyatt, R.G., Cline, W.L., Arrobio, J.L., Brandt, C.D., Rodriguez, W.J., Sack, D.A., Chanock, R.M., and Parrott, R.H., 1976b, Human reovirus-like agent associated with "winter" gastroenteritis, *N. Engl J. Med.* **294**:965.

Kapikian, A.Z., Yolken, R.H., Greenberg, H.B., Wyatt, R.G., Kalica, A.R., Chanock, R.M., and Kim, H.W., 1979, Gastroenteritis viruses, in: *Diagnostic Procedures for Viral, Rickettsial and Chlamydial Infections*, 5th ed. (E.H. Lennette and N.J. Schmidt, eds.), pp. 927–995, American Public Health Association, Washington, D.C.

Kapikian, A.Z., Wyatt, R.G., Greenberg, H.B., Kalica, A.R., Kim, H.W., Brandt, C.D., Rodriguez, W.J., Parrott, R.H., and Chanock, R.M., 1980, Approaches to immunization of infants and young children against gastroenteritis due to rotavirus, *Rev. Infect. Dis.* **2**:459.

Kapikian, A.Z., Cline, W.L., Greenberg, H.B., Wyatt, R.G., Kalica, A.R., Banks, C.E., James, H.D., Jr., Flores, J., and Chanock, R.M., 1981, Antigenic characterization of human and animal rotaviruses by immune adherence hemagglutination assay (IAHA): Evidence for distinctness of IAHA and neutralization antigens, *Infect. Immun.* **33**:415.

Keller, R., Dwyer, J.E., Oh, W., and D'Amodio, M., 1969, Intestinal IgA neutralizing antibodies in newborn infants following poliovirus immunization, *Pediatrics* **43**:330.

Keswick, B.H., and Gerba, C.P., 1980, Viruses in groundwater, *Environ. Sci. Technol.* **14**:1290.

Kim, H.W., Brandt, C.D., Kapikian, A.Z., Wyatt, R.G., Arrobio, J.O., Rodriguez, W.J., Chanock, R.M., and Parrott, R.H., 1977, Human reovirus-like agent infection; Occurrence in adult contacts of pediatric patients with gastroenteritis, *J. Am. Med. Assoc.* **238**:404.

Kimura, T., 1981, Immuno-electron microscopic study on the antigenicity of tubular structures associated with human rotavirus, *Infect. Immun.* **33**:611.

Kogasaka, R., Akihara, M., Horino, K., Chiba, S., and Nakao, T., 1979, A morphological study of human rotavirus, *Arch. Virol.* **61**:41.

Konno, T., Imai, A., Suzuki, H., and Ishida, N., 1975, Mercaptoethanol-sensitive antibody to reovirus-like agents in acute epidemic gastroenteritis, *Lancet* **2**:1312.

Konno, T., Suzuki, H., Imai, A., Kutsuzawa, T., Ishida, N., Katsushima, N., Sakamoto, M., Kitaoka, S., Tsuboi, R., and Adachi, M., 1978a, A long-term survey of rotavirus infection in Japanese children with acute gastroenteritis, *J. Infect. Dis.* **138**:569.

Konno, T., Suzuki, H., Kutsuzawa, T., Imai, A., Katsushima, N., Sakamoto, M., Kitaoka, S., Tsuboi, R., and Adachi, M., 1978b, Human rotavirus infection in infants and young children with intussusception, *J. Med. Virol.* **2**:265.

Koopman, J.S., 1978, Diarrhea and school toilet hygiene in Cali, Colombia, *Am. J. Epidemiol.* **107**:412.

Korolev, M.B., Khaustov, V.I., and Shekoian, L.A., 1981, Morphogenesis of human rotavirus in a cell culture, *Vopr. Virusol.* **3**:309.

Kraft, L.M., 1957, Studies on the etiology and transmission of epidemic diarrhea of infant mice, *J. Exp. Med.* **106**:743.

Kraft, L.M., 1958, Observations on the control and natural history of epidemic diarrhea of infant mice (EDIM), *Yale J. Biol. Med.* **31**:121.

Kurogi, H., Inaba, Y., Takahashi, E., Sato, K., Goto, Y., and Omori, T., 1976, Cytopathic effect of Nebraska calf diarrhea virus (Lincoln strain) on secondary bovine kidney cell monolayer, *Natl. Inst. Anim. Health Q.* **16**:133.

La Bonnardiere, C., de Vaureix, C., l'Haridon, R., and Scherrer, R., 1980, Weak susceptibility of rotavirus to bovine interferon, *Arch. Virol.* **64**:167.

Laemmli, U.K., 1970, Cleavage of structural proteins during the assembly of the head of bacteriophage T4, *Nature (London)* **227**:680.

Lecatsas, G., 1972, Electron microscopic and serological studies on simian virus SA 11 and the "related" O agent, *Onderstepoort J. Vet. Res.* **39**:133.

Lecce, J.G., and King, M.W., 1978, Role of rotavirus (reo-like) in weanling diarrhea of pigs, *J. Clin. Microbiol.* **8**:454.

Lecce, J.G., and King, M.W., 1979, The calf reo-like virus (rotavirus) vaccine: An ineffective immunization agent for rotaviral diarrhea of piglets, *Can. J. Comp. Med.* **43**:90.

Lecce, J.G., King, M.W., and Mock, R., 1976, Reovirus-like agent associated with fatal diarrhea in neonatal pigs, *Infect. Immun.* **14**:816.

Lecce, J.G., King, M.W., and Dorsey, W.E., 1978, Rearing regimen producing piglet diarrhea (rotavirus) and its relevance to acute infantile diarrhea, *Science* **199**:776.

Lewis, H.M., Parry, J.V., Davies, H.A., Parry, R.P., Mott, A., Dourmashkin, R.R., Sanderson, P.J., Tyrrell, D.A.J., and Valman, H.B., 1979, A year's experience of the rotavirus syndrome and its association with respiratory illness, *Arch. Dis. Child.* **54**:339.

L'Haridon, R., and Scherrer, R., 1977, Culture *in vitro* in rotavirus associé aux diarrhées néonatales du veau, *Ann. Rech. Vet.* **7**:373.

Liebermann, H., Dietz, G., Holl, U., Heinrich, H.W., Solisch, P., and Liebermann, H., 1979, Structure of calf rotavirus. 1. Hydrodynamic and electron microscopic parameters of the virus, *Arch. Exp. Veterinaermed.* **33**:367.

Light, J.S., and Hodes, H.L., 1949, Isolation from cases of infantile diarrhea of a filtrable agent causing diarrhea in calves, *J. Exp. Med.* **90**:113.

Linhares, A.C., Pinheiro, F.P., Freitas, R.B., Gabbay, Y.B., Shirley, J.A., and Beard, G.M., 1981, An outbreak of rotavirus diarrhea among a nonimmune, isolated South American Indian community, *Am. J. Epidemiol.* **113**:703.

Lourenco, M.H., Nicolas, J.C., Cohen, J., Scherrer, R., and Bricout, F., 1981, Study of human rotavirus genome by electrophoresis: Attempt of classification among strains isolated in France, *Ann. Virol. (Inst. Pasteur)* **132E**:161.

Lycke, E., Blomberg, J., Berg, G., Eriksoon, A., and Madsen, L., 1978, Epidemic acute diarrhea in adults associated with infantile gastroenteritis virus, *Lancet* **2**:1056.

Madeley, C.R., and Cosgrove, B.P., 1975, Viruses in infantile gastroenteritis, *Lancet* **2**:124.

Mahalanabis, D., Merson, M.H., and Barua, D., 1981, Oral rehydration therapy—recent advances, *World Health Forum* **2**:245.

Maiya, P.P., Pereira, S.M., Mathan, M., Bhat, P., Albert, M.J., and Baker, S.J., 1977, Aetiology of acute gastroenteritis in infancy and early childhood in Southern India, *Arch. Dis. Child.* **52**:482.

Mäki, M., 1981, A prospective clinical study of rotavirus diarrhoea in young children, *Acta Paediatr. Scand.* **70**:107.

Malherbe, H.H., and Strickland-Cholmley, M., 1967, Simian virus SA 11 and the related O agent, *Arch. Gesamte Virusforsch.* **22**:235.

Martin, M.L., Palmer, E.L., and Middleton, P.J., 1975, Ultrastructure of infantile gastroenteritis virus, *Virology* **68**:146.

Mason, B.B., Graham, D.Y., and Estes, M.K., 1980, *In vitro* transcription and translation of simian rotavirus SA 11 gene products, *J. Virol.* **33**:1111.

Mathan, M., Almeida, J.D., and Cole, J., 1977, An antigenic subunit present in rotavirus infected faeces, *J. Gen. Virol.* **34**:325.

Matsuno, S., and Mukoyama, A., 1979, Polypeptides of bovine rotavirus, *J. Gen. Virol.* **43**:309.

Matsuno, S., and Nakajima, K., 1982, RNA of rotavirus: Comparison of RNAs of human and animal rotaviruses, *J. Virol.* **41**:710.

Matsuno, S., Inouye, S., and Kono, R., 1977a, Plaque assay of neonatal calf diarrhea virus and the neutralizing antibody in human sera, *J. Clin. Microbiol.* **5**:1.

Matsuno, S., Inouye, S. and Kono, R., 1977b, Antigenic relationship between human and bovine rotaviruses as determined by neutralization, immune adherence hemagglutination and complement fixation tests, *Infect. Immun.* **17**:661.

Matsuno, S., Hasegawa, A., Kalica, A.R., and Kono, R., 1980, Isolation of a recombinant between simian and bovine rotaviruses, *J. Gen. Virol.* **48**:253.

Matthews, R.E.F., 1979, The classification and nomenclature of viruses, *Intervirology* **11**:133.

McCrae, M.A., and Faulkner-Valle, G.P., 1981, Molecular biology of rotaviruses. I. Characterization of basic growth parameters and pattern of macromolecular synthesis, *J. Virol.* **39**:490.

McCrae, M.A., and Joklik, W.K., 1978, The nature of the polypeptide encoded by each of the 10 double-stranded RNA segments of reovirus type 3, *Virology* **89**:578.

McCrae, M.A., and McCorquodale, J.G., 1982, The molecular biology of rotaviruses. II. Identification of the protein-coding assignments of calf rotavirus genome RNA species, *Virology* **117**:435.

McKimm-Breschkin, J.L., and Holmes, I.H., 1982, Conditions required for induction of interferon by rotaviruses and for their sensitivity to its action. *Infect. Immun.* **36**:857.

McLean, B., and Holmes, I.H., 1980, Transfer of antirotaviral antibodies from mothers to their infants, *J. Clin. Microbiol.* **12**:320.

McLean, B.S., and Holmes, I.H., 1981, Effects of antibodies, trypsin and trypsin inhibitors on susceptibility of neonates to rotavirus infection, *J. Clin. Microbiol.* **13**:22.

McLean, B., Sonza, S., and Holmes, I.H., 1980, Measurement of immunoglobulins A, G and M class rotavirus antibodies in serum and mucosal secretions, *J. Clin. Microbiol.* **12**:314.

McNulty, M.S., 1978, Rotaviruses, *J. Gen. Virol.* **40**:1.

McNulty, M.S., 1979, Morphology and chemical composition of rotaviruses, in: *Viral Enteritis in Humans and Animals* (F. Bricout and R. Scherrer, eds.), pp. 111–140, Editions INSERM, Paris.

McNulty, M.S., Allan, G.M., and McFerran, J.B., 1976a, Isolation of a cytopathic calf rotavirus, *Res. Vet. Sci.* **21**:114.

McNulty, M.S., Allan, G.M., Pearson, G.R., McFerran, J.B., Curran, W.L., and McCracken, R.M., 1976b, Reovirus-like agent (rotavirus) from lambs, *Infect. Immun.* **14**:1332.

McNulty, M.S., Curran, W.L., and McFerran, J.B., 1976c, The morphogenesis of a cytopathic bovine rotavirus in Madin–Darby bovine kidney cells, *J. Gen. Virol.* **33**:503.

McNulty, M.S., Pearson, G.R., McFerran, J.B., Collins, D.S., and Allan, G.M., 1976d, A reovirus-like agent (rotavirus) associated with diarrhoea in neonatal pigs, *Vet. Microbiol.* **1**:55.

McNulty, M.S., Allan, G.M., and McFerran, J.B., 1977, Cell culture with a cytopathic bovine rotavirus, *Arch. Virol.* **54**:201.

McNulty, M.S., Curran, W.L., Allan, G.M., and McFerran, J.B., 1978, Synthesis of coreless, probably defective virus particles in cell cultures infected with rotaviruses, *Arch. Virol.* **58**:193.

McNulty, M.S., Allan, G.M., Todd, D., and McFerran, J.B., 1979, Isolation and cell culture propagation of rotaviruses from turkeys and chickens, *Arch. Virol.* **61**:13.

McNulty, M.S., Allan, G.M., Todd, D., McFerran, J.B., McKillop, E.R., Collins, D.S., and McCracken, R.M., 1980, Isolation of rotaviruses from turkeys and chickens: Demonstration of distinct serotypes and RNA electropherotypes, *Avian Pathol.* **9**:363.

McNulty, M.S., Allan, G.M, Todd, D., McFerran, J.B., and McCracken, R.M., 1981, Isolation from chickens of a rotavirus lacking the rotavirus group antigen, *J. Gen. Virol.* **55**:405.

Mebus, C.A., 1976, Reovirus-like calf enteritis, *Dig. Dis.* **21**:592.

Mebus, C.A., and Newman, L.E., 1977, Scanning electron, light, and immunofluorescent microscopy of intestine of gnotobiotic calf infected with reovirus-like agent, *Am. J. Vet. Res.* **38**:553.

Mebus, C.A., Underdahl, N.R., Rhodes, M.B., and Twiehaus, M.J., 1969, Calf diarrhea (scours): Reproduced with a virus from a field outbreak, *Univ. Nebraska Agric. Exp. Station Res. Bull.* **233**:1.

Mebus, C.A., Kono, M., Underdahl, N.R., and Twiehaus, M.J., 1971a, Cell culture propagation of neonatal calf diarrhea (scours) virus, *Can. Vet. J.* **12**:69.

Mebus, C.A., Stair, E.L., Underdahl, N.R., and Twiehaus, M.J., 1971b, Pathology of neonatal calf diarrhea induced by a reo-like virus, *Vet. Pathol.* **8**:490.

Mebus, C.A., White, R.G., Bass, E.P., and Twiehaus, M.J., 1973, Immunity to neonatal calf diarrhea virus, *J. Am. Vet. Med. Assoc.* **163**:880.

Mebus, C.A., Wyatt, R.G., Sharpee, R.L., Sereno, M.M., Kalica, A.R., Kapikian, A.Z., and Twiehaus, M.J., 1976, Diarrhea in gnotobiotic calves caused by the reovirus-like agent of human infantile gastroenteritis, *Infect. Immun.* **14**:471.

Mebus, C.A., Wyatt, R.G., and Kapikian, A.Z., 1977, Intestinal lesions induced in gnotobiotic calves by the virus of human infantile gastroenteritis, *Vet. Pathol.* **14**:273.

Middleton, P.J., Szymanski, M.T., Abbott, G.D., Bortolussi, R., and Hamilton, J.R., 1974, Orbivirus acute gastroenteritis of infancy, *Lancet* **1**:1241.

Middleton, P.J., Petric, M., and Szymanski, M.T., 1975, Propagation of infantile gastroenteritis virus (orbi-group) in conventional and germfree piglets, *Infect. Immun.* **12**:1276.

Middleton, P.J., Petric, M., Hewitt, C.M., Szymanski, M.T., and Tam, J.S., 1976, Counter-immunoelectro-osmophoresis for the detection of infantile gastroenteritis virus (orbi-group) antigen and antibody, *J. Clin. Pathol.* **29**:191.

Middleton, P.J., Holdaway, M.D., Petric, M., Szymanski, M.T., and Tam, J.S., 1977a, Solid-phase radioimmunoassay for the detection of rotaviruses, *Infect. Immun.* **16**:439.

Middleton, P.J., Szymanski, M.T., and Petric, M., 1977b, Viruses associated with acute gastroenteritis in young children, *Am. J. Dis. Child.* **131**:733.

Moosai, R.B., Gardner, P.S., Almeida, J.D., and Greenaway, M.A., 1979, A simple immunofluorescent technique for the detection of human rotavirus, *J. Med. Virol.* **3**:189.

Much, D.H., and Zajac, I., 1972, Purification and characterization of epizootic diarrhea of infant mice virus, *Infect. Immun.* **6**:1019.

Murphy, F.A., Coleman, P.H., Harrison, A.K., and Gary, G.W., Jr., 1968, Colorado tick fever virus: An electron microscopic study, *Virology* **35**:28.

Murphy, F.A., Borden, E.C., Shope, R.E., and Harrison, A., 1971, Physicochemical and morphological relationships of some arthropod-borne viruses to bluetongue virus—a new taxonomic group: Electron microscopic studies, *J. Gen. Virol.* **13**:273.

Murphy, A.M., Albrey, M.B., and Crewe, E.B., 1977, Rotavirus infections of neonates, *Lancet* **2**:1149.

Murphy, A.M., Grohmann, G.S., and Sexton, M., 1983, Infectious gastroenteritis in Norfolk Island and recovery of viruses from drinking water, *Med. J. Aust.*

Newman, J.F.F., Brown, F., Bridger, J.C., and Woode, G.N., 1975, Characterization of a rotavirus, *Nature (London)* **258**:631.

Nicolaieff, A., Obert, G., and van Regenmortel, M.H.V., 1980, Detection of rotavirus by serological trapping on antibody-coated electron microscope grids, *J. Clin. Microbiol.* **12**:101.

Novo, E., and Esparza, J., 1981, Composition and topography of structural polypeptides of bovine rotavirus, *J. Gen. Virol.* **56**:325.

Obert, G., Gloeckler, R., Burckard, J., and van Regenmortel, M.H., 1981, Comparison of immunoabsorbent electron microscopy, enzyme immunoassay and counter-immunoelectrophoresis for detection of human rotavirus in stools, *J. Virol. Methods* **3**:99.

Ogra, P.L., and Karzon, D.T., 1969, Poliovirus antibody response in serum and nasal secretions following intranasal inoculation with inactivated poliovaccine, *J. Immunol.* **105**:15.

Ørstavik, I., and Haug, K.W., 1976, Virus-specific IgM antibodies in acute gastroenteritis due to a reovirus-like agent (rotavirus), *Scand. J. Infect. Dis.* **8**:237.

Ørstavik, I., Haug, K.W., and Søvde, 1976, Rotavirus-associated gastroenteritis in two adults probably caused by virus reinfection, *Scand. J. Infect. Dis.* **8**:277.

Palmer, E.L., Martin, M.L., and Murphy, F.A., 1977, Morphology and stability of infantile gastroenteritis virus: Comparison with reovirus and bluetongue virus, *J. Gen. Virol.* **35**:403.

Payne, C.M., Ray, C.G., and Yolken, R.H., 1981, The 30- to 54-nm rotavirus-like particles in gastroenteritis: Incidence and antigenic relationship to rotavirus, *J. Med. Virol.* **7**:299.

Pearson, G.R., and McNulty, M.S., 1977, Pathological changes in the small intestine of neonatal pigs infected with a pig reovirus-like agent (rotavirus), *J. Comp. Pathol.* **87**:363.

Pearson, G.R., and McNulty, M.S., 1979, Ultrastructural changes in small intestinal epithelium of neonatal pigs infected with pig rotavirus, *Arch. Virol.* **59**:127.

Petric, M., Szymanski, M.T., and Middleton, P.J., 1975, Purification and preliminary characterization of infantile gastroenteritis virus (orbivirus group), *Intervirology* **5**:233.

Petric, M., Middleton, P.J., Grant, C., Tam, J.S., and Hewitt C.M., 1978, Lapine rotavirus—preliminary studies on epizootology and transmission, *Can. J. Comp. Med.* **42**:143.

Petrie, B.L., Graham, D.Y., and Estes, M.K., 1981, Identification of rotavirus particle types, *Intervirology* **16**:20.

Ramia, S., and Sattar, S.A., 1979, Simian rotavirus SA 11 plaque formation in the presence of trypsin, *J. Clin. Microbiol.* **10**:609.

Ramia, S., and Sattar, S.A., 1980, Proteolytic enzymes and rotavirus SA 11 plaque formation, *Can. J. Comp. Med.* **44**:232.

Ramig, R.F., 1982, Isolation and genetic characterization of temperature-sensitive mutants of simian rotavirus SA 11, *Virology* **120**:93.

Reed, D.E., Daley, C.A., and Shave, H.J., 1976, Reovirus-like agent associated with neonatal diarrhoea in pronghorn antelope, *J. Wildl. Dis.* **12**:488.

Robbins, P.W., Hubbard, S.C., Turco, S.J., and Wirth, D.F, 1977, Proposal for a common oligosaccharide intermediate in the synthesis of membrane glycoproteins, *Cell* **12**:893.

Rodger, S.M., and Holmes, I.H., 1979, Comparison of the genomes of simian, bovine, and human rotaviruses by gel electrophoresis and detection of genomic variation among bovine isolates, *J. Virol.* **30**:839.

Rodger, S.M., Craven, J.A., and Williams, I., 1975a, Demonstration of reovirus-like particles in intestinal contents of piglets with diarrhoea, *Aust. Vet. J.* **51**:536.

Rodger, S.M., Schnagl, R.D., and Holmes, I.H., 1975b, Biochemical and biophysical characteristics of diarrhea viruses of human and calf origin, *J. Virol.* **16**:1229.

Rodger, S.M., Schnagl, R.D., and Holmes, I.H., 1977, Further biochemical characterization, including the detection of surface glycoproteins, of human, calf, and simian rotaviruses, *J. Virol.* **24**:91.

Rodger, S.M., Holmes, I.H., and Studdert, M.J., 1980, Characteristics of the genomes of equine rotaviruses, *Vet. Microbiol.* **5**:243.

Rodger, S.M., Bishop, R.F., Birch, C., McLean, B., and Holmes, I.H., 1981, Molecular epidemiology of human rotaviruses in Melbourne, Australia, from 1973–1979, as determined by electrophoresis of genome ribonucleic acid, *J. Clin. Microbiol.* **13**:272.

Rodriguez, W.J., Kim, H.W., Arrobio, J.O., Brandt, C.D., Chanock, R.M., Kapikian, A.Z., Wyatt, R.G., and Parrott, R.H., 1977, Clinical features of acute gastroenteritis associated with human reovirus-like agent in infants and young children, *J. Pediatr.* **91**:188.

Rodriguez, W.J., Kim, H.W., Brandt, C.D., Yolken, R.H., Arrobio, J.O., Kapikian, A.Z., Chanock, R.M., and Parrott, R.H., 1978, Sequential enteric illnesses associated with different rotavirus serotypes, *Lancet* **2**:37.

Rodriguez, W.J., Kim, H.W., Brandt, C.D., Yolken, R.H., Richard, M., Arrobio, J.O., Schwartz, R.H., Kapikian, A.Z., Chanock, R.M., and Parrott, R.M., 1979, Common exposure outbreak of gastroenteritis due to Type 2 rotavirus with high secondary attack rate within families, *J. Infect. Dis.* **140**:353.

Rodriguez, W.J., Kim, H.W., Brandt, C.D., Bise, B., Kapikian, A.Z., Chanock, R.M., Curlin, G., and Parrott, R.H., 1980, Rotavirus gastroenteritis in the Washington, D.C., area: Incidence of cases resulting in admission to the hospital, *Am. J. Dis. Child.* **134**:777.

Rogers, F.G., Chapman, S., and Whitby, H., 1981, A comparison of lyphogel, ammonium sulphate and ultracentrifugation in the concentration of stool viruses for electron microscopy, *J. Clin. Pathol.* **34**:227.

Roseto, A., Escaig, J., Delain, E., Cohen, J., and Scherrer, R., 1979, Structure of rotaviruses as studied by the freeze–drying technique, *Virology* **98**:471.

Roseto, A., Lema, F., Cavalieri, F., Dianoux, L., Sitbon, M., Ferchal, F., Lasneret, J., and Peries, J., 1980, Electron microscopy detection and characterization of viral particles in dog stools, *Arch. Virol.* **66**:89.

Rubenstein, D., Milne, R.G., Buckland, R., and Tyrrell, D.A.J., 1971, The growth of the virus of epidemic diarrhoea of infant mice (EDIM) in organ cultures in intestinal epithelium, *Br. J. Exp. Pathol.* **52**:442.

Ryder, R.W., McGowan, J.E., Jr., Hatch, M.H., and Palmer, E.L., 1977, Reovirus-like agent as cause of nosocomial diarrhea, *J. Pediatr.* **90**:698.

Ryder, R.W., Oquist, C.A., Greenberg, H., Taylor, D.N., Orskov, F., Orskov, I., Kapikian, A.Z., and Sack, R.B., 1981, Travelers diarrhea in Panamanan tourists in Mexico, *J. Infect. Dis.* **144**:442.

Saif, L.J., Theil, K.W., and Bohl, E.H., 1978, Morphogenesis of porcine rotavirus in porcine kidney cell cultures and intestinal epithelial cells, *J. Gen. Virol.* **39**:205.

Sanekata, T., Yoshida, Y., and Oda, K., 1979, Detection of rotavirus from faeces by reversed passive haemagglutination method, *J. Clin. Pathol.* **32**:963.

Sarkkinen, H.K., Meurman, O.H., and Halonen, P.E., 1979, Solid phase radioimmunoassay of IgA, IgG and IgM antibodies to human rotavirus, *J. Med. Virol.* **3**:281.

Sarkkinen, H.K., Tuokko, H., and Halonen, P.E., 1980, Comparison of enzyme immunoassay and radioimmunoassay for detection of human rotaviruses and adenoviruses in stool specimens, *J. Virol. Methods* **1**:331.

Sato, K., Inaba, Y., Shinozaki, T., Fujii, R., and Matsumoto, M., 1981, Isolation of human rotavirus in cell cultures, *Arch. Virol.* **69**:155.

Saulsbury, F.T., Winkelstein, J.A., and Yolken, R.H., 1980, Chronic rotavirus infection in immunodeficiency, *J. Pediatr.* **97**:61.

Scherrer, R., and Bernard, S., 1977, Application of enzyme-linked immunosorbent assay (ELISA) to the detection of calf rotavirus and rotavirus antibodies, *Ann. Microbiol. (Inst. Pasteur)* **128A**:499.

Schnagl, R.D., and Holmes, I.H., 1976, Characteristics of the genome of human infantile enteritis virus (rotavirus), *J. Virol.* **19**:267.

Schnagl, R.D., Rodger, S.M., and Holmes, I.H., 1981, Variation in human rotavirus electropherotypes occurring between rotavirus gastroenteritis epidemics in Central Australia, *Infect. Immun.* **33**:17.

Schroeder, B.A., Street, J.E., Kalmakoff, J., and Bellamy, A.R., 1982, Sequence relationships between the genome segments of human and animal rotavirus strains, *J. Virol.* **43**:379.

Scott, A.C., Luddington, J., Lucas, M., and Gilbert, F.R., 1978, Rotavirus in goats, *Vet. Rec.* **103**:145.

Shearman, D.J., Parkin, D.M., and McClelland, D.B., 1972, The demonstration and function of antibodies in the gastrointestinal tract, *Gut* **13**:483.

Shepherd, R.W., Truslow, S., Walker-Smith, J.A., Bird, R., Cutting, W., Darnell, R., and Barker, C.M., 1975, Infantile gastroenteritis: A clinical study of reovirus-like agent infection, *Lancet* **2**:1082.

Sheridan, J.F., Aurelian, L., Barbour, G., Santosham, M., Sack, R.B., and Ryder, R.W., 1981, Traveller's diarrhea associated with rotavirus infection: Analysis of virus-specific immunoglobulin classes, *Infect. Immun.* **31**:419.

Shinozaki, T., 1978, Hemagglutinin from human reovirus-like agent, *Lancet* **1**:877.

Shirley, J.A., Beards, G.M., Thouless, M.E., and Flewett, T.H., 1981, The influence of divalent cations on the stability of human rotaviruses, *Arch. Virol.* **67**:1.

Smith, E.M., Estes, M.K., Graham, D.Y., and Gerba, C.P., 1979, A plaque assay for the simian rotavirus SA 11, *J. Gen. Virol.* **43**:513.

Smith, M., and Tzipori, S., 1979, Gel electrophoresis of rotavirus RNA derived from six different animal species, *Aust. J. Exp. Biol. Med. Sci.* **57**:583.

Smith, M.L., Lazdins, I., and Holmes, I.H., 1980, Coding assignments of ds RNA segments of SA 11 rotavirus established by *in vitro* translation, *J. Virol.* **33**:976.

Snodgrass, D.R., and Herring, J.A., 1977, The action of disinfectants on lamb rotavirus, *Vet. Rec.* **101**:81.

Snodgrass, D.R., and Wells, P.W., 1976, Rotavirus infection in lambs: Studies on passive protection, *Arch. Virol.* **52**:201.

Snodgrass, D.R., and Wells, P.W., 1978, Passive immunity in rotaviral infections, *J. Am. Vet. Med. Assoc.* **173**:565.

Snodgrass, D.R., Smith, W., Gray, E.W., and Herring, J.A., 1976, A rotavirus in lambs with diarrhoea, *Res. Vet. Sci.* **20**:113.

Snodgrass, D.R., Angus, K.W., and Gray, E.W., 1977a, Rotavirus infection in lambs: Pathogenesis and pathology, *Arch. Virol.* **55**:263.

Snodgrass, D.R., Madeley, C.R., Wells, P.W., and Angus, K.W., 1977b, Human rotavirus in lambs: Infection and passive protection, *Infect. Immunol.* **16**:268.

Snodgrass, D.R., Angus, K.W., and Gray, E.W., 1979, A rotavirus from kittens, *Vet. Rec.* **104**:222.

Snodgrass, D.R., Fahey, K.J., Wells, P.W., Campbell, I., and Whitelaw, A., 1980, Passive immunity in calf rotavirus infections: Maternal vaccination increases and prolongs immunoglobulin $G_1$ antibody secretion in milk, *Infect. Immun.* **28**:344.

Soenarto, Y., Sebodo, T., Ridho, R., Alrasjid, H., Rohde, J.E., Bugg, H.C., Barnes, G.L., and Bishop, R.F., 1981, Acute diarrhoea and rotavirus infection in newborn babies and children in Yogyakarta, Indonesia, from June 1978 to June 1979, *J. Clin. Microbiol.* **14**:123.

Soike, K.F., Gary, G.W., and Gibson, S., 1980, Susceptibility of nonhuman primate species to infection by simian rotavirus SA 11, *Am. J. Vet. Res.* **41**:1098.

Sonza, S., and Holmes, I.H., 1980, Coproantibody response to rotavirus infection, *Med. J. Aust.* **2**:496.

Spence, L., Fauvel, M., Petro, R., and Bloch, S., 1976, Haemmagglutinin from rotavirus, *Lancet* **2**:1023.

Spence, L., Fauvel, M., Petro, R., and Bloch, S., 1977, Comparison of counterimmunoelectrophoresis and electron microscopy for laboratory diagnosis of human reovirus-like agent-associated infantile gastroenteritis, *J. Clin. Microbiol.* **5**:248.

Spence, L., Fauvel, M., Petro, R., and Babiuk, L.A., 1978, Comparison of rotavirus strains by hemagglutination inhibition, *Can. J. Microbiol.* **24**:353.

Stair, E.L., Mebus, C.A., Twiehaus, M.J., and Underdahl, N.R., 1973, Neonatal calf diarrhea: Electron microscopy of intestines infected with a reovirus-like agent, *Vet. Pathol.* **10**:155.

Stannard, L.M., and Schoub, B.D., 1977, Observations on the morphology of two rotaviruses, *J. Gen. Virol.* **37**:435.

Stoltzfus, C.M., Morgan, M., Banerjee, A.K., and Shatkin, A.J., 1974, Poly (a) polymerase activity in reovirus, *J. Virol.* **13**:1338.

Street, J.E., Croxson, M.C., Chadderton, W.F., and Bellamy, A.R., 1982, Sequence diversity of human rotavirus strains investigated by Northern blot hybridization analysis, *J. Virol.* **43**:369.

Stuker, G., Oshiro, L.S., and Schmidt, N.J., 1980, Antigenic comparisons of two new rotaviruses from rhesus monkeys, *J. Clin. Microbiol.* **11**:202.

Sutmoller, F., Azeredo, R.S., Lacerda, M.D., Barth, O.M., Pereira, H.G., Hoffer, E., and Schatzmayr, H.G., 1982, An outbreak of gastroenteritis caused by both rotavirus and *Shigella sonnei* in a private school in Rio de Janiero, *J. Hyg. (Cambridge)* **88**:285.

Suzuki, H., and Konno, T., 1975, Reovirus-like particles in jejunal mucosa of a Japanese infant with acute infectious non-bacterial gastroenteritis, *Tohoku J. Exp. Med.* **115**:199.

Svennerholm, A.-M., Holmgren, J., Hanson, L.A., Lindblad, B.S., Quereshi, F., and Rahmintoola, J., 1977, Boosting of secretory IgA antibody responses in man by parenteral cholera vaccination, *Scand. J. Immunol.* **6**:1345.

Tam, J.S., Szymanski, M.T., Middleton, P.J., and Petric, M., 1976, Studies on the particles of infantile gastroenteritis virus (orbivirus group), *Intervirology* **7**:181.

Tan, J.A., and Schnagl, R.D., 1981, Inactivation of a rotavirus by disinfectants, *Med. J. Aust.* **1**:19.

Theil, K.W., Bohl, E.H., and Agnes, A.G., 1977, Cell culture propagation of porcine rotavirus (reovirus-like agent), *Am. J. Vet. Res.* **38**:1765.

Theil, K.W., Bohl, E.H., Cross, R.F., Kohler, E.M., and Agnes, A.G., 1978, Pathogenesis of porcine rotaviral infection in experimentally inoculated gnotobiotic pigs, *Am. J. Vet. Res.* **39**:213.

Theil, K.W., McCloskey, C.M., Saif, L.J., Redman, D.R., Bohl, E.H., Hancock, D.D., Kohler, E.M., and Moorhead, P.D., 1981, A rapid, simple method for preparing rotaviral double-stranded ribonucleic acid for analysis by polyacrylamide gel electrophoresis, *J. Clin. Microbiol.* **14**:273.

Thouless, M.E., 1979, Rotavirus polypeptides, *J. Gen. Virol.* **44**:187.

Thouless, M.E., Bryden, A.S., and Flewett, T.H., 1977a, Rotavirus neutralization by human milk, *Br. Med. J.* **2**:1390.

Thouless, M.E., Bryden, A.S., Flewett, T.H., Woode, G.N., Bridger, J.C., Snodgrass, D.R., and Herring, J.A., 1977b, Serological relationships between rotaviruses from different species as studied by complement fixation and neutralization, *Arch. Virol.* **53**:287.

Thouless, M.E., Bryden, A.S., and Flewett, T.H., 1978, Serotypes of human rotavirus, *Lancet* **1**:39.

Todd, D., and McNulty, M.S., 1976, Characterization of pig rotavirus RNA, *J. Gen. Virol.* **33**:147.

Todd, D., and McNulty, M.S., 1977, Biochemical studies on a reovirus-like agent (rotavirus) from lambs, *Virology* **21**:1215.

Todd, D., McNulty, M.S., and Allan, G.M., 1980, Polyacrylamide gel electrophoresis of avian rotavirus RNA, *Arch. Virol.* **63**:87.

Torres-Medina, A., Wyatt, R.G., Mebus, C.A., Underdahl, N.R., and Kapikian, A.Z., 1976, Diarrhea caused in gnotobiotic piglets by the reovirus-like agent of human infantile gastroenteritis, *J. Infect. Dis.* **133**:22.

Tufvesson, B., and Johnsson, T., 1976, Immunoelectroosmophoresis for detection of reo-like virus: Methodology and comparison with electron microscopy, *Acta Pathol. Microbiol. Scand. Sect. B* **84**:225.

Tzipori, S., 1976, Human rotavirus in young dogs, *Med. J. Aust.* **2**:922.

Tzipori, S., and Caple, I.W., 1976, Isolation of a rotavirus from deer, *Vet. Rec.* **99**:398.

Tzipori, S., and Walker, M., 1978, Isolation of rotavirus from foals with diarrhoea, *Aust. J. Exp. Biol. Med. Sci.* **56**:453.

Tzipori, S., and Williams, I.H., 1978, Diarrhoea in piglets inoculated with rotavirus, *Aust. Vet. J.* **54**:188.

Tzipori, S.R., Makin, T.J., Smith, M.L., and Krautil, F.L., 1981, Clinical manifestations of diarrhea in calves infected with rotavirus and enterotoxigenic *Escherichia coli, J. Clin. Microbiol.* **13**:1011.

Urasawa, T., Urasawa, S., and Taniguchi, K., 1981, Sequential passages of human rotavirus in MA-104 cells, *Microbiol. Immunol.* **25**:1025.

Verly, E., and Cohen, J., 1977, Demonstration of size variation of RNA segments between different isolates of calf rotavirus, *J. Gen. Virol.* **35**:583.

Vollet, J.J., Ericsson, C.D., Gibson, G., Pickering, L.K., Du Pont, H.L., Kohl, S., and Conklin, R.H., 1979, Human rotavirus in an adult population with traveler's diarrhoea and its relationship to the location of food consumption, *J. Med. Virol.* **4**:81.

Von Bonsdorff, C.-H., Hovi, T., Makela, P., Hovi, L., and Tevalvoto-Aarn, M., 1976, Rotavirus associated with acute gastroenteritis in adults, *Lancet* **2**:423.

Ward, R.L., and Ashley, C.S., 1980, Effects of wastewater sludge and its detergents on the stability of rotavirus, *Appl. Envir. Microbiol.* **39**:1154.

Welch, A.B., 1971, Purification, morphology and partial characterization of a reovirus-like agent associated with neonatal calf diarrhea, *Can. J. Comp. Med.* **35**:195.

Welch, A.B., and Thompson, T.L., 1973, Physicochemical characterization of a neonatal calf diarrhea virus, *Can. J. Comp. Med.* **37**:295.

Welch, A.B., and Twiehaus, M.J., 1973, Cell culture studies of a neonatal calf diarrhea virus, *Can. J. Comp. Med.* **37**:287.

Wenman, W.M., Hinde, D., Feltham, S., and Gurwith, M., 1979, Rotavirus infections in adults: Results of a prospective family study, *N. Engl. J. Med.* **301**:303.

W.H.O., 1976, Treatment and prevention in diarrheal diseases—a guide for use at the primary level, W.H.O., Geneva.

W.H.O., 1979, The W.H.O. Diarrhoeal Diseases Control Programme, *WHO Weekly Epidemiol. Rec.* **16**:121.

W.H.O. Scientific Working Group, 1980, Rotavirus and other viral diarrhoeas, *Bull, W.H.O.* **58**:183.

Wilsnack, R.E., Blackwell, J.H., and Parker, J.C., 1969, Identification of an agent of epizootic diarrhea of infant mice by immunofluorescent and complement fixation tests, *Am. J. Vet. Res.* **30**:1195.

Wolf, J.L., Cukor, G., Blacklow, J.R., Dambrauskas, R., and Trier, J.S., 1981, Susceptibility of mice to rotavirus infection: Effects of age and administration of corticosteroids, *Infect. Immun.* **33**:565.

Woode, G.N., and Bridger, J.C., 1975, Viral enteritis of calves, *Vet. Rec.* **96**:85.

Woode, G.N., Jones, J., and Bridger, J., 1975, Levels of colostral antibodies against neonatal calf diarrhoea virus, *Vet. Rec.* **97**:148.

Woode, G.N., Bridger, J., Hall, G.A., Jones, J.M., and Jackson, G., 1976a, The isolation of reovirus-like agents (rotaviruses) from acute gastroenteritis of piglets, *J. Med. Microbiol.* **9**:203.

Woode, G.N., Bridger, J.C., Jones, J.M., Flewett, T.H., Bryden, A.S., Davies, H.A., and White, G.B.B., 1976b, Morphological and antigenic relationships between viruses (rotaviruses) from acute gastroenteritis of children, calves, piglets, mice and foals, *Infect. Immunol.* **14**:804.

Woode, G.N., Bew, M.E., and Dennis, M.J., 1978, Studies on cross protection induced in calves by rotaviruses of calves, children and foals, *Vet. Rec.* **103**:32.

Wyatt, R.G., Kapikian, A.Z., Thornhill, T.S., Sereno, M.M., Kim, H.W., and Chanock, R.M., 1974, *In vitro* cultivation in human fetal intestinal organ culture of a reovirus-like agent associated with nonbacterial gastroenteritis in infants and children, *J. Infect. Dis.* **130**:523.

Wyatt, R.G., Gill, V.W., Sereno, M.M., Kalica, A.R., Van Kirk, D.H., Chanock, R.M., and Kapikian, A.Z., 1976a, Probable *in vitro* cultivation of human reovirus-like agent of infantile diarrhoea, *Lancet* **1**:98.

Wyatt, R.G., Sly, D.L., London, W.T., Palmer, A.E., Kalica, A.R., Van Kirk, D.H., Chanock, R.M., and Kapikian, A.Z., 1976b, Induction of diarrhea in colostrum-deprived newborn rhesus monkeys with the human reovirus-like agent of infantile gastroenteritis, *Arch. Virol.* **50**:17.

Wyatt, R.G., Mebus, C.A., Yolken, R.H., Kalica, A.R., James, H.D., Jr., Kapikian, A.Z., and Chanock, R.M., 1979a, Rotaviral immunity in gnotobiotic calves: Heterologous resistance to human virus induced by bovine virus, *Science* **203**:548.

Wyatt, R.G., Yolken, R.H., Urrutia, J.J., Mata, L., Greenberg, H.B., Chanock, R.M., and Kapikian, A.Z., 1979b, Diarrhea associated with rotavirus in rural Guatemala: A longitudinal study of 24 infants and young children, *Am. J. Trop. Med. Hyg.* **28**:325.

Wyatt, R.G., James, W.D., Bohl, E.H., Theil, K.W., Saif, L.J., Kalica, A.R., Greenberg, H.B., Kapikian, A.Z., and Chanock, R.M., 1980, Human rotavirus type 2: Cultivation *in vitro*, *Science* **207**:189.

Wyatt, R.G., Greenberg, H.B., James, W.D., Pittman, A.L., Kalica, A.R., Flores, J., Chanock, R.M., and Kapikian, A.Z., 1982, Definition of human rotavirus serotypes by plaque reduction assay,

Yolken, R.H., Kim, H.W., Clem, T., Wyatt, R.G., Kalica, A.R., Chanock, R.M., and Kapikian, A.Z., 1977, Enzyme linked immunosorbent assay (ELISA) for detection of human reovirus-like agent of infantile gastroenteritis, *Lancet* **2**:263.

Yolken, R.H., Mata, L., Garcia, B., Urrutia, J.J., Wyatt, R.G., Chanock, R.M., and Kapikian, A.Z., 1978a, Secretory antibody directed against rotavirus in human milk—measurement by means of enzyme-linked immunosorbent assay (ELISA), *J. Pediatr.* **93**:916.

Yolken, R.H., Wyatt, R.G., Zissis, G., Brandt, C.D., Rodriguez, W.J., Kim, H.W., Parrott, R.H., Urrutia, J.J., Mata, L., Greenberg, H.B., Kapikian, A.Z., and Chanock, R.M., 1978b, Epidemiology of human rotavirus types 1 and 2 as studied by enzyme-linked immunosorbent assay, *N. Engl. J. Med.* **299**:1156.

Zamotin, B.A., Libüäinen, L.T., Bortnik, F.L., Chernetskaia, E.P., and Enina, Z.I., 1981, Waterborne outbreaks of rotavirus infections, *Zh. Mikrobiol. Epidemiol. Immunobiol.* **11**:100.

Zissis, G., and Lambert, J.P., 1978, Different serotypes of human rotavirus, *Lancet* **1:**38.

Zissis, G., and Lambert, J.P., 1980, Enzyme-linked immunosorbent assays adapted for serotyping of human rotavirus strains, *J. Clin. Microbiol.* **11:**1.

Zissis, G., Lambert, J.P., Kapsenberg, J.G., Enders, G., and Mutanda, L.N., 1981, Human rotavirus serotypes, *Lancet* **1:**944.

## CHAPTER 9

# Cytoplasmic Polyhedrosis Viruses

CHRISTOPHER C. PAYNE AND PETER P. C. MERTENS

## I. INTRODUCTION

The discovery that cytoplasmic polyhedrosis viruses (CPVs) contain a ten-segmented double-stranded RNA (dsRNA) genome confirmed that they have many biochemical characteristics in common with members of the Reoviridae (Kalmakoff *et al.*, 1969; Fujii-Kawata *et al.*, 1970). They are most clearly distinguished from other members of this family in having virus particles that lack a double capsid layer and that are commonly occluded within large proteinaceous inclusion bodies ("polyhedra") during replication (Matthews, 1979). In addition, their host range appears to be restricted to arthropods, and there is only one report of a CPV infection outside the Insecta (Federici and Hazard, 1975). The limited studies that have been carried out also suggest that they are genetically distinct from other genera within the Reoviridae (Black and Knight, 1970; Martinson and Lewandowski, 1974; Kodama and Suzuki, 1973). For these reasons, they have been classified as a distinct genus, but no precise generic name has yet been allocated (Matthews, 1979).

Historically, there have been two main reasons for studying this group of viruses, both linked to the economic importance of their insect hosts. In the first instance, a CPV was recognized as causing disease in the silkworm, *Bombyx mori*, and was associated with a complex of viruses that have caused extensive losses in the silk industry, particularly in Japan (Aruga, 1971). Much research has therefore been carried out on

CHRISTOPHER C. PAYNE • Glasshouse Crops Research Institute, Littlehampton, West Sussex BN16 3PU, United Kingdom.    PETER P. C. MERTENS • Animal Virus Research Institute, Pirbright, Woking, Surrey GU24 0NF, United Kingdom.

the silkworm CPV, which is the type species for the CPV group. This virus has proved a useful model for studies of molecular biology, in particular the investigation of RNA transcription. The detection of an RNA-polymerase associated with purified virus particles (Lewandowski *et al.*, 1969) led to the first discovery of the "blocked" and methylated 5' cap structure in messenger RNA (mRNA) molecules transcribed *in vitro* (Furuichi and Miura, 1975) and the allosteric control of polymerase activity by *S*-adenosyl-L-methionine (Furuichi, 1974, 1978a, 1981; Mertens and Payne, 1978). These are described in detail in Section III.A.1.

A further motivation for CPV research lies in the fact that they are found as pathogens of many pest insect species (particularly Lepidoptera) and have been considered as potential insect-control agents. Although one isolate is commercially available in Japan for control of the pine caterpillar, *Dendrolimus spectabilis* (Katagiri, 1981), CPVs have received less attention in this role than baculoviruses (Payne, 1982). This is partly because they are biochemically similar to viruses that are known to be pathogenic to vertebrates and plants and partly because they often produce chronic disease in the insect host, rather than a lethal infection. The chronic nature of the disease is a consequence of the pronounced tissue tropism of the viruses for the gut epithelial cells of the host, which have an extensive regenerative capacity (Payne, 1981). Since CPVs have no recorded adverse effects on test vertebrates (Katagiri, 1981) and induce sublethal effects in the host that would be significant in the control of pest insect populations (Payne, 1982), it is likely that the use of CPVs as pest-control agents will be reassessed in the future.

CPV isolates have been recorded in approximately 200 species of insects (Martignoni and Iwai, 1977, 1981), with well-documented evidence of infections in Lepidoptera, Hymenoptera, and Diptera. Genotype analysis of the RNA segments of several isolates has revealed considerable genetic heterogeneity among viruses, probably greater than that observed in any other genus of the Reoviridae (Payne and Rivers, 1976). This is not surprising, considering the vast number of host species that are susceptible to CPV infection. The genetic complexity of the group emphasizes, however, that what may be true for one particular CPV isolate does not necessarily apply to all CPVs. Other isolates may differ both structurally and in the biology of the infection process.

Much of the literature, and hence much of this review, concentrates on *B. mori* (type 1) CPV. Other virus isolates will be referred to by the host from which the virus was originally isolated and the virus "type" (where this is known) as described in a provisional classification of CPVs based on the electrophoretic mobilities of the viral RNA genome segments (Payne and Rivers, 1976). The review will examine sequentially the morphology and biochemistry of the structural components of the virus, RNA transcription and translation *in vitro*, and the biology of infection *in vivo*. Other reviews that will be found useful in conjunction with this chapter are those on CPV general biology (Aruga and Tanada,

1971; Wood, 1973; Joklik, 1974; K.M. Smith, 1976; Tinsley and Harrap, 1978), morphology (Payne and Harrap, 1977), pathogenesis (Payne, 1981), and molecular biology (Harrap and Payne, 1979; Miura, 1981).

## II. STRUCTURAL PROPERTIES OF POLYHEDRA AND VIRUS PARTICLES

### A. Morphological and Biophysical Properties

#### 1. Polyhedra

Virus infections in a number of insect virus groups [cytoplasmic polyhedrosis viruses (CPVs), baculoviruses, and entomopoxviruses] generally culminate in the occlusion of between one and several thousand mature virus particles within large proteinaceous inclusion bodies (IBs) or polyhedra. These represent a terminal stage of infection, are highly stable, and serve as the vehicle for the transmission of "packages" of infectious virus from one susceptible host to another. Polyhedra formed in the cytoplasm of infected midgut epithelial cells are a characteristic sign of CPV infection. They were first described by Ishimori (1934) in diseased silkworms, but it was not until the studies of K.M. Smith and Wyckoff (1950) that CPV polyhedra were distinguished from other insect virus IBs.

CPV polyhedra can be readily purified from infected insects by differential centrifugation, and banding in quasi-equilibrium sucrose gradients, in which they have a density of about 1.28 g/cm³ (Martignoni, 1967). They vary considerably in both size and shape (Fig. 1A–C). In infections of Lepidoptera, polyhedra ranging in size from 0.2 to 10 μm in diameter have been recorded (Ignoffo and Adams, 1966; Stairs et al., 1968), although the size range 1–3 μm is more common (Payne, 1981). Polyhedra observed in CPV infections in Hymenoptera and Diptera (particularly in *Simulium* spp.) are generally smaller (Longworth and Spilling, 1970; Bailey et al., 1975; Weiser, 1978).

Thin sections of infected cells or purified polyhedra reveal their main structural features; isometric virus particles are occluded at random within a crystalline protein lattice (Fig. 1D, E). The virus particles are composed of an electron-dense core surrounded by an outer, capsid layer. Arnott et al. (1968) calculated that up to 10,000 virus particles were occluded within a single polyhedron of *Danaus plexippus* CPV, and the lattice structure appeared to be disrupted only slightly, if at all, by the presence of virus particles (Bergold and Suter, 1959; Arnott et al., 1968). The protein matrix ("polyhedrin") appears to be arranged as a face-centered cubic lattice with center-to-center spacing varying between 4.1 and 7.4 nm (Bergold and Suter, 1959; Arnott et al., 1968; Stoltz and Hilsenhoff, 1969; Longworth and Spilling, 1970). No other morphologically recog-

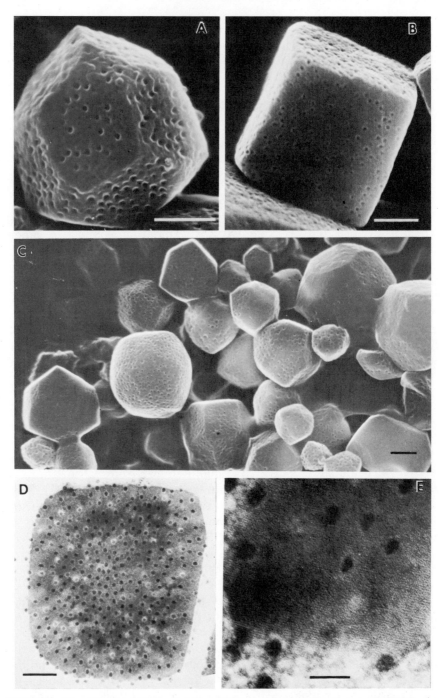

FIGURE 1. Morphology of CPV polyhedra. (A) Scanning electron micrograph (SEM) of a polyhedron of type 2 CPV from *Inachis io* illustrating its dodecahedral shape. Scale bar: 1 μm. (B) SEM of the strain of type 1 CPV from *Bombyx mori* that produces cuboidal polyhedra. Scale bar: 1 μm. (C) SEM of purified polyhedra of type 2 CPV showing the wide variation

nizable structures, such as cellular organelles, appear to be occluded within polyhedra.

Unlike the polyhedra of some other insect virus groups [particularly baculoviruses (Harrap, 1972)], CPV polyhedra are not surrounded by a limiting layer (Hills and Smith, 1959) and scanning electron microscope (SEM) pictures reveal cavities at the surface that are consistent in size with virus particles (Fig. 1A and B). Free virus particles are often found in stored aqueous suspensions of polyhedra, presumably released from the polyhedral surface (Bird, 1965).

While most CPV polyhedra occlude large numbers of virions, very small IBs containing single virions have been observed in CPV infections of anopheline mosquitoes (Anthony et al., 1973), while larger and irregularly shaped IBs that appear to have arisen by a process of fusion of numerous singly occluded particles have been observed in culicines (Clark et al., 1969; Andreadis, 1981) and chironomids (Stoltz and Hilsenhoff, 1969). Major differences in polyhedral shape have also been observed in CPV infections of Lepidoptera. The most frequent distinction is between the polyhedral and cubic shapes shown in Fig. 1A and B. Hukuhara and Hashimoto (1966a) observed, in thin section, three variants of B. mori CPV producing icosahedral, hexahedral (cubic), and triangular polyhedra. These structures were analyzed further by SEM (Rao, 1973), confirming that the icosahedral strain was typically a rhombic dodecahedron, the hexahedron was cubic, and the triangular form was a tetragonal tristetrahedron.

It is still not clear what governs the shape of polyhedra. Certainly the hexahedral and icosahedral strains of B. mori CPV retain their shape after several passages (Aruga, 1971). This would suggest that the shape is controlled by the virus genome. The matrix protein (polyhedrin) is virus-coded (Section III.B.2), and different polyhedra shapes could reflect some structural changes in this protein. On the other hand, there is increasing evidence that different cellular conditions may also affect polyhedron shape (Belloncik and Bellemare, 1980). On the few occasions when CPVs have been successfully grown in cell cultures, it has often been noted that a significant proportion of cubic polyhedra are produced (Grace, 1962; Granados et al., 1974; Quiot and Belloncik, 1977; Belloncik and Bellemare, 1980; Longworth, 1981). Thus, in the infection of a Lymantria dispar cell line with a type 5 CPV of Euxoa scandens, the polyhedra produced were exclusively cubic, while in larvae, the infection produced 99% spherical polyhedra and 1% cubic forms. When the cubic polyhedra from cell-culture infections were used to infect insects, pre-

---

in size of individual inclusion bodies. The sockets at the surfaces of polyhedra represent sites from which virus particles have been lost during purification. Scale bar: 1 μm. (D) Thin section of type 1 CPV polyhedron illustrating the large number of virions embedded within the matrix protein (polyhedrin). Scale bar: 500 nm. (E) Lattice structure of polyhedrin in Anoplonyx destructor CPV polyhedra. Scale bar: 100 nm.

dominantly spherical polyhedra were produced again (Belloncik and Bel-lemare, 1980). In addition, the electrophoretic mobilities of the polypep-tides of polyhedra from insect or cell culture were indistinguishable (Grancher-Barray *et al.*, 1981), suggesting that selection from a mixed population of distinct viruses was unlikely. Such studies suggest that the factors that control polyhedral shape may be complex and indicate that distinctive shapes are unlikely to be conclusive criteria for the recogni-tion of different virus strains, even though they were used for this purpose in early cross-transmission studies (Tanada and Chang, 1962; Tanaka, 1971).

Polyhedra are highly resistant to nonionic or ionic detergents, and they may be purified without apparent harm in solutions containing up to 1% sodium dodecyl sulfate (SDS) (Mertens, 1979). The polyhedra can be dis-solved at high pH, a process that presumably mimics *in vivo* breakdown in the guts of insect larvae during the first stages of infection (Section IV.A). Under conditions of controlled pH and ionic strength, intact virus particles can be released. These virions constitute approximately 2–5% of the total weight of polyhedra (Hukuhara and Hashimoto, 1966b; Payne, 1971). Optimal conditions for the release of virions may vary with the particular virus isolate, but solutions of sodium carbonate, or carbonate–bicarbonate buffers at pH values above pH 10.5, have proved most effec-tive (Hills and Smith, 1959; Hukuhara and Hashimoto, 1966b; Cun-ningham and Longworth, 1968; Hayashi and Bird, 1970). Recently, it has been found that trypsin and chymotrypsin will also completely dissolve polyhedra of *Heliothis armigera* CPV at pHs as low as 8.0 (R. Rubinstein and A. Polson, personal communication), and it may be that a virion-and/or polyhedron-associated protease is also involved in the liberation of virus particles (Section III.A.2.b).

## 2. Virus Particles

Although large numbers of CPV particles may be occluded within polyhedra, many may remain "nonoccluded" even at an advanced stage of infection. Whereas Longworth (1981) observed few nonoccluded par-ticles in a CPV-infected culture of *Spodoptera frugiperda* cells, only 12 and 45% of the virions produced during CPV infections in *Malacosoma disstria* and *B. mori*, respectively, were occluded within polyhedra (Hay-ashi, 1970a; Hayashi *et al.*, 1970). The nonoccluded virions can be readily purified and appear to be almost indistinguishable from the virions that can be released from polyhedra by alkali treatment (Payne and Kalmakoff, 1974a), although rather more apparently empty particles have been ob-served in nonoccluded virus samples (Hayashi and Bird, 1968a). Virions from either source are infectious for insects (Hukuhara and Hashimoto, 1966b; Miyajima and Kawase, 1967; Kawase and Miyajima, 1969).

Some of the morphological and biophysical properties of the virions of type 1 CPV from *B. mori* are summarized in Table I. The virions are

TABLE I. Morphological and Biophysical Properties of Virus Particles of *Bombyx mori* (Type 1) Cytoplasmic Polyhedrosis Virus

| Characteristics | | References |
|---|---|---|
| Morphology | Icosahedral; 55–69 nm diameter<br>Twelve spikes, about 20 nm long, 15–23 nm wide<br>Core: 35 nm<br>Single capsid | Hosaka and Aizawa (1964), Miura *et al.* (1969), Yazaki and Miura (1980), Hatta and Francki (personal communication) |
| Sedimentation coefficient | Intact particles: 400–440 S; "empty" particles: 260 S | Hukuhara and Hashimoto (1966b), Kawase (1967), Kalmakoff *et al.* (1969), Miyajima *et al.* (1969) |
| Buoyant density (CsCl) | Intact particles: 1.44 g/cm³; "empty" particles: 1.30 g/cm³ | Lewandowski and Traynor (1972), Payne and Kalmakoff (1974a) |
| OD 260:280 | 1.7 | Hayashi *et al.* (1970) |
| RNA (%) | 23–30% | Nishimura and Hosaka (1969), Kalmakoff *et al.* (1969) |
| Molecular weight | 54 × 10⁶ | Kalmakoff *et al.* (1969) |

icosahedral with a hexagonal or circular outline (Fig. 2A and B) (Hills and Smith, 1959; Hosaka and Aizawa, 1964). Size measurements for negatively stained *B. mori* CPV virus particles range from 55 nm (T. Hatta and R.I.B. Francki, personal communication) to 69 nm (Hosaka and Aizawa, 1964), and the range is even greater among other isolates, with a mean diameter of 36 nm given for *Pectinophora gossypiella* CPV (Ignoffo and Adams, 1966) and 80–90 nm for *Noctua* (= *Triphaena*) *pronuba* CPV (Lipa, 1970). In laboratories where several CPV isolates have been examined together, the particle dimensions quoted are reasonably uniform (Cunningham and Longworth, 1968), suggesting that much of this variation can be attributed more to different techniques than to fundamental differences in structure among isolates. In thin sections, particles possess an electron-dense core surrounded by an outer shell (Wittig *et al.*, 1960; Xeros, 1966; Arnott *et al.*, 1968; Andreadis, 1981). The size of the central core (about 35 nm) in which the viral RNA is presumably located is similar to that observed in reovirus and orbivirus particles, which also have genomes of comparable size (Joklik, 1974). Some authors have suggested that the diameter of the virion appears to be increased by as much as 30% following occlusion (Arnott *et al.*, 1968; Longworth and Spilling, 1970), but this may simply be attributable to a clearer definition of the virus capsid in thin sections (Payne, 1981).

Although an early model of the structure of CPV virions proposed that particles contain two concentric protein coats or capsids (Hosaka and Aizawa, 1964), electron micrographs show no distinct double-capsid structure of the reovirus type, and empty particles clearly consist of one

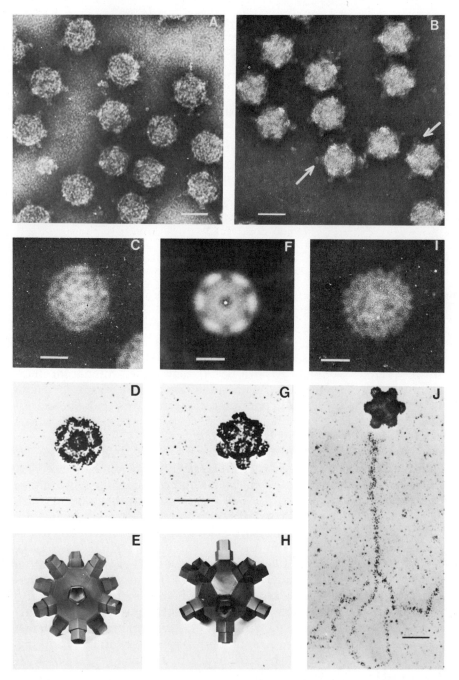

FIGURE 2. Morphology of CPV virus particles. (A, B) Negatively stained virus particles of *Phalera bucephala* and type 1 (*Bombyx mori*) CPV virus particles, respectively, released from polyhedra by treatment with dilute alkali. (→) Possible subunit structure of the virus spike. Scale bar: 50 nm. (C) Type 1 CPV virions arranged on a presumptive five-fold vertex showing apparently hollow morphological subunits, probably positioned at the base of a

capsid layer (Stoltz, 1969; Stoltz and Hilsenhoff, 1969; Lewandowski and Traynor, 1972). In fact, many of the properties of CPV particles are similar to those of subviral particles or "cores" of other reoviruses including reovirus, Fiji disease virus, and leafhopper A virus (LAV) (Joklik, 1974; Payne and Tinsley, 1974; Hatta and Francki, personal communication), and it may be that the polyhedrin in CPVs serves the function of the outer capsid layer that is present in these other viruses.

Probably the most outstanding structural feature of CPV virions are the projections or spikes that can be clearly seen in negatively stained particles (Fig. 2A,B,G). These projections led Yazaki and Miura (1980) to describe the structure of the virion aptly as "a 12-starred pyrotechnic mine," since the spikes are located at each of the 12 vertices of the icosahedral particle. Thus, in particles the central axis of which is situated on a presumptive three-fold vertex, six projections can be clearly seen (Fig. 2G and H). Optical rotation of such particles at 60° intervals illustrates their hexagonal symmetry and indicates that there are discontinuities in the capsid layer at the base of the spikes (Fig. 2F) and that the spikes are hollow, with a central channel about 10 nm wide (Miura *et al.*, 1969). Material from the virus core appears to extend into the spike (Bird, 1965; Lewandowski and Traynor, 1972), and the release of core material, including RNA, from the spike (Fig. 2J) could account for the presence of "tail"-like structures in some particles (Arnott *et al.*, 1968; Lewandowski and Traynor, 1972).

The precise structure of the spikes is unresolved, although they are clearly segmented. Hosaka and Aizawa (1964) suggested that each spike was composed of up to four subunits, each approximately 7.5 nm in length. Miura *et al.* (1969) concluded that the spikes were 20 nm long and segmented, while Asai *et al.* (1972) observed 12-nm spherical particles on top of the spikes. None of these studies differs significantly from the recent structural interpretation of Hatta and Francki (personal communication) illustrated in the models shown in Fig. 2E and H. They observed the close morphological similarity between CPVs and cores of LAV obtained from *Cicadulina bimaculata* (Boccardo *et al.*, 1980) and suggested that the CPV spike consisted of two sections and was similar to that observed for Fijivirus and related plant reoviruses (see Chapter 10). The shorter section or "B spike" attached to the virus capsid was 7–9 nm high

---

spike. Scale bar: 25 nm. (D) Type 1 CPV virions, negatively stained and shadowed, positioned as in (C). Scale bar: 50 nm. (E) Model depicting the same aspect as particles in (C) and (D). (F) Particle on a presumptive three-fold vertex enhanced by $n = 6$ rotations, illustrating the discontinuities in the capsid that probably represent the base of the spike. Scale bar: 25 nm. (G) Type 1 CPV virion, negatively stained and shadowed, positioned as in (F). Scale bar: 50 nm. (H) Model depicting the same aspect as particles in (F) and (G). (I) Type 1 CPV virions showing evidence of 20 peripheral capsomers. Scale bar: 25 nm. (J) Mild disruption of type 1 CPV virions with EDTA. Genome RNA is released from a spike, giving a "tail-like" appearance. Scale bar: 50 nm. Taken from Payne and Harrap (1977) (C, F, I) and from Yazaki and Miura (1980) (D, G, J) with the kind permission of the authors and Academic Press.

and 13–15 nm wide [compared with 23 nm wide Yazaki and Miura (1980)], and superimposed on this was the narrower "A spike," 8–9 nm high and 9–11 nm wide. Such an interpretation, relating the virion structures of insect and plant reoviruses, is an attractive hypothesis.

Although morphological subunits are visible on the surface of CPV virions, it has not been possible to define the number of capsomers and their spatial arrangement with certainty. Particles apparently on their five-fold axis of symmetry can, like reovirus, sometimes be seen with 20 capsomers visible around their circumference (Fig. 2I) (Lewandowski and Traynor, 1972; Luftig et al., 1972; Joklik, 1974; Payne and Harrap, 1977). In this axis, a pentagonal arrangement of hollow, five-sided morphological units can be observed apparently around the base of a spike (Fig. 2C). These units were first described by Hosaka and Aizawa (1964) as being 16–20 nm in diameter with an inner diameter of 3–6 nm. These authors suggested that CPV virions had 12 "capsomeres" composed of a spike and its surrounding morphological units. However, a more likely interpretation would be that the virus is composed of 12 spikes plus 20 hollow morphological units arranged effectively on the 20 faces of an icosahedron.

While the morphology of CPV virions most closely resembles that of the subviral particles of a number of reoviruses, their biophysical properties are also most akin to those of reovirus cores. The sedimentation coefficients measured for B. mori CPV virions range from 400 to 440 S (Table I), while similar values (373–399 S) have been observed for other CPV isolates (Cunningham and Longworth, 1968; Hayashi, 1968; Hayashi and Bird, 1968b; Rubinstein, 1979). "Empty" particles lacking the central core have lower sedimentation coefficients of 215–260 S (Hayashi and Bird, 1968a; Miyajima et al., 1969). Although the first report of CPV virion buoyant density in CsCl was consistent with the density of intact reovirus particles [1.37 g/cm$^3$ (Nishimura and Hosaka, 1969)], values obtained from more recent studies are consistently higher and average 1.44 g/cm$^3$ (Lewandowski and Traynor, 1972; Payne and Kalmakoff, 1974a; Rubinstein, 1979). Less dense virus particles (1.425 g/cm$^3$) lack some of the smaller RNA genome segments, while particles with a density of 1.3 g/cm$^3$ contain no RNA (Lewandowski and Traynor, 1972).

The absorbance ratio of purified particles at 260 : 280 nm ranges from 1.5 to 1.8 (Hayashi and Bird, 1968a, 1968b; Hayashi et al., 1970), and the E$_{260}$ of between 3.7 and 7 (Miyajima et al., 1969; Hayashi and Bird, 1970) reflects the high RNA content of the particles, which has generally been estimated as 23–30% (Hayashi and Bird, 1968a; Nishimura and Hosaka, 1969; Kalmakoff et al., 1969). Values for total virion molecular weights of 43.4–51.5 × 10$^6$ for H. armigera CPV (Rubinstein, 1979) or 54 × 10$^6$ for B. mori CPV (Kalmakoff et al., 1969) are most consistent with an RNA content of about 30% and a total RNA molecular weight of about 15 × 10$^6$ (Section II.C).

CPV particles appear to be extremely stable, their infectivity being retained for several weeks after storage at $-15$, 5, or 25°C (Hukuhara and Hashimoto, 1966b). They are not disrupted by treatment with trypsin, chymotrypsin, ribonuclease A, deoxyribonuclease, or phospholipase C at enzyme concentrations of 50–100 μg/ml (Hayashi and Bird, 1968b; Lewandowski and Traynor, 1972). The particles are also stable in detergents such as sodium deoxycholate at 0.5–1% (Hayashi and Bird, 1968b; Hayashi et al., 1970), but are completely disrupted in 0.5–1% SDS. Sodium carbonate treatment at 0.05 M, which is ideal for releasing virus particles from polyhedra, only disrupts them at elevated temperatures (Hayashi et al., 1970). One or two fluorocarbon treatments do not affect virus infectivity (Miyajima et al., 1969), but ethanol or heat treatment at 60°C for 60 min leads to the release of RNA from the virions (Hayashi and Bird, 1968b; Richards, 1970; Payne and Kalmakoff, 1974b).

## B. Structural Proteins

### 1. Virus Particles

There are relatively few studies of CPV structural proteins and, once again, the majority have concentrated on type 1 CPV from B. mori. Five polypeptides have been consistently observed on SDS–polyacrylamide gels, using phosphate-buffered, "Laemmli," or gradient gels (Fig. 3a) (Lewandowski and Traynor, 1972; Payne and Kalmakoff, 1974a; Payne and Rivers, 1976; Mertens, 1979). The estimated molecular weights of these proteins are shown in Table II. In further discussions, the polypeptides will be referred to by the molecular weights ($\times 10^{-3}$) attributed to them on a phosphate gel system by Payne and Rivers (1976). These values are very similar to the estimates of Lewandowski and Traynor (1972) (Table II).

TABLE II. Virus-Particle Polypeptides of *Bombyx mori* (Type 1) Cytoplasmic Polyhedrosis Virus

| Polypeptide | Molecular weight ($\times 10^{-3}$) | | Proportion of total protein[a] | Number of copies per virion[c] |
| --- | --- | --- | --- | --- |
| | a | b | | |
| 1 | 151[d] | 146 | 36 | 82 |
| 2 | 142 | 138 | 16 | 39 |
| 3 | 130[d] | 125 | 34 | 91 |
| 4 | 67 | 70 | 2 | 10 |
| 5 | 33 | 31 | 10 | 108 |

[a] Data from Lewandowski and Traynor (1972).
[b] Data from Payne and Rivers (1976).
[c] Estimated from a virion molecular weight of $54 \times 10^6$ (Kalmakoff et al., 1969), a total RNA molecular weight of $14.63 \times 10^6$ (Fujii-Kawata et al., 1970), and hence an estimated total protein weight of $33.37 \times 10^6$. The proportion by weight of each polypeptide and their individual molecular weights (Payne and Rivers, 1976) were used to calculate the number of copies per virion.
[d] Identified as surface components (Lewandowski and Traynor, 1972).

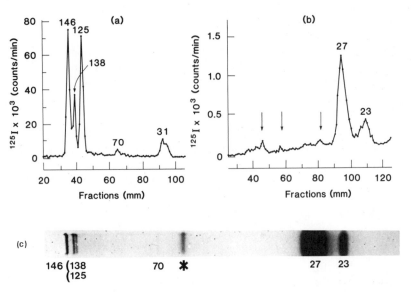

FIGURE 3. Structural polypeptides of type 1 CPV from *Bombyx mori*. (a, b) Electrophoresis on phosphate-buffered SDS–5% polyacrylamide gels (PAGE) of iodinated proteins from purified virions and polyhedra, respectively. Results taken from Lewandowski and Traynor (1972), with the kind permission of the authors and the American Society for Microbiology. The molecular weights ($\times$ 10$^{-3}$) of the structural proteins are taken from Payne and Rivers (1976). (c) SDS-PAGE of the polypeptides of polyhedra separated on 12% Laemmli gels (Laemmli, 1970; Mertens, 1979). It is possible, though not yet certain, that the mobilities of polypeptides 138 and 125 are inverted on this gel system, as are some reovirus proteins (McCrae and Joklik, 1978). The virion polypeptide 31 is masked by the polyhedrin band. (*) Position of a probable dimer of polyhedrin.

    Three of the virion polypeptides have molecular weights in excess of 100,000. Lewandowski and Traynor (1972) iodinated SDS-disrupted virions and calculated the proportion of radioactivity in each, making it possible to estimate the approximate number of copies of each polypeptide in the virion (Table II). When intact virions were iodinated, the relative amounts of $^{125}$I in polypeptides 146 and 125 increased, or remained the same, while radioactivity in the others was lowered, suggesting that these two polypeptides are surface components. Storage of CPV virions leads to degradation of the largest protein (146) and the appearance of another polypeptide with a molecular weight of about 5000 less (Section III.A.2.b) (Lewandowski and Traynor, 1972; Payne and Kalmakoff, 1974a).

    The virion polypeptide composition bears a close resemblance to that of reovirus cores, in which there are three proteins in excess of 100,000, one at 80,000, and one at 38,000, plus a sixth, minor component at 70,000 (Joklik, 1980). It has also been noted that the molecular weights of the structural polypeptides are close to the theoretical sizes of the primary gene products of some of the viral RNA segments (Lewandowski and

Traynor, 1972). The assignment of specific polypeptides to genome RNAs is discussed in detail in Section III.B.2. Several enzyme activities are associated with the virus particles (Section III.A), and virions also hemagglutinate sheep, mouse, and chicken erythrocytes (Miyajima and Kawase, 1969), but the specific locations of the enzymes or hemagglutinin are not yet clear (Section III.A.1.g).

The viral polypeptides of some CPV isolates are distinct from those of type 1 CPV and from other types. One consistent feature is that at least two and more commonly three polypeptides with molecular weights in excess of 100,000 have been resolved (Payne and Rivers, 1976; Mertens, 1979; Grancher-Barray et al., 1981).

## 2. Polyhedra

CPV polyhedra contain at least one major virus-specific polypeptide in addition to the virus particle structural proteins (Fig. 3b) (Lewandowski and Traynor, 1972; Payne and Kalmakoff, 1974a; Payne and Rivers, 1976). The molecular weight of this component may differ for different CPV isolates, but is within the range 25,000–37,000, and represents the alkali-soluble matrix protein fraction (polyhedral protein or polyhedrin) of the IB. This protein appears to be glycosylated as judged by periodic acid–Schiff's reagent staining (Payne and Tinsley, 1974) or [$^{14}$C]glucosamine incorporation (Payne and Kalmakoff, 1974a). Polypeptides that may be dimers of the polyhedrin molecule are also often observed (Fig. 3) (Payne and Rivers, 1976). Both Lewandowski and Traynor (1972) and Mertens (1979) noted a nonvirion component of lower molecular weight (about 19,000–23,000) than the polyhedrin molecule in B. mori CPV (Fig. 3b and c). This has not been detected by other workers, nor has a similar component been found in CPV polyhedra of other isolates (Payne and Rivers, 1976; C. Payne, unpublished observations), and the possibility remains that this protein could be a contaminant or is derived by degradation of the polyhedrin molecule (Section III.A.2.b). In addition to these major structural components, trace amounts of polypeptides (which did not comigrate with virion proteins) were observed within polyhedra of B. mori CPV (arrows in Fig. 3b) (Lewandowski and Traynor, 1972). These may represent virus-coded "non-structural" proteins that are adventitiously incorporated within polyhedra during replication.

Amino acid analysis of the alkali-soluble fraction of B. mori CPV strains demonstrates that about 23% of the residues are aspartic or glutamic acid (Kawase, 1964; Kawase and Yamaguchi, 1974) and that the properties of the protein are similar but not identical to those of polyhedrins obtained from insect baculoviruses (Hukuhara and Hashimoto, 1966b; Harrap and Payne, 1979). Recently, Rohrmann et al. (1980) have compared the amino acid sequence of the polyhedrin of a type 5 CPV isolate from Orgyia pseudotsugata with that of the polyhedrin of a nuclear polyhedrosis virus isolated from the same host. The N-terminal sequences of the two molecules were quite distinct, suggesting that the similar biological properties of the two proteins resulted from independent but parallel evolution. The long-standing question over the viral or host origin of polyhedrins has now been resolved, and it is clear that the polyhedrins of both CPVs and baculoviruses are virus-specific molecules (Van der Beek et al., 1980) (Section III.B.2).

## C. RNA Components

### 1. Double-Stranded RNA

Early studies of the nucleic acid contained within CPV polyhedra showed that only RNA was present (Krieg, 1956; Xeros, 1962), but base composition studies did not reveal possible base-pairing or the double-stranded nature of the viral RNA (Aizawa and Iida, 1963). It was later realized that two categories of RNA can be extracted from B. mori CPV polyhedra. The first is probably in the form of oligonucleotides that are eluted from columns of methylated albumin kieselguhr (MAK) by 0.2 M NaCl and are absent from purified virions. The second is also obtained from purified virions and was eluted from MAK columns by 0.6–0.65 M NaCl (Hayashi and Kawase, 1964, 1965a,b). The significance of the oligonucleotides in polyhedra is still a matter of conjecture. It is not known whether they are virus-specific, but they contain much adenine and could resemble the adenine-rich oligonucleotides and/or the "initiation" oligonucleotides (Section III.A.1.f), which are detected within reovirus particles (Joklik, 1980).

It is now well established that the RNA extracted from virions is double-stranded. Base composition analysis of RNA from strains of type 1 CPV revealed G/C and A/U ratios close to 1.0 and a G + C mole fraction of 42–43% (Hayashi and Kawase, 1964; Miura et al., 1968; Kawase and Yamaguchi, 1974). When treated in standard saline–citrate (SSC) solutions, the extracted RNA of B. mori CPV shows the characteristic sharp hyperchromicity of double-stranded nucleic acids with a $T_m$ of 80°C in 0.01 × SSC (Miura et al., 1968; Kawase and Yamaguchi, 1974). An identical $T_m$ was obtained for RNA from H. armigera (type 5) CPV (Rubinstein et al., 1976). In contrast, Payne and Churchill (1977) observed that RNAs from CPV types 2 and 3 (from Inachis io and Spodoptera exempta, respectively) had lower $T_m$'s, consistent with lower G + C mole fractions of 36–37%.

Double-stranded RNA (dsRNA) from both type 1 and other CPVs is also highly-resistant [unlike single-stranded RNAs (ssRNAs)] to degradation by pancreatic ribonuclease in conditions of high salt concentration, but is degraded after heat denaturation or when the salt concentration is lowered (Miura et al., 1968; Rubinstein et al., 1976). The RNA does not react with formaldehyde, indicating an absence of exposed amino groups on bases that are presumably involved in base-pairing (Miura *et al.*, 1968). A buoyant density in $Cs_2SO_4$ of 1.6–1.65 $g/cm^3$ (Rubinstein *et al.*, 1976; Harley and Rubinstein, 1978) and the circular dichroic and optical rotatory dispersion properties are also consistent with dsRNA structures (Miura *et al.*, 1968; Wada *et al.*, 1971; Wells and Yang, 1974; Zama and Ichimura, 1976).

All these studies have been conducted with RNA following extraction from the virus particle. Recently, Yamakawa *et al.* (1981) have confirmed that both type 1 CPV RNA and Reovirus RNA are double-stranded *in situ* and do not exist within the virion as complementary single strands that could anneal together during extraction. Using the alkylating agent [$^3$H]dimethylsulfate, which methylates ssRNA and dsRNAs in different positions, the methylation patterns of the alkylated viral RNAs were found to be the same irrespective of whether the virus particles or the extracted RNAs were treated.

## 2. Segmented Genome

The discovery that CPV dsRNA was segmented came first from studies of the sedimentation of viral RNA on sucrose gradients, followed by electron-microscopic (EM) analysis and polyacrylamide gel electrophoresis (PAGE). Unlike reovirus RNA, the RNA from *B. mori* CPV sedimented in only two peaks in sucrose gradients with sedimentation coefficients of 15 and 12 S (Miura *et al.*, 1968; Furusawa and Kawase, 1973). These components, referred to as large (*L*) and small (*S*) RNAs by Kalmakoff *et al.* (1969), represented 90 and 10% by weight, respectively, of the total viral RNA (Miura *et al.*, 1968). The viral RNA of *M. disstria* CPV also contained 15 and 12 S components (Hayashi and Donaghue, 1971), even though it was later established by PAGE that the viral RNAs differed from those of *B. mori* CPV (Hayashi and Krywienczyk, 1972).

EM studies also showed a bimodal distribution of RNAs extracted from *B. mori* CPV virions, with average lengths of about 0.4 and 1.3 μm and estimated average molecular weights of about $1 \times 10^6$ and $3 \times 10^6$, respectively (Miura *et al.*, 1968; Nishimura and Hosaka, 1969). However, when virus was treated with urea directly on the EM grid, larger RNA molecules were observed, up to 6.8 μm in length with an estimated molecular weight of $14-18 \times 10^6$ (Nishimura and Hosaka, 1969). The likelihood that molecules of this size represented the complete genome became clear from gel electrophoresis analyses of the viral RNA. When RNA from *B. mori* CPV was electrophoresed on low-concentration polyacrylamide gels, ten equimolar bands were observed (Figs. 4 and 5), ranging in molecular weight from 0.35 to $2.55 \times 10^6$ when compared with the RNA genome segments of reovirus, and totaling about $15 \times 10^6$ (Fujii-Kawata *et al.*, 1970; Lewandowski and Traynor, 1972). The *S* RNA fraction from sucrose gradients contained the five smallest segments and the *L* fraction the remaining larger segments (Kalmakoff *et al.*, 1969).

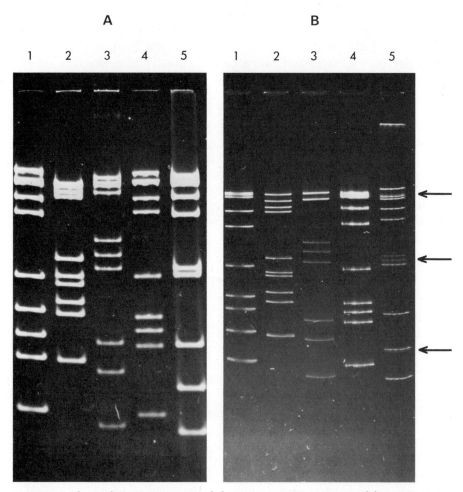

FIGURE 4. Electrophoretic separation of the genome RNA segments of five CPV types illustrating the considerable variation in segment mobilities. (A) Electrophoresis on a 3% polyacrylamide slab gel in a Tris–acetate buffer (Payne and Rivers, 1976); (B) electrophoresis on a 10% polyacrylamide slab gel using a discontinuous Tris-HCl/Tris–glycine buffer (Laemmli, 1970). (1) Type 1 CPV from *Bombyx mori*; (2) type 2 CPV from *Inachis io*; (3) type 7 CPV from *Noctua pronuba*; (4) type 12 CPV from *Autographa gamma*; (5) type 5 CPV from *Orgyia pseudotsugata*. The presence of submolar bands in B5 (←) suggests that the virus isolate is a probable mixture of two electropherotypes (Payne *et al.*, 1983).

The electrophoretic profile obtained with this virus RNA is quite consistent and therefore does not represent random scission of a larger molecule. Endoribonuclease cleavage of an intact RNA chain would be expected to produce molecules with 3′-monophosphate termini, but since the 3′ termini of CPV RNA segments are not phosphorylated, the segments are probably not derived by specific nuclease action (Lewandowski and Leppla, 1972; Furuichi and Miura, 1972). Each particular CPV isolate has a characteristic profile of genome segments, and the molecular weights of RNAs from different viruses may be quite distinctive (Fig. 4) (Payne and Tinsley, 1974; Payne and Rivers, 1976; Rubinstein *et al.*, 1976; Harley and Rubinstein, 1978; Payne *et al.*, 1978). Payne and Rivers (1976) proposed that "major" differences between RNA profiles could be used as a basis for the classification of these viruses into a number of types. This is examined in more detail in Section II.D.

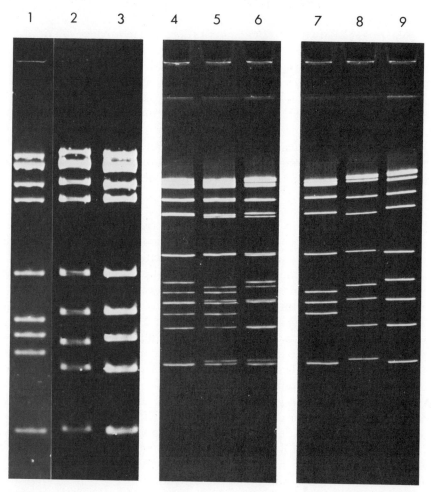

FIGURE 5. Comparison of the genome RNA segments of two electropherotypes of type 1 CPV and type 12 CPV. RNAs of the type 1 electropherotypes show considerable similarities on 3% gels (2, 3), but are clearly distinguishable on 10% gels (6, 8, 9). (1–3) Electrophoresis on 3% polyacrylamide gels in a Tris–acetate buffer (Payne and Rivers, 1976); (4–9) electrophoresis on 10% polyacrylamide gels using a discontinuous buffer (Laemmli, 1970). (1, 7) Type 12 CPV from *Autographa gamma*; (2, 8) type 1 CPV from *Bombyx mori*; (3, 9) type 1 CPV from *Dendrolimus spectabilis*. (4–6) Coelectrophoresis of RNAs from *A. gamma* and *D. spectabilis* CPVs (4), *A. gamma* and *B. mori* CPVs, (5), and *B. mori* and *D. spectabilis* CPVs (6) (Payne *et al.*, 1983).

The detection by EM of long RNA molecules suggests that individual segments are normally linked to one another in the virion. The 6.8-μm molecules may represent the complete viral genome of all ten segments of *B. mori* CPV attached end-to-end with one copy of each segment present within a virion. Larger, and often circular, molecules (contour length 15 μm) observed by Kavenoff *et al.* (1973) may represent RNA aggregates, since the total RNA molecular weight that they represent, of about 35 × 10⁶, is incompatible with data on virion molecular weight and RNA content (Table I).

The linkage between RNA segments does not involve covalent phosphodiester bonds, since studies with [³H]sodium borohydride reduction of oxidized 3′ termini indicated that RNA within the virion had the same number of 3′ termini as the segmented RNA extracted from the virus (Lewandowski and Millward, 1971). Instead, protein links may be involved. Yazaki and Miura (1980) have shown that the viral dsRNA is in intimate association with some part (probably the base) of the virion spikes. RNA gently released from the virus and treated with glutaraldehyde appeared to be supercoiled, and in some examples, two dsRNA molecules appeared to be linked by a protein complex consistent in size with the spike.

It is now clear (as described in Section III.A.1.a) that, like the RNAs of other reoviruses, each genome segment is transcribed by a virus-associated polymerase to produce monocistronic messenger RNAs (mRNAs) that can be translated into virus-specific proteins. Studies of *B. mori* CPV have confirmed that one strand (the "plus" strand) of each genome segment has the same 5′ and 3′ termini as the viral mRNA transcripts. The 3′ termini of viral dsRNA were first identified by [³H]borohydride labeling as a 1:1 mixture of uridine (U) and cytosine (C), unlike reovirus, in which only C termini were found (Furuichi and Miura, 1972). One strand of the duplex of all ten segments terminates in U and the other in C (Lewandowski and Leppla, 1972; Furuichi and Miura, 1973). ³²P-Labeling of the 5′ termini of genome RNA revealed two classes of structure—[³²P]-Gp and [³²P]-AmpGp—the adenylic acid residue being methylated in the latter case (Miura *et al.*, 1974). These and subsequent studies have shown that each segment is a complete duplex in which the 5′ terminus of the plus RNA strand bears the methylated adenylic acid residue and is blocked and "capped" with a methylated guanylic acid residue linked by a pyrophosphate bond (Miura, 1981). Each segment has the following probable molecular structure:

```
         5'                                                 3'
m⁷GpppAmG—U - - - - - - - [G—U—U—A]—G—C—C        plus (+)
         U—C—A[—U—U]- - - - - - - - - - - C—G—G        minus (−)
         3'                                                 5'
```

The sequences shown in brackets on the two strands have been observed in each of the five genome segments of *B. mori* CPV that have so far been examined to this extent (R.E. Smith *et al.*, 1981; Y. Furuichi, personal communication). The significance of the m⁷G cap structure and the common sequences at the 5' and 3' termini, in terms of polymerase recognition sites, is discussed in Section III.A.1.a.

Recent studies by R.E. Smith *et al.* (1981) have shown that it is possible to separate the viral plus and minus RNA strands of the duplex. Using agarose gels containing 7 M urea, denatured *B. mori* CPV RNA was resolved into 18 bands (2 of the larger RNA segments have similar molecular weights and are difficult to resolve). It was confirmed by selective labeling of the 5' terminus of the "uncapped" strand that plus strands migrated faster than the corresponding minus strands. These migration differences disappeared when glyoxal (a G-specific modifier preventing inter- and intramolecular base pairing) was used to eliminate secondary structural differences. This ability to separate strands should facilitate RNA sequencing, cloning, and translation studies and further advance the study of genome RNA structure.

## D. Cytoplasmic Polyhedrosis Virus Classification

### 1. RNA Electropherotypes and Their Use in Studies of Cytoplasmic Polyhedrosis Virus Variation

As mentioned above, the profile of genome RNA segments is consistent for a particular CPV isolate, but profound differences have been observed in the electrophoretic profiles ("electropherotypes") of RNA from a range of CPV isolates (see Fig. 4). It is now certain that each RNA segment codes for a specific protein and that differences in the electrophoretic mobilities of genome segments reflect other differences among viruses. On this premise, "major" differences among electropherotypes have been used to construct a provisional classification of these viruses (Payne and Rivers, 1976).

At present, the interpretation of major differences is partly subjective, but all CPV types defined so far differ from one another in the size of at least three of the genome segments resolved on 3% polyacrylamide gels using a Tris–acetate electrophoresis buffer. In an initial survey of 33 virus isolates, 11 distinct virus types were defined (Payne and Rivers, 1976), and a 12th type has subsequently been identified (Payne *et al.*, 1977). Although many of the CPV isolates had electropherotypes identical to those of the type strains, there was some evidence of further intratype variation, with small differences in the electrophoretic mobility of at least one genome segment [e.g., type 5 CPV (Payne and Rivers, 1976) and type 1 CPV (Payne *et al.*, 1978)]. The molecular weights of the RNA segments of the 12 type strains, calculated by comparison with a standard *B. mori* (type 1) CPV RNA sample, are shown in Table III, and an example of the segment profiles of five types in Fig. 4.

TABLE III. Molecular Weights ($\times$ $10^{-6}$) of the RNA Genome Segments of Cytoplasmic Polyhedrosis Virus Types 1–12[a]

| RNA segment | CPV type | | | | | | | | | | | |
|---|---|---|---|---|---|---|---|---|---|---|---|---|
| | 1[b] | 2[c] | 3[c] | 4[c] | 5[c] | 6[c] | 7[d] | 8[c] | 9[c] | 10[c] | 11[c] | 12[e] |
| 1 | 2.55 | 2.29 | 2.42 | 2.35 | 2.35 | 2.35 | 2.44 | 2.56 | 2.44 | 2.43 | 2.59 | 2.50 |
| 2 | 2.42 | 2.29 | 2.32 | 2.35 | 2.35 | 2.29 | 2.34 | 2.56 | 2.36 | 2.43 | 2.48 | 2.32 |
| 3 | 2.32 | 2.16 | 2.32 | 2.35 | 2.35 | 2.23 | 2.27 | 2.48 | 2.30 | 2.27 | 2.48 | 2.32 |
| 4 | 2.03 | 2.06 | 2.08 | 2.20 | 2.08 | 2.10 | 2.15 | 2.21 | 2.04 | 2.27 | 2.16 | 2.07 |
| 5 | 1.82 | 1.25 | 2.03 | 1.37 | 1.82 | 1.54 | 1.43 | 2.08 | 1.32 | 1.41 | 1.12 | 1.86 |
| 6 | 1.12 | 1.09 | 1.29 | 1.22 | 1.22 | 1.33 | 1.28 | 1.07 | 0.97 | 1.29 | 1.12 | 1.13 |
| 7 | 0.84 | 1.01 | 1.21 | 1.10 | 1.16 | 1.26 | 1.14 | 0.73 | 0.97 | 1.29 | 0.76 | 0.81 |
| 8 | 0.62 | 0.88 | 0.61 | 0.97 | 0.68 | 0.92 | 0.61 | 0.67 | 0.44 | 0.95 | 0.72 | 0.72 |
| 9 | 0.56 | 0.78 | 0.47 | 0.81 | 0.50 | 0.79 | 0.48 | 0.50 | 0.39 | 0.68 | 0.55 | 0.64 |
| 10 | 0.35 | 0.55 | 0.34 | 0.81 | 0.34 | 0.51 | 0.30 | 0.37 | 0.39 | 0.56 | 0.40 | 0.36 |
| TOTAL | 14.63 | 14.36 | 15.09 | 15.53 | 14.82 | 15.32 | 14.44 | 15.23 | 13.62 | 15.58 | 14.38 | 14.73 |

[a] All determinations were made using *B. mori* CPV RNA as a standard, after fractionation of the RNAs by electrophoresis on 3% polyacrylamide gels with a Tris–acetate running buffer.
[b] Fujii–Kawata *et al.* (1970).    [c] Payne and Rivers (1976).
[d] Amended from data in Payne and Rivers (1976) by C. Payne (unpublished observations).
[e] Payne *et al.* (1977).

All CPV types almost certainly contain ten equimolar genome segments, although all ten are rarely clearly resolved. At least four segments have molecular weights in excess of $2 \times 10^6$, and the total genome size ranges only between 13.6 and $15.6 \times 10^6$ (Payne and Rivers, 1976). Harley *et al.* (1977) have pointed out that the relationship between the electrophoretic mobilities of dsRNA segments and their molecular weights deviates from linearity above $1 \times 10^6$, while the molecular weights shown in Table III were calculated assuming a linear relationship throughout the range $0.34–2.59 \times 10^6$. Thus, these values must be considered as approximates only but nonetheless indicative of the range of differences among CPV types.

Recent studies have added to this type classification of CPVs, and an up-to-date summary of 68 virus isolates from a total of 49 species and 14 families of Lepidoptera is shown in Table IV. CPV types 2, 3, 6, and 11 have been most frequently reported in 15, 11, 9, and 8 separate species, respectively. A maximum of five distinct virus types have been isolated from one species (*S. exempta*), indicating how essential it is to classify a virus by its characteristics rather than by the host from which it was originally isolated. Unfortunately, the latter approach has been all too common in insect pathology.

Similar methods of analysis are being used for comparing other reoviruses (Hrdy *et al.*, 1979), particularly rotavirus strains (Rodger and Holmes, 1979; Rodger *et al.*, 1980, 1981; Lourenco *et al.*, 1981). On the whole, the differences among rotavirus electropherotypes seem of lesser magnitude than those observed among CPV types, in which there is often no overall resemblance in segment distribution (as between types 1 and 2 in Fig. 4). Recent studies with electropherotypes of other reoviruses have generally examined segment mobility differences in higher-concentration (e.g., 5–10%) polyacrylamide gels with a discontinuous buffer system (Laemmli, 1970; Sharpe et al., 1978; Rodger and Holmes, 1979). This system has greater resolving power and, when applied to CPV RNAs, has been useful in further assessing inter- and intratype variation (Payne et al., 1983).

When RNAs from five distinct CPV types were electrophoresed on 10% Laemmli gels (see Fig. 4b), the differences among electropherotypes were as pronounced as on 3% Tris–acetate gels (Fig. 4a), although some segments showed different relative mobilities on the two gel systems. In particular, the largest RNA segments of type 2 CPV were clearly resolved into four bands only on Laemmli gels, while the four largest segments of type 7 CPV were resolved only on Tris–acetate gels. Overall, the resolution on Laemmli gels was superior. The basis of the altered relative migration of some segments in the different gel systems is not entirely clear, but seems attributable to the different gel concentrations, rather than to the different buffer systems used (Payne et al., 1983). Such differences have also been observed with reovirus and rotavirus RNAs (Sharpe et al., 1978; Rodger and Holmes, 1979).

The Laemmli system has proved most effective for a closer examination of intratype variation. The type 1 CPVs from D. spectabilis and B. mori have been shown to be antigenically closely related, and appeared to differ only with respect to the mobility of a single RNA genome segment when compared on 3% Tris–acetate gels (Payne et al., 1978) [Fig. 5(2, 3)]. When the RNAs of these viruses, and RNA from type 12 CPV (which differs in the mobility of at least three genome segments on 3% gels), were run on Laemmli gels, much greater differences were found, with relatively few bands comigrating (Fig. 5). These shifts in mobility may be caused only by very small changes in segment size or base composition or both, but they nonetheless indicate that the viruses are more distinct than was first appreciated.

Although differences in the sizes of CPV RNA segments imply that the viruses differ in a range of biochemical and biological properties, viruses the RNA segments of which comigrate may not necessarily be identical. The RNA segments could be the same size but contain sequence differences, which, despite the use of high-resolution Laemmli gels, may be distinguished only by additional studies of RNA homology and antigenic relationship.

In addition to its usefulness in comparing virus isolates, gel electrophoresis of viral RNA also provides a method for the detection of mixed virus infections by the resolution of more than 10 genome segments that may not all be present in equimolar amounts. Mixed CPV infections of quite distinct virus types have been observed (Table IV) (Payne, 1976; Payne and Rivers, 1976; Mertens, 1979), and it is most likely that the

TABLE IV. Summary of the Type Classification of Cytoplasmic Polyhedrosis Viruses Obtained from 49 Insect Species[a]

| Family | Species | CPV type | | | | | | | | | | | |
|---|---|---|---|---|---|---|---|---|---|---|---|---|---|
| | | 1 | 2 | 3 | 4 | 5 | 6 | 7 | 8 | 9 | 10 | 11 | 12 |
| Arctiidae | Arctia caja | | +[b] | +[b] | | | | | | | | | |
| | A. villica | | | | | | | | | | | | |
| Bombycidae | Bombyx mori | A | | | | | | | | | | | |
| Danaidae | Danaus plexippus | | | + | | | | | | | | | |
| Gelechiidae | Pectinophora gossypiella | | | | | | | | | | | | |
| Geometridae | Abraxas grossulariata | | | | | | | | | | | B[c] | |
| | Aplocera (= Anaitis) plagiata | | | +[b] | | | +[b] | | + | | | | |
| | Biston betularia | | + | + | | | + | | | | | | |
| | Operophtera brumata | | | | | | | | | | | | |
| Lasiocampidae | Dendrolimus spectabilis | B[d] | | | | | | | | | | | |
| | Eriogaster lanestris | | +[b,c] | | | | +[b,c] | | | | | | |
| | Gonometa rufibrunnea | | | + | | | | | | | | | |
| | Lasiocampa quercus | | | | | | + | | | | | | |
| | Malacosoma disstria | | +[b,c] | +[b,c] | | | | | +[b,e] | | | | |
| | M. neustria | | + | | | | | | | | | | |
| Lymantriidae | Dasychira pudibunda | | + | | | | | | | | | | |
| | Lymantria dispar | B[d] | | | | | | | | | | +[f] | |
| Noctuidae | Orgyia pseudotsugata | | | | | B/C[c,g] | + | | | | | | |
| | Agrochola (= Anchoscelis) helvola | | | | | | + | | | | | | |
| | A. lychnidis | | | | | | | | | | | | |
| | Agrotis segetum | | | | | | | | | | | | |
| | Aporophyla lutulenta | | | | | | | | | | | | |
| | Autographa gamma | | | | | | | | | + | + | | |
| | Euxoa scandens | | | | | +[h] | | | | | | | |
| | Heliothis armigera | | | | | A1/A2[i] | | | +[j] | | | A[c] | |
| | H. zea | | | | | | | | | | | A[c] | +[c] |

| Family | Species | 1 | 2 | 3 | 4 | 5 | 6 | 7 | 8 | 9 | 10 | 11 | 12 |
|---|---|---|---|---|---|---|---|---|---|---|---|---|---|
| | *Lacanobia oleracea* | +[c] | . | . | . | . | . | . | . | . | . | A[c,f] | +[b,k] |
| | *Mamestra brassicae* | +[f] | . | . | . | . | . | . | +[f] | + | . | . | . |
| | *Noctua (= Triphaena) pronuba* | . | . | . | . | . | . | +[b] | + | . | . | . | . |
| | *Phlogophora meticulosa* | . | . | . | . | . | . | +[b] | . | . | . | . | . |
| | *Polymixis (= Antitype) xanthomista* | . | . | . | . | . | . | . | . | . | . | . | . |
| | *Pseudaletia unipuncta* | . | . | . | . | +[b,f] | . | . | +[b,f] | . | . | B[c] | . |
| | *Spodoptera exempta* | + | . | . | . | . | . | . | . | +[b,f] | . | A[c] | +[l] |
| | *S. exigua* | . | . | . | . | . | . | . | . | . | . | A | . |
| | *Trichoplusia ni* | B/C[c] | . | . | . | . | . | . | . | . | . | . | . |
| Notodontidae | *Phalera bucephala* | + | . | . | . | . | . | . | . | . | . | . | . |
| Nymphalidae | *Aglais urticae* | +[b] | . | . | . | . | +[b] | +[b] | . | . | . | . | . |
| | *Agraulis vanillae* | +[c] | . | . | . | . | . | . | . | . | . | . | . |
| | *Boloria (= Clossiana) dia* | + | . | . | . | . | . | . | . | . | . | . | . |
| | *Inachis io* | + | . | . | . | . | . | . | . | . | . | . | . |
| Papilionidae | *Papilio machaon* | + | . | . | . | . | . | . | . | . | . | . | . |
| Pieridae | *Pieris rapae* | +[c] | + | . | . | . | . | . | . | . | . | . | +[k] |
| Saturniidae | *Actias selene* | . | + | . | . | . | . | . | . | . | . | . | . |
| | *Antheraea mylitta* | . | + | . | . | . | . | . | . | . | . | . | . |
| | *A. pernyi* | . | + | . | . | . | . | . | . | . | . | . | . |
| | *Automeris io* | . | . | + | . | . | . | . | . | +[j] | . | . | . |
| | *Nudaurelia cytherea* | +[c] | . | . | . | . | . | . | . | . | . | . | . |
| Sphingidae | *Hyloicus pinastri* | . | . | +[c] | . | . | . | . | . | . | . | . | . |
| | *Sphinx ligustri* | . | . | +[c] | . | . | . | . | . | . | . | . | . |

[a] Data from Payne and Rivers (1976) with the exceptions noted. (+) Virus of this "type" was recorded in the listed species. Where variant electropherotypes were observed within a virus type (e.g., types 1, 5, and 11), these variants have been differentiated by the capital letters A, B, and C. The terminology for the electropherotypes of the type 5 CPV variants from *Heliothis armigera* (A1 and A2) has been retained (Rubinstein and Harley, 1978). Not all records are "natural" virus isolates, but also include the positive results of cross-transmission tests.

[b] Virus was present as a mixture with another identified or unidentified CPV.

[c] C. Payne (unpublished observations).     [d] Payne *et al.* 1978.     [f] Mertens (1979).

[e] Hayashi and Krywienczyk (1972) and Harrap and Payne (1979).     [i] Rubinstein and Harley (1978).

[g] Rohrmann *et al.* (1980).     [h] Grancher–Barray *et al.* (1981).     [j] Rubinstein and Harley (1978).

[i] Harley and Rubinstein (1978).     [k] Allaway (1982).     [l] Payne *et al.* (1977).

resolution of 16 RNA components in *M. disstria* CPV RNA also arose from a mixture of two viruses (Hayashi and Krywienczyk, 1972; Harrap and Payne, 1979). The existence of mixed infections provides the potential for genetic reassortment of segments, although this has not yet been observed. In fact, the only reported genetic modification in CPVs comes from a study of a virus isolated from *H. armigera*. On three separate occasions, a rapid and reproducible alteration in the genome RNA occurred with no apparent loss of infectivity (Rubinstein and Harley, 1978). After one passage in laboratory stocks of insect larvae, one of the larger RNA genome segments $(2.3 \times 10^6)$ had become submolar, and by the second passage it had disappeared. Its disappearance corresponded with the appearance of another much smaller segment $(0.4 \times 10^6)$. The altered profile remained stable for at least five subsequent passages. The authors suggested that a specific deletion event was the most likely explanation, since the likelihood of spontaneous mutation, or the selection of one virus from the same mixture of two virus strains, was unlikely to have occurred on three independent occasions. Such modifications add both to the genetic complexity of the CPV group and to the difficulty of using electropherotype analysis as the sole basis of CPV classification.

## 2. RNA Homology, Proteins, and Antigenic Properties of Cytoplasmic Polyhedrosis Virus Electropherotypes

The usefulness of the electropherotype classification of CPVs depends on how well it is supported by other distinguishing features of the viruses. Studies of RNA homology and the antigenic properties of the viruses provide useful alternative methods for investigating CPV variation.

### a. RNA Homology

Virus-specific ssRNA of *B. mori* (type 1) CPV synthesized *in vivo* did not anneal to denatured viral RNA of *Arctia caja* CPV (Payne and Kalmakoff, 1973), which was subsequently shown to be a mixture of CPV types 2 and 3 (Payne, 1976). There was also no appreciable annealing of type 1 CPV ssRNA transcripts (synthesized by the viral polymerase *in vitro*) to denatured dsRNA of type 2 CPV or vice versa, while approximately 95% annealing was obtained between transcripts and the homologous viral RNA (Mertens, 1979). These results indicate that at least for CPV types 1, 2, and 3, there is a sound genetic basis for their classification as distinct types. The use of ssRNA transcripts from wound tumor virus (WTV) and rice dwarf virus, or the denatured viral RNA of reovirus and WTV, has also demonstrated that there appears to be no genetic relationship between CPVs and other genera of reoviruses (Black and Knight, 1970; Kodama and Suzuki, 1973; Martinson and Lewandowski, 1974).

In contrast, transcripts of type 1 CPV variants isolated from *B. mori,*
*L. dispar,* or *D. spectabilis* (see Fig. 5) reannealed to all the heterologous
denatured viral RNAs, but to different extents, suggesting that *L. dispar*
and *D. spectabilis* CPVs were identical but shared only 52–76% homology with *B. mori* CPV (Payne *et al.,* 1978). However, as pointed out at
the time, accurate quantification of homology would depend on the presence of equivalent amounts of transcripts of all ten genome segments. It
is now clear that more copies of the transcripts of the smaller genome
segments are produced by CPV in *in vitro* polymerase assays (Section
III.A.1.a) and that sequence homology estimates obtained using transcripts would be biased toward the sequences of the smaller genome segments. Nonetheless, the variants of type 1 CPV that were grouped on the
basis of RNA electropherotype do share a significant degree of sequence
homology.

### b. Proteins and Antigenic Properties

While the proteins of different CPV types are distinct, the molecular
weights of proteins from two isolates of type 5 CPV from *Trichoplusia*
*ni* and *Euxoa scandens* show considerable similarities (Payne and Rivers,
1976; Grancher-Barray *et al.,* 1981), and no major differences were observed among the structural proteins of type 1 CPV isolates that were
distinguishable by RNA differences (Payne *et al.,* 1978). This indicates
that PAGE analysis of structural polypeptides is likely to provide a less
sensitive method of differentiating viruses than electropherotype analysis. Nonetheless, similarities in RNA composition are reflected in similar structural proteins.

Whereas most virus group classifications are based on the antigenic
properties of the viruses, no serotype classification exists for CPVs at
present. The absence of adequate cell-culture systems has inhibited the
use of infectivity neutralization tests, and the available data are restricted
to limited analyses of structural antigens.

The first indication that some CPVs may be antigenically distinct
came from the work of Cunningham and Longworth (1968), who, despite
noting extensive cross-reactions between a group of closely related or
identical CPVs, observed that *I. io* CPV virions (probably type 2 CPV)
did not cross-react with antiserum to *B. mori* CPV in complement-fixation (CF) tests. Mertens (1979) also failed to detect cross-reactions between virions of types 1 and 2 CPVs in double-diffusion tests. Using a
liquid-phase radioimmunoassay (RIA), Payne (1976) observed that virions
of CPV types 2 and 3 did not react with antiserum to the heterologous
virus. Since intact virions were used in this assay, it was possible that
the specificity was a feature of the antigens exposed at the surface of the
virus, as well as of the method used (Harrap and Payne, 1979). Further
evidence of such specificity has come from recent studies using the indirect enzyme-linked immunosorbent assay (ELISA) (Payne *et al.,* 1983).

When intact virions of type 1 (*B. mori*), type 2 (*I. io*), and type 5 (*H. armigera*) CPVs were used as antigens, no cross-reactions were observed, the virions reacting only with the homologous antiserum (Fig. 6). However, when different electropherotypes of type 1 CPV (from *B. mori* and *D. spectabilis*) were compared, the viruses cross-reacted (Fig. 6) (Payne *et al.*, 1978) and also showed a slight reaction at the highest antigen concentration with virions of type 12 CPV isolated from *Autographa gamma*. This last virus appears to differ from *B. mori* CPV by the mobility of only three RNA genome segments on Tris–acetate gels (Payne *et al.*, 1977).

Just as the surface antigens of virions appear to be largely type-specific in ELISA, so there are also no significant cross-reactions among polyhedrins of CPV types 1, 2, and 5 in gel-diffusion or indirect ELISA tests (Fig. 7). Similarly, different variants of CPV types 1, 12, and 5 (Fig. 7) show only intratype cross-reactions, with the exception (noted also for virions above) that there is a slight antigenic relationship between polyhedrins of type 1 and type 12 CPV.

It should be pointed out that using other serological tests, cross-reactions between distinct CPV types have also been reported. Krywien-

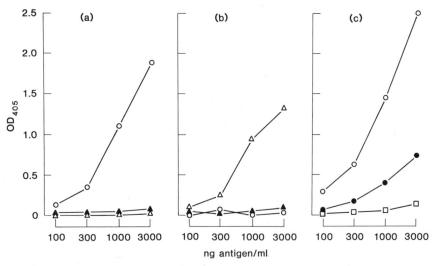

FIGURE 6. Antigenic properties of CPV virions. Distinct virus types show little or no cross-reaction in ELISA with heterologous antibodies at antigen concentrations up to 3 μg/ml. Purified virus particles were used as antigens in an indirect ELISA (Crook and Payne, 1980) with unlabeled homologous and heterologous anti-virion immunoglobulin (IgG) and goat anti-rabbit IgG conjugated with alkaline phosphatase. The extent of the reaction was measured by absorbance at 405 nm after incubation with *p*-nitrophenylphosphate. (a, c) Antibody to type 1 CPV from *Bombyx mori*; (b) antibody to type 2 CPV from *Inachis io*. (○) Type 1 CPV virion antigen from *B. mori*; (●) type 1 CPV virion antigen from *Dendrolimus spectabilis*; (□) type 12 CPV virion antigen from *Autographa gamma*; (▲) type 5 CPV virion antigen from *Heliothis armigera*; (△) type 2 CPV virion antigen from *I. io* (Payne *et al.*, 1983).

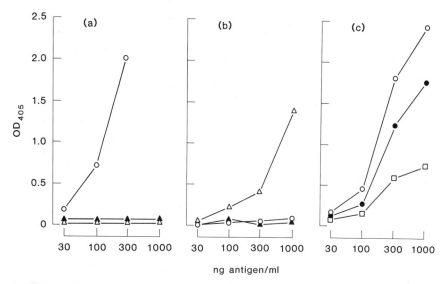

ng antigen/ml

FIGURE 7. Antigenic properties of CPV polyhedrins. Polyhedrins of distinct virus types show little or no cross-reaction in ELISA. The assays were carried out as described in the Fig. 6 caption, using polyhedra dissolved in 0.2 M carbonate–bicarbonate buffer (pH 10.8) at 1 mg/ml. (a, c) Antibody to type 1 CPV polyhedrin from *Bombyx mori*; (b) Antibody to type 2 CPV polyhedrin from *Inachis io*. (O) Type 1 CPV polyhedrin antigen from *B. mori*; (●) type 1 CPV polyhedrin antigen from *Dendrolimus spectabilis*; (□) type 12 CPV polyhedrin antigen from *Autographa gamma*; (▲) type 5 CPV polyhedrin antigen from *Heliothis armigera*; (△) type 2 CPV polyhedrin antigen from *I. io* (Payne *et al.*, 1983).

czyk *et al.* (1969) obtained five precipitin lines in double-diffusion tests with virions of CPVs from *B. mori*, *Malacosoma disstria*, and *Orgyia leucostigma*. Although the virions of *M. disstria* and *O. leucostigma* CPVs were indistinguishable, each shared only one or two antigenic determinants with *B. mori* CPV. Immunoelectrophoresis and CF tests suggested some quantitative differences in the antigenic composition of *M. disstria* and *O. leucostigma* CPVs, a feature that can be more readily interpreted now it is known that the *M. disstria* CPV used in these experiments was a mixture of two CPVs, both of which were distinct from *B. mori* CPV (Hayashi and Krywienczyk, 1972; Harrap and Payne, 1979). The cross-reactions could arise from a CPV group-specific antigen, but because the dsRNA genome of reoviruses is immunogenic, misleading serological cross-reactions can be obtained when comparing CPVs. Krywienczyk and Hayashi (1970, 1971, 1972) noted that antisera to CPV virions cross-reacted with cellular components that contained RNA (including ribosomes from bacteria, invertebrates, and vertebrates) as well as the dsRNA extracted from the virus. Cross-reacting antibodies to a range of viral dsRNAs were also detected in antisera to CPV virions, although the cross-reactions that these could give rise to could be reduced by serum dilution (Payne and Kalmakoff, 1974b) or cross-absorption with the artificial dsRNA polymer poly rI:rC (Payne and Churchill, 1977).

These results suggest that ELISA and perhaps RIA could provide the most useful and sensitive methods for comparing different CPVs. Where the viruses have been sufficiently distinct in their RNA profiles to be classified as virus types, their polyhedrin and virion antigens are only slightly (types 1 and 12) if at all (types 1, 2, 3, and 5) related. In the absence of neutralization tests, these methods could provide an additional basis for virus classification and the distinction of serotypes. In the light of the slight antigenic relationship observed between CPV types 1 and 12, it could be most appropriate to group these within the same serotype, but with this one exception, RNA type and ELISA serotype appear to be synonymous at present. The ELISA has also proved useful for typing human rotaviruses, and distinct electropherotypes ("short" and "long") can be classified into two serotypes by this method (Rodger et al., 1981).

It is interesting that while both the Tris–acetate and the Laemmli gel systems are adequate for discriminating virus types, the Tris–acetate gels have poor resolving power for investigating CPV intratype variation and in this respect may even prove useful in grouping viruses that appear related by other criteria including antigenic relationships. In contrast, the Laemmli system provides the resolution that is required for investigating variation within antigenically related groups. This has been observed in isolates of both type 1 and type 5 CPVs and is expressed in Table IV by listing the individual electropherotypes detected within a type by the letters A–C. Both gel systems should be of great value in future studies of CPV variation and molecular epidemiology.

In conclusion, the classification of CPV types on the basis of major differences among RNA electropherotypes provides at present a subjective but nonetheless valuable method for grouping virus isolates and is supported by the current knowledge of CPV RNA homology and the antigenic properties of different isolates.

## III. *IN VITRO* STUDIES ON THE MECHANISMS OF CYTOPLASMIC POLYHEDROSIS VIRUS REPLICATION

Studies of the biochemistry of CPV replication *in vivo* lag well behind those carried out with other members of the Reoviridae, in particular reovirus. Until recently, adequate cell-culture systems for the study of virus replication have not existed (see Section IV.C). However, *in vitro* studies using the viral genome and the enzymes associated with virus particles have enabled much information to be obtained on the synthesis of messenger RNA (mRNA) and the proteins coded for by the different genome RNA segments.

## A. Virus-Associated Enzymes

The fact that CPVs have a double-stranded RNA (dsRNA) genome that is not, by itself, infectious (Lewandowski et al., 1969; Kawase, 1971), and is unlikely to function as mRNA (Miura and Muto, 1965) or be transcribed efficiently by any cellular enzymes led to the search for, and discovery of, an RNA polymerase (transcriptase) associated with purified virus particles of type 1 CPV from the silkworm, Bombyx mori (Lewandowski et al., 1969). Since then, the enzyme and its in vitro products have been extensively investigated, and there have been significant advances in the understanding of mRNA structure and transcription, in particular the discovery of the blocked 5′ terminus or cap structure (Furuichi and Miura, 1975) (Sections III.A.1.a and f). Such studies have been helped by the relative ease of producing large quantities of highly purified virus and the relative stability of the CPV polymerase (Shimotohno and Miura, 1973a,b, 1974; Furuichi, 1974, 1978a, 1981; Mertens and Payne, 1978; Mertens, 1979; Miura, 1981). During these investigations, there have been recorded in type 1 CPV virions other enzyme activities that are associated with the production and processing of mRNA. These include a nucleotide phosphohydrolase (Storer et al., 1974a,b; Shimotohno and Miura, 1977), guanylyltransferase (Shimotohno and Miura, 1976; Furuichi, 1978a; R.E. Smith and Furuichi, 1982a), and transmethylase activities (Furuichi, 1974, 1978a; 1981; Shimotohno and Miura, 1976; Mertens, 1979; Wertheimer et al., 1980). These enzymes, and their role in the in vitro production of mRNA, are considered in the following sections. CPV virions or polyhedra or both have also been reported to contain a single-stranded RNA (ssRNA) exonuclease (Storer et al., 1974a,b) and alkaline protease activities, respectively (M. Carter and C. Payne, unpublished observations; R. Rubinstein and A. Polson, personal communication).

## 1. Enzymes Involved in Messenger RNA Production and Processing

### a. RNA Polymerase

Although the polymerases associated with CPV virions from unrelated virus types are not identical in all their properties (Mertens, 1979), the enzyme of type 1 CPV represents a useful model. Most of the following discussion deals with results obtained with this virus. The enzyme activity is in many ways similar to that associated with other members of the Reoviridae, since it completely copies one strand only of each segment of the viral genome by a fully conservative mechanism (Lewandowski et al., 1969; Joklik, 1974; Furuichi, 1974; Shimotohno and Miura 1973a,b, 1974; Miura et al., 1974; R.E. Smith and Furuichi, 1980a,b; R.E. Smith et al., 1981). However, unlike other reoviruses, modification of virions by heat shock, $Ca^{2+}$ chelation, or incubation with chymotrypsin is not required for activation of the CPV polymerase (Lewandowski et al., 1969; Furuichi, 1974). This may well be due to the absence of an outer-capsid layer in CPVs that in other reoviruses is removed or modified by such treatment (R.E. Smith et al., 1969; Lewandowski and Traynor, 1972; Cohen et al., 1979; Van Dijk and Huismans, 1980). The release of CPV virions from polyhedra by alkali treatment could be considered as a possible activation step, but the polymerase is also active without pretreatment in "free" virions, which have never been occluded within polyhedra (Donaghue and Hayashi, 1972), although their specific activity may be lower (Mertens, 1979). In contrast, Shimotohno and Miura (1973a,b) and Miura (1981) have reported that pretreatment with dichlorodifluoromethane was essential for the activation of the type 1 CPV polymerase. This has not been found necessary by other workers, and Joklik (1974) suggested that the substituted hydrocarbon may appear to activate transcription by depressing levels of ribonuclease that may often be associated with CPV preparations.

The CPV polymerase is also highly resistant to proteolytic digestion, since it is not affected by the presence of 100 μg/ml of proteinase K in an in vitro assay. Under similar conditions, the polymerase of reovirus is first activated by removal of the outer capsid shell, then inactivated by further digestion of the resulting cores (R.E. Smith and Furuichi, 1980a).

The optimum conditions for the synthesis of ssRNA by purified CPV virions have been studied by in vitro assays in which RNA synthesis is measured by the incorporation of radioactive ribonucleoside triphosphates (rNTPs) into trichloracetic-acid-precipitable macromolecules. The polymerase requires all four rNTPs (ATP, CTP, GTP, and UTP) and divalent cations for efficient ssRNA synthesis. Magnesium ions at a concentration of 12 mM are optimal, and although some synthesis can occur in the presence of manganese ions, little or no activity was observed in the presence of calcium ions (Lewandowski et al., 1969; Shimotohno and Miura, 1973a,b; Furuichi, 1974; Mertens and Payne, 1978; Mertens, 1979). Recent studies have shown that ATP (and to a lesser extent GTP) is of particular importance (Mertens, 1979; Yazaki and Miura, 1980; Furuichi, 1981), suggesting that ATP may serve some function in the polymerase assay apart from merely acting as a substrate for RNA chain elongation. Since the 5'-terminal nucleoside of type 1 CPV mRNA is adenosine (see below) (Miura et al., 1974; Shimotohno and Miura, 1976), the absence of ATP from a reaction mixture could prevent the synthesis of even very short RNA chains. In addition, the virus-associated nucleotide phosphohydrolase (Section III.A.1.b) hydrolyzes ATP faster than any other rNTP (Storer et al., 1974a; Shimotohno and Miura, 1977), and this could deplete the level of ATP in the substrate pool available to the polymerase. However, the $K_m$ value calculated for ATP hydrolysis suggested that this enzyme activity does not interfere with RNA synthesis by excessively depleting levels of rNTPs (Shimotohno and Miura, 1977). It was also first reported that the inclusion of an ATP-regenerating system was required for full enzyme activity, but rather low levels (0.5 μmole) of ATP were used in these assays (Lewandowski et al., 1969; Shimotohno and Miura, 1973a,b). ATP concentrations of 2 mM or higher (Furuichi, 1974; Mertens and Payne, 1978; Mertens, 1979) obviate the need for ATP regeneration in most assays.

The optimum temperature for RNA synthesis is approximately 31°C, although considerable transcription occurs over a wide temperature range from 15 to 40°C (Lewandowski et al., 1969; Shimotohno and Miura, 1973b; Furuichi, 1974; Mertens, 1979; R.E. Smith and Furuichi, 1980a). In contrast, the polymerase of reovirus is active over a higher temperature range and is almost inactive below 30°C (Kapuler, 1970). This difference probably reflects the fact that CPV is adapted for replication in poikilotherms, while reovirus replicates in homeotherms. The CPV polymerase has optimal activity in vitro at pH 8.0–8.2 and, unlike DNA-dependent RNA polymerases, is not significantly inhibited by actinomycin D (Lewandowski et al., 1969; Shimotohno and Miura, 1973b; Furuichi, 1974; Mertens, 1979).

One of the most significant observations of the CPV polymerase, which led ultimately to a study of mRNA methylation and the 5' cap structure of CPV mRNA (see below), was that the enzyme activity of type 1 CPV was stimulated massively (approximately 60-fold) by the addition of the methyl group donor S-adenosyl-L-methionine (AdoMet) to in vitro transcription assays (Furuichi, 1974). The role of AdoMet and related compounds is discussed more fully in Section III.A.1.e, but apparently involves an allosteric modification of the type 1 CPV polymerase (Furuichi, 1981). In contrast, the polymerase of CPV type 2 can synthesize significant amounts of mRNA in the absence of AdoMet, and a much lower (3-fold) stimulation is achieved on its addition (Mertens and Payne, 1978). AdoMet also has little or no effect on the polymerases of reovirus (Shatkin, 1974; Levin and Samuel, 1977) or wound tumor virus (WTV) (Rhodes et al., 1977), although some increase in activity was reported for the bluetongue virus polymerase (Van Dijk and Huismans, 1980).

The relatively stable nature of the CPV polymerase is indicated by the fact that mRNA synthesis is linear with time for at least 10 hr, and activity is also directly proportional to virus concentration provided that sufficient substrate is available (Lewandowski et al., 1969; Shimotohno and Miura, 1973b; Furuichi, 1974; Mertens, 1979; R.E. Smith and Furuichi, 1980a,b). In 26 hr, mRNA equivalent to 30 times the added dsRNA template can be synthesized in vitro (Smith and Furuichi, 1980a).

Polymerase activity is considerably reduced by techniques designed to separate the protein and RNA components of the virus (Storer et al., 1974a,b; Lewandowski and Traynor, 1972), but prolonged storage for 9 months at −20°C with repeated freeze–thawing has little effect on total polymerase activity. Some reduction in the dependence of enzyme activity on the presence of AdoMet was noted under these conditions (Mertens, 1979). Repeated sedimentation by centrifugation also has little effect on the level of polymerase activity, although, once again, it does reduce the enzyme dependence on AdoMet (Furuichi, 1978a, 1981; Wertheimer et al., 1980).

The products of the complete transcription assay include ssRNAs that sediment in the same two main size classes as denatured viral genome RNA (Shimotohno and Miura, 1973a,b). Two classes of oligonucleotides have also recently been detected in reaction mixtures (Furuichi, 1981; R.E. Smith and Furuichi, 1982a). The significance of these oligonucleotides, which include molecules with a structure related to (p)ppApG (corresponding to the 5'-terminal sequence of virus mRNAs) and Gppp(p)N (derived by guanylation), is discussed in Sections III.A.1.c and f).

The macromolecular products of the CPV polymerase in vitro are ssRNAs that do not self-anneal, but hybridize to all ten viral genome segments, giving dsRNAs that then comigrate during electrophoresis with the genome RNA (Furuichi, 1974; Mertens, 1979; R.E. Smith and Furuichi, 1980a). This suggests that they represent complete ssRNA copies of one strand of each of the viral dsRNA segments. Further evidence for this comes from studies of the 5' and 3' termini of the synthesized RNAs. Judging from the 5'-terminal nucleotide of the viral dsRNA of type 1 CPV (Section II.C.2), the starting nucleotide residue at the 5' terminus of the transcript should be either adenylic or guanylic acid. The 5'-terminal structure found to be common for all ten ssRNAs produced

TABLE V. Relative Transcription Frequency of the Single-Stranded RNA Transcripts Produced by Bombyx mori (Type 1) Cytoplasmic Polyhedrosis Virus Polymerase in Vitro

| Genome segment | Mertens (1979) | | R.E. Smith and Furuichi (1980a) | |
| --- | --- | --- | --- | --- |
| | Relative numbers[a] | Relative mass[b] | Relative numbers[a] | Relative mass[b] |
| 1 | 0.09[c] | 0.66 | 0.08 | 0.58 |
| 2 | 0.09[c] | 0.62 | 0.10[c] | 0.69 |
| 3 | 0.09[c] | 0.60 | 0.10[c] | 0.66 |
| 4 | 0.07 | 0.40 | 0.17 | 0.99 |
| 5 | 0.19 | 0.99 | 0.21 | 1.09 |
| 6 | 0.33 | 1.06 | 0.18 | 0.58 |
| 7 | 0.73[c] | 1.75 | 0.42 | 1.01 |
| 8 | 0.73[c] | 1.29 | 0.60 | 1.06 |
| 9 | 0.51 | 0.82 | 0.75 | 1.20 |
| 10 | 1.00 | 1.00 | 1.00 | 1.00 |

[a] Relative numbers of each transcript normalized to segment 10.
[b] Relative mass [from RNA segment molecular weights given by Fujii–Kawata et al. (1970)] of each transcript normalized to segment 10.
[c] Averages for two or more ssRNA species that were not resolved after electrophoresis.

by the CPV polymerase in the absence of AdoMet was first identified as ppApGpUp (Shimotohno and Miura, 1974; Miura, 1981), while the 3′ terminus was −C (Miura, 1981). Thus, it would appear that all ssRNAs have the same 3′ and 5′ termini as one strand of the viral genome RNA. The loss of the γ-phosphate from the 5′-terminal adenylic acid residue was attributed to the virus-associated nucleotide phosphohydrolase activity (Section III.A.1.b).

Subsequent studies have shown that both the ssRNAs and the plus strand of the viral genome RNA are blocked at the 5′ terminus with a cap structure identified as m⁷GpppAmGp Pyp (Furuichi and Miura, 1975) [Section III.A.1.f (m denotes the presence of a methyl group)]. In the transcription process, it appears that the viral RNA minus strand, which does not bear this cap structure, acts as the template for the polymerase-directed synthesis of CPV mRNA (Miura, 1981):

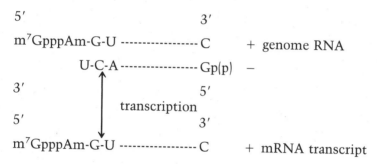

The *in vitro* synthesis of the methylated and guanylated cap structure indicated that virus-associated transmethylase and guanylyltransferase enzymes must also be involved in the production of complete mRNAs.

In some recent, unpublished studies (quoted in R.E. Smith *et al.*, 1981; Y. Furuichi, personal communication), genome segments 4, 5, 8, 9, and 10 have been partially sequenced to approximately 50 nucleotides from the 3′ terminus. Common sequences of –GUUAGCC were observed at the 3′ terminus of the plus strand of each of the five genome RNAs, while the sequence UCAUU– was common to all 3′ termini on the minus strands (Section II.C.2). It has been suggested that these common sequences (which may be present on all ten genome segments) act as initiation signals for the virus-associated polymerase, as well as the polymerase (replicase) that must be involved in the production of viral genome duplex RNA.

The individual ssRNA transcripts from the different genome segments can be readily separated by electrophoresis (Fig. 8). R.E. Smith and Furuichi (1980a) have confirmed that the order of migration of the different products of transcription during electrophoresis in agarose–polyacrylamide gels correlates with the migration of the viral genome segments analyzed by electrophoresis on 3% polyacrylamide gels. This has permitted an analysis of the relative transcription frequencies of the dif-

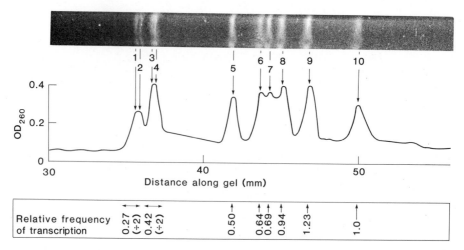

FIGURE 8. Electrophoretic analysis of the polymerase products of type 2 CPV revealing the separation of 9–10 RNA components synthesized in amounts proportional to their molecular weights. Single-stranded RNA was purified from polymerase assays and analyzed by electrophoresis on urea–agarose–polyacrylamide gels. After electrophoresis, the gels were scanned at 260 nm. The relative frequency of transcription of each RNA component was estimated from the amount of RNA in each band, calculated from the total absorbance of each peak (Mertens, 1979).

ferent genome segments. For both CPV types 1 and 2, the transcription frequency of the different genome segments (Table V and Fig. 8) shows some correlation to their molecular weights such that approximately equal amounts by weight of each of the ssRNA species are synthesized under optimal conditions, though these conditions varied with different workers. Whereas R.E. Smith and Furuichi (1980a) found it necessary to include 100–150 mM sodium acetate and high rNTP concentrations to obtain effective transcription, particularly of the larger genome segments of type 1 CPV, Mertens (1979) observed that the ssRNAs of both type 1 and type 2 CPVs were synthesized efficiently without these modifications to a standard polymerase assay. The relatively small differences between the results on transcription frequency of type 1 CPV in the two studies shown in Table V may be accounted for by the different reaction conditions used. Unlike the rotavirus polymerase activity (Bernstein and Hruska, 1981), no RNA transcript appears to be produced consistently at a much greater rate than any other; the slightly lower relative masses of the transcripts of genome segments 1–3 may have arisen by premature termination. Unlike reovirus (Nichols *et al.*, 1972; Joklik, 1974), the relative amounts of the transcripts were also constant regardless of changes in magnesium ion concentration (1–12 mM) or the inclusion of AdoMet and other adenosine analogues (0–1.5 mM) in the assay (Mertens, 1979).

These results suggested that it was most likely that each genome segment was transcribed independently (Mertens, 1979; R.E. Smith and Furuichi, 1980a,b). This conclusion was confirmed by R.E. Smith and

Furuichi (1982b), who demonstrated that initiation of transcription of all ten dsRNA segments of type 1 CPV occurred simultaneously. The times required for the completion of the ten RNA transcripts were proportional to their chain lengths, with elongation rates of six nucleotides per second that were directly comparable to those obtained with reovirus (Joklik, 1974). It is therefore most likely that each of the CPV genome RNA segments undergoes multiple cycles of independent transcription and that CPV genome RNAs are probably not functionally linked within the particle (R.E. Smith and Furuichi, 1982b). It is interesting that active transcriptase complexes that have been isolated recently from virions of type 1 CPV are composed of a single RNA genome segment associated with protein (Dai *et al.*, 1982).

From *in vitro* studies, when the concentration of AdoMet is optimal, no form of transcriptional control other than the rate-limiting step of chain elongation has been observed that would result in the production of larger relative amounts of any given mRNA species (R.E. Smith and Furuichi, 1982b). No studies on transcriptional control *in vivo* have been described. Evidence that the RNA transcripts produced *in vitro* are monocistronic mRNAs that direct the synthesis of virus-specific proteins is considered in Section III.B.

It seems probable, by analogy with reovirus (Joklik, 1974, 1980), that the ssRNAs synthesized by the polymerase also act as templates *in vivo* for the synthesis of complementary minus strands to form the duplex genome RNAs. At present, no studies have been made on the process of duplex RNA formation in CPV infections or on the components of the polymerase (replicase) that would be required.

### b. Nucleotide Phosphohydrolase

Like reovirus cores (Borsa *et al.*, 1970; Kapuler *et al.*, 1970), virions of type 1 CPV contain a nucleotide phosphohydrolase that converts rNTPs to diphosphates (Storer *et al.* 1974a,b). There is no subsequent hydrolysis of the diphosphates to monophosphates. These results have since been confirmed by Shimotohno and Miura (1977). The enzyme is dependent on the presence of magnesium ions, and the rates at which rNTPs are hydrolyzed are in the order $ATP \gg CTP > GTP > UTP$. Unlike the RNA polymerase, the enzyme does not require the presence of viral nucleic acid, since it is active in apparently empty virus particles or in particles disrupted by sonication or degraded by pronase treatment. Storer *et al.* (1974a,b) reported that both the polymerase and the nucleotide phosphohydrolase had a similar dependence on magnesium ions, with similar pH optima, and therefore suggested that they were parts of a multicomponent enzyme complex. However, Shimotohno and Miura (1977) observed significantly higher pH and temperature optima for the phosphohydrolase (pH 9–10 and 45°C) as compared to those for the polymerase (pH 8.1 and 31°C).

As described in Section III.A.1.a, the 5' terminus of the viral mRNA was first identified as ppApGpUp with the terminal adenylic acid lacking the γ-phosphate. It seems likely that one role of the nucleotide phosphohydrolase is the removal of this γ-phosphate from ATP (Shimotohno and Miura, 1977) or from some short oligonucleotide RNA precursor (Furuichi, 1981), rather than from completed RNA Molecules. The 5'-terminal triphosphates in completed RNA chains such as Qβ phage RNA cannot act as substrate for the CPV phosphohydrolase (Shimotohno and Miura, 1977). It is not surprising that the enzyme favors ATP as a substrate when the 5'-terminal residue of nascent RNA chains is adenylic acid. However, the equivalent enzyme in reovirus shows a similar though less marked specificity for ATP (Borsa et al., 1970; Kapuler et al., 1970), even though uncapped reovirus mRNA has a 5'-terminal guanylic acid (Furuichi et al., 1975, 1976a,b; Furuichi and Shatkin, 1976, 1977). It is possible both in reovirus and in CPV that the nucleotide phosphohydrolase serves some other function as well as the modification of the 5' terminus of mRNA. Yazaki and Miura (1980), noting the high ATP consumption by the CPV virion during transcription, suggested that ATP could provide an energy source for movement of the viral RNA genome within the virion during transcription and that the phosphohydrolase could be involved in this. In addition, the phosphohydrolase enzymes of both CPV and reovirus hydrolyze the deoxyribonucleoside triphosphate dATP as effectively as, if not more effectively than, ATP. It has been suggested for reovirus (Joklik, 1974) that this activity may help to account for a reduction in host-cell DNA synthesis observed during reovirus replication, by depletion of the substrate pool available to the host-cell DNA polymerase. There is insufficient information on the biochemistry of CPV replication in vivo to know whether this is likely.

### c. Guanylyltransferase

Virus particles of type 1 CPV, like reovirus cores, contain a guanylyltransferase that forms the blocked cap structure at the 5' terminus of viral mRNA (Furuichi and Miura, 1975; Shimotohno and Miura, 1976; Miura, 1981; R.E. Smith and Furuichi, 1982a). This enzyme adds GMP from GTP to the 5' terminus of the nascent RNA chains, which, as a result of the phosphohydrolase activity end in ppA. . . . This results in the formation of a three-phosphate link (5'—5') between the guanosine residue and the 5' terminus of the ssRNA (GpppA . . .). When first discovered in CPV mRNA, this type of structure, in which two nucleosides link 5' to 5' in a confronting state with two pyrophosphate links, had not been known previously in nucleic acid molecules (Furuichi and Miura, 1975; Miura, 1981).

The step in cap-structure formation catalyzed by the guanylyltransferase is inhibited, and may even be reversed, by high levels of pyrophosphate $(PP_i)$, which is one of the end products of the reaction (Furuichi, 1978a; Furuichi and Shatkin, 1976) [see Fig. 11 (Section III.A.1.f)]. Therefore, pyrophosphatase has been added to some transcription assays to increase the proportion of mRNAs blocked by the addition of a 5′ guanosine residue (Furuichi, 1978a).

The guanylyltransferase of CPV but not of reovirus can also add GMP from GTP to other ribonucleoside triphosphates and diphosphates, resulting in the production of three or four phosphates linked 5′—5′ between guanosine and either guanosine, adenosine, cytosine, or uridine [e.g., G(p)pppG, G(p)pppA (R.E. Smith and Furuichi, 1982a)]. It has been suggested that this class of phosphatase-resistant oligonucleotides, which are synthesized in large amounts in transcriptase assays (Table VI), represents the products of an uncoupled side reaction of the guanylyltransferase rather than intermediates in cap-structure formation (see Fig. 11). As yet, it is not known whether they have any function during CPV replication (R.E. Smith and Furuichi, 1982a).

### d. Transmethylases

Miura et al. (1974) found that one strand of the viral genome RNA [subsequently identified as the plus strand by Shimotohno and Miura (1974)] contained a methylated adenylic acid residue at the 5′ terminus. Furuichi (1974) confirmed that methyl groups from AdoMet were incorporated into CPV mRNA at an early stage of transcription, and further analysis revealed that the 5′ terminus was methylated at both the adenylic acid residue and the 5′-terminal guanylic acid cap (Furuichi and Miura, 1975). The plus strand of genome RNA was subsequently found to be similarly methylated (Miura, 1981).

These observations were among the earliest demonstrations of 5′-terminal methylation of viral and eukaryotic mRNAs and have led to considerable interest in the capping and methylation of mRNA, a subject that is reviewed in detail in Banerjee (1980). They also implied that CPV virions contained enzymes capable of methylating the two 5′-terminal residues.

Methyl groups are incorporated by both CPV types 1 and 2 into all the ssRNA species that can be resolved by electrophoresis of the in vitro–synthesized transcriptase products (Mertens and Payne, 1978; Mertens, 1979), using AdoMet or related compounds, including [³H]methylmethionine, as methyl-group donors (Furuichi, 1974; Wu et al., 1981).

Transmethylase enzymes, including those detected in reovirus cores, often appear to be quite specific in terms of their substrate requirements (Zappia et al., 1969; Deguchi and Barchas, 1971; Furuichi et al., 1975; 1976a,b). Since methyl groups are incorporated at both the 7-position of the blocking guanosine and the 2′-O-position of the 5′-terminal adenosine

(Shimotohno and Miura, 1976; Furuichi, 1978a), it is possible that CPV virions, like vaccinia virus (Martin *et al.*, 1975; Barbosa and Moss, 1978), contain two distinct transmethylase activities.

*e. Interregulation of Enzymes Involved in Cytoplasmic Polyhedrosis Virus Messenger RNA Synthesis*

Since transcription of type 1 CPV RNA appeared to be virtually dependent on the presence of AdoMet, it was originally suggested that transcription was coupled to methylation (Furuichi, 1974). However, several analogues of AdoMet are also effective stimulators of CPV RNA transcription. These include *S*-adenosyl-L-homocysteine [(L-AdoHcy) the end product of the transmethylation reaction], which is incapable of acting as a methyl-group donor and actively inhibits transmethylation (Figs. 9 and 10). This indicates that the increased transcription observed in the presence of these compounds is not simply dependent on methylation of the transcriptase products (Furuichi, 1978a, 1981; Mertens and Payne, 1978; Mertens, 1979; Wertheimer *et al.*, 1980).

The structural features of AdoMet that are important for the increase in RNA synthesis include the terminal carboxyl and amino groups of methionine, the carbon chain length of the amino acid, the 2′- and 3′-hydroxy groups of the adenosine ribose, and an adeninelike structure, in particular the 6-amino group of adenine (Wertheimer *et al.*, 1980). Most recently, Wu *et al.* (1981) have shown that methylmethionine alone is also a very effective stimulator of CPV RNA transcription.

In general, AdoMet analogues that cannot act as methyl donors in-

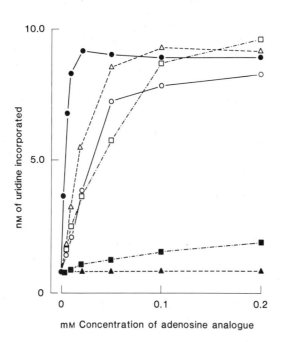

FIGURE 9. Comparison of the stimulation of the RNA polymerase of type 1 CPV virions by adenosine and some of its analogues (Mertens, 1979). (○) AdoMet; (●) L-AdoHcy; (□) *S*-adenosyl-L-ethionine; (■) adenosine; (△) D-AdoHcy; (▲) homocysteine.

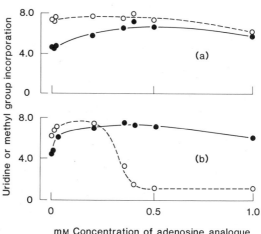

FIGURE 10. Effect of some analogues of adenosine on the methylation of ssRNA synthesized by CPV type 1 virions (Mertens, 1979). (●) Transcription (nmoles uridine incorporated into ssRNA); (○) methylation (pmoles methyl groups incorporated into ssRNA). D-AdoHcy (a) does not significantly reduce methylation, in contrast to L-AdoHcy (b).

hibit methylation of the mRNA, but stimulate transcription. Those analogues that are weak methylation inhibitors are also usually weak stimulators of RNA synthesis (Wertheimer *et al.*, 1980). This is consistent with the idea that AdoMet or its analogues bind to a single site on the virus (e.g., to a methyl transferase) and in so doing stimulate RNA synthesis as well as serve as methyl donors or competitive inhibitors of methylation. However, with such a model, it is difficult to explain the different responses of type 1 CPV to both the D- and L-forms of AdoHcy (Mertens, 1979) and some other AdoMet analogues (Wertheimer *et al.*, 1980). While L-AdoHcy stimulates transcription and inhibits methylation, D-AdoHcy also efficiently stimulates transcription (Fig. 9), but does not significantly reduce methylation (Fig. 10). Other compounds have also been discovered that inhibit CPV mRNA synthesis without reducing methylation. These include compounds such as *S*-guanosyl-L-methionine (GuaMet) and *S*-uridyl-L-methionine (UriMet), in which the adenine of AdoMet is replaced by another base, and AdoHcy-dialdehyde, in which a dialdehyde replaces the *cis*-diol in the ribose moiety (Wertheimer *et al.*, 1980).

The results can best be explained by an alternative model [also considered by Wertheimer *et al.* (1980)] in which molecules of AdoMet or its analogues bind separately to functionally distinct sites, i.e., one to one or more methyl transfer sites, the other to the transcription-control site (Mertens, 1979). While AdoMet and L-AdoHcy may bind efficiently to both sites, D-AdoHcy, GuaMet, UriMet, and AdoHcy-dialdehyde may bind only to the site involved in transcription control. Since it is possible that the CPV virion contains two distinct transmethylases, it seems probable that the polymerase control site is structurally distinct from the active site of at least one of these enzymes.

The 5' termini of mRNAs synthesized in the absence of AdoMet can be guanylated, and therefore the guanylyltransferase is probably not under the direct control of AdoMet—or its analogues (Wertheimer *et al.*, 1980).

It has been suggested that stored or pelleted and resuspended preparations of type 1 CPV virions, which are capable of mRNA synthesis in the absence of AdoMet, may have lost the function of the transcription-control site(s) involved in AdoMet binding (Section III.A.1.a). In contrast, CPV type 2 always shows high transcriptase activity in the absence of AdoMet even when freshly prepared, indicating that different CPVs, like different reoviruses, vary in their response to AdoMet (Mertens and Payne, 1978; Mertens, 1979).

The stimulating effect of AdoMet and its analogues on the transcription of type 1 CPV appears to be linked with initiation, since it is not required for the efficient elongation of nascent ssRNA chains (Furuichi, 1978a, 1981). Virions containing nascent RNA chains that were transferred from a reaction mixture containing AdoMet or AdoHcy to one lacking these components were able to perform RNA chain elongation, but were unable to reinitiate new RNA chains (Furuichi, 1981). This indicated that the effect of AdoMet was not an irreversible modification of the virus-associated enzymes. In assays in which UTP and CTP were excluded, AdoMet stimulated the production of so-called initiation oligonucleotides, including the most abundant product, (p)ppApG, and a variety of closely related, capped and methylated compounds (Table VI)

TABLE VI. Relative Proportions of the Reaction Products of *Bombyx mori* (Type 1) Cytoplasmic Polyhedrosis Virus Transcription *in Vitro*

| Species | Relative amounts synthesized[a] | | |
|---|---|---|---|
| | Complete reaction (ATP, GTP, CTP + UTP)[b] | Incomplete reaction (no CTP or UTP) | |
| | | b | c |
| mRNA | 0.31 | — | — |
| pppApG | | | 7.7 |
| ppApG | } 7.3 | 51 | { 43.3 |
| GpppApG | | | 2.4 |
| m⁷GpppApG | } 0.33 | 6.2 | { 2.8 |
| ppApGpN | — | — | 3.7 |
| GpppApGpN | — | — | 0.1 |
| m⁷GpppApGpN | — | — | 0.3 |
| m⁷GpppAᵐpGpN | — | — | 0.2 |
| GppppApG | 0.33 | 1.1 | — |
| GppppG | 3.6 | 8.7 | — |
| GppppA | 5.9 | 11.8 | — |
| GppppC | | — | — |
| GppppU | } 14.4 | — | — |

[a] Picomoles synthesized during 1-hr transcription (R.E. Smith and Furuichi, 1982a).
[b] Data from R.E. Smith and Furuichi (1982a).
[c] Hypothetical figures calculated from data in Furuichi (1981) after equalizing the amount of (p)ppApG produced in the two studies of R.E. Smith and Furuichi (1982a) and Furuichi (1981).

(Furuichi, 1981; R.E. Smith and Furuichi, 1982a). These compounds, which do not substitute for AdoMet in promoting mRNA synthesis, correspond to the 5′-terminal sequence of virus mRNAs and are apparently produced during repeated initiation by the viral polymerase. In the absence of AdoMet, little RNA and few if any such oligonucleotides were produced (Furuichi, 1981).

Subsequent experiments have shown that efficient mRNA synthesis is controlled by the concentrations of AdoMet, the initiating nucleotide ATP, and to a lesser extent GTP. It appears most likely that RNA initiation is governed by some interaction between AdoMet and the polymerase, producing a conformational change that in turn lowers the apparent $K_m$ for the initiating nucleotide ATP. The definition of such an allosteric mechanism for the stimulation of the type 1 CPV polymerase may extend to other RNA polymerases and could be general for eukaryotic transcription systems (Furuichi, 1981).

### f. Mechanism of Cap Formation and Chain Elongation

Although the functions of the different CPV-associated enzymes in the formation of methylated cap structures is clear, the precise order of the reaction sequence is less certain. Two different sequences have been proposed (Shimotohno and Miura, 1976; Furuichi, 1981). The following sequence was originally suggested by Shimotohno and Miura (1976) and Yamaguchi *et al.* (1976): GTP + ATP (or ADP) → GpppA → m$^7$GpppApG → m$^7$GpppAmpGpUp. . . . This sequence, according to Furuichi (1981), does not take into consideration the relatively large amount of initiation oligonucleotides, particularly (p)ppApG, produced in transcriptase reaction mixtures (Table VI). The production of (p)ppApG and the ability of the virus particles to reutilize it as a substrate for incorporation into fully capped and methylated CPV mRNA (unpublished results quoted in R.E. Smith and Furuichi, 1982a) clearly indicate that the formation of a phosphodiester bond to yield pppApG takes place before guanylation and methylation. Thus, the most likely progression is that shown in Fig. 11 (Furuichi, 1981).

After the formation of the first phosphodiester bond (—pppApG), a further step in cap formation seems essential for the continuation of chain elongation. This is suggested from studies with γ–β-imido analogues of ATP, which, if added to assays before transcription starts, inhibit mRNA initiation. If these compounds are added after initiation, they do not prevent RNA chain elongation. It is known that such ATP analogues cannot be cleaved by the nucleotide phosphohydrolase, and it is likely that this reaction or, more probably, the subsequent guanylation (Fig. 11) represents the critical step controlling elongation (Furuichi, 1978a; R.E. Smith and Furuichi, 1982b). Abortion at this step would account for the presence of the oligonucleotide (p)ppApG in assay mixtures (Table VI). It is clear from studies on the methylation of added GpppA that after guanylation

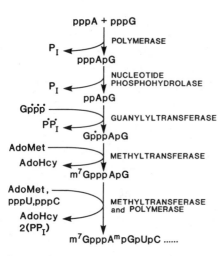

FIGURE 11. Proposed mechanism for the formation of single-stranded capped RNA by the enzymes associated with type 1 CPV virions.

of ppApG, methylation at the 7-position of the blocking guanosine residue is carried out first and that the ribose moiety in the adenosine residue is not methylated until a few nucleotides are linked to it (Shimotohno and Miura, 1976).

The oligonucleotide GpppApG and its methylated derivative $m^7$GpppApG, which are found in some quantities in *in vitro* polymerase assays (Table VI), could have some additional role in virus replication other than acting as intermediates in the pathway of cap formation. 7-Methylguanosine and related compounds inhibit the formation of ribosome complexes with capped and methylated mRNAs (Section III.B.1). If these oligonucleotides are produced by CPV virions during replication *in vivo*, they could inhibit host-cell protein synthesis by competitive inhibition of ribosome binding by capped mRNAs.

The potential role of the cap in complete CPV mRNAs, and of similar structures found in eukaryotic and viral mRNAs, has excited much interest (Banerjee, 1980; Miura, 1981) and is discussed more fully in relation to protein synthesis in Section III.B.1. The importance of capping in mRNA chain elongation explains why high levels of GTP (as well as ATP) are required for efficient mRNA synthesis (Furuichi, 1981).

The dependence of completion of RNA transcripts on capping could be significant *in vivo* in ensuring that a high proportion of the transcripts are capped and can function effectively as mRNA. However, in recent studies of reovirus replication, it has been reported that ssRNA transcripts synthesized late in infection by progeny subviral particles are predominantly uncapped, unlike the material produced earlier by parental cores (Skup and Millward, 1980a; Zarbl *et al.*, 1980; Skup *et al.*, 1981). This change is mirrored by a modification of the host-cell translation mechanism such that uncapped, rather than capped, mRNA is preferentially translated (Skup and Millward, 1980b; Sonenberg *et al.*, 1981). This change could leave the capped transcripts derived from parental cores free to act as template for the production of genome dsRNA (Skup *et al.*, 1981). Although there are no detailed data on CPV replication *in vivo*, it is known that the guanylation step in mRNA capping is reversible in the presence of excess $PP_i$ [an end product of guanylation (Fig. 11)]. This reversed reaction could account for the many uncapped ppA . . . 5' termini observed on mRNA synthesized *in vitro* in the presence of AdoHcy (Furuichi, 1978a) and could produce uncapped termini *in vivo*.

### g. Location of RNA-Processing Enzymes within the Virion

Although the virus-associated enzyme activities and their products are well characterized, little progress has been made in assigning specific enzyme functions to specific structural polypeptides. As mentioned above, the polymerase activity of type 1 CPV was considerably reduced by any technique that separates the protein and RNA components of the virion (Storer *et al.*, 1974a; Lewandowski and Traynor, 1972). This may well result from loss of the dsRNA template when the virion structure is damaged. It is also possible that the transcriptase can function only as part of a multicomponent enzyme complex that is disrupted by these techniques and cannot be reconstituted into an enzymatically active form. In this context, it has been noted that there are some similarities in the molecular weights of the polypeptides of DNA-dependent RNA polymerase complexes of nonviral origin and those of the structural proteins of CPV virions and reovirus cores (Joklik *et al.*, 1970; Payne, 1971; Lewandowski and Traynor, 1972).

Relatively few protein species are present in the virions of CPV type 1 [five have been observed most consistently (Section II.B.1)], yet the virions contain a minimum of four distinct enzyme functions. It is therefore possible that the majority of the virion structural proteins have some enzymatic function and could be involved in transcription. Results from both electron-microscopic and biochemical studies indicate that virions of type 1 CPV, like those of reovirus, are capable of synthesizing several ssRNA molecules simultaneously from different genome segments (Yazaki and Miura, 1980; R.E. Smith and Furuichi, 1982b). It therefore seems probable that each virion contains several copies of the polymerase and capping enzymes and that there is at least one complex of enzymes for each of the ten segments of the viral genome. This conclusion is supported by the recent isolation (following UV irradiation of virions) of at least nine different active transcriptase complexes from type 1 CPV, each of which contained one of the genome segments (Dai *et al.*, 1982). This is the first report to indicate that the RNA polymerase can still be active in some subviral components.

As described in Section III.A.1.a, the different mRNAs are produced *in vitro* in molar amounts inversely related to their molecular weights and are simultaneously initiated at the onset of transcription. These results are consistent with the independent transcription of each genome segment by a recirculating polymerase rather than the random binding of the polymerase to the genome segments, a smaller number of transcriptase complexes than RNA segments, or recirculation of all ten segments through one enzyme complex. Yazaki and Miura (1980) recently demonstrated that the spikes of the CPV virion are the probable sites of the transcriptase enzymes. When the genome dsRNA was released from the virion, "knob-shaped" proteins were often found associated with the end of an RNA strand in glutaraldehyde-fixed preparations (Fig. 12). The dimensions of this protein were similar to those of the spike. When virions were incubated in a transcriptase assay mixture, the spikes swelled and became deformed. When these virions were disrupted, the proteins were attached not only to the end of the dsRNA strand, but also at different positions along the strand (Fig. 12). Yazaki and Miura (1980) concluded that the dsRNA molecules were associated with the spike and, during transcription, passed through part of the projection in a similar manner to a tape moving through a tape-recorder head, while the ssRNA copies passed out through the tubular spike.

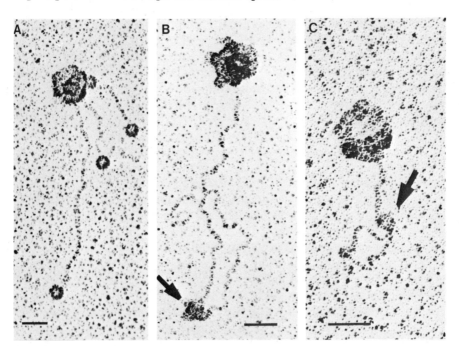

FIGURE 12. Association of CPV genome RNA with virion spikes. (A) Double-stranded genome RNA released from virions of type 1 CPV by EDTA. A protein particle with dimensions similar to those of part of the virus spike is attached to the edge of each released strand. (B, C) Virions incubated in a polymerase assay medium before fixation and disruption. Loops of dsRNA are seen. Protein particles (→) can be seen at different positions along the released dsRNA strand. Single-stranded RNA was not visualized under the conditions used. Scale bars: 50 nm. Reproduced from Yazaki and Miura (1980), by kind permission of the authors and Academic Press.

The active subviral transcriptase complexes that have recently been isolated from the CPV virion (Dai *et al.*, 1982) may be the same subunits as described by Yazaki and Miura (1980). In addition to the RNA polymerase, these complexes also contain the virus-associated transmethylase activities (Dai *et al.*, 1982). The transmethylase(s) may well be the last enzyme(s) in the pathway for cap-structure formation and will be dependent on the action of other capping enzymes (see Fig. 11). It is therefore possible that these complexes also contain the nucleotide phosphohydrolase and guanylyltransferase enzymes. In a recent paper, R.E. Smith and Furuichi (1982a) presented some preliminary observations that suggested that $[^{32}P]$-GMP became covalently attached to a virus structural protein with a molecular weight of 120,000 (perhaps equivalent to the protein with a molecular weight of 125,000 described in Section II.B.1). The possibility that this may be the guanylyltransferase provides at present the only potential assignment of an enzyme function to a specific viral polypeptide.

Although there are only 10 genome RNA segments in CPVs and 12 spikes, it is interesting that the maximum number of RNA genome segments among members of the Reoviridae does not exceed 12, e.g., WTV (Kalmakoff *et al.*, 1969) and Colorado tick fever virus (Knudson, 1981). This is the maximum number that would be expected if each RNA segment is associated with a separate transcriptase complex located at the vertex of an icosahedral particle. It could be that a 12-segment genome represents the ancestral type of reoviruses from which others have diverged.

## 2. Other Virus-Associated Enzymes

### a. Exoribonuclease

Storer *et al.* (1974a) suggested that an exoribonuclease that degrades ssRNA was associated with purified virions of type 1 CPV. This enzyme, like the RNA polymerase and nucleotide phosphohydrolase, was apparently dependent on the presence of ATP and magnesium ions for its activity. Since one of the major enzyme functions of CPV virions is the synthesis of ssRNA, it seems somewhat improbable that virus particles would contain an enzyme of this type, although Storer *et al.* (1974a) suggested that the nuclease might play some role in the formation of the dsRNA segments. In view of the ubiquity of ribonucleases, it seems likely that low levels of the enzyme found in CPV preparations could be the result of contamination, e.g., from the sucrose gradients used in virus purification. Nonetheless, capped and methylated CPV mRNA is relatively resistant to exoribonuclease digestion (Section III.B.1), and the possible involvement of such an enzyme in some viral function cannot be totally excluded.

### b. Protease

Insect viruses that produce proteinaceous inclusion bodies appear to have alkaline proteases associated with them (Payne and Kalmakoff, 1978; Harrap and Payne, 1979). Although there are no published reports that clearly demonstrate the presence of proteases in CPV polyhedra, Payne and Tinsley (1974) observed that the polyhedrin of type 2 CPV was extensively degraded during prolonged exposure to alkali. At the time, this degradation was attributed to breakdown following the loss of carbohydrate groups from the protein, but it could equally have arisen from endogenous protease activity. In addition, Lewandowski and Traynor (1972) and Payne and Kalmakoff (1974a) observed some modification in the polypeptide components of type 1 CPV virions during storage. An additional high-molecular-weight polypeptide appeared, probably as a result of the breakdown of the largest viral polypeptide. Once again, this could be attributed to endogenous protease activity.

More recently, M. Carter and C.C. Payne (unpublished observations) have confirmed the existence of alkaline protease activity in purified preparations of type 1 CPV polyhedra. The enzyme degraded polyhedrin (molecular weight 27,000) progressively into components ranging in molecular weight from 4600 to 15,500. The enzyme was inactivated by incubating polyhedra at 80°C for 30 min, and activity was optimal at pH 10.8. Purified virions were also proteolytically active, but attempts to locate the active component gave inconclusive results. Proteases were also detected in polyhedra of type 5 CPV from *H. armigera* (R. Rubinstein and A. Polson, personal communication), but the similarity in the molecular weights of these and insect proteases, particularly trypsin, suggested that they may represent the nonspecific occlusion of insect-derived enzymes. Nonetheless, purified virions of this virus also contained proteolytically active polypeptides of different molecular weights.

The nature of the CPV-associated alkaline protease requires further investigation. In particular, there is no clear indication that it is virus-directed. The role of such an enzyme could be similar to that proposed for the proteases observed in other occluded insect viruses, i.e., the release of virions from the surrounding polyhedrin, in the high-pH conditions prevailing in the midgut of many insect larvae (Payne and Kalmakoff, 1978), prior to infection of the insect gut epithelial cells.

## B. Translation of Cytoplasmic Polyhedrosis Virus RNA and Control of *in Vitro* Protein Synthesis

In the absence of effective cell-culture systems for CPV replication, the study of the translation of CPV mRNA *in vivo* has been greatly restricted. In particular, the exact numbers and biochemical properties of the virus nonstructural proteins are not yet resolved. The differential control of translation of different protein species is also largely a matter of speculation. However, as described above, the virion-associated polymerase and enzyme complex synthesize intact plus-strand copies of all ten viral genome segments *in vitro*. These transcripts, and the plus strand of viral genome RNA, have provided the starting material for *in vitro* studies on CPV RNA translation.

## 1. Translation of Cytoplasmic Polyhedrosis Virus Messenger RNA Transcripts

The first studies on CPV RNA translation indicated that the transcripts produced *in vitro* were capable of stimulating amino acid incorporation into protein in a wheat germ translation system (Shimotohno *et al.*, 1977; Mertens, 1979; Sun *et al.*, 1981). However, effective protein synthesis was dependent on the presence of complete methylated cap structures at the 5' terminus of the RNA molecule. Thus, mRNAs synthesized in the presence of AdoMet, which are predominantly fully capped and methylated (Furuichi, 1978a), increased amino acid incorporation much more than mRNAs synthesized in the presence of AdoHcy (Mertens, 1979), which have 5'-terminal structures in the forms of GpppA (37%) and ppA (63%) (Furuichi, 1978a). In addition, when the $m^7Gp$ cap is removed by tobacco phosphodiesterase (Shinshi *et al.*, 1976), the translation-template activity of the mRNA is reduced to 10–20% of that of capped mRNA (Shimotohno *et al.*, 1977). It is known that the methylated cap structure has a stabilizing effect on CPV mRNA, probably by increasing resistance to degradation by exoribonucleases (Furuichi and Miura, 1975; Furuichi *et al.*, 1977; Shimotohno *et al.*, 1977; Miura, 1981). However, the cap structure is also involved in ribosome binding, since the removal of $m^7Gp$ was associated with a loss of 80 S ribosome–RNA protein synthesis–initiation complexes (Shimotohno *et al.*, 1977). Such complexes are also strongly inhibited by the addition of free $m^7Gp$ (Miura *et al.*, 1979), which presumably competes with mRNA for ribosome-binding sites. The stimulation of translation produced by unmethylated CPV or reovirus mRNA can be increased by adding AdoMet to the wheat germ translation system (Muthukrishnan *et al.*, 1975; Mertens, 1979). This effect is probably due to the ability of this system to methylate exogenous mRNA (Muthukrishnan *et al.*, 1975; Levin and Samuel, 1977).

CPV mRNA synthesized in the presence of $S$-adenosyl-L-ethionine contains partially ethylated 5' caps (Furuichi, 1978b). Such mRNA is even less effective than uncapped and unmethylated RNA in inducing protein synthesis, and translation is not increased by adding AdoMet to a wheat germ translation system (Mertens, 1979). It therefore seems probable that methylated (rather than ethylated) cap structures are required at the 5' terminus of CPV mRNAs to promote efficient translation. Similar conclusions have been reached with other viral and eukaryotic mRNAs, and these studies are summarized in reviews by Banerjee (1980) and Miura (1981).

When intact CPV mRNA transcripts were used as templates for *in vitro* translation in a reticulocyte lysate system, several of the synthesized polypeptides comigrated with virus structural proteins during gradient PAGE (Mertens, 1979) (Fig. 13A). It appeared that none of the products was produced in great molar excess relative to the remainder, suggesting that, *in vitro*, the mRNAs are translated with comparable though not identical frequency (McCrae and Mertens, 1983). This in turn suggests that any form of translational control that results in the production of significantly larger amounts of polyhedrin *in vivo* is not effective in this *in vitro* translation system (McCrae, 1982).

To discover which of the genome RNA segments encodes which virus-specific polypeptide, the different mRNA species must be separated and subsequently translated independently. It is theoretically possible to do this with the mRNA products from the CPV polymerase, but there are

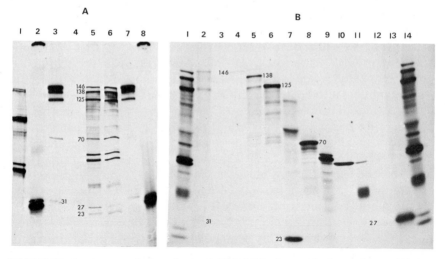

FIGURE 13. *In vitro* translation of type 1 CPV RNAs in a reticulocyte lysate. (A) Autoradiograph of 5–15% polyacrylamide gradient slab gel comparing [35S]methionine-labeled proteins synthesized in response to added RNAs. (1) Proteins synthesized after addition of denatured reovirus type 3 dsRNA; (2, 8) iodinated proteins of type 1 CPV polyhedra; (3, 7) iodinated proteins of type 1 CPV virions; (4) endogenous products of reticulocyte lysate; (5) proteins synthesized after addition of the ssRNA transcripts of type 1 CPV polymerase; (6) proteins synthesized after addition of denatured dsRNA of type 1 CPV. (B) Autoradiograph of 5–15% slab gel comparing the protein products of the separated denatured genome RNA segments of type 1 CPV. (1, 14) Protein products of total denatured RNA; (2) iodinated proteins of type 1 CPV virions; (3) endogenous products of the reticulocyte lysate; (4–12) protein products of RNA genome segments 1–10, respectively (genome segments 2 and 3 were not separated, and the products are shown in gel well 5); (13) iodinated proteins of type 1 CPV polyhedra. The structural proteins of virions and polyhedra are marked by their molecular weights ($\times 10^{-3}$). Results from Mertens (1979), McCrae (1982) and McCrae and Mertens (1983) included by kind permission of the authors.

no full reports that indicate that it has yet been achieved, although R.E. Smith and Furuichi (1980b) briefly mention unpublished preliminary studies of the translation products of separated mRNAs. Instead, an alternative and technically less demanding approach has been used, and this is described in the next section.

## 2. Translation of Denatured Genome RNAs

McCrae and Joklik (1978) demonstrated that the plus strands of reovirus genome dsRNA were capable of being translated *in vitro* and produced recognizable virus-specific polypeptides. The separation of the different reovirus and CPV dsRNA genome segments can be carried out more effectively than that of the ssRNA transcripts. This is due partly to the greater resolution of dsRNA molecules during electrophoresis and the greater stability of dsRNAs during the subsequent reextraction procedures. Because of the mechanism of CPV genome RNA production, each segment contains a fully capped plus strand apparently identical to the equivalent mRNA transcript (Section III.A.1.a) (Miura, 1981), and like reovirus RNA, this can be translated *in vitro*. The proteins produced do not appear to differ from those translated from the polymerase-derived ssRNA transcripts (Fig. 13A).

The plus and minus strands of RNA have been separated on a preparative scale by electrophoresis in denaturing gels (R.E. Smith *et al.*, 1981). However, this is not essential for successful translation, since it is only necessary to denature the dsRNA molecules to separate the two strands using conditions such as 90% dimethylsulfoxide at 50°C (McCrae and Joklik, 1978) or 10 mM methyl mercury (Sangar *et al.*, 1981). The minus strands appear not to form complexes with ribosomes (perhaps because they do not bear the 5'-terminal cap structure) and are therefore inactive in protein synthesis (R.E. Smith *et al.*, 1981).

The genome segments of type 1 CPV have been fractionated, denatured individually, and used as templates for protein synthesis in a reticulocyte lysate (Mertens, 1979; McCrae, 1982; McCrae and Mertens, 1983). Electrophoretic analysis of the polypeptides synthesized (Fig. 13B) has made it possible to construct a provisional genetic map (Fig. 14) of this virus genome. The recognition of specific polypeptides at present relies solely on their comigration in SDS–polyacrylamide gels with known virus structural proteins. Thus, the genetic map cannot be regarded as definitive until more conclusive evidence (such as peptide maps) is available to confirm the identity of each of the *in vitro* translation products. However, studies with other dsRNA viruses have indicated that comigration of polypeptides forms a sound basis for the provisional assignment of viral proteins to genome segments (McCrae and Joklik, 1978; Mertens and Dobos, 1982).

The translation of genome segment 1 of type 1 CPV produced a polypeptide that comigrated with V146 (see Fig. 13B). The major polypeptide

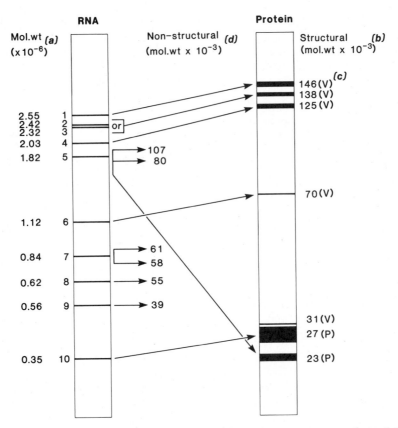

FIGURE 14. Provisional coding assignments of the genome RNA segments of type 1 CPV. (a) RNA molecular weights from Fujii-Kawata *et al.* (1970). (b) Structural-protein molecular weights from Payne and Rivers (1976). (c) Structural polypeptides of purified virions (V) and nonvirion structural polypeptides of polyhedra (P). (d) Nonstructural-protein molecular weights calculated from data in Fig. 13, using virion structural-protein molecular weights as standards (Payne and Rivers, 1976).

produced from genome segments 2 and 3 (which were not separated before addition to the reticulocyte lysate) produced one major polypeptide that comigrated with V138 and minor bands that comigrated with V146 and V125 which may indicate contamination of the RNA primer with segments 1 and 4. Alternatively, either segment 2 or 3 may code for another polypeptide with the same electrophoretic mobility as V146, V138, or V125. The major product from genome segment 4 comigrates with V125. Although translation of segment 5 produced two large proteins (107,000 and 80,000), the major translation product was much smaller (23,000) and may comigrate with P23, a protein that has been reported to be a polyhedral structural protein by some workers (Lewandowski and Traynor, 1972; Mertens, 1979) or that may represent a nonstructural protein

(McCrae, 1982). Provisional peptide maps of the two larger products from segment 5 show peptide similarities, and the smaller of these may represent a premature termination or proteolytic cleavage of the larger (McCrae, 1982). The same may also be true of the smallest, and most intense, polypeptide.

Translation of genome segment 6 produced a protein that comigrated with V70, while the products from segments 7, 8, and 9 did not comigrate with any of the virion- or polyhedron-associated polypeptides and are probably nonstructural proteins. While the major products of segments 8 and 9 were well-defined polypeptides with molecular weights of approximately 55,000 and 39,000 translation of segment 7 produced at least two protein species, with molecular weights of 61,000 and 58,000. These have peptides in common (McCrae and Mertens, 1983), and the smaller may arise by premature termination or proteolytic cleavage.

Translation of genome segment 10 produced a protein that comigrated with P27, the major component of polyhedra (polyhedrin). This provided the first direct evidence that CPV polyhedrin is a virus-coded protein and is consistent with the observation that type 1 CPV virions lacking the smaller genome segment can infect insects, but cannot produce polyhedra (C.C. Payne and J. Kalmakoff, unpublished observations). Apart from the production of the two smaller polypeptides from segment 5, these provisional assignments of CPV type 1 proteins are in good agreement with other preliminary translation results (R.E. Smith and Furuichi, 1980b).

At least two problems emerge from these protein–RNA assignments. First, there does not appear to be any major translation product that comigrates with V31. This, or a protein of similar molecular weight, has been quite consistently observed in preparations of CPV virions (Lewandowski and Traynor, 1972; Payne and Kalmakoff, 1974a; Payne and Rivers, 1976; Mertens, 1979). It is possible that V31 is produced by a posttranslational modification (e.g., cleavage, glycosylation) that is inoperative in the reticulocyte lysate. Similar results have been observed during *in vitro* translation with one of the proteins of bluetongue virus (D. Sangar and P.P.C. Mertens, unpublished observations) and with infectious pancreatic necrosis virus (Mertens and Dobos, 1982).

The second problem concerns the relative rates of translation of the different proteins. It had been anticipated that the RNA segment that codes for polyhedrin might be translated more efficiently than most other RNAs, since this protein is produced in vast quantities during infection. In fact, assuming that virus particles represent only 5% by weight of the polyhedron (Hukuhara and Hashimoto, 1966b), approximately 30 times more polyhedrin by weight is present in polyhedra than the total of all other viral polypeptides. The existing evidence (Fig. 13A) does not indicate that the polyhedrin mRNA is more efficiently translated, and an improved understanding of the control of translation awaits an effective cell-culture system for the investigation of CPV replication. Some possible mechanisms of control are discussed below.

### 3. Control of Cytoplasmic Polyhedrosis Virus Protein Synthesis

R.E. Smith and Furuichi (1982b) have mentioned some unpublished results that suggest that viral polypeptide synthesis in CPV-infected cells of *B. mori* is regulated to a considerable extent at the transcriptional or translational level, or both. Some inherent control may occur if (as *in vitro*) the genome segments are transcribed with frequencies inversely related to their molecular weights (Section III.A.1.a). Thus, at the same elongation rate, approximately seven mRNA transcripts of segment 10 (coding for polyhedrin) will be produced for every one transcript of segment 1, and this could lead to a higher level of polyhedrin production. In reovirus, transcription appears to be regulated *in vivo*, but not *in vitro* (Joklik, 1974), and this may yet be found in CPV infections. It is clear from the results obtained using unfractionated RNA transcripts or denatured dsRNA (Fig. 13A) that some proteins are produced in larger amounts than others, suggesting that in the reticulocyte lysate there is some translational control of gene expression (McCrae, 1982). However, this is unlikely to reflect accurately the control present *in vivo*.

Variations in the 5'-terminal structures of CPV mRNAs affect their relative abilities to form initiation complexes with ribosomes and may thereby affect their relative translation frequency. However, CPV mRNA synthesized *in vitro* in the presence of AdoMet contains almost exclusively m$^7$GpppAm caps that are distributed evenly among the different transcripts (Furuichi, 1978a; Mertens, 1979). Thus, unless some differential control of capping by the virion-associated enzymes exists *in vivo*, but is not evident *in vitro*, it is unlikely that methylation and capping are responsible for differential expression of the genome segments. Nonetheless, differential ribosome binding may also occur with different mRNAs, mediated by the base sequence between the 5'-terminal cap and the AUG initiation codon (Nakashima *et al.*, 1980) or by variations in the secondary structure of the mRNA (R.E. Smith and Furuichi, 1982a). Such possibilities await further investigation.

### IV. CYTOPLASMIC POLYHEDROSIS VIRUS REPLICATION *IN VIVO*

With several recent exceptions (see Section IV.C), most studies of CPV replication have been carried out in the larval stages of insects, rather than in cell culture. CPV infection and pathogenesis in larvae have been reviewed recently by Payne (1981), and the following description summarizes the main features of the biology of virus infection and host response.

## A. Mode of Infection and Virus Morphogenesis

### 1. Virus Entry

Virus infection of larvae under natural conditions follows the ingestion of food contaminated with CPV polyhedra. The polyhedrin is rapidly dissolved in the alkaline gut juice of the insect and is probably degraded by endogenous or gut proteases (Vago, 1959; Vago et al., 1959) (Section III.A.2.b). The released virus particles are then free to infect susceptible cells. Infection can also be artificially induced by the injection of virions into the hemocoel (Faust and Cantwell, 1968), but regardless of the mechanism of virus application, replication is confined almost exclusively to cells of the gut epithelium (Section IV.B.1 ).

The method of entry of virions into susceptible cells is unclear. Virus attachment to cell membranes may be mediated by the viral hemagglutinin (Miyajima and Kawase, 1969). By analogy with reovirus, it is likely that the retention of the structural integrity of the virus is important for infection so that the viral polymerase remains active. Kobayashi (1971) has suggested that infection proceeds following the injection of core material into the cell through a spike. From what is now known of the probable enzymatic functions associated with the spikes. this mode of entry seems unlikely. Both Bird (1965) and Quiot and Belloncik (1977) have observed large numbers of vacuoles or lysosomes containing virions, in infected larvae and an *L. dispar* cell line, respectively, at an early stage of infection. Perhaps, like reovirus (Silverstein and Dales, 1968), CPV virion uptake also occurs by viropexis. Unlike reovirus, however (Silverstein *et al.*, 1972), subsequent modifications of virus structural proteins to activate the viral polymerase may not be necessary, since the CPV enzyme is, in effect, already activated by the release of virions from polyhedra.

### 2. Assembly of Virus Particles

#### *a. RNA and Protein Synthesis*

*In vivo* studies of viral RNA synthesis in CPV-infected larvae are limited, but have clearly demonstrated that virus-specific single-stranded RNA (ssRNA) molecules are produced that are consistent in size and other properties with the well-studied *in vitro* products of the polymerase. Two size classes (15 and 22 S) of ssRNAs were synthesised in CPV-infected larvae of *O. leucostigma* and *M. disstria* when host RNA synthesis was inhibited by actinomycin D (Hayashi, 1970b; Hayashi and Kawarabata, 1970; Hayashi and Donaghue, 1971). Similar results were obtained in infected *B. mori* larvae (Furusawa and Kawase, 1971, 1973; Kawase and Furusawa, 1971). It was confirmed that these RNAs did not self-anneal but reannealed specifically to denatured viral RNA (Furusawa and

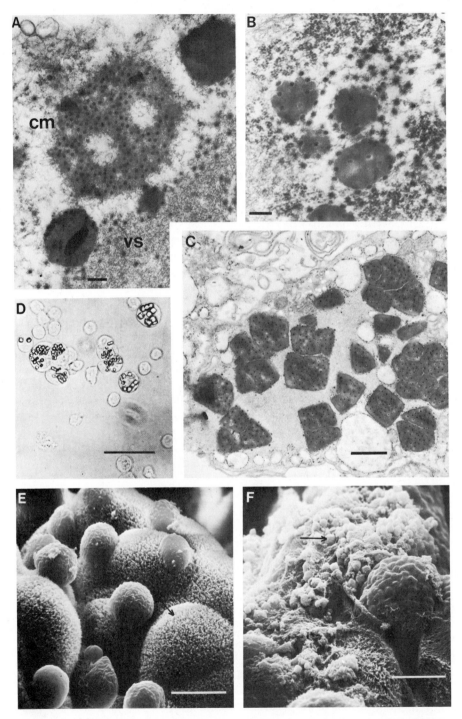

FIGURE 15. Morphogenesis of CPVs. (A) EM revealing the virogenic stroma (vs) and crystallogenic matrix (cm) in a CPV-infected gut epithelial cell of *Bupalus piniarius*. Scale bar:

Kawase, 1973; Payne and Kalmakoff, 1973) and probably represented the plus-strand products of the viral polymerase.

Newly synthesized viral genome RNA has also been detected during infection (Kawase and Kawamori, 1968). The precise machanism of its synthesis is unknown, although it seems most likely that the process will be analogous to that observed in reovirus infections, where plus-strand RNAs synthesized by the viral polymerase serve as templates for minus-strand production within a subviral replicative complex composed of virus structural and nonstructural proteins (Joklik, 1980). The double-stranded RNAs (dsRNAs) synthesized in CPV-infected larvae fall in the same size classes (15S and 12S) as genome RNA, and at the height of infection, this dsRNA amounted to as much as 12% of the total RNA in CPV-infected midguts of *B. mori* (Hayashi and Kawase, 1965d). Several authors have suggested that the nucleus is the main site of viral RNA synthesis. Viral dsRNA amounted to 16, 8.1, and 5.2%, respectively, of the total RNA in nuclear, mitochondrial, and microsomal fractions of midguts of CPV-infected *B. mori* (Hayashi and Kawase, 1965c). Autoradiographic studies have also indicated that much radiolabeled uridine is incorporated within nuclei during infection (Watanabe, 1967b), even in the presence of sufficient actinomycin D to suppress most host-cell RNA synthesis (Hayashi and Retnakaran, 1970). It is perhaps surprising that the nucleus may be involved, since all other aspects of replication appear to be confined to the cytoplasm and neither subviral particles nor mature virions have been detected in the nuclei of infected cells (Kobayashi, 1971). Further work is needed to directly characterize the nuclear involvement.

Although virus-specific antigens can be detected in the cytoplasm 4–8 hr after infection (Kawase and Miyajima, 1969; Granados, 1975), the first distinctive morphological feature observed in electron-microscopic (EM) studies is the development of viroplasms or "virogenic stromata" within the cytoplasm (Xeros, 1956; Sohi *et al.*, 1971). These stromata have a micronet appearance (Fig. 15A) and lack cellular components such as mitochondria and endoplasmic reticulum (Xeros, 1956; Arnott *et al.*, 1968; Longworth and Spilling, 1970; Sohi *et al.*, 1971). They first form near the brush border of the gut epithelial cells and subsequently extend to the perinuclear region and base of the cell later in infection (Xeros, 1966; Kawase and Miyajima, 1969; Granados, 1975). The early stages of these stromata lack cytochemically detectable RNA, although late in

---

200 nm. (B) Developing polyhedra in *B. piniarius*. Scale bar: 200 nm. (C) Gut epithelial cell of *Bombyx mori* filled with CPV polyhedra. Scale bar: 1 μm. (D) Light micrograph showing cuboidal *Chrysodeixis eriosoma* CPV polyhedra in a *Spodoptera frugiperda* cell culture. Scale bar: 30 μm. (E, F), Scanning EMs of the microvillar surface (→) of healthy (E) and CPV-infected (F) gut epithelium of *Hyphantria cunea*. Polyhedra (→) are visible at the surface of the infected gut epithelium. Scale bar: 10 μm. Reproduced from Payne and Harrap (1977) (A–C), Longworth (1981) (D), and Boucias and Nordin (1978) (E, F), by kind permission of the authors and Academic Press.

infection, much RNA is present (Xeros, 1966). This is consistent with the stromata not being the main sites of viral RNA synthesis, but they are certainly active in the synthesis or accumulation of viral proteins, or both, as measured by autoradiography (Watanabe, 1968a) and immuno-fluorescence (Kawase and Miyajima, 1969; Granados, 1975).

There is little information on the production of specific viral proteins. The synthesis of polyhedrin is detected later than virion proteins and may represent a late function in virus replication (Kawase and Miyajima, 1969). Nothing is known of the biochemical processes of virus assembly, including the mechanism by which the ten RNA genome segments are assembled into each virion.

### b. Virion Morphogenesis

There are two main theories of CPV virion morphogenesis based on EM observations of infected gut cells in both infected larvae and cell culture. The observations of Bird (1965) and Kobayashi (1971) are consistent with the interpretation that viral core material is synthesized separately and subsequently encapsidated. Kobayashi (1971) observed particles resembling the core of the virion developing from, or associated with, electron-dense circular bodies formed near the virogenic stromata. These particles appeared to diffuse into the stromata, where they were encapsidated and the spikes were added. Bird (1965) observed concurrent synthesis of core material and capsids, with encapsidation occurring when a partially formed capsid enclosed a core. He suggested that fully formed capsids remained empty. In contrast, Arnott et al. (1968) and Andreadis (1981) have interpreted the numerous capsids or empty particles as the first stage of morphogenesis, with these subsequently being filled with core material. Andreadis (1981) suggested that virus assembly began in areas of sparse virogenic stromata with the production of capsids and particles with incomplete cores. Areas of dense stromata contained only the complete particles, which were subsequently occluded within polyhedra. These different interpretations may be resolved only by an infection time–course study carried out in synchronously infected cell cultures, together with an improved understanding of the biochemical events involved in virus replication.

### 3. Assembly of Polyhedra

The first polyhedra appear only about 15 hr postinfection or later, although polyhedrin synthesis may be detected as early as 9 hr after infection at sites near the brush border of the gut epithelial cells (Kawase and Miyajima, 1969). The virogenic stromata are the main sites for polyhedron synthesis and virus occlusion (Andreadis, 1981). In CPV-infected Lepidoptera, polyhedron formation starts at several sites with the crystallization of protein around groups of mature virions (Fig. 15A and B), often at the periphery of the stromata (Xeros, 1956, 1966; Kobayashi, 1971; Granados et al., 1974; Quiot and Belloncik, 1977; Payne and Harrap, 1977). Only mature virions are occluded (Sohi et al., 1971). In some cases, many tightly packed fibrils can be seen adjacent to developing polyhedra (Fig. 15A). This structure, referred to as a "crystallogenic matrix," probably represents fibers of polyhedrin that crystallize only in the presence of virus particles (Arnott et al., 1968; Kobayashi, 1971; Payne and Harrap, 1977). A different process of polyhedron assembly has been noted in CPV infections of some dipterans, where the virions are first occluded singly rather than in groups, and these small inclusion bodies appear to fuse subsequently (Clark et al., 1969; Stoltz and Hilsenhoff, 1969; Federici, 1974; Payne and Harrap, 1977; Andreadis, 1981).

The precise mechanism of polyhedron formation is unknown. It is likely that there is some specific recognition (such as between enzyme and substrate) between virions and polyhedrin. In a mixed infection of A. caja by CPV types 2 and 3, polyhedrin characteristic of only type 3 CPV is produced (Payne, 1976). Judged by the RNAs from these polyhedra, far more virions of type 3 CPV were occluded than type 2 virus. In contrast, the nonoccluded virions contained proportionally more type 2 viral RNA than type 3. This observation is consistent with some specificity in the process of virus occlusion. The unusually large amount of non-occluded virus observed in M. disstria CPV infections (Hayashi, 1970a) may also have arisen by a similar exclusion of a proportion of the virions from polyhedra, since this infection is now known to arise from a mixture of two CPV strains (Section II.C.2).

Factors that govern the shapes and size of polyhedra are still open to speculation (Section II.A.1). At the final stages of infection, many polyhedra may be found in a single cell (Fig. 15C and D). In lepidopteran cell cultures, normal average yields of polyhedra range between 3.7 and 6.25 polyhedra per infected cell (Belloncik and Arella, 1981; Longworth, 1981). In thin sections of infected lepidopteran larvae, 20 polyhedra have been observed in a single cell (Arnott et al., 1968), compared with up to 100 smaller polyhedra in an infected cell of the hymenopteran Anoplonyx destructor (Longworth and Spilling, 1970). By this stage, the cell cytoplasm is filled with polyhedra, the stromata have regressed, and cell lysis may occur, although this appears rare in infected cell cultures (Section IV.C).

## B. Effect of Virus Infection on Cell and Host

### 1. Cellular Pathology

Virus infection is generally restricted in larvae to the columnar epithelial cells of the midgut, although goblet cells may also be infected (Martignoni et al., 1969; Iwashita, 1971). Only Stoltz (1969) has observed

CPV replication in tissues other than the alimentary tract, in this example, in the fat body of *Chironomus plumosus*. Nonetheless, in cell cultures, both primary cells and cell lines derived from ovarian, tracheal, or gut tissue are susceptible to infection (Grace, 1962; Vago and Bergoin, 1963; Kobayashi, 1971; Sohi *et al.*, 1971; Granados *et al.*, 1974; Granados, 1976; Quiot and Belloncik, 1977; Longworth, 1981).

In the infected larva, virus infection spreads throughout the midgut region, although in some species the entire gut is rarely infected (Bailey *et al.*, 1975; Weiser, 1978). The huge numbers of polyhedra produced give the infected gut a characteristically creamy-white appearance. Within the cell, the endoplasmic reticulum is progressively degraded (Xeros, 1956; Arnott *et al.*, 1968; Kobayashi, 1971; Andreadis, 1981), mitochondria enlarge, and the cytoplasm becomes highly vacuolated. In contrast, relatively few pathological changes have been observed in the nucleus (Payne, 1981). One exception is the discovery of a CPV strain infecting *B. mori* that produces within the nucleus cubic inclusions that superficially resemble polyhedra, although they do not contain virions. In such infections, polyhedron formation in the cytoplasm is absent or infrequent (Hukuhara and Yamaguchi, 1973; Kawase *et al.*, 1973).

At the final stages of infection, cellular hypertrophy is common (Grace, 1962; Harpaz *et al.*, 1965; Xeros, 1966; Granados *et al.*, 1974). The microvilli protruding into the gut lumen from the surface of the epithelial cells are reduced or completely absent (Fig. 15E and F) Boucias and Nordin, 1978). This undoubtedly reduces the adsorptive capacity of the gut cells. Apical-cell membranes eventually break down and polyhedra are released into the gut lumen and excreted in large numbers in the feces of diseased larvae (Ishimori, 1934; K.M. Smith, 1963; Watanabe, 1968b; Boucias and Nordin, 1978). The gut pH is lowered during infection (Vago *et al.*, 1959), perhaps as a consequence of the buildup of uric acid (Watanabe, 1971a, 1972), and this prevents the dissolution of the polyhedra in the gut fluid.

## 2. Host Pathology

### a. Effect of Cytoplasmic Polyhedrosis Virus Infection on Host Development

The pathological consequences of CPV infection in insect larvae are manifested by the symptoms of starvation that result from the changes in gut cell structure and physiology. CPV-infected larvae rapidly lose their appetite and stop feeding as early as 2 days after infection (Arnott *et al.*, 1968). Larval body size and weight are reduced and diarrhea is common, the feces being highly contaminated with polyhedra (Payne, 1981). The duration of the larval stage is increased considerably. In one case, CPV-infected larvae of *P. gossypiella* remained in the first instar for periods exceeding the normal generation time (Ignoffo and Adams, 1966), and time delays in larval development extending to about 1½ times the normal generation time are not unusual (Bellemare and Belloncik, 1981; Payne, 1981). On occasion, increased numbers of larval molts occur both in infected larvae and in uninfected, starved insects (Magnoler, 1974).

Following the reductions in larval size, pupae produced from infected individuals are also smaller (Neilson, 1965; Ignoffo and Adams, 1966; Bird, 1969; Vail et al., 1969; Vail and Gough, 1970; Bullock et al., 1970; Simmons and Sikorowski, 1973; Magnoler, 1974; Bell and Kanavel, 1976; Bellemare and Belloncik, 1981) and, in common with diseased larvae, contain smaller reserves of lipid and protein (Bell, 1977; Bell and Kanavel, 1977; Sikorowski and Thompson, 1979; Thompson and Sikorowski, 1980).

The considerable sublethal effects of CPV infection are particularly pronounced in the adult stage, in which, as in the pupa, infection still continues in the gut epithelium. The majority of diseased adults are malformed, and they may be unable to emerge satisfactorily or are rendered flightless (Atger, 1962; Neilson, 1965; Vail and Gough, 1970; Magnoler, 1974; Bellemare and Belloncik, 1981). Mating is therefore affected, and the fecundity is further limited by reduced egg-laying by the smaller females (Neilson, 1965; Vail and Gough, 1970; Simmons and Sikorowski, 1973). The effects can be so dramatic that the reproductive capacity of the host is reduced to almost nil (Neilson, 1965).

The virus can be transmitted from infected or contaminated females on the surface of the egg and produces high levels of infection in the subsequent larval generation (Bellemare and Belloncik, 1981). This surface contamination is probably an important method for the vertical transmission of the disease in nature. There is no good evidence that CPV is transmitted from generation to generation within the egg. No virus development has been detected in ovarian tissue or developing oocytes, nor has transovum transmission been observed when the surface of the egg is satisfactorily disinfected (Sikorowski et al., 1973).

### b. Host Susceptibility and Virus Virulence

The majority of CPV infections produce a chronic disease, often without extensive larval mortality, with many individuals reaching the adult stage even though heavily diseased (Bird and Whalen, 1954; Stairs et al., 1968; Bird, 1969; Magnoler, 1974; Bell and Kanavel, 1976; Bailey, 1977; Payne, 1981). There are exceptions, particularly for infections of young larvae, and the CPV of O. pseudotsugata appears highly virulent, with an $LD_{50}$ of only 8 polyhedra per neonate larva (Martignoni et al., 1969). In contrast, the $LD_{50}$ for P. gossypiella CPV in neonate larvae was approximately 1000 times greater than the dose required merely to infect half the treated insects (Bell and Kanavel, 1976). As with some other insect virus infections, older larvae are less susceptible to CPV infection (Aruga and Watanabe, 1964; Bird, 1969; Aruga, 1971; Boucias and Nordin,

1979; Bellemare and Belloncik, 1981; Payne, 1982). Interference may also occur between different CPV strains, and this has been seen as one way by which the cross-protection induced by a virus that was first administered to a larva may be advantageous in reducing disease incidence following application of a second virus. Thus, Aruga and Watanabe (1970) found that when a UV-inactivated CPV was fed to B. mori larvae, protection from a challenge CPV was extensive. No clear explanation has been provided for such interference phenomena, but Quiot et al. (1980b) have observed that filtered supernatant from cell cultures infected with CPV may restrict virus replication (in this example a baculovirus) in cells to which the filtrate is added.

The selection of silkworm strains for their resistance to CPV has been an important aspect of sericultural research. A large difference (about 1000-fold) among the median infective doses of virus has been observed in a number of insect strains tested (Watanabe, 1966a). In addition, high selection pressures, in excess of 60% mortality, resulted in the development of silkworm strains that were approximately 10 times more resistant to CPV infection (Watanabe, 1967a). Host resistance to peroral infection by CPV probably depends on inhibitors in the gut lumen (Watanabe, 1966b). Antiviral substances in the gut have been found to reduce CPV infectivity, perhaps by virion agglutination (Aratake et al., 1974; Briese, 1981). In other papers, it was suggested that different characteristics of cell-membrane receptors, the larval peritrophic membrane (Watanabe, 1971b), or changes in gut pH (Payne, 1981) may also be significant.

Finally, larvae can recover from a CPV infection because the gut epithelium has a considerable regenerative capacity. Infected columnar cells are discharged into the gut lumen at larval molt, and new cells develop in their place (Inoue and Miyagawa, 1978). These new cells may become reinfected, but this does not always occur. In Hyphantria cunea larvae, regenerated columnar cells were not susceptible to reinfection with B. mori CPV (Yamaguchi, 1979).

*c. Virus Host Range*

CPVs have been isolated only from arthropods, and attempts to infect vertebrates or vertebrate cell lines have proved negative (Sections IV.C and V.B). Virus replication is inhibited at the usual temperatures used for the culture of vertebrate cells. Thus, at 35°C and above, even insect larvae treated with virus do not develop infections (Tanada and Chang, 1968; Aruga, 1971), since the early stages of virus infection, including RNA synthesis, are restricted at these temperatures (Kobayashi and Kawase, 1980).

Cross-transmission tests within the Insecta (reviewed by K.M. Smith, 1963; Ignoffo, 1968a, 1973; Tanaka, 1971) have suggested that CPVs have a relatively wide host range in comparison with some other groups of insect viruses, particularly the baculoviruses. There are several reports of successful cross-transmission of some CPV isolates between different families within the Lepidoptera. Unfortunately, the published data are open to the criticism that the identity of the virus isolates used has not been established by any satisfactory criteria. At best, distinctive polyhedral shapes have been used for virus identification (Tanaka, 1971). Nonetheless, in recent studies in which the viruses were identified by their RNA electropherotype, type 2 and type 12 CPV isolates were shown to be cross-infectious for species from both the Pieridae and Noctuidae families of Lepidoptera (Allaway, 1982; S. Wilson and C. Payne, unpublished observations).

The data shown in Table IV (Section II.D.1) probably provide the best indication of the range of species that may be infected by a specific virus "type," but it should not be assumed without further experimentation that one species will be infected by all variants of a particular CPV "type." Even minor differences in electropherotype could lead to differences in viruses that would modify their host range.

## C. Cytoplasmic Polyhedrosis Virus Infection in Cell Cultures

Progress toward a detailed understanding of CPV replication would be greatly improved by the development of a productive cell-culture system for virus growth. Recent studies have made encouraging progress toward this end.

The replication of CPVs in cell cultures was first reported in primary cells from ovarian, tracheal, and gut tissue (Grace, 1962; Vago and Bergoin, 1963; Kobayashi, 1971; Sohi et al., 1971). Kawarabata and Hayashi (1972) were the first to report CPV development in an established line of Aedes aegypti cells. Although polyhedra were not observed, replication was measured by the incorporation of [$^3$H]uridine into components with the same sedimentation coefficient as virions. The identity of the cells used by these workers has subsequently been questioned, and it appears most likely that they were of lepidopteran and not dipteran origin (Krywienczyk and Sohi, 1973). Since this study, complete CPV morphogenesis, including polyhedron formation, has been observed in several lepidopteran and dipteran cell lines (Table VII). With the questionable exception of M. disstria CPV (Kawarabata and Hayashi, 1972), all CPVs isolated from Lepidoptera have failed to replicate in dipteran cell lines.

Successful infection in insect cells has been achieved using nonoccluded virus from the guts of infected larvae as inoculum (Granados et al., 1974; Quiot and Belloncik, 1977; Longworth, 1981). Relatively low levels of cell infection (1–40%) have been obtained, with the exception of E. scandens CPV in L. dispar cells, in which more than 90% of cells produced polyhedra (Quiot and Belloncik, 1977).

TABLE VII. Cytoplasmic Polyhedrosis Virus Replication in Insect Cell Lines

| Virus inoculum | Cell line | Infection[a] | Reference |
|---|---|---|---|
| Aedes sollicitans CPV | Aedes albopictus | + | C.M. Barry and J.L. Fowler (unpublished observations quoted in Granados, 1976) |
| Chrysodeixis eriosoma CPV | A. aegypti | − | Longworth, 1981 |
| | Spodoptera frugiperda | + | |
| | Trichoplusia ni | + | |
| Euxoa scandens CPV | Bombyx mori | + | Quiot et al. (1980b) |
| | Choristoneura fumiferana | + | Arella et al. (1981) |
| | Lymantria dispar | + | Quiot and Belloncik (1977) |
| | Spodoptera frugiperda | + | C. Payne (unpublished observations) |
| Inachis io CPV (type 2 CPV) | Aedes aegypti[b] | ± | Kawarabata and Hayashi (1972) |
| Malacosoma disstria CPV | A. aegypti | − | Granados (1978) |
| Trichoplusia ni CPV | Anopheles stephensi | − | |
| | A. gambiae | − | |
| | Culex quinquefasciatus | − | |
| | Estigmene acrea | + | Granados (1975, 1976, 1978) |
| | Lymantria dispar | + | |
| | Spodoptera frugiperda | + | |
| | Trichoplusia ni | + | Granados et al. (1974) |

[a] (+) Polyhedra observed, (±) probable replication, but no polyhedra detected; (−) no replication observed.
[b] The nature of this cell line is not certain. More recent evidence suggests that the cells used were of lepidopteran origin (Krywienczyk and Sohi, 1973).

Two common features have emerged from several of these studies:

1. Many of the polyhedra produced in infected cell cultures are cuboidal (Fig. 15D) and differ from the shape of polyhedra produced by the same virus isolate in larvae (the significance of this is discussed in Section II.A.1).

2. With the exception of *Aedes sollicitans* CPV infection [C.M. Barry and J.L. Fowler, unpublished observations quoted in Granados (1976)], the cell-culture supernatant fractions contain very little virus, and this has led to difficulties in passaging. Three features have been noted that help account for this: First, infected cells in culture do not lyse (Granados *et al.*, 1974). Second, "free" virions may be present in very low levels, since the vast majority may be occluded (Longworth, 1981). Third, individual virus particles are not released from the cell. Instead, infected cells may bud off large membrane-bound areas containing polyhedra, virions, and viroplasm such that 7–8 days after infection, the culture supernatant contains large numbers of these detached structures (Quiot *et al.*, 1980a). Infective virus may be released from cells by freeze–thawing or, more effectively, sonication (Belloncik and Chagnon, 1980).

The production of polyhedra in infected cells provides a readily recognizable cytopathic effect that has been used as the basis for assay of infectious centers and median tissue-culture infectious doses, and the dose–response obtained is consistent with a single particle being able to initiate infection (Belloncik and Chagnon, 1980; Longworth, 1981).

Polyhedra produced *in vitro* are infectious for the original insect host, although when a quantitative bioassay was conducted, the $LC_{50}$ for *E. scandens* CPV polyhedra produced in cell culture was approximately 5 times less than for polyhedra produced in larvae. This would be accounted for if fewer virions were occluded in cell culture, but could also be influenced by the fact that polyhedra from larvae dissolved more readily in high-pH conditions similar to those that might prevail in the insect gut (Belloncik and Bellemare, 1980). Further observations on virus development in cell culture should prove most profitable.


# V. CYTOPLASMIC POLYHEDROSIS VIRUS INFECTIONS IN INSECT POPULATIONS

## A. Natural Incidence

The recognition of different virus electropherotypes provides a sound basis for studies of CPV molecular epidemiology. It has been noted that type 1 CPV isolates may have originated from Japan, type 4 isolates have all been obtained from the same locality in Asia, and types 2, 3, and 6 shared a predominantly European or Palearctic geographic distribution (Payne and Rivers, 1976; Payne et al., 1977, 1978). However, no detailed epidemiological studies have been undertaken.

In the relatively few examples in which the incidence of CPV infection has been monitored, the disease has normally been found to be enzootic and only infrequently reaches epizootic proportions. The viruses appear to persist even at low host density, a feature of a pathogen that has relatively low virulence for its host but that has stable infectious forms and is transmitted efficiently (Anderson and May, 1981). In surveys of members of the Simuliidae, the incidence of CPV infections generally ranged from 1 to 6.5% of the larvae (Bailey, 1977; Weiser, 1978), although up to 54% of individuals of the species *Cnephia mutata* were infected (Bailey, 1977). The CPV of the pine caterpillar, *D. spectabilis*, was present at a low level in almost all Japanese populations examined (Koyama and Katagiri, 1968; Katagiri, 1969), while a 1964 survey showed that *B. mori* CPV was established at enzootic or even epizootic levels in almost all districts of Japan (Aruga, 1971).

The production of large numbers of polyhedra can lead to prolonged virus persistence in the environment. Thus, *Lymantria fumida* CPV may persist on trees for at least 1 year (Katagiri, 1981), while the CPVs of several alfalfa pests appeared to persist in field soils throughout the winter (Tanada and Omi, 1974), where the polyhedra are probably adsorbed to soil particles (Hukuhara, 1972, 1977; Hukuhara and Wada, 1972). The virus is also effectively spread and transmitted by infected or contaminated adults, probably by surface contamination of the egg by excreta containing polyhedra. Tanada et al. (1964) observed that adults of *Colias eurytheme* collected from five separate field populations laid eggs that gave rise to CPV-infected larvae. The viruses may also survive within alternative hosts. CPVs isolated from several lepidopterous pests of alfalfa appeared to be cross-infectious (Tanada and Omi, 1974), while type 2 CPV will infect many lepidopterous pests of *Brassica* spp. (Payne, 1982). The presence of several susceptible host species may therefore provide a reservoir for the maintenance of a CPV within a particular ecosystem.

Given the appropriate conditions of persistence, spread, and host and virus density, a CPV infection may reach epizootic proportions and dramatically reduce the population of the host. Such natural epizootics seem rare, but Deseo et al. (1980) concluded that a CPV of *Lobesia botrana* was exerting some natural control, while Jahn (1978) observed a population collapse of *L. dispar* in Austria, brought about by a CPV infection.

For these viruses to be successfully used as biological control agents (Section V.B), the aim is to manipulate the virus so that a high incidence of infection occurs in the host population with consequent reduction in pest damage. In contrast, in sericulture, the aim is to prevent disease, and much research has been carried out on the factors responsible for the development of virus epizootics in silkworms. This work has been reviewed by Aruga (1971) and centers largely on the belief that the CPV can exist in a latent form within the insect population. The evidence for this is not convincing. Several treatments have been used to convert suspected "latent" CPV infections from a state wherein virus is not demonstrable into an actively multiplying one. The treatments include different food quality, temperature fluctuations (Aruga and Nagashima, 1962), and other stresses including crowding (Steinhaus and Dineen, 1959), chemicals, and other pathogens, including viruses (K.M. Smith and Rivers, 1956; Tanada et al., 1964; Aruga, 1971; Jurkovicova, 1979; Mertens, 1979). Certainly these effects may alter the susceptibility of larvae and allow a more frank expression of a chronic infection produced by a small contaminating inoculum of virus. The likelihood of contamination should not be underestimated. CPVs are among the most prevalent of insect microbial pathogens, and polyhedra can frequently be detected in rearing facilities (Aruga, 1971; Mery and Dulmage, 1975). It is particularly interesting that when B. mori larvae are reared under aseptic conditions on an artificial diet, they show no evidence of "latent" infections, since they do not respond to chemical or temperature stressors by producing an active CPV infection (Kurata, 1971).

## B. Cytoplasmic Polyhedrosis Viruses as Biological Control Agents

With the increasing problems of insect pest resistance to chemical pesticides in agriculture and forestry, there will be a need for selective biological methods of control in future pest-management systems. Virus diseases of insects and mites provide a large reservoir of potential control agents. Of these, the baculovirus group has received most attention partly because it appears unique among insect viruses in sharing no overt structural or biochemical similarities with viruses in vertebrates and higher plants, and partly because baculoviruses are responsible for most of the natural virus epizootics observed in insects and mites (Summers et al., 1975; Entwistle, 1978; Summers and Kawanishi, 1978; Tinsley, 1979; Payne, 1982). CPVs, although often less pathogenic than many baculoviruses, also possess characteristics that can be useful in biological-control programs. The nature of CPV infection, virus transmission and environmental persistence is such that the viruses could best be used in a stable environment where both a residual insect population and a certain degree of crop damage can be tolerated (Payne, 1982). For such reasons, most field trials with these viruses have been carried out against insect pests of forests (Morris, 1980; Katagiri, 1981).

The first CPV to be used as an insecticide was directed against the pine processionary caterpillar, Thaumetopoea pityocampa, in the Ventoux region of France, and the application seemed successful, with only 3–4% of the initial larval population surviving to pupation (Biliotti et al.,

1959; Grison *et al.*, 1959; Martouret and Dusaussoy, 1959). More recently, a CPV of *D. spectabilis* was registered in 1974 as a microbial insecticide in Japan (Katagiri, 1981). The developmental work had shown that spray applications of $1$–$3 \times 10^{11}$ polyhedra per hectare were most effective in reducing the larval population. This is equivalent to the amount (per hectare) of virus produced by 200–600 fully-grown larvae (Koyama and Katagiri, 1968; Katagiri, 1969, 1981). Although it was necessary to apply virus before the larval population reached its peak, the best results were achieved when virus application was delayed until the penultimate instar, even though the larvae were not at their most susceptible at this age. This delayed treatment permitted the virus to add to the control naturally exerted on young larvae by other diseases, parasitoids, and predators (Katagiri, 1969). CPVs have also been used in field trials against other forest pests, including *Dendrolimus punctatus* (Ying, 1970; Hsiao, 1981). *L. fumida,* and *L. dispar* (Katagiri, 1969, 1981), and also against the cabbage looper, *T. ni* (Granados, 1978).

In general, the problems associated with CPV treatments are similar to those outlined for baculoviruses (Tinsley, 1979; Payne, 1982). For example, should virus application be directed against the more susceptible younger stages, or older, more resistant larvae? In a productive infection of older larvae, more virus is released into the environment and could result in a virus load within the ecosystem that could provide long-term, self-sustaining control of pest populations. Applications against younger larvae will almost certainly require periodic "topping-up." Like other insect viruses, spray formulations of CPV polyhedra are liable to inactivation by UV light on treated foliage. Ignoffo *et al.* (1977) have demonstrated that CPV polyhedra from *Heliothis virescens* have a half-life of only 2.2 hr (comparable to that of some baculoviruses) when exposed to combined short- and long-wavelength UV.

The potential environmental safety of these viruses must also be given the most careful consideration. The limited safety tests on vertebrates that were conducted prior to registration of *D. spectabilis* CPV revealed that the virus appeared to have no significant pathogenic effects on mice, rats, hamsters, rabbits, chick embryos, or fish, when injected or applied orally (Katagiri, 1981). Earlier tests on *B. mori* CPV had failed to detect replication (as measured by polyhedron formation) in turkey embryos (Cantwell *et al.*, 1968) or several vertebrate cell lines (Ignoffo, 1968a, 1973). Nonetheless, these tests were carried out at a time when little or no replication of CPVs had been reported in insect cell lines and would therefore merit closer investigation now. More recently, Granados (1978) reported that *T. ni* CPV, which has been shown to replicate in a number of insect cell lines (Table VII), did not infect HeLa or mouse fibroplast L-929 cells. Certainly, CPV replication even in insect hosts is inhibited at temperatures equivalent to those naturally found in homeotherms (Section IV.B.2.c).

The potential ecological hazard of CPV infections is not restricted to vertebrates. Although the use of CPVs against Lepidoptera does not appear to harm plants or beneficial arthropods [parasitoids and predators (Ignoffo, 1968b)], the relatively broad host range of at least some CPVs warrants further study. On the one hand, this extended host range may be useful among a number of pest Lepidoptera species on a single crop (Tanada and Omi, 1974; Payne, 1982). On the other hand, potential effects on nonpest Lepidoptera could be damaging. It may seem somewhat surprising that *D. spectabilis* CPV is available for use as a microbial insecticide in Japan, when it is known to infect the silkworm, *B. mori* (Yamamasu and Kawakita, 1962).

The means of producing a CPV for use as an insecticide is at present restricted to growth in larvae. Cell-culture techniques could provide further improvements in virus production. Belloncik and Bellemare (1980) calculated that $10^9$ polyhedra of *E. scandens* CPV could be produced per 25-cm$^2$ cell-culture flask, compared with $10^7$ polyhedra per larva. Attempts to improve yields by cobalt-60 irradiation of cells increased the numbers of polyhedra per cell, but fewer cells were infected and the total yield was not increased (Belloncik and Arella, 1981). Production, quality control, safety testing, virus identification, and assay are all features that need further study before CPVs can be made more freely available for use as part of a new generation of microbial insecticides.

# VI. CONCLUSIONS

In the 12 years since the appearance of a book devoted to the silkworm CPV (Aruga and Tanada, 1971), considerable progress has been made on the biochemical properties of the CPV group. Studies of the genome RNAs have provided methods for analyzing and classifying the considerable variation that exists among different virus isolates. The enzymatic properties of type 1 CPV virions, and their RNA products, have provided a means of investigating the molecular basis of RNA transcription. Nonetheless, there is still little information on the intracellular replication cycle of these virions. In particular, it is to be hoped that future developments on virus multiplication in insect cell cultures will provide the basis for studies of transcriptional and translational control *in vivo*, the identity of viral nonstructural proteins, and the mechanisms of genome RNA production and virus assembly.

ACKNOWLEDGMENTS. We are particularly grateful to Drs. Y. Furuichi, K. Miura, Y. Sun, and S. Belloncik for their cooperation in supplying reprints or preprints or both during the preparation of this review. We would also like to thank Drs. Malcolm McCrae and Riva Rubinstein for helpful discussions and for their permission to include details of unpublished work. Drs. K. Miura, G. Boucias, J.F. Longworth, and L.J. Lewandowski kindly gave their permission to include figures taken from their publications. Maurice Bone, Andrew Smith, Peter Fiske, and Margaret Arnold gave their usual careful attention to other diagrams and photographs. Sue Bewsey coped magnificently with our handwriting to produce the typescript. Brian Underwood, Norman Crook, and Graham Allaway made helpful comments on the draft manuscript.

## REFERENCES

Aizawa, K., and Iida, S., 1963, Nucleic acids extracted from the virus polyhedra of the silkworm, *Bombyx mori* (Linnaeus), *J. Insect Pathol.* **5**:344.

Allaway, G.P., 1982, Infectivity of some occluded insect viruses, Ph.D. thesis, University of London.

Anderson, R.M., and May, R.M., 1981, The population dynamics of microparasites and their invertebrate hosts, *Philos. Trans. R. Soc. London Ser. B.* **291**:451.

Andreadis, T.G., 1981, A new cytoplasmic polyhedrosis virus from the saltmarsh mosquito, *Aedes cantator* (Diptera: Culicidae), *J. Invertebr. Pathol.* **37**:160.

Anthony, D.W., Hazard, E.I., and Crosby, S.W., 1973, A virus disease in *Anopheles quadrimaculatus*, *J. Invertebr. Pathol.* **22**:1.

Aratake, T., Kayamura, T., and Watanabe, H., 1974, Inactivation of a cytoplasmic-polyhedrosis virus by gut-juice of the silkworm, *Bombyx mori* L. *J. Seric. Sci. Jpn.* **41**:41 (in Japanese; English summary).

Arella, M., Belloncik, S., Barray, S. and Devauchelle, G., 1981, Dual infection of a lepidopteran cell line with the cytoplasmic polyhedrosis virus and the *Chilo* iridescent virus, Abstracts of the 5th International Congress for Virology, Strasbourg, p. 427.

Arnott, H.J., Smith, K.M., and Fullilove, S.L., 1968, Ultrastructure of a cytoplasmic polyhedrosis virus affecting the monarch butterfly *Danaus plexippus*, *J. Ultrastruct. Res.* **24**:479.

Aruga, H., 1971, Cytoplasmic polyhedrosis of the silkworm—historical, economical and epizootiological aspects, in: *The Cytoplasmic-Polyhedrosis Virus of the Silkworm* (H. Aruga and Y. Tanada, eds.), pp. 3–21, University of Tokyo Press, Tokyo.

Aruga, H., and Nagashima, E., 1962, Generation to generation transmission of the cytoplasmic polyhedrosis virus of *Bombyx mori* Linnaeus, *J. Insect Pathol.* **4**:313.

Aruga, H., and Tanada, Y. (eds.), 1971, *The Cytoplasmic-Polyhedrosis Virus of the Silkworm*, University of Tokyo Press, Tokyo.

Aruga, H., and Watanabe, H., 1964, Resistance to *per os* infection with cytoplasmic-polyhedrosis virus in the silkworm, *Bombyx mori* (Linnaeus), *J. Insect Pathol.* **6**:387.

Aruga, H., and Watanabe, H., 1970, Interference between UV-inactivated and active cytoplasmic-polyhedrosis viruses in the silkworm, *Bombyx mori* L. II. Silkworm strain and time interval of inoculation, *J. Seric. Sci. Jpn.* **39**:382 (in Japanese; English summary).

Asai, J., Kawamoto, F., and Kawase, S., 1972, On the structure of the cytoplasmic-polyhedrosis virus of the silkworm, *Bombyx mori*, *J. Invertebr. Pathol.* **19**:279.

Atger, P., 1962, Virose intestinale chez la noctuelle du chou *Mamestra brassicae* L. (Lepidoptera), *Ann. Epiphytol. (Paris)* **13**:263.

Bailey, C.H., 1977, Field and laboratory observations on a cytoplasmic polyhedrosis virus of blackflies (Diptera: Simuliidae), *J. Invertebr. Pathol.* **29**:69.

Bailey, C.H., Shapiro, M., and Granados, R.R., 1975, A cytoplasmic polyhedrosis virus from the larval blackflies, *Cnephia mutata* and *Prosimulium mixtum* (Diptera: Simuliidae), *J. Invertebr. Pahol.* **25**:273.

Banerjee, A.K., 1980, 5′-Terminal cap structure in eukaryotic messenger ribonucleic acids, *Microbiol. Rev.* **44**:175.

Barbosa, E., and Moss, B., 1978, mRNA (nucleoside-2′-)-methyltransferase from vaccinia virus: Characteristics and substrate specificity, *J. Biol. Chem.* **253**:7698.

Bell, M.R., 1977, Pink bollworm: Effect of infection by a cytoplasmic polyhedrosis virus on diapausing larvae, *Ann. Entomol. Soc. Am.* **70**:675.

Bell, M.R., and Kanavel, R.F., 1976, Effect of dose of cytoplasmic polyhedrosis virus on infection, mortality, development rate, and larval and pupal weights of the pink bollworm, *J. Invertebr. Pathol.* **28**:121.

Bell, M.R., and Kanavel, R.F., 1977, The effect of a cytoplasmic polyhedrosis virus on lipid and protein content of pupae of the pink bollworm (Lepidoptera: Gelechiidae), *J. Kans. Entomol. Soc.* **50**:359.

Bellemare, N., and Belloncik, S., 1981, Études au laboratoire des effets d'une polyédrose cytoplasmique sur le ver gris blanc *Euxoa scandens* (Lepidoptère: Noctuidae Agrotinae), *Ann. Soc. Entomol. Que.* **26**:28.

Belloncik, S., and Arella, M., 1981, Production of cytoplasmic polyhedrosis virus (CPV) polyhedra in a gamma irradiated *Lymantria dispar* cell line, *Arch. Virol.* **68**:303.

Belloncik, S., and Bellemare, N., 1980, Polyèdres du CPV d'*Euxoa scandens* (Lep: Noctuidae) produits *in vivo* et sur cellules cultivées *in vitro*: Etudes comparatives, *Entomophaga* **25**:199.

Belloncik, S., and Chagnon, A., 1980, Titration of a cytoplasmic polyhedrosis virus by a tissue microculture assay: Some applications, *Intervirology* **13**:28.

Bergold, G.H., and Suter, J., 1959, On the structure of cytoplasmic polyhedra of some Lepidoptera, *J. Insect Pathol.* **1**:1.

Bernstein, J.M., and Hruska, J.F., 1981, Characterization of RNA polymerase products of Nebraska calf diarrhea virus and SA 11 rotavirus, *J. Virol.* **37**:1071.

Biliotti, E., Grison, P., Maury, R., and Vago, C., 1959, Emploi d'une poudre à base de virus spécifique contre la chenille processionaire du pin dans le massif du Ventoux, *C. R. Acad. Agric. Fr.* **45**:407.

Bird, F.T., 1965, On the morphology and development of insect cytoplasmic-polyhedrosis virus particles, *Can. J. Microbiol.* **11**:497.

Bird, F.T., 1969, Infection and mortality of spruce budworm *Choristoneura fumiferana*, and forest tent caterpillar, *Malacosoma disstria* caused by nuclear and cytoplasmic polyhedrosis viruses, *Can. Entomol.* **101**:1269.

Bird, F.T., and Whalen, M.M., 1954, A nuclear and a cytoplasmic polyhedral virus disease of the spruce budworm, *Can. J. Zool.* **32**:82.

Black, D.R., and Knight, C.A., 1970, Ribonucleic acid transcriptase activity in purified wound tumor virus, *J. Virol.* **6**:194.

Boccardo, G., Hatta, T., Francki, R.I.B., and Grivell, C.J., 1980, Purification and some properties of reovirus-like particles from leafhoppers and their possible involvement in wallaby ear disease of maize, *Virology* **100**:300.

Borsa, J., Grover, J., and Chapman, J.D., 1970, Presence of nucleoside triphosphate phosphohydrolase activity in purified virions of reovirus, *J. Virol.* **6**:295.

Boucias, D.G., and Nordin, G.L., 1978, A scanning electron microscope study of *Hyphantria cunea* CPV-infected midgut tissue, *J. Invertebr. Pathol.* **32**:229.

Boucias, D.G., and Nordin, G.L., 1979, Susceptibility of *Hyphantria cunea* to a cytoplasmic polyhedrosis virus, *J. Kans. Entomol. Soc.* **52**:641.

Briese, D.T., 1981, Resistance of insect species to microbial pathogens, in: *Pathogenesis of Invertebrate Microbial Diseases*, (E.W. Davidson, ed.), pp. 511–545, Allanheld, Osmun, Totowa, New Jersey.

Bullock, H.R., Martinez, E., and Stuermer, C.W. Jr., 1970, Cytoplasmic-polyhedrosis virus and the development and fecundity of the pink bollworm, *J. Invertebr. Pathol.* **15**:109.

Cantwell, G.E., Faust, R.M., and Poole, H.K., 1968, Attempts to cultivate insect viruses in avian eggs, *J. Invertebr. Pathol.* **10**:161.

Clark, T.B., Chapman, H.C., and Fukuda, T., 1969, Nuclear-polyhedrosis and cytoplasmic-polyhedrosis virus infection in Louisiana mosquitoes, *J. Invertebr. Pathol.* **14**:284.

Cohen, J., Laporte, J., Charpilienne, A., and Scherrer, R., 1979, Activation of rotavirus RNA polymerase by calcium chelation, *Arch. Virol.* **60**:177.

Crook, N.E., and Payne, C.C., 1980, Comparison of three methods of ELISA for baculoviruses, *J. Gen. Virol.* **46**:29.

Cunningham, J.C., and Longworth, J.F., 1968, The identification of some cytoplasmic polyhedrosis viruses, *J. Invertebr. Pathol.* **11**:196.

Dai, R., Wu, A., Shen, X., Qian, L., and Sun, Y., 1982, Isolation of genome–enzyme complex from cytoplasmic polyhedrosis virus of silkworm, *Bombyx mori, Sci. Sin.* **25**:29.

Deguchi, T., and Barchas, J., 1971, Inhibition of transmethylations of biogenic amines by *S*-adenosyl-homocysteine: Enhancement of transmethylation by adenosylhomocysteinase, *J. Biol. Chem.* **246**:3175.

Deseo, K.V., Brunelli, A., Marani, F., and Bertaccini, A., 1980, Il ruolo delle malattie nella dinamica della popolozioni di *Lobesia botrana* Den. E. Schiff. (Lepidoptera; Tortricidae) e la loro importanza pratica, *Atti Giorn. Fitopatol.* **3**:441.

Donaghue, T.P., and Hayashi, Y., 1972, Cytoplasmic polyhedrosis virus (CPV) of *Malacosoma disstria*: RNA polymerase activity in purified free virions, *Can. J. Microbiol.* **18**:207.

Entwistle, P.F., 1978, Microbial control of insects and other pests, in: *Crop Protection*, pp. 72–96. British Association for the Advancement of Science, Annual meeting Bath University 1978, University of Bath, United Kingdom.

Faust, R.M., and Cantwell, G.E., 1968, Inducement of cytoplasmic polyhedrosis by intrahemocoelic injection of freed viral particles into the silkworm *Bombyx mori, J. Invertebr. Pathol.* **11**:119.

Federici, B.A., 1974, Virus pathogens of mosquitoes and their potential use in mosquito control, in: *Le contrôle des Moustiques/Mosquito Control* (A. Aubin, J.-P. Bourassa, S. Belloncik, M. Pellisier, and E. Lacoursière, eds.), pp. 93–135, Les Presses de l'Université du Quebec.

Federici, B.A., and Hazard, E.I., 1975, Iridovirus and cytoplasmic polyhedrosis virus in the fresh water daphnid *Simocephalus expinosus, Nature* (*London*) **254**:327.

Fujii-Kawata, I., Miura, K., and Fuke, M., 1970, Segments of genome of viruses containing double-stranded RNA, *J. Mol. Biol.* **51**:247.

Furuichi, Y., 1974, Methylation-coupled transcription by virus associated transcriptase of cytoplasmic polyhedrosis virus containing double stranded RNA, *Nucleic Acids Res.* **1**:809.

Furuichi, Y., 1978a, Pretranscriptional capping in the biosynthesis of cytoplasmic polyhedrosis virus mRNA, *Proc. Natl. Acad. Sci. U.S.A.* **75**:1086.

Furuichi, Y., 1978b, Stimulation of cytoplasmic polyhedrosis virus mRNA synthesis by *S*-adenosylmethionine and its derivatives, Abstracts of the 4th International Congress for Virology, The Hague, p. 332, Centre for Agricultural Publishing and Documentation, Wageningen.

Furuichi, Y., 1981, Allosteric stimulatory effect of *S*-adenosyl methionine on the RNA polymerase in cytoplasmic polyhedrosis virus: A model for the positive control of eukaryotic transcription, *J. Biol. Chem.* **256**:483.

Furuichi, Y., and Miura, K.I., 1972, The 3' termini of the genome RNA segments of silkworm cytoplasmic polyhedrosis virus, *J. Mol. Biol.* **64**:619.

Furuichi, Y., and Miura, K.I., 1973, Identity of the 3'-terminal sequences in ten genome segments of silkworm cytoplasmic polyhedrosis virus, *Virology* **55**:418.

Furuichi, Y., and Miura, K., 1975, A blocked structure at the 5' terminus of mRNA from cytoplasmic polyhedrosis virus, *Nature* (*London*) **253**:374.

Furuichi, Y., and Shatkin, A.J., 1976, Differential synthesis of blocked and unblocked 5'-termini in reovirus mRNA: Effect of pyrophosphate and pyrophosphatase, *Proc. Natl. Acad. Sci. U.S.A.* **73**:3448.

Furuichi, Y., and Shatkin, A.J., 1977, 5'-Termini of reovirus mRNA: Ability of viral cores to form caps post-transcriptionally, *Virology* **77**:566.

Furuichi, Y., Morgan, M., Muthukrishnan, S., and Shatkin, A.J., 1975, Reovirus messenger RNA contains a methylated blocked 5'-terminal structure: $m^7G(5')ppp(5')G^mpCp$-, *Proc. Natl. Acad. Sci. U.S.A.* **72**:362.

Furuichi, Y., Muthukrishnan, S., Tomasz, J., and Shatkin, A.J., 1976a, Caps in eukaryotic mRNAs: Mechanism of formation of reovirus mRNA 5' terminal $^7mGpppGm$-C, *Prog. Nucleic Acid Res. Mol. Biol.* **19**:3.

Furuichi, Y., Muthukrishnan, S., Tomasz, J., and Shatkin, A.J., 1976b, Mechanism of formation of reovirus mRNA 5'-terminal blocked and methylated sequence, m⁷GpppGᵐpC, *J. Biol. Chem.* **251**:5043.

Furuichi, Y., La Fiandra, A., and Shatkin, A.J., 1977, 5'-Terminal structure and mRNA stability, *Nature (London)* **266**:235.

Furusawa, T., and Kawase, S., 1971, Synthesis of ribonucleic acid resistant to actinomycin-D in silkworm midguts infected with the cytoplasmic-polyhedrosis virus, *J. Invertebr. Pathol.* **18**:156.

Furusawa, T., and Kawase, S., 1973, Virus-specific RNA synthesis in the midgut of silkworm, *Bombyx mori* infected with cytoplasmic-polyhedrosis virus, *J. Invertebr. Pathol.* **22**:335.

Grace, T.D.C., 1962, The development of a cytoplasmic polyhedrosis in insect cells grown *in vitro*, *Virology* **18**:33.

Granados, R.R., 1975, Multiplication of a cytoplasmic polyhedrosis virus (CPV) in insect tissue cultures, Abstracts of the 3rd International Congress for Virology, Madrid, p. 98.

Granados, R.R., 1976, The infection and replication of insect pathogenic viruses in tissue culture, *Adv. Virus Res.* **20**:189.

Granados, R.R., 1978, Biology of cytoplasmic polyhedrosis viruses and entomopoxviruses, in: *Viral Pesticides: Present Knowledge and Potential Effects on Public and Environmental Health* (M.D. Summers and C.Y. Kawanishi, eds.), pp. 89–101, U.S. Environmental Protection Agency, Research Triangle Park, North Carolina.

Granados, R.R., McCarthy, W.J., and Naughton, M., 1974, Replication of a cytoplasmic polyhedrosis virus in an established cell line of *Trichoplusia ni* cells, *Virology* **59**:584.

Grancher-Barray, S., Boisvert, J., and Belloncik, S., 1981, Electrophoretic characterization of proteins and RNA of cytoplasmic polyhedrosis virus (CPV) from *Euxoa scandens*, *Arch. Virol.* **70**:55.

Grison, P., Vago, C. and Maury, R., 1959, La lutte contre le processionaire du pin *Thaumetopoea pityocampa* Schiff, dans le massif du Ventoux: Essai d'utilisation pratique d'un virus spécifique, *Rev. For. Fr. (Nancy)* **5**:353.

Harley, E.H., and Rubinstein, R., 1978, The multicomponent genome of a different cytoplasmic polyhedrosis virus isolated from *Heliothis armigera*, *Intervirology* **10**:351.

Harley, E.H., Rubinstein, R., Losman, M., and Lutton, D., 1977, Molecular weights of the RNA genome segments of a cytoplasmic polyhedrosis virus determined by a new comparative approach, *Virology* **76**:210.

Harpaz, I., Zlotkin, E., and Ben Shaked, Y., 1965, On the pathology of cytoplasmic and nuclear polyhedrosis of the Cyprus processionary caterpillar, *Thaumetopoea wilkinsoni* Tams., *J. Invertebr. Pathol.* **7**:15.

Harrap, K.A., 1972, The structure of nuclear polyhedrosis viruses. I. The inclusion body, *Virology* **50**:114.

Harrap, K.A., and Payne, C.C., 1979, The structural properties and identification of insect viruses, *Adv. Virus Res.* **25**:273.

Hayashi, Y., 1968, Constitution of the ribosomal fraction from the forest tent caterpillar (*Malacosoma disstria*) midgut infected with a CPV, *J. Invertebr. Pathol.* **12**:468.

Hayashi, Y., 1970a, Occluded and free virions in midgut cells of *Malacosoma disstria* infected with cytoplasmic polyhedrosis virus (CPV), *J. Invertebr. Pathol.* **16**:442.

Hayashi, Y., 1970b, RNA in midgut of tussock moth, *Orgyia leucostigma*, infected wth cytoplasmic-polyhedrosis virus, *Can. J. Microbiol.* **16**:1101.

Hayashi, Y., and Bird, F.T., 1968a, The use of sucrose gradients in the isolation of cytoplasmic polyhedrosis virus particles, *J. Invertebr. Pathol.* **11**:40.

Hayashi, Y., and Bird, F.T., 1968b, Properties of a cytoplasmic polyhedrosis virus from the white-marked tussock moth, *J. Invertebr. Pathol.* **12**:140.

Hayashi, Y., and Bird, F.T., 1970, The isolation of cytoplasmic polyhedrosis virus from the white-marked tussock moth *Orgyia leucostigma* (Smith), *Can. J. Microbiol.* **16**:695.

Hayashi, Y., and Donaghue, T.P., 1971, Cytoplasmic polyhedrosis virus: RNA synthesized *in vivo* and *in vitro* in infected midgut, *Biochem. Biophys. Res. Commun.* **42**:214.

Hayashi, Y., and Kawarabata, T., 1970, Effect of actinomycin on synthesis of cell RNA and replication of insect cytoplasmic-polyhedrosis viruses in vivo, *J. Invertebr. Pathol.* **15**:461.

Hayashi, Y., and Kawase, S., 1964, Base pairing in ribonucleic acid extracted from the cytoplasmic polyhedra of the silkworm, *Virology* **23**:611.

Hayashi, Y., and Kawase, S., 1965a, Studies on the RNA in the cytoplasmic polyhedra of the silkworm, *Bombyx mori* L. I. Specific RNA extracted from cytoplasmic polyhedra, *J. Seric. Sci. Jpn.* **34**:83 (in Japanese; English summary).

Hayashi, Y., and Kawase, S., 1965b, Studies on the RNA in CPV of *Bombyx mori*. III. Comparison between icosahedral and hexagonal (polyhedra) RNA, *J. Seric. Sci. Jpn.* **34**:167 (in Japanese; English summary).

Hayashi, Y., and Kawase, S., 1965c, Studies on the RNA in the cytoplasmic polyhedra of the silkworm, *Bombyx mori* L. IV. Subcellular distribution, *J. Seric. Sci. Jpn.* **34**:171 (in Japanese; English summary).

Hayashi, Y., and Kawase, S., 1965d, Studies on the RNA in the cytoplasmic polyhedra of the silkworm, *Bombyx mori* L. V. Changes in fractionated RNA of the midgut, *J. Seric. Sci. Jpn.* **34**:244 (in Japanese; English summary).

Hayashi, Y., and Krywienczyk, J., 1972, Electrophoretic fractionation of cytoplasmic polyhedrosis virus genome, *J. Invertebr. Pathol.* **19**:160.

Hayashi, Y., and Retnakaran, A., 1970, The site of RNA synthesis of a cytoplasmic-polyhedrosis virus (CPV) in *Malacosoma disstria*, *J. Invertebr. Pathol.* **16**:150.

Hayashi, Y., Kawarabata, T., and Bird, F.T., 1970, Isolation of a cytoplasmic-polyhedrosis virus of the silkworm, *Bombyx mori*, *J. Invertebr. Pathol.* **16**:378.

Hills, G.J. and Smith, K.M., 1959, Further studies on the isolation and crystallization of insect cytoplasmic viruses, *J. Insect Pathol.* **1**:121.

Hosaka, Y., and Aizawa, K., 1964, The fine structure of the cytoplasmic polyhedrosis virus of the silkworm, *Bombyx mori* (Linnaeus), *J. Insect. Pathol.* **6**:53.

Hrdy, D.B., Rosen, L., and Fields, B.N., 1979, Polymorphism of the migration of double-stranded RNA genome segments of reovirus isolates from humans, cattle and mice, *J. Virol.* **31**:104.

Hsiao, K.-J., 1981, The use of biological agents for the control of the pine defoliator, *Dendrolimus punctatus* (Lepidoptera, Lasiocampidae) in China, *Prot. Ecol.* **2**:297.

Hukuhara, T., 1972, Demonstration of polyhedra and capsules in soil with scanning electron microscope, *J. Invertebr. Pathol.* **20**:375.

Hukuhara, T., 1977, Purification of polyhedra of a cytoplasmic polyhedrosis virus from soil using Metrizamide, *J. Invertebr. Pathol.* **30**:270.

Hukuhara, T., and Hashimoto, Y., 1966a, Studies of two strains of cytoplasmic-polyhedrosis virus, *J. Invertebr. Pathol.* **8**:184.

Hukuhara, T., and Hashimoto, Y., 1966b, Serological studies of the cytoplasmic and nuclear polyhedrosis viruses of the silkworm, *Bombyx mori*, *J. Invertebr. Pathol.* **8**:234.

Hukuhara, T., and Wada, H., 1972, Adsorption of polyhedra of a cytoplasmic-polyhedrosis virus by soil particles, *J. Invertebr. Pathol.* **20**:309.

Hukuhara, T., and Yamaguchi, K., 1973, Ultrastructural investigation on a strain of a cytoplasmic-polyhedrosis virus with nuclear inclusions, *J. Invertebr. Pathol.* **22**:6.

Ignoffo, C.M., 1968a, Specificity of insect viruses, *Bull. Entomol. Soc. Am.* **14**:265.

Ignoffo, C.M., 1968b, Viruses—living insecticides, *Curr. Top. Microbiol. Immunol.* **42**:129.

Ignoffo, C.M., 1973, Effects of entomopathogens on vertebrates, *Ann. N.Y., Acad. Sci.* **217**:141.

Ignoffo, C.M., and Adams, J.R., 1966, A cytoplasmic-polyhedrosis virus, *Smithiavirus pectinophorae* sp. n. of the pink bollworm, *Pectinophora gossypiella* (Saunders), *J. Invertebr. Pathol.* **8**:59.

Ignoffo, C.M., Hostetter, D.L., Sikorowski, P.P., Sutter, G., and Brooks, W.M., 1977, Inactivation of representative species of entomopathogenic viruses, a bacterium, fungus, and protozoan by an ultra-violet light source, *Environ. Entomol.* **6**:411.

Inoue, H., and Miyagawa, M., 1978, Regeneration of midgut epithelial cell in the silkworm, *Bombyx mori*, infected with viruses, *J. Invertebr. Pathol.* **32**:373.

Ishimori, N., 1934, Contribution a l'étude de la Grasserie du ver a soie, *C. R. Seances Soc. Biol. Fil.* **116**:1169.

Iwashita, Y., 1971, Histopathology of cytoplasmic polyhedrosis, in: *The Cytoplasmic-Polyhedrosis Virus of the Silkworm* (H. Aruga and Y. Tanada, eds.), pp. 79–101, University of Tokyo Press, Tokyo.

Jahn, E., 1978, Uber das Auftreten einer cytoplasmatischen Polyedrose beim Zusammenbruch der Gradation von *Lymantria dispar* (Lepidoptera; Lymantriidae) im Leithagebirge von Osterreich 1973, *Z. Angew. Zool.* **66**:9.

Joklik, W.K., 1974, Reproduction of Reoviridae, in: *Comprehensive Virology*, Vol. 2 (H. Fraenkel-Conrat and R.R. Wagner, eds.), pp. 231–334, Plenum Press, New York.

Joklik, W.K., 1980, The structure and function of the reovirus genome, in: *Genetic Variation of Viruses* (P. Palese and B. Roizman eds.), *Ann. N. Y. Acad. Sci.* **354**:107.

Joklik, W.K., Skehel, J.J., and Zweerink, H.J., 1970, The transcription of the reovirus genome, *Cold Spring Harbor Symp. Quant. Biol.* **35**:791.

Jurkovicova, M., 1979, Activation of latent virus infections in larvae of *Adoxophyes orana* (Lepidoptera: Tortricidae) and *Barathra brassicae* (Lepidoptera: Noctuidae) by foreign polyhedra, *J. Invertebr. Pathol.* **34**:213.

Kalmakoff, J., Lewandowski, L.J., and Black, D.R., 1969, Comparison of the ribonucleic acid subunits of reovirus, cytoplasmic polyhedrosis virus, and wound tumor virus, *J. Virol.* **4**:851.

Kapuler, A.M., 1970, An extraordinary temperature dependence of the reovirus transcriptase, *Biochemistry* **9**:4453.

Kapuler, A.M., Mendelsohn, N., Klett, H., and Acs, G., 1970, Four base-specific nucleoside 5′-triphosphatases in the subviral core of reovirus, *Nature (London)* **225**:1209.

Katagiri, K., 1969, Review on microbial control of insect pests in forests in Japan, *Entomophaga* **14**:203.

Katagiri, K., 1981, Pest control by cytoplasmic polyhedrosis viruses, in: *Microbial Control of Pests and Plant Diseases 1970–1980* (H.D. Burges, ed.), pp. 433–440. Academic Press, London and New York.

Kavenoff, R., Klotz, L.C., and Zimm, B.H., 1973, On the nature of chromosome-sized DNA molecules, *Cold Spring Harbor Symp. Quant. Biol.* **38**:1.

Kawarabata, T., and Hayashi, Y., 1972, Development of a cytoplasmic polyhedrosis virus in an insect cell line, *J. Invertebr. Pathol.* **19**:414.

Kawase, S., 1964, The amino acid content of viruses and their polyhedron proteins of the polyhedroses of the silkworm *Bombyx mori* (Linnaeus), *J. Insect Pathol.* **6**:156.

Kawase, S., 1967, Ribonucleic acid extracted from the cytoplasmic-polyhedrosis virus of the silkworm, *Bombyx mori*, *J. Invertebr. Pathol.* **9**:136.

Kawase, S., 1971, Chemical nature of the cytoplasmic-polyhedrosis virus, in: *The Cytoplasmic-Polyhedrosis Virus of the Silkworm* (H. Aruga and Y. Tanada, eds.), pp. 37–59, University of Tokyo Press, Tokyo.

Kawase, S., and Furusawa, T., 1971, Effect of actinomycin D on RNA synthesis in the midguts of healthy and CPV-infected silkworms, *J. Invertebr. Pathol.* **18**:33.

Kawase, S., and Kawamori, I., 1968, Chromatographic studies on RNA synthesis in the midgut of the silkworm, *Bombyx mori*, infected with a cytoplasmic-polyhedrosis virus, *J. Invertebr. Pathol.* **12**:395.

Kawase, S., and Miyajima, S., 1969, Immunofluorescence studies on the multiplication of cytoplasmic polyhedrosis virus of the silkworm, *Bombyx mori*, *J. Invertebr. Pathol.* **13**:330.

Kawase, S., and Yamaguchi, K., 1974, A polyhedrosis virus forming polyhedra in midgut-cell nucleus of silkworm, *Bombyx mori*. II. Chemical nature of the virion, *J. Invertebr. Pathol.* **24**:106.

Kawase, S., Kawamoto, F., and Yamaguchi, K., 1973, Studies on the polyhedrosis virus forming polyhedra in the midgut-cell nucleus of the silkworm, *Bombyx mori*. I. Purification procedure and form of the virion, *J. Invertebr. Pathol.* **22**:266.

Knudson, D.L., 1981, Genome of Colorado tick fever virus, *Virology* **112**:361.

Kobayashi, M., 1971, Replication cycle of cytoplasmic-polyhedrosis virus as observed with the electron microscope, in: *The Cytoplasmic-Polyhedrosis Virus of the Silkworm* (H. Aruga and Y. Tanada, eds.), pp. 103–128, University of Tokyo Press, Tokyo.

Kobayashi, M., and Kawase, S., 1980, Absence of detectable accumulation of cytoplasmic polyhedrosis viral RNA in the silkworm *Bombyx mori*, reared at a supraoptimal temperature, *J. Invertebr. Pathol.* **35**:96.

Kodama, T., and Suzuki, N., 1973, RNA polymerase activity in purified rice dwarf virus, *Ann. Phytopathol. Soc. Jpn.* **39**:251.

Koyama, R., and Katagiri, K., 1968, Use of cytoplasmic polyhedrosis virus for the control of pine caterpillar, *Dendrolimus spectabilis* Butler (Lepidoptera: Lasiocampidae), Proceedings of the joint United States–Japan seminar on microbial control of insect pests, Fukuoka, April 21–23 1967, pp. 63–69, Institute of Biological Control, Kyushu University, Japan.

Krieg, A., 1956, Uber die Nucleinsäuren der Polyeder-Viren, *Naturwissenschaften* **43**:537.

Krywienczyk, J., and Hayashi, Y., 1970, Serological comparison of ribosomal and viral components, *J. Invertebr. Pathol.* **15**:165.

Krywienczyk, J., and Hayashi, Y., 1971, Specificity of serological cross-reactions between insect viruses and ribosomes, *J. Invertebr. Pathol.* **17**:321.

Krywienczyk, J., and Hayashi, Y., 1972, Serological investigations of subcellular fractions from *Malacosoma disstria* larvae infected with cytoplasmic polyhedrosis, *J. Invertebr. Pathol.* **20**:150.

Krywienczyk, J., and Sohi, S.S., 1973, Serologic characterization of a *Malacosoma disstria* Hübner(Lepidoptera: Lasiocampidae) cell line, *In Vitro (Rockville)* **8**:459.

Krywienczyk, J., Hayashi, Y., and Bird, F.T., 1969, Serological investigations of insect viruses. I. Comparison of three highly purified cytoplasmic-polyhedrosis viruses, *J. Invertebr. Pathol.* **13**:114.

Kurata, K., 1971, On symptoms of silkworms, *Bombyx mori*, infected with a cytoplasmic-polyhedrosis virus under aseptic conditions, *J. Seric. Sci. Jpn.* **40**:32 (in Japanese; English summary).

Laemmli, U.K., 1970, Cleavage of structural proteins during the assembly of the head of bacteriophage T4, *Nature (London)* **227**:680.

Levin, K.H., and Samuel, C.E., 1977, Biosynthesis of reovirus-specified polypeptides: Effect of methylation on the efficiency of reovirus genome expression *in vitro*, *Virology* **77**:245.

Lewandowski, L.J., and Leppla, S.H., 1972, Comparison of the 3′ termini of discrete segments of the double-stranded ribonucleic acid genomes of cytoplasmic polyhedrosis viruses, wound tumor virus and reovirus, *J. Virol.* **10**:965.

Lewandowski, L.J., and Millward, S., 1971, Characterization of the genome of cytoplasmic polyhedrosis virus, *J. Virol.* **7**:434.

Lewandowski, L.J., and Traynor, B.L., 1972, Comparison of the structure and polypeptide composition of three double-stranded ribonucleic acid-containing viruses (Diplornaviruses): Cytoplasmic polyhedrosis virus, wound tumor virus, and reovirus, *J. Virol.* **10**:1053.

Lewandowski, L.J., Kalmakoff, J., and Tanada, Y., 1969, Characterization of a ribonucleic acid polymerase activity associated with purified cytoplasmic polyhedrosis virus of the silkworm, *Bombyx mori*, *J. Virol.* **4**:857.

Lipa, J.J., 1970, A cytoplasmic polyhedrosis virus of *Triphaena pronuba* (L) (Lepidoptera, Noctuidae), *Acta Microbiol. Pol. Ser. B.* **2**:237.

Longworth, J.F., 1981, The replication of a cytoplasmic polyhedrosis virus from *Chrysodeixis eriosoma* (Lepidoptera: Noctuidae) in *Spodoptera frugiperda* cells, *J. Invertebr. Pathol.* **37**:54.

Longworth, J.F., and Spilling, C.R., 1970, A cytoplasmic polyhedrosis of the larch sawfly, *Anoplonyx destructor*, *J. Invertebr. Pathol.* **15**:276.

Lourenco, M.H., Nicolas, J.C., Cohen, J., Scherrer, R., and Bricout, F., 1981, Study of human rotavirus genome by electrophoresis: Attempt of classification among strains isolated in France, *Ann. Virol. (Paris)* **132**:161.

Luftig, R.B., Kilham, S., Hay, A.J., Zweerink, H.J., and Joklik, W.K., 1972, An ultrastructural study of virions and cores of reovirus type 3, *Virology* **48**:170.

Magnoler, A., 1974, Effects of a cytoplasmic polyhedrosis on larval and post larval stages of the gypsy moth, *Porthetria dispar, J. Invertebr. Pathol.* **23:**263.

Martignoni, M.E., 1967, Separation of two types of viral inclusion bodies by isopycnic centrifugation, *J. Virol.* **1:**646.

Martignoni, M.E., and Iwai, P., 1977, A catalog of viral diseases of insects and mites, USDA Forest Service General Technical Report, PNW-40.

Martignoni, M.E., and Iwai, P.J., 1981, A catalogue of viral diseases of insects, mites and ticks, in: *Microbial Control of Pests and Plant Diseases 1970–1980* (H.D. Burges, ed.), pp. 897–911, Academic Press, London and New York.

Martignoni, M.E., Iwai, P.J., Hughes, K.M., and Addison, R.B., 1969, A cytoplasmic polyhedrosis of *Hemerocampa pseudotsugata, J. Invertebr. Pathol.* **13:**15.

Martin, S.A., Paoletti, E., and Moss, B., 1975, Purification of mRNA guanylyl-transferase and mRNA (guanine-7-)methyl transferase from vaccinia virus, *J. Biol. Chem.* **250:**9322.

Martinson, H.G., and Lewandowski, L.J., 1974, Sequence homology studies between the double-stranded RNA genomes of cytoplasmic polyhedrosis virus, wound tumor virus and reovirus strains 1, 2 and 3, *Intervirology* **4:**91.

Martouret, D., and Dusaussoy, G., 1959, Multiplication et extraction des corps d'inclusion de la virose intestinale de *Thaumetopoea pityocampa* Schiff., *Entomophaga* **4:**253.

Matthews, R.E.F., 1979, Classification and nomenclature of viruses, *Intervirology* **12:**131.

McCrae, M.A., 1982, Coding assignments for the genes of a cytoplasmic polyhedrosis virus, Proceedings of the IIIrd International Colloquium on Invertebrate Pathology, pp. 20–24.

McCrae, M.A., and Joklik, W.K., 1978, The nature of the polypeptide encoded by each of the 10 double-stranded RNA genome segments of reovirus type 3, *Virology* **89:**578.

McCrae, M.A., and Mertens, P.P.C., 1983, *In vitro* translation studies on and RNA coding assignments for cytoplasmic polyhedrosis viruses, Proceedings of the International Double-Stranded RNA Virus Conference, Virgin Islands, October 1982, Elsevier, Amsterdam (in press).

Mertens, P.P.C., 1979, A study of the transcription and translation (*in vitro*) of the genomes of cytoplasmic polyhedrosis viruses types 1 and 2, D.Phil. thesis, University of Oxford.

Mertens, P.P.C., and Dobos, P., 1982, The messenger RNA of infectious pancreatic necrosis virus is polycistronic, *Nature (London)* **297:**243.

Mertens, P.P.C., and Payne, C.C., 1978, S-Adenosyl-L-homocysteine as a stimulator of viral RNA synthesis by two distinct cytoplasmic polyhedrosis viruses, *J. Virol.* **26:**832.

Mery, C., and Dulmage, H.T., 1975, Transmission, diagnosis and control of cytoplasmic polyhedrosis virus in colonies of *Heliothis virescens, J. Invertebr. Pathol.* **26:**75.

Miura, K.-I., 1981, The cap structure of eukaryotic messenger RNA as a mark of a strand carrying protein information, *Adv. Biophys.* **14:**205.

Miura, K.I., and Muto, A., 1965, Lack of messenger RNA activity of a double-stranded RNA, *Biochim. Biophys. Acta* **108:**707.

Miura, K., Fujii, I., Sakaki, T., Fuke, M., and Kawase, S., 1968, Double-stranded ribonucleic acid from cytoplasmic polyhedrosis virus of the silkworm, *J. Virol.* **2:**1211.

Miura, K., Fujii-Kawata, I., Iwata, H., and Kawase, S., 1969, Electron-microscopic observation of a cytoplasmic polyhedrosis virus from the silkworm, *J. Invertebr. Pathol.* **14:**262.

Miura, K., Watanabe, K., and Sugiura, M., 1974, 5'-Terminal nucleotide sequences of the double-stranded RNA of silkworm cytoplasmic polyhedrosis virus, *J. Mol. Biol.* **86:**31.

Miura, K.-I., Kodama, Y., Shimotohno, K., Fukui, T., Ikehara, M., Nakagawa, I., and Hata, T., 1979, Inhibitory effect of methylated derivatives of guanylic acid for protein synthesis with reference to the functional structure of the 5'-cap in viral messenger RNA, *Biochim. Biophys. Acta* **564:**264.

Miyajima, S., and Kawase, S., 1967, Different infection response of silkworm larvae to the locality of CPV virus injection, *J. Invertebr. Pathol.* **9:**441.

Miyajima, S., and Kawase, S., 1969, Hemagglutination with cytoplasmic polyhedrosis virus of the silkworm, *Bombyx mori, Virology* **39:**347.

Miyajima, S., Kimura, I., and Kawase, S., 1969, Purification of a cytoplasmic-polyhedrosis virus of the silkworm, *Bombyx mori, J. Invertebr. Pathol.* **13:**296.

Morris, O.N., 1980, Entomopathogenic viruses: Strategies for use in forest insect pest management, *Can. Entomol.* **112:**573.

Muthukrishnan, S., Both, G.W., Furuichi, Y., and Shatkin, A.J., 1975, 5' Terminal 7-methyl guanosine in eukaryotic mRNA required for translation, *Nature (London)* **255**:33.

Nakashima, K., Darzynkiewicz, E., and Shatkin, A.J., 1980, Proximity of mRNA 5'-region and 18S rRNA in eukaryotic initiation complexes, *Nature (London)* **286**:226.

Neilson, M.M., 1965, Effects of a cytoplasmic polyhedrosis on adult Lepidoptera, *J. Invertebr. Pathol.* **7**:306.

Nichols, J.L., Hay, A.J., and Joklik, W.K., 1972, 5'-Terminal nucleotide sequence of reovirus mRNA synthesized *in vitro, Nature (London) New Biol.* **235**:105.

Nishimura, A., and Hosaka, Y., 1969, Electron microscopic study on RNA of cytoplasmic polyhedrosis virus of the silkworm, *Virology* **38**:550.

Payne, C.C., 1971, Properties and replication of some occluded insect viruses, D.Phil. thesis, University of Oxford.

Payne, C.C., 1976, Biochemical and serological studies of a cytoplasmic polyhedrosis virus from *Arctia caja:* A naturally-occurring mixture of two virus types, *J. Gen. Virol.* **30**: 357.

Payne, C.C., 1981, Cytoplasmic polyhedrosis viruses, in: *Pathogenesis of Invertebrate Microbial Diseases* (E.W. Davidson, ed.), pp. 61–100, Allanheld, Osmun, Totowa, New Jersey.

Payne, C.C., 1982, Insect viruses as control agents, in: *Parasites as Biological Control Agents* (R.M. Anderson and E.U. Canning, eds.), *Parasitology.*

Payne, C.C., and Churchill, M.P., 1977, The specificity of antibodies to double-stranded (ds) RNA in antisera prepared to three distinct cytoplasmic polyhedrosis viruses, *Virology* **79**:251.

Payne, C.C., and Harrap, K.A., 1977, Cytoplasmic polyhedrosis viruses, in: *The Atlas of Insect and Plant Viruses* (K. Maramorosch, ed.), pp. 105–129, Academic Press, New York.

Payne, C.C., and Kalmakoff, J., 1973, The synthesis of virus-specific single-stranded RNA in larvae of *Bombyx mori* infected with a cytoplasmic polyhedrosis virus, *Intervirology* **1**:34.

Payne, C.C., and Kalmakoff, J., 1974a, Biochemical properties of polyhedra and virus particles of the cytoplasmic polyhedrosis virus of *Bombyx mori, Intervirology* **4**:354.

Payne, C.C., and Kalmakoff, J., 1974b, A radioimmunoassay for the cytoplasmic polyhedrosis virus from *Bombyx mori, Intervirology,* **4**:365.

Payne, C.C., and Kalmakoff, J., 1978, Alkaline protease associated with virus particles of a nuclear polyhedrosis virus: Assay, purification, and properties, *J. Virol.* **26**:84.

Payne, C.C., and Rivers, C.F., 1976, A provisional classification of cytoplasmic polyhedrosis viruses based on the sizes of the RNA genome segments, *J. Gen. Virol.* **33**:71.

Payne, C.C., and Tinsley, T.W., 1974, The structural proteins and RNA components of a cytoplasmic polyhedrosis virus from *Nymphalis io* (Lepidoptera: Nymphalidae), *J. Gen. Virol.* **25**:291.

Payne, C.C., Piasecka-Serafin, M., and Pilley, B., 1977, The properties of two recent isolates of cytoplasmic polyhedrosis viruses, *Intervirology* **8**:155.

Payne, C.C., Mertens, P.P.C., and Katagiri, K., 1978, A comparative study of three closely-related cytoplasmic polyhedrosis viruses, *J. Invertebr. Pathol.* **32**:310.

Payne, C.C., Rubinstein, R., Crook, N.E., and Mertens, P.P.C., 1983, Serological and molecular studies of variation in cytoplasmic polyhedrosis viruses, Proceedings of the International Double-Stranded RNA Virus Conference, Virgin Islands, October 1982, Elsevier, Amsterdam (in press).

Quiot, J.M., and Belloncik, S., 1977, Caractérisation d'une polyédrose cytoplasmique chez le lépidoptère *Euxoa scandens,* Riley (Noctuidae, Agrotinae): Etudes *in vivo* et *in vitro, Arch. Virol.* **55**:145.

Quiot, J.-M., Vago, C., and Belloncik, S., 1980a, Reaction antiviral par bourgeonnement cellulaire: Etude en culture de cellules de Lépidoptère infectée par un reovirus de polyédrose cytoplasmique, *C. R. Acad. Sci. Paris, Ser. D* **291**:481.

Quiot, J.-M., Vago, C., and Tchoukchry, M., 1980b, Etude experimentale sur culture de tissus de Lépidoptères, de l'interaction de deux virus d'invertebres, *C. R. Acad. Sci. Paris, Ser. D.* **290**:199.

Rao, C.B.J., 1973, Surface topography and shapes of polyhedral inclusion bodies of the cytoplasmic polyhedrosis virus, *J. Ultrastruct. Res.* **42**:582.

Rhodes, D.P., Reddy, D.V.R., MacLeod, R., Black, L.M., and Banerjee, A.K., 1977, *In vitro* synthesis of RNA containing 5'-terminal structure $^7$mG(5')ppp(5') Ap$^m$ . . . by purified wound tumor virus, *Virology* **76**:554.

Richards, W.C., 1970, Disruption of a cytoplasmic polyhedrosis virus by ethanol. *J. Invertebr. Pathol.* **15**:457.

Rodger, S.M., and Holmes, I.H., 1979, Comparison of the genomes of simian, bovine and human rotaviruses by gel electrophoresis and detection of genomic variation among bovine isolates, *J. Virol.* **30**:839.

Rodger, S.M., Holmes, I.H., and Studdert, M.J., 1980, Characteristics of the genomes of equine rotaviruses, *Vet. Microbiol.* **5**:243.

Rodger, S.M., Bishop, R.F., Birch, C., McLean, B., and Holmes, I.H., 1981, Molecular epidemiology of human rotaviruses in Melbourne, Australia, from 1973 to 1979 as determined by electrophoresis of genome ribonucleic acid, *J. Clin. Microbiol.* **13**:272.

Rohrmann, G.F., Bailey, T.J., Becker, R.R., and Beaudreau, G.S., 1980, Comparison of the structure of C- and N-polyhedrins from two occluded viruses pathogenic for *Orgyia pseudotsugata*, *J. Virol.* **34**:360.

Rubinstein, R., 1979, Some physical characteristics of a cytoplasmic polyhedrosis virus of *Heliothis armigera*, *Intervirology* **12**:340.

Rubinstein, R., and Harley, E.H., 1978, Reproducible alterations of cytoplasmic polyhedrosis virus double-stranded RNA genome patterns on laboratory passage, *Virology* **84**:195.

Rubinstein, R., Harley, E.H., Losman, M., and Lutton, D., 1976, The nucleic acids of viruses infecting *Heliothis armigera*, *Virology* **69**:323.

Sangar, D.V., Taylor, J., and Gorman, B.M., 1981, The identification of genome segments coding for bluetongue virus polypeptides, Abstracts of the 5th International Congress for Virology, Strasbourg, p. 428.

Sharpe, A.H., Ramig, R.F., Murtoc, T.A., and Fields, B.N., 1978, A genetic map of reovirus. 1. Correlation of genomic RNAs between serotypes 1, 2, 3, *Virology* **84**:63.

Shatkin, A.J., 1974, Methylated messenger RNA synthesis *in vitro* by purified reovirus, *Proc. Natl. Acad. Sci. U.S.A.* **71**:3204.

Shimotohno, K., and Miura, K., 1973a, Transcription of double-stranded RNA in cytoplasmic polyhedrosis virus *in vitro*, *Virology* **53**:283.

Shimotohno, K., and Miura, K.-I., 1973b, Single-stranded RNA synthesis *in vitro* by the RNA polymerase associated with cytoplasmic polyhedrosis virus containing double-stranded RNA, *J. Biochem. (Tokyo)* **74**:117.

Shimotohno, K., and Miura, K.-I., 1974, 5'-Terminal structure of messenger RNA transcribed by the RNA polymerase of silkworm cytoplasmic polyhedrosis virus containing double-stranded RNA, *J. Mol. Biol.* **86**:21.

Shimotohno, K., and Miura, K., 1976, The process of formation of the 5'-terminal modified structure in messenger RNA of cytoplasmic polyhedrosis virus, *FEBS Lett.* **64**:204.

Shimotohno, K., and Miura, K.-I., 1977, Nucleoside triphosphate phosphohydrolase associated with cytoplasmic polyhedrosis virus, *J. Biochem. (Tokyo)* **81**:371.

Shimotohno, K., Kodama, Y., Hashimoto, J., and Miura, K.-I., 1977, Importance of 5'-terminal blocking structure to stabilize mRNA in eukaryotic protein synthesis, *Proc. Natl. Acad. Sci. U.S.A.* **74**:2734.

Shinshi, H., Miura, M., Sugimura, T., Shimotohno, K., and Miura, K., 1976, Enzyme cleaving the 5'-terminal methylated blocked structure of messenger RNA, *FEBS Lett.* **65**:254.

Sikorowski, P.P., and Thompson, A.C., 1979, Effects of cytoplasmic polyhedrosis virus on diapausing *Heliothis virescens*, *J. Invertebr. Pathol.* **33**:66.

Sikorowski, P.P., Andrews, G.L., and Broome, J.R., 1973, Trans-ovum transmission of a cytoplasmic polyhedrosis virus of *Heliothis virescens* (Lepidoptera: Noctuidae), *J. Invertebr. Pathol.* **21**:41.

Silverstein, S.C., and Dales, S., 1968, The penetration of reovirus RNA and initiation of its genetic function in L-strain fibroblasts, *J. Cell. Biol.* **36**:197.

Silverstein, S.C., Astell, C., Levin, D.H., Schonberg, M., and Acs, G., 1972, The mechanisms of reovirus uncoating and gene activation *in vivo*, *Virology* **47**:797.

Simmons, C.L., and Sikorowski, P.P., 1973, A laboratory study of the effects of cytoplasmic polyhedrosis virus on *Heliothis virescens* (Lepidoptera: Noctuidae), *J. Invertebr. Pathol.* **22**:369.

Skup, D., and Millward, S., 1980a, mRNA capping enzymes are masked in reovirus progeny subviral particles, *J. Virol.* **34**:490.

Skup, D., and Millward, S., 1980b, Reovirus-induced modification of cap-dependent translation in infected L cells, *Proc. Natl. Acad. Sci. U.S.A.* **77**:152.

Skup, D., Zarbl, H., and Millward, S., 1981, Regulation of translation in L-cells infected with reovirus, *J. Mol. Biol.* **151**:35.

Smith, K.M., 1963, Cytoplasmic polyhedroses, in: *Insect Pathology: An Advanced Treatise,* Vol. 1 (E.A. Steinhaus, ed.), pp. 457–497, Academic Press, New York.

Smith, K.M., 1976, *Virus–Insect Relationships,* Longman, London.

Smith, K.M., and Rivers, C.F., 1956, Some viruses affecting insects of economic importance, *Parasitology* **46**:235.

Smith, K.M., and Wyckoff, R.W.G., 1950, Structure within polyhedra associated with insect virus diseases, *Nature (London)* **166**:861.

Smith, R.E., and Furuichi, Y., 1980a, Gene mapping of cytoplasmic polyhedrosis virus of silkworm by the full-length mRNA prepared under optimized conditions of transcription *in vitro, Virology* **103**:279.

Smith, R.E., and Furuichi, Y., 1980b, Separation of full length transcripts and genome RNA plus and minus strands from cytoplasmic polyhedrosis virus of *Bombyx mori,* in: *Animal Virus Genetics* (B.N. Fields and R. Jaenisch, eds.), pp. 391–400, Vol. 23, *ICN-UCLA Symposium on Molecular and Cellular Biology* (C.F. Fox ed.), Academic Press, New York.

Smith, R.E., and Furuichi, Y., 1982a, A unique class of compound, guanosine-nucleoside tetraphosphate G(5')pppp(5')N, synthesized during the *in vitro* transcription of cytoplasmic polyhedrosis virus of *Bombyx mori, I. Biol. Chem.* **257**:485.

Smith, R.E., and Furuichi, Y., 1982b, Segmented CPV genome dsRNAs are independently transcribed, *J. Virol.* **41**:326.

Smith, R.E., Zweerink, H.J., and Joklik, W.K., 1969, Polypeptide components of virions, top component and cores of reovirus type 3, *Virology* **39**:791.

Smith, R.E., Morgan, M.A., and Furuichi, Y., 1981, Separation of the plus and minus strands of cytoplasmic polyhedrosis virus and human reovirus double-stranded genome RNAs by gel electrophoresis, *Nucleic Acids Res.* **9**:5269.

Sohi, S.S., Bird, F.T., and Hayashi, Y., 1971, Development of *Malacosoma disstria* cytoplasmic polyhedrosis virus in *Bombyx mori* ovarian and tracheal tissue cultures, Proceedings of the International Colloquium on Insect Pathology, College Park Maryland, 1970, pp. 340–351.

Sonenberg, N., Skup, D., Trachsel, H., and Millward, S., 1981, *In vitro* translation in reovirus- and poliovirus-infected cell extracts, *J. Biol. Chem.* **256**:4138.

Stairs, G.R., Parrish, W.B., and Allietta, M., 1968, An histopathological study involving a cytoplasmic polyhedrosis virus of the salt-marsh caterpillar, *Estigmene acrea, J. Invertebr. Pathol.* **12**:359.

Steinhaus, E.A., and Dineen, J.P., 1959, A cytoplasmic polyhedrosis of the alfalfa caterpillar, *J. Insect Pathol.* **1**:171.

Stoltz, D.B., 1969, Observations on naturally occurring viruses in larvae of the midge, *Chironomus plumosus,* PhD. thesis, McMaster University, Hamilton, Ontario.

Stoltz, D.B., and Hilsenhoff, W.L., 1969, Electron-microscopic observations on the maturation of a cytoplasmic-polyhedrosis virus, *J. Invertebr. Pathol.* **14**:39.

Storer, G.B., Shepherd, M.G., and Kalmakoff, J., 1974a, Enzyme activities associated with cytoplasmic polyhedrosis virus from *Bombyx mori.* I. Nucleotide phosphohydrolase and nuclease activities, *Intervirology* **2**:87.

Storer, G.B., Shephard, M.G., and Kalmakoff, J., 1974b, Enzyme activities associated with cytoplasmic polyhedrosis virus from *Bombyx mori.* II. Comparative studies, *Intervirology* **2**:193.

Summers, M.D., and Kawanishi, C.Y., 1978, *Viral pesticides: Present Knowledge and Potential Effects on Public and Environmental Health*, U.S. Environmental Protection Agency, Research Triangle Park, North Carolina.

Summers, M.D., Engler, R., Falcon, L.A., and Vail, P.V., 1975, *Baculoviruses for Insect Pest Control: Safety Considerations*, American Society for Microbiology, Washington D.C.

Sun, Y., Wu, A., Dai, R., and Shen, X., 1981, Synthesis of structural proteins in a cell free system directed by silkworm cytoplasmic polyhedrosis virus mRNA synthesized *in vitro, Sci. Sin.* **24**:684.

Tanada, Y., and Chang, G.Y., 1962, Cross-transmission studies with three cytoplasmic polyhedrosis viruses, *J. Insect. Pathol.* **4**:361.

Tanada, Y., and Chang, G.Y., 1968, Resistance of the alfalfa caterpillar, *Colias eurytheme*, at high temperatures to a cytoplasmic-polyhedrosis virus and thermal inactivation point of the virus, *J. Invertebr. Pathol.* **10**:79.

Tanada, Y., and Omi, E.M., 1974, Persistence of insect viruses in field populations of alfalfa insects, *J. Invertebr. Pathol.* **23**:360.

Tanada, Y., Tanabe, A.M., and Reiner, C.E., 1964, Survey of the presence of a cytoplasmic-polyhedrosis virus in field populations of the alfalfa caterpillar, *Colias eurytheme* Boisduval, in California, *J. Insect Pathol.*, **6**:439.

Tanaka, S., 1971, Cross transmission of cytoplasmic-polyhedrosis viruses, in: *The Cytoplasmic-Polyhedrosis Virus of the Silkworm* (H. Aruga and Y. Tanada, eds.), pp. 201–207, University of Tokyo Press, Tokyo.

Thompson, A.C., and Sikorowski, P.P., 1980, Fatty acid and glycogen requirement of *Heliothis virescens* infected with cytoplasmic polyhedrosis virus, *Comp. Biochem. Physiol. B* **66**:93.

Tinsley, T.W., 1979, The potential of insect pathogenic viruses as pesticidal agents, *Annu. Rev. Entomol.* **24**:63.

Tinsley, T.W., and Harrap, K.A., 1978, Viruses of invertebrates, in: *Comprehensive Virology*, Vol. 12 (H. Fraenkel-Conrat and R.R. Wagner, eds.), pp. 1–101, Plenum Press, New York.

Vago, C., 1959, Sur le mode d'infection de la virose intestinale de *Thaumetopoea pityocampa* Schiff., *Entomophaga* **4**:311.

Vago, C., and Bergoin, M., 1963, Développement des virus a corps d'inclusion du Lépidoptère *Lymantria dispar* en cultures cellulaires, *Entomophaga* **8**:253.

Vago, C., Croissant, O., and Lepine, P., 1959, Processus de l'infection à virus à partir des corps d'inclusion de "polyédrie cytoplasmique" ingérés par le Lépidoptère sensible, *Mikroskopie* **14**:36.

Vail, P.V., and Gough, D., 1970, Effects of cytoplasmic-polyhedrosis virus on adult cabbage loopers and their progeny, *J. Invertebr. Pathol.* **15**:397.

Vail, P.V., Hall, I.M., and Gough, D., 1969, Influence of a cytoplasmic polyhedrosis virus on various developmental stages of the cabbage looper, *J. Invertebr. Pathol.* **14**:237.

Van der Beek, C.P., Saaijer-Riep, J.D., and Vlak, J.M., 1980, On the origin of the polyhedral protein of *Autographa californica* nuclear polyhedrosis virus, *Virology* **100**:326.

Van Dijk, A.A., and Huismans, H., 1980, The *in vitro* activation and further characterization of the bluetongue virus associated transcriptase, *Virology* **104**:347.

Wada, A., Kawata, I., and Miura, K.-I., 1971, Flow-dichroic spectra of double-stranded RNA, *Biopolymers* **10**:1153.

Watanabe, H., 1966a, Relative virulence of polyhedrosis viruses and host-resistance in the silkworm, *Bombyx mori* L. (Lepidoptera: Bombycidae), *Appl. Entomol. Zool.* **1**:139.

Watanabe, H., 1966b, Some aspects on the mechanism of resistance to peroral infection by cytoplasmic-polyhedrosis virus in the silkworm, *Bombyx mori* L., *J. Seric., Sci. Jpn.* **35**:411 (in Japanese; English summary).

Watanabe, H., 1967a, Development of resistance in the silkworm, *Bombyx mori* to peroral infection of a cytoplasmic-polyhedrosis virus, *J. Invertebr. Pathol.* **9**:474.

Watanabe, H., 1967b, Site of viral RNA synthesis within the midgut cells of the silkworm, *Bombyx mori*, infected with cytoplasmic-polyhedrosis virus, *J. Invertebr. Pathol.* **9**:480.

Watanabe, H., 1968a, Light radioautographic study of protein synthesis in the midgut epithelium of the silkworm, *Bombyx mori*, infected with a cytoplasmic-polyhedrosis virus, *J. Invertebr. Pathol.* **11**:310.

Watanabe, H., 1968b, Pathogenic changes in the faeces from the silkworm, *Bombyx mori* L., after peroral inoculation with a cytoplasmic-polyhedrosis virus, *J. Seric. Sci. Jpn.* **37**:385 (in Japanese; English summary).

Watanabe, H., 1971a, Pathophysiology of cytoplasmic polyhedrosis in the silkworm, in: *The Cytoplasmic-Polyhedrosis Virus of the Silkworm* (H. Aruga and Y. Tanada, eds.), pp. 151–167, University of Tokyo Press, Tokyo.

Watanabe, H., 1971b, Resistance of the silkworm to cytoplasmic-polyhedrosis virus, in: *The Cytoplasmic-Polyhedrosis Virus of the Silkworm* (H. Aruga and Y. Tanada, eds.), pp. 169–184, University of Tokyo Press, Tokyo.

Watanabe, H., 1972, Pathophysiology of nitrogen catabolism in the midgut of silkworm, *Bombyx mori* L. (Lepidoptera: Bombycidae), infected with a cytoplasmic-polyhedrosis virus, *Appl. Entomol. Zool.* **6**:163.

Weiser, J., 1978, A new host, *Simulium argyreatum* Meig., for the cytoplasmic polyhedrosis virus of blackflies in Czechoslovakia, *Folia Parasitol. (Prague)* **25**:361.

Wells, B.D., and Yang, J.T., 1974, A computer probe of the circular dichroic bands of nucleic acids in the ultraviolet region. II. Double-stranded ribonucleic acid and deoxyribonucleic acid, *Biochemistry* **13**:1317.

Wertheimer, A.M., Chen, S.Y., Borchardt, R.T., and Furuichi, Y., 1980, *S*-Adenosyl methionine and its analogs: Structural features correlated with synthesis and methylation of mRNAs of cytoplasmic polyhedrosis virus, *J. Biol. Chem.* **255**:5924.

Wittig, G., Steinhaus, E.A., and Dineen, J.P., 1960, Further studies of the cytoplasmic polyhedrosis virus of the alfalfa caterpillar, *J. Insect Pathol.* **2**:334.

Wood, H.A., 1973, Viruses with double-stranded RNA genomes, *J. Gen. Virol.* **20**(Suppl.):61.

Wu, A., Dai, R., Shen, X., and Sun, Y., 1981, [$^3$H-methyl]methionine as possible methyl donor for formation of 5′-terminus of *in vitro* synthesized mRNA of cytoplasmic polyhedrosis virus of silkworm, *Bombyx mori*, *Sci. Sin.* **24**:1737.

Xeros, N., 1956, The virogenic stroma in nuclear and cytoplasmic polyhedrosis, *Nature (London)* **178**:412.

Xeros, N., 1962, The nucleic acid content of the homologous nuclear and cytoplasmic polyhedrosis viruses, *Biochim. Biophys. Acta* **55**:176.

Xeros, N., 1966, Light microscopy of the virogenic stromata of cytopolyhedroses, *J. Invertebr. Pathol.* **8**:79.

Yamaguchi, K., 1979, Natural recovery of the fall webworm, *Hyphantria cunea* to infection by a cytoplasmic-polyhedrosis virus of the silkworm, *Bombyx mori*, *J. Invertebr. Pathol.* **33**:126.

Yamaguchi, K., Nakagawa, I., Hata, T., Shimotohno, K, Hiruta, M., and Miura, K.-I., 1976, Chemical and biological synthesis of confronted nucleotide structure at 5′-terminus of messenger RNA of CP virus, *Nucleic Acids Res.* **S2**:S151.

Yamakawa, M., Shatkin, A.J., and Furuichi, Y., 1981, Chemical methylation of RNA and DNA viral genomes as a probe of *in situ* structure, *J. Virol.* **40**:482.

Yamamasu, Y., and Kawakita, T., 1962, Studies on the grasserie of the silk-producing insects. IV. On the cytoplasmic polyhedrosis of pine moth, *Dendrolimus spectabilis* Butler., *Bull. Fac. Text. Sci. Kyoto Univ.* **3**:421 (in Japanese; English summary).

Yazaki, K., and Miura, K.-I., 1980, Relation of the structure of cytoplasmic polyhedrosis virus and the synthesis of its messenger RNA, *Virology* **105**:467.

Ying, S.L., 1970, Application of *Isaria* sp., cytoplasmic polyhedrosis virus and *Bacillus thuringiensis* against the pine caterpillar, *Dendrolimus punctatus* Walker (Lasiocampidae: Lepidoptera), *Q. J. Chinese For.* **4**:51.

Zama, M., and Ichimura, S., 1976, Induced circular dichroism of acridine orange bound to double-stranded RNA and transfer RNA, *Biopolymers* **15**:1693.

Zappia, V., Zydek-Cwick, C.R. and Schlenk, F., 1969, The specificity of *S*-adenosylmethionine derivatives in methyl transfer reactions, *J. Biol. Chem.* **244**:4499.

Zarbl, H., Skup, D., and Millward, S., 1980, Reovirus progeny subviral particles synthesize uncapped mRNA, *J. Virol.* **34**:497.

CHAPTER 10

# The Plant Reoviridae

R.I.B. FRANCKI AND GUIDO BOCCARDO

## I. INTRODUCTION

Several of the leafhopper- and planthopper-borne diseases we now know to be caused by viruses that belong to the Reoviridae were studied for many years without their etiological agents being recognized (Lyon, 1910; Fukushi, 1931; L.N. Black, 1944; Biraghi, 1949; Fenaroli, 1949). It was only after the development of suitable purification procedures and other techniques that the viruses were isolated and characterized. The first plant Reoviridae to be purified and studied were wound tumor virus (WTV) in America and rice dwarf virus (RDV) in Japan. It was shown that both had polyhedral particles about 70 nm in diameter (Brakke *et al.*, 1954; Fukushi and Kimura, 1959). More sophisticated electron-microscopic studies established the similarity of WTV and RDV particles to those of reovirus (Bils and Hall, 1962; Fukushi *et al.*, 1962; Vasquez and Tournier, 1962). The similarity was further highlighted when it was demonstrated that these viruses all contained double-stranded RNAs (dsRNAs) (L.M. Black and Markham, 1963; Gomatos and Tamm, 1963a,b; Miura *et al.*, 1966). Subsequent studies revealed that a number of other viruses with reoviruslike particles from vertebrates, insects, and plants contained dsRNA. Verwoerd (1970) recommended that they be classified in one group, for which he suggested the name diplornaviruses. However, it soon became clear that this name was inappropriate (Wood, 1973), since a number of viruses with dsRNA the particles of which were unlike those of reovirus were isolated from fungi and bacteria (Matthews, 1982). Furthermore, small viruslike polyhedral particles, symptomlessly carried by

R. I. B. FRANCKI • Department of Plant Pathology, Waite Agricultural Research Institute, The University of Adelaide, Adelaide 5064, South Australia.    GUIDO BOCCARDO • Istituto di Fitovirologia Applicata del C.N.R., 10135 Torino, Italy.

plants and referred to as cryptic viruses, have also recently been shown to contain dsRNA (Lisa *et al.*, 1981).

In the 1960s, several other viruses from plants, such as maize rough dwarf virus (MRDV) and Fiji disease virus (FDV), were shown to have reoviruslike particles (Gerola *et al.*, 1966; Teakle and Steindl, 1969) and to contain dsRNA (Redolfi and Pennazio, 1972; Francki and Jackson, 1972). Subsequent observations established that the particle fine structure of MRDV and FDV was similar but distinguishable from that of WTV and RDV (Milne *et al.*, 1973; Hatta and Francki, 1977). Furthermore, whereas MRDV and FDV particles contained 10 segments of dsRNA, those of WTV and RDV contained 12 (Reddy *et al.*, 1974, 1975a). These differences are the basis for dividing the plant-infecting Reoviridae into two genera, with WTV and FDV as the type species of *Phytoreovirus* and *Fijivirus*, respectively (Matthews, 1982).

The research progress on the plant-infecting Reoviridae can be traced from reviews that have appeared at regular intervals (L.M. Black, 1959, 1972, 1982; Maramorosch *et al.*, 1969; Harpaz, 1972; Wood, 1973; Joklik, 1974a; Milne and Lovisolo, 1977). The understanding of the biochemistry of the plant-infecting members of the family cannot rival that of the viruses that infect vertebrates (Joklik, 1974a). The reasons are not only that much greater effort has gone into research on viruses of medical and veterinary interest, but also that there are technical difficulties associated with studies of the viruses in plants (Matthews, 1981). The plant Reoviridae are expecially difficult experimental subjects because they usually replicate only in vascular or neoplastic tissues and because they cannot be transmitted by mechanical inoculation. Their transmission is dependent on insect vectors, in which they also multiply (see Section V.A). The establishment of cell cultures from leafhoppers that could be infected with WTV (Chiu and Black, 1967, 1969) holds out hopes of a better system for studying virus replication. Their potential has not yet been fully exploited (L.M. Black, 1979).

In this chapter, we have endeavored to summarize the present knowledge of the two genera of plant Reoviridae, discussing in more detail information reported subsequent to the review by Milne and Lovisolo (1977). Where appropriate, we have placed some emphasis on the comparison between the properties of the Reoviridae that infect plants and those that infect other organisms.


## II. MEMBERS OF THE REOVIRIDAE THAT INFECT PLANTS

### A. *Phytoreovirus*

The three viruses that belong to this genus (Table I), wound tumor virus (WTV), rice dwarf virus, (RDV), and rice gall dwarf virus (RGDV), have particles similar to viruses in the genus *Reovirus* that infect only vertebrates (Matthews, 1982). However, the phytoreoviruses have genomes consisting of 12 double stranded RNA (dsRNA) segments (Fig. 1) instead of 10, as do the reoviruses. All three phytoreoviruses are transmitted by leafhoppers (Cicadellidae, Jassoidea). Both WTV and RDV have been studied for many years, and data on them are documented in a long list of publications. On the other hand, RGDV has only recently been isolated for the first time in Thailand (Omura *et al.*, 1980), but its properties are sufficiently well known for the virus to be included in the genus *Phytoreovirus* with confidence (Inoue and Omura, 1982; Omura, personal communication).

## B. *Fijivirus*

The name of this genus is derived from Fiji disease virus (FDV), the type member (Matthews, 1982). Fiji disease was recognized early in this century as a serious disease of sugarcane in Fiji (Lyon, 1910), but virus particles were not observed until nearly 60 years later (Giannotti *et al.*, 1968; Teakle and Steindl, 1969). Several other viruses are now included in the genus by virtue of their particle structure and dsRNA genome of ten segments. Maize rough dwarf virus (MRDV) and oat sterile dwarf virus (OSDV) are both serologically unrelated to each other and to FDV. FDV, MRDV, and OSDV are thus type species of three different serological

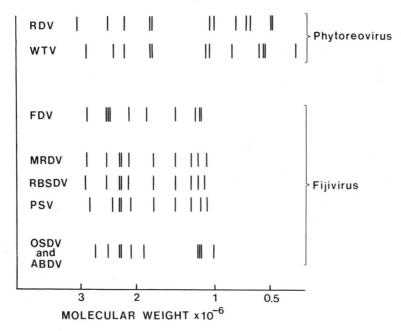

FIGURE 1. Diagramatic representation of the relative sizes of the individual RNA genome segments of phytoreoviruses and fijiviruses.

TABLE I. Members of the Reoviridae That Infect Plants

| Virus | Host(s) | Vector(s) | Geographic distribution | References |
|---|---|---|---|---|
| Genus *Phytoreovirus* | | | | |
| 1. Wound tumor (WTV) | *Trifolium incarnatum* *Melilotus alba* *M. officinalis* *Rumex acetosa* | *Agallia constricta* *A. quadripunctata* *Agalliopsis novella* | — | L. M. Black (1970), Reddy and Black (1974, 1977), Reddy and MacLeod (1976), Reddy et al. (1977), Nuss and Peterson (1980) |
| 2. Rice dwarf (RDV) | *Oryza sativa* *Echinocloa crus-galli* | *Nephotettix cincticeps* *N. apicalis* *Recilia (= Inazuma) dorsalis* | Japan, China, Korea | Iida et al. (1972), Kodama and Suzuki (1973), Nakata et al. (1978) |
| 3. Rice gall dwarf (RGDV) | *Oryza sativa* | *Nephotettix nigropictus* *N. cincticeps* *N. malayanus* *N. virescens* *Recilia (= Inazuma) dorsalis* | Thailand | Omura et al. (1980), Inoue and Omura (1981) Morinaka et al. (1982) |
| Genus *Fijivirus*[a] | | | | |
| 1. Fiji disease (FDV) | *Saccharum officinarum* | *Perkinsiella saccharicida* *P. vastatrix* *P. vitiensis* | Australia, Fiji, Madagascar, New Britain, New Guinea, New Hebrides, Samoa | Hutchinson and Francki (1973), Ikegami and Francki (1974, 1975, 1976), Hatta and Francki (1976, 1977, 1981) |
| 2. Maize rough dwarf (MRDV) | *Zea mays* *Digitaria sanguinalis* *Echinocloa crus-galli* *Triticum aestivum* *Avena sativa* | *Laodelphax striatellus (= Calligypona marginata)* *Delphacodes propinqua* *Sogatella vibrix* *Iavesella (= Calligypona) pellucida* | Czechoslovakia, France, Italy, Israel, Spain, Switzerland, Yugoslavia | Lovisolo (1971), Harpaz (1972), Milne and Lovisolo (1977), Milne and Luisoni (1977), Boccardo and Milne (1981) |

| | | | | |
|---|---|---|---|---|
| 2a. Cereal tillering disease (CTDV) | Hordeum vulgare, Avena sativa, Triticum aestivum, Zea mays | Laodelphax striatellus, Dicranotropis hamata | Sweden | Lindsten et al. (1973), Milne et al. (1974) |
| 2b. Rice black-streaked dwarf (RBSDV) | Oryza sativa, Avena sativa, Triticum aestivum, Zea mays | Laodelphax striatellus, Unkanodes albifascia, U. sapporona | Japan, China | Shikata (1974), Reddy et al. (1975b), Shikata and Kitagawa (1977) |
| 2c. Pangola stunt (PSV) | Digitaria decumbens | Sogatella furcifera | Brazil, Guyana, Fiji, Peru, Taiwan | Milne (1977a), Boccardo et al. (1979) |
| 3. Oat sterile dwarf (OSDV) | Avena sativa, Arrhenatherum elatius, Lolium spp. | Javesella pellucida, J. discolor, J. dubia | Britain, Czechoslovakia, Finland, Germany, Norway, Poland, Sweden | Luisoni et al. (1979), Boccardo and Milne (1980a) |
| 3a. Arrhenatherum blue dwarf (ABDV) | Arrhenatherum elatius | Javesella pellucida | East Germany | Mühle and Kempiak (1971), Milne and Lesemann (1978) |
| 3b. Lolium enation (LEV) | Lolium spp. | Javesella pelluccida | West Germany | Huth (1975), Milne and Lesemann (1978) |
| Unclassified Rice ragged stunt (RRSV) | Oryza spp. | Nilaparvata lugens | Bangla-desh, China, India, Indonesia, Japan, Malaysia, Philippines, Taiwan, Thailand | Hibino et al. (1977, 1979), Ling et al. (1978), Milne et al. (1982) |

[a] Numbers denote viruses that are serologically unrelated (possibly distinct species); letter suffixes denote viruses that may be strains of better-known viruses.

subgroups within the genus. No other virus has so far been found related to FDV, while cereal tillering disease virus (CTDV), rice black-streaked dwarf virus (RBSDV), and pangola stunt virus (PSV) are serologically closely related to MRDV, and *Arrhenatherum* blue dwarf virus (ABDV) and *Lolium* enation virus (LEV) appear to be antigenically indistinguishable from OSDV (Table I). In addition to similar particle structure and genome size (Fig. 1), all the fijiviruses infect only monocotyledons (Gramineae), in which they induce neoplastic growths that have been referred to as vein swellings, galls, or enations by various authors. All fijiviruses are transmitted by planthoppers (Delphacidae, Fulguroidea).

## C. Unclassified: Rice Ragged Stunt Virus

Rice ragged stunt virus (RRSV) causes a serious disease of rice in Asia (Hibino *et al.*, 1977; Ling *et al.*, 1978) and has reoviruslike particles (Hibino *et al.*, 1979; Shikata *et al.*, 1979). However, the particles differ in structure from those of both phytoreoviruses and fijiviruses (Milne, 1981) and appear to contain only eight segments of dsRNA. The total molecular weight of the RRSV genome was estimated to be only approximately three quarters that of phytoreoviruses or fijiviruses (Boccardo and Milne, 1980b). Examination of RRSV particles in purified preparations and sections of infected plant cells failed to find evidence for the presence of a particle outer shell (Milne, 1981). Hence, it seems that the complete RRSV particle corresponds to a structure more like a fijivirus subviral particle (SVP) or core (see Section 3.3.2), which could account for the smaller virus genome. Milne (1981) also observed that the projections of RRSV particles were different from the B spikes of fijivirus SVPs; they appeared more like truncated cones than the pentagonal cylinders characteristic of fijiviruses.

Some properties of RRSV are similar to those of fijiviruses. RRSV produces stunting, leaf-twisting, excessive tillering, and vein swellings on the lower surfaces of leaf blades and culms of infected rice plants (Hibino *et al.*, 1977; Ling *et al.*, 1978). The swellings appear to be caused by the disorganized proliferation of vascular tissues (Hibino *et al.*, 1977; Hatta *et al.*, 1982). The virus is transmitted by the brown rice planthopper (Table I) in a persistent manner (Ghosh and John, 1980) and appears to replicate in the tissues of the vector (Hibino *et al.*, 1979). Plant and insect cells infected with RRSV contain viroplasms, characteristic of infection by plant Reoviridae (Hibino *et al.*, 1979; Milne, 1981).

## D. Viruses Possibly Involved with Plant Diseases of Uncertain Etiology

There are a number of plant diseases that at one time or another have been claimed to be caused by, or to be associated with, reolike viruses or virus particles but the etiology of which remains obscure. We will briefly discuss the evidence for and against any possible role of these viruses in the etiology of the diseases.

## 1. Maize Wallaby Ear Disease

Maize wallaby ear disease (MWED) was first described in Australia by Tryon (1910) and later attributed to a virus because of the symptoms developed on maize plants colonized by the leafhopper *Cicadulina* (*Cicadula*) bipunctella bimaculata (Schindler, 1942). The plants developed enations on their lower leaf surfaces and became stiff and stunted. Maramorosch *et al.* (1961) observed a similar disease in the Philippines induced on maize and rice colonized by *C. bipunctella* in the apparent absence of any virus. They concluded that the disease was induced by an insect toxin. However, no data were presented that would convincingly demonstrate that neither plants nor insects were infected with virus.

Grylls (1975) concluded that a reolike virus was responsible for MWED; his evidence was as follows: (1) Polyhedral viruslike particles about 85 nm in diameter with cores about 50 nm across were observed in thin sections of salivary glands of *C. bimaculata* capable of inducing MWED, but not in those of insects from colonies that did not produce the disease. (2) Isometric particles about 45 nm in diameter, thought to be the SVPs of the 85-nm particles observed in insects, were detected in partially purified preparations from diseased maize and grasses, and from insects capable of inducing MWED. (3) MWED symptoms were transmitted from maize to maize by a side-approach graft. (4) MWED symptoms could also be induced in maize and grasses by another leafhopper, *Nesoclutha pallida*. It was also later shown that preparations of viruslike particles from *C. bimaculata* and *N. pallida* contained dsRNA that was resolved into ten components by polyacrylamide gel electrophoresis (PAGE) (Reddy *et al.*, 1976).

Boccardo *et al.* (1980) confirmed that reolike virus particles, which they referred to as leafhopper A virus (LAV), were present in *C. bimaculata* from colonies able to induce MWED, but not in those that produced no symptoms. They also detected rhabdoviruslike particles in insects from the former but not the latter colonies. However, attempts to demonstrate that either of these viruslike particles were able to replicate in maize plants were unsuccessful. Furthermore, when insects that induced MWED in maize were removed from plants showing early symptoms, the plants recovered and produced normal growth. These observations cast serious doubts on the conclusions reached by Grylls (1975) that MWED is caused by a leafhopper-borne virus. Boccardo *et al.* (1980) suggested that MWED may be caused by an insect toxin as concluded by Maramorosch *et al.* (1961 ), but that toxin production by the insects was dependent on their being infected with virus. However, more recent studies have shown that some *C. bimaculata* colonies in which no LAV could be detected by immuno-electron-microscopic tests could also induce

MWED symptoms in maize (F. A. Ofori and R.I.B. Francki, unpublished results). These observations fail to support that there is any connection of LAV with MWED.

LAV particles are structurally similar to those of the fijiviruses, although no serological relationship could be traced to FDV, MRDV, PSV, OSDV, or RRSV (Boccardo et al., 1980). LAV was also shown to contain ten segments of dsRNA (Boccardo et al., 1980). However, although it shares many properties with members of the genus *Fijivirus*, LAV cannot be included in the genus because it has not been shown to infect plants. It would appear that reolike viruses also infect other insect species. Herold and Munz (1967), while studying the planthopper *Peregrinus maidis* infected with maize mosaic virus, observed reoviruslike particles in tissues of the insect. Maillet and Folliot (1967) and R.G. Milne, R. Lenzi, and M. Conti (private communication), observed similar particles in the tissues of the leafhoppers *Thyphlocyba douglasi* and *Laodelphax striatellus* apparently unconnected with any plant disease. Reoviruslike particles have also been detected in the tissues of a number of other insects: in *Frankiniella fusca*, a thrips vector of tomato spotted wilt virus (Paliwal, 1979); in the common housefly, *Musca domestica* (Moussa, 1978, 1980, 1981); and in *Drosophila melanogaster* cell lines (Haars et al., 1980; Alatortsev et al., 1981). Therefore, there may be a number of Reoviridae that will not fit into any of the currently recognized genera.

## 2. Rugose Leaf Curl Disease

Rugose leaf curl disease (RLCD) was first reported by Grylls (1954) from Lucerne and was characterized by extreme rugosity and leaf curling. It was shown to be transmitted by the leafhopper *Austroagallia torrida* and to infect 20 *Dicotyledon* species in eight families. Later, Grylls et al., (1974) observed reoviruslike particles about 75 nm in diameter in thin sections of *A. torrida* that induced RLCD symptoms in *Trifolium pratense*. Such particles were not detected in diseased plants; however, viruslike particles about 42 nm in diameter were detected in partially purified preparations from diseased plants, but not in those from healthy ones. Since very similar particles were also present in *A. torrida* extracts, these investigators concluded that the purified particles were the SVPs of the 75-nm particles seen in sections of insect cells.

Behncken and Gowenlock (1976) observed numerous rickettsialike bodies in the phloem of clovers suffering from RLCD. Similar bodies were not detected in the tissues of healthy plants. Furthermore, these workers demonstrated remission of the symptoms in clover plants treated with penicillin. These observations support the view that the disease may be caused by a rickettsialike organism and that the reoviruslike particles observed in *A. torrida* (Grylls et al., 1974) may be unconnected with the plant disease. Further research will be needed to establish beyond doubt the etiology of RLCD.

### 3. Corn White Leaf Disease

Corn white leaf disease (CWLD), a disease that causes stunting and the development of yellow to white continuous leaf stripes on maize, has been reported by Trujillo *et al.* (1974) in Venezuela. The disease was shown to be transmitted by the planthopper *Peregrinus maidis* to maize and to the grass *Rottboellia exaltata*. Trujillo *et al.* (1974) detected polyhedral viruslike particles, 55–60 nm in diameter, in crude extracts from infected plants. Stained in phosphotungstic acid, the particles appear as similarly treated *Fijivirus* particles that degrade to SVPs devoid of B spikes (Milne and Lovisolo, 1977). More detailed investigations are required to substantiate the viral etiology of CWLD, remembering that reoviruslike particles have been observed in thin sections of *P. maidis*, but are apparently unable to induce any disease in maize in Venezuela (Herold and Munz, 1967).

## III. PROPERTIES OF VIRUS PARTICLES

### A. Virus Purification

In developing virus-purification procedures, assay methods that will allow virus losses to be assessed at each step are of the utmost importance (Francki, 1972). This is usually done by infectivity assays when the properties of the virus are not well known. Assays are straightforward if the virus is sap-transmissible, and reasonably accurate results are obtained when the assay plants are local lesion hosts of the virus. Since plant Reoviridae are not sap-transmissible, infectivity can be determined only by injecting preparations into virus-free insect vectors that are then tested for their ability to transmit the virus to susceptible plants. Such assays are both laborious and inaccurate and hence of little value for the quantitative determination of virus concentration (L.M. Black, 1979). It is probably true to say that the lack of accurate biological assays for the plant Reoviridae has been the most serious impediment to their eventual purification. Once antisera became available, they could be used to monitor various purification steps. Another serious problem has been the difficulty of obtaining sufficient virus. Virus particles are restricted to the neoplastic tissues of plants infected by most of the plant Reoviridae (see Section IV.B). Hence, obtaining large quantities of plant tissues containing virus as starting material for purification is usually difficult.

### 1. Phytoreoviruses

Wound tumor virus (WTV) has been purified satisfactorily from root tumors of sweet clover grown under defined conditions that favor large-tumor production; a photoperiod of 16 hr in the light at 20°C and 8 hr in the dark at 17°C appears to be very satisfactory (L.M. Black, 1965). Although the virus has been purified by a number of methods (Table II), some appear to be better than others. Early methods (Brakke et al., 1954; L.M. Black et al., 1967) were rather inefficient in that only 2–5% of the virus particles extracted from tissue were recovered in the final preparations (Gamez and Black, 1967). An improved method in which virus was concentrated by precipitation with polyethylene glycol 6000 (PEG), rather than by ultracentrifugation, yielded 3–5 times more virus with a 50-fold higher specific infectivity (Reddy and Black, 1973a). Furthermore, Reddy and Black (1973a) subsequently demonstrated that the PEG and NaCl concentrations and the pH used for virus precipitation were critical for optimal virus yield and specific infectivity. However, conditions for optimal yields were not identical to those for maximum specific infectivity. It has been shown that even mild organic solvents such as fluorocarbons and carbon tetrachloride used for clarifying plant extracts decreased the infectivity of WTV (Reddy and Black, 1973a), as did CsCl used for isopycnic density-gradient centrifugation (Reddy and MacLeod, 1976).

A number of methods have also been used for the purification of rice dwarf virus (RDV) (Table II), but their relative merits do not appear to have been compared in such detail as for WTV. However, it has been shown that like WTV, RDV can be recovered in higher yields and specific infectivity when concentrated by precipitation with PEG than by ultracentrifugation (Kimura, 1976a). Like WTV, RDV was also shown to be partially degraded in CsCl (Nakata et al., 1978).

RGDV has also recently been purified by a method involving PEG precipitation and quasi-equilibrium sucrose density-gradient centrifugation (Omura, personal communication).

## 2. Fijiviruses

Purification of viruses belonging to the genus *Fijivirus* is more difficult than that of the phytoreoviruses because of the ease with which the particles lose their outer protein shells (Milne et al., 1973; Ikegami and Francki, 1974). This is especially so with Fiji disease virus (FDV), with which, after extraction and a single sedimentation, all the intact particles (IPs), about 75 nm in diameter, were converted to cores or subviral particles (SVPs), about 55 nm in diameter, with projections (B spikes) (Hatta and Francki, 1977). Consequently, most purified preparations of fijiviruses studied contain largely or entirely SVPs. However, maize rough dwarf virus (MRDV) prepared by Milne et al. (1973) contained 60–70% IPs, and the preparations were shown to be infectious. Nevertheless, the IPs were readily degraded to SVPs by a few days' storage at 4°C or by treatments such as heating for 10 min at 50°C, repeated sedimentations, or contact with chloroform. More drastic treatments such as staining with 1% phosphotungstic acid or contact with CsCl or *n*-butanol also removed the B spikes from the SVPs (Milne *et al.*, 1973; Boccardo and Milne, 1975).

The approach taken to the purification of FDV has been somewhat different from that for the remaining fijiviruses. MRDV and most of the other viruses have been purified from root tissues in which the viruses appear to reach higher concentrations and from which host materials are more readily eliminated. Organic solvents have been useful for the initial clarification of these tissue extracts (Table II). Use of the same organic solvents for the clarification of gall extracts of FDV-infected sugarcane led to serious losses of virus (Ikegami and Francki, 1974). The most satisfactory preparations of FDV SVPs have been obtained when a nonionic detergent was used for the removal of host impurities (Ikegami and Francki, 1974; van der Lubbe *et al.*, 1979).

Methods for the purification of FDV were devised with the aid of serological assays for double-stranded RNA (dsRNA), assuming that dsRNA concentrations were a reflection of virus content (Francki and Jackson, 1972). These studies suggested that only the galls on leaves and stems of FDV-infected sugarcane were a source of the virus. This was later confirmed by electron-microscopic observations (Hatta and Francki, 1976) and serological assays using an antiserum to FDV SVPs (Rohozinski *et al.*, 1981). Because of the difficulty in obtaining FDV preparations containing IPs, attempts have been made to stabilize the particles after extraction. Addition of glutaraldehyde (GA) to a final concentration of 1% to freshly prepared extracts of gall tissues was shown to stabilize FDV IPs to a certain extent. However, such extracts yielded purified preparations that still contained SVPs as well as IPs (Hatta and Francki, 1977).

Only SVPs have been obtained in purified preparations of oat sterile dwarf virus (OSDV) and pangola stunt virus (PSV) (Luisoni *et al.*, 1979; Boccardo *et al.*, 1979). Though Shikata and Kitagawa (1977) reported that purified rice black-streaked dwarf virus (RBSDV) was infectious, it is not clear which kind of particles were present in their preparations.

## B. Virus Composition

### 1. Nucleic Acid

#### a. Phytoreoviruses

WTV and RDV have been reported to contain 22 and 11% RNA with G + C contents of 39 and 44%, respectively (L.M. Black and Markham, 1963; Toyoda *et al.*, 1965; Miura *et al.*, 1966; Kalmakoff *et al.*, 1969). The discrepancy in the RNA content of the two viruses seems excessive in view of the similarity of their particles (see Section III.C.1) and should be reinvestigated. The RNAs of both viruses have been shown to be seg-

TABLE II. Purification of Some Plant Reoviridae

| Virus[a] | Extraction buffers | Clarifying agents | Concentration and purification | Final products[b] | References |
|---|---|---|---|---|---|
| Genus *Phytoreovirus* | | | | | |
| WTV | 0.1 M Glycine–0.01 M MgCl$_2$ | CHCl$_3$; CCl$_4$; fluorocarbon (Freon 113) | Differential and sucrose density-gradient centrifugation, electrophoresis | IPs | Brakke *et al.* (1954), L. M. Black *et al.* (1967) |
| | | | PEG precipitation, quasi-equilibrium density-gradient centrifugation | IPs | Reddy and Black (1973a) |
| | | | CsCl isopycnic centrifugation | IPs + SVPs | Reddy and MacLeod (1976) |
| RDV | 0.03 M Phosphate, pH 6.8 | CHCl$_3$; Freon 113; lytic enzymes | Differential and sucrose-density gradient centrifugation | IPs | Fukushi *et al.* (1962) |
| | 0.1 M Phosphate, pH 7.2; 6.5 | | DEAE–cellulose chromotography | IPs | Toyoda *et al.* (1965) |
| | | | PEG, quasi-equilibrium density-gradient centrifugation | IPs | Kimura (1976a) |
| | | | CsCl isopycnic centrifugation | IPs + SVPs | Nakata *et al.* (1978) |
| RGDV | | CCl$_4$ | PEG, sucrose density-gradient centrifugation | IPs | Omura, private communication |

| Genus *Fijivirus* | | | | |
|---|---|---|---|---|
| FDV | 0.1 M Glycine–5 mM EDTA, pH 8.5, 0.16 M phosphate, pH 7.4 | Nonidet P-40 | Differential and sucrose density-gradient centrifugation | SVPs<br>IPs + SVPs | Ikegami and Francki (1974), van der Lubbe et al. (1979), Hatta and Francki (1976) |
| MRDV | 0.03 M Na$_2$PO$_4$–0.01 M Na$_2$SO$_3$–1 mM EDTA<br>0.3 M Glycine–0.03 M MgCl$_2$–Tris-Cl, pH 7.5<br>0.05 M Na$_2$HPO$_4$–5 mM EDTA–10 mM Na$_2$SO$_3$ | CHCl$_3$; celite<br>None<br>Freon 113; CCl$_4$ | Differential and sucrose density-gradient centrifugation<br>Sucrose | SVPs<br>IPs + SVPs | Wetter et al. (1969), Milne et al. (1973), Redolfi and Boccardo (1974), Boccardo and Milne (1981) |
| RBSDV | 0.1 M Ammonium acetate, pH 6–9 | CHCl$_3$; CCl$_4$; n-butanol | Differential and sucrose density-gradient centrifugation | —[c] | Shikata and Kitagawa (1977) |
| PSV | 0.05 M Phosphate, pH 7.0, + 5 mM EDTA–10 mM Na$_2$SO$_3$ | Freon 113 | Differential and Cs$_2$SO$_4$ density-gradient centrifugation | SVPs | Boccardo et al. (1979) |
| OSDV | 0.4 M phosphate, pH 7.0, + 5 mM EDTA–10 mM Na$_2$SO$_3$ | Freon 113 | Differential and Cs$_2$SO$_4$ density-gradient centrifugation | SVPs | Luisoni et al. (1979) |

[a] (WTV) Wound tumor virus; (RDV) rice dwarf virus; (RGDV) rice gall dwarf virus; (FDV) Fiji disease virus; (MRDV) maize rough dwarf virus; (RBSDV) rice black-streaked dwarf virus; (PSV) pangola stunt virus; (OSDV) oat sterile dwarf virus.
[b] (IP) Intact particle; (SVP) subviral particle.
[c] The quality of the electron micrographs does not allow confident interpretation.

mented, and although there had been some controversies over the numbers of segments (Kalmakoff *et al.*, 1969; Fujii-Kawata *et al.*, 1970; Wood and Streissle, 1970; Lewandowski and Leppla, 1972; Reddy and Black, 1973b), it is now generally accepted that both viruses each contain 12 segments (Reddy *et al.*, 1974) (Fig. 1), a property considered to be one of the basic features of the genus *Phytoreovirus* (Matthews, 1982)).

Electron-microscopic examination of WTV and RDV showed that the segments vary from 0.1 to 1.5 μm in length (Kleinschmidt *et al.*, 1964; Fujii-Kawata *et al.*, 1970). The double-strandedness of the RNAs has been confirmed by X-ray diffraction studies (Tomita and Rich, 1964; Sato *et al.*, 1966 ). The X-ray data also indicate that the ribose-phosphate chains from the two intertwined helices are 13 Å apart along their axes. Furthermore, the orientation of the $PO_2^-$ groups on the polynucleotide chains appears to be different from that of the same groups in dsDNA molecules. This is also supported by infrared dichroism and optical rotatory dispersion data (Sato *et al.*, 1966; Samejima *et al.*, 1968).

The biological properties of *Phytoreovirus* genomes are as would be expected for dsRNA; they have no affinity for ribosomes, messenger RNA (mRNA) activity, or infectivity (Miura and Muto, 1965; L.M. Black, 1970; Iida *et al.*, 1972). Also, as expected, the RNAs of both WTV and RDV have been shown to be efficient inducers of interferon in animal cells (D.R. Black *et al.*, 1972; Takehara and Suzuki, 1973).

The 3' termini of WTV RNA have been investigated by labeling with tritiated borohydride and analyzing the tri-alcohols released after alkali hydrolysis (Lewandowski and Leppla, 1972). It was shown that equal amounts of $C_{OH}$ and $U_{OH}$ were present in the viral RNA.

Although the WTV genome consists of 12 dsRNA segments or parts of segments (Reddy and Black, 1974, 1977), Reddy and Black (1974) observed that virus isolates that had been maintained for long periods by vegetative propagation in crimson clover partially or totally lost their ability to be transmitted by vectors or to replicate in vector-cell monolayers. These isolates were shown to be deletion mutants. Reddy and Black (1974) detected 16 subvectorial or exvectorial mutants after maintaining the virus in vegetatively propagated plants for 2 years. Ten, nine, four, and one of the mutants were identified to have deletions on RNA segments 5, 1, 2, and 7, respectively. Mutants that had lost segments 2 or 5 completely were shown to be able to replicate in sweet clover and to produce characteristic root tumors (Reddy and Black, 1977).

### b. Fijiviruses

The dsRNA genomes of the fijiviruses have not been investigated as extensively as those of the phytoreoviruses. First indications that FDV and MRDV contained dsRNA came from immunochemical data and from thermal denaturation and enzymatic studies, respectively (Francki and Jackson, 1972; Redolfi and Pennazio, 1972). Confirmatory evidence for

FDV RNA came from physical and enzymatic studies (Ikegami and Francki, 1975). The genomic dsRNAs of MRDV and FDV were resolved into nine segments by PAGE (Redolfi and Boccardo, 1974; Ikegami and Francki, 1975). This was later corrected by Reddy *et al.* (1975a), who resolved the RNAs of each virus into ten segments. Similarly, it has been established that RBSDV, PSV, OSDV, and *Arrhenatherum* blue dwarf virus (ABDV) all have genomes consisting of ten dsRNA segments (Reddy *et al.*, 1975b; Boccardo *et al.*, 1979; Luisoni *et al.*, 1979).

From the data summarized in Fig. 1, it can be seen that the sizes of the genomic RNAs of the fijiviruses are quite distinct from those of the phytoreoviruses. Furthermore, it seems that the *Fijivirus* dsRNA genomes fall into three groups; FDV alone; MRDV, RBSDV, and PSV; and the indistinguishable OSDV and ABDV.

## 2. Capsid Proteins

### a. Phytoreoviruses

Reddy and MacLeod (1976) separated seven polypeptides by PAGE from WTV preparations purified from infected plants, insect vectors, and vector-cell monolayers (Table III). Two of the polypeptides (II and IV) were not detected in virus preparations incubated with trypsin or chymotrypsin or both, but the virus remained fully infective. This suggested that these polypeptides were located on the surface of the particles. Furthermore, peptides VI and VII were removed from WTV particles during sedimentation in CsCl gradients. Earlier, Lewandowski and Traynor (1972) had detected only four polypeptides with estimated molecular weights of 156,000, 122,000, 63,000, and 44,000 in their WTV preparations. However, they used CsCl density-gradient centrifugation in their purification

TABLE III. Structural Polypeptides of Some Plant Reoviridae[a]

| Polypeptide | WTV | | RDV[d] | MRDV[e] |
|---|---|---|---|---|
| | b | c | | |
| I | 160 | 155 | 193 | 139 |
| II | 131 | 130 | 152 | 126 |
| III | 118 | 108 | 131 | 123 |
| IV | 96 | ~74 | 110 | 111 |
| V | 58 | 57 | 62 | 97 |
| VI | 36 | 42 | 46 | 64 |
| VII | 35 | 41.5 | 45 | — |

[a] Figures are molecular weight $\times 10^3$.    [b] Reddy and Macleod (1976).
[c] Nuss and Peterson (from *in vitro* and *in vivo* transcription experiments).
[d] Nakata *et al.* (1978).    [e] Boccardo and Milne (1975).

procedure. It seems that the first three of these correspond to polypeptides I, III, and V, respectively, in Table III. The fourth may correspond to either polypeptide VI or VII, or perhaps both, since they are similar in size and may not have been resolved under the electrophoretic conditions used by Lewandowski and Traynor (1972).

Nakata *et al.* (1978) also detected seven polypeptides in preparations of RDV with molecular weights resembling those reported for WTV (Table III). As in WTV, polypeptides II and IV of RDV could be removed by chymotrypsin. Furthermore, these two polypeptides, together with polypeptides VI and VII, were lost from particles sedimenting through CsCl. Thus, it would appear that both RDV and WTV particles have stable cores, composed of polypeptides I, III, and V, containing the genomic dsRNA. Nakata *et al.* (1978) observed that polypeptide V was released from the particles by freezing and thawing. This led them to suggest that this peptide may be located internally and associated with the viral RNA, a suggestion that requires more substantiating data.

The available information indicates that WTV and RDV capsids are both composed of seven polypeptides. However, it must be remembered that this is based only on PAGE studies. Some of the larger polypeptides detected could be aggregates of smaller ones, or some of the smaller ones could be partial digests of larger ones. These possibilities should be considered in any future research on the proteins of the phytoreoviruses. Any such research should always be combined with high-resolution electron-microscopic studies in an endeavor to determine which proteins are associated with which structures.

### b. Fijiviruses

Boccardo and Milne (1975) separated six polypeptides by PAGE from MRDV preparations containing high proportions of IPs, three (I, II, and III) from preparations of MRDV SVPs with B spikes, and only two from SVPs devoid of the spikes (I and II) (Table 3). This allowed a partial assignment of polypeptides to the various structures of the MRDV particle. Although peptides II, III, and VI appeared to be present in larger amounts than peptides I, IV, and V, a reliable estimate of their molar ratios was not possible due to the heterogeneity of the virus preparations and the poor resolution of some of the protein bands in the polyacrylamide gels.

Van der Lubbe *et al.* (1979) separated three polypeptides from FDV SVP preparations, but the protein composition of IPs could not be investigated because of their instability during purification (Hatta and Francki, 1977). The three SVP polypeptides had molecular weights of 145,000–150,000, 125,000–129,000, and 36,000–39,000. The two larger polypeptides could correspond to two of the polypeptides I–III of MRDV (Table III), but the third appears to be very much smaller than any of the six polypeptides isolated from MRDV.

Currently available information on the proteins of the fijiviruses is rather fragmentary, and a more sophisticated approach to the problem is needed for a clearer picture to emerge.

## 3. Virus-Associated Enzyme Activities

### a. Phytoreoviruses

D.R. Black and Knight (1970) detected an RNA-dependent RNA polymerase (transcriptase) in purified preparations of WTV. Transcriptional products were shown to consist of single-stranded RNA (ssRNA) molecules that annealed to WTV genomic RNA. Reddy et al. (1977) showed that transcriptase activity was not stimulated by $Cs^+$, $K^+$, $Li^+$, or $Na^+$, but it was by $Mg^{2+}$ and $Mn^{2+}$, though the latter had inhibitory effect at concentrations higher than 5 mM. The reaction was also shown to be partially dependent on the energy-regeneration system (2.5 mM phosphoenolpyruvate and 12 μg/ml phosphoenolpyruvate-kinase). Annealing of the transcription products to denatured WTV dsRNA and PAGE analysis of the hybrids showed that 12 ssRNA molecules of the same length as the genomic dsRNA segments were synthesized (Reddy et al., 1977). Nuss and Peterson (1981a) resolved, by PAGE, 11 discrete mRNA species transcribed in vitro from the WTV dsRNA genome and demonstrated that each mRNA hybridized selectively to one genome segment.

Rhodes et al. (1977) and Nuss and Peterson (1981b) demonstrated that purified WTV also contained another enzyme activity that, in the presence of the donor methyl [$^3$H]$S$-adenosyl-L-methionine, catalyzed methylation of the 12 ssRNA transcripts. Methylation was shown to occur exclusively at the 5′ termini in the "blocked" structure $^7mG(5')ppp-(5')A_p^m$. This indicated that only the strands that contain $U_{OH}$ at their 3′ termini are transcribed (Rhodes et al., 1977). Both WTV transcriptase and methylase activities appeared to be located in the inner cores of the virus particles (Reddy et al., 1977; Rhodes et al., 1977).

Purified RDV was also shown to contain a transcriptase (Kodama and Suzuki, 1973). Like the WTV enzyme, it was dependent on the presence of all the four ribonucleoside triphosphates and $Mg^{2+}$, and was stimulated by an energy-regenerating system. Its location in the virus particle, however, was not elucidated (Kodama and Suzuki, 1973). The transcriptional products appeared to consist of ssRNA Molecules that were immediately released from the virus particles on completion of transcription. They could be annealed to RDV genomic RNA. The inhibitory effect of pyrophosphate but not orthophosphate on the reaction was taken to indicate the presence in the purified virus of base-specific ribonucleotidyl-transferase activity (Kodama and Suzuki, 1973). However, this needs to be substantiated experimentally.

## b. Fijiviruses

Ikegami and Francki (1976) failed to detect transcriptase activity in purified preparations of FDV SVPs. However, the enzyme was detected in freshly prepared extracts from galls of FDV-infected sugarcane leaves, but not in tissue extracts from healthy plants. The enzyme activity was short-lived in that it was no longer detectable after 15–20 min of incubation. The products annealed to FDV genomic RNA, but consisted of a heterogeneous mixture of low-molecular-weight ssRNA molecules. This was not altogether surprising, since the reaction took place in relatively crude leaf extracts that must have contained ribonucleases and other degradative enzymes. The transcriptase from FDV-infected sugarcane leaves sedimented at the same rate as the SVPs, indicating that the enzyme was associated with them. Their lack of activity after purification was probably due to their instability. It has been shown that FDV SVPs readily extrude their dsRNA *in vitro* (see Section III.C.2), a process that could well be expected to result in a loss of their ability to function as transcription complexes. Current problems associated with preparing highly purified IPs of the fijiviruses (see Section III.A.2) must be solved to facilitate meaningful investigations of the *in vitro* transcription of their genomes.

## C. Virus-Particle Structure

### 1. Phytoreoviruses

Like reovirus, WTV and RDV particles have double shells of protein with icosahedral symmetry. Streissle and Granados (1968) studied the morphology of WTV particles in uranyl-acetate-stained preparations and remarked on their similarity to the particles of reovirus. However, the WTV particles appeared a little smaller than those of reovirus and more angular with hexagonal outlines. They measured about 78 nm in diameter from vertex to vertex and about 72 nm from side to side. Reddy and MacLeod (1976) reported that WTV particles stained in phosphotungstic acid were about 73 nm in diameter and observed that slightly smaller particles were present in virus preparations that had been treated with chymotrypsin or CsCl, with diameters of 63 and 59 nm, respectively. These treatments were shown to remove some of the structural peptides (see Section III.B.3.a). In earlier studies, Bils and Hall (1962) reported that WTV particles were only about 60 nm in diameter. In retrospect, it seems that they may have been studying particles that had lost their outer shells during purification (see Section III.B.2.a). However, large particles like those of Reoviridae may undergo flattening during negative staining to make their apparent diameters greater. This possibility is supported by measurements on uranyl-acetate-stained and freeze–dried and shadowed particles of FDV, which were reported to be about 71 and 67 nm in diameter, respectively (Hatta and Francki, 1977).

Particles about 70 nm in diameter have been observed in preparations of RDV (Fukushi *et al.*, 1962; Kimura and Shikata, 1968). Kimura, in a personal communication to Shikata (1977), remarked on the similarity of RDV and WTV particles (Fig. 2), for which a structural model consisting of 32 morphological subunits with $T = 3$ lattice symmetry has been proposed (Kimura and Shikata, 1968).

### 2. Fijiviruses

Reoviruslike particles have been observed in negatively stained crude extracts of plants infected with all the known fijiviruses [see Table I

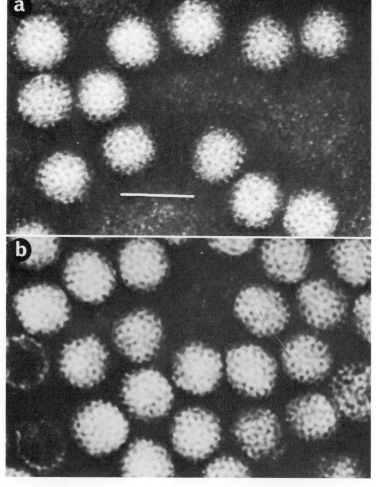

FIGURE 2. Electron micrographs of the phytoreoviruses WTV (a) and RDV (b). Scale bar: 100 nm. Courtesy of Dr. I. Kimura.

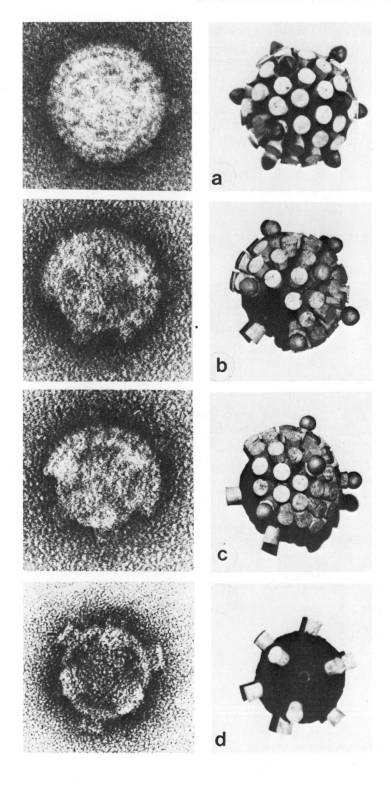

TABLE IV. Dimensions of Structures of Fiji
Disease Virus and Maize Rough Dwarf
Virus Particles in Uranyl Acetate Stain

| Structure | Dimension (nm) | FDV[a] | MRDV[b] |
|-----------|----------------|--------|---------|
| IPs | Diameter | 70.8 | 63–67 |
| A spikes | Width | 10.6 | 11 |
| | Height | 9.3 | 11 |
| SVPs | Diameter | 54 | 50–57 |
| B spikes | Width | 13.6 | 11 |
| | Height | 8.0 | 8 |

[a] Hatta and Francki (1977).    [b] Milne et al. (1973).

(Section II)]. Many IPs can easily be observed in preparations stained in uranyl acetate; however, when stained in phosphotungstic acid, they readily degrade to SVPs, unless the extracts are previously fixed (Milne and Lovisolo, 1977). From these observations, it seems that all *Fijivirus* particles (IPs and SVPs) have very similar morphology.

The fine structures of only MRDV and FDV particles (IPs and SVPs) in purified preparations have been studied in any detail (Milne et al., 1973; Boccardo and Milne, 1975, 1981; Hatta and Francki, 1977; van der Lubbe et al., 1979). It was concluded that both viruses consist of inner SVPs with B spikes surrounded by an outer protein shell with A spikes. On the basis of observation of negatively stained MRDV particles (Milne et al., 1973), Milne and Lovisolo (1977) proposed a structural model for the virus particle (Fig. 3). Later, Hatta and Francki (1977) studied the structure of FDV particles in negatively stained and shadowed preparations (Figs. 4 and 5), from which they also derived a model (Fig. 6). The reported dimensions of the structural components of FDV and MRDV particles are in good agreement, considering that the measurements were done in different laboratories (Table IV). Hatta and Francki (1977) were not prepared to reach any definite conclusion about the capsomeric structure of FDV from their micrographs (Figs. 4 and 5). Milne et al. (1973) and Milne and Lovisolo (1977) suggested that the outer shell of the MRDV particle may consist of 92 morphological subunits arranged with $T = 9$ icosadeltahedral symmetry (Fig. 3). The two proposed models (Figs. 3 and 6) are compatible, and it would appear that the particle structure of FDV and MRDV is very similar if not the same. However, their stabilities seem to differ in that the outer shell of FDV appears to be removed much more easily (Hatta and Francki, 1977).

The protein shells of the MRDV and FDV SVPs appear to be very

←————————————————————————————————————————

FIGURE 3. Selected electron micrographs of MRDV in various stages of disassembly (*left*) with appropriate models (*right*). (a) Particle is intact, showing its A spikes; (d) particle has lost all its A spikes and outer protein shell; (b,c) particles are partially disassembled. Courtesy of Dr. R.G. Milne.

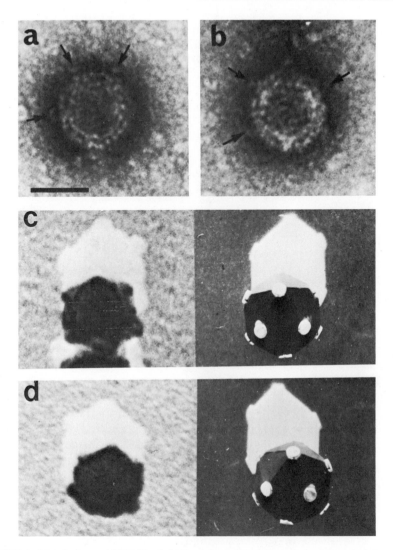

FIGURE 4. Morphology of FDV IPs. (a,b) Particles stained in uranyl acetate. (→) A spikes. (c,d) Particles have been freeze–dried and shadowed with platinum–carbon. At right of (c) and (d) are appropriate models photographed in light from an angle similar to that used for shadowing the particles in (c) and (d). Scale bar: 50 nm. After Hatta and Francki (1977).

stable (Milne *et al.*, 1973; Boccardo and Milne, 1975; Hatta and Francki, 1977). The B spikes are shorter and wider than the A spikes, which are attached to the B spikes (Figs. 3 and 6). Each B spike consists of a pentagonal structure with five subunits leaving a central hole about 4.5 nm in diameter (Fig. 5). Each B spike appears to be attached to a baseplate that is part of the inner protein shell (Milne *et al.*, 1973). It has been shown that the SVP shell of FDV remains intact while the dsRNA extrudes (Fig. 7). This extrusion may take place through the central holes of the B spikes (van der Lubbe *et al.*, 1979).

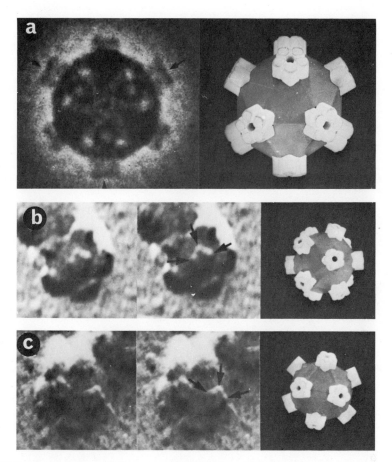

FIGURE 5. Morphology of FDV cores on SVPs. (a) The particle was stained in uranyl acetate and its image was positively printed and rotated three times through 120°; its B spikes (←) appear slightly longer than the other three, indicating that the particle's positioning is like that of the model at right. (b,c) The particles were freeze–dried and shadowed with platinum–carbon. (←) Pentamer clusters of the B-spike subunits. Models at aspects corresponding to the particles are at right. After Hatta and Francki (1977).

FIGURE 6. Scale model of a particle of FDV showing its A and B spikes (labeled A and B). A part of the particle outer shell (O) and one of the A spikes has been removed to expose the inner particle core (C) and structure of the B spikes. Although it is considered that the A spikes, outer shell, and core are constructed from morphological subunits, their arrangement remains obscure. After Hatta and Francki (1977).

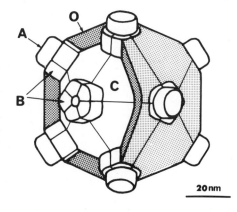

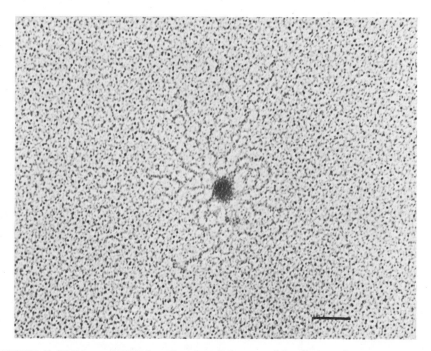

FIGURE 7. FDV core on SVP (see Figs. 5 and 6) prepared by the cytochrome $c$ spreading and rotary shadowing technique (van der Lubbe *et al.*, 1979). Scale bar: 100 nm. Courtesy of Dr. T. Hatta.

## D. Antigenic Properties of the Viruses

### 1. Phytoreoviruses

Both WTV and RDV appear to be reasonably efficient immunogens, and antisera to the viruses have been used extensively (L.M. Black and Brakke, 1954; Iida *et al.*, 1972), but no serological relationships between the two have been detected (Iida *et al.*, 1972). It has also been shown that rice gall dwarf virus is antigenically distinct from WTV and RDV (Omura, private communication).

WTV and RDV antisera have been especially useful for the preparation of fluorescein-conjugated antibodies used in detecting their respective viruses in plant tissues, smears from insect vectors, and cultured vector-cell monolayers (Nagaraj *et al.*, 1961; Sinha and Reddy, 1964; Reddy and Black, 1966; Iida *et al.*, 1972; Kimura and Miyajima, 1976).

It has been observed that ultracentrifugal supernatants from extracts of WTV-infected plants and insects contained antigen that was assumed to be a protein smaller than the virus (L.M. Black and Brakke, 1954; Whitcomb and Black, 1961). This antigen has been referred to as wound-tumor soluble antigen (WTSA) by L.M. Black (1959), but does not appear to have been characterized in any detail. As far as we are aware, its antigenic relationship to WTV remains obscure. The possibility that WTSA is the viral dsRNA as suggested by Luisoni et al. (1975) has not been tested. Kimura and Miyajima (1976) also reported the presence of a soluble antigen in RDV-infected rice plants, and antibodies specific to dsRNA have been detected in anti-RDV serum (Ikegami and Francki, 1973). However, the possibility that the soluble antigen may be the dsRNA has not been investigated.

An antigenic relationship between WTV and reovirus was detected by Streissle and Maramorosch (1963) in complement-fixation tests. However, this relationship could not be confirmed in subsequent studies using neutralization, hemagglutination-inhibition, and passive hemagglutination tests (Gomatos and Tamm, 1963b; Streissle et al., 1967; Gamez et al., 1967). The most obvious explanation for the original claim that the two viruses were related is that the rabbit used for immunization with WTV carried antireovirus antibodies. However, the possibility that antibodies specific to dsRNA were responsible for the apparent relationship between WTV and reovirus should also be considered.

## 2. Fijiviruses

Antisera prepared to fijiviruses have been used to establish interrelationships among members of the genus. On the basis of their antigenic properties, the fijiviruses have been divided into three unrelated groups (Fig. 8). The first group contains FDV, which does not appear to be related to any other member of the genus; the second contains MRDV, cereal tillering disease virus (CTDV), PSV, and RBSDV, all of which are interrelated; and the third, OSDV, ABDV, and Lolium enation virus (LEV), among which no antigenic differences have been detected using an anti-OSDV serum. These relations are all based on reactions of antibodies to SVP protein antigens, since none of the antisera used in these studies appears to have contained antibodies to the outer shells of the virus particles (Milne et al., 1973; Luisoni et al., 1973; 1975, 1979; Ikegami and Francki, 1974; Milne and Luisoni, 1977; Milne and Lesemann, 1978; van der Lubbe et al., 1979; Boccardo et al., 1979). Most of the antisera, however, did contain antibodies to dsRNA, and since such antibodies show very little specificity to different double-stranded polyribonucleotides (Stollar, 1973; Luisoni et al., 1975; Ikegami and Francki, 1977), care had to be exercised in interpreting the results of tests in which they were used. Immunodiffusion and immuno-electron-microscopic decoration (Fig. 9) are especially suitable tests for use with antisera containing antibodies to dsRNA, since any RNA-specific reactions can easily be identified (Ikegami and Francki, 1974; Luisoni et al., 1975; Milne and Luisoni, 1977; van der Lubbe et al., 1979). Recent studies on the use of the enzyme-linked immunosorbent assay (ELISA) indicate that it is a sensitive method for detecting FDV protein antigens without involving any reactions of dsRNA using sera containing antibodies to both antigens (Rohozinski et

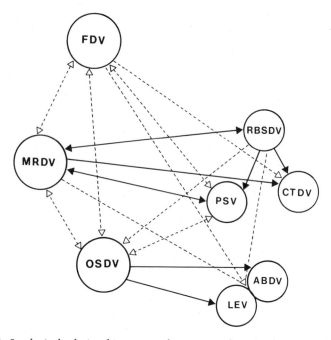

FIGURE 8. Serological relationships among fijiviruses indicating that the viruses fall into three serogroups. FDV does not appear to be related to any other virus (Group 1); MRDV is related to RBSDV, PSV, and CTDV (Group 2); and OSDV to ABDV and LEV. (———) Reported positive serological reactions; (– – – –) reports that no reactions were detected. Arrowheads at both ends of a line indicate that antisera to both viruses were used, whereas only a single arrow indicates that antiserum to only one of the viruses was used, the one from which the arrowhead points.

*al.,* 1981). This method may therefore be useful for studying the serological reactions of fijiviruses involving their protein antigens without interference from the viral dsRNAs.

It is not surprising that anti-*Fijivirus* sera have usually been found not to contain detectable amounts of antibodies to antigens of the particle outer shells, since it is known that these structures are readily lost to expose the SVPs (see Section III.C.2). It would seem that even if preparations used as immunogens did contain numerous IPs, these would degrade to SVPs soon after injection into the animals. The only anti-FDV sera in which antibodies to outer-shell antigens have been detected are those from rabbits that had been immunized with preparations containing FDV IPs fixed with GA (T. Hatta and R.I.B. Francki, unpublished results). Although GA-fixed MRDV IPs failed to induce antibodies to outer shells of the IPs, virus preparations stabilized with the cross-linking agent dithiobis-(succinimidyl)-propionate elicited high titers (Boccardo and Milne, 1981). Using these antisera, preliminary investigations suggest that the antigenic relationships among the fijiviruses established by tests with antisera containing antibodies only to SVP antigens may be similar to those observed when antigens specific to the outer shells of the particles are taken into consideration (Boccardo *et al.*, 1980; Boccardo and Milne, 1981).

*Fijivirus* SVPs have been found to be relatively efficient in eliciting antibodies to dsRNA. For example, when FDV SVPs were injected into mice, plasma with antibody titers in gel-diffusion tests of 1:256 and 1:64 to FDV protein and dsRNA, respectively, was obtained (Ikegami and Francki, 1974). Relatively high serum antibody titers to dsRNA were also obtained from rabbits immunized with FDV SVPs (van der Lubbe *et al.*, 1979). This was surprising, since the dsRNAs of the Reoviridae are considered to be located internally to their two protein shells, and hence one would not expect them to be exposed so as to act as immunogens. If the SVPs were so unstable in the body fluids of the immunized animals as to disrupt and release the dsRNA, the released molecules would not be expected to be efficient immunogens, since double-stranded polyribonucleotides are very poor in eliciting antibodies unless complexed with proteins (Stollar, 1973, 1975). Van der Lubbe *et al.* (1979) demonstrated that after purification from plant tissues, FDV SVPs consist of the characteristic B-spiked structures from which the viral dsRNA is partially extruded (see Section III.C.2), as seen in micrographs of the particles that have been spread on cytochrome *c* and shadowed at a low angle (Fig. 7). If they retain their integrity in the immunized animals, these structures could be expected to be efficient elicitors of antibodies to both dsRNA and protein by virtue of being relatively large complexes of the two components, both of which are exposed.

Attempts to prepare antisera to dsRNAs from a number of fijiviruses by immunizing rabbits with the antigens complexed to methylated bovine serum albumin as recommended by Schwartz and Stollar (1969) gave very disappointing results. The highest-titered antiserum obtained was to MRDV RNA, with a titer of 1:8 in immunodiffusion tests, whereas rabbits immunized with FDV, PSV, and OSDV RNAs failed to elicit detectable amounts of antibodies (G. Boccardo and R.I.B. Francki, unpublished results). This suggests that the SVPs of the fijiviruses are more efficient immunogens for the production of antibodies to dsRNA than the isolated RNAs coupled with methylated bovine serum albumin.

### 3. Rice Ragged Stunt Virus

Antisera prepared against RRSV have been used to test for possible relationships of the virus to SVPs of the fijiviruses FDV, MRDV, PSV, and OSDV by immunoelectron microscopy. However, these and reciprocal tests failed to establish any relationships (Milne *et al.*, 1979; Boccardo *et al.*, 1980). Antibodies to dsRNA were detected in the anti-RRSV serum by immunoelectron miscroscopy (Boccardo *et al.*, 1980), but not by immunodiffusion tests (Milne *et al.*, 1979). Recently, anti-RRSV serum has been used successfully for the detection of the virus in individual

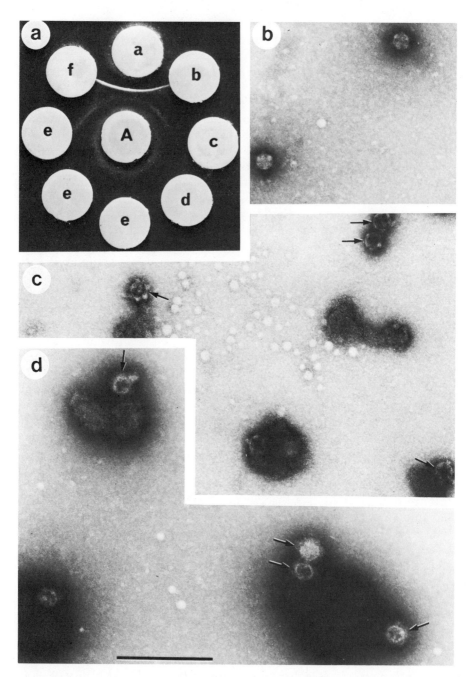

FIGURE 9. Serological reactions of FDV. (a) A two-dimensional immunodiffusion test is shown with an antiserum to FDV cores (SVPs) loaded in the center well (A); the outer wells contained antigens as follows: (a) FDV SVPs (homologous antigen); (b) FDV nucleic acid; (c) host plant nucleic acid; (d) juice from healthy plants; (f) poly(I):poly(c); (e) wells were not used. Data after Ikegami and Francki (1974). (b–d) Immuno-electron-microscopic tests

insect vectors by ELISA (Hibino and Kimura, 1982). Similar tests have also been used to detect MRDV (E. Luisoni *et al.*, 1982) and FDV (J. Rohozinski and R.I.B. Francki, unpublished results) in individual leafhoppers.

## IV. PLANT–VIRUS RELATIONSHIPS

### A. Diseases Caused by the Plant Reoviridae and Their Transmission

All the plant-infecting Reoviridae are transmitted only by their insect vectors (see Section II.A). None of the viruses is spread by mechanical transmission, although Harpaz (1959) did occasionally transmit maize rough dwarf virus (MRDV) experimentally by needle pricking. However, transmission was so erratic as to be considered of no value for experimental work. Brakke *et al.* (1954) reported five cases of mechanical transmission of wound tumor virus (WTV), but were unable to repeat the experiment. Later, Kimura and Black (1972) suggested that the original success may have been the result of accidental hopper transmission.

The economic importance of the plant diseases caused by the Reoviridae varies from serious, as in the case of the devastating damage in crops of rice and sugarcane infected with rice dwarf virus (RDV) and Fiji disease virus (FDV), respectively (Ishii and Yoshimura, 1973; Egan *et al.*, 1982), to nil in the case of WTV (L.M. Black, 1970). WTV was accidentally discovered in leafhoppers of the species *Agalliopsis novella* while L.M. Black (1944) was investigating the field biology of another virus disease, caused by potato yellow dwarf virus, now known to be a member of the Rhabdoviridae (Matthews, 1982). Since its discovery, WTV has been confined to the laboratory and has never been found in the field. Interest in WTV has probably been maintained because of its tumor-inducing properties (L.M. Black, 1965, 1972), its ability to replicate in both plant and insect cells (L.M. Black, 1959, 1969, 1979), and as the prototype member of the Reoviridae that infect plants.

### 1. Phytoreoviruses

The three members of the genus [see Table I (Section II)] are quite different in that WTV infects only dicotyledonous plants, causing neoplastic growths (L.M. Black, 1970); RDV infects only Gramineae, in which

---

of FDV SVPs. (b) The particles have been reacted with anti-potato virus x serum as a control; (c) with anti-poly(I):poly(c) serum; (d) with anti-FDV serum (homologous antiserum). Arrows point to FDV SVPs in (c) and (d); they are decorated with antibody in (d) but not in (c). Material extruding from FDV SVPs (see Fig. 7) is decorated in both (c) and (d) but not in (b). Scale bar: 200 nm. Data after van de Lubbe *et al.* (1979).

it induces mosaic symptoms and dwarfing but no neoplasia (Iida *et al.*, 1972); and rice gall dwarf virus (RGDV) has been observed to infect only rice, in which it induces neoplastic growth (Omura *et al.*, 1980).

### a. Wound Tumor Virus

The virus can be transmitted experimentally to 43 plant species in 20 different dicotyledonous families (L.M. Black, 1945). Usually, it induces prominent enlargements on leaf and stem veins and root tumors such as those seen in Fig. 10. The root tumors usually originate at the bases of emerging lateral roots and vary considerably in size and number on different hosts (L.M. Black, 1945, 1965, 1972).

Streissle and Maramorosch (1969) successfully grew virus-induced tumors *in vitro*, but found that the WTV concentration became depleted with time, so that the virus was no longer detectable after 5–7 months of culture.

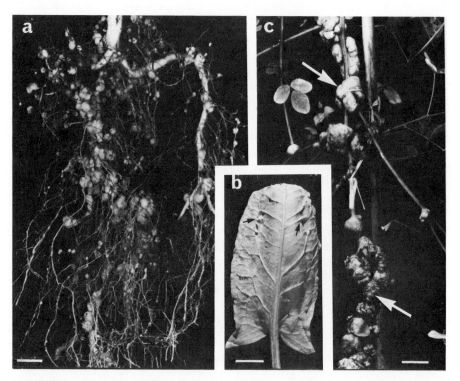

FIGURE 10. Symptoms of WTV infection on the roots of sweet clover (*Melilotus alba*), showing severe galling (a); on dock (*Rumex acetosa*) leaves, which have small tumors, the most prominent of which are marked with arrows (b); and on the stems of yellow sweet clover (*Melilotus officinalis*), on which the largest galls are marked with arrows (c). Scale bars: 1 nm. Courtesy of Dr. L.M. Black.

L.M. Black (1969) observed that vegetatively propagating WTV-infected plants resulted in some virus isolates that had partially or totally lost their ability to be insect-transmitted (subvectorial and exvectorial mutants, respectively). Changes in the composition of the double-stranded RNA (dsRNA) genomes that accompany these mutations are discussed in Section III.B.1.a.

### b. Rice Dwarf Virus

In Japan, the disease later shown to be caused by RDV was first recorded in Shiga Prefecture as early as 1883 (Fukushi, 1969). It was later observed also in Korea (Iida et al., 1972) and mainland China (Anonymous, 1978). It was initially thought to be caused by leafhopper colonization, but was later shown to be a virus disease transmitted by the leafhoppers (Fukushi, 1931). It was also established that RDV was transmissible through the insect eggs (Fukushi, 1933, 1940). Fukushi also argued that the long latent period of the virus in leafhopper vectors indicated that the virus was capable of multiplying in the insects, a controversial and hotly debated idea at the time (Bawden, 1939). Like WTV, RDV has also been observed to lose its ability to be vector-transmitted after prolonged maintenance by vegetative propagation (Kimura, 1976b).

Rice plants infected with RDV are severely stunted (Fig. 11) and develop numerous small tillers, giving the plants a rosettelike appearance. The leaves develop chlorotic specks (Fukushi, 1934). The severity of the disease is dependent on the age at which the plants become infected. Infection of young seedlings results in early symptom appearance, and although the plants may survive, they produce no grain (Shinkai, 1962). There appear to be a number of rice varieties that have some resistance to the virus and others to the principal natural vector, *Nephotettix cincticeps* (Shinkai, 1962; Ling, 1972). RDV can infect 16 other Gramineae (Iida et al., 1972).

### c. Rice Gall Dwarf Virus

RGDV has recently been isolated from diseased rice plants (Omura et al., 1980). The infected plants were dark green and stunted and developed small neoplastic growths on the lower surfaces of the leaves. These symptoms are reminiscent of those produced by the fijiviruses on their hosts (see Section IV.A.2). RGDV can be transmitted by four *Nephotettix* spp. (see Table I) in a propagative manner, and was shown to be transovarially carried to 87% of the progeny in *N. nigropictus* eggs (Inoue and Omura, 1981). So far, the disease seems to be localized in central Thailand, but it might become widespread and important due to the great population density of the green rice leafhoppers in the sub- and northern tropical regions (Inoue and Omura, 1981).

FIGURE 11. Symptoms on rice plants infected with RDV. (a) The plants at right are infected, whereas those at left are healthy controls. (b) The leaf from an infected plant (*left*) is compared to that of a healthy one (*right*). Courtesy of Dr. I. Kimura.

## 2. Fijiviruses

Unlike the phytoreoviruses, fijiviruses all induce rather similar symptoms in their hosts, which are all confined to the Gramineae. They usually include stunting of the plants, shortening of the shoot internodes, and excessive tillering. In addition, the leaves assume a darker green color and develop neoplastic growths (enations, galls) along the veins on stems and lower leaf surfaces. The size of these growths varies considerably with the virus and the host variety.

Fijiviruses have been shown to be transovarially transmitted with various degrees of efficiency. Vacke (1966) found only 0.2% oat sterile dwarf virus (OSDV) transmission through the eggs of *Javesella pellucida*, but 4 and 17% *Perkinsiella saccharicida* eggs have been reported to transmit pangola stunt virus (PSV) (Chang, 1977). Harpaz (1972) reported the transmission of MRDV through about 4% of *Laodelphax striatellus* eggs, but this could not be confirmed by M. Conti (cited as unpublished results by Milne and Lovisolo, 1977).

The principal plant hosts and insect vectors of the fijiviruses, as well as their geographical distributions, are presented in Table I. More detailed information has been reviewed by Milne and Lovisolo (1977).

### a. Fiji Disease Virus

The disease caused by FDV was first described by Lyon (1910) in Fiji, from whence the virus got its name and where it caused considerable damage. It also occurs in some other sugarcane-growing regions (Table I), where it can cause serious losses and is expensive to control by extensive roguing and planting of virus-free setts (Egan *et al.*, 1983). Although FDV can be transmitted experimentally to maize and sorghum (Hutchinson *et al.*, 1972), sugarcane appears to be its only natural host (Fig. 12a). This and the availability of virus-resistant varieties of sugarcane provide approaches to effective control measures (Egan *et al.*, 1983).

Leaf galls can also be produced on sugarcane by causes other than FDV, such as in the case of pseudo-Fiji disease, which appears to be induced by an insect toxin (Antoine, 1959; Sheffield, 1969).

### b. Maize Rough Dwarf Virus and Related Viruses

Following the introduction of high-yielding North American maize hybrids, a new disease called "nanismo" (= dwarfing) was observed in northern Italy (Biraghi, 1949; Fenaroli, 1949). The same disease also appeared ten years later in Israel when the maize hybrids were introduced there (Harpaz *et al.*, 1958). MRDV, the causal agent of these outbreaks, had probably been present in the regions, but was not noticed because the local maize varieties were tolerant or resistant (Harpaz, 1972; Milne and Lovisolo, 1977). Similar events may have led to the recent identification of what appears to be MRDV in Argentina (Bradfute *et al.*, 1981; Nomé *et al.*, 1981). In addition to maize, wheat, oats, and the grasses *Cynodon dactylon, Digitaria sanguinalis*, and *Echinocloa crusgalli* are all natural hosts of MRDV (Klein and Harpaz, 1969; Luisoni and Conti, 1970; Conti, 1972; Conti and Milne, 1977). In addition to typical *Fijivirus* symptoms on maize leaves infected with the virus, the roots develop swellings, resulting in their frequent splitting (Fig. 12c and d).

Lindsten and Gerhardson (1971) recognized a severe disease of barley and oats in Sweden, which they named cereal tillering. The disease was readily distinguished from the one caused by OSDV occurring in the same

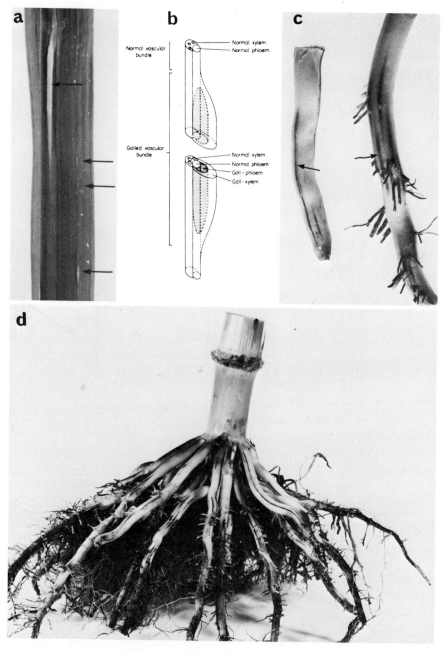

FIGURE 12. Symptoms of plants infected by fijiviruses. (a) Typical galls (←) on the under-
sides of leaves of a sugarcane infected with FDV; (b) distribution of abnormal tissues of a
leaf gall; (c,d) typical symptoms of root swellings and cracking (←) on maize plants infected
with MRDV.

area, by difference in host range, symptomatology, and vector specificity (Lindsten *et al.*, 1973). However, the symptoms of the diseased plants were indistinguishable from those infected with MRDV and cereal tillering disease virus (CTDV), the causal agent of cereal tillering disease, and reacted with antisera to MRDV. It seems likely that CTDV and MRDV may be the same virus or closely related strains of the same virus (Milne *et al.*, 1975; Milne and Luisoni, 1977). However, their exact relationship will remain obscure until CTDV is fully characterized.

In Japan, Kuribayashi and Shinkai (1952) recognized a serious disease of rice that was later found to be caused by rice black-streaked dwarf virus (RBSDV), which has been shown to be related to MRDV (Luisoni *et al.*, 1973, 1975). The virus has been transmitted experimentally to 25 species of Gramineae (Shinkai, 1962) and also occurs naturally in wheat and maize, causing serious diseases of these crops (Ishii and Yoshimura, 1973). The virus has also been reported from mainland China (Anonymous, 1974). Symptoms caused by RBSDV on common hosts are indistinguishable from those caused by MRDV and CTDV.

PSV has been observed to cause severe diseases in pastures of pangola grass (*Digitaria decumbens*), especially in South America, where it was first recorded (Dirven and van Hoof, 1960). The virus has a host range restricted to a dozen *Digitaria* spp. It is not known to infect any of the major hosts of MRDV, CTDV, or RBSDV, but is serologically related to MRDV (Milne and Luisoni, 1977; Boccardo *et al.*, 1979). The availability of *Digitaria* lines tolerant to the virus provides an approach to disease control (Hunkar *et al.*, 1974).

### c. Oat Sterile Dwarf Virus and Related Viruses

Diseased oats with similar symptoms were simultaneously observed in Czechoslovakia and Sweden (Průša, 1958; Průša *et al.*, 1959; Lindsten, 1959). The causal agent was shown to be a virus and named oat dwarf tillering virus in Sweden (Lindsten, 1961) and OSDV in Czechoslovakia (Brčák *et al.*, 1972). OSDV, the generally accepted name of the virus, can experimentally infect 30 species of Gramineae belonging mostly to the tribe Poaceae and has been reported to cause various degrees of damage to the infected crops in different countries (Boccardo and Milne, 1980a). Two viruses similar to OSDV, *Arrhenatherum* blue dwarf virus (ABDV) from eastern Germany (Mühle and Kempiak, 1971; Milne *et al.*, 1974) and *Lolium* enation virus (LEV) from western Germany (Huth, 1975; Lesemann and Huth, 1975), are serologically closely related to OSDV )Milne and Lesemann, 1978). Also, the genomes of OSDV and ABDV [see Fig. 1 (Section II.A)] have dsRNA segments of indistinguishable electrophoretic mobilities (Luisoni *et al.*, 1979). However, further work will be required to establish whether OSDV, ABDV, and LEV are identical or related strains of the same virus.

## B. Histopathology and Cytopathology of Infected Plants

Particles of the plant Reoviridae are large and characteristic enough to be detected and identified in infected cells by electron microscopy. Small variations reported in the particle size of different plant Reoviridae are probably a reflection more of differences in preparative and electron-microscopic techniques used than of any intrinsic size differences. Infected cells also contain large inclusion bodies detectable by light as well as electron microscopy; they are usually referred to as viroplasms (Milne and Lovisolo, 1977; Shikata, 1977, 1981), but earlier workers referred to them as x-bodies (McWhorter, 1922). The presence of these structures has made infected plants amenable to histological and cytological investigations using both light and electron microscopy (Fig. 13). These studies have been reviewed by Milne and Lovisolo (1977) and Shikata (1977, 1981).

The particles of all plant Reoviridae have a similar appearance in thin sections of infected cells (Fig. 14a). They are about 70 nm in diameter, with densely staining inner cores due to the presence of dsRNA (Hatta and Francki, 1978). Except for FDV, some of the 70-nm virus particles may be found enclosed in long tubules (Fig. 14b and c) of unknown origin and function. However, the majority of these particles are usually scattered in the cytoplasm or aggregated into crystalline inclusions (Fig. 14a). Again, FDV seems to be an exception in that no regularly arranged particle aggregates have been observed (Hatta and Francki, 1981).

Viroplasms in plant cells infected by all plant Reoviridae show a number of common features. They occur in the cytoplasm, vary in size, and consist of fibrillar and amorphous material unbounded by any membrane (Milne and Lovisolo, 1977; Shikata, 1977, 1981). Electron-microscopic observations led Shikata (1977, 1981) to conclude that viroplasms in cells infected with phytoreoviruses tend to be smaller and more spherical than those induced by fijiviruses. However, inspection of published light micrographs indicates that no significant differences exist, at least in the case of FDV- and RDV-infected cells (Kunkel, 1924; Fukushi, 1934).

It is generally considered that viroplasms are the sites of virus synthesis. Originally, the suggestion was made from cytological studies (Shikata and Maramorosch, 1967). However, much more convincing evidence comes from the work of Favali et al. (1974) and Favali and Lotti (1981), who showed by [³H]uridine labeling and autoradiography that RNA was synthesized in the viroplasms of MRDV- and OSDV-infected cells, respectively, and then incorporated into mature virus particles.

### 1. Phytoreoviruses

RDV differs from all other plant Reoviridae in that it does not induce neoplastic growths. RDV-infected plants are stunted and develop patches of chlorotic tissue like Gramineae infected by viruses belonging to a va-

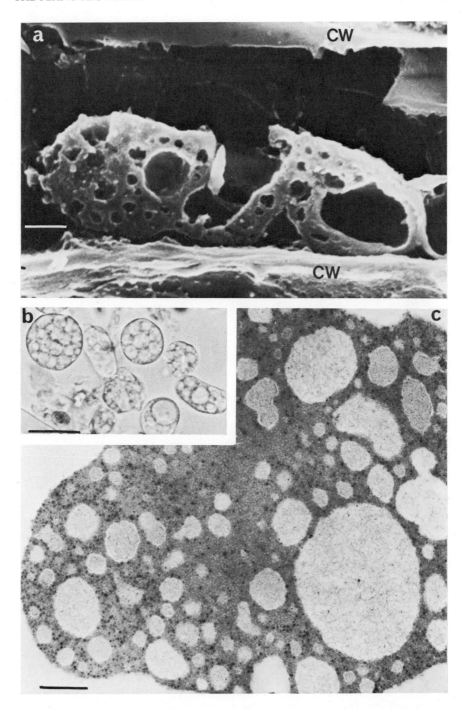

FIGURE 13. Viroplasms in leaf gall cells of sugarcane infected with FDV as seen by scanning electron microscopy (a), phase-contrast light microscopy (b), and transmission electron microscopy (c). (CW) Cell wall. Scale bars: (a,c) 1 μm; (b) 20 μm. Courtesy of Dr. T. Hatta.

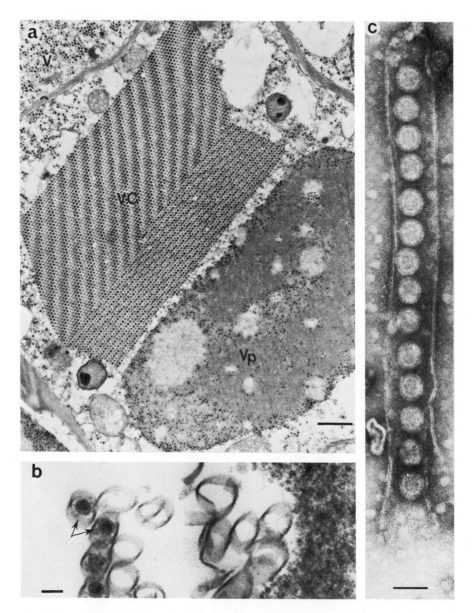

FIGURE 14. Cytopathological structures associated with the multiplication of some fiji-viruses. (a) A cell infected with MRDV, showing a large crystal of virus particles (VC) and a viroplasm (Vp); virus particles (V) are also seen scattered in the cytoplasm of an adjoining cell. (b) Characteristic cytoplasmic tubes cut obliquely in the cytoplasm of a cell infected with OSDV; some are empty, whereas others contain virus particles. (c) A tube similar to those seen in (b) but isolated from an MRDV-infected maize plant and negatively stained with phosphotungstate. Scale bars: (a) 1 μm; (b,c) 100 nm. Courtesy of Drs. A. Appiano (a) and R.G. Milne (b,c).

riety of taxonomic groups. The cells in the chlorotic areas contain virus particles and viroplasms (Fukushi, 1934; Fukushi et al., 1962; Hirai et al., 1964; Shikata, 1966).

WTV appears to be confined to neoplastic growths, which are produced irregularly as infection spreads systemically. Most of the studies have been on infected sweet clover (Melitotus alba) and M. officinalis, which develop severe symptoms and are easily propagated vegetatively. Stem tumors are sometimes produced, and their number can be greatly increased by artificial wounding, hence the name of the virus (L.M. Black, 1965, 1970, 1972). Root tumors appear to originate at the bases of emerging lateral roots and contain scattered zones of cells surrounded by numerous abnormal phloemlike cells and some distorted tracheids (Littau and Black, 1952). Many of the tumor cells were seen to contain inclusions by light microscopy (Littau and Black, 1952). These are probably the same structures found to stain by fluorescent antibodies to WTV (Nagaraj and Black, 1961) and identified in electron micrographs as viroplasms. The cells containing viroplasms were also shown to contain numerous virus particles (Shikata and Maramorosch, 1965). RGDV particles have been detected in phloem cells of diseased rice (Omura et al., 1980).

The fine structure of plant cells infected with the phytoreoviruses is not well known. Although RDV and WTV were among the first viruses with isometric particles to be studied in infected plant cells (Fukushi et al., 1962; Shikata et al., 1964), the electron-microscopic techniques then available were still rather primitive. Hence, the published micrographs are not very informative, and it would seem that the cytopathology of all three phytoreoviruses could well be reinvestigated by the more sophisticated methods now available.

### 2. Fijiviruses

The neoplastic growths induced by all the fijiviruses appear to be histologically similar (Hatta et al., 1982). The neoplastic tissues consist principally of modified phloem cells, although some deformed tracheids can also be detected, especially in the galls on sugarcane infected with FDV (Hatta and Francki, 1976). The virus appears to be confined to the gall tissues (Hatta and Francki, 1976; Rohozinski et al., 1981). It has been shown that leaf meristems in FDV-infected sugarcane supporting virus synthesis (Hatta and Francki, 1981) proliferate and differentiate into the abnormal cells of leaf galls [see Fig. 12b (Section IV.A.1.a)]. No dividing cells were ever observed in galls on expanded green leaves. It seems that FDV can replicate only in very young cells, transforming them into neoplastic tissue; however, no information is available on the mechanism of this process. It was also observed that the virus reached the maximum concentration in infected cells at about the same time as the leaf tissue started to green, and it decreased thereafter (Hatta and Francki, 1981).

MRDV and OSDV have been observed to induce cell-wall prolifer-ations, sometimes enclosing virus particles (Bassi et al., 1974; Appiano and Lovisolo, 1979; Brčák and Králík, 1980; Favali and Lotti, 1981). Elec-tron microscopy coupled with energy-dispersion spectrometry has shown that MRDV-induced enations contain less Ca and Si but more P, K, and S than normal tissues (Favali et al., 1980). However, the significance of these observations remains to be elucidated.

Plant cells infected by fijiviruses all contain very similar viroplasms, and earlier observations on them have been reviewed by Milne and Lov-isolo (1977) and Martelli and Russo (1977). Viroplasms induced by fiji-viruses appear to differ from those induced by phytoreoviruses in that they are differentiated into two distinct zones when viewed in thin sec-tions. However, this difference may only be apparent, and further studies on phytoreovirus-induced cytopathology may help to elucidate this point. Lighter-staining zones consist of intertwined strands about 7 nm wide when observed in thin sections. The darker-staining zones contain elec-tron-dense particles, 30–35 nm in diameter, and less heavily stained ring structures, about 55 nm in diameter (Giannotti and Milne, 1977; Hatta and Francki, 1981). Hatta and Francki (1981) have studied the fine struc-ture of viroplasms in FDV-infected cells by light and electron microscopy in conjunction with cytochemical experiments. They confirmed previous observations that the viroplasms consist mainly of protein and RNA (Giannotti and Monsarrat, 1968; Bassi and Favali, 1972). Freeze–etching studies indicate that the 7-nm-wide filaments seen in thin sections are helical filaments about 13 nm in diameter. They are similar to the fila-ments seen in uranyl-acetate-stained crude extracts of gall tissues from plants infected by several fijiviruses (Milne, 1977b) and in partially pur-ified preparations of FDV subviral particles (SVPs) (Hatta and Francki, 1977). The suggestion (Hatta and Francki, 1977) that the helical filaments may be aggregated protein subunits of FDV particle outer shells now seems untenable. The darkly stained zones of viroplasms viewed in thin sections of FDV-infected cells appear to contain the bulk of the viroplasm-associated RNA, which was identified as dsRNA (Hatta and Francki, 1981). It was also concluded that these zones contained strands similar to those in the lighter-staining zones. However, among them were nu-merous SVPs, some of which contained dsRNA and others of which did not. These structures correspond to the dense cores and ringlike struc-tures, respectively, as seen in thin sections.

The work of Hatta and Francki (1981) indicates that although much more biochemical work is needed to elucidate the mechanisms of fijivirus synthesis and assembly, the viroplasms must play a central role in these processes, as suggested earlier by Favali et al. (1974). Hatta and Francki (1981) tentatively suggested that the empty and RNA-containing SVPs in viroplasms are virus particles in the process of assembly. The high concentration of ribosomes in the cytoplasm adjacent to these zones may be involved in the synthesis of virus-specified polypeptides. The large number of complete virus particles in the cytoplasm and their absence in viroplasms suggests that they may be released immediately after assembly. Alternatively, the final stages of assembly may take place at the periphery of the lighter-stained viroplasm zones. The function of the strands that make up the bulk of their mass is obscure, but it is possible that they may play a key role in the assembly of the virus particles.

FDV-infected cells differ from those infected by all the other fijiviruses in that tubules containing virus particles or accumulations of large crystalline aggregates have not been observed. There appears to be disagreement about the cytopathic effects of OSDV in that ordered arrays of particles located in tubules have been observed by some authors (Boccardo and Milne, 1980a), but could not be detected by others (Brčák and Králík, 1980). The tubules are very characteristic and prominent features of cells infected by the majority of fijiviruses, but their function and significance remain obscure. In some instances, the tubules may be incompletely closed, thus forming scrolls (Martelli and Russo, 1977; Milne and Lovisolo, 1977; Giannotti and Milne, 1977; Shikata and Kitagawa, 1977; Appiano and Lovisolo, 1979; Milne, 1980; Brčák and Králík, 1980). The tubules in MRDV-infected cells have been shown by cytochemical methods to consist of protein (Bassi and Favali, 1972). In transverse section, their walls appear to consist of three dense layers sandwiching two light areas, and their surfaces in uranyl-acetate-stained preparations appear to have a square lattice repeating every 4 nm (Milne and Lovisolo, 1977; Milne, 1980). The tubules from MRDV- and OSDV-infected cells failed to react with their respective anti-SVP sera (Milne and Lovisolo, 1977; Boccardo and Milne, 1980a). It remains to be determined whether the tubule protein or proteins are virus- or host-specified.

## V. VIRUS–VECTOR RELATIONSHIPS

All the Reoviridae that infect plants are transmitted by their vectors in a persistent (propagative) manner and multiply in them. The duration of the acquisition, incubation (latent), and inoculation periods appears to vary with the virus, insect, and environmental conditions (Maramorosch, 1950; Klein and Harpaz, 1970). It seems that nymphs of all instars and adults are capable of virus transmission, although the relative efficiencies can vary (Harris, 1979).

Although the viruses multiply in their vectors, no neoplastic growths like those in plants are produced. However, cytopathological effects similar to those induced in plant cells have been observed in all the organs of the various vectors, with the exception of the testes (Harris, 1979). The multiplication of phytoreoviruses has been studied in intact insects (Sinha, 1968; Maramorosch et al., 1969), as well as in vector-cell monolayers (L.M. Black, 1979), wound tumor virus (WTV) being the most extensively studied. As yet, no planthopper cells have been cultured *in vitro,* and only the multiplication of maize rough dwarf virus (MRDV) in its vector appears to have been studied in any detail (Vidano, 1970). Phytoreoviruses and fijiviruses do not appear to have easily recognizable effects on their insect hosts, and the pertaining observations are variable. For example, Fukushi (1969) observed that rice dwarf virus (RDV) infection significantly reduced the fecundity of the leafhoppers, whereas Maramorosch (1975) reported that WTV infection did not affect either the fecundity or the longevity of its vector. It has also been observed that MRDV-infected planthoppers laid fewer eggs with impaired viability (Harpaz, 1972).

## A. Virus Multiplication in Insects

### 1. Phytoreoviruses

Demonstration of WTV multiplication in an insect vector was first achieved by L.M. Black and Brakke (1952) by serial injections of infected *Agallia constricta* extracts into virus-free leafhoppers through seven passages. Titration of virus at each passage showed that there was no WTV titer decrease with each passage when the insects were maintained on plants immune to WTV. Further evidence was provided by Whitcomb and Black (1961), who demonstrated an increase of virus titer in injected leafhoppers by serological assays. Confirmatory evidence was also obtained using infectivity assay by means of injection of vectors (Reddy and Black, 1966), by virus-particle counts (Gamez and Black, 1968), and finally and most precisely by bioassay on vector-cell monolayers (Reddy and Black, 1972).

Sinha (1965) used the fluorescent-antibody technique to detect WTV antigens in various organs of infected leafhoppers and to follow the sequence in which the tissues were infected. Following 1 day's aquisition feeding on infected plants, WTV antigens were detected after 4 days in one corner of the filter chamber; after 7 days, in the entire filter chamber and part of the ventriculum; after 12 days, in the hemolymph; after 14 days, in the fat body, brain, and Malpighian tubules; and in the salivary glands only after the incubation period of 17 days, that is, at the start of transmissions. These results indicate that the primary site of infection is the intestinal filter chamber, from whence the virus spreads systemically. Electron-microscopic examination of thin sections of WTV-infected insects has confirmed the systemic spread of the virus (Maramorosch *et al.,* 1969; Maramorosch, 1975). The cytopathological changes in infected vector cells resembled those observed in plant cells (see Section IV.B.1) (Shikata and Maramorosch, 1967).

Studies on the multiplication of RDV in insects have produced a picture very similar to that of WTV. The initial indication that RDV multiplies in its vector came from the painstaking experiments done by

Fukushi (1940). He demonstrated that *Nephotettix cincticeps* was capable of carrying virus transovarially for up to five generations, starting with insects from individual infected eggs. The insects did not have access to a source of virus from plants, since they were individually transferred daily to new rice plants. Fukushi (1940) argued that since the amount of virus in a single egg must be very small, multiplication must have occurred to explain his results. This conclusion was confirmed by insect injection experiments (Fukushi and Kimura, 1959). Electron-microscopic observations on thin sections of RDV-infected leafhoppers and their eggs also support the conclusion that the virus replicates in insects, inducing cytopathological effects similar to those caused by WTV (Fukushi *et al.*, 1960, 1962; Fukushi and Shikata, 1963; Nasu, 1965). The detailed work of Nasu (1965) has provided some insight into the events that lead to infection of the eggs.

## 2. Fijiviruses

Electron-microscopic observations on thin sections of viruliferous insects provide the only evidence that fijiviruses such as Fiji disease virus (Francki and Grivell, 1972), MRDV (Gerola *et al.*, 1966; Vidano, 1970), rice black-streaked dwarf virus (Shikata and Kitagawa, 1977), pangola stunt virus (Kitajima and Costa, 1970), and oat sterile dwarf virus (Brčák *et al.*, 1970) replicate in most tissues of their planthopper vectors. The cytopathological effects induced by the viruses appear very similar to those produced in plant cells (see Section IV.B.2). However, there are some minor differences such as the formation of crystalline aggregates of FDV particles in insect vectors that are absent in plant cells (Francki and Grivell, 1972).

Vidano (1970) proposed a sequence of events involving eight phases during the invasion of *Laodelphax striatellus* by MRDV. However, he observed that the insects also contained bacilliform viruslike particles, which could mean that they were infected by a second virus. Hence, it is difficult to be certain which of the observed pathological effects were caused by MRDV.

## B. Virus Multiplication in Insect Cell Cultures

There are obvious technical problems in studying the multiplication of plant Reoviridae at the biochemical level in whole insects. Similarly, many approaches to such studies in plant cells are frustrated by the fact that the only method of infecting plants is by vectors. Attempts to establish WTV in plant tissue cultures failed to provide a suitable system for meaningful experiments, since the virus did not replicate and was soon eliminated from the cells (Streissle and Maramorosch, 1969). However, the pioneering work of Black and his colleagues opened up new avenues for studying the replication of phytoreoviruses by establishing vector-cell monolayer cultures that can be grown indefinitely. Their work centered on cultures of *Agallia constricta* cells that could be infected with WTV (Chiu and Black, 1967; L.M. Black, 1969, 1979). Previously, Chiu *et al.* (1966) and Mitsuhashi and Nasu (1967) had been able to inoculate primary cultures of leafhopper cells with WTV and RDV, respectively, and to show that the viruses multipled in them.

It has been observed that WTV can multiply in cell lines from some leafhoppers that are not vectors (Hirumi and Maramorosch, 1967; Chiu and Black, 1969). It may be possible to exploit this phenomenon for investigating the mechanism of virus-vector specificity. The techniques used in culturing leafhopper vector cells in monolayers have been reviewed in some detail (Reddy, 1977; L.M. Black, 1979; Maramorosch, 1979).

Phase-contrast microscopy failed to detect any differences between normal and WTV-infected *A. constricta* cells in monolayers. Chiu and Black (1967) used fluorescein-conjugated antibodies to detect infected cells. Using this technique, they established a linear relationship between virus concentration in the inoculum and the number of infection foci in the monolayers (Chiu and Black, 1969). Kimura and Black (1971) defined the optimal conditions for inoculating the monolayers and later demonstrated that only 3–6 virus particles were required for establishing each focus of infection. This approaches the value for the theoretical infection unit of 1 virus particle (Kimura and Black, 1972). These experiments point to the dramatic difference in sensitivity of this assay method compared to the ones used for assaying most plant viruses. For example, it has been estimated that between $10^4$ and $10^5$ tobacco mosaic virus particles must be inoculated into a local-lesion host plant to produce one lesion (Matthews, 1981).

It has been observed that prolonged maintenance of WTV in insect cell cultures can cause the virus to lose its ability to infect plant hosts (Martinez-Lopez and Black, 1979). This appears to be a phenomenon similar to that observed when the virus is grown in plants propagated vegetatively, resulting in the loss of infection toward insects (see Section IV.A.1.a).

Nuss and Peterson (1980) have made use of the *A. constricta* cell monolayer system for investigating the synthesis of WTV-specified polypeptides. Using [$^{35}$S]methionine as label, they detected 12 such polypeptides that corresponded well to those expected to be encoded by each of the virus genome segments. The polypeptides also had the same electrophoretic mobilities as those synthesized *in vitro* by the messenger RNAs transcribed from the viral genome by the virus-associated transcriptase. Six and possibly seven of the polypeptides were identified as virus capsid proteins. The function of the remaining five needs to be determined.

Peterson and Nuss (1982) observed that WTV infection does not appear to affect the synthesis of normal cellular proteins in vector-cell monolayers. The cells synthesized virus-specified polypeptides at a high rate between 8 and 60 hr after infection. Thereafter, the rate of synthesis declined and continued very slowly. This synthesis was persistent and continued at the low rate even after extensive subculturing. WTV was detected with fluorescein-conjugated antibodies in more than 95% of the persistently infected subcultured cells. The virus-specific polypeptide synthesis rate could not be stimulated in such cells by superinfection.

Cells supporting WTV multiplication never show any obvious effects of infection other than the appearance of viroplasms and virus particles in their cytoplasm (L.M. Black, 1979). The mechanism by which the synthesis of WTV is regulated remains to be elucidated. It would seem that a similar regulatory mechanism may operate in WTV-infected whole insects. It has been observed that the amount of WTV increases continuously for 3–4 weeks after acquisition by feeding and remains more or less constant thereafter, without inducing any obvious pathological effects (Reddy and Black, 1972). It is not known whether the apparently regulatory mechanism is host- or virus-specified. However, it is clearly of great advantage to the survival of the virus, since it ensures that the vehicle by which it is transferred from plant to plant is not in any way harmed.

## VI. CLASSIFICATION OF AND POSSIBLE RELATIONSHIPS AMONG THE REOVIRIDAE

### A. Classification Problems

The majority of viruses with isometric particles and multisegmented double-stranded RNA (dsRNA) genomes have been assigned to the six genera of the family Reoviridae approved by the International Committee on Taxonomy of Viruses (Matthews, 1982). Their host ranges, physical properties, and cytopathological effects are summarized in Table V. According to the accepted criteria, however, there are some viruses the taxonomic position of which is not clear. For example, a virus such as the one detected in the leafhopper *Cicadulina bimaculata*, leafhopper A virus (LAV), the physical properties and induced cytopathological effects of which are such as to qualify it for inclusion in the genus *Fijivirus*, does not appear to be capable of infecting plants (Boccardo *et al.*, 1980). A second virus that seems not to fit into any of the currently recognized genera is rice ragged stunt virus (RRSV). It infects rice plants, producing symptoms reminiscent of those of Gramineae infected with fijiviruses, and it is vectored by the planthopper *Nilaparvata lugens*. However, the structure of the virus particles, which contain only 8 dsRNA segments (Boccardo and Milne, 1980b; Milne, 1981), precludes its assignment to the genus *Fijivirus* or to any of the other currently accepted genera [see Table I (Section II) and Table V]. The housefly virus described by Moussa (1978, 1980, 1981) presents yet another taxonomic enigma by virtue of its unusual structure.

TABLE V. Biological, Structural, and Cytopathological Properties of the Reoviridae

| Properties | Genus | | | | | |
|---|---|---|---|---|---|---|
| | Reo-virus | Orbi-virus | Rota-virus | Cypo-virus | Phytoreo-virus | Fiji-virus |
| Hosts[a] | | | | | | |
| Vertebrates | + | + | + | − | − | − |
| Insects | − | + | − | +[q] | + | + |
| Plants | − | − | − | − | + | + |
| Particle structure | | | | | | |
| Diameter (nm) | | | | | | |
| Intact particles | 76.5[b] | 69[c] | 75[d] | 60–65[e] | 73[f] | 67–70[g] |
| Core or subviral particles | 52[b] | 63[c] | 65[d] | ?[e] | 59[f] | 55[g] |
| Outer shell | + | + | + | − | + | +[g] |
| Spikes | | | | | | |
| A | − | − | − | + | − | +[g] |
| B | + | − | − | + | − | +[g] |
| Number | | | | | | |
| RNA segments | 10[h] | 10[i] | 10[j] | 10[e] | 12[k] | 10[l] |
| Viral proteins | 9[h] | 7[i] | 8[m] | 5[e] | 7[f] | 6(?)[n] |
| Cytopathology | | | | | | |
| Viroplasms | +[o] | +[o] | +[p] | +[e,r] | + | + |
| Tubules | +[o] | +[o] | +[p] | − | + | +[s] |
| Matrix protein | − | − | − | +[e] | − | − |

[a–p] Data from: [a] Matthews (1982); [b] Luftig et al. (1972); [c] Martin and Zweerink (1972); [d] Esparza and Gil (1980); [e] Payne and Harrap (1977); [f] Reddy and MacLeod (1976); [g] Milne et al. (1973) and Hatta and Francki (1977); [h] McCrea and Joklik (1978); [i] Verwoerd et al. (1972); [j] Obijieski et al. (1977); [k] Reddy et al. (1974); [l] Reddy et al. (1975a,b), Luisoni et al. (1979), and Boccardo et al. (1979); [m] Matsuno and Mukoyama (1979); [n] Boccardo and Milne (1975); [o] Joklik (1974a); [p] Holmes et al. (1975).
[q] Aside from various species of insects, one virus of the genus has been isolated from the crustacean Simocephalus expinosus (Payne and Harrap, 1977).
[r] Referred to as "virogenic stroma" (Payne and Harrap, 1977).
[s] Except for the type member, Fiji disease virus (see Section IV.B.2).

The problem of our inability to assign newly described viruses, obviously belonging to the Reoviridae, to existing genera is unlikely to abate in the future. Already, there are a number of reports of viruslike particles in cells of various insect species that have not been characterised and that may or may not be involved in diseases of plants (see Section II.D). Their structure suggests that they will be shown to belong to the Reoviridae, but it seems likely that at least some of them will be difficult to accomodate in the existing genera (Table V).

## B. Possible Relationships among Existing Genera

### 1. Reoviridae That Infect Plants

Although both *Phytoreovirus* and *Fijivirus* particles conform to the general Reoviridae structural plan of a double-shelled capsid about 70 nm in diameter containing segmented dsRNA, the details are quite different

(Table V and Section III.C). *Fijivirus* particles have A spikes on their outer capsid shells and B spikes on their subviral particles (SVPs). *Phytoreovirus* particles apparently have no projections on their surfaces, and although some of the capsid proteins may be removed by various treatments (see Section III.A.1), there is no evidence of a B-spiked SVP. However, it must be remembered that these comparisons are based on micrographs from various laboratories in which sometimes quite different preparative and electron-microscopic techniques have been used. Truly informative data will be obtained only from comparative studies using exactly the same techniques, preferably done in the same laboratory. At present, we are aware of only two such studies, where the structure of *Reovirus* particles has been compared to those of wound tumor virus (WTV) and Fiji disease virus (FDV) by Streissle and Granados (1968) and Hatta and Francki (1977), respectively.

The genomes of phytoreoviruses and fijiviruses differ in the number of dsRNA segments [see Fig. 1 (Section II.A)]. However, this does not necessarily indicate remote relationship, since it has been demonstrated that WTV can lose as many as two genome segments and still retain infectivity (see Section III.B.1.a).

Biological properties of viruses belonging to the two genera also show some differences. Whereas the phytoreoviruses are all transmitted by leafhoppers, all fijiviruses have planthopper vectors. The significance of this is not clear and may be only of minor importance in determining the degree of relationship of the viruses. The properties of some plant viruses in other taxa suggest that vector specificity may not be a good indicator of relationship. For example, within the Geminivirus group, some members are transmitted by leafhoppers and others by whiteflies (Goodman, 1981). The range of hosts infected by the three different phytoreoviruses and the symptoms they induce are quite different (see Section IV.A.1). On the other hand, the type of host and symptoms induced are common for all recognized fijiviruses (see Section IV.A.2). However, "typical" *Fijivirus* symptoms are also characteristic of rice gall dwarf virus and RRSV infections. Therefore, it would seem unlikely that any conclusions about the relationships of the plant Reoviridae can be implied from the plant hosts they infect and the symptoms they induce.

## 2. Reoviridae That Infect Insects

Four of the six Reoviridae genera include viruses that multiply in insects. However, only members of the genus *Cypovirus* are pathogenic to insects and have no alternative plant or vertebrate hosts (Matthews, 1982). The other three genera of insect-infecting viruses also have vertebrate (*Orbivirus*) or plant (*Phytoreovirus* and *Fijivirus*) hosts, and are usually referred to as vertebrate or plant viruses, a designation that is not completely logical. Their ability to replicate in insect hosts without causing any indication of disease suggests a long association and may be indicative of evolution from common insect-infecting ancestors. Although viruses within the various Reoviridae genera display significant differences, they share many basic properties. Their architecture and modes of replication are essentially similar (Table V) and so distinctive when compared to those of any other virus that it seems extremely unlikely that any of the Reoviridae originated independently (Joklik, 1974b). If their evolutionary root indeed reaches insect-infecting ancestors, some probably aquired the ability to infect vertebrates or plants relatively recently. In the case of *Reovirus* and *Rotavirus*, this may have been followed by loss of their capacity to infect insects. Such an event is not difficult to envisage, since it has been demonstrated experimentally that continuous passage of WTV through only one of its alternate hosts may result in the virus losing its ability to replicate in the other (see Sections IV.A.1.a and V.B).

A casual glance suggests that the cypoviruses differ greatly from all other Reoviridae in their particle structure (Table V). However, recent studies have revealed that the particle structure of the *Bombyx mori* cytoplasmic polyhedrosis virus (CPV) is remarkably similar to that of the SVPs of FDV and LAV (Hatta and Francki, 1982). It seems that the only significant structural difference between CPV and LAV and FDV is that CPV lacks an outer capsid shell (Fig. 15). This supports the view that cypoviruses and fijiviruses may be closely related, having evolved from a common ancestor relatively recently. This putative ancestral stock may have developed along two divergent lines. One of the virus-specified polypeptides may have evolved to become either the protein of the polyhedra (*Cypovirus*) or the outer shell coat protein (*Fijivirus* and LAV). Both these types of proteins appear to have the same function, that of protecting the genome transcriptional enzyme complex. This suggestion finds some support from data on the protein composition of these viruses. It appears that *Fijivirus* and *Cypovirus* particles consist of six and five structural proteins, respectively, but the CPV genome also encodes for the protein of the polyhedra (Table V) (Harrap and Payne, 1979).

Judging from the similarity of their particles, it may well be that viruses such as LAV are closely related to the fijiviruses. LAV and similar

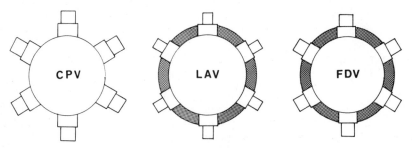

FIGURE 15. Comparative morphology of some plant and insect Reoviridae, CPV, LAV, and FDV [see the more detailed three-dimensional model of FDV in Fig. 6 (Section III.C.2)].

insect viruses may never have aquired the ability to infect plants. Alternatively, they may have aquired and later lost it.

ACKNOWLEDGMENTS. We thank Drs. L.M. Black and R.G. Milne for reading the manuscript and offering many critical suggestions; Drs. A. Appiano, L.M. Black, T. Hatta, I. Kimura, and R.G. Milne for supplying material for illustrations; Drs. L.M. Black, M.A. Favali, H. Hibino, H. Inone, I. Kimura, M. Lotti, R.E.F. Matthews, D.L. Nuss, T. Omura, and A.J. Peterson for sending us manuscripts prior to publication; Mrs. L. Wichman for art work; and Miss L. Castree and Mr. C.J. Grivell for checking the manuscript. We are indebted to all these colleagues for help with the preparation of the manuscript, but the responsibility for any errors of fact or interpretation must be solely ours. The work in the senior author's laboratory was supported by generous grants from the Australian Research Grants Committee, the Rural Credits Development Fund of the Reserve Bank of Australia, the Bureau of Sugar Experiment Stations, Brisbane, and a Special Research Grant of the Australian Commonwealth Department of Primary Industry. The junior author was also the recipient of travel grants from the C.N.R. Progelto Finalizzato Virus, Rome.

# REFERENCES

Alatortsev, V.E., Ananiev, E.V., Gushchina, E.A., Grigoriev, V.B., and Gushchin, B.V., 1981, A virus of the Reoviridae in established cell lines of *Drosophila melanogaster*, *J. Gen. Virol.* **54**:23.

Anonymous, 1974, Studies on the pathogenesis of rice black-streaked dwarf disease, *Sci. Sin.* **17**(2):273.

Anonymous, 1978, The pathogens of some virus diseases of cereals in China. IV. Serological determination of the percentage of active individuals transmitting rice dwarf disease in a population of insect vectors, *Acta Biochim. Biophys. Sin.* **10**:357.

Antoine, R., 1959, Cane diseases, *Mauritius Sugar Ind. Res. Inst. Annu. Rep.* **59**.

Appiano, A., and Lovisolo, O., 1979, Ultrastructure of maize roots infected with maize rough dwarf virus and presence of virus particles in vacuoles with lysosomal activity, *Microbiologica* **2**:37.

Bassi, M., Favali, M.A., 1972, Electron microscopy of maize rough dwarf virus assembly sites in maize: Cytochemical and autoradiographic observations, *J. Gen. Virol.* **16**:153.

Bassi, M., Favali, M.A., and Appiano, A., 1974, An autoradiographic study of the cell wall modifications induced by maize rough dwarf virus in the host cells, *Riv. Patol. Veg.* **10**:19.

Bawden, F.C., 1939, *Plant Viruses and Virus Diseases*, Chronica Botanica, Leiden, 272 pp.

Behncken, G.M., and Gowenlock, D.H., 1976, Association of a bacterium-like organism with rugose leaf curl disease of clover, *Aust. J. Biol. Sci.* **29**:137.

Bills, R.F., and Hall, C.E., 1962, Electron microscopy of wound tumor virus, *Virology* **17**:123.

Biraghi, A., 1949, Reperti istologici su piante di mais affette da "nanismo," *Notiz. Mal. Piante* **7**:1.

Black, D.R., and Knight, C.A., 1970, Ribonucleic acid transcriptase activity in purified wound tumor virus, *J. Virol.* **6**:194.

Black, D.R., Eckstein, F., Hobbs, J.B., Sternbach, H., and Merigan, T.C., 1972, The antiviral activity of certain thiophosphate and 2'-chloro substituted polynucleotide homopolymer duplexes, *Virology* **48**:537.

Black, L.M., 1944, Some viruses transmitted by agallian leafhoppers, *Proc. Am. Philos. Soc.* **88**:132.

Black, L.M., 1945, A virus tumor disease of plants, *Am. J. Bot.* **32**:408.

Black, L.M., 1959, Biological cycles of plant viruses in insect vectors, in: *The Viruses*, Vol. 2, (F.M. Burnet and W.M. Stanley, eds.), pp. 157–185, Academic Press, New York.

Black, L.M., 1965, Physiology of virus-induced tumors in plants, in: *Handbuch der Pflanzenphysiologie*, Vol. 15, part 2 (W. Ruhland *et al.*, eds.), pp. 236–266, Springer–Verlag, Berlin.

Black, L.M., 1969, Insect tissue cultures as tools in plant virus research, *Annu. Rev. Phytopathol.* **7**:73.

Black, L.M., 1970, Wound tumor virus, *C.M.I./A.A.B. Descriptions of Plant Viruses*, No. 34.

Black, L.M., 1972, Plant tumors of viral origin, *Prog. Exp. Tumor Res.* **15**:110.

Black, L.M., 1979, Vector cell monolayers and plant viruses, *Adv. Virus Res.* **25**:191.

Black, L.M., 1982, Wound tumor disease, in: *Molecular Biology of Plant Tumors* (G. Kahl and J. Schell, eds.), pp. 69–105, Academic Press, New York.

Black, L.M., and Brakke, M.K., 1952, Multiplication of wound-tumor virus in an insect vector, *Phytopathology* **42**:269.

Black, L.M., and Brakke, M.K., 1954, Serological reactions of a plant virus transmitted by leafhoppers, *Phytopathology* **44**:482.

Black, L.M., and Markham, R., 1963, Base paring in the ribonucleic acid of wound tumor virus, *Neth. J. Plant Pathol.* **69**:215.

Black, L.M., Reddy, D.V.R., and Reichmann, M.E., 1967, Virus inactivation by moderate forces during quasi-equilibrium zonal density gradient centrifugation, *Virology* **31**:713.

Boccardo, G., and Milne, R.G., 1975, The maize rough dwarf virion. I. Protein composition and distribution of RNA in different viral fractions, *Virology* **68**:79.

Boccardo, G., and Milne, R.G., 1980a, Oat sterile dwarf virus, *C.M.I./A.A.B. Descriptions of Plant Viruses*, No. 217.

Boccardo, G., and Milne, R.G., 1980b, Electrophoretic fractionation of the double-stranded RNA genome of rice ragged stunt virus, *Intervirology* **14**:57.

Boccardo, G., and Milne, R.G., 1981, Enhancement of the immunogenicity of the maize rough dwarf virus outer shell using the cross-linking reagent dithiobis (succinimidyl) proprinate, *J. Virol. Methods* **3**:109.

Boccardo, G., Milne, R.G., and Luisoni, E., 1979, Purification, serology and nucleic acid of pangola stunt virus subviral particles, *J. Gen. Virol.* **45**:659.

Boccardo, G., Hatta, T., Francki, R.I.B., and Grivell, C.J., 1980, Purification and some properties of reovirus-like particles from leafhoppers and their possible involvement in wallaby ear disease of maize, *Virology* **100**:300.

Bradfute, O.E., Teyssandier, E., Marino, E., and Dodd, J.L., 1981, Reolike virus associated with maize Rio Cuarto disease in Argentina, *Phytopathology* **71**:205.

Brakke, M.K., Vatter, A.E., and Black, L.M., 1954, Size and shape of wound-tumor virus, *Brookhaven Symp. Biol.* No. 6, p. 137.

Brčák, J., and Králík, O., 1980, Cytopathic effects of oat sterile dwarf virus in enation cells of oat leaves, *Acta Virol.* **24**:346.

Brčák, J., Králík, O., and Vacke, J., 1970, Virions of the oat sterile dwarf virus in the midgut cells of its vector, *Javesella pellucida*, *Int. Symp. Plant Pathol.*, New Dehli, Abstract 93.

Brčák, J., Králík, O., and Vacke, J., 1972, Virus origin of the oat sterile dwarf disease, *Biol. Plant.* **14**:302.

Chang, V.C.S., 1977, Transovarial transmission of the Fiji disease virus in *Perkinsiella saccharicida* Kirk, *Sugarcane Pathol. Newsl.* **18**:22.

Chiu, R.-J., and Black, L.M., 1967, Monolayer cultures of insect cell lines and their inoculation with a plant virus, *Nature (London)* **215**:1076.

Chiu, R.-J., and Black, L.M., 1969, Assay of wound tumor virus by the fluorescent cell counting technique, *Virology* **37**:667.

Chiu, R.-J., Reddy, D.V.R., and Black, L.M., 1966, Inoculation and infection of leafhopper tissue cultures with a plant virus, *Virology* **30**:562.

Conti, M., 1972, Investigations on the epidemiology of maize rough dwarf virus. I. Overwintering of virus in its planthopper vector, *Acta III Congr. Un. Fitopat. Medit., Oeiras 22–28 Outubro 1972*, 11.

Conti, M., and Milne, R.G., 1977, Some new natural hosts of maize rough dwarf virus (MRDV), *Ann. Phytopathol.* **9**:255.

Dirven, J.G.P., and van Hoof, H.A., 1960, A destructive virus disease of pangolagrass, *Tijdschr. Plantenziekten* **66**:344.

Egan, B.T., Ryan, C.C., and Francki, R.I.B., 1983, Fiji disease, in: *Diseases of Sugarcane* (A.G. Gillaspie, C. Ricaud, and B.T. Egan, eds.), Elsevier, Amsterdam (in press).

Esparza, J., and Gil, F., 1978, A study on the ultrastructure of human rotavirus, *Virology* **91**:141.

Favali, M.A., and Lotti, M., 1981, Oat leaves infected with oat sterile dwarf virus (OSDV): An ultrastructural and autoradiographic study, *Microbiologica* **4**:435.

Favali, M.A., Bassi, M., and Appiano, A., 1974, Synthesis and migration of maize rough dwarf virus in the host cell: An autoradiographic study, *J. Gen. Virol.* **24**:563.

Favali, M.A., Barbieri, N., Bianchi, A., Bonecchi, R., and Conti, M., 1980, X-ray microanalysis of leaf tumors from maize plant experimentally infected with maize rough dwarf virus: Scanning and transmission electron microscopic study, *Virology* **103**:357.

Fenaroli, L., 1949, Il "nanismo" del mais, *Notiz. Mal. Piante* **3**:38.

Francki, R.I.B., 1972, Purification of viruses, in: *Principles and Techniques in Plant Virology* (C. Kado and H.O. Agrawal, eds.), pp. 295–335, Van Nostrand Reinhold, New York.

Francki, R.I.B., and Grivell, C.J., 1972, Occurrence of similar particles in Fiji disease virus-infected sugar cane and insect vector cells, *Virology* **48**:305.

Francki, R.I.B., and Jackson, A.O., 1972, Immunochemical detection of double-stranded ribonucleic acid in leaves of sugar cane infected with Fiji disease virus, *Virology* **48**:275.

Fujii-Kawata, I., Miura, K.-I., and Fuke, M., 1970, Segments of genomes of viruses containing double-stranded ribonucleic acid, *J. Mol. Biol.* **51**:247.

Fukushi, T., 1931, On the intracellular bodies associated with dwarf disease of rice plant, *Trans. Sapporo Nat. Hist. Soc.* **12**:35.

Fukushi, T., 1933, Transmission of the virus through the egg of an insect vector, *Proc. Imp. Acad. (Tokyo)* **9**:457.

Fukushi, T., 1934, Studies on the dwarf disease of rice plants, *J. Fac. Agric. Hokkaido Imp. Univ.* **37**:41.

Fukushi, T., 1940, Further studies on the dwarf disease of rice plant, *J. Fac. Agric. Hokkaido Imp. Univ.* **45**:83.

Fukushi, T., 1969, Relationships between propagative rice viruses and their vectors, in: *Viruses, Vectors and Vegetation* (K. Maramorosch, ed.), pp. 279–301, Interscience, New York.

Fukushi, T., and Kimura, I., 1959, On some properties of the rice dwarf virus, *Proc. Jpn. Acad.* **35**:482.

Fukushi, T., and Shikata, E., 1963, Localization of rice dwarf virus in its insect vector, *Virology* **21**:503.

Fukushi, T., Shikata, E., Kimura, I., and Nemoto, M., 1960, Electron microscopic studies on the rice dwarf virus, *Proc. Jpn. Acad.* **36**:352.

Fukushi, T., Shikata, E., and Kimura, I., 1962, Some morphological characters of rice dwarf virus, *Virology* **18**:192.

Gamez, R., and Black, L.M., 1967, Application of particle-counting to a leafhopper-borne virus, *Nature (London)* **215**:173.

Gamez, R., and Black, L.M., 1968, Particle counts of wound tumor virus during its peak concentration in leafhoppers, *Virology* **34**:444.

Gamez, R., Black, L.M., and MacLeod, R., 1967, Reexamination of the serological relationship between wound tumor virus and reovirus, *Virology* **32**:163.

Gerola, F.M., Bassi, M., Lovisolo, O., and Vidano, C., 1966, Virus-like particles in both maize plants infected with maize rough dwarf virus and the vector *Laodelphax striatellus* Fallèn, *Phytopathol. Z.* **56**:97.

Ghosh, A., and John, V.T., 1980, Rice ragged stunt virus disease in India, *Plant Dis.* **64**:1032.

Giannotti, J., and Milne, R.G., 1977, Pangola stunt virus in thin sections and in negative stain, *Virology* **80**:347.

Giannotti, J., and Monsarrat, P., 1968, Etude histologique et histochimique des tumeurs foliaires des cannes à sucre atteintes de la maladie de Fidji, *Ann. Epiphytol.* **19**:707.

Giannotti, J., Monsarrat, P., and Vago, C., 1968, Structures des corps X des tumeurs foliaires de cannes a sucre atteintes de la maladie de Fidji, *Ann. Epiphytol.* **19**:31.

Gomatos, P.J., and Tamm, I., 1963a, The secondary structure of reovirus RNA, *Proc. Natl. Acad. Sci. U.S.A.* **49**:707.

Gomatos, P.J., and Tamm, I., 1963b, Animal and plant viruses with double-helical RNA, *Proc. Natl. Acad. Sci. U.S.A.* **50**:878.

Goodman, R.M., 1981, Geminiviruses, in: *Handbook of Plant Virus Infections: Comparative Diagnosis* (E. Kurstak ed), pp. 879–910, Elsevier/North-Holland, Amsterdam.

Grylls, N.E., 1954, Rugose leaf-curl—a new virus disease transovarially transmitted by the leafhopper *Austroagallia torrida*, *Aust. J. Biol. Sci.* **7**:47.

Grylls, N.E., 1975, Leafhopper transmission of a virus causing maize wallaby ear disease, *Ann. Appl. Biol.* **79**:283.

Grylls, N.E., Waterford, C.J., Filshie, B.K., and Beaton, C.D., 1974, Electron microscopy of rugose leaf curl virus in red clover, *Trifolium pratense* and in the leafhopper vector *Austroagallia torrida*, *J. Gen. Virol.* **23**:179.

Haars, R., Zentgraf, H., Gateff, E., and Bautz, F.A., 1980, Evidence for endogenous reovirus-like particles in a tissue culture cell line from *Drosophila melanogaster*, *Virology* **101**:124.

Harpaz, I., 1959, Needle transmission of a new maize virus, *Nature (London)* **184**:77.

Harpaz, I., 1972, *Maize Rough Dwarf: A Planthopper Virus Disease Affecting Maize, Rice, Small Grain and Grasses*, Israel Universities Press, Jerusalem, 251 pp.

Harpaz, I., Minz, G., and Nitzany, F., 1958, Dwarf disease of maize, *FAO Plant Prot. Bull.* **7**:43.

Harrap, K.A., and Payne, C.C., 1979, The structural properties and identification of insect viruses, *Adv. Virus Res.* **25**:273.

Harris, K.F., 1979, Leafhoppers and aphids as biological vectors: Vector–virus relationships, in: *Leafhopper Vectors and Plant Disease Agents* (K. Maramorosch and K.F. Harris, eds.), pp. 217–308, Academic Press, New York.

Hatta, T., and Francki, R.I.B., 1976, Anatomy of virus-induced galls on leaves of sugar cane infected with Fiji disease virus and the cellular distribution of virus particles, *Physiol. Plant Pathol.* **6**:321.

Hatta, T., and Francki, R.I.B., 1977, Morphology of Fiji disease virus, *Virology* **76**:797.

Hatta, T., and Francki, R.I.B., 1978, Enzyme cytochemical identification of single-stranded and double-stranded RNAs in virus-infected plant and insect cells, *Virology* **88**:105.

Hatta, T., and Francki, R.I.B., 1981, Development and cytopathology of virus-induced galls on leaves of sugar cane infected with Fiji disease virus, *Physiol. Plant Pathol.* **19**:337–346.

Hatta, T., and Francki, R.I.B., 1982, Similarity in the structure of cytoplasmic polyhedrosis virus, leafhopper A virus and Fiji disease virus particles, *Intervirology* **18**:203.

Hatta, T., Boccardo, G., and Francki, R.I.B., 1982, Anatomy of leaf galls induced by some Reoviridae and by wallaby ear disease, *Physiol. Plant. Pathol.* **20**:43.

Herold, F., and Munz, K., 1967, Virus particles in apparently healthy *Peregrinus maidis*, *J. Virol.* **1**:1028.

Hibino, H., and Kimura, I., 1982, Detection of rice ragged stunt virus in insect vectors by enzyme-linked immunosorbent assay, *Phytopathology* (in press).

Hibino, H., Roechan, M., Sudarisman, S., and Tantera, D.M., 1977, A virus disease of rice (kerdil hampa) transmitted by the brown planthopper *Nilaparvata lugens* Stol, in Indonesia, *Contr. Centr. Res. Inst. Agric. (Bogor)* **35**:1.

Hibino, H., Saleh, N., and Roechan, M., 1979, Reovirus-like particles associated with rice ragged stunt diseased rice and insect vector cells, *Ann. Phytopathol. Soc. Jpn.* **45:**228.

Hirai, T., Suzuki, N., Kimura, I., Nakazawa, M., and Kashiwagi, Y., 1964, Large inclusion bodies associated with virus diseases of rice, *Phytopathology* **54:**367.

Hirumi, H., and Maramorosch, K., 1967, Electron microscopy of wound tumor virus in cultured embryonic cells of the leafhopper *Macrosteles fascifrons, Second Int. Coll. Invert. Tissue Culture (Proceedings)*, pp. 203–217, Istituto Lombardo, Academia di Scienze E Lettre, Pavia.

Holmes, I.H., Ruck, B.J., Bishop, R.F., and Davidson, G.P., 1975, Infantile enteritis viruses: Morphogenesis and morphology, *J. Virol.* **16:**937.

Hunkar, A.E.S., Hung, A.T.A., Dulder, I.G., Schank, S.C., Holder, N., and Edwards, C., 1974, Assessment of field resistance to pangola stunt virus (PSV) of *Digitaria* introduction in Surinam and Guyana, *Trop. Agric. (Trinidad)* **52**(1):75.

Hutchinson, P.B., and Francki, R.I.B., 1973, Sugar cane Fiji disease virus, *C.M.I./A.A.B. Descriptions of Plant Viruses*, No. 119.

Hutchinson, P.B., Forteath, G.N.R., and Osborn, A.W., 1972, Corn, sorghum and Fiji disease, *Sugarcane Pathol. Newslett.* **9:**12.

Huth, W., 1975, Eine für die Bundesrepublik Deutschland neue Virose bei Weidelgräsern. *Nachrichtenbl. Dtsch. Pflanzenschutzdienst (Braunschweig)* **27:**49.

Iida, T.T., Shinkai, A., and Kimura, T., 1972, Rice dwarf virus, *C.M.I.,/A.A.B. Descriptions of Plant Viruses*, No. 102.

Ikegami, M., and Francki, R.I.B., 1973, Presence of antibodies to double-stranded RNA in sera of rabbits immunized with rice dwarf and maize rough dwarf viruses, *Virology* **56:**404.

Ikegami, M., and Francki, R.I.B., 1974, Purification and serology of virus-like particles from Fiji disease virus-infected sugar cane, *Virology* **61:**327.

Ikegami, M., and Francki, R.I.B., 1975, Some properties of RNA from Fiji disease subviral particles, *Virology* **64:**464.

Ikegami, M., and Francki, R.I.B., 1976, RNA dependent RNA polymerase associated with subviral particles of Fiji disease virus, *Virology* **70:**292.

Ikegami, M., and Francki, R.I.B., 1977, Antigenic variation among double-stranded RNAs from virus and a synthetic polyribonucleotide, *Ann. Phytopathol. Soc. Jpn.* **43:**59.

Inoue, H., and Omura, T., 1982, Transmission of rice gall dwarf virus by the green rice leafhopper, *Plant. Dis.* **66:**57.

Ishii, M., and Yoshimura, S., 1973, Epidemiological studies on rice black-streaked dwarf virus in Kanto-Tosan district of Japan, *J. Cent. Agric. Exp. Sta.* **17:**16.

Joklik, W.K., 1974a, Reproduction of the reoviridae, in: *Comprehensive Virology*, Vol. 2 (H. Fraenkel-Conrat and R.R. Wagner, eds.), pp. 231–334, Plenum Press, New York,

Joklik, W.K., 1974b, Evolution of viruses, *Symp. Soc. Gen. Microbiol.* **24:**293.

Kalmakoff, J., Lewandowski, L.J., and Black, D.R., 1969, Comparison of the ribonucleic acid subunits of reovirus, cytoplasmic polyhedrosis virus, and wound tumor virus, *J. Virol.* **4:**851.

Kimura, I., 1976a, Improved purification of rice dwarf virus by the use of polyethylene glycol, *Phytopathology* **66:**1470.

Kimura, I., 1976b, Loss of vector transmissibility in an isolate of rice dwarf virus, *Ann. Phytopathol. Soc. Jpn.* **42:**322.

Kimura, I., and Black, L.M., 1971, Some factors affecting infectivity assay of wound-tumor virus on cell monolayers from an insect vector, *Virology* **46:**266.

Kimura, I., and Black, L.M., 1972, The cell-infecting unit of wound tumor virus, *Virology* **49:**549.

Kimura, I., and Miyajima, N., 1976, Serological detection of virus antigen in the rice plants infected with rice dwarf virus, *Ann. Phytopathol. Soc. Jpn.* **42:**266.

Kimura, I., and Shikata, E., 1968, Structural model of rice dwarf virus, *Proc. Jpn. Acad.* **44:**538.

Kitajima, E.W., and Costa, A.S., 1970, Electron microscopy of pangola stunt virus-infected plants and viruliferous planthopper vectors, 7èm Congr. Int. micr. électr. Crenoble, Vol. 3, p. 323.

Klein, M., and Harpaz, I., 1969, Changes in resistance of graminaceous plants to delphacid planthoppers induced by maize rough dwarf virus (MRDV), Z. Angew. Entomol. **64**:39.

Klein, M., and Harpaz, I., 1970, Heat suppression of plant virus propagation in the insect vector's body, Virology **41**:72.

Kleinschmidt, A.K., Dunnebacke, T.H., Splendove, R.S., Schaffer, F.L., and Whitcomb, R.F., 1964, Electron microscopy of RNA from reo-virus and wound tumor virus, J. Mol. Biol. **10**:282.

Kodama, T., and Suzuki, N., 1973, RNA polymerase activity in purified rice dwarf virus, Ann. Phytopathol. Soc. Jpn. **39**:251.

Kunkel, L.O., 1924, Histological and cytological studies on the Fiji disease of sugarcane, Bull Exp. Sta. Hawaiian Sugar Planters Assoc. **3**:99.

Kuribayashi, K., and Shinkai, A., 1952, On the rice black-streaked dwarf disease, a new disease of rice in Japan, Ann. Phytopathol. Soc. Jpn. **16**:41.

Lesemann, D.-E., and Huth, W., 1975, Nachweis von maize rough dwarf virus—ähnlichen Partikeln in Enationen von Lolium-Pflanzen aus Deutschland, Phytopathol. Z. **82**:246.

Lewandowski, L.J., and Leppla, S.H., 1972, Comparison of the 3' termini of discrete segments of the double-stranded ribonucleic acid genomes of cytoplasmic polyhedrosis virus, wound tumor virus and reovirus, J. Virol. **10**:965.

Lewandowski, L.J., and Traynor, B.L., 1972, Comparison of the structure and polypeptide composition of three double-stranded ribonucleic acid-containing viruses (diplornaviruses): Cytoplasmic polyhedrosis virus, wound tumor virus and reovirus, J. Virol. **10**:1053.

Lindsten, K., 1959, A preliminary report of virus diseases of cereals in Sweden, Phytopathol. Z. **35**:420.

Lindsten, K., 1961, Studies on virus diseases of cereals in Sweden. II. On virus diseases transmitted by the leafhopper, Calligypona pellucida (F.), K. Lantdrhögsk. Annlr. **27**:199.

Lindsten, K., and Gerhardson, B., 1971, Stråsädens bestockningssjuka—en ny och svårartad viros som under 1971 påträffats i Östergötland, Vaxtskyddsnotiser **35**:66.

Lindsten, K., Gerhardson, B., and Pettersson, J., 1973, Cereal tillering disease in Sweden and some comparisons with oat sterile dwarf and maize rough dwarf, Statens Vaxtskyddanst. Medd. (Natl. Swed. Inst. Plant Prot.) **15**:151, 375.

Ling, K.C., 1972, Rice Virus Diseases, International Rice Research Institute, Los Baños, Philippines, 134 pp.

Ling, K.C., Tiongco, E.R., and Aguiero, V.M., 1978, Rice ragged stunt, a new virus disease, Plant Dis. Rep. **62**:701.

Lisa, V., Boccardo, G., and Milne, R.G., 1982, Double-stranded ribonucleic acid from carnation cryptic virus, Virology (in press).

Littau, V.C., and Black, L.M., 1952, Spherical inclusions in plant tumors caused by a virus, Am. J. Bot. **39**:87.

Lovisolo, O., 1971, Maize rough dwarf virus, C.M.I./A.A.B. Descriptions of Plant Viruses, No. 72.

Luftig, R.B., Kilham, S.S., Hay, A.J., Zweerink, H.J., and Joklik, W.K., 1972, An ultrastructural study of virions and cores of reovirus type 3, Virology **48**:170.

Luisoni, E., and Conti, M., 1970, Digitaria sanguinalis (L) Scop., a new natural host of maize rough dwarf virus, Phytopathol. Mediterr. **9**(2–3):102.

Luisoni, E., Lovisolo, O., Kitagawa, Y., and Shikata, E., 1973, Serological relationship between maize rough dwarf virus and rice black-streaked dwarf virus, Virology **52**:281.

Luisoni, E., Milne, R.G., and Boccardo, G., 1975, The maize rough dwarf virion. II. Serological analysis, Virology **68**:86.

Luisoni, E. Milne, R.G., and Roggero, P. (1982) Diagnosis of rice ragged stunt virus by enzyme-linked immunosorbent assay and immunosorbent electron microscopy, Plant Dis. **66**:929.

Luisoni, E., Boccardo, G., Milne, R.G., and Conti, M., 1979, Purification, serology and nucleic acid of oat serile dwarf virus subviral particles, *J. Gen. Virol.* **45**:651.

Lyon, H.L., 1910, A new cane disease now epidemic in Fiji, *Hawaii. Plant. Rec.* **3**:200.

Maillet, P.L., and Folliot, R., 1967, Sur la presence d'un virus dans le testicule chez un insecte homoptere, *C. R. Acad. Sci. Ser. D* **264**:2828.

Maramorosch, K., 1950, Influence of temperature on incubation and transmission of the wound-tumor virus, *Phytopathology* **40**:1071.

Maramorosch, K., 1975, Infection of arthropod vectors by plant pathogens, in: *Invertebrate Immunity* (K. Maramorosch and R.E. Shope, eds.), pp. 49–53, Academic Press, New York.

Maramorosch, K., 1979, Leafhopper tissue culture, in: *Leafhopper Vectors and Plant Disease Agents* (K. Maramorosch and K.R. Harris, eds.), pp. 485–511, Academic Press, New York.

Maramorosch, K., Calica, C.A., Agati, J.A., and Pableo, G., 1961, Further studies on the maize and rice leaf galls induced by *Cicadulina bipunctella, Entomol. Exp. Appl.* **4**:86.

Maramorosch, K., Shikata, E., Hirumi, H., and Granados, R.R., 1969, Multiplication and cytopathology of a plant tumor virus in insects, *Natl. Cancer Inst. Monogr.* **31**:493.

Martelli, G.P., and Russo, M., 1977, Plant virus inclusion bodies, *Adv. Virus Res.* **21**:175.

Martin, S.A., and Zweerink, H.J., 1972, Isolation and characterization of two types of blue-tongue virus particles, *Virology* **50**:495.

Martinez-Lopez, G., and Black, L.M., 1979, Is infectivity for plants lost during continuous passage in V.C.M.? *Adv. Virus Res.* **25**:191.

Matsuno, S., and Mukoyama, A., 1979, Polypeptides of bovine rotavirus. *J. Gen. Virol.* 43, 309.

Matthews, R.E.F., 1981, *Plant Virology*, 2nd ed. Academic Press, New York, 858 pp.

Matthews, R.E.F., 1982, Classification and nomenclature of viruses: Fourth report of the International Committee on Taxonomy of Viruses, *Intervirology* **17**:1.

McCrae, M.A., and Joklik, W.K., 1978, The nature of the polypeptide encoded by each of the 10 double-stranded RNA segments of reovirus type 3, *Virology* **89**:578.

McWhorter, F.P., 1922, The nature of the organism found in the Fiji galls of sugarcane, *Philipp. Agric.* **11**:103.

Milne, R.G., 1977a, Pangola stunt virus, *C.M.I./A.A.B. Descriptions of Plant Viruses*, No. 175.

Milne, R.G., 1977b, Structure of the viroplasm of the maize rough dwarf-like viruses, *Ann. Phytopathol.* **9**:333.

Milne, R.G., 1980, Electron microscopy of thin sections of Italian ryegrass infected with both ryegrass cryptic virus and oat sterile dwarf virus, *Microbiologica* **3**:333.

Milne, R.G., 1981, Does rice ragged stunt virus lack the typical double shell of the Reoviridae? *Intervirology* **14**:331.

Milne, R.G., and Lesemann, D.-E., 1978, An immunoelectron microscopic investigation of oat sterile dwarf and related viruses, *Virology* **90**:299.

Milne, R.G., and Lovisolo, O., 1977, Maize rough dwarf and related viruses, *Adv. Virus Res.* **21**:267.

Milne, R.G., and Luisoni, E., 1977, Serological relationships among maize rough dwarf-like viruses, *Virology* **80**:12.

Milne, R.G., Conti, M., and Lisa, V., 1973, Partial purification, structure and infectivity of complete maize rough dwarf virus particles, *Virology* **53**:130.

Milne, R.G., Kempiak, G., Lovisolo, O., and Mühle, E., 1974, Viral nature of *Arrhenatherum* blue dwarf (Blauverzwergung von *Arrhenatherum elatius*),1*Phytopathol. Z.* **79**:315.

Milne, R.G., Lindsten, K., and Conti, M., 1975, Electron microscopy of the particles of cereal tillering disease virus and oat sterile dwarf virus, *Ann. Appl. Biol.* **79**:371.

Milne, R.G., Luisoni, E., and Ling, K.C., 1979, Preparation and use of an antiserum to rice ragged stunt virus subviral particles, *Plant Dis. Rep.* **63**:445.

Milne, R.G., Luisoni, E., Boccardo, G., and Ling, K.C., 1982, Rice ragged stunt virus, *C.M.I./A.A.B. Descriptions of Plant Viruses*, No. 248.

Mitsuhashi, J., and Nasu, S., 1967, An evidence for the multiplication of rice dwarf virus in the vector cell cultures inoculated *in vitro*, *Appl. Entomol. Zool. Jpn.* **2**:113.

Miura, K.-I., and Muto, A., 1965, Lack of messenger RNA activity of a double-stranded RNA, *Biochim. Biophys. Acta* **108**:707.

Miura, K.-I., Kimura, I., and Suzuki, N., 1966, Double-stranded ribonucleic acid from rice dwarf virus, *Virology* **28**:571.

Moussa, A.Y., 1978, A new virus disease in the housefly, *Musca domestica* (Diptera), *J. Invertebr. Pathol.* **31**:204.

Moussa, A.Y., 1980, The housefly virus contains double-stranded RNA, *Virology* **106**:173.

Moussa, A.Y., 1981, Studies of the housefly virus structure and disruption during purification procedures, *Micron* **12**:131.

Morinaka, T., Putta, M., Chattanachit, D., Parejarearn, A., Disthaporn,S., Omura, T., and Inoue, H., 1982, Transmission of rice gall dwarf virus by cicadellid leafhoppers *Recilia dorsalis* and *Nephoteltix nigropictus* in Thailand, *Plant Dis.* **66**:703.

Mühle, E., and Kempiak, G., 1971, Zur Geschichte, Ätiologie und Symptomatologie der Blauverzwergung des Glatthafers [*Arrhenatherum elatius* (L.), I. et C. Presl.], *Phytopathol. Z.* **72**:269.

Nagaraj, A.N., and Black, L.M., 1961, Localization of wound-tumor virus antigen in plant tumors by use of fluorescent antibodies, *Virology* **15**:289.

Nagaraj, A.N., Sinha, R.C., and Black, L.M., 1961, A smear technique for detecting virus antigen in individual vectors by the use of fluorescent antibodies, *Virology* **15**:205.

Nakata, M., Fukunaga, K., and Suzuki, N., 1978, Polypeptide components of rice dwarf virus, *Ann. Phytopathol. Soc. Jpn.* **44**:288.

Nasu, S., 1965, Electron microscopic studies on transovarial passage of rice dwarf virus, *Appl. Entomol. Zool. Jpn.* **9**:225.

Nomé, S.F., Lenardon, S.L., Raju, B.C., Laguna, I.G., Lowe, S.K., and Docampo, D., 1981, Association of reovirus-like particles with "Enfermedad de Rio IV" of maize in Argentina, *Phytopathol. Z.* **101**:7.

Nuss, D.L., and Peterson, A.J., 1980, Expression of wound tumor virus gene products *in vivo* and *in vitro*, *J. Virol.* **34**:532.

Nuss, D.L., and Peterson, A.J., 1981a, Resolution and genome assignment of mRNA transcripts synthesized *in vitro* by wound tumor virus, *Virology* **114**:399.

Nuss, D.L., and Peterson, A.J., 1981b, *In vitro* synthesis and modification of mRNA by exvectorial isolates of wound tumor virus, *J. Virol.* **39**:954.

Obijeski, J.F., Palmer, E.L., and Martin, M.L., 1977, Biochemical characterization of infantile gastroenteritis virus (IGV), *J. Gen. Virol.* **34**:485.

Omura, T., Inoue, H., Morinaka, T., Saito, Y., Chattanachit, D., Putta, M., Parejarearn, A., and Disthaporn, S., 1980, Rice gall dwarf, a new virus disease, *Plant. Dis.* **64**:795.

Paliwal, Y.C., 1979, Occurrence and localization of spherical viruslike particles in tissues of apparently healthy tobacco thrips, *Frankliniella fusca*, a vector of tomato spotted wilt virus, *J. Invertebr. Pathol.* **33**:307.

Payne, C.C., and Harrap, K.A., 1977, Cytoplasmic polyhedrosis viruses, in: *The Atlas of Insect and Plant Viruses* (K. Maramorosch, ed.), pp. 105–129, Academic Press, New York.

Peterson, A.J., and Nuss, D.L., 1982, Polypeptide synthesis in cultured vector cells initially and persistently infected with an insect-transmitted plant virus, *Virology* (in press).

Průša, V., 1958, Die sterile Verzwergung des Hafers in der Tschechoslowakischen Republik, *Phytopathol. Z.* **33**:99.

Průša, V., Jermoljev, E., and Vacke, J., 1959, Oat sterile-dwarf virus disease, *Biol. Plant.* **1**:223.

Reddy, D.V.R., 1977, Techniques of invertebrate tissue culture for the study of plant viruses, *Methods Virol.* **6**:393.

Reddy, D.V.R., and Black, L.M., 1966, Production of wound-tumor virus and wound-tumor soluble antigen in the insect vector, *Virology* **30**:551.

Reddy, D.V.R., and Black, L.M., 1972, Increase of wound tumor virus in leafhoppers as assayed on vector cell monolayers, *Virology* **50:**412.

Reddy, D.V.R., and Black, L.M., 1973a, Estimate of absolute specific infectivity of wound tumor virus purified with polyethylene glycol. *Virology* **54:**150.

Reddy, D.V.R., and Black, L.M., 1973b, Electrophoretic separation of all components of the double-stranded RNA of wound tumor virus, *Virology* **54:**557.

Reddy, D.V.R., and Black, L.M., 1974, Deletion mutants of the genome segments of wound tumor virus, *Virology* **61:**458.

Reddy, D.V.R., and Black, L.M., 1977, Isolation and replication of mutant populations of wound tumor virions lacking certain genome segments, *Virology* **80:**336.

Reddy, D.V.R., and MacLeod, R., 1976, Polypeptide components of wound tumor virus, *Virology* **70:**274.

Reddy, D.V.R., Kimura, I., and Black, L.M., 1974, Co-electrophoresis of dsRNA from wound tumor and rice dwarf viruses, *Virology* **60:**293.

Reddy, D.V.R., Boccardo, G., Outridge, R., Teakle, D.S., and Black, L.M., 1975a, Electrophoretic separation of dsRNA genome segments from Fiji disease and maize rough dwarf viruses, *Virology* **63:**287.

Reddy, D.V.R., Shikata, E., Boccardo, G., and Black, L.M., 1975b, Co-electrophoresis of double-stranded RNA from maize rough dwarf and rice black-streaked dwarf viruses, *Virology* **67:**279.

Reddy, D.V.R., Grylls, N.E., and Black, L.M., 1976, Electrophoretic separation of dsRNA genome segments from maize wallaby ear virus and its relationship to other phytoreoviruses, *Virology* **73:**36.

Reddy, D.V.R., Rhodes, D.P., Lesnaw, J.A., MacLeod, R., Banerjee, A.K., and Black, L.M., 1977, *In vitro* transcription of wound tumor virus RNA by virion-associated RNA transcriptase, *Virology* **80:**356.

Redolfi, P., and Boccardo, G., 1974, Fractionation of the double-stranded RNA of maize rough dwarf virus subviral particles, *Virology* **59:**319.

Redolfi, P., and Pennazio, S., 1972, Double-stranded ribonucleic acid from maize rough dwarf virus, *Acta Virol. (Prague)* **16:**369.

Rhodes, D.P., Reddy, D.V.R., MacLeod, R., Black, L.M., and Banerjee, A.K., 1977, *In vitro* synthesis of RNA containing 5'-terminal structure $^7mG(5')ppp(5')A_p^m$ . . . by purified wound tumor virus *Virology* **76:**554.

Rohozinski, J., Francki, R.I.B., and Chum P.W.G., 1981, Detection and identification of Fiji disease virus in infected sugarcane in immunodiffusion, immuno-osmophoretic and enzyme-linked immunosorbent assays, *J. Virol. Methods* **3:**177.

Samejima, T., Hashizume, H., Imahori, K., Fujii, I., and Miura, K.-I., 1968, Optical rotatory dispersion and circular dichroism of the rice dwarf virus ribonucleic acid, *J. Mol. Biol.* **34:**39.

Sato, T., Kyogoku, Y., Higuchi, S., Mitsui, Y., Iitaka, Y., Tsuboi, M., and Miura, K.-I., 1966, A preliminary investigation on the molecular structure of rice dwarf virus ribonucleic acid, *J. Mol. Biol.* **16:**180.

Schindler, A.J., 1942, Insect transmission of "wallaby ear" disease of maize, *Aust. Inst. Agric. Sci. J.* **8:**35.

Schwartz, E.F., and Stollar, B.D., 1969, Antibodies to polyadenylate–polyuridylate copolymers as reagents for double strand RNA and DNA–RNA hybrid complexes, *Biochem. Biophys. Res. Commun.* **35:**115.

Sheffield, F.M.L., 1969, The cause of leaf galls (pseudo-Fiji disease), *Sugarcane Pathol. Newslett.* **3:**25.

Shikata, E., 1966, Electron microscopic studies in plant viruses, *J. Fac. Agric. Hokkaido Univ.* **55:**1.

Shikata, E., 1974, Rice black-streaked dwarf virus, *C.M.I./A.A.B. Descriptions of Plant Viruses*, No. 135.

Shikata, E., 1977, Plant reovirus group, in: *The Atlast of Insect and Plant Viruses* (K. Maramorosch, ed.), pp. 377–404, Academic Press, New York.

Shikata, E., 1981, Reoviruses, in: *Handbook of Plant Virus Infections: Comparative Diagnosis* (E. Kurstak, ed.), pp. 423–454, Elsevier/North-Holland, Amsterdam.

Shikata, E., and Kitagawa, Y., 1977, Rice black-streaked dwarf virus: Its properties, morphology and intracellular localization, *Virology* 77:826.

Shikata, E., and Maramorosch, K., 1965, An electron microscopic study of plant neoplasia induced by wound tumor virus, *J. Natl. Cancer Inst.* 36:97.

Shikata, E., and Maramorosch, K., 1967, Electron microscopy of wound tumor virus assembly sites in insect vectors and plants, *Virology* 32:363.

Shikata, E., Orenski, S.W., Hirumi, H., Mitsuhashi, J., and Maramorosch, K., 1964, Electron micrographs of wound-tumor virus in an animal host and in a plant tumor, *Virology* 23:441.

Shikata, E., Senboku, T., Kamjaipai, K., Chou, T.-G., Tiongco, E.R., and Ling, K.C., 1979, Rice ragged stunt virus, a new member of plant reovirus group, *Ann. Phytopathol. Soc. Jpn.* 45:436.

Shinkai, A., 1962, Studies on the insect transmission of rice virus diseases, *Rep. Natl. Agric. Sci. (Jpn.)* C-14:1.

Sinha, R.C., 1965, Sequential infection and distribution of wound-tumor virus in the internal organs of a vector after ingestion of virus, *Virology* 26:673.

Sinha, R.C., 1968, Recent work on leafhopper-transmitted viruses, *Adv. Virus Res.* 13:181.

Sinha, R.C., and Reddy, D.V.R., 1964, Improved fluorescent smear technique and its application in detecting virus antigens in an insect vector, *Virology* 24:626.

Stollar, B.D., 1973, Nucleic acid antigens, in: *The Antigens* Vol. 1 (M. Sela, ed.), p. 1, Academic Press, New York.

Stollar, B.D., 1975, The specificity and applications of antibodies to helical nucleic acids, *Crit. Rev. Biochem.* 3:45.

Streissle, G., and Granados, R.R., 1968, The fine structure of wound tumor virus and reovirus, *Arch. Gesamte Virusforsch.* 25:369.

Streissle, G., and Maramorosch, K., 1963, Reovirus and wound-tumor virus: Serological cross reactivity, *Science* 140:996.

Streissle, G., and Maramorosch, K., 1969, Cultivation *in vitro* of wound tumor virus infected and healthy tissue from sweet clover, *Phytopathology* 59:403.

Streissle, G., Rosen, L., and Tokumitsu, T., 1967, Host specificity of wound tumor virus and reoviruses and their serological relationship, *Arch. Gesamte Virusforsch.* 22:409.

Takehara, M., and Suzuki, N., 1973, Interfereon induction by rice dwarf virus RNA, *Arch. Gesamte Virusforsch,* 40:291.

Teakle, D.S., and Steindl, D.R.L., 1969, Virus-like particles in galls on sugarcane plants affected by Fiji disease, *Virology* 37:139.

Tomita, K., and Rich, A., 1964, X-ray diffraction investigations of complementary RNA, *Nature (London)* 201:1160.

Toyoda, S., Kimura, I., and Suzuki, N., 1965, Purification of rice dwarf virus, *Ann. Phytopathol. Soc. Jpn.* 30:225.

Trujillo, G.E., Acosta, J.M., and Pinero, A., 1974, A new corn disease found in Venezuela, *Plant Dis. Rep.* 58:122.

Tryon, H., 1910, Report of the Department of Agriculture and Stock, Queensland, 1909–1910, p. 81.

Vacke, J., 1966, Study of transovarial passage of the oat sterile-dwarf virus, *Biol. Plant* 8:127.

Van der Lubbe, J.L.M., Hatta, T., and Francki, R.I.B., 1979, Structure of the antigen from Fiji disease virus particles eliciting antibodies specific to double-stranded polyribonucleotides, *Virology* 95:405.

Vasquez, C., and Tournier, P., 1962, The morphology of reovirus, *Virology* 17:503.

Verwoerd, D.W., 1970, Diplornaviruses: A newly recognized group of double-stranded RNA viruses, *Prog. Med. Virol.* 12:192.

Verwoerd, D.W., Els, H.J., De Villiers, E.-M., and Huismans, H., 1972, Structure of bluetongue virus capsid, *J. Virol.* 10:783.

Vidano, C., 1970, Phases of maize rough dwarf virus multiplication in the vector *Laodelphax striatellus* (Fallen), *Virology* **41**:218.

Wetter, C., Luisoni, E. Conti, M., and Lovisolo, O., 1969, Purification and serology of maize rough dwarf virus from plant and vector, *Phytopathol. Z.* **66**:197.

Whitcomb, R.F., and Black, L.M., 1961, Synthesis and assay of wound-tumor soluble antigen in an insect vector, *Virology* **15**:136.

Wood, H.A., 1973, Viruses with double-stranded RNA genomes, *J. Gen. Virol.* **20**(Suppl.):61.

Wood, H.A., and Streissle, G., 1970, Wound tumor virus: Purification and fractionation of the double-stranded ribonucleic acid, *Virology* **40**:329.

# Index